Texts and Monographs in
Symbolic Computation

A Series of the
Research Institute for Symbolic Computation,
Johannes-Kepler-University, Linz, Austria

Edited by
B. Buchberger and G. E. Collins

A. Ben-Israel
R. Gilbert

Computer-Supported Calculus

Springer-Verlag Wien GmbH

Dr. Adi Ben-Israel
Rutgers Center for Operations Research and Department of Mathematics
Rutgers University, New Brunswick, New Jersey, U.S.A.

Dr. Robert Gilbert
Department of Mathematical Science and Computer and Informational Sciences
University of Delaware, Newark, Delaware, U.S.A.

Originally published by Springer-Verlag Wien New York in 2002
Softcover reprint of the hardcover 1st edition 2002

Data conversion by HD Ecker: TeXtservices, Bonn
Printed on acid-free and chlorine-free bleached paper

With 191 Figures

Library of Congress Cataloging-in-Publication Data

Ben-Israel, Adi.
    Computer-supported calculus / A. Ben-Israel, R. Gilbert
      p.   cm. — (Texts and monographs in symbolic computation, ISSN 0943-853X)
    Includes index.
    ISBN 978-3-7091-7230-8   ISBN 978-3-7091-6146-3 (eBook)
    DOI 10.1007/978-3-7091-6146-3
    1. Calculus—Data processing. 2. MACSYMA. I. Gilbert, Robert P., 1932– II. Title.
III. Series.
QA303.5.D37B46 2001
515′.0285—dc21   2001049370

ISSN 0943-853X
  ISBN 978-3-7091-7230-8

Preface

This is a new type of calculus book: Students who master this text will be well versed in calculus and, in addition, possess a useful working knowledge of one of the most important mathematical software systems, namely, MACSYMA. This will equip them with the mathematical competence they need for science and engineering and the competitive workplace. The choice of MACSYMA is not essential for the didactic goal of the book. In fact, any of the other major mathematical software systems, e.g., AXIOM, MATHEMATICA, MAPLE, DERIVE, or REDUCE, could have been taken for the examples and for acquiring the skill in using these systems for doing mathematics on computers. The symbolic and numerical calculations described in this book will be easily performed in any of these systems by slight modification of the syntax as soon as the student understands and masters the MACSYMA examples in this book. What is important, however, is that the student gets all the information necessary to design and execute the calculations in at least one concrete implementation language as this is done in this book and also that the use of the mathematical software system is completely integrated with the text.

In these times of globalization, firms which are unable to hire adequately trained technology experts will not prosper. For corporations which depend heavily on science and engineering, remaining competitive in the global economy will require hiring employees having had a traditionally rigorous mathematical education. Educators do students an injustice when they offer oversimplified material under the deceptive guise of calculus reform. We direct this book to those teachers and students who share our commitment to excellence in mathematical preparation.

The instructor who uses this book should schedule student work on the computer at least one day per week. Lectures should include either live demonstrations (using the MACSYMA code given in this book or equivalent code in any other mathematical software system) or interactive sessions, with both input and output lines projected on a screen for student viewing.

When we were teaching with this text, we have typically combined three lectures with one laboratory session and ensured students access to the computer laboratory for special projects.

By selecting the appropriate sections of the book, the text can be used for a broad spectrum of course schedules in the various curricula of mathematics, computer science, engineering, business, health sciences, etc.

Contents

Advice to the student on using this book

Begin with the introduction to MACSYMA in Appendix A, which contains enough information to get you started (but is no substitute for the MACSYMA User Guide).

Use MACSYMA to plot the graphs of all functions encountered here (you will find in this book fewer drawings than in traditional calculus texts because we expect you to draw your own). Use MACSYMA to compute and to check your hand calculations, but do not rely on it. Remember that software is no substitute for thinking.

This text has many problems and exercises, some routine (testing your knowledge of basic concepts) and other more challenging. Do as many as you can of each type.

Do not believe anything you read, even in a calculus book. This is why you should demand to see proofs. Simple proofs are given throughout the book, but complicated proofs are kept in separate and easily avoidable sections.

Equations are numbered on the right side.

Computer work lends itself to working in teams where members compare notes, check each other's work, and suggest new ideas. A congenial study group is a good start.

Functions, limits, and continuity

1 Functions

1.1 Introduction

The concept of function is central to mathematics. A function is a rule assigning values to certain objects. If a function is called f, the value it assigns to x is denoted by $f(x)$. A well-defined function f assigns to each such x a single value $f(x)$. However, several objects $x_1, x_2, \ldots, x_n$ may get the same value $f(x_1) = f(x_2) = \cdots = f(x_n)$. Here are some examples.

Example 1.1. $f(x) := x^2$ is the function which assigns to each x the value x^2. For example $f(3) = 9$, $f(-3) = 9$.

Example 1.2. $g(x) := \sqrt{x}$ assigns to each *nonnegative* x the value $\sqrt{x}$. For example, $g(9) = 3$, but $g(-9)$ is undefined.

MACSYMA-Session 1.1 (Defining and evaluating functions with MACSYMA). The function $f(x) := x^2$ is defined in MACSYMA by typing the input on the command line c1 below, followed by the <ENTER>. This causes the definition of the function to echo on the output display line d1.

c1. `f(x):= x^2 <ENTER>`
d1. $f(x) := x^2$

From now on we will omit the <ENTER>, which is always understood after an input line. Also, if an echo serves no purpose, we suppress it by ending the input line with a $. When it comes to function names and variable names, MACSYMA prefers (i.e., uses as default) lowercase letters. If you insist on the function being called F and not f, type \f, for example,

c2. `\f(x):= x^3`

gives the desired uppercase:

d2. $F(x) := x^3$

To evaluate $f(x)$ for a particular value of x, such as $x = 3$, just type

c3. `f(3)`

to get the correct answer

d3.   9

Similarly,

c4. `\f(3)`
d4.   27

The square-root function is defined as

c5. `h(x):= sqrt(x)$`

c6. h(3)

d6. $\sqrt{3}$

Unlike the function $g(x)$ in Example 1.2, the MACSYMA function sqrt(x) is defined also for negative x, using the *imaginary number* $i := \sqrt{-1}$, written in MACSYMA as %i. For example,

c7. h(-9)

d7. $3\,i$

Alternatively, we can use here h(x) := x^(1/2) instead of h(x):=sqrt(x).
Finally, clean up:

c8. remfunction(f, \f, h)$

Example 1.3. Consider the process of assigning social-security numbers to individuals. We can regard it as a function

$$SSN(person) := \text{social-security number}$$

which is well-defined, mathematically, if no person gets more than one number.

Example 1.4. A calculus professor uses the following procedure for giving final grades. Each student is first given a numerical grade x, computed on the basis of examinations, homework assignments, and quizzes throughout the semester. The numerical grade satisfies $0 \le x \le 100$. A final grade is then determined by the function

$$grade(x) := \begin{cases} A & \text{if } 91 \le x \\ B & \text{if } 75 \le x < 91 \\ C & \text{if } 60 \le x < 75 \\ D & \text{if } 50 \le x < 60 \\ F & \text{if } \quad x < 50 \end{cases}$$

which assigns to every number $0 \le x \le 100$ one of the letters A, B, C, D, and F.

MACSYMA-Session 1.2. The professor in Example 1.4 may use MACSYMA to compute the function *grade(x)*. It is expressed in MACSYMA as follows.

```
c1. grade(x) := if x<50 then "F" else
               if x<60 then "D" else
                   if x<75 then "C" else
                       if x<91 then "B" else "A"$
```

using the function if... then... else in nested form (see Appendix A, A.16). For example, the third condition, $x < 75$, is checked only after the previous conditions, $x < 60$ and $x < 50$, tested negative. To compute the grade corresponding to $x = 63$:

c2. grade(63)

d2. C

a fair grade, but the computer has no heart:

c3. grade(90.9999)

d3. B

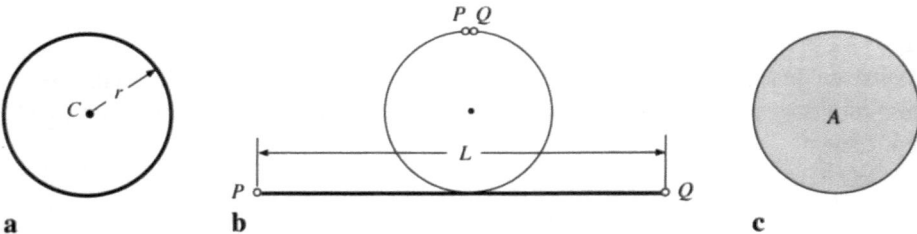

Fig. 1.1 a–c. A circle, its circumference and area. **a** A circle with center C and radius r; **b** the circumference L; **c** the area A

Example 1.5. A *circle* is the set of points in the plane which have the same distance (called radius) from a given point, called the center of the circle (Fig. 1.1 a). The length of the circle (Fig. 1.1 b) is called its circumference. You may recall that the circumference L of the circle is proportional to its radius r, $L = 2\pi r$, where $\pi \approx 3.14159$ ("$\approx$" means approximately equal). This relation gives L as a function of r,

$$L(r) := 2\pi r \ . \tag{1.1}$$

For example, the circumference of a circle with a radius of 25 [cm] is $L(25) = 2\pi 25 \approx 157.08$ [cm].

You also recall that the *area* A enclosed by the circle (Fig. 1.1 c) is

$$A(r) := \pi r^2 \ , \tag{1.2}$$

a function of the radius r. For example, the area of a circle with a radius of 25 [cm] is $A(25) = \pi 25^2 \approx 1963.49$ [cm^2].

Example 1.6. A *sphere* is a set of points in 3-dimensional space which have the same distance (the radius again) from a given center.[1] The *surface area* S of the sphere is related to its radius r by the formula

$$S(r) := 4\pi r^2 \ . \tag{1.3}$$

The *volume* V enclosed by the sphere is given by

$$V(r) := \frac{4}{3}\pi r^3 \ . \tag{1.4}$$

Both surface area S and volume V are functions of the radius r. For a sphere with a radius of 25 [cm], $S(25) = 4\pi 25^2 \approx 7853.98$ [cm^2] and $V(25) = \frac{4}{3}\pi 25^3 \approx 65449.85$ [cm^3].

MACSYMA-Session 1.3. The functions in Examples 1.5 and 1.6 are now expressed as MACSYMA functions, and combined in one list:

```
c1. [circumference(r):=2*%pi*r, area(r):=%pi*r^2,
            surface(r):=4*%pi*r, volume(r):=4/3*%pi*r^3]$
```

---

1 The sphere is a generalization of the circle to 3-dimensional space.

These functions can be evaluated for different values of r:

c2. [circumference(3), area(a), surface(1), volume(r)]

d2. $\left[6\pi, \pi a^2, 4\pi, \dfrac{4\pi r^3}{3} \right]$

There was no need to define the functions in c1, since MACSYMA has these functions (and thousands of others) built in. The corresponding MACSYMA functions are found in the file geofunct, which must first be loaded,

c3. load(geofuncts)$

c4. [circumference_circle(r), area_circle(r), area_sphere(r),
         vol_sphere(r)]

d4. $\left[2\pi r, \pi r^2, 4\pi r^2, \dfrac{4\pi r^3}{3} \right]$

MACSYMA also knows the surface and volume measure of spheres in n-dimensional spaces. In the following line, 3 stands for 3 dimensions (these formulas should be familiar to you), and 4 stands for 4 dimensions.

c4. [sphere_surface(3,r), sphere_volume(3,r), sphere_surface(4,r),
         sphere_volume(4,r)]

d4. $\left[4\pi r^2, \dfrac{4\pi r^3}{3}, 2\pi^2 r^2, \dfrac{\pi^2 r^4}{2} \right]$

Formulas (1.1)–(1.4) can be derived and verified mathematically. For example, we can prove that for any $r > 0$, the volume of a sphere with radius r is given by (1.4). These are examples of "theoretical" functions. The next example illustrates functions which arise "experimentally", giving a table of values from which an analytical formula may be deduced.

Example 1.7. A biological experiment requires counting certain bacteria in a given test area. The bacteria are counted every 5 minutes. At time 0 min there are 400,000 bacteria. The next measurement, at time 5 min, gives a count of 488,000 bacteria. The results of the first 50 minutes are listed in Table 1.1. While the number of bacteria $N(t)$ at time t is an unknown function, we can (after making some reasonable assumptions on how bacteria multiply) approximate it by the *exponential function*

$$N(t) = N(0)\exp(\lambda t), \quad t \geq 0, \tag{1.5}$$

where $N(0)$ is the number of bacteria at time 0, $\lambda > 0$ is a given constant (describing the rate of growth of the bacteria population[2]) and exp x is the exponential function that we will introduce in Chap. 2.

Here the initial population $N(0)$ is given as 4.0×10^5. From Table 1.1 we can estimate (by methods to be described later) the rate of growth λ as 0.04. The results of Table 1.1 can then be approximated by the function

$$N(t) = 4.00 \times 10^5 \exp(0.04t), \quad t \geq 0. \tag{1.6}$$

---

2 Not all bacteria populations grow at the same rate, so λ is specific to the population in question.

Table 1.1. Results of the experiment in Example 1.7

Time [min]	No. of bacteria	Time [min]	No. of bacteria
0	4.00×10^5	30	1.335×10^6
5	4.88×10^5	35	1.571×10^6
10	5.98×10^5	40	2.054×10^6
15	7.32×10^5	45	2.355×10^6
20	8.79×10^5	50	2.970×10^6
25	1.070×10^6		

MACSYMA-Session 1.4. How well does the function (1.6) represent the numerical results of Table 1.1? To check this, we enter Eq. (1.6) into MACSYMA.

c1. n(t):=4.0e5*exp(0.04*t)

d1. $n(t) := 400000.0\,\exp(0.04t)$

We can now compute $N(t)$ at any time t

c2. n(45)

d2.   2419860.0

The 11 values $N(0)$, $N(5)$, ..., $N(45)$, $N(50)$ can be computed simultaneously by

c3. makelist(N(5*t), t, 0, 10)

d3.   [400000.0, 488561.0, 596729.5, 728847.2, 890216.0, 1087310.0, 1087310.0, 1328050.0, 1622080.5, 1981210.5, 2419860.0, 2955620.0]

which can be compared with the values of N in Table 1.1. For larger times t, the function $N(t)$ returns higher and higher values. For example after 10 hours,

c4. n(600)

d4. $1.05956 \cdot 10^{16}$

MACSYMA's ordinary single-precision floating-point numbers can represent numbers with absolute value up to about 10^{38}. Since $N(t)$ increases exponentially, we resort to double-float numbers, which can represent numbers up to about 10^{306}, and define a double-float version of $N(t)$:

c5. nd(t) := 4.0d5*exp(0.04d0*t)

d5. $n\,d(t) := 400000.0\,\exp(0.04t)$

After 100 hours, we get the absurdly large number

c6. nd(6000)

d6. $6.80355105427028 \cdot 10^{109}$

Since resources necessary to sustain the bacteria (e.g., food, space) are finite, such large a population cannot be sustained. We conclude that the mathematical model (1.6) is not valid (i.e., does not describe the bacteria population as a function of time) if t is very large. You can plot the function $N(t)$ by the command plot(n(t),t,t_min,t_max). For example, $N(t)$ for $0 \le t \le 1000$:

c7. plot(n(t),t,0,1000,"Time","Population")$

Try this and other values of t_min, t_max.

Example 1.8. A ball is released, at time 0, from the top of a tower, 200 meters high, and falls to the ground, see Fig. 1.2a. The distance s traveled by the ball is a *quadratic function* of time,

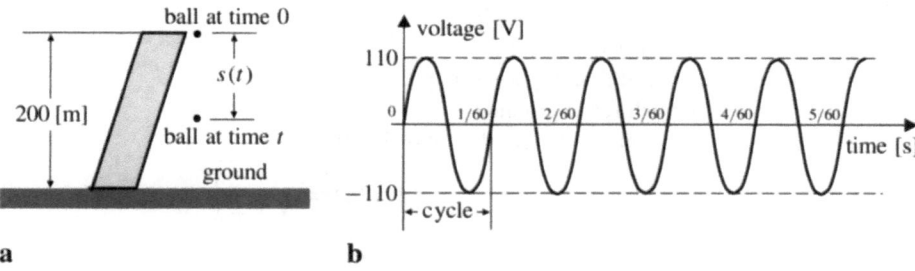

Fig. 1.2 a, b. Illustrations of Example 1.8, ball dropped from a tower (**a**), and Example 1.9, alternating voltage (**b**)

$$s(t) = (g/2)t^2 \, ,$$

where the time t is measured in seconds [s], the distance s is measured in meters [m], and $g \approx 9.81$ m/s^2 is the *gravity acceleration*.

Example 1.9. The alternating current, supplied by your friendly electrical power company, has voltage V – measured in volt [V] – which is a *sinusoidal function* of time. In the US, domestic voltage is given by

$$V(t) = 110\sin(120\pi t) \, , \tag{1.7}$$

where time t is measured in seconds. This function is depicted in Fig. 1.2b. See also Exercise 2.13.

Exercises

1.1 The grading system described in Example 1.4 has been changed and new grades were added. The new rule is

$$\text{new\_grade}(x) := \begin{cases} A & \text{if } 95 \le x \\ A- & \text{if } 91 \le x < 95 \\ B+ & \text{if } 85 \le x < 91 \\ B & \text{if } 80 \le x < 85 \\ B- & \text{if } 75 \le x < 80 \end{cases}$$

$$\text{new\_grade}(x) := \begin{cases} C+ & \text{if } 70 \le x < 75 \\ C & \text{if } 65 \le x < 70 \\ C- & \text{if } 60 \le x < 65 \\ D & \text{if } 50 \le x < 60 \\ F & \text{if } \qquad x < 50 \end{cases}$$

Adapt the function **grade(x)** of MACSYMA-Session 1.2 to compute the new grades.

1.2   In Example 1.8, where will the ball be
      (a) 1 s   (b) 6 s   (c) 10 s   (d) 100 s
      after its release? Recall that the tower in question is 200 m tall.
      (e) When will the ball reach ground?

1.3   Find $f(a)$, $f(\sqrt{a})$, and $f(1/a)$ for the following functions

(a) $f(x) := \dfrac{1}{x^2}$ (b) $f(x) := \sqrt{\dfrac{1-x}{1+x}}$

(c) $f(x) := x^2 + \dfrac{1}{x^2}$ (d) $f(x) := \sqrt{x^2 + 2x + 1}$

(e) $f(x) := \dfrac{x-a}{x+a}$ (f) $f(x) := (x+a)^4$

1.4   Express the area of a square in terms of its circumference.

1.5   Express the area of a circle in terms of its circumference.

1.6   Express the total surface area of a cube in terms of its volume.

1.7   Express the perimeter of an equilateral triangle in terms of its area.

1.8   The volume of a right circular cone is given by $V = \frac{\pi}{3} r^2 h$, and the lateral surface
      area is $A = \pi r h$, where h is the altitude of the cone and r the radius of the base. If
      the altitude is twice the diameter of the base, find the volume as a function of the
      area.

1.9   In the previous problem consider the total surface area S of the cone, that is, lateral
      area plus area of the base. Find the volume V as a function of S.

1.10 A pyramid with square base of sides $2a$ and altitude h has a volume $V = \frac{4}{3} h a^2$. The
      area of one of the lateral sides is given by $A = a\sqrt{h^2 + a^2}$. If the altitude is $1/2$ a
      side of the base, find A as a function of V. Repeat the problem when the ratio of
      altitude to a base-side is 1.

1.11 An isosceles triangle is inscribed in a circle of radius 1. If one of the equal sides of
      the triangle is a and the other side is b, find b as a function of a.

1.2 Functions and their graphs

The examples in Sect. 1.1 showed some special functions. In this section we give
formal definitions and study general properties which apply to all functions.

Definition 1.10. A function f is a rule which assigns (single) values $f(x)$ to
certain elements x. We say then that f is *defined* at x. This assignment is denoted
$x \mapsto f(x)$, or

$$y := f(x) , \qquad\qquad (1.8)$$

where
- x is called the *argument* or *independent* variable of the function f, and
- y is called the *dependent* variable of f.

The value $f(\xi)$ assigned by f to a particular ξ is called the *value* of f at ξ.[3]
A *real function* is a function f which has real arguments x and real values $f(x)$.

---

3 ξ, pronounced "ksi", is the Greek letter for "x". We use ξ here for particular values
of the variable x.

The function *SSN* of Example 1.3 has arguments which are not real numbers. The function *grade* of Example 1.4 has values which are not real numbers. Both these functions "flunk" the test of a "real function". All other functions in Sect. 1.1 are real functions.

It is important to distinguish between a *function*, denoted by f or $f(x)$ (where x is variable), and particular *values* such as $f(2.3)$ or $f(a)$ (where a is constant). The names used for the function variables are immaterial. For example,
-   $f(x) := x^2$ is the same function as $f(u) := u^2$ (independent variable denoted by u),
-   $y := x^2$ represents the same function as $v := u^2$.

Definition 1.11. The *domain* (also domain of definition) of a function f is the set of values over which it is defined.[4] We denote the domain of f by $D(f)$.

For example the domain $D(SSN)$ of the function *SSN* (Example 1.3) consists of all persons who were issued a social-security number.

Remark 1.12. Two functions, f and g, are called *equal*, denoted by $f = g$, if they have the same domain

$$D(f) = D(g) \, ,$$

and if $f(x) = g(x)$ for any x in their domain.

The function $g(x)$ of Example 1.2 and the MACSYMA function `sqrt(x)` (see MACSYMA-Session 1.1) have different domains,[5]

$$D(g) = [0, \infty) \ \text{(nonnegative numbers)} \, ,$$

$$\text{but} \quad D(\text{sqrt}) = (-\infty, \infty) \ \text{(all real numbers)} \, ,$$

and are considered different functions, even though $g(x) = \text{sqrt}(x)$ whenever x is nonnegative.

Definition 1.13. The *range* of a function f, denoted $R(f)$, is the set of values which f assumes over its domain,

$$R(f) := \{ \, y \mid y = f(x) \text{ for some } x \in D(f) \, \} \, . \tag{1.9}$$

Table 1.2 presents some real functions, their (natural) domains, and ranges. As usual, we denote by **R** the set of real numbers, $(-\infty, \infty)$. Check the statements in Table 1.2. Why is the point 0 excluded from both domain and range in row g?

---

4 That is, those values of x which have been assigned a value $f(x)$.

5 The square-root function in MACSYMA is even defined for complex numbers.

Table 1.2. Selection of real functions

Function f	Name	Domain $D(f)$	Range $R(f)$
a c (a real constant)	constant	$\mathbf{R}$	$\{c\}$
b x	identity	$\mathbf{R}$	$\mathbf{R}$
c x^2	square	$\mathbf{R}$	$[0, \infty)$
d x^3	cubic	$\mathbf{R}$	$\mathbf{R}$
e $+\sqrt{x}$	square root	$[0, \infty)$	$[0, \infty)$
f $-\sqrt{x}$	negative square root	$[0, \infty)$	$(-\infty, 0]$
g $\dfrac{1}{x}$	reciprocal	$(-\infty, 0) \cup (0, \infty)$	$(-\infty, 0) \cup (0, \infty)$
h $\|x\| := \begin{cases} x & \text{if } x \geq 0 \\ -x & \text{if } x < 0 \end{cases}$	absolute value	$\mathbf{R}$	$[0, \infty)$
i $\mathrm{sign}(x) := \begin{cases} 1 & \text{if } x > 0 \\ -1 & \text{if } x < 0 \end{cases}$	sign	$(-\infty, 0) \cup (0, \infty)$	$\{-1\} \cup \{1\}$

Definition 1.14. The *graph* of a real function f is the set of points

$$\mathrm{graph}(f) := \{ (x, y) \mid y = f(x), \, x \in D(f) \} .$$

Figure 1.3 a shows the graph of a real function f whose domain $D(f)$ is the interval $[a, b]$, and whose range $R(f)$ is $[c, d]$. Several of the real functions in Table 1.2 are illustrated in Fig. 1.4.

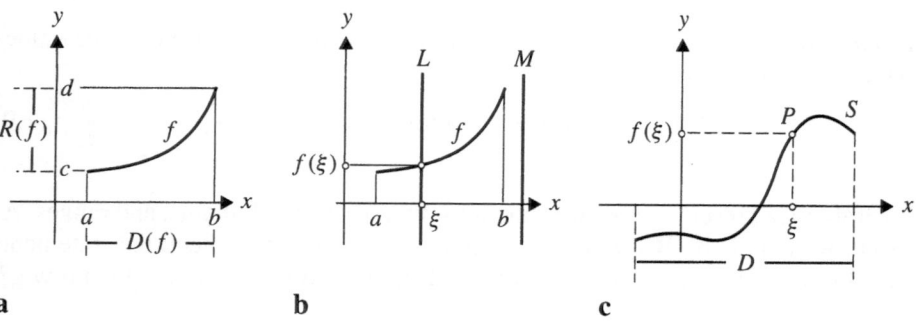

a   **b**   **c**

Fig. 1.3 a–c. Graph, domain, range, and the "vertical-line test". **a** The graph of f, its domain $D(f)$, and its range $R(f)$. **b** A vertical line intersects the graph of f in at most one point. **c** A set S satisfying the vertical-line test is the graph of a function f

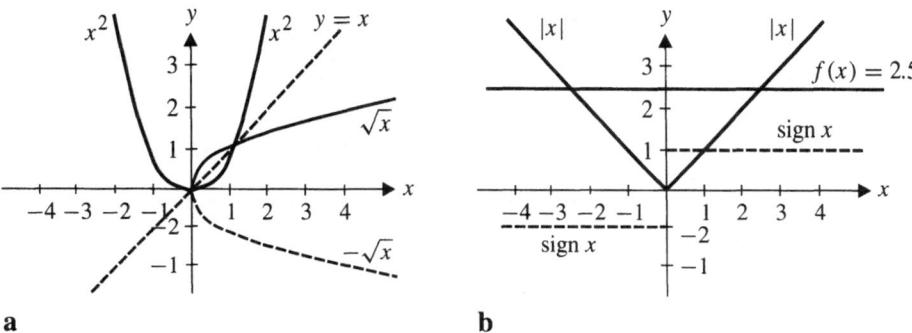

Fig. 1.4 a, b. Graphs of several real functions. **a** The identity function $y = x$, the square function $y = x^2$, and the two square root functions; **b** the constant function $f(x) = 2.5$, the absolute-value function $|x|$, and the sign function

Vertical-line test

A real function f assigns *at most one* value for each real ξ (one value if $\xi \in D(f)$, *none* if $\xi \notin D(f)$). This shows that any vertical line

$$x := \xi$$

meets the graph of f at most once. If $\xi \in D(f)$, then the vertical line $x := \xi$ meets the graph of f at the point $(\xi, f(\xi))$; see, e.g., the line L in Fig. 1.3 b. If $\xi \notin D(f)$, then the vertical line through $(\xi, 0)$ does not meet the graph, see, e.g., the line M in Fig. 1.3 b.

Definition 1.15. If a set of points S in the plane meets each vertical line at most once, we say that S satisfies the vertical-line test.

Such a set is illustrated in Fig. 1.3 c.
    We can now answer the question: Given a set of points S in the xy-plane, when is it the graph of a function $y = f(x)$?

Proposition 1.16. A nonempty set of points S in the xy-plane is a graph of a real function f

$$S = \{ (x, f(x)) \mid x \in D(f) \} \tag{1.10}$$

if and only if S satisfies the vertical-line test.

Proof. If S is a graph of a function f, then, as shown above, it satisfies the vertical-line test. Conversely, let a set S pass the vertical-line test, and define

$$D := \{ \xi \in \mathbf{R} \mid \text{the vertical line } x = \xi \text{ meets } S \} .$$

The set D is the domain of a function f satisfying (1.10). Indeed, for each $\xi \in D$, the vertical line $x = \xi$ intersects S at exactly one point P, whose y-coordinate is defined as $f(\xi)$, see Fig. 1.3 c. □

Table 1.3. Selected operations on functions (Definition 1.17)

	Name of function	Notation	Defined, for all $x \in D(f) \cap D(g)$, by
a	sum of f, g	$f + g$	$(f + g)(x) := f(x) + g(x)$
b	product of f, g	fg or $f \cdot g$	$(fg)(x) := f(x)g(x)$
c	quotient of f, g	$\dfrac{f}{g}$ or (f/g)	$\left(\dfrac{f}{g}\right)(x) := \dfrac{f(x)}{g(x)}$, provided $g(x) \neq 0$
d	maximum of f, g	$\max\{f, g\}$	$\max\{f, g\}(x) := \max\{f(x), g(x)\}$
e	minimum of f, g	$\min\{f, g\}$	$\min\{f, g\}(x) := \min\{f(x), g(x)\}$

Operations on functions

We can operate on real functions f, g to form new functions. Some of the more useful combinations are listed in Table 1.3.

Definition 1.17. Let f and g be real functions, with respective domains $D(f)$ and $D(g)$. We define the sum, product, etc., of f and g as in Table 1.3.

For a combined function, such as $f + g$ or fg, to be defined at x, both f and g must be defined at x. Therefore, all the functions in Table 1.3 have domain $D(f) \cap D(g)$, the intersection[6] of $D(f)$ and $D(g)$. If that intersection is empty, the definition is vacuous, i.e., no function has been defined.

MACSYMA-Session 1.5. The functions in Table 1.3 can be computed with MACSYMA. For example, the sum of the functions f and g
```
c1. f(x)  := x^2$
c2. g(x)  := sqrt(x)$
```
can be computed as
```
c3. s(x)  := f(x)+g(x)$
```
In particular,
```
c4. s(4)
d4.     18
```
The way the function s is defined above, f and g are determined at the time when s is applied to its argument, x. For example, if we redefine the function f right now, s will use the new function f instead of the one which existed when s was defined.
```
c5. f(x)  := x^3$
c6. f(4)
d6.     64
```
If we want the sum to be defined in terms of the functions f and g which are defined at the time when the sum is defined, then we must use the MACSYMA command `define`.
```
c7. define(ss(x), f(x)+g(x))
```
$$d7. \qquad ss(x) := x^3 + \sqrt{x}$$
```
c8. ss(4)
d8.     66
```

---

6 $D(f) \cap D(g) := \{x \mid x \in D(f) \text{ and } x \in D(g)\}$

Changing the definition of f after ss is defined does not affect ss.
c9. f(x):=x^2$
c10. ss(4)
d10.   66
The other operations in Table 1.3 can be carried out in MACSYMA for any two functions f and g as follows.
c11. p(x):= f(x)*g(x)$
c12. p(4)
d12.   32
c13. d(x):= f(x)/g(x)$
c14. d(4)
d14.   8
c15. mx(x):= max(f(x),g(x))$
c16. mx(4)
d16.   16
c17. mn(x):= min(f(x),g(x))$
c18. mn(4)
d18.   2
The absolute-value function can be defined in terms of the maximum function:
c19. a(x):= max(x,-x)$
c20. a(4)
d20.   4
c21. a(-4)
d21.   4

Another important operation on functions, not listed in Table 1.3, is given next.

Definition 1.18 (Composition). Let f and g be functions such that the range of f has some points contained in the domain of g. Then the *composition* of g and f, denoted by $(g \circ f)$ or $g(f)$, is defined by

$$(g \circ f)(x) := g(f(x)) .$$

For $g(f(x))$ to be defined it is necessary that f be defined at x, and g defined at $f(x)$. Therefore

$$D(g \circ f) = \{ x \in D(f) \mid f(x) \in D(g) \} , \tag{1.11}$$

in words: the domain of $(g \circ f)$ consists of those points $x \in D(f)$ such that $f(x) \in D(g)$.

Example 1.19. Let $f(x) := \sqrt{x}$, $g(x) := \sqrt{1 - x^2}$. Then

$$(g \circ f)(x) := g(f(x)) = \sqrt{1 - f^2(x)} = \sqrt{1 - (\sqrt{x})^2} = \sqrt{1 - x} .$$

With $D(f) = [0, \infty)$ and $D(g) = [-1, 1]$, it follows from (1.11) that

$$D(g \circ f) = [0, 1] \quad \text{(verify!)}$$

Note that the domain of the function $h(x) := \sqrt{1-x}$ is $(-\infty, 1]$, and therefore the functions h and $g \circ f$ are considered different.

MACSYMA-Session 1.6 (Composition of functions). The MACSYMA function `compose_functions(g,x,f) := limit(g,x,f)` builds the composite function $g \circ f$, assuming that g depends on the variable x. Here, `limit(g,x,a)` is the MACSYMA command for taking the limit of $g(x)$ as the variable x approaches the value a. We use `compose_functions` with the functions from Example 1.19. Indeed,
c1. `compose_functions(sqrt(1-x^2),x,sqrt(x))`
yields the function $\sqrt{1-x}$ again,
d1. $\quad \sqrt{1-x}$
c2. `remfunction(compose_functions)`

Implicitly defined functions

Two variables x and y are sometimes related by an equation like

$$F(x, y) = 0 , \qquad (1.12)$$

called an *implicit relation* of x and y.

The graph of the implicit relation (1.12) is the set of points which satisfy it

$$\text{graph}(F) := \{ (x, y) \mid F(x, y) = 0 \} .$$

We assume that graph(F) is nonempty, i.e., (1.12) holds for at least one pair (x, y).

MACSYMA-Session 1.7. The unit circle (see Fig. 1.6a)

$$x^2 + y^2 = 1 \qquad (1.13)$$

is the graph of the implicit relation $F(x, y) := x^2 + y^2 - 1 = 0$. Given a point $P(x, y)$ in the plane, does it belong to the unit circle? We can check this by substituting the coordinates (x, y) in the expression $F(x, y) := x^2 + y^2 - 1$. Define the function of two variables
c1. `f(x,y) := x^2 + y^2 - 1$`
The point $(1/2, 3/2)$ is on the unit circle because
c2. `f(1/2,sqrt(3)/2)`
d2. $\quad 0$
The point $(1/2, 1/2)$ is not on the unit circle because
c3. `f(1/2,1/2)`
d3. $\quad -\dfrac{1}{2}$

Given an implicit relation $F(x, y) = 0$ and a particular value $x := \xi$, it is often necessary to find a value (or values) η such that[7]

$$F(\xi, \eta) = 0 . \qquad (1.14)$$

---

7 η, pronounced "ate-uh," is the seventh letter in the Greek alphabet.

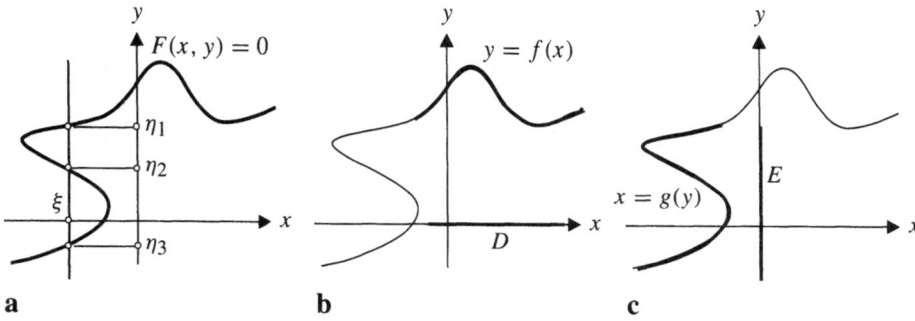

Fig. 1.5 a–c. Illustration of Definition 1.20. **a** The solutions of $F(x, y) = 0$ for y at $x := \xi$; **b** function $y = f(x)$ defined by $F(x, y) = 0$ in D; **c** function $x = g(y)$ defined by $F(x, y) = 0$ in E

Definition 1.20. Such η is denoted by $y(\xi)$ and is called a *solution* of $F(x, y) = 0$ *at $x = \xi$.* We say then that $F(x, y) = 0$ is solvable for (or with respect to) y at $x = \xi$. If there is a unique solution $y(\xi)$, we call $F(x, y) = 0$ *uniquely solvable* for y at $x = \xi$.

If $F(x, y) = 0$ is solvable for y at *all* points x in some set D, we denote such solution(s) by $y(x)$. If the solution is unique in D, we get a function $y = f(x)$ (1.8) defined in D. The points $\{(x, f(x)) \mid x \in D\}$ satisfy $F(x, y) = 0$, i.e., lie on graph(F) (see Fig. 1.5 b).

The same terminology applies when we solve $F(x, y) = 0$ "for x" at a particular $y = \eta$, giving a solution $x = \xi$ satisfying (1.14), or when we solve "for x" at all points y in some set E, giving a function of y, $x = g(y)$ (1.15) as illustrated in Fig. 1.5 c.

Definition 1.21. A function $y := f(x)$ is *defined implicitly* by $F(x, y) = 0$, or *implicit* in F, if the points $(x, f(x))$ satisfy the implicit relation (1.12).

Similarly, a function

$$x := g(y) \tag{1.15}$$

with points $(g(y), y)$ on graph(F) is the function of y *defined implicitly* by $F(x, y) = 0$.

Remark 1.22 (Geometric interpretation). For any ξ, a solution $y(\xi)$ of $F(x, y) = 0$ is the y-coordinate of a point where the vertical line $x = \xi$ intersects the graph of $F(x, y) = 0$ (see Fig. 1.5 a). This graph defines a function $y = f(x)$ in a set D if and only if it intersects exactly once with every vertical line $x := \xi$, for $\xi \in D$. See also Proposition 1.16.

The unit circle (1.13) "flunks" the vertical-line test (see line L in Fig. 1.6 b), and is therefore not the graph of an explicit function f. Indeed, solving (1.13) for y we get two solutions

$$y = \pm\sqrt{1 - x^2}, \quad -1 \le x \le 1,$$

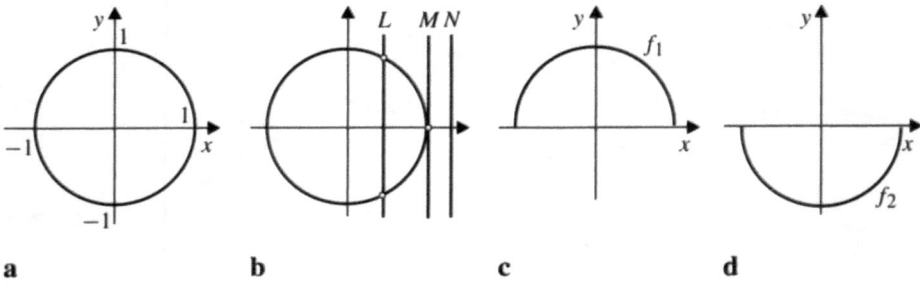

a   **b**   **c**   **d**

Fig. 1.6 a–d. Unit circle defining two explicit functions $y = f_1(x)$ and $y = f_2(x)$. **a** The unit circle is the graph of $x^2 + y^2 = 1$. **b** This graph fails the vertical-line test. **c** Upper semicircle. **d** Lower semicircle

and no solutions for $|x| > 1$. The "positive solution" is an explicit function

$$y = f_1(x) := +\sqrt{1 - x^2}, \quad -1 \le x \le 1,$$

whose graph is the upper semicircle (Fig. 1.6c). Similarly, the lower semicircle (Fig. 1.6d) is the graph of the "negative solution"

$$y = f_2(x) := -\sqrt{1 - x^2}, \quad -1 \le x \le 1.$$

The unit circle is the union of the graphs of f_1 and f_2.

Example 1.23. The solutions at x for the implicit relations $x^2 + y^2 - 1 = 0$ and $xy - 1 = 0$ with $x = 1/\sqrt{2}, 1, 2$ and $x = 2, -2, 0$, respectively, are presented in Table 1.4.

MACSYMA-Session 1.8. MACSYMA can be used to solve implicit relations $F(x, y) = 0$ by use of the `solve` command, as in `solve(F(x,y),y)`, which gives the solution(s) y as function(s) $y = y(x)$.
For example, we solve $F(x, y) := x^2 + y^2 - 1 = 0$ for y by using:
c1. `f(x,y):=x^2+y^2-1=0$`
c2. `y(x):=solve(f(x,y),y)$`

Table 1.4. Solutions of implicit relations of Example 1.23

Implicit relation	x	Solution at x
$x^2 + y^2 - 1 = 0$	$x := 1/\sqrt{2}$	$\pm 1/\sqrt{2}$ (2 solutions)
	$x := 1$	0
	$x := 2$	none
$xy - 1 = 0$	$x := 2$	$1/2$
	$x := -2$	$-1/2$
	$x := 0$	none

c3. y(x)

d3. $[y = -\sqrt{1-x^2}, y = \sqrt{1-x^2}]$

The solutions for $x = 1/\sqrt{2}$ are then obtained as

c4. y(1/sqrt(2))

d4. $\left[y = -\dfrac{1}{\sqrt{2}}, y = \dfrac{1}{\sqrt{2}}\right]$

which agrees with Example 1.23. If there are no real solutions, MACSYMA may yield complex solutions. For example, $F(x, y) := x^2 + y^2 - 1 = 0$ has no real solutions for y at $x = 2$, but we have

c5. y(2)

d5. $[y = -\sqrt{3}i, y = \sqrt{3}i]$

which are complex numbers. To solve $F(x, y) = 0$ for x as a function(s) of y, we use solve(F(x,y),x). For example, to solve $F(x, y) := xy - 1 = 0$

c6. f(x,y) := x*y-1 = 0$

c7. solve(f(x,y),x)

d7. $\left[x = \dfrac{1}{y}\right]$

we could also write

c8. solve(x*y-1,x)

d8. $\left[x = \dfrac{1}{y}\right]$

since the expression $x * y - 1$ is treated as $x * y - 1 = 0$ by solve.

Exercises

1.12 For each of the following functions $f(x)$, determine the domain $D(f)$ and the range $R(f)$.

(a) $f(x) = x^3 - x^2 - x + 1$ (b) $f(x) = \sqrt{1 + x^2}$

(c) $f(x) = \sqrt{1 - x^2}$ (d) $f(x) = \dfrac{1}{1 + x^2}$ (e) $f(x) = \dfrac{1}{1 - x^2}$

(f) $f(x) = \sqrt{9 - \sqrt{x}}$ (g) $f(x) = \dfrac{x}{x^2 + 1}$ (h) $f(x) = \dfrac{1}{\sqrt{x}}$

(i) $f(x) = \sqrt{1 + \dfrac{1}{x^2}}$ (j) $f(x) = \dfrac{1}{|x - 1|}$ (k) $f(x) = \dfrac{x}{1 - x^2}$

(l) $f(x) = \dfrac{1}{9 - \sqrt{x}}$

1.13 Use MACSYMA to plot the graphs of the functions in the previous exercise.

1.14 Use the MACSYMA command solve to find those values of x where the functions below take on the values 0, 6, and a.

(a) $f(x) := x^4 - 10$ (b) $f(x) := x^3 + 6x + 10$ (c) $f(x) := \sqrt{x^2 - 100}$

(d) $f(x) := x + \dfrac{1}{x}$ (e) $f(x) := \dfrac{x^2 - 1}{x^2 + 1}$ (f) $f(x) := x^3 + 2x^2 + 4x$

1.15 Using MACSYMA, plot the functions in the previous exercise. Verify the values of x where $f(x) = 0$.

1.16 For the following functions find those values of x such that $f(x) = 0$

(a) $f(x) := 4x - 7$ (b) $f(x) := 2x^2 + 3x + 10$

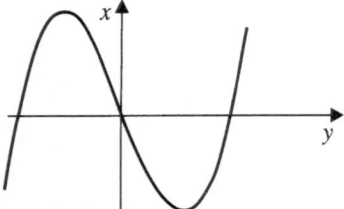

Fig. 1.7. Solution of (1.16) for x as function of y

(c) $f(x) := \sqrt{x^2 - 25}$ (d) $f(x) := 5 + \dfrac{1}{x}$ (e) $f(x) := \dfrac{x - 1}{x + 1}$

(f) $f(x) := x^3 + 2x^2 + x$

1.17 Use MACSYMA to plot the graphs of the functions:

(a) $x^2 + \sqrt{x}$ (b) $\max\{x^2, \sqrt{x}\}$ (c) $\dfrac{\sqrt{x}}{x^2 - 1}$

In each case, specify the domain and range.

1.18 Use MACSYMA to plot the graph of the function $V(t)$ of Example 1.9 with appropriate scaling.

1.19 A set S in the plane satisfies the horizontal-line test if every horizontal line intersects it at most once. Prove that the set S is the graph of a function $x = g(y)$ (1.15) if and only if it satisfies the horizontal-line test.

1.20 For each of the following implicit relations

(a) $2x + 3y - 1 = 0$ (b) $xy - 1 = 0$ (c) $x - y^2 = 0$

(d) $(x - y)^2 - 1 = 0$

determine if the graph satisfies

   i. the vertical-line test (in which case give the implicit function $y = f(x)$ defined by the graph),

   ii. the horizontal-line test (if "yes", give the implicit $x = g(y)$ defined by the graph).

1.21 Consider the implicit relation

$$y^3 - 3y - x = 0 . \tag{1.16}$$

   a. Does the graph satisfy the horizontal-line test?

   b. Solve for x as a function of y, and plot the graph with MACSYMA. You will get a picture like Fig. 1.2. Note that you now have the y-axis pointing to the right, and the x-axis pointing up. How will this graph look in the usual xy-coordinates system?

   c. Does the graph satisfy the vertical-line test?

   d. Use MACSYMA to solve (1.16) for y as a function of x. You will get three solutions (the MACSYMA output seems quite mysterious, and will make sense later, so accept it for now on faith). Plot all three solutions, and compare with the plot in b.

   e. Solve (1.16) for y, as a function of x, with the condition

$$y(0) = 0 .$$

What is the domain of the resulting function?

Hint: The condition $y(0) = 0$ is satisfied by exactly one of the three solutions in d.

1.22 Find the real solutions of the equation $\sqrt{x + 3 - 4\sqrt{x - 1}} + \sqrt{x + 8 - 6\sqrt{x - 1}} = 1$.

Hint: Determine the domain of the function $f(x) := \sqrt{x+3-4\sqrt{x-1}} + \sqrt{x+8-6\sqrt{x-1}} - 1$ and use MACSYMA to plot its graph.

1.23 Plot the graph of the implicit relation $||x| + |y-3| - 3| = 1$.

1.24 Consider the implicit relation (1.12),

$$F(x, y) = 0,$$

and substitute x/a for x where $a > 0$ is a given constant.
a. Show that the graph of (1.12) is "stretched" along the x-axis if $a > 1$.
b. What happens if $a < 1$?
c. Describe the behavior of the graph if we substitute y/b for b, where $b > 0$.
d. Plot the unit circle (see Appendix A, A.17).

1.25 Find the zeros of the following expression $\sqrt{x-1} + 1 - \sqrt{2\sqrt{x-1} + x}$.

1.3 Polynomials

Given a nonnegative integer n, and $n+1$ real numbers $a_0, a_1, \ldots, a_n$ with $a_n \neq 0$, the expression

$$p(x) = a_0 + a_1 x + a_2 x^2 + \cdots + a_n x^n = \sum_{k=0}^{n} a_k x^k, \qquad (1.17)$$

is called a (real) polynomial in the variable x (we use in (1.17) the convention $x^0 := 1$). The real numbers $a_0, a_1, \ldots, a_n$ are called the *coefficients* of the polynomial p, and the highest exponent n (with $a_n \neq 0$) is called its *degree*. The polynomial (1.17) is said to be of the nth *degree*. Some special cases are listed in Table 1.5.

Remark 1.24. Every polynomial p is defined on the entire real line, so $D(p) = \mathbf{R}$. Two polynomials

$$p(x) = a_0 + a_1 x + a_2 x^2 + \cdots + a_n x^n$$
$$\text{and} \quad q(x) = b_0 + b_1 x + b_2 x^2 + \cdots + b_n x^n$$

are therefore *equal* (see Remark 1.12) if $p(x) = q(x)$ for all $x \in \mathbf{R}$, see also Remark 1.31. It can be shown that polynomials are equal if and only if their

Table 1.5. Special cases of polynomials

Degree	Polynomial	Name
$n = 0$	$p(x) = a_0$	constant function, in particular, the zero polynomial $p(x) = 0$
$n = 1$	$p(x) = a_0 + a_1 x$	linear function
$n = 2$	$p(x) = a_0 + a_1 x + a_2 x^2$	quadratic polynomial
$n = 3$	$p(x) = a_0 + a_1 x + a_2 x^2 + a_3 x^3$	cubic polynomial

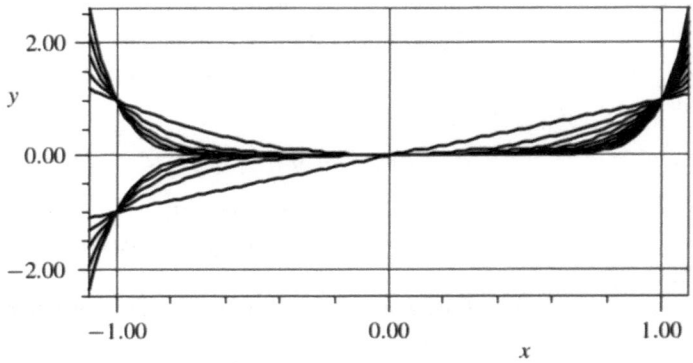

Fig. 1.8. First ten monomials for $-1.1 \leq x \leq 1.1$

coefficients agree, i.e.,

$$a_0 = b_0, \; a_1 = b_1, \; \ldots, \; a_n = b_n \; .$$

Polynomials of different degrees cannot be equal.

MACSYMA-Session 1.9. A *monomial* is a polynomial with a single term $p(x) = x^n$. To plot the first ten monomials we first make a list of them:

c1. `makelist(x^n,n,1,10)`

d1. $[x, x^2, x^3, x^4, x^5, x^6, x^7, x^8, x^9, x^{10}]$

then

c2. `plot2(''%,x,-1.1,1.1)`

to plot the 10 monomials for $-1.1 \leq x \leq 1.1$. See the result in Fig. 1.8.

Note the use of `''%` which means to read in the previous result, which is stored by the name `%`, into the current calculation. We must force the evaluation at read time by use of `''` (two single quotes) because the plotting commands do not evaluate their first argument. The graphs of all of these monomials have the common points $(0, 0)$ and $(1, 1)$. The higher the degree n, the quicker the growth of x^n for $x > 1$ and the "flatter" the graph near $x = 0$. Compare the graphs of the monomials x^n for n even and odd.

Example 1.25 (Evaluation of polynomials). The polynomial (1.17),

$$p(x) = a_0 + a_1 x + a_2 x^2 + \cdots + a_n x^n \; ,$$

can be rewritten as

$$p(x) = ((\ldots(a_n x + a_{n-1})x + a_{n-2})\ldots)x + a_2)x + a_1)x + a_0 \quad (1.18)$$

requiring n multiplications and n additions (check!). The original representation (1.17) requires more multiplications (how many?). For this reason, we prefer to use (1.18) when evaluating polynomials, specially if the degree n is large.

For example, the polynomial

$$p(x) := 11 - 8x + 7x^2 - 4x^3 + 7x^4 + 2x^5$$

is the same as

$$p(x) = ((((2x+7)x-4)x+7)x-8)x+11 .$$

Evaluating p at any point, say $x := \sqrt{3}$, requires 5 multiplications:

$$p(\sqrt{3}) = ((((2\sqrt{3}+7)\sqrt{3}-4)\sqrt{3}+7)\sqrt{3}-8)\sqrt{3}+11 = 95 - 2\sqrt{3} .$$

MACSYMA-Session 1.10. The representation (1.22) is computed (recursively) by the MACSYMA function

```
c1. polynomial(a,x) := if length(a) = 1 then part(a,1)
                       else part(a,1)+x*polynomial(rest(a),x)$
```

where:

a is the *list of coefficients* $[a_0, a_1, \ldots, a_n]$,

length(a) is the number of elements in the list a,

part(a,n) is the nth term in the list a,

rest(a,n) gives what is left of the list a after the first n elements are removed. Here $n = 1$ is the default value.

Note that checking the length on each pass is not very efficient. We can overcome that by testing for an empty list [ ] instead:

```
c2. polynomial(a,x) := if a=[ ] then 0
                       else part(a,1)+x*polynomial(rest(a),x)$
```

The polynomial in Example 1.25 has coefficients $ia := [11, -8, 7, -4, 7, 2]$ and is evaluated by means of:

```
c3. polynomial([11,-8,7,-4,7,2],x)
```
d3. $x(x(x(x(2x+7)-4)+7)-8)+11$

while the value at $x = \sqrt{3}$ is computed as

```
c4. polynomial([11,-8,7,-4,7,2],sqrt(3))
```
d4. $\sqrt{3}(\sqrt{3}(\sqrt{3}(\sqrt{3}(2\sqrt{3}+7)-4)+7)-8)+11$

which can be expanded:

```
c5. expand(%)
```
d5. $95 - 2\sqrt{3}$

where % refers to the previous result. If there is just one coefficient, the polynomial is constant. However, the coefficient must still appear as a list with one element, a_0, and not as the single number a_0 in our function. For example, polynomial([2],x) gives the constant polynomial $p(x) := 2$, but polynomial(2,x) does not work with our definition. (Why?)

Try this: the 10th-degree polynomial

$$p(x) := \sum_{k=0}^{10} (-1)^k x^k / k! = 1 - x + x^2/2! - x^3/3! + \cdots + -x^9/9! + x^{10}/10!$$

is computed by:

c6. `polynomial(makelist((-1)^k/k!,k,0,10),x)`

d6. $x \left(x \left(x \left(x \left(x \left(x \left(x \left(x \left(\left(\frac{x}{3628800} - \frac{1}{362880} \right) x + \frac{1}{40320} \right) - \frac{1}{5040} \right) + \frac{1}{720} \right) - \frac{1}{120} \right) + \frac{1}{24} \right) - \frac{1}{6} \right) + \frac{1}{2} \right) - 1 \right) + 1$

Roots and factors

Definition 1.26. A number ξ is a *root* (or zero) of the polynomial p if $p(\xi) = 0$.

We associate the name "root" with vegetables which are good for you (such as carrots) but are also hard to get (you extract them by pulling hard or by digging them out). This impression is correct also for roots of polynomials: They are necessary but usually hard to extract.

The simplest case is that of quadratic polynomials, where the roots are easily computed. Even if you are familiar with quadratic equations, you may still benefit from reading the next example.

Example 1.27 (Quadratic equations). The roots of a quadratic polynomial written as

$$q(x) = ax^2 + bx + c, \quad \text{with } a \neq 0, \tag{1.19}$$

are the solutions of the quadratic equation

$$ax^2 + bx + c = 0. \tag{1.20}$$

Completing (1.19) to a square

$$q(x) = ax^2 + bx + c = a\left(x^2 + \frac{b}{a}x\right) + c$$

$$= a\left(x^2 + \frac{b}{a}x + \frac{b^2}{4a^2}\right) + c - \frac{b^2}{4a} \tag{1.21}$$

$$= a\left(x + \frac{b}{2a}\right)^2 + \left(c - \frac{b^2}{4a}\right),$$

we rewrite (1.20) as

$$a\left(x + \frac{b}{2a}\right)^2 + \left(c - \frac{b^2}{4a}\right) = 0, \quad \text{or} \quad \left(x + \frac{b}{2a}\right)^2 = \left(\frac{b^2 - 4ac}{4a^2}\right).$$

Taking the square root, assuming $(b^2 - 4ac)$ is nonnegative,

$$x + \frac{b}{2a} = \pm\sqrt{\frac{b^2 - 4ac}{4a^2}} = \pm\frac{\sqrt{b^2 - 4ac}}{2a},$$

gives the two solutions,

$$x_1 = \frac{-b - \sqrt{b^2 - 4ac}}{2a}, \quad x_2 = \frac{-b + \sqrt{b^2 - 4ac}}{2a}, \tag{1.22}$$

illustrated in Fig. 1.9a.

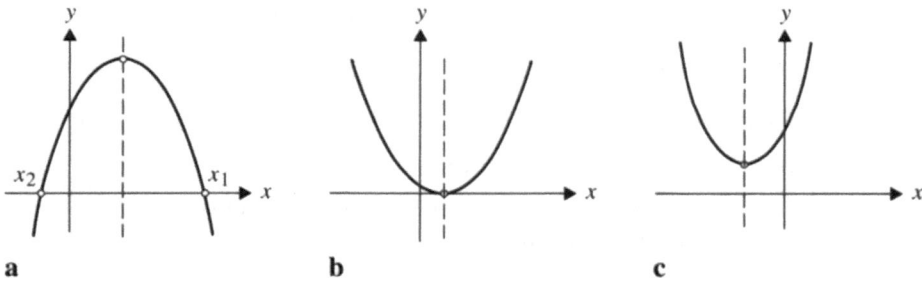

Fig. 1.9 a–c. Roots of a quadratic equation. **a** Two real roots; **b** one real root; **c** no real root

If $b^2 - 4ac = 0$, then the two roots (1.22) coincide (Fig. 1.9b).

If $b^2 - 4ac < 0$, there are no real roots, and the graph of (1.19) does not intersect the x-axis (Fig. 1.9c). However, the complex roots of (1.20) are

$$x_1 = -\frac{b}{2a} - i\frac{\sqrt{4ac - b^2}}{2a}, \quad x_2 = -\frac{b}{2a} + i\frac{\sqrt{4ac - b^2}}{2a}. \tag{1.23}$$

Definition 1.28. If $q(x)$ is a real polynomial of degree $n - 1$ and ξ is any real number, then

$$p(x) = (x - \xi)q(x) \tag{1.24}$$

is a real polynomial of degree n, and $(x - \xi)$ is called a *(linear) factor* of p.

In Definition 1.28 the polynomial p is "constructed" of q and the factor $(x - \xi)$. Often we have the opposite situation: A real polynomial $p(x) = a_0 + a_1 x + a_2 x^2 + \cdots + a_n x^n$ is given, and we need to factor it, that is, to write it as (1.24) for some number ξ and polynomial $q(x)$.

Factorization is important because it is equivalent to finding the roots of a polynomial.

Proposition 1.29. Given a real polynomial $p(x)$ of degree n, a number ξ is a root of p if and only if $p(x) = (x - \xi)q(x)$ (1.24), where $q(x)$ is a polynomial of degree $n - 1$.

Proof. If: Let $(x - \xi)$ be a linear factor of $p(x)$. Substituting $x := \xi$ in (1.24) shows that $p(\xi) = 0$.

Only if: Let ξ be a root of $p(x)$. We use the identity

$$x^k - \xi^k = (x - \xi)(x^{k-1} + \xi x^{k-2} + \cdots + \xi^{k-2} x + \xi^{k-1}) \tag{1.25}$$

which holds for all $x, \xi \in \mathbf{R}$ and all integers $k \geq 1$.

We conclude that

$$x^k - \xi^k = (x - \xi)q_{k-1}(x) \tag{1.26}$$

where $q_{k-1}(x)$ is a polynomial of degree $k - 1$.

Since ξ is a root of p, we subtract $p(\xi) = 0$ from $p(x)$ to get

$$p(x) = p(x) - p(\xi)$$
$$= (a_0 + a_1 x + a_2 x^2 + \cdots + a_n x^n) - (a_0 + a_1 \xi + a_2 \xi^2 + \cdots + a_n \xi^n)$$
$$= (a_0 - a_0) + a_1(x - \xi) + a_2(x^2 - \xi^2) + \cdots + a_n(x^n - \xi^n)$$
$$= (x - \xi)(a_1 + a_2 q_1(x) + a_3 q_2(x) + \cdots + a_n q_{n-1}(x)) ,$$
$$\text{using (1.26) for each } k ,$$
$$= (x - \xi)q(x) .$$

Here $q(x) := q_1(x) + q_2(x) + \cdots + q_{n-1}(x)$ is a polynomial of degree $n - 1$, the same degree as $q_{n-1}(x)$.   □

We can use Proposition 1.29 to prove

Lemma 1.30. Let $p(x)$ be a polynomial of degree $\leq n$. Then p has $n + 1$ distinct roots if and only if $p(x)$ is the zero polynomial, $p(x) \equiv 0$.

Proof. If: If p is the zero polynomial, it is zero everywhere, in particular at any $n + 1$ points.

Only if: We prove this by induction on n.

The statement is clearly true for $n = 1$ since a line $y = a_0 + a_1 x$ with two distinct roots must have $a_0 = a_1 = 0$.

We assume the statement true for $n - 1$. Let p be a polynomial of degree n with distinct roots $\xi_1, \xi_2, \ldots, \xi_{n+1}$. Then, by Proposition 1.29,

$$p(x) = (x - \xi_1)q(x), \quad \text{where } q \text{ is a polynomial of degree less than or equal to } n - 1.$$

The n roots $\xi_2, \ldots, \xi_{n+1}$ of p must also be roots of $q(x)$ (why?). Therefore q is the zero polynomial. This proves that p is zero.   □

Remark 1.31. Another way of stating Lemma 1.30 is:

Two polynomials $p(x)$ and $q(x)$ of degrees $\leq n$ are equal if and only if they agree (have same values) at $n + 1$ different points $\xi_1, \xi_2, \ldots, \xi_{n+1}$

$$p(\xi_k) = q(\xi_k), \quad k = 1, 2, \ldots, n + 1 .$$

Any $n + 1$ points $\{ (x_k, y_k) \mid k = 1, 2, \ldots, n + 1 \}$ with different x_k determine a unique polynomial of degree $\leq n$ passing through them.

If we factor the polynomial $q(x)$ in (1.24), we get

$$p(x) = (x - \xi_1)(x - \xi_2)r(x)$$

where $r(x)$ is a polynomial of degree $n - 2$. Continuing in this manner we finally write the polynomial (1.17),

$$p(x) = a_0 + a_1 x + a_2 x^2 + \cdots + a_n x^n ,$$

as a product of factors,

$$p(x) = a_n(x - \xi_1)(x - \xi_2) \ldots (x - \xi_n) ,$$

giving n roots $\xi_1, \xi_2, \ldots, \xi_n$ of p. The following example shows that this is not always possible.

Example 1.32. The quadratic equation

$$x^2 + x + 1 = 0 \qquad (1.27)$$

has no real roots, see Example 1.27. In other words, the quadratic polynomial $q(x) := x^2 + x + 1$ cannot be factored as

$$x^2 + x + 1 = (x - a)(x - b), \quad \text{with real } a \text{ and } b .$$

Such quadratic factors are called *irreducible*.

However, the quadratic equation (1.27) has complex roots

$$x_1 = -\frac{1}{2} - i\frac{\sqrt{3}}{2}, \quad x_2 = -\frac{1}{2} + i\frac{\sqrt{3}}{2}, \quad \text{by (1.23)},$$

giving the *complex factorization*

$$x^2 + x + 1 = (x - x_1)(x - x_2) = \left(x + \frac{1}{2} + i\frac{\sqrt{3}}{2}\right)\left(x + \frac{1}{2} - i\frac{\sqrt{3}}{2}\right).$$

It can be shown that every real polynomial $p(x)$ of degree n can be factored as the product

$$\begin{aligned} p(x) = a_n(x - \xi_1)(x - \xi_2) \ldots (x - \xi_l)(x^2 + b_1 x + c_1) \cdot \\ \cdot (x^2 + b_2 x + c_2) \ldots (x^2 + b_m x + c_m) \end{aligned} \qquad (1.28)$$

of *linear factors* $(x - \xi_1)$, $(x - \xi_2)$, $\ldots$, $(x - \xi_l)$,
and *irreducible quadratic factors* $(x^2 + b_1 x + c_1)$, $(x^2 + b_2 x + c_2)$, $\ldots$, $(x^2 + b_m x + c_m)$.
The degree of the polynomial (1.28) is $n = l + 2m$ (check!). Each of the irreducible quadratic factors $(x^2 + b_k x + c_k)$ in (1.28) can be factored further as

$$x^2 + b_k x + c_k = (x - (\alpha_k - i\beta_k))(x - (\alpha_k + i\beta_k))$$

where $\alpha_k - i\beta_k$ and $\alpha_k + i\beta_k$ are the *complex roots* of the quadratic equation

$$x^2 + b_k x + c_k = 0 .$$

This gives the *complex factorization* of the polynomial $p(x)$,

$$p(x) = \prod_{j=1}^{l}(x - \xi_j) \prod_{k=1}^{m}(x - (\alpha_k - i\beta_k))(x - (\alpha_k + i\beta_k))$$

listing all roots of p:

the l real roots $\xi_1, \xi_2, \ldots, \xi_l$ and

the $2m$ complex roots $\alpha_1 \pm i\,\beta_1$, $\alpha_2 \pm i\,\beta_2$, $\ldots$, $\alpha_m \pm i\,\beta_m$.

A real polynomial of degree n therefore has n roots, including the complex roots of its irreducible quadratic factors. This result is known as the "Fundamental Theorem of Algebra."

Obtaining a polynomial by multiplying its factors is called *expansion*.

MACSYMA-Session 1.11. MACSYMA can expand as well as factor. For example, factor the polynomial $x^2 - 1$ to get the factored form $(x - 1)(x + 1)$:

c1. `factor(x^2-1)`

d1. $(x - 1)(x + 1)$

giving the roots $x = 1$ and $x = -1$ of $x^2 = 1$. Expanding the last result, %, gives back the original polynomial:

c2. `expand(%)`

d2. $x^2 - 1$

The polynomial $x^3 - 3x^2 + 3x - 1$ is similarly factored to give

c3. `factor(x^3-3*x^2+3*x-1)`

d3. $(x - 1)^3$

showing that $x = 1$ is a triple root of $x^3 - 3x^2 + 3x - 1 = 0$.

By default, factorization is performed over rational numbers. Many polynomials fail to factorize completely when only rational numbers are permitted. For example, $x^2 + x - 1$ is irreducible over the rationals, since

c4. `p:=x^2+x-1`

d4. $x^2 + x - 1$

c5. `factor(p)`

d5. $x^2 + x - 1$

fails to find any factors.

There is an alternative way to factor a polynomial. We first use `solve` to find the roots of p,

c6. `solve(p)`

d6. $$\left[x = -\frac{\sqrt{5}+1}{2},\ x = \frac{\sqrt{5}-1}{2} \right]$$

map the right-hand sides (`rhs`) to get the list

c7. `map('rhs,%)`

d7. $$\left[-\frac{\sqrt{5}+1}{2},\ \frac{\sqrt{5}-1}{2} \right]$$

subtract that list from x to get list of factors of p,

c8. `x-%`

d8. $$\left[x + \frac{\sqrt{5}+1}{2},\ x - \frac{\sqrt{5}-1}{2} \right]$$

and finally multiply the factors

c9. `apply("*",%)`

d9. $$\left(x - \frac{\sqrt{5}-1}{2} \right)\left(x + \frac{\sqrt{5}+1}{2} \right)$$

which is the sought factored form of p. This approach is preferred to using `factor` alone, because `solve` tries `factor` first, and if `factor` fails to give a full factorization in linear terms, then `solve` switches to other appropriate methods.

Another example is

c10. `factor(x^3-1)`

d10. $(x-1)(x^2+x+1)$

which gives a partial factorization, while

c11. `solve(x^3-1)`

d11. $\left[x = \dfrac{\sqrt{3}\,i-1}{2}, \; x = -\dfrac{\sqrt{3}\,i+1}{2}, \; x = 1 \right]$

c12. `map('rhs,%)`

d12. $\left[\dfrac{\sqrt{3}\,i-1}{2}, \; -\dfrac{\sqrt{3}\,i+1}{2}, \; 1 \right]$

c13. `x-%`

d13. $\left[x - \dfrac{\sqrt{3}\,i-1}{2}, \; x + \dfrac{\sqrt{3}\,i+1}{2}, \; x - 1 \right]$

c14. `apply("*",%)`

d14. $(x-1)\left(x - \dfrac{\sqrt{3}\,i-1}{2} \right)\left(x + \dfrac{\sqrt{3}\,i+1}{2} \right)$

yields a full factorization. Note that `factor` and `solve` on high-order polynomials can be very time consuming, and will not always yield a useful result, such as `solve(x^6+4*x+1)`. With univariate polynomials, the `allroots` command may be used to find the approximate (complex) numerical roots as in

c15. `solve(x^6+4*x+1)`

d15. $[0 = x^6 + 4x + 1]$

c16. `allroots(x^6+4*x+1)`

d16. $\big[x = -0.25006052753622, \; x = -1.26254009243826,$
$x = 1.26075392527414\,i - 0.35695100744472,$
$x = -1.26075392527414\,i - 0.35695100744472,$
$x = 0.77814583055394\,i + 1.11325131743196,$
$x = 1.11325131743196 - 0.77814583055394\,i \big]$

c17. `apply("*",x-map('rhs,%))`

d17. $(x + 0.25006052753622)(x + 1.26254009243826)$
$(x - 1.26075392527414\,i + 0.35695100744472)$
$(x - 0.77814583055394\,i - 1.11325131743196)$
$(x + 0.77814583055394\,i - 1.11325131743196)$
$(x + 1.26075392527414\,i + 0.35695100744472)$

where the last step shows the factorization.

Exercises

1.26 Find the real roots of the following polynomials. Check your results by plotting the curves with MACSYMA,

(a) $x^4 + 4x^3 + 6x^2 + 4x + 1$ (b) $x^4 + 4x^3 + 6x^2 + 4x + 5$

(c) $x^4 + 4x^3 + 6x^2 + 4x - 15$ (d) $x^3 - 6x^2 + 12x - 8$

(e) $x^3 - 6x^2 + 12x - 12$ (f) $x^3 - 6x^2 + 12x$

1.27 Find any complex roots for the polynomials in the preceding problems. Recall that you have already found the real roots and use these to reduce the order of the given polynomial.

1.28 Sketch the graphs of the polynomials below. Check your results with MACSYMA.

(a) $p(x) := x^3$ (b) $p(x) := x^4$ (c) $p(x) := x^5$

(d) $p(x) := (x - 1)^6$ (e) $p(x) := x^2(x - 1)^2$

(f) $p(x) := (x - 1)x(x + 1)$ (g) $p(x) := (x - 1)^2 x(x + 1)^2$

(h) $p(x) := (x - 1)x^2(x + 1)$ (i) $p(x) := (x - 1)^2 x^2(x + 1)^3$

1.29 Use MACSYMA to plot the graph of $p(x) := x^2 - x - 3/4$. Determine from the graph the x-intervals for which the inequalities $p(x) < 0$ and $p(x) > 0$ hold. Solve these inequalities with MACSYMA: solve the expressions $x^2 - x - 3/4 < 0$ and $x^2 - x - 3/4 > 0$ and interpret the results.

1.30 In MACSYMA use the command product((x-1/k),k,1,10) and expand to get a polynomial of the 10th degree. What are its roots? Factor the polynomial to get back the factors $(x - 1/k)$, $k = 1, \ldots, 10$.

1.31 Explain the MACSYMA program polynomial(a,x) in Session 1.10.

1.32 Verify the identity (1.25), by expanding the right-hand side.

1.33 Find two *different* polynomials of the 4th degree which pass through the points

$$(-2, 3), \ (0, 3), \ (4, -2), \ (5, 1) \ .$$

Does this contradict Remark 1.31?

1.34 The graph of the quadratic polynomial (1.19), with $a \neq 0$,

$$y = ax^2 + bx + c \tag{1.29}$$

$$= a(x + b/2a)^2 + (c - b^2/4a), \quad \text{by (1.21)},$$

which suggests changing to new variables

$$\bar{x} := x + b/2a, \quad \text{a translation of } b/2a \ ,$$

$$\bar{y} := y - (c - b^2/4a), \quad \text{a translation of } b^2/4a - c \ ,$$

giving a "simpler" graph $\bar{y} = a\bar{x}^2$. We conclude that the general quadratic graph (1.29) is a translation of the graph of $y = ax^2$, to the point

$$\left(-\frac{b}{2a}, \ c - \frac{b^2}{4a} \right), \tag{1.30}$$

the origin of the $\bar{x}\bar{y}$-coordinate system. Therefore the graph of (1.29) is a *parabola* with axis parallel to the y-axis, and with vertex at the point (1.30) (see also Appendix D, D.3). This parabola "opens up" if $a > 0$ (see the graph of x^2 in Fig. 1.4 a), and "opens down" if $a < 0$ (see Fig. 1.9 a).

     Use MACSYMA to plot $y = ax^2 + bx + c$ for different values of a, b, and c. In each case verify that the vertex of the parabola is at (1.30).

1.35 We saw that the graph of $y = ax^2 + bx + c$ behaves like the graph of its "leading term" $y = ax^2$.

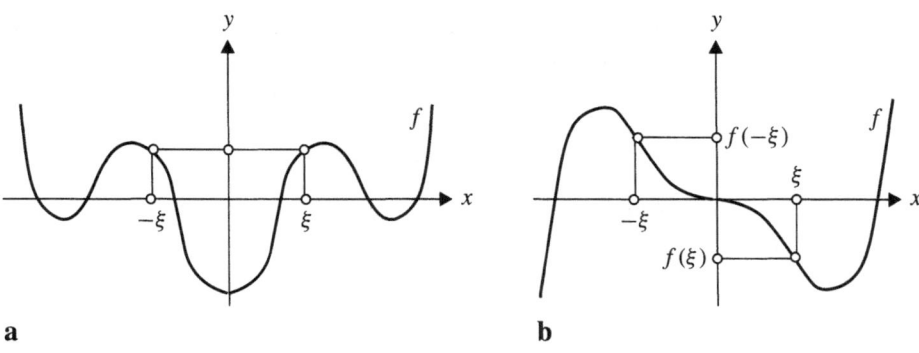

Fig. 1.10 a, b. Even and odd functions. **a** An even function; **b** an odd function

a. Is the same true for polynomials of higher degree? In other words, are the graphs of

$$y = a_n x^n + a_{n-1} x^{n-1} + \cdots + a_2 x^2 + a_1 x + a_0 x^n, \quad \text{with } (a_n \neq 0),$$
and $\quad y = a_n x^n$

similar? Show that for x sufficiently far from the origin the answer is "yes" in the sense that

$$\frac{a_n x^n + a_{n-1} x^{n-1} + \cdots + a_2 x^2 + a_1 x + a_0}{a_n x^n}$$

gets closer and closer to 1 as $|x|$ increases (in either direction).

b. Illustrate this for the 3rd-degree polynomial $0.1x^3 + 2x^2 - 5x + 1$ and its leading term $0.1x^3$. Use MACSYMA to compute their ratio for $x = 1000$ and $x = -1000$.

c. Use MACSYMA to plot the graphs of these polynomials sufficiently "far from the origin". This requires varying x-min and x-max in the command plot(expr,x, x-min,x-max).

1.36 We saw in MACSYMA-Session 1.9 that if n is even, the graph of $f(x) := x^n$ is *symmetric* with respect to the *y-axis*. We express this analytically as

$$f(-x) = f(x), \quad \text{for all } x . \tag{1.31}$$

Indeed, $(-x)^n = (-1)^n x^n = x^n$ if n is even. We call any function (not just a monomial) which satisfies (1.31) an *even function*.

If n is odd, the monomial $f(x) = x^n$ satisfies

$$f(-x) = -f(x), \quad \text{for all } x . \tag{1.32}$$

Any function f satisfying (1.32) is called *odd*. The graph of an odd function is *symmetric* with respect to the *origin* in the following sense: If the point $(\xi, f(\xi))$ is on the graph, so is its "reflection" with respect to the origin $(-\xi, -f(\xi))$.

Even and odd functions are illustrated in Fig. 1.10.

For each of the following functions determine if it is even, odd, or neither.

(a) $\sin x$ (b) $\cos x$ (c) $\tan x$ (d) $\exp x$

1.37 An arbitrary function f with domain[8] $D(f) = \mathbf{R}$ can be written as the sum of an even function f_{even} and an odd function f_{odd}. These two functions are called the *even* and the *odd part* of f and are constructed, for each x, as follows

$$f_{\text{even}}(x) := \frac{f(x) + f(-x)}{2}, \quad f_{\text{odd}}(x) := \frac{f(x) - f(-x)}{2}.$$

This construction will be used later, e.g., to define the hyperbolic functions.
 Find the even and odd parts of the following functions:

(a) $f(x) := 1 - 5x + x^2$

(b) $f(x) := \sum_{k=0}^{6} (-1)^k x^k$

(c) $f(x) := \dfrac{1}{1+x^2}$

(d) $f(x) := a_0 + a_1 x + a_2 x^2$

(e) $f(x) := \sum_{k=0}^{n} a_k x^k$

(f) $f(x) := \dfrac{1}{1-x}$

Plot the functions c, d, e, and f and their odd and even parts.

1.38 Show that if x_0 is a zero of the polynomial $p(x) = a_0 + a_1 x + \cdots + a_n x^n$ then $p(x)/(x - x_0)$ is a polynomial of degree $n - 1$.

1.39 Examine the sum, difference, product, and ratio of even and odd functions. What kind of symmetry have the resulting functions?

1.40 Show that the function $\sum_{k=0}^{n} a_k x^k \cdot \sum_{k=0}^{n} (-1)^k a_k x^k$ is even.

1.41 Use MACSYMA to plot the function $f(x) := x(x - 1)(x - 2)(x + 1)(x + 2)$. What kind of symmetry does f have? Can you see this from the factored definition of f? Or from its expanded form?

1.42 Often we can guess a factor $(x - \xi_1)$ of a given polynomial $p(x)$. Dividing, we get $q(x) = p(x)/(x - \xi_1)$ and compute the factorization (1.24). Try to factor
(a) $p(x) = x^3 - x^2 - x + 1$

(b) $p(x) = x^3 - 2x^2 + x$
(c) $p(x) = x^4 + 2x^2 - 4x + 1$

(d) $p(x) = 12x^3 + 36x^2 + 15x - 18$
(e) $p(x) = x^4 + x^3 - 2x^2 - 6x - 4$

(f) $p(x) = x^3 + 4x^2 + 5x + 2$

1.43 Use MACSYMA to factor the polynomials in Exercise 1.42.

1.44 Show (using completion of a square, as in Example 1.27) that the equation

$$x^2 + ax + y^2 + by + c = 0 \tag{1.33}$$

represents a *circle* (in particular, a point which is a circle with a radius of 0) or has no real graph. Show that the radius r of the circle is given by

$$r = \frac{\sqrt{a^2 + b^2 - 4c}}{2}.$$

Test if the following equations correspond to circles, calculate their center and radii and plot them with MACSYMA.
(a) $x^2 + x + y^2 + y = 0$

(b) $2x^2 - x + 2y^2 - 2y - 3 = 0$

---

8 Instead of $\mathbf{R}$, the domain $D(f)$ may be an interval I which is symmetric with respect to 0.

(c) $x^2 + 4x + y^2 - 4y - 3 = 0$ (d) $3x^2 - 20x + 3y^2 - 4y + 3 = 0$

(e) $x^2 + y^2 = 0$ (f) $x^2 - 2x + y^2 + 2y + 2 = 0$

1.45 The graph of the implicit equation $y^2/b^2 + x^2/a^2 = 1$ gives a stretched circle, called an *ellipse*. Plot the ellipses corresponding to the equation for $b := 1$, and the values $a := 1/2, 1, 2,$ and 3.

1.46 Produce an equation analogous to Eq. (1.33) for a general ellipse generated by translation in the direction of the axes.

1.47 Use MACSYMA to find the numbers a, b, and c of the general parabola $ax^2 + bx + c$ that are given by the condition that the three points (x_1, y_1), (x_2, y_2), and (x_3, y_3) lie on its graph.

     Use the result to calculate the parabola whose graph meets the points

(a) $(-1, 1)$, $(0, 0)$, and $(1, 1)$ (b) $(-1, 0)$, $(0, 0)$, and $(1, 1)$

(c) $(-1, -1)$, $(0, 0)$, and $(1, 1)$ (d) $(0, 0)$, $(k, 0)$, and $(1, 1)$,

                                               for $k \in \mathbf{R}\backslash\{1\}$.

1.48 Factor the expressions

(a) $n^4 + 4$ (b) $a^{10} + a^5 + 1$

1.49 Give approximations of the real zeros of

(a) $x^3 + 2x^2 + 10x - 20$ (b) $x^3 - 5x^2 - 10x + 1$

1.4 Rational functions

Polynomials are constructed by the arithmetic operations of addition, subtraction, and multiplication. Allowing also division, one gets the (real) rational functions $r(x)$, which are ratios

$$r(x) := \frac{p(x)}{q(x)} \tag{1.34}$$

of two real polynomials $p(x)$ and $q(x)$. We denote the degree of the numerator $p(x)$ by n and the degree of the denominator $q(x)$ by m.

     Factors common to the numerator and denominator can be cancelled. For example,

$$\begin{aligned} r(x) = \frac{p(x)}{q(x)} &= \frac{x^7 - 3x^6 + 3x^5 - 4x^4 + 5x^3 - 3x^2 + 3x - 2}{x^4 - 2x^3 + x^2 - 4x + 4} \\ &= \frac{(x-1)^2(x-2)(x^2+1)(x^2+x+1)}{(x-1)(x-2)(x^2+x+2)} \\ &= \frac{(x-1)(x^2+1)(x^2+x+1)}{x^2+x+2} . \end{aligned}$$

Also, if $n > m$, then we may divide p by q to get

$$r(x) = p_0(x) + \frac{p_1(x)}{q_1(x)} , \tag{1.35}$$

where $p_0(x)$, $p_1(x)$ and $q_1(x)$ are polynomials, and the degree of p_1 is less than that of $q_1(x)$.

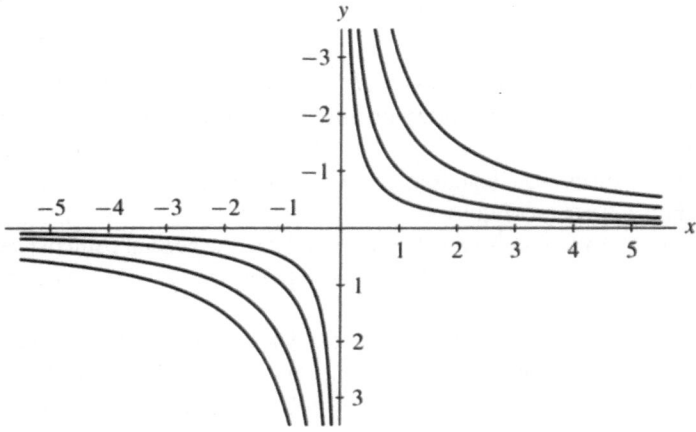

Fig. 1.11. Graphs of $f(x) = a/x$ for $a = 1/2, 1, 2$, and 3

For example,

$$r(x) = \frac{(x-1)(x^2+1)(x^2+x+1)}{x^2+x+2}$$

$$= x^3 - x^2 + 1 - \frac{x+3}{x^2+x+2} = p_0(x) + \frac{p_1(x)}{q_1(x)},$$
(1.36)

where p_1 is of degree 1 and the degree of q_1 is 2.

We may therefore assume
- the degree of the numerator n is less than the degree m of the denominator,
- there are no common factors in p and q.

The *degree* of the rational function p/q is defined as the degree of the polynomial q. While polynomials are defined for all $x \in \mathbf{R}$, rational functions are not defined at the roots of the denominator $q(x)$, which are called *poles* of the rational function.

Example 1.33. Consider the rational function

$$r(x) := \frac{ax+b}{cx+d},$$
(1.37)

where a, b, c, and d are real numbers, and $c \neq 0$. Then

$$r(x) = \frac{ax+b}{cx+d} = \frac{a}{c} + \frac{b/c - ad/c^2}{x+d/c} = A + \frac{B}{x-C}$$
(1.38)

has a pole at $x = C = -d/c$. The graph of (1.38) is a hyperbola centered at $(C, -A) = (-d/c, -a/c)$.

Figure 1.11 shows the behavior of hyperbolae $r(x) = a/x$ near the pole $x := 0$.

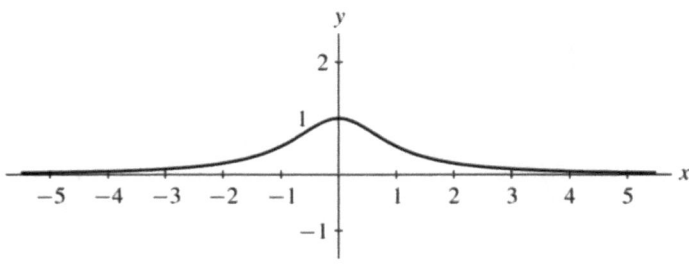

Fig. 1.12. Graph of the rational function (1.39)

Example 1.34. **The rational function**

$$r(x) = 1/(1 + x^2) \tag{1.39}$$

has no poles and is consequently defined throughout **R**. Its graph is shown in Fig. 1.12. Compare with the graphs of the hyperbolae in Fig. 1.11.

Next we see how MACSYMA handles rational functions.

MACSYMA-Session 1.12. Consider the expression $(1 + x)/((1 - x)(2 + x)(x - 3))$:

c1. `(1+x)/((1-x)*(2+x)*(x-3))`

d1. $\dfrac{x + 1}{(1 - x)(x - 3)(x + 2)}$

Performing a partial fractions decomposition of this gives

c2. `partfrac(%,x)`

d2. $\dfrac{1}{15(x + 2)} + \dfrac{1}{3(x - 1)} - \dfrac{2}{5(x - 3)}$

representing the function as the sum of lower-degree rational functions. Partial fractions decompositions are further studied in Sect. 10.3, where they are applied in integration. Now plot the rational function `(1+x)/((1-x)*(2+x)*(x-3))`. There are poles at $x := -2, x := 1$, and $x := 3$. c3. `plot2((1+x)/((1-x)*(2+x)*(x-3)),x,-10,10)`

g3.   2D Graphic - Bounds: -10. < X < 10.0 -3.5 < Y < 5.8

d3.   done

Exercises

1.50 Sketch the graphs of the following rational functions.

(a) $r(x) := \dfrac{x - 1}{x + 1}$

(b) $r(x) := \dfrac{(x - 1)^2}{(x + 1)^2}$

(c) $r(x) := \dfrac{x(x - 1)}{x + 1}$

(d) $r(x) := \dfrac{x - 1}{x(x + 1)(x - 2)}$

(e) $r(x) := \dfrac{(x - 1)^2}{x + 1}$

(f) $r(x) := \dfrac{x^2}{x^2 - 1}$

(g) $r(x) := \dfrac{x^4 - 4x^3 + 6x^2 - 4x + 1}{x^4 + 1}$

(h) $r(x) := \dfrac{x^2 - 1}{x^4 - 4x^3 + 6x^2 - 4x + 1}$

(i) $r(x) := \dfrac{x^3 - 1}{x^2 + 1}$

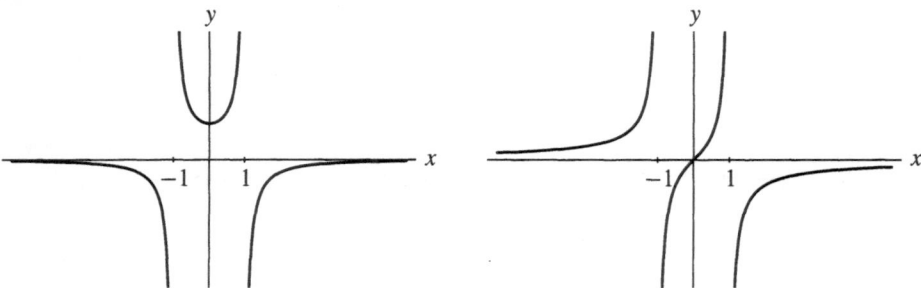

Fig. 1.13. What rational functions have these graphs?

1.51 Use the MACSYMA command `partfrac` to obtain the partial fractions decompositions for the rational functions above. Use this information to once more sketch the
functions, and compare with the previous results.

1.52 Use MACSYMA to see the graphs of the rational monomials $r_n(x) := 1/x^n$ for
$n = 1, \ldots, 10$. Note that they all have $(1, 1)$ as a common point.

1.53 Do the rational functions $r_n(x) := 1/(x - a)^n$, where $n = 1, 2, \ldots$ pass through a
common point? If so, what are the coordinates of this point?

1.54 Plot the graph of $f(x) = (x^2 + 2x + 1)/(x^2 - 1)$. Note that the denominator of f
vanishes at the two points -1 and 1. Why does f only have one pole, at $x = 1$?

1.55 Suppose $x = 1$. Then obviously $x = (1 - x) + 1$. Dividing by $(1 - x)$ we get

$$\frac{x}{1 - x} = \frac{1 - x}{1 - x} + \frac{1}{1 - x} = 1 + \frac{1}{1 - x},$$

and by subtracting $x/(1 - x)$ from both sides,

$$0 = 1 + \frac{1}{1 - x} - \frac{x}{1 - x} = 1 + \frac{1 - x}{1 - x} = 2,$$

"proving" that $0 = 2$. What is wrong with this "proof"?

1.56 For the rational function $f(x) = (1 + x^2)/(x^3 - 2x - 1)$,
   a. compute the poles, and guess the shape of the graph of f,
   b. plot the graph,
   c. compute the partial fractions decomposition.

1.57 Verify Eq. (1.38) (a) with paper and pencil, (b) with MACSYMA.

1.58 Consider the graphs of two, unknown, rational functions in Fig. 1.13. Guess these
rational functions, then use MACSYMA to plot your guesses. Repeat if necessary,
until you solve the mystery.

1.59 Use the MACSYMA function `if` to write a function symmetry(f, x) that gives
   –   the answer "odd" if f is odd with respect to x,
   –   the answer "even" if f is even with respect to x, and
   –   the answer "unknown symmetry" if none of the above symmetries is found for f
       with respect to x.

   Use the function to determine the symmetry of several rational examples. *Hint:* You
   can nest one `if` function inside another.

1.5 Inverse functions

Recall the function *SSN* of Example 1.3. If no number is assigned to more than one person, then the social-security numbers can be used to identify individuals. In this case we have an inverse function

$$SSN^{-1}(social\text{-}security\ number) := \text{person}$$

which identifies the individual with a given number. The domain of SSN^{-1} is the set of numbers already assigned, which is the range of the function *SSN*. This works fine if, for all persons and numbers,

$$SSN^{-1}(SSN(person)) = \text{same person} ,$$

$$SSN(SSN^{-1}(number)) = \text{same number} .$$

In general, given a function f with domain $D(f)$ and range $R(f)$, it is sometimes possible to find a function $\phi(x)$ such that $\phi(f(x)) = x$ for all $x \in D(f)$. Such a function provides an *inverse mapping* to f, as it takes the point $f(x) \in R(f)$ back onto the point $x \in D(f)$:

$$x \mapsto f(x) \mapsto \phi(f(x)) = x .$$

To make these ideas precise we need

Definition 1.35. The functions f and ϕ are *inverses* of each other, a fact denoted by $f = \phi^{-1}$ and $\phi = f^{-1}$, if

$$D(f) = R(\phi), \quad R(f) = D(\phi) , \tag{1.40a}$$
$$\phi(f(x)) = x, \quad x \in D(f) , \tag{1.40b}$$
$$\text{and} \quad f(\phi(y)) = y, \quad y \in R(f) . \tag{1.40c}$$

A function f for which an inverse function exists, is called *injective*.

> The inverse function f^{-1} is not the same as the reciprocal function $1/f(x)$. The inverse of $y = f(x)$, where x is the independent variable, is a function $f^{-1}(y)$, with y being the independent variable.
> Since the names of the variables are immaterial, we often write the inverse as $y = f^{-1}(x)$, with x being the independent variable.

How is the inverse function computed? If the function f has an inverse f^{-1}, we compute it by solving

$$f(x) = y, \quad \text{for all } y \in R(f) , \tag{1.41}$$

to get a unique solution

$$x := \phi(y) . \tag{1.42}$$

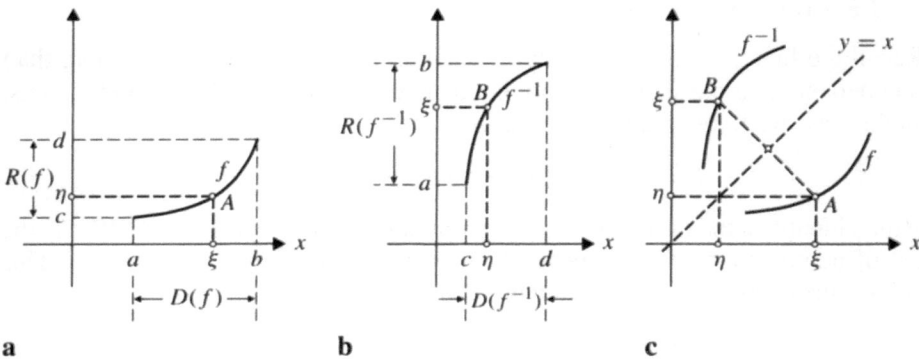

Fig. 1.14 a–c. Graphs of f and f^{-1} being symmetric with respect to the line $y = x$. **a** The graph of f, its domain $D(f)$, and its range $R(f)$. **b** Point B on the graph of f^{-1} corresponds to point A on the graph of f. **c** Point B is the reflection of point A with respect to the line $y = x$

The function ϕ is the sought inverse function. To prove this we must check every item in Definition 1.35,

$$R(\phi) = D(f) \quad \text{and} \quad D(\phi) = R(f) \quad \text{by Definitions 1.11 and 1.13}$$
$$\phi(f(x)) = x \quad \text{and} \quad f(\phi(y)) = y \quad \text{by (1.41) and (1.42)}$$

This is illustrated in Fig. 1.14 a, where we solve (1.41) for the point $y := \eta \in R(f)$, to get the solution $x := \xi \in D(f)$. This gives a point $B = (\eta, \xi)$ on the graph of f^{-1}, i.e., $\xi = f^{-1}(\eta)$ (see Fig. 1.14 b).

If for some $y \in R(f)$ Eq. (1.41) does not have a unique solution x, then f does not have an inverse.

Remark 1.36. The point $B = (\eta, \xi)$ and the point $A = (\xi, \eta)$ (on the graph of f) are reflections of each other with respect to the line $y = x$ (see Fig. 1.14 c). We conclude:

The graphs of f and f^{-1} are symmetric with respect to the line $y = x$.

MACSYMA-Session 1.13. Rewriting (1.41) as an implicit relation $f(x) - y = 0$, we can compute the inverse f^{-1} by solving this equation for x as a function of y, as in Session 1.8. Therefore the function

```
c1. inverses(f,x,y) := solve(f-y,x)$
```

gives the solution x as function(s) of y. If there is only one solution, it is the sought inverse. For example, the inverse of $f(x) = ax + b$, (with $a \neq 0$), is:

```
c2. inverses(a*x+b,x,y)
```

$$\text{d2.} \quad \left[x = \frac{y - b}{a} \right]$$

The inverse of $y = x^3$ gives the three solutions

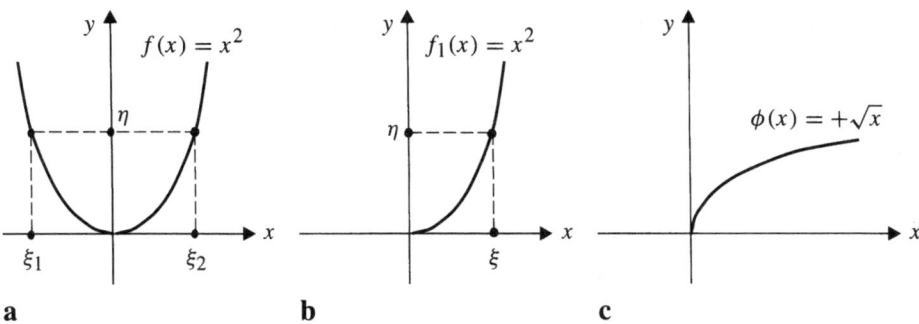

Fig. 1.15 a–c. The need to restrict the domain. **a** Function $f(x) = x^2$ in its natural domain; **b** restricting the domain to nonnegative x; **c** inverse function of f_1

c3. `inverses(x^3,x,y)`

d3. $\left[x = \dfrac{(\sqrt{3}\,i - 1)\sqrt[3]{y}}{2}, \; x = -\dfrac{(\sqrt{3}\,i + 1)\sqrt[3]{y}}{2}, \; x = \sqrt[3]{y} \right]$

the last of which is the sought inverse function $x = y^{1/3}$, and the first two (complex) solutions are irrelevant for our purpose. What is the inverse of $f(x) = x^2$? Computing

c4. `inverses(x^2,x,y)`

d4. $\left[x = -\sqrt{y}, \; x = \sqrt{y} \right]$

we get the two solutions. In fact the function $f(x) := x^2$ does not have an inverse, since solving $\xi^2 = \eta$, for any $\eta > 0$, gives two solutions $\xi_1 = -\sqrt{\eta}$ and $\xi_2 = +\sqrt{\eta}$ (see Fig. 1.15). For positive η, however, both of these are so-called *local inverses* of f (see also Example 1.37).

Since `solve(f-y,x)` gives a list of solutions, we may decide to select always the first one. The function

c5. `inverse_1(f,x,y):=first(solve(f-y,x))$`

does this. Some examples of its use:

c6. `inverse_1(a*x+b,x,y)`

d6. $x = \dfrac{y - b}{a}$

which has only one solution

c7. `inverse_1(x^3,x,y)`

d7. $x = \dfrac{(\sqrt{3}\,i - 1)\sqrt[3]{y}}{2}$

the first of the three solutions is selected

c8. `inverse_1(x^2,x,y)`

d8. $x = -\sqrt{y}$

The last two results show that the first solution (or for that matter any arbitrarily selected solution) is not always a natural choice.

The MACSYMA command `taylor_solve` is useful for finding a power series solution of the inverse, as in

c9. `inverse(f,x,y):=taylor_solve(f,x,y,0,5)$`

which uses powers through y^5 to approximate $x(y)$. Try this function on the previous three examples.

Try all three inverse functions `inverses`, `inverse_1`, and `inverse` on such favorites as $f(x) = \sin(x)$, $f(x) = \text{asin}(x)$, and $f(x) = \text{atan}(x)$. Note any warning messages.

Example 1.37. The functions $f(x) = x^2$ and $\phi(x) = +\sqrt{x}$ are not inverses, since

$$D(f) = \mathbf{R} \neq R(\phi) = [0, +\infty) \, .$$

We can correct the situation by restricting the domain $D(f)$. Consider the function

$$f_1(x) := x^2, \quad \text{with domain } D(f_1) = [0, +\infty)$$

(see Fig. 1.15b). Now for every $\eta \geq 0$ there is a unique solution ξ of

$$f_1(\xi) = \eta \, .$$

The inverse of f_1 is the function $\phi(x) := +\sqrt{x}$ (see Fig. 1.15c). Note that the graphs of f_1 and ϕ are symmetric with respect to $y = x$.

We recall (Exercise 1.19) that a set S in the plane satisfies the horizontal-line test, if any horizontal line intersects the set S at most once. This test gives a simple rule to determine if a function f has an inverse.

Proposition 1.38. A function f has an inverse f^{-1} if and only if its graph satisfies the horizontal-line test.

Indeed, the graphs of f and f^{-1} are symmetric with respect to the line $y := x$, see Remark 1.36. This symmetry means that:

$$\text{graph } f \text{ satisfies the horizontal-line test}$$

$$\text{if and only if} \quad \text{graph } f^{-1} \text{ satisfies the vertical-line test.}$$

Definition 1.39 (Monotone functions). A function f defined in an interval $[a, b]$ is called

- nondecreasing if $f(u) \leq f(v)$
- increasing if $f(u) < f(v)$
- nonincreasing if $f(u) \geq f(v)$
- decreasing if $f(u) > f(v)$

whenever $a \leq u < v \leq b$.

A function is called monotone in $[a, b]$ if it is one of the above four types. A function is *strictly monotone* in $[a, b]$ if it is either increasing or decreasing there.

Example 1.40. If f is strictly monotone (increasing or decreasing) throughout its domain, then its graph satisfies the horizontal-line test, and f has an inverse. Examples of functions that are strictly monotone throughout $\mathbf{R}$:
(a) $f(x) := x^3$ (b) $f(x) := \sqrt[3]{x}$ (c) $f(x) := e^x$ (d) $f(x) := e^{-x}$

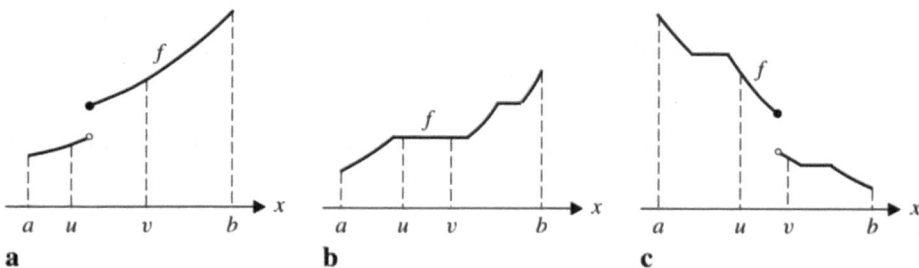

Fig. 1.16 a–c. Illustration of monotone functions. **a** Increasing function; **b** nondecreasing function; **c** nonincreasing function

Example 1.41. Does the inverse of $f(x) := x^m$, for m being a nonnegative integer, exist? The answer depends on whether m is even or odd.

If m is odd, then x^m is increasing throughout $\mathbf{R}$ (check!), and its inverse exists.

If m is even, then every horizontal line $y = \eta$ with $\eta > 0$ intersects its graph twice (check!). As in Example 1.37 we define a new function

$$f_1(x) = x^m, \quad \text{restricted to } [0, \infty), \quad \text{where } x^m \text{ is increasing}$$

which has an inverse.

Either way, the inverse function is called the *m*th *root*, and denoted by $\sqrt[m]{x}$ or $x^{1/m}$, with the understanding that the domain

$$D(x^{1/m}) = \begin{cases} \mathbf{R} & \text{if } m \text{ is odd,} \\ [0, \infty) & \text{if } m \text{ is even.} \end{cases}$$

Particular examples are the *square* root $x^{1/2}$ (denoted simply by $\sqrt{x}$), and the *cube* root $x^{1/3}$ or $\sqrt[3]{x}$.

Definition 1.42 (The power x^a). We define x^a for positive x and any real number a. The definition has three stages.

a. Integer a, say $a = m$. If m is nonnegative, the monomial x^m is defined for all x. If $m < 0$, we define

$$x^m := \frac{1}{x^{-m}}, \quad \text{except at the pole } x := 0 \,. \tag{1.43}$$

b. Rational a, say $a = m/n$, for integers m and n. We define

$$x^{m/n} := (x^{1/n})^m = (\sqrt[n]{x})^m \tag{1.44}$$

restricted to positive x.

Indeed, if n is even, then the definition of $\sqrt[n]{x}$ requires x be nonnegative. However, the point $x = 0$ may be a pole, see (1.43). Therefore, the restriction to positive x is essential if $x^{m/n}$ is to be defined for all integers m and n.

c. Real a. Given a positive x and a rational number $r = m/n$, we defined x^r. We can now define x^a for any real a, using the fact that real numbers can be approximated arbitrarily close by rationals (see Appendix B, B.2).

Any real number a can be approximated by nested intervals with rational endpoints $[r_k, s_k]$

$$r_1 < r_2 < \cdots < r_n < r_{n+1} < \cdots \quad \cdots < a < \cdots < s_{n+1} < \tag{1.45}$$
$$< s_n < \cdots < s_2 < s_1$$

with $s_k - r_k \to 0$ as $k \to \infty$. Moreover, we can assume that all numbers in (1.45) have the same sign, positive or negative.
Therefore, if $a > 0$,

$$x^{r_1} < x^{r_2} < \cdots < x^{r_n} < x^{r_{n+1}} < \cdots \quad \cdots < x^{s_{n+1}} < x^{s_n} < \cdots < x^{s_2} < x^{s_1}$$

and we define x^a as the intersection of the nested intervals $[x^{r_k}, x^{s_k}]$.
Similarly, if $a < 0$,

$$x^{s_1} < x^{s_2} < \cdots < x^{s_n} < x^{s_{n+1}} < \cdots \quad \cdots < x^{r_{n+1}} < x^{r_n} < \cdots < x^{r_2} < x^{r_1}$$

and x^a is defined as the intersection of the nested intervals $[x^{s_k}, x^{r_k}]$.

See also Exercise 1.66.

MACSYMA-Session 1.14. The power $32^{3/5}$ is computed by
c1. `32^(3/5)`
d1.   8
which is $(32^{1/5})^3$ or $(2)^3$ which is 8, or $((32)^3)^{1/5}$ which is $(32768)^{1/5}$ which is 8. To compute $(-32)^{3/5}$ which is $(-32768)^{1/5}$ which is -8, we do
c2. `(-32)^(3/5)`
d2.   8
We can get the principal, rather than the real, branch of rational exponents by resetting some switches. `m1pbranch:false` and `domain:real` can be set to `true` and `complex`, respectively, to obtain the principal branch:
c3. `(m1pbranch:true, domain:complex, (-32)^(3/5))`
d3. $8\,e^{3i\pi/5}$
The power $\pi^{\sqrt{2}}$ is computed as
c4. `%pi^sqrt(2)`
d4. $\pi^{\sqrt{2}}$
and can be numerically approximated as
c5. `%,numer`
d5.   5.04749

Given a real number a, the function

$$f(x) := x^a, \quad \text{defined for all positive } x ,$$

is called a *power function*.

Proposition 1.43 (Power rules). For any positive x and y, and any real a and b,

$$x^a \cdot x^b = x^{a+b} , \qquad (1.46a)$$

$$x^{a \cdot b} = (x^a)^b , \qquad (1.46b)$$

$$x^a \cdot y^a = (x \cdot y)^a . \qquad (1.46c)$$

Proposition 1.44. For any real a, the inverse of the function $f(x) = x^a$ (defined in $(0, \infty)$) is

$$f^{-1}(x) = x^{1/a} .$$

Exercises

1.60 Give an example of a function f, defined on $[0, 1]$, whose graph satisfies the horizontal-line test (and therefore f^{-1} exists, by Proposition 1.38), but f is not strictly monotone in $[a, b]$, see Definition 1.39.

1.61 An implicit relation

$$F(x, y) := \sum_{j=0}^{m} \sum_{k=0}^{n} a_{jk} x^j y^k = 0 , \qquad (1.47)$$

with real coefficients a_{jk} is called *algebraic*. Note that $F(x, y)$ is here a polynomial in both variables x and y.

A function $y = f(x)$ implicitly defined by (1.47) is called an *implicit* algebraic function of x. To get such a function in an interval D, Eq. (1.47) must be uniquely solvable for y at all $x \in D$.

Note that all functions we already have met are algebraic. A rational function r of the variable x is the solution of the equation $q(x)y - p(x) = 0$ for some polynomials p and q. The square root function is one of the particular solutions of the equation $y^2 - x = 0$.

1.62 Use MACSYMA to plot the graphs of the following algebraic functions.
(a) $x^3 + y^3 = 3xy$ (b) $x^2 + y^4 - y^2 = 1$
(c) $y^m = x$ for $m = 2, \ldots, 5$ (d) $(x^2 + y^2)^2 = 9(x^2 - y^2)$
(e) $x^4 - 4x^3 + 3x^2 + 2x^2 y^2 = y^2 + 4xy^2 - y^4$

1.63 Given below are implicit algebraic relations $F(x, y) = 0$. For each relation, attempt to find a function $y = f(x)$ implicitly defined by it. Such a function may be defined differently for different intervals on the x-axis.
(a) $x^2 - y^2 = 1$ (b) $y^2 = x^2$ (c) $y^2 = x^4$
(d) $y^2 + x^3 = 1$ (e) $x^2 - xy + y^2 = 1$ (f) $x^2 - 2xy + y^2 = 1$

1.64 Prove Proposition 1.43 if a and b are rational.

1.65 Prove Proposition 1.44.

1.66 The irrational number $\sqrt{2}$ is approximated by the two rational numbers $1.414 < \sqrt{2} < 1.415$.
     a. Use MACSYMA to verify that $\pi^{1.414} < \pi^{\sqrt{2}} < \pi^{1.415}$.
     b. Find two rational numbers r and s such that $r < \sqrt{2} < s$, and $\pi^s - \pi^r < 10^{-6}$.
Conclude that $\pi^{\sqrt{2}}$ is approximated well by either π^s or π^r.

1.67 Determine which pairs of functions below are inverses, and for which respective domains they are inverses. Here a, b, c, and d are nonzero constants. In part i assume that $ad \neq bc$.

(a) $f(x) := ax$, $g(x) := \dfrac{1}{a}x$

(b) $f(x) := ax + 1$, $g(x) := \dfrac{1}{a}x - 1$

(c) $f(x) := a(x + 1)$, $g(x) := \dfrac{1}{a}x - 1$

(d) $f(x) := x^4$, $g(x) := x^{1/4}$

(e) $f(x) := x^3$, $g(x) := x^{1/3}$

(f) $f(x) := \sqrt{x^3}$, $g(x) := x^{2/3}$

(g) $f(x) := \dfrac{1-x}{1+x}$, $g(x) := \dfrac{1+x}{1-x}$

(h) $f(x) := \dfrac{1-x}{1+x}$, $g(x) := \dfrac{1-x}{1+x}$

(i) $f(x) := \dfrac{ax + b}{cx + d}$, $g(x) := \dfrac{-dx + b}{cx - a}$

1.68 For the following functions determine whether there exists an inverse function defined on some interval. If there is an inverse, determine its range and domain.

(a) $f(x) := \dfrac{a-x}{b+x}$

(b) $f(x) := x^5$

(c) $f(x) := x^2 + 1$

(d) $f(x) := x^2 + 2x + 1$

(e) $f(x) := x^2 - x + 3$

(f) $f(x) := x^3 - 4x - 15$

(g) $f(x) := \dfrac{1-x^2}{1+x^2}$

(h) $f(x) := x^3 + 3x^2 + 3x + 1$

(i) $f(x) := x^4 - 4x^3 + 6x^2 - 4x + 1$

(j) $f(x) := x + \dfrac{1}{x}$

(k) $f(x) := x^2 + \dfrac{1}{x^2}$

(l) $f(x) := \left(x^2 + 1\right)^{1/2}$

(m) $f(x) := x^2 + x + 1$

(n) $f(x) := x^2 - x$

(o) $f(x) := \left(x^2 + x\right)^{1/3}$

1.69 Consider the functions in the previous exercise for which an inverse exists, and use MACSYMA to plot each such function and its inverse. By Remark 1.36, these graphs are reflections of one another w.r.t. the line $y = x$. Check if this is actually so, and if not, explain.

2 Elementary functions used in calculus

This chapter is a brief introduction to the exponential, trigonometric, and hyperbolic functions, and their inverses. These functions, together with the polynomial and rational functions of Chap. 1, are used throughout calculus.

You may already be familiar with some, or all, of these functions. Either way, you can skip this chapter at first, and return to it when you need its results.

2.1 Exponential and logarithmic functions

In this section we recall two elementary functions, namely, the exponential function and the logarithmic function. We assume you are somewhat familiar with these functions and give here a brief summary.

In Definition 1.42 we defined x^a for all positive x and real a. Reversing the roles of a and x, we similarly define a^x for all positive a and all $x \in \mathbf{R}$. For a fixed positive $a \neq 1$, we thus have

$$f(x) := a^x, \quad \text{defined for all } x \in \mathbf{R}, \tag{2.1}$$

called the *exponential function to base a*.

❙ The exponential function a^x should not be confused with the power function x^a.

The function $f(x) := a^x$ has domain $D(f) = \mathbf{R}$ and range $R(f) = (0, \infty)$ (check!). It is *strictly monotone* throughout $\mathbf{R}$ (strictly increasing if $a > 1$, strictly decreasing if $0 < a < 1$). Therefore its inverse $f^{-1}(x)$ exists. By Definition 1.35, the inverse function satisfies

$$f(f^{-1}(y)) = a^{f^{-1}(y)} = y, \quad \text{for all } y \in R(f) = [0, \infty).$$

For positive y, the value x such that $a^x = y$ is called the logarithm of y to the base a and denoted by $\log_a y$. Consequently, the inverse function of a^x is called the *logarithmic function to the base a* and denoted by $\log_a(x)$ or $\log_a x$.

The basic exponentiation rules, and the corresponding rules for logarithms, are listed in Table 2.1.

Proposition 2.1. For any positive a, u, v and real x, y, the basic exponent and logarithm rules are those presented in Table 2.1.

If we know the logarithm $\log_a x$ to some base a, we can compute the logarithm

Table 2.1. Exponent and logarithm rules for any positive a, u, v and real x, y (Proposition 2.1)

	Exponent rule	Logarithm rule *
a	$a^0 = 1$	$\log_a 1 = 0$
b	$a^1 = a$	$\log_a a = 1$
c	$a^x \cdot a^y = a^{x+y}$	$\log_a(uv) = \log_a u + \log_a v$
d	$a^x/a^y = a^{x-y}$	$\log_a(u/v) = \log_a u - \log_a v$
e	$(a^x)^y = a^{x \cdot y}$	$\log_a(u^y) = y \log_a(u)$

* Each logarithm rule is obtained from the corresponding exponent rule by using $a^x = u$, $a^y = v$, or equivalently, $x = \log_a u$, $y = \log_a v$.

$\log_b x$ to *any other* base b as follows

$$\log_b x = \log_b a \ \log_a x .$$ (2.2)

Indeed

$$x = a^{\log_a x} = \left(b^{\log_b a}\right)^{\log_a x}$$
$$= b^{\log_b a \cdot \log_a x}, \quad \text{proving (2.2)} .$$

While logarithms $\log_a x$ can be defined for any positive base $a \neq 1$, the only bases used in practice are the three listed in Table 2.2.

The number[1] e is called the *base of the natural logarithms* or the *Euler*[2] *constant*. It is a transcendental number (see Appendix B, B.2), and therefore has a representation as a nonrepeating decimal. For many applications the 6-digit approximation

$$e \approx 2.71828$$ (2.3)

suffices.

Table 2.2. Logarithms defined for base $a = 10, e, 2$

a	Name of logarithms $\log_a$	Notation	User
10	common logarithms	$\log_{10}$	everyone
e	natural logarithms	ln, $\log_e$, or just log	mathematicians, scientists, engineers
2	logarithm to base 2	$\log_2$	computer scientists, see, e.g., Exercise 2.4

---

[1] The significance of this number can be appreciated by reading the story in Example 3.30.

[2] Leonhard Euler (1707–1783).

MACSYMA-Session 2.1. The number e can be entered in MACSYMA as %e.
The first 50 digits of e are computed with *bigfloat numbers*, and setting the number of significant digits to 50,[3]
c1. bfloat(%e), bfprecision:50
d1.   2.7182818284590452353602874713526624977572470937

If the base a in a^x or $\log_a x$ is unspecified, it is taken as the natural base e. Thus *exponential function* (without specifying base) always means e^x, which is also denoted by

$$\exp(x) := e^x .$$

Similarly, *logarithm* (without specifying base) always means the natural logarithm $\ln x$, also denoted $\log x$.

MACSYMA-Session 2.2. We use MACSYMA to reproduce the various rules in Proposition 2.1. Many of the simplification rules are applied by default. For example,
c1. a^0
d1.   1
c2. a^x * a^y
d2. a^{y+x}
c3. a^x / a^y
d3. $\dfrac{1}{a^{y-x}}$
We can change the display of this result by setting the MACSYMA option variable exptdispflag to false
c4. exptdispflag:false$
c5. a^x / a^y
d5. a^{x-y}
c6. reset(exptdispflag)$
To get rule e of Table 2.1, we need to tell MACSYMA that $a > 0$
c7. assume(a>0)$
c8. (a^x)^y
d8. $a^{x\,y}$
The corresponding logarithmic transformation is performed by default, even when we do not explicitly tell MACSYMA that $a > 0$.
c9. log(a^x)
d9. $\log(a)\,x$
We can suppress this automatic logarithmic transformation by setting the MACSYMA option variable logexpand:false.
c10. log(a^x), logexpand:false
d10. $\log a^x$
MACSYMA works internally with natural logarithms, written as $\log(x)$. To specify a logarithm to the base a, enter logb(a,x). This logarithm simplifies as follows.

---

3 The commands for big float, double float, and single float when executed print the exponents in the forms "d5" and "b5" for 10^5 in double-float and big-float notation. We have eliminated this when it does not seem to be necessary.

c11. `logb(a,x)`

d11. $\dfrac{\log x}{\log a}$

c12. `logb(a,1)`

d12.   0

MACSYMA expands $\log(xy)$ to $\log(x) + \log(y)$ only when we tell it to do so, by setting the MACSYMA option variable `logexpand:all`.[4]

c13. `log(x*y), logexpand:all`

d13. $\log y + \log x$

The expansion $\log(x/y) = \log(x) - \log(y)$ is applied by MACSYMA only if we set the option variable `logexpand:super`.

c14. `log(x/y), logexpand:super`

d14. $\log x - \log y$

These expansions are reversed by the MACSYMA command `logcontract`.

c15. `logcontract(log(x) + log(y))`

d15. $\log(xy)$

c16. `logcontract(log(x) - log(y))`

d16. $\log\left(\dfrac{x}{y}\right)$

By the way, it is often helpful to display a list of all the assumed facts which we have added to the MACSYMA computation environment.

c17. `facts()`

d17. $\{a > 0\}$

It is also often convenient to know how to remove assumptions from the MACSYMA computation environment. This is done with the `forget` command.

c18. `forget(x > 0, y > 0, a > 0)`

d18. $\{\text{not\_assumed}, \text{not\_assumed}, a > 0\}$

Exercises

2.1 Consider the mathematical model (1.5) for the number of bacteria $N(t)$ at time t.

   a. Let $T > 0$ be a constant. At any time t, we compare the population at times t and $t + T$ by $N(t + T)/N(t)$. Prove that, for any t, $N(t + T)/N(t) = e^{\lambda T}$, independently of t!

   b. Conclude that there is a constant $T_2 > 0$ (called the doubling time) such that $N(t + T_2)/N(t) = 2$, for all t, and compute T_2.

2.2 Find the following logarithms (here $\log_b a$ denotes the logarithm of a to the base b).

   (a) $\log_2 8$ (b) $\log_2(1/16)$ (c) $\log_2 1024$

   (d) $\log_{10} 1000$ (e) $\log_{1000} 10$ (f) $\log_{49} 7$

2.3 Why is the number $b = 1$ useless as a base of logarithms?

2.4 Given a positive integer n, its *binary representation* is

$$n = b_k\, b_{k-1}\, \ldots\, b_1\, b_0 \qquad\qquad (2.4)$$

where the b_i's are 0 or 1 (called binary digits, or bits), and $n = \sum_{i=0}^{k} b_i\, 2^i$. For

---

4 The two normal settings `logexpand:false` and `logexpand:true` do not cause this expansion to occur.

example, $19 = 1 \times 2^4 + 0 \times 2^3 + 0 \times 2^2 + 1 \times 2^1 + 1 \times 2^0$, and its binary representation is therefore:
$$10011.$$

Prove: The number of bits needed to represent a positive integer n is the smallest integer which is greater than $\ln_2 n$.

Note: This may explain why logarithms to base 2 are popular in computer science.

2.5   The representation to base 3 of a positive integer n is

$$n = t_m\, t_{m-1} \ldots t_1\, t_0, \quad \text{where } t_i = 0, 1 \text{ or } 2, \text{ and } n = \sum_{i=0}^{m} t_i\, 3^i . \qquad (2.5)$$

For example, the base 3 representation of 19 is: 2 0 1.

a. How many digits t_i are needed in the representation (2.5) of a positive integer n?

b. Approximate the ratio $\dfrac{\text{number of digits in (2.5)}}{\text{number of bits in (2.4)}}$ for large n.

2.6   Recalling that $\ln x$ is only defined for $x > 0$, whereas $\exp x$ is defined for all real values of x, simplify the following composite functions and state where the result is valid. Check your work with MACSYMA.

(a) $e^{\ln x^3}$ (b) $e^{\ln(-x)}$ (c) $\ln(1/x^3)$

(d) $\exp(\ln(1 + x))$ (e) $\ln(1 - \exp(x^2))$

2.7   In Proposition 2.1, verify each logarithm rule from the corresponding exponent rule, and vice versa.

2.8   Give the real solutions of the equations

(a) $e^{2x} + e^x - \ln(e^2) = 0$ (b) $u^{x-2} = v^{x+3}$ for $u, v \in \mathbf{R}^+$

Give the result for $u := 100$, $v := 10$.

2.9   Use MACSYMA to plot the even and odd parts of the exponential function $\exp x$. These functions will be considered in Sect. 2.4. Can you imagine how their inverses will look? We consider these inverses in Sect. 2.5.

2.10 Determine whether the following functions have inverses if defined on a suitable domain. When this is the case, give the domains and ranges for the original function and its inverse.

(a) $f(x) := e^{1/x}$ (b) $f(x) := e^{x^2}$ (c) $f(x) := e^{1/x^2}$

(d) $f(x) := e^{x+1}/x$ (e) $f(x) := \ln(1/x)$ (f) $f(x) := \ln x^2$

(g) $f(x) := \frac{1}{2}(e^x - e^{-x})$ (h) $f(x) := \ln x^2$ (i) $f(x) := \ln\dfrac{x-1}{x+1}$

(j) $f(x) := \frac{1}{2}(e^x + e^{-x})$ (k) $f(x) := \dfrac{e^x - e^{-x}}{e^x + e^{-x}}$ (l) $f(x) := \dfrac{1}{e^x - e^{-x}}$

2.11 Consider the functions, in the previous exercise, which have an inverse in some suitable domain. Use MACSYMA to plot each such function, and its inverse. Are the graphs the reflections of each other w.r.t. the line $y = x$?

2.2 Trigonometric functions

We assume that you are familiar with the trigonometric functions, $\sin\theta$, $\cos\theta$ and $\tan\theta$. These are defined for $0 \le \theta \le 2\pi$ in Appendix C, C.4, by the unit circle (see Fig. C.13). The reciprocals of these functions are often used and deserve their own names:

$$\cot \theta := \frac{1}{\tan \theta}, \quad \text{the cotangent}, \tag{2.6a}$$

$$\sec \theta := \frac{1}{\cos \theta}, \quad \text{the secant}, \tag{2.6b}$$

$$\csc \theta := \frac{1}{\sin \theta}, \quad \text{the cosecant}. \tag{2.6c}$$

These reciprocals are added to the list of trigonometric functions, see also Exercise 2.12.

Angles are measured in radians [rad] or degrees [°], where (see Appendix C, C.3)

$$2\pi \,[\text{rad}] = 360° . \tag{2.7}$$

Given a central angle θ defined by a point $P(x, y)$ on the unit circle (see Fig. 2.1), the 6 trigonometric functions are therefore:

$$\cos \theta = x, \quad \sin \theta = y, \quad \tan \theta = z = y/x \text{ (see Exercise 2.24)},$$
$$\sec \theta = 1/x, \quad \csc \theta = 1/y, \quad \cot \theta = 1/z = x/y.$$

Having defined the trigonometric functions $f(\theta)$ for all[5] $0 \leq \theta \leq 2\pi$, we extend these functions to all real θ by the definition

$$f(\theta) := f(\theta + 2\pi) . \tag{2.8}$$

Explanation: A full rotation (2π [rad] or 360°) around the circle brings us to the same point. For this reason we cannot distinguish between the angles θ and $\theta + n2\pi$, the latter represents n full rotations (counterclockwise if n is positive, clockwise if n is negative).

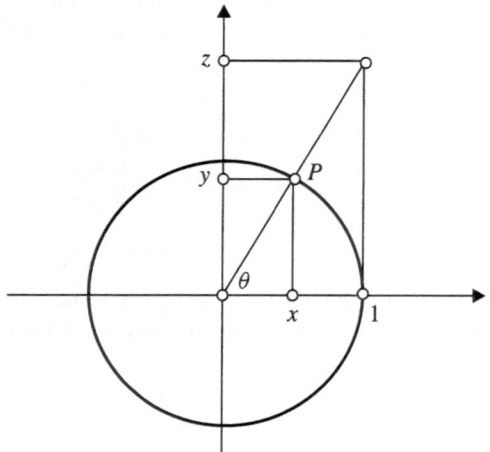

Fig. 2.1. Unit circle and a central angle

---

5 This is true for sin and cos. Each of the other trigonometric functions has poles where it "blows up". For example, the function tan is undefined at $\pi/2$ and $(3\pi)/2$.

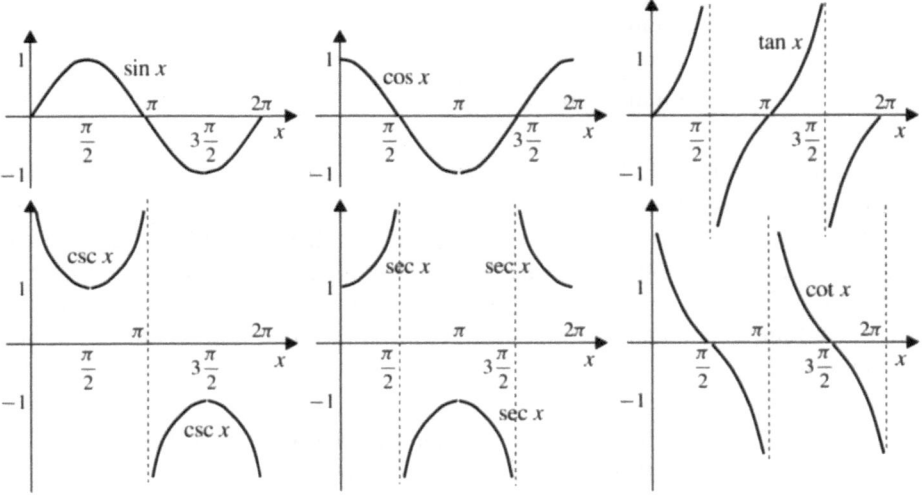

Fig. 2.2. Trigonometric functions in the interval $[0, 2\pi]$

To emphasize that the trigonometric functions are no longer restricted to the angles $0 \le \theta \le 2\pi$, we denote their arguments by letters x, y, etc.

A function f is called *periodic*, with period T, if

$$f(x) = f(x + T), \quad \text{for all } x .\tag{2.9}$$

The trigonometric functions are periodic. The functions sin, cos, sec, and csc have period 2π. The functions tan and cot have period π. See the graphs of these functions in Fig. 2.2, with sin, cos, tan on top, and their respective reciprocals csc, sec, cot under them.

Relations satisfied by trigonometric functions throughout their domains, are called *trigonometric identities*. Some of these are proved in Appendix C, Proposition C.4 and Exercise C.10. We list next some important trigonometric identities.

Proposition 2.2. For all x and y:

(a) $\sin^2 x + \cos^2 x = 1$ (b) $\sin\left(x + \dfrac{\pi}{2}\right) = \cos x$ (c) $\cos\left(x + \dfrac{\pi}{2}\right) = -\sin x$

(d) $\sin(x + \pi) = -\sin x$ (e) $\cos(x + \pi) = -\cos x$

(f) $\tan\left(x + \dfrac{\pi}{2}\right) = -\dfrac{1}{\tan x}$

(g) $\sin(x + y) = \sin x \cos y + \cos x \sin y$

(h) $\sin(x - y) = \sin x \cos y - \cos x \sin y$

(i) $\cos(x + y) = \cos x \cos y - \sin x \sin y$

(j) $\cos(x - y) = \cos x \cos y + \sin x \sin y$

(k) $\tan(x + y) = \dfrac{\tan x + \tan y}{1 - \tan x \, \tan y}$,   if $\tan x \, \tan y \neq 1$

(l) $\tan(x - y) = \dfrac{\tan x - \tan y}{1 + \tan x \, \tan y}$,   if $\tan x \, \tan y \neq -1$

Part a is a restatement of Pythagoras's theorem, see, e.g., the right triangle $\triangle(OAB)$ of Fig. C.13a (Appendix C). The expressions $\sin^2 x$ and $\cos^2 x$ are abbreviations for the squares $(\sin x)^2$ and $(\cos x)^2$, respectively. It is possible to use just the sine and cosine functions, since the remaining four trigonometric functions can be expressed in terms of sin and cos (see Exercise 2.12a). However, we will use the other four functions for convenience, specially when they result in simpler expressions.

MACSYMA-Session 2.3. In MACSYMA the arguments of trigonometric functions are measured in radians by default. You can define degree by
c1. `degree:=%pi/180`

d1. $\dfrac{\pi}{180}$

and then 360 degrees is as expected
c2. `360*degree`

d2. 2π

To determine the number of degrees in a radian, we compute 1/degree
c3. `1/degree`

d3. $\dfrac{180}{\pi}$

and approximate it numerically by
c4. `%,numer`

d4.   57.2958

The number %pi can be approximated to, say, 50 digits, by
c5. `bfloat(%pi),bfprecision:50`

d5.   3.1415926535897932384626433832795028841971693993751

Here is the computation of sin(30 degrees):
c6. `sin(30*degree)`

d6. $\frac{1}{2}$

or alternatively,
c7. `sin(%pi/6)`

d7. $\frac{1}{2}$

both simplifying to $1/2$. The expression $\cos(\pi)$ simplifies to -1
c8. `cos(%pi)`

d8. -1

We can use MACSYMA to reproduce trigonometric identities. For example, $\sin(x + y)$ can be subjected to the command `trigexpand` which uses the sum and/or product rules for angles to convert to sums, products, and powers of trigonometric functions
c9. `trigexpand(sin(x+y))`

d9. $\cos x \, \sin y + \sin x \, \cos y$

in agreement with Proposition 2.2g. Similarly, $\cos(x + y)$ becomes c10. `trigexpand(cos(x+y))`

d10. $\cos x\, \cos y - \sin x\, \sin y$

while the command `trigreduce` collects products and powers of trigonometric functions into those of sums and products of angles.

c11. `trigreduce(sin(x)*cos(y))`

d11. $\dfrac{\sin(y+x)}{2} - \dfrac{\sin(y-x)}{2}$

We can also compute the powers $(\sin x)^m$ and $(\cos x)^m$ for fixed integer m. For example, to compute $\sin(x)^5$ in terms of multiple angles, we do

c12. `trigreduce(sin(x)^5)`

d12. $\dfrac{\sin(5x) - 5\sin(3x) + 10\sin x}{16}$

We can expand the multiple angle $\sin(5x)$

c13. `trigexpand(sin(5*x))`

d13. $\sin^5 x - 10\cos^2 x\, \sin^3 x + 5\cos^4 x\, \sin x$

To force only sines to appear in the answer, we can substitute for all powers and multiples of $(\cos x)^2$ using

c14. `ratsubst(1-sin(x)^2,cos(x)^2,%)`

d14. $16\sin^5 x - 20\sin^3 x + 5\sin x$

which says to replace $(\cos x)^2$ by $1 - (\sin x)^2$ whenever $(\cos x)^2$ appears in a subexpression. Thus the $(\cos x)^2$ and $(\cos x)^4 = ((\cos x)^2)^2$ are replaced in terms of sines. If $(\cos x)^3$ appeared, it would be viewed as $(\cos x)(\cos x)^2$ and then $(\cos x)(1 - (\sin x)^2)$ would result. Now try $\sin(10x)$

c15. `trigexpand(sin(10*x))`

d15. $10\cos x\, \sin^9 x - 120\cos^3 x\, \sin^7 x + 252\cos^5 x\, \sin^5 x - 120\cos^7 x\, \sin^3 x + 10\cos^9 x\, \sin x$

c16. `ratsubst(1-sin(x)^2,cos(x)^2,%)`

d16. $512\cos x\, \sin^9 x - 1024\cos x\, \sin^7 x + 672\cos x\, \sin^5 x - 160\cos x\, \sin^3 x + 10\cos x\, \sin x$

c17. `factor(%)`

d17. $2\cos x\, \sin x\,(4\sin^2 x - 2\sin x - 1)(4\sin^2 x + 2\sin x - 1)(16\sin^4 x - 20\sin^2 x + 5)$

MACSYMA can handle some rather complicated symbolic expressions. Consider for example the sum of cosines

c18. `1/2+sum(cos(k*x),k,1,n)`

d18. $\displaystyle\sum_{k=1}^{n} \cos(kx) + \dfrac{1}{2}$

which can be summed via

c19. `closedform(%)`

d19. $\dfrac{e^{inx+ix}}{2(e^{ix}-1)} - \dfrac{e^{-inx}}{2(e^{ix}-1)} - \dfrac{e^{ix}}{2(e^{ix}-1)} + \dfrac{1}{2(e^{ix}-1)} + \dfrac{1}{2}$

We next derive the formula for $\tan(2x)$ in terms of $\sin x$ and $\cos x$,

c20. `ratsubst(1-cos(x)^2, sin(x)^2, subst(sin(x)/cos(x), tan(x),`
      `trigexpand(tan(2*x))))`

d20. $\dfrac{2\cos x\, \sin x}{2\cos^2 x - 1}$

Similarly, we express $\tan(x/2)$ in terms of $\sin x$ and $\cos x$, c21. `tan(x/2), halfangles:true`

d21. $\dfrac{1 - \cos x}{\sin x}$

c22. `%*(1-cos(x)^2)/(1-cos(x))/(1+cos(x))`

d22. $\dfrac{1 - \cos^2 x}{(\cos x + 1)\sin x}$

c23. `subst(1-sin(x)^2,cos(x)^2,%)`

d23. $\dfrac{\sin x}{\cos x + 1}$

Exercises

2.12 Consider the trigonometric functions, $\sin x$, $\cos x$, $\tan x$, $\cot x$, $\sec x$, and $\csc x$. It is convenient to have these 6 functions, but not all of them are really necessary.
   a. Express the other 4 functions in terms of $\sin x$ and $\cos x$.
   b. Express the other 4 functions in terms of $\tan x$ and $\sin x$.
   c. Can just one of the trigonometric functions be used to give the other five?

2.13 Consider the alternating voltage $V(t)$ in Example 1.9. The voltage $V(t)$ is a periodic function meaning that there is a positive T such that

$$V(t + T) = V(t), \quad \text{for all } t . \tag{2.10}$$

The smallest such T is called the cycle (or period) of the voltage (see Fig. 1.2b). The frequency of a voltage is the number of cycles per second or hertz [Hz]. Finally, the amplitude of $V(t)$ is its highest value.
   What are the (a) amplitude, (b) cycle, and (c) frequency of the voltage (1.7)? (d) In many countries, alternating voltage has amplitude 220 V and frequency 50 Hz. Find its function, analogous to (1.7).

2.14 Use MACSYMA to verify the identities in Proposition 2.2.

2.15 Prove the following identities and check them with MACSYMA
   (a) $\sin x + \sin y = 2 \sin((x + y)/2) \cos((x - y)/2)$
   (b) $\cos x + \cos y = 2 \cos((x + y)/2) \cos((x - y)/2)$
   (c) $\tan(x + y) = \dfrac{\tan x + \tan y}{1 - \tan x \tan y}$ (d) $\cot(x + y) = \dfrac{\cot x \cot y - 1}{\cot x + \cot y}$
   (e) $\sin 3x = 3 \sin x - 4 \sin^3 x$ (f) $\cos 3x = 4 \cos^3 x - 3 \cos x$

2.16 Expand the expressions $\sin mx$ and $\cos mx$ for $m := 2, 3, 4, 5$ with MACSYMA.

2.17 Collect the expressions $\sin^m x$ and $\cos^m x$ for $m := 2, 3, 4, 5$ with MACSYMA. Verify the MACSYMA result by checking it against your own calculations.

2.18 Use MACSYMA to check the following identities for various values of n
   (a) $\sin nx = \binom{n}{1} \sin x \, \cos^{n-1} x - \binom{n}{3} \sin^3 x \, \cos^{n-3} x + \binom{n}{5} \sin^5 x \, \cos^{n-5} x - \ldots$
   (b) $\cos nx = \binom{n}{0} \cos^n x - \binom{n}{2} \cos^{n-2} x \, \sin^2 x + \binom{n}{4} \cos^{n-4} x \, \sin^4 x - \ldots$

2.19 Prove the half-angle formulae for the sine and cosine functions

$$\cos(x/2) = \pm\sqrt{(1 + \cos x)/2}, \quad \sin(x/2) = \pm\sqrt{(1 - \cos x)/2} .$$

What are the correct signs in these expressions?

2.20 Use MACSYMA to find closed-form expressions (not containing a sum or product sign)

(a) $\displaystyle\sum_{k=0}^{n} \sin(kx)$ (b) $\displaystyle\sum_{k=0}^{n} \cos(kx)$

and prove the results by induction.

2.21 Calculate $\sin x$ from the equation $\sin^{10} x + \cos^{10} x = 1/5$.

2.22 By definition all trigonometric functions have period 2π. Show that $\tan x$ and $\cot x$ have a smaller period. What is it?

2.23 Prove the following two versions of Pythagoras's theorem of Proposition 2.2 a.

(a) $\sec^2 x - \tan^2 x = 1$ (b) $\csc^2 x - \cot^2 x = 1$

2.24 In Fig. 2.1, show (by a geometric argument) that $z = y/x$.

2.3 Inverse trigonometric functions

In this section we consider the inverse trigonometric functions. The trigonometric functions were initially defined for angles, or arcs (see Appendix C), e.g., $\sin(\pi/6\,[\mathrm{rad}]) = 1/2$. The inverse functions therefore assign angles, or arcs, to numbers

$$\text{inverse sine } (1/2) = \pi/6\,[\mathrm{rad}] .$$

For this reason, the inverse trigonometric functions have the prefix *arc*. For example, *arcsin* is the inverse sine function, *arccos* is the inverse cosine function, etc.

Here we abbreviate $\arcsin x$ by $\operatorname{asin} x$, $\arccos x$ by $\operatorname{acos} x$, $\arctan x$ by $\operatorname{atan} x$, etc. This is in agreement with the notation `asin(x)`, `acos(x)`, `atan(x)`, etc., used by MACSYMA.[6]

| Do not confuse $\operatorname{asin} x$ with $a \times \sin x$, etc.

The trigonometric functions are periodic, and therefore their graphs "flunk" the horizontal-line test. Indeed, every horizontal line which intersects the graph of a periodic function has infinitely many points of intersection with that graph. As in Example 1.37, in order to guarantee inverse functions, we have to restrict the domains of the functions in question.

Example 2.3 (Inverse sine function). To define the inverse sine function, restrict the domain of the sine function to the interval $[-\pi/2, \pi/2]$, to get the graph in Fig. 2.3 a. Using symmetry with respect to $y = x$ (see Remark 1.36), we get the graph in Fig. 2.3 b as the graph of the inverse sine function. We see that its domain and range are

$$D(\operatorname{asin}) = [-1, 1] = R(\sin), \quad R(\operatorname{asin}) = [-\pi/2, \pi/2] = D(\sin) .$$

Since the sine function is odd (check!), its inverse asin is also odd (see Exercise 2.29).

---

6 Other calculus books use the notation $\sin^{-1} x$, $\cos^{-1} x$, etc. However, MACSYMA interprets `sin^(-1)` x as $1/\sin x$, in the same way that it takes `sin^2` x to mean $\sin^2 x$.

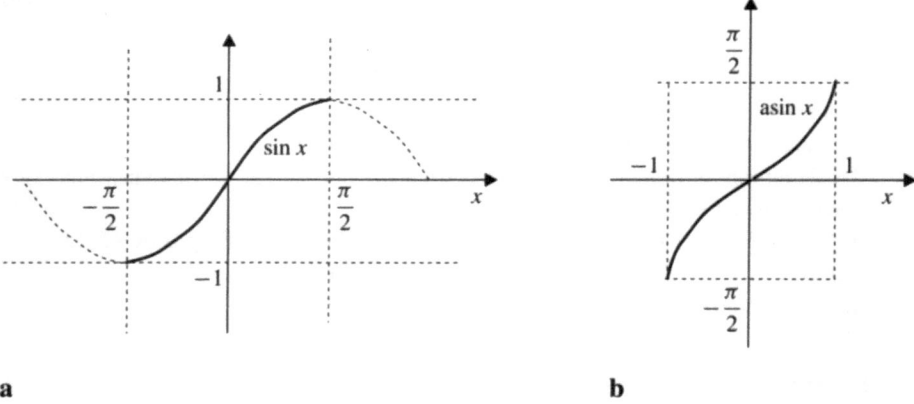

Fig. 2.3. (Restricted) sine function (**a**) and its inverse (**b**)

To illustrate how the inverse function asin x combines with other trigonometric functions, consider the expression $\cos(\text{asin}\, x)$, over the domain $[-1, 1]$ of the function asin.

Using the Pythagorean theorem $\cos^2 y + \sin^2 y = 1$, for $y = \text{asin}\, x$, and the fact that $\sin(\text{asin}\, x) = x$, we get

$$\cos(\text{asin}\, x) = \pm\sqrt{1 - \sin^2(\text{asin}\, x)} = \pm\sqrt{1 - x^2}\ .$$

The minus sign does not apply here, since $\cos y$ is nonnegative for y in the range $[-\pi/2, \pi/2]$ of asin x. Therefore

$$\cos(\text{asin}\, x) = \sqrt{1 - x^2}$$

(see also Fig. 10.1).

Example 2.4 (Inverse cosine function). To define an inverse cosine function or arccosine function, we restrict the domain[7] of $\cos x$ to the interval $[0, \pi]$, where cos is monotone decreasing (see Fig. 2.4).

The inverse function acos x then has domain and range

$$D(\text{acos}) = [-1, 1] = R(\cos), \quad R(\text{acos}) = [-\pi/2, \pi/2] = D(\cos)\ .$$

Example 2.5 (Inverse tangent function). The inverse tangent function or arctangent function is defined after restricting the domain of $\tan x$ to the open interval[8] $(-\pi/2, \pi/2)$, see the graph in Fig. 2.5 a.

---

7 Why can't we restrict the domain of $\cos x$ the same way we restricted the domain of $\sin x$ in Example 2.3?

8 The function $\tan x$ is not defined for $x = \pm\pi/2$.

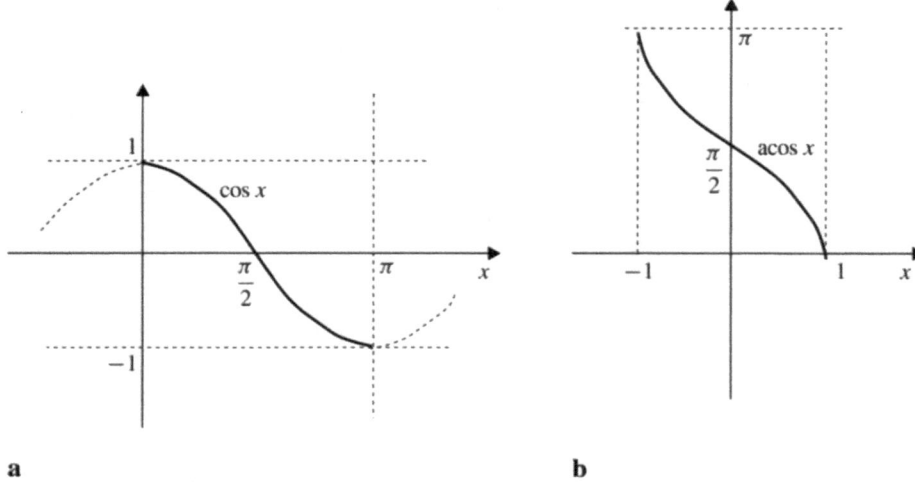

Fig. 2.4. (Restricted) cosine function (**a**) and its inverse (**b**)

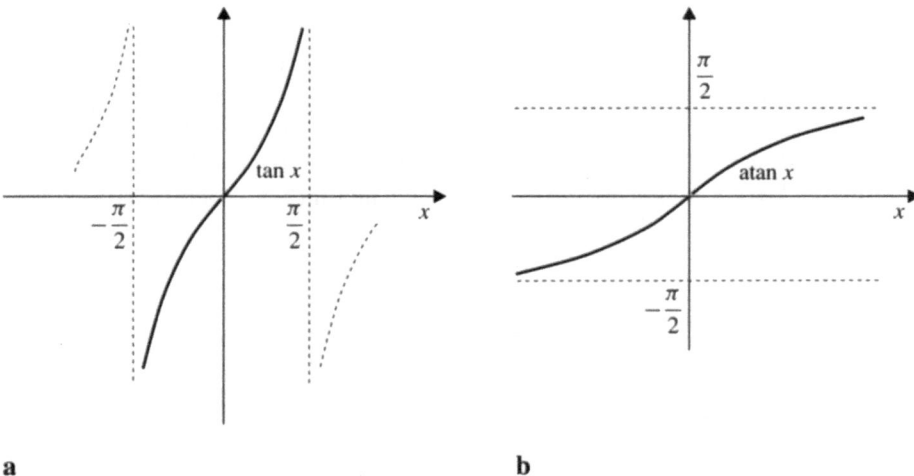

Fig. 2.5. (Restricted) tangent function (**a**) and its inverse (**b**)

The function atan x has domain and range

$$D(\text{atan}) = (-\infty, \infty) = R(\tan), \quad R(\text{atan}) = (-\pi/2, \pi/2) = D(\tan) \,.$$

The inverse function atan is odd, see its graph in Fig. 2.5b.

To see how this function combines with other trigonometric functions consider the expression $\sec^2(\text{atan}\, x)$. Using Pythagoras's theorem in the form (see Exercise 2.23)

$$\sec^2 y = 1 + \tan^2 y$$

with $y := \operatorname{atan} x$, and using the fact $\tan(\operatorname{atan} x) = x$, we get the pretty expression

$$\sec^2(\operatorname{atan} x) = 1 + x^2 .$$

Similarly,

$$\csc^2(\operatorname{acot} x) = 1 + x^2 .$$

The inverse cotangent function or arccotangent function $\operatorname{acot} x$ is similarly defined, once we restrict the domain of $\cot x$ to the open interval $(0, \pi)$.

Example 2.6 (Inverse secant function). The inverse secant or arcsecant function $\operatorname{asec} x$ is a bit more complicated. From Fig. 2.6 we see that in order to define a one-to-one mapping for $\sec x$, we may restrict the domain of $\sec x$ to the union of the intervals $[0, \pi/2)$ and $(\pi/2, \pi]$. So the inverse function $\operatorname{asec} x$ is defined in the union of the intervals $(-\infty, -1]$ and $[1, \infty)$.

Similarly the inverse cosecant function or arccosecant function $\operatorname{acsc} x$ is the inverse of the restriction of $\csc x$ to the union of the intervals $[-\pi/2, 0)$ and $(0, \pi/2]$. The domain of $\operatorname{acsc} x$ is the union of intervals $(-\infty, -1]$ and $[1, \infty)$.

Exercises

2.25 Prove the identity $\operatorname{atan} x + \operatorname{atan} y = \operatorname{atan}((x + y)/(1 - xy))$.

2.26 Evaluate the following expressions:

    (a) $\cos(\operatorname{atan} \pi/4)$ (b) $\cos(2 \operatorname{acos}(3/5))$ (c) $\sin(2 \operatorname{asin}(4/5))$

    (d) $\tan(2 \operatorname{acsc}(3/2))$ (e) $\operatorname{acot}(\sin(\pi/2))$ (f) $\cos(2 \operatorname{asin}(3/5))$

    (g) $\cos(\operatorname{asin}(1/3) + \operatorname{acos}(2/3))$ (h) $\tan(\operatorname{asin}(2/3) + \operatorname{acos}(1/3))$

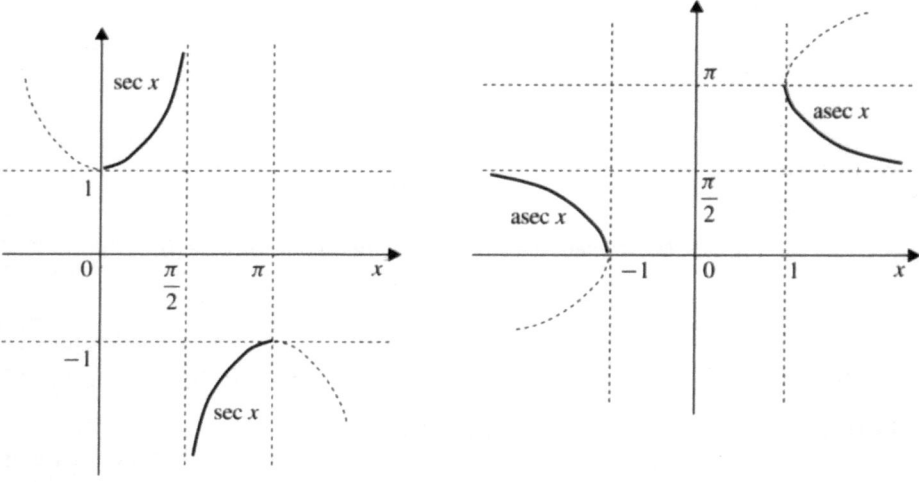

a   b

Fig. 2.6. (Restricted) secant function (**a**) and its inverse (**b**)

2.27 Use Pythagoras's theorem to evaluate the following expressions:
    (a) $\sin(\text{acos}\, x)$ (b) $\cot(\text{atan}\, x)$ (c) $\tan(\text{asec}\, x)$
    (d) $\cos(\text{acsc}\, x)$ (e) $\sec(\text{acot}\, x)$ (f) $\cot(\text{asec}\, x)$

2.28 Plot the expression $2\,\text{acos}(\sqrt{3}/3) + \text{acos}(1/3) - \pi$. What do you conjecture? Is MACSYMA able to prove this conjecture?

2.29 Prove: If the function $f : [-a, a] \to \mathbf{R}$ has an inverse, and if f is odd, then f^{-1} is odd, too.

2.30 For which values $x \in \mathbf{R}$ does the equality $\text{acot}\, x = \text{atan}(1/x)$ hold? Which relation is true for the rest of the real line?

2.31 For which values $x \in \mathbf{R}$ do the following identities hold?
    (a) $\text{asin}\, x + \text{acos}\, x = \pi/2$ (b) $\text{atan}\, x + \text{acot}\, x = \pi/2$

2.32 Use MACSYMA to plot the functions $\text{asin}(\sin x)$, $\text{acos}(\cos x)$, $\text{atan}(\tan x)$, and $\text{acot}(\cot x)$. Explain the graphs.

2.33 Determine whether you can find for some domain $D(f)$ for which the function f given below has an inverse function. Determine the domain of its inverse function. Use MACSYMA to plot the function and its inverse.
    (a) $f(x) := \sin\sqrt{x}$ (b) $f(x) := \sin\sqrt{x^2 + 1}$ (c) $f(x) := \sin(x + 1/x)$
    (d) $f(x) := \text{asin}\, e^x$ (e) $f(x) := \text{atan}(1/x)$ (f) $f(x) := \sin e^x$
    (g) $f(x) := \text{atan}\left(\dfrac{e^x - e^{-x}}{2}\right)$ (h) $f(x) := \text{asin}\left(\dfrac{e^x + e^{-x}}{2}\right)$
    (i) $f(x) := \text{atan}(e^x)$

2.34 Simplify the following expressions. Check your results with MACSYMA.
    (a) $\text{acos}(\sin x)$ (b) $\text{asin}(\cos x)$ (c) $\text{acos}\, x + \text{asin}\, x$
    (d) $\text{atan}\, x + \text{acot}\, x$ (e) $\text{asec}\, x + \text{acsc}\, x$ (f) $\text{asec}(-x)$

2.4 Hyperbolic functions

For each trigonometric function (sine, cosine, tangent, etc.) there is a corresponding function called hyperbolic (hyperbolic sine, hyperbolic cosine, hyperbolic tangent, etc.). The hyperbolic functions play an important role in applications to science and engineering.

The *hyperbolic sine*, denoted by $\sinh x$,[9] is the odd part of the exponential function,

$$\sinh x := \frac{\exp x - \exp(-x)}{2}. \tag{2.11}$$

The *hyperbolic cosine*, denoted by $\cosh x$,[10] is the even part of $\exp x$,

$$\cosh x := \frac{\exp x + \exp(-x)}{2}. \tag{2.12}$$

By adding and subtracting (2.11) and (2.12) we can express the exponential func-

---

9 The word sinh is commonly pronounced sinch, or cinch, which is slang for something sure or easy.

10 Pronounced cosh.

tions in terms of the hyperbolic functions,

$$\exp x = \cosh x + \sinh x \ ,$$
$$\exp(-x) = \cosh x - \sinh x \ .$$

The remaining hyperbolic functions are defined in analogy with the trigonometric functions:

$$\tanh x := \frac{\sinh x}{\cosh x}, \quad \coth x := \frac{1}{\tanh x}, \quad \text{sech}\, x := \frac{1}{\cosh x}, \quad \text{csch}\, x := \frac{1}{\sinh x} \ .$$
$$(2.13)$$

| Do not confuse $\sinh x$ with $\sin(hx)$, etc.

The hyperbolic functions satisfy certain identities which are analogous to trigonometric identities. For example, in analogy with the Pythagorean identity $\cos^2 x + \sin^2 x = 1$, we have

$$\cosh^2 x - \sinh^2 x = 1, \quad \text{for all } x \ . \tag{2.14}$$

This may be proved using the expressions for $\sinh x$ and $\cosh x$ in terms of the exponential functions to obtain

$$\cosh^2 x - \sinh^2 x = \tfrac{1}{4}(\exp x + \exp(-x))^2 - \tfrac{1}{4}(\exp x - \exp(-x))^2$$
$$= \tfrac{1}{4}(\exp(2x) + 2 + \exp(-2x) - \exp(2x) + 2 - \exp(-2x))$$
$$= 1 \ .$$

Similarly, the sine addition formula (Proposition 2.2g) has its counterpart in the following hyperbolic identity

$$\sinh(x + y) = \sinh x \ \cosh y + \cosh x \ \sinh y \ ,$$

proved as follows

$$\sinh(x + y) = \tfrac{1}{2}(e^{x+y} - e^{-x-y}) = \frac{e^x e^y - e^{-x} e^{-y}}{2}$$
$$= \tfrac{1}{2}((\cosh x + \sinh x)(\cosh y + \sinh y)$$
$$- (\cosh x - \sinh x)(\cosh y - \sinh y))$$
$$= \sinh x \ \cosh y + \cosh x \ \sinh y \ .$$

An analogous addition formula holds for the $\cosh x$; indeed, one has

$$\cosh(x + y) = \cosh x \ \cosh y + \sinh x \ \sinh y \ ,$$

see Exercise 2.36.

Exercises

2.35 Prove, by (2.14), (a) $1 - \tanh^2 x = \operatorname{sech}^2 x$, and (b) $\coth^2 x - 1 = \operatorname{csch}^2 x$.

2.36 Show the following properties of the hyperbolic functions
   (a) $\cosh x + \sinh x = e^x$ (b) $\cosh x - \sinh x = e^{-x}$
   (c) $\cosh(x + y) = \cosh x \, \cosh y + \sinh x \, \sinh y$

2.37 Establish the hyperbolic function analogues of the double-angle formulae.
   (a) $\cosh 2x = 2 \sinh^2 x + 1$ (b) $\sinh 2x = 2 \sinh x \, \cosh x$

2.38 Establish the identities
   (a) $\tanh(x \pm y) = \dfrac{\tanh x \pm \tanh y}{1 \pm \tanh x \, \tanh y}$ (b) $\cosh \dfrac{x}{2} = \sqrt{\dfrac{1}{2}(\cosh x + 1)}$

   (c) $\sinh \dfrac{x}{2} = \pm \sqrt{\dfrac{1}{2}(\cosh x - 1)}$

   (d) $\sinh x + \sinh y = 2 \sinh\left(\dfrac{x+y}{2}\right) \cosh\left(\dfrac{x-y}{2}\right)$

   (e) $\cosh x + \cosh y = 2 \cosh\left(\dfrac{x+y}{2}\right) \cosh\left(\dfrac{x-y}{2}\right)$

2.39 Plot the functions
   (a) $f(x) := e^{-x}$ (b) $f(x) := \cosh x - 1$
   and give approximations of the real solutions of the equations $x = f(x)$.

2.5 Inverse hyperbolic functions

The inverse hyperbolic functions get here the same terminology, and notation, as the inverse trigonometric functions of Sect. 2.3. For example, the inverse hyperbolic sine function is called arc sine hyperbolic and denoted by $\operatorname{asinh} x$.[11]

Some inverse hyperbolic functions are even simpler than their trigonometric counterparts.

Example 2.7 (Inverse hyperbolic sine). The hyperbolic sine

$$\sinh x := \tfrac{1}{2}(\exp x - \exp(-x))$$

is monotone increasing throughout **R** (check!), and its range is $(-\infty, \infty)$. Therefore the inverse hyperbolic sine, denoted by $\operatorname{asinh} x$, is well-defined throughout **R**. The domain and range of the inverse hyperbolic sine are

$$D(\operatorname{asinh}) = \mathbf{R} = R(\sinh), \quad R(\operatorname{asinh}) = \mathbf{R} = D(\sinh) .$$

The inverse hyperbolic sine can be expressed explicitly as

$$\operatorname{asinh} x = \ln\left(x + \sqrt{x^2 + 1}\right) . \tag{2.15}$$

---

11 A common pronunciation of asinh is arcsinch.

Proof. Since $y = \operatorname{asin} x$ is equivalent to

$$x = \sinh y = \frac{\exp y - \exp(-y)}{2}$$

we get a quadratic expression in $\exp y$, namely

$$(\exp y)^2 - 2x \exp y - 1 = 0 .$$

Solving this equation for $\exp y$ we obtain two roots

$$\exp y = x \pm \sqrt{x^2 + 1} .$$

We notice that for all $x \in \mathbf{R}$ the relation $\sqrt{x^2 + 1} > x$ holds; however, as $\exp y > 0$ for all $y \in \mathbf{R}$, we may disregard the solution with the minus sign as being spurious. So $\exp y = x + \sqrt{x^2 + 1}$, and by taking the natural logarithm we get

$$y = \ln\left(x + \sqrt{x^2 + 1}\right) = \operatorname{asinh} x . \qquad \square$$

MACSYMA-Session 2.4. MACSYMA recognizes the inverse hyperbolic functions and uses the same notation. For example, we reproduce (2.15) by means of

c1. `asinh(x)`

d1. $\operatorname{asinh}(x)$

c2. `logarc(%)`

d2. $\log\left(\sqrt{x^2 + 1} + x\right)$

Example 2.8 (Inverse hyperbolic cosine). The graph of the hyperbolic cosine

$$\cosh x := \tfrac{1}{2}(\exp x + \exp(-x)) ,$$

shown in Fig. 2.7a, does not satisfy the horizontal-line test. We restrict the domain of the hyperbolic cosine to the semi-infinite interval $[0, \infty)$, where it is monotone increasing. The hyperbolic cosine maps $[0, \infty)$ onto the semi-infinite interval $[1, \infty)$.

The inverse hyperbolic cosine function, denoted by $\operatorname{acosh} x$, is therefore defined with domain and range

$$D(\operatorname{acosh}) = [1, \infty) = R(\cosh), \quad R(\operatorname{acosh}) = [0, \infty) = D(\cosh) .$$

Its graph is shown in Fig. 2.7b.

The inverse hyperbolic cosine function, $\operatorname{acosh} x$, can be written explicitly, in analogy with (2.15), as

$$\operatorname{acosh} x = \ln\left(x + \sqrt{x^2 - 1}\right) . \tag{2.16}$$

The other inverse hyperbolic functions can be easily established (Table 2.3). Plot the hyperbolic functions and their inverses with MACSYMA.

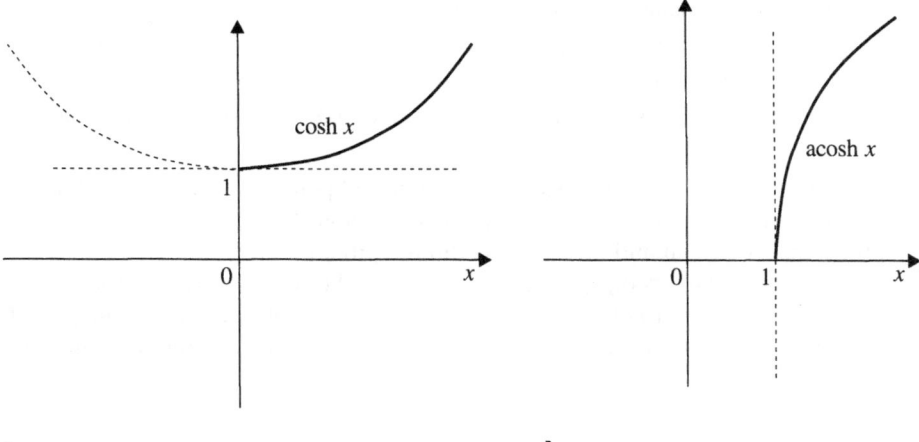

Fig. 2.7. Hyperbolic cosine function (**a**) and its inverse (**b**)

Table 2.3. Inverse hyperbolic functions

Name	Notation	Domain of definition
Inverse hyperbolic sine	asinh x	**R**
Inverse hyperbolic cosine	acosh x	$[1, \infty)$
Inverse hyperbolic tangent	atanh x	**R**
Inverse hyperbolic cotangent	acoth x	the union of the intervals $(-\infty, 0)$ and $(0, \infty)$
Inverse hyperbolic secant	asech x	$[0, \infty)$
Inverse hyperbolic cosecant	acsch x	the union of the intervals $(-\infty, 0)$ and $(0, \infty)$

Exercises

2.40 Prove (2.16).

2.41 Prove the following relations

(a) $\text{atanh } x = \dfrac{1}{2} \ln \dfrac{1+x}{1-x} \quad (-1 < x < 1)$

(b) $\text{acoth } x = \dfrac{1}{2} \ln \dfrac{x+1}{x-1} \quad (|x| > 1)$ (c) $\text{asech } x = \ln\left(\dfrac{1+\sqrt{1-x^2}}{x}\right)$

(d) $\text{acsch } x = \ln\left(\dfrac{\sqrt{x^2+1}+1}{x}\right) \quad (x > 0)$

2.42 Establish the following identities

(a) $\text{asech } x = \text{acosh}(1/x) \quad (0 < x \le 1)$

(b) $\text{acsch } x = \text{asinh}(1/x) \quad (x \ne 0)$

(c) $\text{atanh } x = \text{acoth}(1/x) \quad (|x| < 1)$

2.43 Show the identity $\ln(\sqrt{1+x^2} - x) + \ln(\sqrt{1+x^2} + x) = 0$.

2.44 Show that the expressions that MACSYMA returns for `acosh(x)` and `asech(x)` are equivalent to (2.16) and Exercise 2.41 a, respectively.

2.45 By restricting the domains of the functions given below, find their inverse functions.

What are the domains of the inverse functions?

(a) $f(x) := \ln(\cosh x)$ (b) $f(x) := \sqrt{\sinh x}$ (c) $f(x) := \ln(\operatorname{asinh} x)$

(d) $f(x) := \operatorname{asinh}(1/x)$ (e) $f(x) := \exp(\operatorname{asec} x)$ (f) $f(x) := \operatorname{asinh}(\sinh x)$

(g) $f(x) := \operatorname{asinh}(\cosh x)$ (h) $f(x) := \sinh(\operatorname{acosh} x)$ (i) $f(x) := \operatorname{asinh}(\tanh x)$

When you compute an expression with paper and pencil, and with MACSYMA, the two results may not look the same, even though they may be identical (assuming that both you and MACSYMA did the job correctly).

It is important not to become overly dependent on the computer (and the symbolic software). From time to time check the MACSYMA results with paper and pencil. Symbolic computation is a labor-saving device but not a substitute for thinking.

3 Limits and continuity

The most basic concept in calculus is the limit. It is used in the study of continuity, derivatives, integrals, and all other important topics in calculus. Indeed, one cannot use calculus intelligently without understanding limits.

We denote by

$$\lim_{x \to x_0} f(x) = L \quad \text{(the limit of } f \text{ as } x \to x_0)$$

the fact that the values $f(x)$ approach L (in a sense to be explained below) as x gets closer and closer to x_0. First some examples.

Example 3.1. The rational function $f(x) := (x^2 - 4)/(x - 2)$ is

$$f(x) = \frac{x^2 - 4}{x - 2} = \frac{(x - 2)(x + 2)}{x - 2} = \begin{cases} x + 2 & \text{if } x \neq 2, \\ ? & \text{if } x = 2. \end{cases}$$

Indeed, attempting to compute the value of f at $x := 2$, we get

$$f(2) = \frac{2^2 - 4}{2 - 2} = \frac{0}{0},$$

which is undefined. However, as x gets closer and closer to 2, the values of $f(x)$ get arbitrarily close to 4. We denote this by

$$\lim_{x \to 2} \frac{x^2 - 4}{x - 2} = 4.$$

Example 3.2. Consider the function

$$f(x) := \begin{cases} \sin(1/x) & \text{if } x \neq 0, \\ 0 & \text{if } x = 0. \end{cases}$$

Although this function is defined everywhere, its behavior near $x = 0$ is quite wild, see the graph in Fig. 3.3.

For any integer n,

$$\sin \frac{1}{x} = \begin{cases} 1 & \text{if } x = \dfrac{2}{\pi(4n + 1)}, \\ -1 & \text{if } x = \dfrac{2}{\pi(4n - 1)}. \end{cases}$$

This shows that for any x (no matter how close to 0) there are points even closer to 0, where the function takes on the values 1, −1 and any value in between.

We say that the limit $\lim_{x \to 0} \sin(1/x)$ does not exist.

Example 3.3. Consider the area A of the unit circle, which is $A = \pi \approx 3.14159$. We approximate π by expressing A as the limit of areas of regular polygons. First some terms:

- a *regular* polygon is a polygon with equal sides,
- an *n*-polygon (or *n*-gon) has n vertices,
- an *inscribed* *n*-gon has its n vertices on the unit circle (and is therefore contained in it),
- a *circumscribed* *n*-gon contains the unit circle and is tangent to it at n points.

Figure 3.1 a and c illustrates inscribed and circumscribed regular hexagons.

We denote by
I_n the area of an inscribed regular n-gon,
C_n the area of a circumscribed regular n-gon;
then clearly, $I_n < A < C_n$, for all $n \geq 3$. Furthermore, as n increases, the areas I_n increase while the areas C_n decrease, giving

$$I_3 < I_4 < \cdots < I_n < I_{n+1} < \cdots < A < \cdots < C_{n+1} < C_n < \cdots < C_4 < C_3$$
$$(3.1)$$

which is a useful approximation of π if we can compute I_n, C_n and if we can show that they get arbitrarily close to π.

Computation of I_n: The inscribed regular n-gon is made of n equal triangles OAB (Fig. 3.1 b). Therefore

$$I_n = \frac{n}{2} \sin \frac{2\pi}{n}, \quad n = 3, 4, \ldots$$

Computation of C_n: The circumscribed regular n-gon is made of n equal triangles OCD (Fig. 3.1 d). Its area is

$$C_n = n \tan\left(\frac{\pi}{n}\right), \quad n = 3, 4, \ldots$$

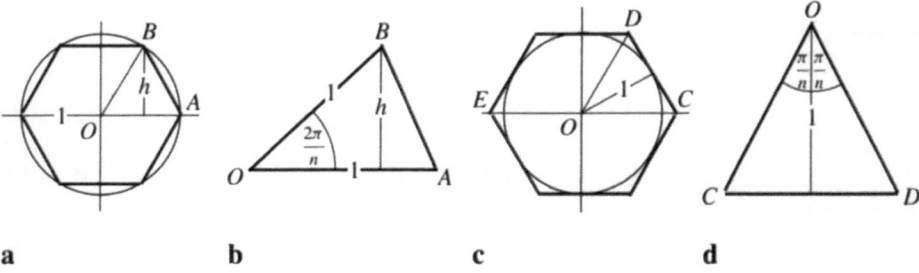

a　　　　　**b**　　　　　**c**　　　　　**d**

Fig. 3.1 a–d. Approximating the area of the unit circle by regular polygons. **a** Inscribed regular hexagon. **b** Inscribed regular n-gon is made of n triangles OAB, each has area $(1/2) \sin(2\pi/n)$. **c** Circumscribed regular hexagon. **d** Circumscribed regular n-gon is made of n triangles OCD, each has area $\tan(\pi/n)$

Therefore

$$\frac{n}{2} \sin \frac{2\pi}{n} < \pi < n \tan\left(\frac{\pi}{n}\right), \quad n = 3, 4, \ldots$$

You will be able to show later (Exercise 4.19) that

$$\lim_{n \to \infty} \frac{n}{2} \sin \frac{2\pi}{n} = \pi, \quad \text{and} \quad \lim_{n \to \infty} n \tan\left(\frac{\pi}{n}\right) = \pi,$$

so the areas of the inscribed and circumscribed n-gons both converge to π.

MACSYMA-Session 3.1. We use MACSYMA to illustrate how the numbers $\{I_n\}$ and $\{C_n\}$ approximate π as $n \to \infty$.

The numbers $\{I_n\}$ and $\{C_n\}$ converge slowly, so we compute them in steps of 100.

To compute the 10 values $I_3, I_{103}, I_{203}, \ldots, I_{1003}$, define the MACSYMA function I(n)[1]

c1. I(n) := n*sin(2*%pi/n)/2

d1. $\quad I(n) := \dfrac{n \sin\left(\frac{2\pi}{n}\right)}{2}$

We construct a list of the values $\{3, 103, 203, \ldots, 903, 1003\}$ of n by specifying the first number, 3, the increment, 100 and the last number, 1003,

c2. list_of_n:(3..100..1003)

d2. $\quad [3, 103, 203, 303, 403, 503, 603, 703, 803, 903, 1003]$

and map the function $I(n)$ over the list

c3. map(I, list_of_n)

d3. $\quad \left[\dfrac{3\sqrt{3}}{4}, \dfrac{103 \sin\left(\frac{2\pi}{103}\right)}{2}, \dfrac{203 \sin\left(\frac{2\pi}{203}\right)}{2}, \ldots, \dfrac{903 \sin\left(\frac{2\pi}{903}\right)}{2}, \dfrac{1003 \sin\left(\frac{2\pi}{1003}\right)}{2}\right]$

and convert the above to single-float numbers (see Appendix A, A.5),

c4. sfloat(%)

d4. $\quad$ [1.29904, 3.13964, 3.14109, 3.14137, 3.14146, 3.14151,
$\qquad$ 3.14153, 3.14155, 3.14156, 3.14156, 3.14157]

Similarly, define the function C(n)

c5. C(n) := n * tan(%pi/n)

d5. $\quad C(n) := n \tan\left(\dfrac{\pi}{n}\right)$

Again, we map the function $C(n)$ over the list of values of n,

c6. map(C, list_of_n)

d6. $\quad \left[3\sqrt{3}, 103 \tan\left(\dfrac{\pi}{103}\right), 203 \tan\left(\dfrac{\pi}{203}\right), \ldots, 903 \tan\left(\dfrac{\pi}{903}\right), 1003 \tan\left(\dfrac{\pi}{1003}\right)\right]$

and convert these answers to single-float numbers

c7. sfloat(%)

d7. $\quad$ [5.19615, 3.14256, 3.14184, 3.1417, 3.14165, 3.14163,
$\qquad$ 3.14162, 3.14161, 3.1416, 3.1416, 3.1416]

The convergence upward of the sequence $I(n)$ and the convergence downward of the sequence $C(n)$ illustrate the convergent limiting process. We can repeat the process with

---

1 MACSYMA uses round parentheses in the definitions of functions; square brackets in MACSYMA denote subscripted variables, such as array elements.

larger values of n by using double-precision floating-point numbers (see Appendix A, A.5) as follows:

c8. dfloat(map(I, (1003..10000..101003)))

d8.   [3.1415721062482, 3.14159248284946, 3.1415926067305, 3.14159263208422,
        3.14159264129482, 3.14159264564346, 3.14159264803514,
        3.14159264948958, 3.14159265043946, 3.14159265109378,
        3.14159265156356]

c9. dfloat(map(C, (1003..10000..101003)))

d9.   [3.14160292732106, 3.14159273895996, 3.14159267701944,
        3.14159266434258, 3.14159265973726, 3.14159265756296,
        3.1415926563671, 3.14159265563988, 3.14159265516496, 3.1415926548378,
        3.1415926546029]

We can repeat these computations with bigfloats (with the default precision of 20 digits). For example,

c10. bfloat(map(I, (100000003..5000000000..10100000003)))

d10.   [3.1415926535897911714, 3.1415926535897932377,
        3.1415926535897932383]

Similarly, for $C(n)$,

c11. bfloat(map(C, (100000003..5000000000..10100000003)))

d11.   [3.1415926535897942272, 3.1415926535897932389, 3.1415926535897932386]

As an exercise, repeat these computations with larger integers n and higher bigfloat precision, say 50 or 100 digits.

Finally, we compute the limits of I_n and of C_n as n approaches infinity,

c12. limit(I(n),n,inf)

d12. π

c13. limit(C(n),n,inf)

d13. π

Example 3.4. The function $f(x) := x^x$ is well-defined for all positive x. How does it behave when x gets closer and closer to zero? We recall that for every positive x,

$$x^0 = 1, \quad \text{and} \quad 0^x = 0 .$$

Does the function x^x act more like x^0, or more like 0^x, as x approaches 0 (from the right, that is through positive values)?

Exercises

3.1 The perimeter of a polygon is the sum of lengths of its sides. In Example 3.3 compute:
   a. the perimeter l_n of the inscribed regular n-gon,
   b. the perimeter L_n of the circumscribed regular n-gon.
   c. Conclude that

   $$l_3 < l_4 < \cdots < l_n < l_{n+1} < \cdots < 2\pi < \cdots < L_{n+1} < L_n < \cdots < L_4 < L_3$$

   and use these inequalities, with the help of MACSYMA, to approximate π.

3.2 The diameter of a set is the maximum distance between any two of its points. For example, the diameter of the circumscribed hexagon in Fig. 3.1 c is the distance from C to E.

    a. In Example 3.3 compute the diameter D_n of the circumscribed regular n-gon, for $n \geq 3$.

    b. What can you say about the diameters $\{D_n\}$, as $n \to \infty$?

3.3 Consider the function $f(x) := x^x$ in Example 3.4. Use MACSYMA to (a) plot the graph of x^x, (b) compute x^x for $x := 1/n$, n large. How would you define x^x at $x := 0$?

3.1 Limits

We study here the behavior of a function f near a point $x := c$. As x approaches c the values of $f(x)$ may converge (i.e., get arbitrarily close) to some value L, which is called the limit of f as $x \to c$. To make this precise, we agree that

"x approaches c" means $|x - c| < \delta$,

                                          for sufficiently small $\delta > 0$;

"the values of $f(x)$ converge to L" means $|f(x) - L| < \varepsilon$,

                                          for sufficiently small $\varepsilon > 0$.

    For example, $|x - 5| < \delta$ means

$$
\begin{array}{ll}
4.99 < x < 5.01 & \text{if } \delta = 0.01, \\
4.999 < x < 5.001 & \text{if } \delta = 0.001, \\
4.9999 < x < 5.0001 & \text{if } \delta = 0.0001,
\end{array}
$$

illustrating how the selection of δ "forces" x to be arbitrarily close to the value 5. Similarly, $|f(x) - 9| < \varepsilon$ means

$$
\begin{array}{ll}
8.99 < f(x) < 9.01 & \text{if } \varepsilon = 0.01, \\
8.999 < f(x) < 9.001 & \text{if } \varepsilon = 0.001, \\
8.9999 < f(x) < 9.0001 & \text{if } \varepsilon = 0.0001.
\end{array}
$$

    Here is then the formal definition of a limit.

Definition 3.5. L is called the *limit* of $f(x)$ as x approaches[2] c, if for any $\varepsilon > 0$, there exists a number $\delta > 0$ such that

$$|f(x) - L| < \varepsilon , \tag{3.2a}$$
$$\text{for all } x \neq c \text{ satisfying} \quad |x - c| < \delta . \tag{3.2b}$$

We denote this by

$$\lim_{x \to c} f(x) = L \quad \text{or} \quad f(x) \to L \text{ for } x \to c . \tag{3.3}$$

Remark 3.6.
a. Definition 3.5 does not depend on the value of f at c, as the point $x := c$ is excluded from consideration. The value $f(c)$ may violate the inequality (3.2a) for

---

    2 Also, the values $f(x)$ *tend* (or converge) to L as x *tends* to c.

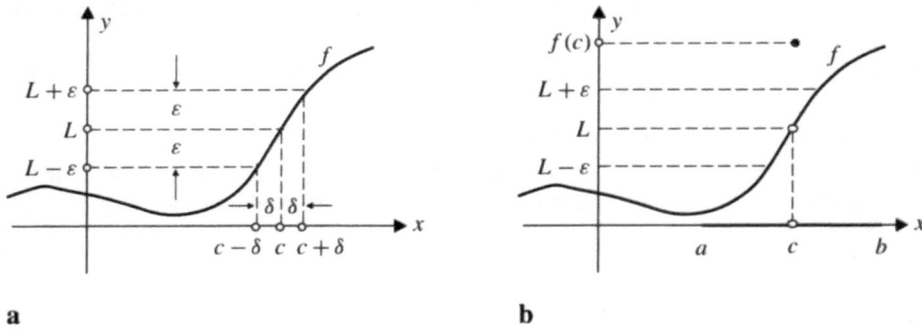

Fig. 3.2 a, b. Definition of limit for a particular ε. **a** If $x \neq c$ and $|x - c| < \delta$, then $|f(x) - L| < \varepsilon$. **b** The value of f at c is immaterial, in fact f does not have to be defined at c

some ε (see Fig. 3.2b). The conclusion $\lim_{x \to c} f(x) = L$ does not change if you redefine the function f at $x := c$ as say $f(c) = 10^6$. The function f does not even have to be defined at $x := c$. For instance, the function $f(x) := (x^2 - 4)/(x - 2)$ is undefined at $x := 2$, where its limit exists (see Example 3.1).

  b. Definition 3.5 uses the values of $f(x)$ as $x \to c$, which means that the function f must be defined at all other points sufficiently close to c. It suffices for f to be defined in some interval (a, c) ("immediately to the left" of c) and in some interval (c, b) ("immediately to the right" of c).

  c. Given the function f and the value L, the inequality (3.2a) means $L - \varepsilon < f(x) < L + \varepsilon$, and therefore ε can be regarded as the desired accuracy of approximating L, i.e., how close to L one wants to get.

  d. For any such ε, Definition 3.5 guarantees the accuracy (3.2a) provided x is sufficiently close to c, but $x \neq c$, i.e., x is in one of the intervals $(c - \delta, c)$ and $(c, c + \delta)$. This is illustrated, for a given ε, in Fig. 3.2a.

  e. The number δ is not uniquely determined by ε. You can always "take a smaller δ" in the sense that if, for a given ε,

$$0 < |x - c| < \delta_1 \quad \Longrightarrow \quad |f(x) - L| < \varepsilon \,,$$

then, for any $0 < \delta < \delta_1$,

$$0 < |x - c| < \delta \quad \Longrightarrow \quad |f(x) - L| < \varepsilon \,.$$

The following example illustrates the use of Definition 3.5 to verify limits.

Example 3.7. a. In Example 3.1 we concluded that

$$\lim_{x \to 2} \frac{x^2 - 4}{x - 2} = 4 \,.$$

For any $\varepsilon > 0$, the condition

$$\left| \frac{x^2 - 4}{x - 2} - 4 \right| < \varepsilon, \quad \text{or} \quad 4 - \varepsilon < \frac{x^2 - 4}{x - 2} < 4 + \varepsilon$$

is equivalent, for $x \neq 2$, to

$$4 - \varepsilon < (x + 2) < 4 + \varepsilon, \quad \text{or} \quad 2 - \varepsilon < x < 2 + \varepsilon$$

showing that at $x := 2$ we can take $\delta = \varepsilon$.

b. At any point $x := c > 0$, the function $f(x) := \sqrt{x}$ has the limit $\lim_{x \to c} \sqrt{x} = \sqrt{c}$. Indeed, if $\varepsilon > 0$ then

$$|\sqrt{x} - \sqrt{c}| < \varepsilon, \quad \text{or} \quad \sqrt{c} - \varepsilon < \sqrt{x} < \sqrt{c} + \varepsilon$$

becomes, by squaring,

$$c - 2\sqrt{c}\varepsilon + \varepsilon^2 < x < c + 2\sqrt{c}\varepsilon + \varepsilon^2.$$

For the given c and ε we can therefore take $\delta := 2\sqrt{c}\varepsilon + \varepsilon^2$.

c. The limit $\lim_{x \to 0} \sin(1/x)$ (see Example 3.2) does not exist. Recall that arbitrarily close to $x := 0$ we have $f(x) = 1$ as well as $f(x) = -1$. Therefore, for any $\varepsilon < 1$ there are no L and δ such that

$$x \neq 0 \quad \text{and} \quad |x - 0| < \delta \quad \Longrightarrow \quad |\sin(1/x) - L| < \varepsilon.$$

The graph of $\sin(1/x)$, for $-2 < x < 2$ and $x \neq 0$, is plotted in Fig. 3.3.

d. For any c, the limit $\lim_{x \to c}(1/(x - c))$ does not exist. Indeed, as x approaches c from the left, the values $1/(x - c)$ become arbitrarily large and negative. If x approaches c from the right, the values $1/(x - c)$ become arbitrarily large and positive. Therefore, for any $\delta > 0$, there are no finite L and ε such that

$$(x \neq c \quad \text{and} \quad |x - c| < \delta) \quad \Longrightarrow \quad L - \varepsilon < 1/(x - c) < L + \varepsilon.$$

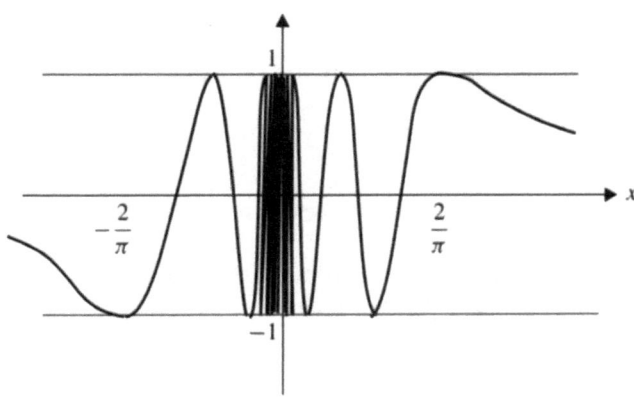

Fig. 3.3. Graph of $\sin(1/x)$

Definition 3.5 by itself may not help to find an actual limit. However, it allows us to establish rules that are useful to compute limits.[3]

Lemma 3.8 (Limit laws). a. If k is constant, then

$$\lim_{x \to a} k = k . \tag{3.4}$$

b. If $\lim_{x \to a} f(x) = L$ and $\lim_{x \to a} g(x) = M$, then

$$\lim_{x \to a} (f(x) \pm g(x)) = L \pm M . \tag{3.5}$$

In words: The limit of a sum (difference) is the sum (difference) of the limits.
  c. For products of functions we have

$$\lim_{x \to a} (f(x)g(x)) = LM , \tag{3.6}$$

i.e., the limit of a product is the product of the limits.
  d. Finally, if $M \neq 0$, then we have the following quotient rule

$$\lim_{x \to a} \left(\frac{f(x)}{g(x)} \right) = \frac{L}{M} . \tag{3.7}$$

Proof. We prove part b. (The other parts are proved similarly. The proof is rather technical and can be skipped at first reading.)
  For any $\varepsilon > 0$, there are positive δ_1 and δ_2 such that

$$x \neq a \quad \text{and} \quad |x - a| < \delta_1 \quad \Longrightarrow \quad |f(x) - L| < \varepsilon/2 ,$$
$$x \neq a \quad \text{and} \quad |x - a| < \delta_2 \quad \Longrightarrow \quad |g(x) - M| < \varepsilon/2 .$$

Let $\delta := \min\{\delta_1, \delta_2\}$. Then, by the triangle inequality (Appendix B, Lemma B.7c), it follows that

$$|(f(x) \pm g(x)) - (L \pm M)| \leq |f(x) - L| + |g(x) - M| \leq \varepsilon/2 + \varepsilon/2 = \varepsilon ,$$

whenever $x \neq a$ and $|x - a| < \delta$. $\qquad\square$

Example 3.9 (Limits of polynomials). If $p(x)$ is a polynomial and a is any point, then

$$\lim_{x \to a} p(x) = p(a) . \tag{3.8}$$

Proof. Clearly, $\lim_{x \to a} x = a$, so the result holds for the identity polynomial $p(x) := x$.
  Since $x^2 = x \cdot x$, it follows from the product rule that $x^2 \to a^2$ as $x \to a$. More generally, if n is any positive integer, then $x^n \to a^n$ as $x \to a$. Combining the product rule with the sum rule, we prove the result for arbitrary polynomials. $\qquad\square$

---

3 We shall occasionally use other letters than c to indicate the point where the limit is being taken. Here we have opted to use a as the generic point.

For example, $\lim_{x\to a}(3x^2 + 2x - 7) = 3a^2 + 2a - 7$.

Example 3.10 (Limits of rational functions). **If** $r(x) := p(x)/q(x)$ **is a rational function, and** a **is any point which is not a pole of** r**, then** $\lim_{x\to a} r(x) = r(a)$.

Proof. Follows from Example 3.9 and the quotient rule (3.7). ∎

Consider for example the rational function

$$\frac{x^2 + 6}{x + 2} \quad \text{and the limit} \quad \lim_{x\to 2} \frac{x^2 + 6}{x + 2}.$$

Since $x = 2$ is not a pole, we write

$$\lim_{x\to 2} \frac{x^2 + 6}{x + 2} = \frac{2^2 + 6}{2 + 2} = \frac{5}{2}.$$

Another example: The rational function $(x^2 - 4)/(x - 2)$ has a pole at $x = 2$. However, for all $x \neq 2$ the quotient has the value $x + 2$, and therefore

$$\lim_{x\to 2} \frac{x^2 - 4}{x - 2} = \lim_{x\to 2}(x + 2) = 4,$$

see Example 3.1.

MACSYMA-Session 3.2. We can compute a limit of $f(x)$ as x approaches a in MACSYMA by `limit(f,x,a)`. One may add an optional fourth argument of plus or minus to denote a one-sided limit. For example,

```
c1. limit((4*x^2-1)/(4*x^2+8*x+3),x,-1/2)
```
d1. -1

The limit of $|4x - 1|$ as $x \to -3$ is computed as

```
c2. limit(abs(4*x-1),x,-3)
```
d2.   13

Now we try a more difficult example, the limit, as x approaches 0, of $(1/x)(1/(x+4) - 1/4)$, which for $x = 0$ takes on the form $0/0$. For $x \neq 0$, the expression simplifies to $-1/(4(x + 4))$, and the limit as $x \to 0$ is $-1/16$, which is verified by

```
c3. limit(1/x*(1/(x+4)-1/4),x,0)
```
d3. $-\dfrac{1}{16}$

Using MACSYMA we can also handle the limits of Example 3.3. Both expressions are equal to π.

```
c4. limit(n*sin(2*%pi/n)/2,n,inf)
```
d4. π
```
c5. limit(n*tan(%pi/n),n,inf)
```
d5. π

The next two lemmas give useful rules for the calculation of limits.

Lemma 3.11 (Substitution rule). Assume $\lim_{x \to a} f(x) = L$ and $\lim_{y \to L} g(y) = g(L)$. Then

$$\lim_{x \to a} g(f(x)) = g(L) . \tag{3.9}$$

Proof. Let $\varepsilon > 0$ be given. Since $\lim_{y \to L} g(y) = g(L)$, there is some $\delta_1 > 0$ such that

$$y \neq L \quad \text{and} \quad |y - L| < \delta_1 \quad \Longrightarrow \quad |g(y) - g(L)| < \varepsilon .$$

Also, since $\lim_{x \to a} f(x) = L$, there is some $\delta > 0$ such that

$$|x - a| < \delta \underset{x \neq a}{} \quad \Longrightarrow \quad |f(x) - L| < \delta_1 .$$

Therefore

$$|x - a| < \delta \underset{x \neq a}{} \quad \Longrightarrow \quad |f(x) - L| = |y - L| < \delta_1$$

$$\Longrightarrow \quad |g(f(x)) - g(L)| = |g(y) - g(L)| < \varepsilon ,$$

proving that $\lim_{x \to a} g(f(x)) = g(L)$. $\square$

Lemma 3.12 (Pinching rule). Suppose that the inequality

$$f(x) \leq g(x) \leq h(x)$$

holds for all x in some interval around a, except perhaps at $x := a$. If

$$\lim_{x \to a} f(x) = L = \lim_{x \to a} h(x) = L ,$$

then also

$$\lim_{x \to a} g(x) = L .$$

Proof. This proof is easy, and is left to the reader (Exercise 3.11). $\square$

Example 3.13.

$$\lim_{\theta \to 0} \sin \theta = 0, \quad \text{and} \quad \lim_{\theta \to 0} \cos \theta = 1 . \tag{3.10}$$

Proof. Let θ be measured in radians,[4] and consider the angle θ as a central angle in a circle with a radius of 1 (Fig. 3.4). Then
- the area of the circular sector OAB is $\theta/2$ (this is so because the radius is 1 and the angle θ is measured in radians);[5]

---

4   The result is true also if θ is measured in degrees (0 radians is the same as 0 degrees), but the proof is a little more complicated.

5   In particular, if $\theta = 2\pi$, then the area is π.

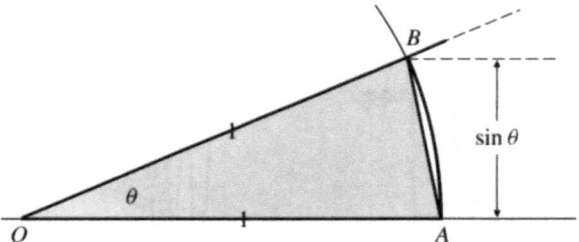

Fig. 3.4. Angle θ and its sine

–   the area of the triangle[6] OAB is $(\sin\theta)/2$ (since its base is 1 and height is $\sin\theta$).
Therefore

$$0 \le \frac{\sin\theta}{2} \le \frac{\theta}{2},$$

"pinching" $\sin\theta$ between 0 and θ, both of which approach 0 as $\theta \to 0$, proving the first limit in (3.10).

From the Pythagorean relation $\sin^2\theta + \cos^2\theta = 1$ we get moreover

$$\lim_{\theta\to 0} \cos^2\theta = \lim_{\theta\to 0} (1 - \sin^2\theta) = 1,$$

but $\lim_{\theta\to 0} \cos^2\theta = (\lim_{\theta\to 0}\cos\theta)^2$, and therefore $\lim_{\theta\to 0}\cos\theta$ is $+1$ or -1. The negative sign is eliminated, since $\cos\theta$ is positive near $\theta = 0$. □

Example 3.14. We use the pinching rule to prove

$$\lim_{x\to 0} x\sin(1/x) = 0.$$

The function $x\sin(1/x)$ is bounded above and below by $|x|$ and $-|x|$, so that

$$-|x| \le x\sin(1/x) \le |x|$$

(see Fig. 3.5a). As $x \to 0$, we also have $|x| \to 0$, and therefore

$$0 \le \lim_{x\to 0} (x\sin(1/x)) \le 0,$$

proving the claimed limit.

Example 3.15. We can similarly show that

$$\lim_{x\to 0} x^2\sin(1/x) = 0,$$

---

6 Note that the triangle OAB is inside the circular sector OAB, and the two are not the same.

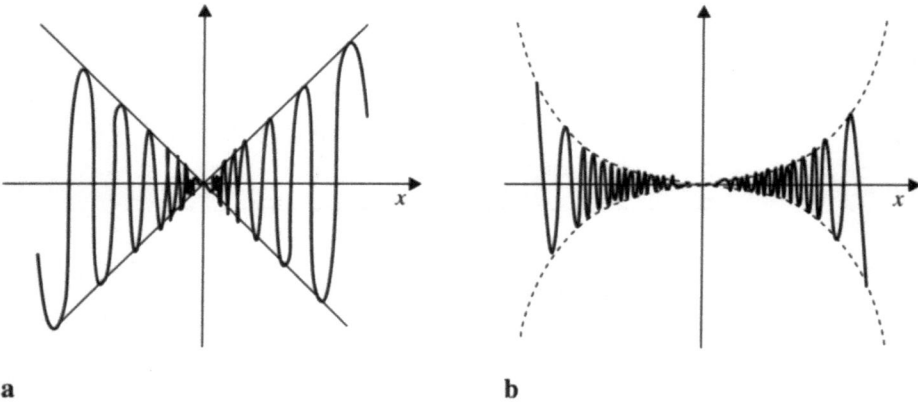

Fig. 3.5 a, b. Two graphs near $x = 0$. **a** Graph of $x \sin(1/x)$ pinched between $\pm x$, **b** graph of $x^2 \sin(1/x)$ between $\pm x^2$

since
$$-x^2 \leq x^2 \sin(1/x) \leq x^2, \quad \text{for all } x \neq 0$$
(see Fig. 3.5 b).

Example 3.16. The limit of the exponential function $\exp(x)$, as $x \to 0$, is
$$\lim_{x \to 0} e^x = 1. \tag{3.11}$$
This follows from the inequalities
$$1 + x < e^x < 1 + x + x^2, \quad \text{whenever } -\infty < x < 1.79328 \tag{3.12}$$
showing that $\exp(x)$ is pinched, near 0, between the functions $1 + x$ and $1 + x + x^2$, both having limit 1 as $x \to 0$ (see Fig. 3.6). The limit (3.11) follows[7] then from the pinching rule Lemma 3.12.

Using (3.12) we can similarly calculate
$$\lim_{x \to 0} \frac{e^x - 1}{x} = 1. \tag{3.13}$$

MACSYMA-Session 3.3. We can verify the above limits with MACSYMA.
```
c1. limit(exp(x),x,0)
d1.    1
c2. limit((exp(x)-1)/x,x,0)
d2.    1
```

---

7 This is not a proof, since it is based on the claim (3.12) which was not proved. Use MACSYMA to plot the three curves $1 + x$, $\exp(x)$, and $1 + x + x^2$ to get Fig. 3.6, and observe (3.12).

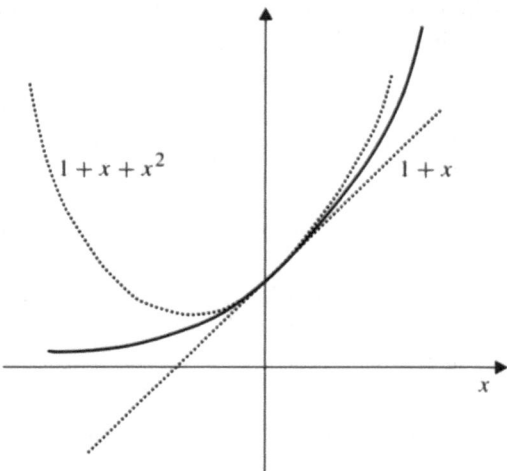

Fig. 3.6. Exponential function pinched between $1 + x$ and $1 + x + x^2$

Example 3.17. We use the substitution rule to compute

$$\lim_{x \to 0} (x \sin(1/x))^3 = 0 .$$

This follows from Lemma 3.11 by using $f(x) := x \sin(1/x)$ and $g(y) = y^3$.

Exercises

3.4   In the following examples determine which limits exist (by the limit laws given above). Check your results with MACSYMA.

(a) $\lim\limits_{x \to 0} (x^4 + 12x^2 - 4)$ (b) $\lim\limits_{x \to 2} \dfrac{x^2 - 4}{x - 2}$ (c) $\lim\limits_{x \to -1} \dfrac{x + 1}{x^2 - x - 2}$

(d) $\lim\limits_{x \to 3} \dfrac{x - 3}{x^4 - 81}$ (e) $\lim\limits_{x \to 3} (4x^3 - 5x^2 + 18)$ (f) $\lim\limits_{x \to 2} \dfrac{1/x - 1/2}{x - 2}$

(g) $\lim\limits_{x \to -1} \left| \dfrac{x^3 + 2x + 1}{x + 1} \right|$ (h) $\lim\limits_{x \to y} \dfrac{x^3 - y^3}{x - y}$ (i) $\lim\limits_{x \to y} \dfrac{x^n - y^n}{x - y}$

(j) $\lim\limits_{x \to 4} \dfrac{2 - \sqrt{x}}{4 - x}$ (k) $\lim\limits_{x \to 0} \left| \dfrac{1 - \sqrt{1 - x^2}}{x} \right|$ (l) $\lim\limits_{x \to 0} \dfrac{(a + x)^{1/3} - a^{1/3}}{x}$

3.5   Use MACSYMA to find the following limits:

(a) $\lim\limits_{x \to 1} \dfrac{1 - \sqrt{x}}{1 - x}$ (b) $\lim\limits_{x \to 0} \dfrac{1 - \sqrt{1 - x^2}}{x^2}$ (c) $\lim\limits_{x \to 0} \dfrac{1 - \sqrt{1 - x^3}}{x^3}$

(d) $\lim\limits_{x \to 5} \dfrac{\sqrt{x + 4} - 3}{x - 5}$ (e) $\lim\limits_{x \to 2} \dfrac{x^2 + 4x + 4}{x^2 - 4}$ (f) $\lim\limits_{x \to 1} \dfrac{x^3 - 1}{x - 1}$

(g) $\lim\limits_{x \to 2} \dfrac{4 - x^2}{x - 2}$ (h) $\lim\limits_{x \to 2} \dfrac{(4 - x^2)^5}{x - 2}$ (i) $\lim\limits_{x \to 0} \left| \dfrac{\sqrt{x^2 + x^4} - |x|}{x} \right|$

3.6 Verify the above limits by hand calculation, using the results of this section.

3.7 Use MACSYMA to find the following limits:

(a) $\lim\limits_{x\to 0} \dfrac{\sin mx}{x}$ (b) $\lim\limits_{x\to 0} \dfrac{\sin^2 3x}{x^2}$ (c) $\lim\limits_{h\to 0} \dfrac{\sin(x+h) - \sin x}{h}$

(d) $\lim\limits_{h\to 0} \dfrac{\cos(x+h) - \cos x}{h^2}$ (e) $\lim\limits_{x\to 0} \dfrac{x \tan x}{1 - \cos x}$

(f) $\lim\limits_{x\to 0} \dfrac{x + \sin^4 x}{4x}$ (g) $\lim\limits_{x\to y} \dfrac{\tan x - \tan y}{x - y}$ (h) $\lim\limits_{x\to y} \dfrac{\sinh x - \sinh y}{x - y}$

(i) $\lim\limits_{x\to y} \dfrac{e^x - e^y}{x - y}$ (j) $\lim\limits_{x\to 0} \dfrac{e^{\sin x} - 1}{x}$ (k) $\lim\limits_{x\to 0} \dfrac{\ln(\cos x)}{x}$

(l) $\lim\limits_{x\to 0} \dfrac{\sqrt{\cos x} - |x|}{x}$

3.8 Consider Lemma 3.11. Why is it necessary to assume that $\lim_{y\to L} g(y) = g(L)$? We may be tempted to substitute for Lemma 3.11 the following.

Claim: Assume $\lim_{x\to a} f(x) = L$ and $\lim_{y\to L} g(y) = M$. Then $\lim_{x\to a} g(f(x)) = M$.

Show that the claim is false by the example

$$f(x) := 0 \text{ for all } x; \quad g(x) := \begin{cases} 1 & \text{if } x \neq 0, \\ 0 & \text{if } x = 0; \end{cases} \quad \text{and the point } a := 0.$$

3.9 Let $f(x) = [x]$ be the greatest integer or *floor function*, that is

$$\text{floor}(x) = [x] := \max\{n \in \mathbf{Z} \mid n \le x\}. \tag{3.14}$$

The floor function is graphed in Fig. 3.7 a. Note that for integer values $k \in \mathbf{N}$ the floor function $[k]$ always equals k, and so takes the *higher* of the two neighboring values. In the graph this is represented by the small circles: a filled circle means that this value is taken, whereas an empty circle means that this value is not taken.

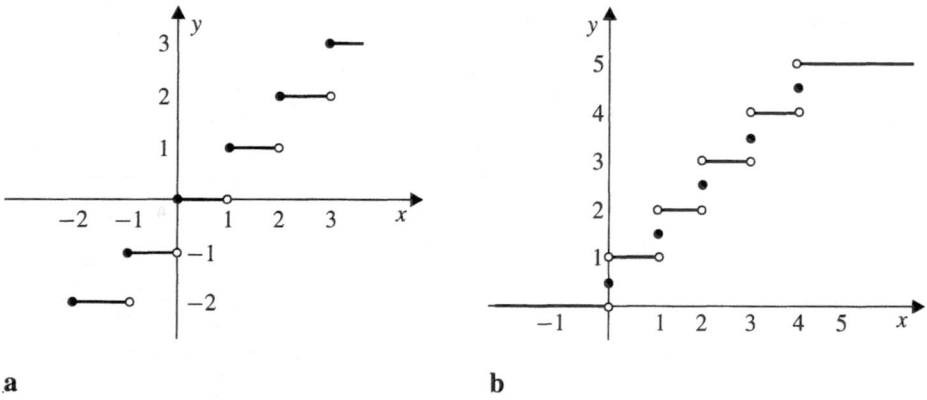

a   **b**

Fig. 3.7. a, b. Floor and staircase functions. **a** Greatest integer or floor function (• value taken, ○ value not taken); **b** staircase function S_4 with $m + 1 = 5$ steps

a. For which values of a does $\lim_{x \to a} f(x)$ exist?

b. We define the MACSYMA function[8]

```
our_floor(x):=if 0 <=x and x <1 then 0
                 else if x>=1 then our_floor(x-1)+1
                       else our_floor(x+1)-1)),
```

see also Appendix A, A.16.

Use MACSYMA to plot it, and calculate its values at the points $k/2$, ($k = 1, 2, \ldots, 10$).

3.10 Plot the following functions by MACSYMA

(a) $f(x) := 1 + [x] + [-x]$ (b) $f(x) := 1/[x] - x$

Does the limit of $f(x)$ exist at the points $x = 0, \pm 1, \pm 2, \ldots$? What happens in the interval $[0, 1]$?

3.11 Prove the pinching rule, i.e., Lemma 3.12.

3.12 Use MACSYMA to compute the following limits:

(a) $\lim\limits_{x \to 1} e^{1/x}$ (b) $\lim\limits_{x \to 1} e^{x^2 - 1}$ (c) $\lim\limits_{x \to 0} e^{\tan x}$

(d) $\lim\limits_{x \to 0} e^{\sinh x}$ (e) $\lim\limits_{x \to 1} \ln\left(\dfrac{1}{x}\right)$ (f) $\lim\limits_{x \to -1} \ln x^2$

(g) $\lim\limits_{x \to 0} \dfrac{x}{e^x - e^{-x}}$ (h) $\lim\limits_{x \to 0} \ln(x^2 + 1)$ (i) $\lim\limits_{x \to 1} \ln\left(\dfrac{x - 1}{x + 1}\right)$

(j) $\lim\limits_{x \to 0} \sin \sqrt{x^2 + \pi^2}$ (k) $\lim\limits_{x \to 0} \sin\left(x + \dfrac{1}{x}\right)$ (l) $\lim\limits_{x \to 0} a\sin e^x$

(m) $\lim\limits_{x \to 0} a\tan \dfrac{1}{x}$ (n) $\lim\limits_{x \to 1} \sin \ln x$ (o) $\lim\limits_{x \to 0} a\tan(e^x - e^{-x})$

(p) $\lim\limits_{x \to 0} a\sin \dfrac{e^x + e^{-x}}{2}$ (q) $\lim\limits_{x \to 0} a\tan(e^x)$

3.2 One-sided limits

The limit (3.3), $\lim_{x \to c} f(x) = L$, in Definition 3.5 is a *two-sided* limit, since the variable x approaches the point c from both sides. In this section we study one-sided limits, where the variable x approaches the point c on one side. This is necessary if the function is defined only on one side of the point in question, or if approaching the point from different sides gives different limits. We use the following terminology:

– "*x approaches c from the right*", also "*x approaches c from above*", denoted by $x \to c^+$ or $x \downarrow c$, means
$c < x < c + \delta$ for sufficiently small $\delta > 0$;

– "*x approaches c from the left*", also "*x approaches c from below*", denoted by $x \to c^-$ or $x \uparrow c$, means
$c - \delta < x < c$ for sufficiently small $\delta > 0$.

Definition 3.18 (One-sided limits). a. L is called the *right limit* of f at c, or the *limit* of $f(x)$ as x approaches c from the *right* (or from *above*), denoted by

---

8 The name floor is reserved in MACSYMA for this function; we therefore use a different name.

$$\lim_{x \to c^+} f(x) = L \quad \text{or} \quad \lim_{x \downarrow c} f(x) = L \,,$$

if for any $\varepsilon > 0$, there exists a number $\delta > 0$ such that

$$c < x < c + \delta \quad \Longrightarrow \quad |f(x) - L| < \varepsilon \,.$$

b. L is called the *left limit* of f at c, or the *limit* of $f(x)$ as x approaches c from the *left* (or from *below*), denoted by

$$\lim_{x \to c^-} f(x) = L \quad \text{or} \quad \lim_{x \uparrow c} f(x) = L \,,$$

if for any $\varepsilon > 0$, there exists a number $\delta > 0$ such that

$$c - \delta < x < c \quad \Longrightarrow \quad |f(x) - L| < \varepsilon \,.$$

Remark 3.19. Comparing Definitions 3.5 and 3.18, we see that if the left limit and the right limit exist and are equal

$$\lim_{x \to c^-} f(x) = \lim_{x \to c^+} f(x) = L \,,$$

then the limit of f at c exists and equals the same value

$$\lim_{x \to c} f(x) = L \,.$$

Example 3.20. The function $\sqrt{x}$ is defined only for $x \geq 0$. As x approaches 0 from the right, the values of $\sqrt{x}$ converge to 0, $\lim_{x \to 0^+} \sqrt{x} = 0$.

Example 3.21. The function

$$\text{sign}(x) := \frac{x}{|x|} = \begin{cases} 1 & \text{if } x > 0, \\ -1 & \text{if } x < 0, \end{cases}$$

does not have a limit at $x := 0$; however, the two one-sided limits exist

$$\lim_{x \to 0^+} \text{sign}(x) = 1 \quad \text{and} \quad \lim_{x \to 0^-} \text{sign}(x) = -1 \,.$$

Example 3.22 (Step and staircase functions). Another interesting function is the *step function*, defined as

$$\text{step}(x) := \begin{cases} 0 & \text{if } x < 0, \\ 1/2 & \text{if } x = 0, \\ 1 & \text{if } x > 0, \end{cases} \tag{3.15}$$

which, for $x \neq 0$, can be expressed as $\text{step}(x) = (1/2)(1 + \text{sign}(x))$. The step function has one-sided limits at 0,

$$\lim_{x \to 0^+} \text{step}(x) = 1, \quad \text{and} \quad \lim_{x \to 0^-} \text{step}(x) = 0 \,.$$

The translated step function $\text{step}(x - a)$ has its step at the point a, where it has the two-sided limits

$$\lim_{x \to a^+} \text{step}(x - a) = 1, \quad \text{and} \quad \lim_{x \to a^-} \text{step}(x - a) = 0 .$$

A *staircase function* is a function with several steps, for example,

$$S_m(x) := \sum_{n=0}^{m} \text{step}(x - n) ,$$

illustrated in Fig. 3.7b for $m = 4$. At each step-point, the staircase function has a left limit and a right limit which are different, and also not equal to the value of the function S_m at that point. At all other points the left limits and the right limits coincide, and therefore the two-sided limits exist.

Example 3.23 (Monotone limits exist). If a function f is monotone and bounded in (a, b), then the one-sided limits $\lim_{x \to b^-} f(x)$ and $\lim_{x \to a^+} f(x)$ exist. To prove the first statement, let us assume that f increases. The other cases are treated similarly. We consider the set

$$S := \{ f(x) \in \mathbf{R} \mid x \in (a, b) \} .$$

By definition S is bounded, and so by the least-upper-bound property of the real numbers (see Appendix B, B.5) $u := \sup S \in \mathbf{R}$. We shall prove that $\lim_{x \to b^-} f(x) = u$. Therefore we observe that by the definition of the least upper bound $f(x) \le u$ for all $x \in (a, b)$, and further for each $\varepsilon > 0$ there must be a point $\xi \in (a, b)$ with $f(\xi) \ge u - \varepsilon$. Since f increases, we have $f(x) \ge u - \varepsilon$ for all $x \in (\xi, b)$, and we get finally $|f(x) - u| \le \varepsilon$ for all $x \in (\xi, b)$ which implies that the limit exists (with the choice $\delta := b - \xi$).

MACSYMA-Session 3.4. MACSYMA can compute left limits as x approaches c from below, and right limits as x approaches c from above, by adding a fourth argument of *minus* or *plus* to the limit function. For example,
c1. `limit(signum(x),x,0,plus)`
d1.   1
c2. `limit(signum(x),x,0,minus)`
d2. -1
A more difficult example is $\pi/2 - \text{atan}(1/x)$, which has limit 0 if x approaches 0 from above, and a limit π if x approaches 0 from below.
c3. `limit(%pi/2-atan(1/x),x,0,plus)`
d3.   0
c4. `limit(%pi/2-atan(1/x),x,0,minus)`
d4. π

Example 3.24. We consider the last example in detail. Let

$$k(x) := \pi/2 - \text{atan}(1/x)$$

(Fig. 3.8). We calculate its one-sided limits at $x := 0$.
    a. The function $\text{atan}\, x$ is the inverse function of the tangent function $\tan x$

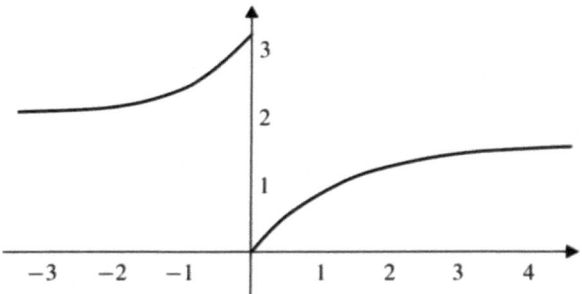

Fig. 3.8. Function $\pi/2 - \text{atan}(1/x)$.

in the interval $(-\pi/2, \pi/2)$. The function $\tan x$ is negative for $-\pi/2 < x < 0$, and positive for $0 < x < \pi/2$. The same is true for the inverse tangent function. Therefore the values of $k(x)$ are strictly larger than $\pi/2$ for $-\pi/2 < x < 0$. In the limit, as $x \uparrow 0$, the argument $1/x \downarrow -\infty$, and therefore

$$\text{atan}\,\frac{1}{x} \downarrow -\frac{\pi}{2} \quad \text{so that finally} \quad \lim_{x \to 0^-} \left(\frac{\pi}{2} - \text{atan}\,\frac{1}{x}\right) = \pi.$$

b. Similarly one may see that $\lim_{x \to 0^+} (\pi/2 - \text{atan}(1/x)) = 0$.

Exercises

3.13 Use MACSYMA to find the following limits:

(a) $\displaystyle\lim_{x \to 1^-} \frac{|1 - x|}{1 - x}$

(b) $\displaystyle\lim_{x \to 1^+} \frac{|1 - x|}{1 - x}$

(c) $\displaystyle\lim_{x \to -3^+} \sqrt{6 - x - x^2}$

(d) $\displaystyle\lim_{x \to -4^-} \frac{16 - x^2}{\sqrt{16 - x^2}}$

(e) $\displaystyle\lim_{x \to 0^-} \frac{x}{x - |x|}$

(f) $\displaystyle\lim_{x \to 0^+} \ln\left(\frac{|x|}{x}\right)$

(g) $\displaystyle\lim_{x \to 1^+} (x - \sqrt{x^2 - 1})$

(h) $\displaystyle\lim_{x \to 0^+} \exp\left(\frac{1}{1 + x^2}\right)$

(i) $\displaystyle\lim_{x \to 0^+} \exp\left(\frac{-1}{x}\right)$

3.14 Compute the following limits by hand:

(a) $\displaystyle\lim_{x \to 2} \frac{\sqrt{x^2 - 4}}{3x - 6}$

(b) $\displaystyle\lim_{x \to 0} \frac{x^2 - 3x}{\sqrt{x}}$

(c) $\displaystyle\lim_{x \to -3^+} \sqrt{6 - x - x^2}$

(d) $\displaystyle\lim_{x \to -2^-} \frac{4 - x^2}{\sqrt{4 - x^2}}$

(e) $\displaystyle\lim_{x \to 0^-} \frac{x^2}{(x - |x|)^2}$

(f) $\displaystyle\lim_{x \to 0^+} \exp\left(\frac{|x|}{x}\right)$

(g) $\displaystyle\lim_{x \to 1^-} \exp\left(\frac{-1}{1 - x^2}\right)$

(h) $\displaystyle\lim_{x \to 0^-} \exp\left(\frac{1}{x}\right)$

(i) $\displaystyle\lim_{x \to 0^-} \frac{x(1 - \csc x)}{(1 + \tan x)^2}$

3.15 Use MACSYMA to find the following limits:

(a) $\displaystyle\lim_{x \to 0^+} \ln\left(\frac{|x|}{x}\right)$

(b) $\displaystyle\lim_{x \to 1^+} \frac{|\sin(1 - x)|}{\sin(1 - x)}$

(c) $\displaystyle\lim_{x \to \pi^+} \sqrt{\cos x}$

(d) $\displaystyle\lim_{x \to 0^-} \frac{\text{asin}\,x}{x}$

(e) $\displaystyle\lim_{x \to 0^+} \ln\left(\frac{\text{acos}\,x}{e^x}\right)$

(f) $\displaystyle\lim_{x \to 0^+} \exp\left(\frac{1}{1 + \sin^2 x}\right)$

(g) $\displaystyle\lim_{x \to 0^+} \exp\left(\frac{-1}{\text{acos}\,x}\right)$

3.3 Infinite limits

"Infinity", ∞, is a mathematical symbol, and not a number which is subject to arithmetic operations.[9] The following expressions involving infinity are used in this section:

$$\begin{array}{llll}
\text{"}x \to +\infty\text{"} & \text{means} & x > M, \text{ for all } M > 0, \\
\text{"}x \to -\infty\text{"} & \text{means} & x < -M, \text{ for all } M > 0, \\
\text{"}f(x) \to +\infty\text{"} & \text{means} & f(x) > M, \text{ for all } M > 0, \\
\text{"}f(x) \to -\infty\text{"} & \text{means} & f(x) < -M, \text{ for all } M > 0.
\end{array}$$

For example, if $x := c$ is a pole of the rational function $f(x) = p(x)/q(x)$, then the values of f become unbounded (i.e., $f(x) \to +\infty$ or $f(x) \to -\infty$) as $x \to c$. This situation is covered in the following.

Definition 3.25 (Infinite limits). The function f has the right limit ∞ at c, denoted by $\lim_{x \to c^+} f(x) = +\infty$, if for any $M > 0$ there is some $\delta > 0$ such that

$$f(x) > M \quad \text{whenever } c < x < c + \delta.$$

The function f has the right limit $-\infty$ at c, denoted by $\lim_{x \to c^+} f(x) = -\infty$, if for any $M > 0$ there is some $\delta > 0$ such that

$$f(x) < -M \quad \text{whenever } c < x < c + \delta.$$

The left limits

$$\lim_{x \to c^-} f(x) = +\infty, \qquad \lim_{x \to c^-} f(x) = -\infty,$$

and the two-sided limits

$$\lim_{x \to c} f(x) = +\infty, \qquad \lim_{x \to c} f(x) = -\infty,$$

are defined analogously.

Example 3.26 (Rational functions). All the infinite limits in Definition 3.25 can be realized in poles of rational functions.

   a. The function

$$f(x) := 1/x^2$$

is positive, and increases indefinitely, as $x \to 0$ from either side. We thus have the two-sided limit

$$\lim_{x \to 0} (1/x^2) = +\infty.$$

   b. The function

$$g(x) := 1/x$$

---

9 For example, $\infty + \infty = \infty$, but $\infty - \infty$ is undefined.

behaves differently at the pole $x := 0$,

$$\lim_{x \to 0^-} (1/x) = -\infty, \quad \text{but} \quad \lim_{x \to 0^+} (1/x) = +\infty \quad \text{(verify!)} \, .$$

Example 3.27. A function can have, at a point, a finite one-sided limit as the point is approached from one side, and an infinite one-sided limit as the point is approached from the other side. For example, the function

$$h(x) := \begin{cases} 0 & \text{if } x \le 0, \\ 1/x & \text{if } x > 0, \end{cases}$$

clearly has $\lim_{x \to 0^-} h(x) = 0$ and $\lim_{x \to 0^+} h(x) = \infty$.

Limits as $x \to +\infty$ or as $x \to -\infty$ are called *limits at infinity*, not to be confused with the infinite limits of Definition 3.25. We have:

Definition 3.28 (Limits at infinity). The number L is the limit of $f(x)$ as x approaches ∞, denoted by $\lim_{x \to \infty} f(x) = L$, if for any $\varepsilon > 0$, there exists a number $M > 0$ such that

$$|f(x) - L| < \varepsilon \quad \text{for all } x > M \, .$$

The limit at $-\infty$, $\lim_{x \to -\infty} f(x) = L$, is defined analogously.

Example 3.29 (A rational example). The function

$$f(x) := \frac{1 - x^2}{1 + x + x^2} \, ,$$

has the following limits at infinity

$$\lim_{x \to -\infty} \frac{1 - x^2}{1 + x + x^2} = -1, \quad \lim_{x \to +\infty} \frac{1 - x^2}{1 + x + x^2} = -1 \, .$$

This is illustrated in Fig. 3.9. As x approaches infinity, the graph of f gets nearer to the horizontal line $y := -1$.

MACSYMA-Session 3.5. MACSYMA can handle infinite limits as well as limits at infinity. The (two-sided) `limit` function can return one of the following results

- $+\infty$ or inf, real positive infinity,
- $-\infty$ or minf, real negative infinity,
- ∞ or infinity, complex infinity,
- ind, for indefinite, used when the answer is indefinite in magnitude but still bounded,
- und, for undefined, used when both one-sided limits exist and are real, at least one of them is finite, but the limiting values are not equal.

Here are some examples.

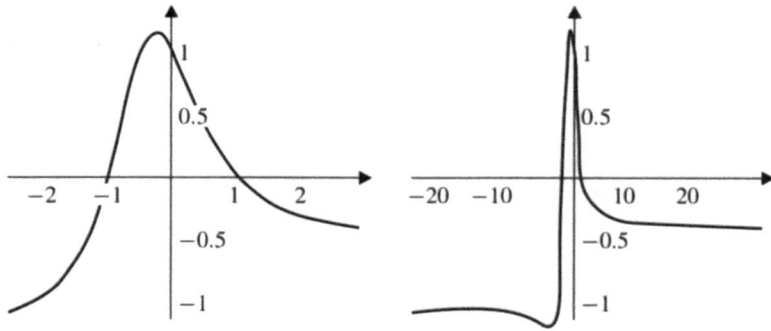

Fig. 3.9. Function $(1 - x^2)/(1 + x + x^2)$ with different axes ratios

c1. `limit(1/x^2,x,0)`
d1. $+\infty$
c2. `limit(-1/sin(x^2),x,0)`
d2. $-\infty$
c3. `limit(1/x,x,0)`
d3. ∞
a complex infinity (the two-sided limit as $x \to 0$ along the real axis does not exist).
c4. `limit(sin(1/x),x,0)`
d4.   ind
indefinite, since $\sin(1/x)$ does not tend to any definite limit as $x \to 0$ (it oscillates
between -1 and 1) but is still bounded.
c5. `limit(atan(1/x),x,0)`
d5.   und
undefined, since $\mathrm{atan}(1/x)$ has different limits as x approaches 0 from above and below,
and both are finite.

Example 3.30. The number e, the base of the natural logarithms, is itself a limit
at infinity,
$$e := \lim_{n \to \infty} (1 + 1/n)^n . \qquad (3.16)$$
To illustrate this limit consider the function
$$f(n) := (1 + 1/n)^n$$
and some of its values

n	1	2	3	4	10	50	100	365	1000	10000
$f(n)$	2	2.25	2.37037	2.44140	2.59374	2.69158	2.70481	2.71456	2.71692	2.71816

With this limit usually goes the following story. You deposit one dollar (the capital)
at a bank where the annual interest rate is 100%.[10] At the end of the year you get
back the capital and the annual interest, a total sum of $1 + 1$ dollars.

---

10 If you find such a bank, let us know.

Suppose now that interest is *compounded quarterly*. This means that the interest is added to the capital every 3 months, and from then on the interest is itself earning interest. After the first quarter your capital has grown to $(1 + 1/4)$ (the interest for $1/4$ year is $100\%/4 = 25\%$). Three months later your capital has grown to $(1 + 1/4)^2$ (the interest earned in the first quarter is now capital, and earning interest). At the end of the year, your capital has grown to

$$(1 + 1/4)^4 = f(4) = 2.44140 \, [\$] \, .$$

Similarly, if interest is compounded n times a year, then the accumulated capital at year's end is

$$f(n) = (1 + 1/n)^n \, ;$$

in particular, if interest is compounded daily,

$$f(365) = (1 + 1/365)^{365} = 2.71456 \, [\$] \, .$$

Let us use MACSYMA to compute the capital, after one year, if $n \to \infty$ (which means that interest is compounded continuously),

```
c1. limit((1+1/n)^n,n,inf)
d1.    e
c2. sfloat(%)
d2.    2.71828
```

Example 3.31. Consider the above story with the bank paying, after one year, $1+x$ dollars for each dollar deposited at the beginning of the year. The number x can be positive (gain) or negative (loss). If interest is compounded n times a year, then the accumulated capital at year's end is

$$f(n) = (1 + x/n)^n \, ,$$

and in the limit

$$e^x = \lim_{n \to 0} (1 + x/n)^n \, .$$

If x is negative, then $f(n)$ may be negative for $n = 1, 3, 5$, etc., but the limit $\lim_{n \to 0} f(n) = e^x \geq 0$.
    Similarly, for any real λ,

$$e^{\lambda x} = \lim_{n \to 0} (1 + \lambda x/n)^n \, . \tag{3.17}$$

Exercises

3.16 Use MACSYMA to find the following limits:

(a) $\displaystyle\lim_{x \to \infty} \frac{x + \sqrt{x}}{x}$

(b) $\displaystyle\lim_{x \to \infty} \frac{1 + \sqrt{1 + x^2}}{5x}$

(c) $\displaystyle\lim_{x \to \infty} \frac{1 + \sqrt{1 + x^3}}{x^3}$

(d) $\displaystyle\lim_{x \to \infty} \frac{\sqrt{x}}{x}$

(e) $\displaystyle\lim_{x \to \infty} \frac{x}{\sqrt{x}}$

(f) $\displaystyle\lim_{x \to \infty} \frac{x^2 + 2x + 1}{x^2}$

(g) $\displaystyle\lim_{x\to\infty} \frac{x^3 + 3x^2 + 3x}{x}$ (h) $\displaystyle\lim_{x\to\infty} \frac{x + \sqrt{x}}{x - \sqrt{x}}$ (i) $\displaystyle\lim_{x\to\infty} \frac{x^{10} + x}{4x^9 + 10^6}$

(j) $\displaystyle\lim_{x\to\infty} \sqrt{x^2 + x} - x$

3.17 Compute the above limits by hand.
3.18 Calculate the following limit (assume $a_n \neq 0$ and $b_m \neq 0$, and distinguish all possible cases)

$$\lim_{x\to\infty} \frac{a_n x^n + \cdots + a_0}{b_m x^m + \cdots + b_0}.$$

3.19 Compute the following limits by hand.

(a) $\displaystyle\lim_{x\to\infty} \frac{\sin x}{x}$ (b) $\displaystyle\lim_{x\to\infty} \sin \frac{1}{x}$ (c) $\displaystyle\lim_{x\to\infty} \left(\frac{1}{x} - \frac{1}{x^2}\right)$

(d) $\displaystyle\lim_{x\to\infty} \frac{x^2(1 + \sin x)}{(1 + \sin x)^2}$ (e) $\displaystyle\lim_{x\to-\infty} \frac{\operatorname{atan} x}{x^2}$

(f) $\displaystyle\lim_{x\to\infty} \frac{1 + \sin(1/x)}{\cos(1/x) + \sin^2(1/x^2)}$ (g) $\displaystyle\lim_{x\to\infty} \frac{|x|}{x}$

(h) $\displaystyle\lim_{x\to\infty} \frac{\sinh x}{e^x}$ (i) $\displaystyle\lim_{x\to\infty} \frac{\cosh x}{\sinh x}$ (j) $\displaystyle\lim_{x\to\infty} \frac{\operatorname{atan} x}{x}$

(k) $\displaystyle\lim_{x\to\infty} \frac{\sqrt{2x^4 + x}}{x^2 - 3x}$ (l) $\displaystyle\lim_{x\to\infty} \frac{(x^{10} + x)^{1/2}}{4x^5 + 10^3}$ (m) $\displaystyle\lim_{x\to\infty} \sqrt{x^2 + 2x} - x$

3.20 Prove:
(a) $\displaystyle\lim_{x\to 0^+} f(1/x) = \lim_{x\to\infty} f(x)$ (b) $\displaystyle\lim_{x\to 0^-} f(1/x) = \lim_{x\to-\infty} f(x)$

(c) $\displaystyle\lim_{x\to\infty} f(x) = \lim_{x\to-\infty} f(-x),$

whenever the limits exist.
3.21 It can be shown that

$$(1 + 1/n)^n < e < (1 + 1/n)^{n+1}, \quad n = 1, 2, \ldots \tag{3.18}$$

Demonstrate these inequalities, using MACSYMA, for selected values of n.

3.4 Continuous functions

Of the many bad things that can happen to a function f at a point c we mention three: (a) the function f is not defined at c, (b) $\lim_{x\to c} f(x)$ does not exist, (c) f is defined at c, the limit $\lim_{x\to c} f(x)$ exists, but $f(c) \neq \lim_{x\to c} f(x)$. Continuity means that none of these happens.

Definition 3.32 (Continuity). A function f is said to be *continuous* at the point c, if the following three conditions are satisfied:
a. f is defined at c,
b. $\lim_{x\to c} f(x)$ exists (so f must be defined in an interval containing c), and
c. the above limit equals the value of f at c,

$$\lim_{x\to c} f(x) = f(c). \tag{3.19}$$

If f is not continuous at a point c in the domain of f, it is said to be *discontinuous* at c.

If f is continuous at every point of (a, b), then f is called *continuous in the interval* (a, b).

We can restate Definition 3.32, using $\varepsilon\delta$-notation, as follows.

Definition 3.33. The function f is continuous at the point c if:
a. f is defined in some open interval (a, b) containing c, and
b. for any $\varepsilon > 0$ there is a $\delta > 0$ such that

$$|f(x) - f(c)| < \varepsilon \quad \text{whenever} \quad |x - c| < \delta . \tag{3.20}$$

Example 3.34. The function $f(x) := x^2$ is continuous at $x = 0$. We show this by Definition 3.33.

Let an $\varepsilon > 0$ be prescribed. We now determine a δ so that

$$|f(x) - f(0)| = x^2 < \varepsilon \quad \text{whenever} \quad |x - 0| < \delta .$$

Since

$$x^2 < \varepsilon \quad \text{whenever} \quad |x| < \sqrt{\varepsilon}$$

we conclude that $\delta := \sqrt{\varepsilon}$ will do.

Use this argument to show that $f(x) := x^2$ is continuous at any point x.

The above definition of continuity required the function to be defined on both sides of the point in question. If we use one-sided limits instead of the two-sided limit in Definition 3.32 we can define *one-sided* continuity as follows.

Definition 3.35. The function f is continuous from the *right* at c, if (a) f is defined at c, (b) $\lim_{x \to c^+} f(x)$ exists, and (c) $\lim_{x \to c^+} f(x) = f(c)$.

Similarly, f is continuous from the *left* at c if left limits are used above.

A function f is continuous at the point c if and only if it is continuous from the left, and from the right, at c.

Example 3.36 (Step function). The step function $\text{step}(x)$, defined in Example 3.22, is discontinuous at $x = 0$. It is also discontinuous at $x = 0$ from the right and from the left, since $\text{step}(0) = 1/2$.

Defining the function differently at 0, we get the *Heaviside function*,[11] which is important in applications,

---

11 Oliver Heaviside (1850–1925).

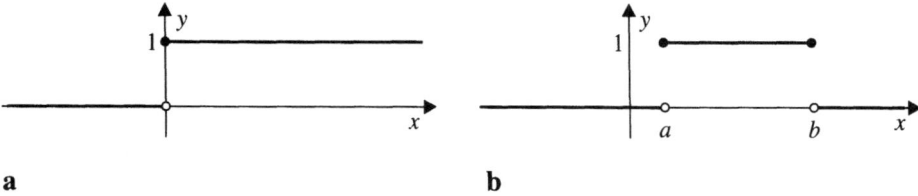

Fig. 3.10 a, b. Heaviside and indicator functions. **a** Heaviside function; **b** indicator function for a closed interval $[a, b]$

$$H(x) := \begin{cases} \text{step}(x) & \text{if } x \neq 0, \\ 1 & \text{if } x = 0. \end{cases}$$

This function is continuous from the right at 0^+, since

$$\lim_{x \to 0^+} H(x) = 1 = H(0)$$

(Fig. 3.10a).

Example 3.37 (Indicator function). Let an interval $I \subset \mathbf{R}$ be given. Then the function[12]

$$\chi_I(x) := \begin{cases} 1 & \text{if } x \in I, \\ 0 & \text{otherwise,} \end{cases}$$

is called the indicator function[13] of the interval I. The function χ_I is continuous for all x, except at the endpoints a and b of the interval I.

For a closed interval $I = [a, b]$, the characteristic function $\chi_I(x)$ can be represented, for $x \neq a, b$, by

$$chi(a, x, b) := \chi_{[a,b]}(x) = \tfrac{1}{2}(\text{sign}(x - a) - \text{sign}(x - b)) . \tag{3.21}$$

MACSYMA-Session 3.6. We define the indicator function $chi(a, x, b)$ by
c1. `chi(a,x,b):=1/2*(signum(x-a)-signum(x-b))$`
Figure 3.10b can be plotted by:
c2. `plot2(chi(1/2,x,3),x,-2,5)$`
Try it.

The word "step" is reserved by MACSYMA, so we rename the function step(x) using the German word for "step",
c3. `schritt(x):=(1+signum(x))/2$`
and the staircase function
c4. `stair(n):=sum(schritt(x-m),m,0,n)$`
Thus the S_4 function of Fig. 3.7b is

---

12 The letter χ is the Greek letter "chi", see also (3.21).
13 This function is also called characteristic function.

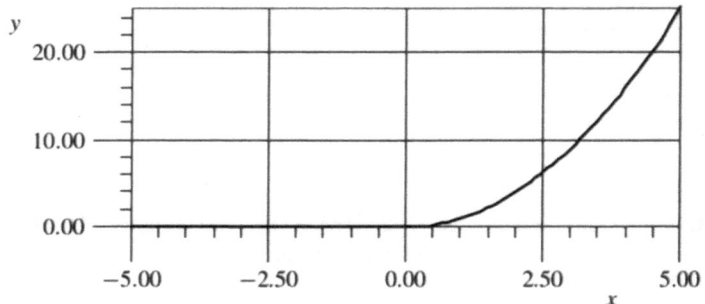

Fig. 3.11. Coincidence of the two graphs in Session 3.6, line c9

c5. `stair(4)`

d5. $\dfrac{\text{signum}(x)+1}{2} + \dfrac{\text{signum}(x-1)+1}{2} + \dfrac{\text{signum}(x-2)+1}{2} + \dfrac{\text{signum}(x-3)+1}{2} + \dfrac{\text{signum}(x-4)+1}{2}$

c6. `plot2(''%,x,-2,6)$`

where `''%` is again used to read in the expression to the plotting function so it will be evaluated the correct number of times. Try this plot.

Functions with "jumps", such as *schritt* or *chi*, can also be defined by the if-then-else special form of MACSYMA. For example, the results of

c7. `f(x):=if x<0 then 0 else x^2 $`

c8. `g(x):=x^2*schritt(x) $`

are two functions which yield the same values for $x < 0$, at $x = 0$, and for $x > 0$, so they are the same function expressed by different means. Plot both of them to see. For which values of x is the function continuous?

c9. `plot2([f(x),g(x)],x,-5,5)$`

Open intervals are natural for the definition of continuity[14]; however, we often need a continuous function in a closed interval $[a, b]$. This means that f is continuous in (a, b), and has one-sided continuity (see Definition 3.35) at the endpoints.

Definition 3.38. A function f is continuous in a closed interval $[a, b]$, if it is (a) defined in $[a, b]$, (b) continuous in the open interval (a, b), and (c) satisfies, at the endpoints a and b,

$$\lim_{x \to a^+} f(x) = f(a), \quad \text{and} \quad \lim_{x \to b^-} f(x) = f(b) . \tag{3.22}$$

Example 3.39. The function

$$f(x) := \begin{cases} x\sin(1/x) & \text{if } 0 < x < \infty, \\ 0 & \text{if } x = 0, \end{cases}$$

---

14 Because two-sided limits are used in Definition 3.32, requiring the function to be defined to the left, and to the right, of every point where it has a limit.

is continuous on the interval $[0, \infty)$. Indeed, this function is continuous on the open interval $(0, \infty)$ (verify!) and we have to verify (3.22) only at the endpoint $x := 0$,

$$\lim_{x \to 0^+} x \sin(1/x) = 0 = f(0) .$$

In the remainder of this section we check the continuity of several functions studied in Chaps. 1 and 2.

Example 3.40 (Polynomials and rational functions). It follows from (3.8) that polynomials (1.17),

$$p(x) = a_0 + a_1 x + a_2 x^2 + \cdots + a_n x^n ,$$

are continuous at all x. Let $r(x) = p(x)/q(x)$ be a rational function with no factors common to the numerator $p(x)$ and the denominator $q(x)$. Using the quotient rule Lemma 3.8d we conclude that the rational function $r(x)$ is continuous at all points x which are not poles. The rational function $r(x)$ "blows up" at each zero of the denominator (which is not a zero of the numerator since p and q have no common factors).

Example 3.41 (Trigonometric functions). The functions $\sin x$ and $\cos x$ are defined on the entire x-axis. In Example 3.13 we proved

$$\lim_{x \to 0} \sin x = 0 = \sin 0, \quad \lim_{x \to 0} \cos x = 1 = \cos 0 ,$$

showing that the functions $\sin x$ and $\cos x$ are continuous at $x := 0$. We show now that these functions are continuous at all points c, $-\infty < c < \infty$.

a. Let c be arbitrary. To compute $\lim_{x \to c} \sin x$ we use the sum formula

$$\sin x = \sin((x - c) + c) = \sin(x - c) \cos c + \cos(x - c) \sin c$$

so that

$$\lim_{x \to c} \sin x = \cos c \left(\lim_{x \to c} \sin(x - c) \right) + \sin c \left(\lim_{x \to c} \cos(x - c) \right)$$

$$= \cos c \left(\lim_{u \to 0} \sin u \right) + \sin c \left(\lim_{u \to 0} \cos u \right), \quad \text{substituting } u := x - c$$

$$= \sin c, \quad \text{by (3.10)}$$

proving $\sin x$ is continuous at c.

b. The continuity of $\cos x$ at an arbitrary point c can be shown similarly using the sum formula for cosines, or using the fact that $\cos x = \sin(x + \pi/2)$, and the continuity of $\sin x$ at the point $c + \pi/2$.

Other trigonometric functions can be checked using the quotient rule Lemma 3.8(d).

c. The function $\tan x = \sin x / \cos x$ is continuous at all points except the zeros of $\cos x$, i.e., $\tan x$ is continuous at all $x \neq \pi/2 \pm n\pi$, $n = 0, 1, \ldots$ At the exceptional points, $\tan x$ is not defined, so it does not even have one-sided continuity. The one-sided limits at $x := \pi/2$ are

$$\lim_{x \to (\pi/2)^-} \tan x = +\infty, \qquad \lim_{x \to (\pi/2)^+} \tan x = -\infty .$$

d. The function $\csc x = 1/\sin x$ is continuous at all points except the zeros of $\sin x$, i.e., at all points $x \neq \pm n\pi, n = 0, 1, \ldots$

We conclude that all trigonometric functions are continuous in their natural domains.

Example 3.42 (Exponential and hyperbolic functions). Consider the exponential function $\exp(x)$ which is defined for all x. Since $\exp(0) = 1$, it follows from (3.11),

$$\lim_{x \to 0} \exp(x) = 1 ,$$

that $\exp(x)$ is continuous at $x = 0$.

As for the sine function, continuity at 0 suffices to show continuity of $\exp(x)$ at any point c. Indeed,

$$\exp(x) = \exp((x - c) + c) = \exp(x - c) \exp(c) ,$$

and therefore

$$\lim_{x \to c} \exp(x) = \exp(c) \lim_{x \to c} \exp(x - c) = \exp(c) \lim_{u \to 0} \exp(u)$$
$$= \exp(c), \quad \text{by (3.11)}$$

which shows that $\exp(x)$ is continuous at any point $x := c$.

The hyperbolic functions (which are defined in terms of the exponential function) are continuous in their natural domains. For example, the hyperbolic cotangent

$$\coth x = \frac{\cosh x}{\sinh x} = \frac{e^x + e^{-x}}{e^x - e^{-x}}$$

is continuous at all points other than $x := 0$, where it is undefined.

MACSYMA-Session 3.7. The way MACSYMA handles limits depends on setting the option `limsubst`. If `limsubst` is false, which is the default setting, limits are evaluated as usual, for example,

c1. `limit(h(x),x,a)`

d1. $\quad \lim_{x \to a} h(x)$

where $h(x)$ is an unspecified function.

Setting `limsubst` to `true` causes the following substitution,

c2. `limit(h(x),x,a),limsubst:true`

d2. $\quad h(a)$

This setting assumes that all unspecified functions are continuous, with limits equal to function values. If a function is not continuous at a point where the limit is taken, the setting `limsubst:true` will give wrong results.

The next two theorems are useful for checking continuity in numerous instances.

Theorem 3.43 (Composition of continuous functions). If the function $f(x)$ is continuous at $x := a$, and the function $g(y)$ is continuous at $y := f(a)$, then their composition $(g \circ f)(x)$ is continuous at $x := a$.

In words: The composition of two continuous functions is continuous.

Proof. This result follows immediately from the substitution law for limits (Lemma 3.11) as

$$\lim_{x \to a} g(f(x)) = g\left(\lim_{x \to a} f(x)\right) = g(f(a)) . \qquad \square$$

To use this theorem for checking the continuity of a function $h(x)$, it is necessary to write h as a composition $h(x) = g(f(x))$ of functions satisfying the conditions of Theorem 3.43.

Theorem 3.44 (Continuity of the inverse function). Let $f(x)$ be continuous in an open interval (a, b), where it has an inverse function $f^{-1} := \phi$. Let (c, d) be an open interval, contained in the set $\{ f(x) \mid a < x < b \}$. Then the inverse function ϕ is continuous in (c, d).

In words: The inverse of a continuous function is continuous.

The proof of this theorem is somewhat past the scope of this text. It is usually proved in a course on intermediate analysis. We outline a proof, for the special case that f is strictly monotone in $[a, b]$, in Sect. 3.6.

Meantime we will conclude from Theorem 3.44 the continuity of the inverses of some common functions. For example, the inverse trigonometric and the hyperbolic functions are continuous by this theorem.

Example 3.45 (Continuity of roots). We apply Theorem 3.44 to check the continuity of the root function $g(x) := x^{1/n}$, the inverse of the power function $f(x) := x^n$.

If n is odd, we take $(-\infty, \infty)$ as the domain of definition for f. Since $f(-\infty, \infty) = (-\infty, \infty)$ for n odd, its inverse g is defined on $(-\infty, \infty)$. For n even, we take the domain of definition of f to be $(0, \infty)$. Then $f(0, \infty) = (0, \infty)$ and the domain of definition of g is also $(0, \infty)$. As f is known to be continuous on $(0, \infty)$, we have that

$$\lim_{x \to a} x^{1/n} = a^{1/n} ,$$

where we must take $a > 0$ if n is even.

Example 3.46 (Continuity of the inverse trigonometric functions). The trigonometric functions were shown in Example 3.41 to be continuous wherever they are defined. Theorem 3.44 then settles the continuity of the inverse trigonometric functions, wherever they are defined and bounded.

Example 3.47 (Continuity of the logarithmic functions). The inverse of the exponential function $\exp(x)$, shown to be continuous in Example 3.42, is the natural logarithm $\ln x$. By Theorem 3.44 it follows that $\ln x$ is continuous in $(0, \infty)$. Now logarithms to other bases can be represented by the natural logarithm, using (2.2),

$$\log_b x = \log_b e \ln x .$$

Therefore, all logarithmic functions are continuous in their natural domain $(0, \infty)$.

The following result says that a continuous function cannot change too abruptly.

Lemma 3.48. Let the function f be continuous and positive at a point c. Then $f(x)$ is positive in some interval containing c as an interior point.

Proof. We use Definition 3.33, the $\varepsilon\delta$-definition of continuity. Assume $f(c) = \gamma > 0$, and take $\varepsilon := \gamma/2$. Then there is a $\delta > 0$ such that $|f(x) - \gamma| \leq \gamma/2$ for all $x \in [c - \delta, c + \delta]$. Therefore, $x \in [c - \delta, c + \delta]$ implies

$$f(x) = \gamma - (\gamma - f(x)) \geq \gamma - |\gamma - f(x)| \geq \gamma - \gamma/2 = \gamma/2 > 0 . \qquad \square$$

Example 3.49 (Dirichlet function). Here we consider a somewhat pathological example: the Dirichlet[15] function f which is defined by

$$f(x) := \begin{cases} 1 & \text{if } x \text{ irrational,} \\ 0 & \text{if } x \text{ rational.} \end{cases}$$

We will show that the Dirichlet function is *nowhere* continuous. This follows from the fact that for each rational number a there are irrational numbers which are arbitrarily near to a, e.g., $a + \sqrt{2}/n$, with $n \in \mathbf{N}$; so f is discontinuous at each rational point. Similarly for each irrational number a there are rational numbers which are arbitrarily near to a. Therefore, f is discontinuous at each irrational point.

Exercises

3.22 Determine where each of the following functions is continuous.

(a) $f(x) := x^2 + \sqrt{x}$ (b) $f(x) := \dfrac{1}{x^2 + 1}$ (c) $f(x) := \dfrac{x}{x^2 - 1}$

(d) $f(x) := \sqrt{\dfrac{x + 1}{x - 1}}$ (e) $f(x) := \dfrac{\sin x}{|\sin x|}$ (f) $f(x) := \sqrt{\dfrac{x^2 + 1}{x^2 - 1}}$

(g) $f(x) := \sqrt{x - 1/x}$ (h) $f(x) := \tan^{1/3} x$

3.23 Determine where each of the following functions is continuous and describe any *discontinuities* in terms of left and right limits.

(a) $f(x) := \dfrac{x \sin(1/x)}{\sin x}$ (b) $f(x) := \dfrac{x - \cos(x - 1)}{x - 1}$

(c) $f(x) := \dfrac{\sin(x - \pi)}{x^2 - \pi^2}$ (d) $f(x) := \dfrac{\sin x}{\sqrt{x}}$ (e) $f(x) := \dfrac{1 - \cos x}{\sqrt{1 - x^2}}$

---

15 Peter Gustav Lejeune Dirichlet (1805–1859).

(f) $f(x) := \dfrac{1}{\sin x}$ (g) $f(x) := \dfrac{1}{\sinh x}$ (h) $f(x) := \dfrac{1}{1 - \ln x}$

3.24 Find a function which is discontinuous at $x = 1, 1/2, 1/3, \ldots$ but continuous at all other points on $(-\infty, \infty)$. Repeat this problem for the set of discontinuities $x = 0, \pm 1, \pm 2, \ldots$

3.25 Determine where the following functions are continuous.[16]

(a) $f(x) = [x]$ (b) $f(x) = x - [x]$ (c) $f(x) = \sqrt{x - [x]}$

(d) $f(x) = [1/x]$ (e) $f(x) = 1/([x] - x)$

(g) $f(x) := \begin{cases} 0 & \text{if } x \text{ is irrational,} \\ 1/q & \text{if } x = p/q \text{ is rational (expressed in lowest terms).} \end{cases}$

3.26 If $\lim_{x \to c} f(x)$ exists, but the function f is discontinuous at c, then it can be redefined as

$$F(x) := \begin{cases} f(x) & \text{if } x \neq c, \\ \lim_{x \to c} f(x) & \text{if } x = c, \end{cases}$$

and the new function $F(x)$ is continuous at c, by Definition 3.32. The discontinuity of f at c is called *removable*, for obvious reasons.

For the following functions f and points c, determine if f is continuous at c. If not, decide if the discontinuity is removable, in which case redefine the function to make the discontinuity go away.

(a) $f(x) := \tan x, \quad c := \pi/2$ (b) $f(x) := \begin{cases} \sqrt[3]{x} & \text{if } x \neq 0, \\ -1 & \text{if } x = 0, \end{cases} \quad c := 0$

(c) $f(x) := \begin{cases} \dfrac{x^2 - 4}{x - 2} & \text{if } x \neq 2, \\ 2 & \text{if } x = 2, \end{cases} \quad c := 2$

(d) $f(x) := \begin{cases} \dfrac{x^2 - 4}{x - 2} & \text{if } x \neq 2, \\ 4 & \text{if } x = 2, \end{cases} \quad c := 2$

(e) $f(x) := \begin{cases} \dfrac{-4 + 16x - 9x^2 - x^3 + x^4}{-4 + 8x - 5x^2 + x^3} & \text{if } x \neq 2, \\ 9 & \text{if } x = 2, \end{cases} \quad c := 0$

(f) $f(x) := \begin{cases} x^2 \sin(1/x) & \text{if } x \neq 0, \\ 0 & \text{if } x = 0, \end{cases} \quad c := 0$

(g) $f(x) := \begin{cases} 1 & \text{if } x \text{ is rational,} \\ 0 & \text{if } x \text{ is irrational,} \end{cases} \quad c := 0$

3.27 Given a function f which is defined at isolated points, it is required here to find a function $\phi(x)$ which coincides with $f(x)$ at the given points and is continuous everywhere.

(a) $f(x) := x$, when x is an integer (b) $f(x) := x$, when $1/x$ is an integer

---

16 A rational number p/q is said to be in lowest terms if p and q have no common factors, and $q > 0$.

(c) $f(x) := |x|$, when $\sin(x/\pi) = 1$ (d) $f(x) := 0$, when x is irrational

(e) $f(x) := x$, when x is irrational   (f) $f(x) := x - |x|$, when x is irrational

3.28 Prove that Definition 3.33 is equivalent to Definition 3.32.

3.29 Restate Definition 3.35 in an $\varepsilon\delta$-form, as in Definition 3.33.

3.30 Given below are functions f, points c and values ε. Each function f is continuous at the given point c. In each case, find a number $\delta > 0$ satisfying (3.20).

(a) $f(x) := x^2$, $\varepsilon := 1/2$, $c := 1$ (b) $f(x) := x^2$, $\varepsilon := 1/2$, c arbitrary

(c) $f(x) := x^3$, $\varepsilon := 1/3$, $c := 0$ (d) $f(x) := \dfrac{x^2 + 1}{x}$, $\varepsilon := 1/10$, $c := -1$

(e) $f(x) := \sqrt{x + 1}$, $\varepsilon := 1/4$, $c := 0$ (f) $f(x) := x^2$, ε arbitrarily > 0, $c := 1$

(g) $f(x) := x^2 + x$, ε arbitrarily > 0, $c := 0$

(h) $f(x) := \dfrac{x^2 + 1}{x - 1}$, ε arbitrarily > 0, $c := 0$

(i) $f(x) := \sqrt{x^2 + 1}$, ε arbitrarily > 0, $c := 0$

Hint: Use the argument of Example 3.34.

3.31 Prove that $\sqrt[3]{x} : \mathbf{R} \to \mathbf{R}$ is continuous using the $\varepsilon\delta$-characterization of continuity given in Definition 3.33.

3.32 Let

$$f(x) := \begin{cases} -2\sin x & \text{if } x \leq -\pi/2, \\ A\sin x + B & \text{if } -\pi/2 < x < \pi/2, \\ \cos x & \text{if } x \geq \pi/2. \end{cases}$$

Choose the numbers $A, B \in \mathbf{R}$ such that f is continuous in $\mathbf{R}$, and plot the graph of f.

3.33 Prove that the function $f : [0, 1] \to \mathbf{R}$ defined by

$$f(x) := \begin{cases} x & \text{if } x \text{ is rational}, \\ 1 - x & \text{otherwise}, \end{cases}$$

is continuous only at $x = 1/2$ and takes each value between $f(1) = 0$ and $f(0) = 1$.

3.34 Use a nested application of MACSYMA's `if` statement[17] to define the step function (3.15) in MACSYMA, and plot it.

3.35 (Monotonicity and the existence of inverse function)

  a. Let f be a continuous function with domain $D(f) = [a, b]$ and range $R(f) = [c, d]$. Then f has an inverse function if and only if f is strictly monotone (increasing or decreasing) on $[a, b]$.

  b. Let f in (a) be strictly increasing. Show that f^{-1} is strictly increasing on $[c, d]$.

  c. Show: If f is not continuous on $[a, b]$, then f^{-1} may exist even if f is not monotone.

  d. Show: If the domain of f consists of two or more intervals $[a_k, b_k]$, then f^{-1} may exist even if f is not monotone.

---

17 The statement takes the following form *if* condition *then* expression1 *else* expression2.

3.5 Continuous functions on closed intervals

In this section we study some important properties of continuous functions on closed intervals (see Definition 3.38).

Lemma 3.50 (Continuity implies boundedness). Let the function f be continuous on $[a, b]$. Then f is bounded on $[a, b]$.

Proof. Suppose f is not bounded on $[a, b]$. We show that there exists a point $c \in [a, b]$ where $\lim_{x \to c} f(x) \neq f(c)$. This will contradict the continuity of f at the point c.

Let $I_0 := [a, b]$. Bisect this interval into two closed intervals $[a, (a+b)/2]$ and $[(a + b)/2, b]$. The function f is unbounded on at least one of these intervals, call it I_1, and bisect it to get a closed interval I_2 of length $|b - a|/4$, where f is again unbounded. Continuing in this fashion we get a sequence of nested closed intervals $\{ I_k \mid k = 0, 1, 2, \dots \}$ whose lengths are $|b - a|/2^k$. These nested intervals have a unique common point[18], say c. Thus, $\lim_{x \to c} f(x) \neq f(c)$, so that f is not continuous at c. This contradiction implies that f is bounded in $[a, b]$. $\square$

The method of proof used above is called *proof by contradiction*.

Example 3.51. The function

$$f(x) := \begin{cases} 1/x & \text{if } 0 < x \leq 1, \\ 0 & \text{if } x = 0, \end{cases}$$

is continuous in the open interval $(0, 1)$ (verify!) but is unbounded there. Therefore, in Lemma 3.50 it is essential that the interval $[a, b]$ is closed.

Since $f(x)$ is unbounded in $(0, 1)$, it cannot be continuous in any closed interval containing $(0, 1)$, such as $[0, 1]$.

The previous property can be considerably strengthened. We can show that a function f continuous on $[a, b]$ is not only bounded there, but actually achieves its maximum (finite) value on $[a, b]$.

Theorem 3.52 (Continuous functions attain their extremal values). Let the function f be continuous on the closed interval $[a, b]$. Then there exist:
a. a number $c \in [a, b]$ where f achieves its maximum, i.e., $f(x) \leq f(c)$ for all $x \in [a, b]$, and
b. a number $d \in [a, b]$ where f achieves its minimum, i.e., $f(x) \geq f(d)$ for all $x \in [a, b]$.

---

18 This is a fundamental property of real numbers, that nested closed intervals, whose lengths decrease to zero, have a unique common point, see Appendix B, B.5.

Proof. Consider the image of $[a, b]$ under the mapping f, i.e., the set

$$S := f([a, b]) := \{ f(x) \mid a \leq x \leq b \} .$$

From our previous discussion we know that S is bounded above (and below). Let the least upper bound of the set S be $u := \sup S \in \mathbf{R}$. Its existence is guaranteed by the least-upper-bound property of the real numbers (see Appendix B, B.5).

We show that the maximum of f in $[a, b]$ is the least upper bound u of S, and, moreover, the maximum occurs for some point $c \in [a, b]$ where $f(c) = u$. We do this by using, once more, the bisection method. Since u is the least upper bound of f on $I_0 := [a, b]$, it must also be the least upper bound on at least one of the intervals $[a, (a+b)/2]$ or $[(a+b)/2, b]$. Continuing with the subdivision procedure we obtain, as before, a nested sequence of closed intervals I_n with $n \in \mathbf{N}_0$, which have a single common point, say c. At this point the function f takes on its least upper bound, i.e., $f(c) = u$.

We can similarly show that there is a point d in $[a, b]$ where the minimum of f on $[a, b]$ is attained.   □

Remark 3.53. The sequence of nested intervals in the proof of Theorem 3.52 has a unique common point, but there may be more than one point in $[a, b]$ where f attains its maximum value. Consider for example the constant function $f(x) := 3$. Then f evaluated at any point in $[a, b]$ is a maximum, as well as a minimum.

Example 3.54. The function

$$f(x) := \begin{cases} 1/x & \text{if } 0 < x \leq 1, \\ 0 & \text{if } x = 0, \end{cases}$$

is continuous in the open interval $(0, 1)$. Since f is unbounded from above (see Example 3.51) its maximum is not attained in $(0, 1)$. The function f is bounded below in $(0, 1)$, and its greatest lower bound in $(0, 1)$ is 1, but this value is not attained in this interval since the point 1 is "missing". For any $\alpha > 1$ we can find x in $(0, 1)$, such that $f(x) < \alpha$, showing that while the greatest lower bound of f cannot be attained, any higher (arbitrarily close) value is attained by f in $(0, 1)$.

Lemma 3.50 allows us to conclude that an unbounded function on a closed interval $[a, b]$ is *not continuous*; it does not help in case the function is bounded since there are bounded functions which are discontinuous. The sign function (Appendix B, B.10) is such an example.

The next theorem implies that the graph of a function, continuous on a closed interval $[a, b]$, has no gaps. This is a geometric interpretation of continuity.

Theorem 3.55 (Intermediate-value theorem). Let f be continuous on the closed interval $[a, b]$ with $f(a) \neq f(b)$, and let k be any value between $f(a)$ and $f(b)$. Then there exists a point $c \in [a, b]$ such that $f(c) = k$.

Proof. The proof again uses the bisection method. Assume $f(a) < f(b)$ (the case $f(a) > f(b)$ is handled similarly). We begin with the interval $I_0 := [a, b]$, and subdivide to obtain $[a, (a + b)/2]$ and $[(a + b)/2, b]$. Then we define the next interval

$$I_1 := \begin{cases} [a, (a+b)/2] & \text{if } f(a) < k < f((a+b)/2), \\ [(a+b)/2, b] & \text{if } f((a+b)/2) < k < f(b) \end{cases}$$

and stop if $k = f((a + b)/2)$. Continuing in this fashion, we obtain either a point $c \in [a, b]$ such that $f(c) = k$, or a sequence of nested closed intervals $I_n := [a_n, b_n]$ so that $f(a_n) < k < f(b_n)$. These nested intervals have a unique common point c, satisfying

$$f(c) = \lim_{n \to \infty} f(a_n) = k$$

by the continuity of f. $\qquad\square$

See also Exercise 3.40.

Example 3.56. Continuity on the *entire* interval is necessary for the intermediate-value property. This is illustrated by the Heaviside function

$$H(x) := \begin{cases} 0 & \text{if } x < 0, \\ 1 & \text{if } x \geq 0. \end{cases}$$

We see that $H(-1) = 0$ and $H(1) = 1$, but no value between 0 and 1 is attained by H in $[-1, 1]$. For example, there is no point x in $[-1, 1]$ where $H(x) = 1/2$. This proves that, indeed, the function $H(x)$ is not continuous on $[-1, 1]$.

The following corollary is the special case $k = 0$ in Theorem 3.55. It gives a simple condition for a continuous function to have a zero in an interval.

Corollary 3.57. Let f be continuous in $[a, b]$, and let[19]

$$f(a) f(b) < 0 . \tag{3.23}$$

Then there is a point $c \in (a, b)$ where $f(c) = 0$.

For strictly monotone functions (see Definition 1.39) the intermediate-value theorem can be stated as follows.

Corollary 3.58. If f is continuous and strictly monotone in $[a, b]$, then f assumes any value between $f(a)$ and $f(b)$ exactly once in $[a, b]$.

---

19 This means either $f(a) < 0 < f(b)$ or $f(a) > 0 > f(b)$.

Example 3.59 (The bisection algorithm). The intermediate-value theorem (Theorem 3.55) provides an algorithmic approach to approximate a zero of a continuous function f in a closed interval $[a, b]$, where $f(a) f(b) < 0$ (3.23).

Consider the function $f(x) := x^2 - 2$, whose positive zero is $\sqrt{2}$. Since $f(0) = -2$ and $f(2) = 2$, the function f has a zero in the interval $[0, 2]$. We approximate it using the bisection method.

The following is a list of the intervals $I_n := [a_n, b_n], n = 1, 2, \ldots, 20$, generated by the bisection method, such that $f(a_n) < 0$ and $f(b_n) > 0$, i.e., $\sqrt{2} \in I_n$ for all n.

$I_1 = [1.00000, 2.00000]$ $I_2 = [1.00000, 1.50000]$ $I_3 = [1.25000, 1.50000]$
$I_4 = [1.37500, 1.50000]$ $I_5 = [1.37500, 1.43750]$ $I_6 = [1.40625, 1.43750]$
$I_7 = [1.40625, 1.42187]$ $I_8 = [1.41406, 1.42187]$ $I_9 = [1.41406, 1.41797]$
$I_{10} = [1.41406, 1.41602]$ $I_{11} = [1.41406, 1.41504]$ $I_{12} = [1.41406, 1.41455]$
$I_{13} = [1.41406, 1.41431]$ $I_{14} = [1.41418, 1.41431]$ $I_{15} = [1.41418, 1.41425]$
$I_{16} = [1.41418, 1.41422]$ $I_{17} = [1.41420, 1.41422]$ $I_{18} = [1.41421, 1.41422]$
$I_{19} = [1.41421, 1.41422]$ $I_{20} = [1.41421, 1.41422]$

So within the current accuracy we have $1.41421 < \sqrt{2} < 1.41422$, obtained in the 18th iteration.

MACSYMA-Session 3.8. The bisection algorithm can be implemented by the programs `bisection_aux(f,x,a,b)` and `bisection(f,x,a,b)` below.

```
c1. bisection_aux(f,x,a,b) := block([mid:(a+b)/2],
        if a=mid or b=mid or limit(f,x,mid)=0 then return(mid)
        else if limit(f,x,mid)*limit(f,x,a) < 0 then
                bisection_aux(f,x,a,mid)
            else bisection_aux(f,x,mid,b))$
c2. bisection(f,x,a,b) := block([prederror:false, numer:true],
        case(is(limit(f,x,a)*limit(f,x,b) < 0),
            unknown, "Boundary conditions could not be verified",
            false, "Bisection not applicable when function has
                    same sign at endpoints",
            true, bisection_aux(f,x,a,b)))$
```

The function `bisection(f,x,a,b)` determines if f has opposite signs at $x = a$ and $x = b$. There are three possible outcomes:

- unknown, i.e., the condition $f(a) f(b) < 0$ cannot be determined. An example where this happens is bisection($x + u, x, -1, 1$), where the extra variable u prevents the determination of signs. In this case the bisection algorithm is not applicable.
- false, i.e., $f(a) f(b) \geq 0$, and again bisection is not applicable. An example of this case is bisection($x^2 - 2, x, 2, 8$).
- true, i.e., $f(a) f(b) < 0$, and the function `bisection_aux` is invoked to find an interval containing a sign change and then to shrink that interval to a floating-point number.

The program `bisection_aux(f,x,a,b)` binds mid as the midpoint of a and b, and returns "mid" if $a = $ mid or $b = $ mid or $\lim_{x \to \text{mid}} f(x) = 0$. If none of these three conditions is met, `bisection_aux` checks to see if the root lies in $[a, \text{mid}]$ or $[\text{mid}, b]$

and then calls itself recursively on the subinterval containing the sign change. The use of `numer:true` in `bisection` causes floating-point numbers to be used, which serves to reduce the size of expressions which appear in the calculations. Here are some examples.

c3. `bisection(x^2-2,x,0,2)`

d3.   1.41421

c4. `bisection(cos(x),x,0,2)`

d4.   1.5708

c5. `bisection(1/x,x,-2,-3)`

d5.   Bisection not applicable when function has same sign at endpoints

A consequence of Theorem 3.55 is the following.

Corollary 3.60 (Real polynomials with odd degree have a real zero). If n is odd and the coefficients $(a_k)_{k=0}^{n-1}$ are real, then the polynomial equation

$$x^n + a_{n-1}x^{n-1} + \cdots + a_1 x + a_0 = 0$$

has at least one real zero.

This is proved in Sect. 3.6.

Example 3.61 (Odd polynomials). To illustrate the above corollary consider the polynomial

$$p(x) := x^5 + 2x^3 - x^2 - 20 .$$

Clearly $p(0) = -20$ and $p(2) = 24$; hence, by the intermediate-value theorem there is a value $x = x_0$ where $p(x_0) = 0$. However, Corollary 3.60 tells us we need not evaluate the polynomial to find that it has a change of sign over an interval. The proof of this corollary makes use of the idea that for very large positive values of x the polynomial becomes positive because x^5 is the dominant term in the polynomial. On the other hand, for very large negative values of x the polynomial must become negative as $(-x)^5 = -x^5$ is negative and dominates the other terms in the polynomial. To illustrate this point more clearly note that if $x = 100$, then $x^5 = 100^5 = 10^{10} = 10,000,000,000$, which is much larger than all the other terms in the polynomial combined! On the other hand, if $x = -100$, then $x^5 = -100^5 = -10^{10}$, which is more negative than all the other terms combined. This is the idea behind Corollary 3.60.

Exercises

3.36 Determine which of the following functions are bounded above and below on the prescribed interval I, and which take on their maxima and minima there.

(a) $f(x) := \cot x$, $I = (0, \pi/2)$ (b) $f(x) := \sec x$, $I = [0, \pi/2]$

(c) $f(x) := \dfrac{1}{1+x^2}$, $I = [-1, 1]$ (d) $f(x) := \sqrt{1+\cot^2 x}$, $I = (\pi/2, \pi)$

(e) $f(x) := \sqrt{\dfrac{1}{x+1}}$, $I = [-1, 1]$ (f) $f(x) := \sqrt{\dfrac{x-1}{x+1}}$, $I = [0, 1]$

3.37 Determine a region where the following functions are bounded above and below.

(a) $f(x) := (x+1)^{100}$ (b) $f(x) := \dfrac{x+1}{x-1}$ (c) $f(x) := |x^2 - 1000|$

(d) $f(x) := e^{1/(x^2+1)}$ (e) $f(x) := e^{1/|x+1|}$ (f) $f(x) := \ln\left(\dfrac{x^2+1}{x^2+4}\right)$

3.38 Determine which of the following functions are bounded above and below, and which take on their maxima and minima on the indicated interval I.

(a) $f(x) := x^3$, $I = (-1, 1)$ (b) $f(x) := x$, $I = \mathbf{R}$

(c) $f(x) := 1/(x^2 + 1)$, $I = \mathbf{R}$ (d) $f(x) := 1/(1 - x^2)$, $I = (-1, 1)$

(e) $f(x) := \begin{cases} x^2 & \text{if } x \le 1, \\ 1 & \text{if } x > 1, \end{cases}$ $I = \mathbf{R}$ (f) $f(x) := [x]$, $I = [-\pi, \pi]$

(g) $f(x) := \begin{cases} x^2 & \text{if } x \le a, \\ 2a + 1 & \text{if } x > a, \end{cases}$ $I = \mathbf{R}$ (consider several different values for a)

The following problems use the interval $I = [-1, 1]$.

(h) $f(x) := \mathrm{atan}(1/(1 - x^2))$ (i) $f(x) := \begin{cases} 0 & \text{if } x \text{ is irrational,} \\ 1/q & \text{if } x = p/q \text{ is rational} \end{cases}$

(j) $f(x) := \begin{cases} x & \text{if } x \text{ is irrational,} \\ 0 & \text{if } x = p/q \text{ is rational} \end{cases}$

(k) $f(x) := \begin{cases} 1 & \text{if } x \text{ is irrational,} \\ 1/q & \text{if } x = p/q \text{ is rational} \end{cases}$

(l) $f(x) := \begin{cases} 1 & \text{if } x \text{ is irrational,} \\ p & \text{if } x = p/q \text{ is rational} \end{cases}$

3.39 Determine whether the functions below are continuous.

(a) $f(x) := \begin{cases} x\sin(1/x) & \text{if } 0 < x \le 1, \\ 0 & \text{if } x = 0 \end{cases}$

(b) $f(x) := \begin{cases} x\cot x & \text{if } 0 < x \le \pi/2, \\ 1 & \text{if } x = 0 \end{cases}$

3.40 Theorem 3.55 says that a function f continuous in a closed interval $[a, b]$ attains there any value between $f(a)$ and $f(b)$. Prove the following related results.

a. Let f be continuous on the closed interval $[a, b]$ where its minimum value is m and its maximum is M. If k is any value between m and M, then there exists a point $c \in [a, b]$ such that $f(c) = k$.

b. Let f be continuous on the open interval (a, b) where its greatest lower bound is l and its least upper bound is U. If $l < k < U$, then there exists a point $c \in (a, b)$ such that $f(c) = k$.

3.41 Suppose f and g are continuous on $[a, b]$. Moreover, assume $f(a) > g(a)$, but $f(b) < g(b)$. Prove there is a point $c \in (a, b)$ where $f(c) = g(c)$.

3.42 Is the function defined by

$$f(x) := \begin{cases} 0 & \text{if } x \text{ is irrational,} \\ x/q & \text{if } x = p/q \text{ is rational,} \end{cases}$$

continuous at $x = 0$?

3.43 Show that the function $f: [0, 2] \to \mathbf{R}$ defined by

$$f(x) := \begin{cases} x & \text{if } 0 \le x < 1, \\ 1 + x & \text{if } 1 \le x \le 2, \end{cases}$$

does not have the intermediate-value property, and thus must be discontinuous. Give another example.

3.44 Using Lemma 3.50, we can conclude that a function f which is unbounded in an interval $[a, b]$ is not continuous in that interval.

Check if the following functions f are continuous in the indicated intervals I.

(a) $f(x) := \begin{cases} x/\sin^2 x & \text{if } -\pi/2 \le x \le \pi/2, \\ 1 & \text{if } x = 0, \end{cases}$ $\quad I = [-\pi/2, \pi/2]$

(b) $f(x) := \begin{cases} \dfrac{1}{x}\sin\dfrac{1}{x} & \text{if } 0 < x \le 1, \\ 0 & \text{if } x = 0, \end{cases}$ $\quad I = [0, 1]$

(c) $f(x) := \begin{cases} x^2\sin^3 x & \text{if } 0 < x \le 1, \\ 0 & \text{if } x = 0, \end{cases}$ $\quad I = [0, 1]$

3.45 Use Corollary 3.57 to prove that the functions f below have a root in the indicated interval I.

(a) $f(x) := x^2 - 7, \quad I = [2, 3]$

(b) $f(x) := x^4 - x^2 + x - 2, \quad I = [1, 2]$

(c) $f(x) := x^3 - 3, \quad I = [1, 2]$

(d) $f(x) := x^5 + 2x^3 - 7x - 10, \quad I = [1, 2]$

(e) $f(x) := \sin x + 2\cos x, \quad I = [0, \pi]$

(f) $f(x) := e^{-x} - x, \quad I = [0, 1]$

(g) $f(x) := \tan x - x, \quad I = [-\pi/4, \pi/4]$

(h) $f(x) := 2\sinh x + \cosh x, \quad I = [-2, 0]$

3.46 Use the MACSYMA function bisection(f,x,a,b) from MACSYMA-Session 3.8 to find a zero of f in the interval I.

(a) $f(x) := x^3 + 2x^2 + 10x - 20, \quad I = [1, 2]$

(b) $f(x) := x^3 + 2x^2 + 10x, \quad I = [1, 2]$

(c) $f(x) := x^3 - 5x^2 - 10x + 1, \quad I = [-3, 3]$

(d) $f(x) := x^3 - 5x^2 - 10x + 1, \quad I = [-3, 0]$

(e) $f(x) := x^3 - 5x^2 - 10x + 1, \quad I = [0, 3]$

(f) $f(x) := \cos x - x, \quad I = [0, 1]$

3.47 Give numerical solutions of the equations (a) $x = e^{-x}$ and (b) $x = \cosh x - 1$.

3.6 Proofs

Proof of Theorem 3.44. We prove the continuity of the inverse f^{-1} of a continuous function, in the special case that the function f is strictly monotone in the interval $[a, b]$. Then by Corollary 3.58, the function f assumes every value between $f(a)$ and $f(b)$ exactly once.

Clearly the inverse function

$$x = f^{-1}(y)$$

is also strictly monotone, in the interval with endpoints $f(a)$ and $f(b)$. Because of the monotonicity, if $x_1 \neq x_2$, then

$$|f(x_1) - f(x_2)| = |y_1 - y_2| > 0 .$$

For any $\varepsilon > 0$, consider the function $|f(x+\varepsilon) - f(x)|$ on $[a, b-\varepsilon]$. This function takes on its minimum value, call it $m(\varepsilon)$, at some point $x^* \in [a, b - \varepsilon]$. Therefore, for all x_1, x_2 in $[a, b-\varepsilon]$, such that $|x_1 - x_2| \geq \varepsilon$, it follows (from the monotonicity of f) that

$$|f(x_1) - f(x_2)| \geq m(\varepsilon) .$$

We conclude that

$$|f(x_1) - f(x_2)| = |y_1 - y_2| < m(\varepsilon) \quad \Longrightarrow \quad |x_1 - x_2| < \varepsilon .$$

The continuity of the inverse function now follows from the $\varepsilon\delta$-definition (Definition 3.33), with $\delta = m(\varepsilon)$. $\square$

Proof of Corollary 3.60. Let us consider the function

$$\begin{aligned} f(x) &:= x^n + a_{n-1}x^{n-1} + \cdots + a_1 x + a_0 \\ &= x^n(1 + a_{n-1}x^{-1} + \cdots + a_1 x^{-n+1} + a_0 x^{-n}) . \end{aligned}$$

To prove the theorem, we need merely find an $x_1 < 0$ and an $x_2 > 0$ where

$$|a_{n-1}x_j^{-1} + \cdots + a_0 x_j^{-n}| \leq 1/2, \quad j = 1, 2 ,$$

so that

$$1 + a_{n-1}x_j^{-1} + \cdots + a_1 x_j^{-n+1} + a_0 x_j^n \geq 1 - 1/2 > 0, \quad j = 1, 2 ,$$

as then

$$f(x_1) = x_1^n\left(1 + a_{n-1}x_1^{-1} + \cdots + a_1 x_1^{-n+1} + a_0 x_1^{-n}\right) < 0 ,$$

and

$$f(x_2) = x_2^n\left(1 + a_{n-1}x_2^{-1} + \cdots + a_1 x_2^{-n+1} + a_0 x_2^{-n}\right) > 0 ,$$

where we used $x_1^n < 0$ and $x_2^n > 0$, which follow from n odd, $x_1 < 0$, and $x_2 > 0$.

Let us see how this is done. If we choose x_j so that it satisfies $|x_j| \geq \max\{1, 2n|a_n|, \ldots, 2n|a_1|\}$, then $|x_j^k| \geq |x_j|$, and for all $k = 1, \ldots, n$

$$\frac{|a_k|}{|x_j|^k} \leq \frac{|a_k|}{|x_j|} \leq \frac{|a_k|}{2n|a_k|} = \frac{1}{2n}, \quad \text{for } j = 1, 2.$$

From this we may get the upper bound

$$\left| \frac{a_{n-1}}{x_j^1} + \frac{a_{n-2}}{x_j^2} + \cdots + \frac{a_1}{x_j^{n-1}} + \frac{a_0}{x_j^n} \right|$$

$$\leq \frac{1}{2n} + \frac{1}{2n} + \cdots + \frac{1}{2n} = \frac{1}{2}, \qquad \text{for } j = 1, 2.$$

Corollary 3.57 now establishes that the function must have a zero between x_1 and x_2. $\qquad\square$

Derivatives

4 Differentiation

Two functions f and g are tangent at a point x_0 if their graphs almost coincide, near x_0, in the sense of Definition 4.1 below. A function f can have, at a given point x_0, at most one tangent which is a linear function, say

$$l(x) = f(x_0) + m(x - x_0) ,$$

in which case
- l is called the *tangent line*, or simply the tangent, of f at x_0,
- the slope m of l is called the *slope*, or *derivative*, of f at x_0.

Calculus is concerned mainly with the computation, properties, and applications of derivatives. The derivative and its computation – a process called differentiation – are the subjects of this chapter.

4.1 Tangency

Definition 4.1 (Tangency). Let f and g be functions with graphs $\mathcal{F}$ and $\mathcal{G}$, respectively, and let the point $P = (x_0, y_0)$ be common to $\mathcal{F}$ and $\mathcal{G}$,

$$f(x_0) = g(x_0) = y_0 . \tag{4.1}$$

Then we call
- the *graphs $\mathcal{F}$ and $\mathcal{G}$ tangent* at P, or
- the *functions f and g tangent* at x_0, if

$$\lim_{x \to x_0} \frac{f(x) - g(x)}{x - x_0} = 0 . \tag{4.2}$$

For the two-sided limit in (4.2) to make sense, it is necessary that both f and g be defined in some open interval (a, b) containing x_0, i.e., $a < x_0 < b$. We then say that f and g are defined in a *neighborhood* of x_0.

Statement (4.2) means that near x_0 the graphs $\mathcal{F}$ and $\mathcal{G}$ approach each other faster than x approaches x_0. An equivalent (and sometimes more convenient) statement is

$$\lim_{x \to x_0} \frac{|f(x) - g(x)|}{|x - x_0|} = 0 . \tag{4.3}$$

The statement

$$\lim_{x \to x_0} (f(x) - g(x)) = 0 , \tag{4.4}$$

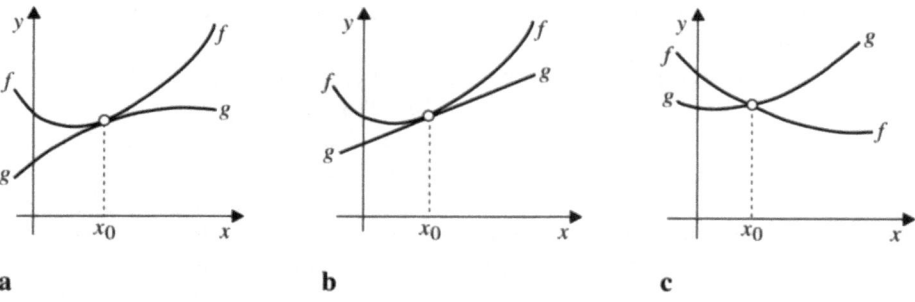

Fig. 4.1 a–c. Tangency. **a** f and g are tangent at x_0; **b** g is linear tangent of f at x_0; **c** f and g are not tangent at x_0

only says that the graphs $\mathcal{F}$ and $\mathcal{G}$ approach each other as x approaches x_0. Therefore (4.4) is weaker than (4.2); see Fig. 4.1c for an illustration of (4.4) without tangency at x_0.

Remark 4.2. Let the function f be defined in a neighborhood of a point x_0. Then f has infinitely many tangent functions at x_0. Indeed, for any real α and any $\beta > 1$, the function

$$g(x) := f(x) + \alpha(x - x_0)^\beta \tag{4.5}$$

is tangent to f at x_0.

Proof.

$$\lim_{x \to x_0} \frac{|f(x) - g(x)|}{|x - x_0|} = \lim_{x \to x_0} \frac{|\alpha(x - x_0)^\beta|}{|x - x_0|}$$
$$= |\alpha| \lim_{x \to x_0} |x - x_0|^{\beta - 1} = 0 \quad \text{since } \beta > 1 \,. \qquad \square$$

Example 4.3. The function $f(x) := x^2$ is tangent, at any point x_0, to the line

$$y = x_0^2 + 2x_0(x - x_0) \,. \tag{4.6}$$

This follows from Remark 4.2 by taking $\alpha = -1$ and $\beta = 2$ in (4.5), giving

$$g(x) = x^2 - (x - x_0)^2 = 2x_0 x - x_0^2$$
$$= x_0^2 + 2x_0(x - x_0) \,,$$

whose graph is a line tangent to x^2 at x_0. For example, a line tangent to x^2 at $x := -2$ is

$$y = 4 - 4(x + 2) = -4 - 4x \,.$$

MACSYMA-Session 4.1. The tangency condition (4.2) can be checked with the
MACSYMA function

```
c1. test_tangency(f,g,x,x0)  := case(limit((f-g)/(x-x0),x,x0),
                                     [0, "Tangent"],
                                     [und,"Limit was undefined"],
                                     [ind,"Limit was not definite"],
                                     [otherwise,"Not tangent"])$
```

The MACSYMA function case(key, [indicator1,form1], [indicator2,form2],
..., [indicatorN,formN]) evaluates the expression "key" in its first argument and
compares it to each indicator, until it finds a match, say indicator j, in which case the
function evaluates form j and returns that value.

In the function test_tangency, we use the case macro to compare
$\lim_{x \to x_0}((f(x) - g(x))/(x - x_0)$ with the given indicators, where
-   if the limit equals indicator 0, the string "Tangent" is returned;
-   if the limit is undefined (und) or indefinite (ind), the function returns the appropriate
    strings; and
-   otherwise returns "Not tangent"; the default result when none of the previous
    indicators matched the key.

Some examples.

```
c2. test_tangency(x^3,x^2+x-1,x,1)
d2.    Tangent
```

The functions x^3 and $x^2 + x - 1$ are thus tangent at $x = 1$, which can be illustrated by
plot([x^3,x^2+x-1],x,0,2). Similarly, we can verify that x^3 is tangent at $x = 1$ to
the line $y = 1 + 3(x - 1)$.

The functions $f(x) = x^2 \sin(1/x)$ and $g(x) = x^2$ are tangent at $x = 0$.

```
c3. test_tangency(x^2*sin(1/x),x^2,x,0)
d3.    Tangent
```

The functions $1 + x$, e^x, and $1 + x + x^2$ are all tangent at $x = 0$.

```
c4. test_tangency(1+x+x^2,%e^x,x,0)
d4.    Tangent
c5. test_tangency(1+x,%e^x,x,0)
d5.    Tangent
```

see also Fig. 3.6.

The functions $f(x) = |x|$ and $g(x) = x$ are not tangent at $x = 0$

```
c6. test_tangency(abs(x),x,x,0)
d6.    Not tangent
```

because the limit of $(|x| - x)/x$ as $x \to 0$ does not exist.

Similarly, the functions

$$f(x) := \begin{cases} 0 & \text{if } x = 0, \\ x * \sin(1/x) & \text{otherwise,} \end{cases}$$

and $g(x) := x$ are not tangent at $x = 0$. Note that the function test_tangency does not
verify that the functions even meet at $x = x_0$. To correct this oversight, we rewrite the
function as follows

```
c2. test_tangent(f,g,x,x0) := case(limit(f-g,x,x0),
      0, case(limit((f-g)/(x-x0),x,x0),
              0, "Tangent",
              und,"Limit was undefined",
              ind,"Limit was not definite",
              otherwise,"Not tangent"),
      otherwise, "Unable to establish that the point is in common")$
```

Here $\lim_{x \to x_0}(f(x) - g(x))$ is first tested

- if the limit is 0, then f and g intersect at x_0, and `test_tangency` proceeds to check tangency as before,
- otherwise the function returns "Unable to establish ..."

The following are easy consequences of the definition of tangency.

Lemma 4.4 (Properties of tangent functions). Let f, g be functions tangent at a point x_0.

a. If g is tangent at x_0 to a third function h, then f and h are also tangent[1] at x_0.
b. If g is continuous at x_0, then f is also continuous at x_0.
c. If f and g are linear functions, then $f = g$.

Proof. a. For any x, we have

$$\lim_{x \to x_0} \frac{f(x) - h(x)}{x - x_0} = \lim_{x \to x_0} \frac{f(x) - g(x)}{x - x_0} + \lim_{x \to x_0} \frac{g(x) - h(x)}{x - x_0} = 0 .$$

b. Let g be continuous at x_0, i.e.,

$$g(x_0) = y_0 = \lim_{x \to x_0} g(x) . \tag{4.7}$$

Then, for all x,

$$|f(x) - y_0| \le |f(x) - g(x)| + |g(x) - y_0| ,$$

and by (4.4) and (4.7),

$$\lim_{x \to x_0} f(x) = y_0 .$$

c. Let f and g be linear functions, passing through $P = (x_0, y_0)$, say

$$f(x) = y_0 + m(x - x_0), \quad g(x) = y_0 + n(x - x_0) .$$

Then

$$\lim_{x \to x_0} \frac{f(x) - g(x)}{x - x_0} = m - n .$$

Since $f(x_0) = g(x_0)$, it follows that f and g are tangent at x_0 if and only if $f = g$. □

---

1 That is, tangency is transitive.

We saw (Remark 4.2) that a function has infinitely many tangents at any point x_0. Can it have more than one linear tangent at x_0? The answer is "no".

Corollary 4.5. A function f can have at most one linear tangent at any point x_0.

Proof. Let f have two linear tangents, l_1 and l_2, at x_0. Then l_1 and l_2 are themselves tangent at x_0, by Lemma 4.4a, and therefore $l_1 = l_2$ by Lemma 4.4c. ☐

Example 4.6. The function $f(x) := |x|$ has no linear tangent at $x = 0$.

Proof. If l is a linear tangent of $|x|$ at $x = 0$, then l is a line through the origin, say $l(x) = mx$ where m is the slope. But the limit

$$\lim_{x \to 0} \frac{|x| - mx}{x} = \lim_{x \to 0} \frac{|x|}{x} - m$$

does not exist, for any m. ☐

Exercises

4.1 Show that tangency, defined in Definition 4.1, implies (4.4),

$$\lim_{x \to x_0} (f(x) - g(x)) = 0 .$$

Hint: If the limit in (4.4) does not exist, or is nonzero, then (4.2) cannot hold.

4.2 Use the function `test_tangency` in MACSYMA-Session 4.1 to test which of the following functions is tangent to $\cos x$ at $x = 0$.
   (a) $1 + x^2$ (b) $1 - x^2$ (c) 1 (horizontal line)
   (d) $1 + 1000x^2$ (e) $1 + 1{,}000{,}000x^2$ (f) $1 + |x|$
   (g) $0.999 + 1000x^2$ (h) $1 + 0.0001x + 1{,}000{,}000x^2$
   (i) $\sec x$

   Draw the graphs of $\cos x$ and of the functions in a–i, whenever possible, and use the graphs to illustrate "tangency" or "nontangency".

4.3 Compute the tangent line of $f(x) := x^3$ at any point x_0.
   Hint: Use the trick of Example 4.3 twice. First get a parabola g which is tangent to x^3 and then get a line l which is tangent to g. The steps are:
   a. The function $g(x) := x^3 - (x - x_0)^3$ is a parabola tangent to x^3 at x_0.
   b. For any α, the function $l(x) := g(x) - \alpha(x - x_0)^2$ is tangent to g at x_0.
   c. Determine α such that l is a line.

4.4 (*n*th-order tangency). Let n be a nonnegative integer. The functions f and g are said to be nth-order tangent at x_0 if $f(x_0) = g(x_0)$ and

$$\lim_{x \to x_0} \frac{f(x) - g(x)}{(x - x_0)^n} = 0 . \tag{4.8}$$

Special cases of n:
 – for $n = 0$, (4.8) reduces to (4.4), which can be called 0*th-order* tangency.

- for $n = 1$, (4.8) reduces to the tangency condition (4.2), which we rename
  first-order tangency.

Prove:

a. If f and g are nth-order tangent at x_0, they are also mth-order tangent, for any
   $0 \leq m \leq n$. Compare with Exercise 4.1.

b. If at x_0 the functions f, g are nth-order tangents and the functions g, h are
   mth-order tangents, then f and h are kth-order tangent, where $k = \min\{m, n\}$.
   Compare with Lemma 4.4 a.

c. A function f can have, at any point x_0, at most one nth-order tangent g which is a
   polynomial of the nth degree: $g(x) = f(x_0) + \sum_{k=1}^{n} c_k (x - x_0)^k$. Compare with
   Corollary 4.5.

4.5 Verify that the MACSYMA function

```
test_tangency_order(f,g,x,x0,default1):= case(limit((f-g)/
                    ((x-x0)^default1),x,x0),
                            [0, "tangent"],
                            [und, "limit was not defined"],
                            [ind, "limit was not definite"],
                            [otherwise, "not tangent"])$
```

tests nth-order tangency, where the fifth argument `default1` is used for $n > 1$.

a. For $n = 1$ the argument `default1` can be omitted, in which case it would
   automatically be assigned its default value of 1. Thus
   `test_tangency_order(cos(x),sec(x),x,0)` suffices to verify that $\cos x$ and
   $\sec x$ are (first-order) tangents at $x = 0$.

b. The functions $\sin x$ and $x - x^3/6$ are 4th-order tangents at $x = 0$.
   c2. `test_tangency_order(sin(x),x-x^3/6,x,0,4)`
   d2.   tangent
   Are these functions also 3rd-order tangent at $x = 0$? 5th-order tangent at $x = 0$?

c. Explain the answer to
   `test_tangency_order(sin(x),cos(x),x,pi/4,0)`

4.6 Write a MACSYMA function `find_tangency_order(f,g,x,x0)` that calculates
the maximal order of tangency so that f and g are nth-order tangents at x_0.

For the following triples $\{f, g, x_0\}$, find the maximal order of tangency.

(a) $f(x) = \sin x$, $g(x) = x - \dfrac{x^3}{3!} + \dfrac{x^5}{5!}$, $x_0 = 0$

(b) $f(x) = |x|$, $g(x) = |x| + x^3$, $x_0 = 0$

(c) $f(x) = |x|$, $g(x) = |x + x^3|$, $x_0 = 0$

(d) $f(x) = \sin x$, $g(x) = \tan x$, $x_0 = 0$

(e) $f(x) = (x - 1)^{10}$, $g(x) = (x - 1)^9$, $x_0 = 1$

4.2 Differentiability

Definition 4.7 (Slope, derivative). If the function f has, at the point x_0, a linear tangent[2]

$$l(x) = f(x_0) + m(x - x_0) \,,$$

then the slope m is called the slope, or derivative, of f at x_0 and is denoted by $f'(x_0)$.

If the derivative $f'(x_0)$ exists, it can be computed by taking a limit. To show this, write the linear tangent of f at x_0 as

$$l(x) = f(x_0) + f'(x_0)(x - x_0) \,. \tag{4.9}$$

Then

$$0 = \lim_{x \to x_0} \frac{f(x) - l(x)}{x - x_0}, \quad \text{by (4.2)} \,,$$

$$= \lim_{x \to x_0} \frac{f(x) - f(x_0)}{x - x_0} - \lim_{x \to x_0} \frac{f'(x_0)(x - x_0)}{x - x_0}, \quad \text{by (4.9)} \,,$$

$$= \lim_{x \to x_0} \frac{f(x) - f(x_0)}{x - x_0} - f'(x_0), \quad \text{since } f'(x_0) \text{ is a constant} \,.$$

Therefore, the derivative of f at x_0 is

$$f'(x_0) = \lim_{x \to x_0} \frac{f(x) - f(x_0)}{x - x_0} \,. \tag{4.10}$$

We often have to calculate the derivative $f'(x)$ of f at a *general point* x. We then use the limit (4.10) with x_0 replaced by x. But x is already used in (4.10) for the variable which tends to x_0, so we have to rename that variable too. Since ξ is the Greek for x, it is appropriate to denote by it a general point on the x-axis.

The derivative of f at the point x is therefore

$$f'(x) := \lim_{\xi \to x} \frac{f(\xi) - f(x)}{\xi - x} \,, \tag{4.11}$$

where $(x, f(x))$ is the point of tangency, and ξ is the variable which tends to x.

Definition 4.8 (Differentiability). a. A function f is called differentiable at the point x_0 if it has a derivative at x_0, i.e., if the limit (4.10) exists. Thus f is differentiable at x_0 if and only if it has a tangent line at x_0, in which case the slope of that line equals the derivative $f'(x_0)$; see Definition 4.7.

b. If the function f is differentiable at every point of an interval I, then f is called differentiable in I.

---

2 Which, by Corollary 4.5, is unique.

Example 4.9 (Absolute-value function). In Example 4.6 we saw that $f(x) = |x|$ does not have a linear tangent at $x = 0$, and is therefore nondifferentiable at $x = 0$. We now show that $|x|$ is differentiable for all $x \neq 0$.

Consider first the case $x > 0$ and use the limit (4.11). We can assume that the variable ξ is also positive, because $\xi \to x$, i.e., ξ comes arbitrarily close to x.

$$f'(x) = \lim_{\xi \to x} \frac{f(\xi) - f(x)}{\xi - x} = \lim_{\xi \to x} \frac{|\xi| - |x|}{\xi - x} = \lim_{\xi \to x} \frac{\xi - x}{\xi - x} = 1 \, .$$

For $x < 0$ we get similarly

$$f'(x) = \lim_{\xi \to x} \frac{f(\xi) - f(x)}{\xi - x} = \lim_{\xi \to x} \frac{|\xi| - |x|}{\xi - x} = \lim_{\xi \to x} \frac{-\xi + x}{\xi - x} = -1 \, ,$$

where we used the assumption $\xi < 0$.

We now give a graphical interpretation of the limit (4.10). Let $\mathcal{F}$ be the graph of a function f, and let $P = (x_0, f(x_0))$ be a point on $\mathcal{F}$. It is required to find the tangent line l of f at the point P, called the *point of tangency*. For example, in Fig. 4.2 the parabola $\mathcal{F}$ is the graph of $f(x) = x^2$, and $P := (1/2, 1/4)$ is the point of tangency.

We show that the tangent line at P is the limit of lines PQ, where the points Q are also on the graph $\mathcal{F}$, and $Q \to P$.

Let Q be another point on $\mathcal{F}$. Then the line PQ (which intersects $\mathcal{F}$ at P and Q) is called a *secant line*. Several secant lines are plotted in Fig. 4.2. We call Q the *moving point*, and denote it by $Q = (\xi, f(\xi))$. If Q is not far from P, then the secant line PQ is an approximation of the tangent line at P, and the approximation gets better as Q gets closer to P.

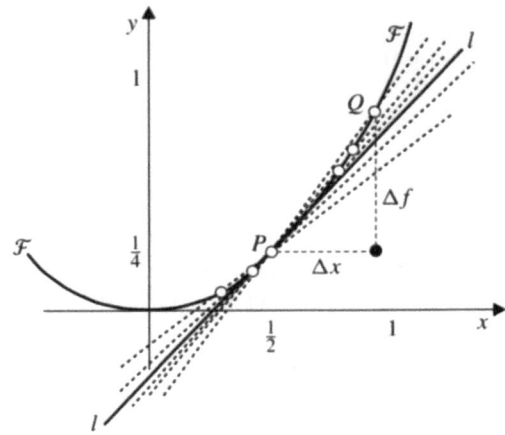

Fig. 4.2. Tangent line l as the limit of secant lines PQ

The secant line PQ has the slope

$$m_{PQ} = \frac{\Delta f}{\Delta x} = \frac{f(\xi) - f(x_0)}{\xi - x_0} \,. \tag{4.12}$$

The limit (4.10),

$$f'(x_0) = \lim_{\xi \to x_0} \frac{f(\xi) - f(x_0)}{\xi - x_0} \,,$$

is therefore

$$f'(x_0) = \lim_{Q \to P} m_{PQ} \,, \tag{4.13}$$

i.e., the (slope of the) tangent line of f at P is the limit of the (slopes m_{PQ} of the) secant lines PQ, as $Q \to P$.

Example 4.10. Return now to $f(x) = x^2$ and the point of tangency $P = (1/2, 1/4)$. Let $Q = (\xi, \xi^2)$ be a moving point on the parabola $\mathcal{F}$. Then the slope of the secant line PQ is

$$m_{PQ} = \frac{\xi^2 - 1/4}{\xi - 1/2} \,.$$

For the points $P = (1/2, 1/4)$ and $Q = (1, 1)$ shown in Fig. 4.2, the slope of the secant line PQ is

$$m_{PQ} = \frac{1 - 1/4}{1 - 1/2} = \frac{3}{2} \,.$$

The derivative of $f(x) = x^2$ at $x = 1/2$ is, by (4.13),

$$f'(1/2) = \lim_{Q \to P} m_{PQ} = \lim_{\xi \to 1/2} \frac{\xi^2 - 1/4}{\xi - 1/2} = \lim_{\xi \to 1/2} \frac{(\xi + 1/2)(\xi - 1/2)}{\xi - 1/2}$$
$$= \lim_{\xi \to 1/2} (\xi + 1/2) = 1 \,.$$

The equation of the tangent line l of $f(x) = x^2$ at $P = (x_0, f(x_0)) = (1/2, 1/4)$ can now be written by (4.9),

$$y = f(x_0) + f'(x_0)(x - x_0) = 1/4 + (x - 1/2) = x - 1/4 \,,$$

in agreement with (4.6), for $x_0 = 1/2$.

MACSYMA-Session 4.2. a. Plot the graphs of the parabola $y = x^2$ and the line $y = x - 1/4$, and verify that they are tangent at the point $P = (1/2, 1/4)$.

c1. `plot2([x^2,x-1/4],x,0,1)$`

b. Use the following MACSYMA function

```
c2. line(x,x1,y1,x2,y2) := block([deltax:x1-x2],
                 if deltax=0 then 0
                 /* we can't divide by 0, so just return 0 */
                 else (y1-y2)*x/deltax + (x1*y2-x2*y1)/deltax)$
```

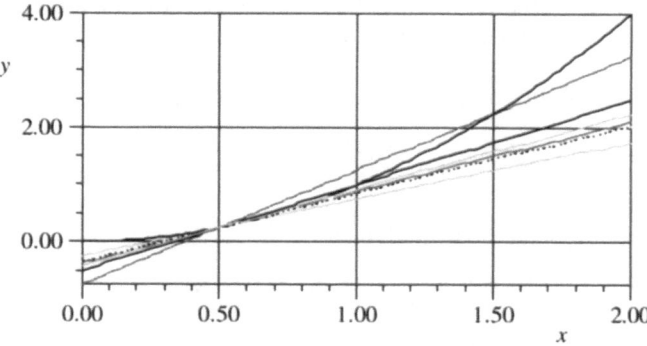

Fig. 4.3. Graphs of Session 4.2, line c5

to plot the sequence of secant lines through P and the points
c3. `makelist(line(x,1/2,1/4,1/2+1/n,(1/2+1/n)^2),n,1,5)`
d3. $[2*x - (3/4), ((3*x)/2) - (1/2), ((4*x)/3) - (5/12), ((5*x)/4) - (3/8), ((6*x)/5) - (7/20)]$
place the functions x^2 and $x - 1/4$ at head of the list of secant lines
c4. `append([x^2,x-1/4],%)$`
and then plot them (to distinguish between the curves, use different line styles),
c5. `plot2(''%,x,0,2,[0,1,2,3,4,5,6])$`

For general functions f the computation of the limit (4.10) may be quite difficult. Here MACSYMA comes to the rescue.

MACSYMA-Session 4.3. The following MACSYMA function computes the derivative of $f(x)$ at $x = x_0$
c1. `deriv(f,x,x0):=limit((f-subst(x0,x,f))/(x-x0),x,x0)$`
Some examples.
a. The derivative of x^3 at $x = 1$ is 3. To confirm, use `deriv(x^3,x,1)`.
b. In Example 4.6 we saw that the function $|x|$ has no linear tangent at $x_0 = 0$, i.e., $|x|$ is nondifferentiable at $x_0 = 0$ and `deriv(abs(x),x,0)` is not defined.
c. The derivative of $x^{2/3}$ at $x = 10^{-6}$ is 200/3, as can be verified by `deriv(x^(2/3),x,10^(-6))`. Similarly, the derivative at $x = -10^{-6}$ is $-200/3$. How does the derivative of $x^{2/3}$ behave as x gets closer to 0? Is $x^{2/3}$ differentiable at the point $x = 0$? Does the function $x^{2/3}$ have a linear tangent at $x = 0$? Draw the graph of $x^{2/3}$ and explain.
d. The function $x \sin(1/x)$ is not differentiable at $x = 0$. This is confirmed by `deriv(x*sin(1/x),x,0)` returning ind, for indefinite, a finitely bounded quantity which fails to approach a limit as x tends to 0.

The last three functions, $|x|$, $x^{2/3}$ and $x \sin(1/x)$ are continuous, but not differentiable, at the point $x := 0$. We show now that the reverse is impossible.

Theorem 4.11 (Differentiability implies continuity). Let the function f be differentiable at a point x_0. Then f is continuous at x_0.

Proof.[3] Differentiability of f at x_0 means that the limit

$$\lim_{x \to x_0} \frac{f(x) - f(x_0)}{x - x_0} = L$$

exists. This equation can be rewritten as

$$\lim_{x \to x_0} (f(x) - f(x_0)) = L \lim_{x \to x_0} (x - x_0) = 0, \quad \text{i.e.,} \quad \lim_{x \to x_0} f(x) = f(x_0) \,,$$

and so f is continuous at the point x_0. $\square$

| If f is differentiable in an open interval I, then it is continuous in I.

Suppose the function f is differentiable at the point x_0, and we know that its derivative is positive, i.e., $f'(x_0) > 0$. What does this tell us about the function in a neighborhood of x_0?

Lemma 4.12. a. If the derivative $f'(x_0) > 0$, then there is a positive δ such that

$$f(v) < f(x_0) < f(u), \quad \text{for all } x_0 - \delta < v < x_0 < u < x_0 + \delta, \quad (4.14)$$

in which case we say that f is *increasing* at the point x_0.

b. Conversely, if (4.14) holds for some positive δ, then the derivative of f, if it exists, is nonnegative at x_0, i.e., $f'(x_0) \geq 0$.

c. If the function f is increasing and differentiable in an interval I, then $f'(x)$ is nonnegative in I.

Proof. The differentiability of f at x_0 means, in light of Definition 3.5 (the $\varepsilon\delta$-definition of limits), that for any $\varepsilon > 0$ there is a $\delta > 0$ such that

$$f'(x_0) - \varepsilon < \frac{f(x) - f(x_0)}{x - x_0} < f'(x_0) + \varepsilon \quad \text{for all } x_0 - \delta < x < x_0 + \delta \,.$$

a. If $f'(x_0) > 0$, select $\varepsilon := f'(x_0)/2$. Then there is a $\delta > 0$ such that

$$\varepsilon < \frac{f(x) - f(x_0)}{x - x_0} < 3\varepsilon \quad \text{for all } x_0 - \delta < x < x_0 + \delta \,. \quad (4.15)$$

If u is in the interval $(x_0, x_0 + \delta)$, i.e., $x_0 < u < x_0 + \delta$, then, from (4.15),

$$\varepsilon < \frac{f(u) - f(x_0)}{u - x_0} \quad \text{or} \quad f(u) > f(x_0) + \varepsilon(u - x_0) > f(x_0) \,.$$

---

3 See also Exercise 4.7.

If v is in the interval $(x_0 - \delta, x_0)$, i.e., $x_0 - \delta < v < x_0$, then, from (4.15),

$$\frac{f(v) - f(x_0)}{v - x_0} < 3\varepsilon \quad \text{or} \quad f(v) < f(x_0) + 3\varepsilon(v - x_0) < f(x_0) \quad (\text{since } v < x_0)$$

proving part a.

b. From (4.14) we can only conclude that $f'(x_0) \geq 0$.

c. If f is increasing in an interval I (according to Definition 1.39), then it is increasing at any point $x_0 \in I$, and the result follows from part b. $\square$

Remark 4.13. The following statements can be proved similarly.

a. If $f'(x_0) < 0$, then the function f is *decreasing* at the point x_0, meaning that there is a $\delta > 0$ such that

$$f(v) > f(x_0) > f(u), \quad \text{for all } x_0 - \delta < v < x_0 < u < x_0 + \delta . \quad (4.16)$$

b. If (4.16) holds for some positive δ, and if f is differentiable at x_0, then $f'(x_0) \leq 0$.

c. If the function f is decreasing and differentiable in an interval I, then $f'(x)$ is nonpositive in I.

Example 4.14. The function $f(x) := x^3$ is increasing at the point $x := 0$ (and at all points). However, $f'(0) = 0$, illustrating why we cannot conclude, in part b of Lemma 4.12, that $f'(x_0)$ is positive.

Example 4.15. If a function f is increasing at a point x_0, it does not follow that there is an interval I around the point x_0 such that f is increasing in I. For example, the function

$$f(x) := \begin{cases} x + 2x^2 \sin(1/x) & \text{if } x \neq 0, \\ 0 & \text{if } x = 0, \end{cases}$$

is increasing at the point $x = 0$ (at $x = 0$, f' does not exist), but not in any interval I around 0.

A function f is not differentiable at the point x_0, if at x_0 the function has no linear tangent or, equivalently, if the limit (4.10),

$$f'(x_0) = \lim_{\xi \to x_0} \frac{f(\xi) - f(x_0)}{\xi - x_0} ,$$

does not exist. In this case, we may consider the one-sided limits

$$\lim_{\xi \to (x_0)+} \frac{f(\xi) - f(x_0)}{\xi - x_0}, \qquad \lim_{\xi \to (x_0)-} \frac{f(\xi) - f(x_0)}{\xi - x_0} ,$$

and the corresponding one-sided derivatives.

Definition 4.16 (One-sided derivatives). Let $f : I \to \mathbf{R}$ be a real function in an interval I, x_0 a point in I.

a. If the limit

$$f'_+(x_0) := \lim_{\xi \to (x_0)+} \frac{f(\xi) - f(x_0)}{\xi - x_0} \tag{4.17}$$

exists, it is called the *right* derivative of f at x_0, and f is called *right* differentiable at x_0. If f is right differentiable at x_0, then the line

$$l_+(x) = f(x_0) + f'_+(x_0)(x - x_0) \tag{4.18}$$

is the *right* tangent of f at x_0, i.e., the limit of secant lines, as the moving point approaches the tangency point from the right.

b. The *left* derivative, and the *left* differentiability, of f at x_0 are analogously defined by the limit from the left

$$f'_-(x_0) := \lim_{\xi \to (x_0)-} \frac{f(\xi) - f(x_0)}{\xi - x_0} . \tag{4.19}$$

The *left* tangent of f at x_0 is

$$l_-(x) = f(x_0) + f'_-(x_0)(x - x_0) .$$

Comparing with Definition 4.8, we see that f is differentiable at x_0 if and only if the following three conditions hold
a. f is right differentiable at x_0,
b. f is left differentiable at x_0, and
c. $f'_+(x_0) = f'_-(x_0)$.

Example 4.17 (Absolute-value function). The function $f(x) = |x|$ is differentiable for all $x \neq 0$ (Example 4.9). At the point $x = 0$, the function f is not differentiable (Example 4.6). However, when we examine Fig. 4.4a, we see that the graph of $|x|$ has a *corner* at $x = 0$, where it has infinitely many "supporting lines", like the two shown in Fig. 4.4a, but no tangent line, according to Definition 4.1.
The right derivative of $f(x) = |x|$ at $x := 0$ is

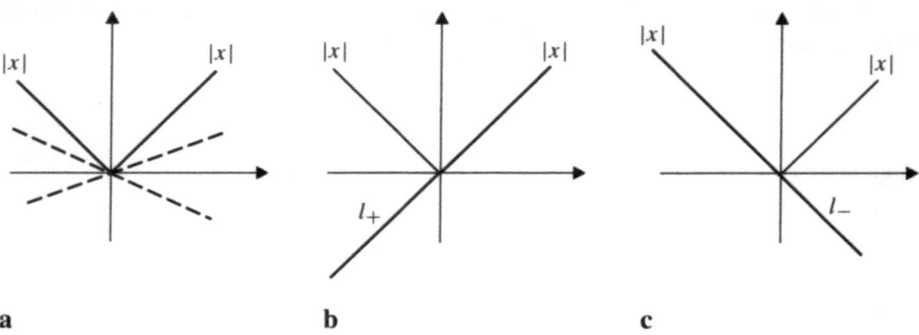

Fig. 4.4. Absolute-value function with two supporting lines (**a**) and its right (**b**) and left (**c**) tangents at 0

$$f'_+(0) = \lim_{\xi \to 0+} \frac{f(\xi) - f(0)}{\xi - 0} = \lim_{\xi \to 0+} \frac{|\xi| - |0|}{\xi - 0} = 1 \,,$$

and the right tangent, by (4.18), is $l_+(x) = x$ (see Fig. 4.4b).
Similarly,

$$f'_-(0) = \lim_{\xi \to 0-} \frac{f(\xi) - f(0)}{\xi - 0} = \lim_{\xi \to 0-} \frac{|\xi| - |0|}{\xi - 0} = -1 \,,$$

giving the left tangent $l_-(x) = -x$ (see Fig. 4.4c).
However, since $f'_+(0) \neq f'_-(0)$, the two-sided limit

$$\lim_{\xi \to 0} \frac{|\xi| - |0|}{\xi - 0}$$

does not exist.

Example 4.18. The function $f(x) := x \sin(1/x)$ has neither right derivative nor left derivative at $x = 0$.

MACSYMA-Session 4.4. The right derivative of f at $x = x_0$ is calculated by the following MACSYMA function.

c1. `dif_right(f,x,x0):=limit((f-limit(f,x,x0))/(x-x0), x, x0, plus)$`

Note that `dif_right(f,x,x0)` substitutes `limit(f,x,x0)` for $f(x_0)$; indeed, the function f need not even be defined at x_0.

The left derivative is calculated similarly.

c2. `dif_left(f,x,x0):=limit((f-limit(f,x,x0))/(x-x0), x, x0, minus)$`

The `plus` and `minus` designators in the `limit` functions cause the point $x = x_0$ to be approached from above and from below, respectively.

The right and left derivatives of $f(x) := |x|$ at $x = 0$ are different

c3. `dif_right(abs(x),x,0)`

d3.   1

c4. `dif_left(abs(x),x,0)`

d4. -1

confirming the fact that $|x|$ is not differentiable at $x = 0$.

Exercises

4.7   Prove Theorem 4.11 by Lemma 4.4b, and the fact that a linear function is continuous.

4.8   Show that if a function $f : I \to \mathbf{R}$ is differentiable at some point $x \in I$, then the function $g : I \to \mathbf{R}$ defined by

$$g(\xi) := \begin{cases} \dfrac{f(\xi) - f(x)}{\xi - x} & \text{if } \xi \neq x, \\ f'(x) & \text{if } \xi = x, \end{cases}$$

is continuous at the point $\xi = x$. Moreover, if f is continuous in $I \setminus \{x\}$, then g is continuous in all of I.

4.9   Consider the following functions f at the given point x_0. Use MACSYMA to determine the derivatives $f'(x_0)$, $f'_+(x_0)$ and $f'_-(x_0)$, the right tangent and left tangent at x_0, whenever they exist.

(a) $f(x) := x^2 + 5x - 8$, $\quad x_0 := 2$ $\qquad$ (b) $f(x) := 1/(x - 2)$, $\quad x_0 := 3$

In the following problems, use $x_0 := 0$.

(c) $f(x) := \begin{cases} x^2 \sin(1/x) & \text{if } x \neq 0, \\ 0 & \text{if } x = 0 \end{cases}$ $\qquad$ (d) $f(x) := \begin{cases} x \sin(1/x) & \text{if } x > 0, \\ |x| & \text{if } x \leq 0 \end{cases}$

(e) $f(x) := x^{1/3}$ $\qquad\qquad\qquad\qquad\qquad$ (f) $f(x) := x^{2/3}$

(g) $f(x) := \begin{cases} x^2 & \text{if } x > 0, \\ 0 & \text{if } x \leq 0 \end{cases}$ $\qquad$ (h) $f(x) := \max\{(x-1)^2, (x+1)^2\}$

4.10 Use MACSYMA to compute $f'_+(x)$ and $f'_-(x)$ for the following functions at $x := 0$.

(a) $f(x) := |\sin x|$ $\qquad$ (b) $f(x) := |x^2|$ $\qquad$ (c) $f(x) := e^{|x|}$

(d) $f(x) := x^2/|x|$ $\qquad$ (e) $f(x) := \dfrac{x^2 - 1}{x - 1}$ $\qquad$ (f) $f(x) := e^{-1/|x|}$

(g) $f(x) := x^2 \sin(1/|x|)$ $\quad$ (h) $f(x) := \ln(1 + |x|)$ $\quad$ (i) $f(x) := \ln\left(\dfrac{1-x}{1+x}\right)$

4.3 Derivative function

In Sect. 4.2 we defined the derivative $f'(x)$ of f at a general point x, (4.11),

$$f'(x) = \lim_{\xi \to x} \frac{f(\xi) - f(x)}{\xi - x} .$$

Since the point x is general, the derivative $f'(x)$ is a *new function* associated with f.

Definition 4.19 (Derivative function). The derivative of a function f at a general point x, defined by the limit (4.11), is called the *derivative* (or derivative function) of f and is denoted by f' (pronounced "f prime"). The domain of f' consists of all x where f is differentiable.

For many functions f the derivative f' can be computed by formal rules (see Chap. 5) avoiding the limit (4.11). Once the derivative function $f'(x)$ is known, the derivative $f'(x_0)$ at any special point x_0 can be computed by substituting $x := x_0$ in $f'(x)$. This is easier than calculating the limit (4.10),

$$f'(x_0) = \lim_{\xi \to x_0} \frac{f(\xi) - f(x_0)}{\xi - x_0} .$$

Example 4.20 (Derivative of x^2). Let $f(x) = x^2$. Then, by (4.11),

$$f'(x) = \lim_{\xi \to x} \frac{\xi^2 - x^2}{\xi - x} = \lim_{\xi \to x} \frac{(\xi + x)(\xi - x)}{\xi - x} = \lim_{\xi \to x} (\xi + x) = 2x . \qquad (4.20)$$

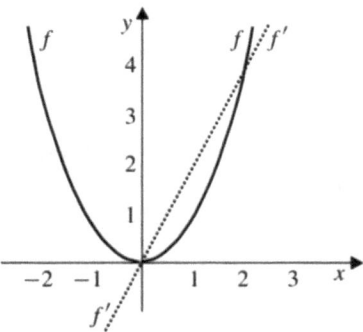

Fig. 4.5. Function $f(x) = x^2$ and its derivative $f'(x) = 2x$

We conclude that the function x^2 is differentiable everywhere. At any special point x_0, the slope of the tangent is therefore $f'(x_0) = 2x_0$, in agreement with Example 4.3. The functions f and f' are plotted in Fig. 4.5. The derivative f' (which is the slope of the parabola x^2) is monotone increasing, going from negative values (for $x < 0$) to positive values (for $x > 0$). The derivative f' equals zero for $x = 0$, which is where the parabola f has a horizontal tangent $y = 0$.

In Sect. 4.2 we expressed the slope $f'(x_0)$ of the tangent line of f at x_0, as the limit of slopes m_{PQ} of the secant lines PQ, where $P = (x_0, f(x_0))$ and $Q = (\xi, f(\xi)) \to P$. The same holds for the derivative $f'(x)$ at a general point x.

Let $P = (x, f(x))$ be the point of tangency, and let $Q = (x + \Delta x, f(x + \Delta x))$ be a nearby point, i.e., $|\Delta x|$ is small[4]. The secant line PQ (see Fig. 4.6 a) is then a good approximation to the tangent line through P (Fig. 4.6 b). The slope of the

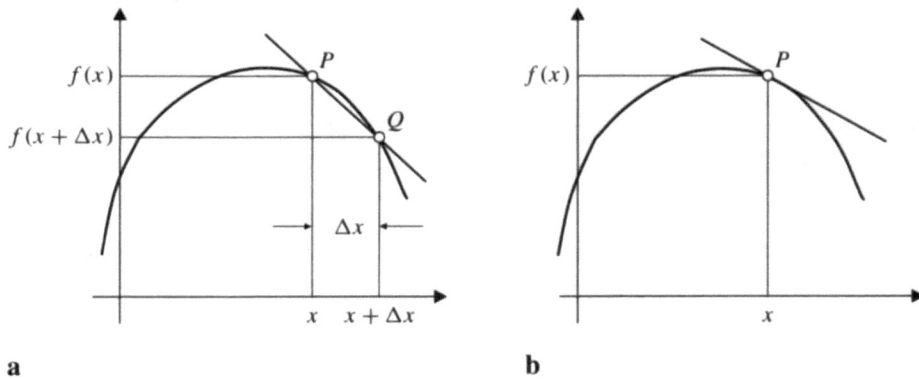

Fig. 4.6 a, b. Tangent as limit of secant lines. **a** Secant line through the points x, $f(x)$ and $(x + \Delta x, f(x + \Delta x))$. **b** Tangent line through the point $(x, f(x))$

---

4 Δx can be positive or negative.

secant line PQ is, as in (4.12),

$$m_{PQ} = \frac{f(x + \Delta x) - f(x)}{\Delta x} = \frac{\Delta f(x)}{\Delta x}$$

where $\Delta f(x)$, abbreviated Δf, is the *difference*[5] of the values of f as its argument changes from x to $x + \Delta x$, i.e., $\Delta f := f(x + \Delta x) - f(x)$.

In the limit as $Q \to P$, i.e., as $\Delta x \to 0$, we get the slope of the tangent at x,

$$f'(x) = \lim_{\Delta x \to 0} \frac{f(x + \Delta x) - f(x)}{\Delta x} = \lim_{\Delta x \to 0} \frac{\Delta f}{\Delta x} .$$

Consider now the tangent line at $(x, f(x))$. If we move to the point $x + dx$, then the vertical change measured along the tangent line is

$$df(x) = f'(x)\, dx . \tag{4.21}$$

We call $df(x)$ (abbreviated df) and dx *differentials*. See Fig. 4.7b for illustration of differentials. The differentials df and dx should not be confused with the differences Δx and Δf.

The differences Δx and Δf measure the distances, along the x- and y-coordinate axes, respectively, between the two points $(x, f(x))$ and $(x + \Delta x, f(x + \Delta x))$ of the graph of f (see Fig. 4.7a).

The differentials dx and df correspond to the x- and y-coordinate distances when the graph of f is replaced by its tangent line at the point $(x, f(x))$ (see Fig. 4.7b). The slope of that tangent is $df \div dx$.

As one moves from a point x_0 to $x_0 + \Delta x$, the difference

$$\Delta f := f(x_0 + \Delta x) - f(x_0) \tag{4.22}$$

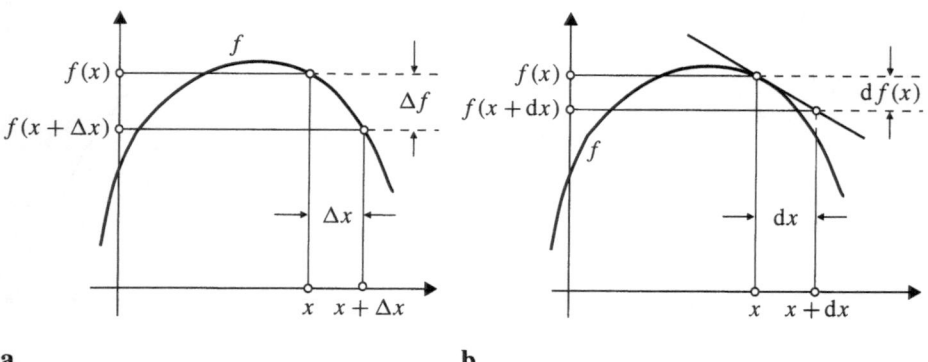

a   **b**

Fig. 4.7 a, b. Differences and differentials. **a** Difference Δf is measured along the graph of f. **b** Differential df is measured along the tangent

---

5 Δf is sometimes called *increment*, but this suggests that an increase takes place, whereas Δf can in fact be negative, as in Fig. 4.7.

is the true change in the value of f. The differential

$$\mathrm{d}f := f'(x_0)\Delta x , \tag{4.23}$$

measured along the tangent, is just an approximation of Δf. What makes this approximation useful is that in many cases where the derivative f' is known, the differential is easy to calculate. For application of differentials see Sect. 4.6.

Remark 4.21 (On notation). There are alternative notations for writing the derivative

$$f'(x) = \lim_{\xi \to x} \frac{f(\xi) - f(x)}{\xi - x} = \lim_{\Delta x \to 0} \frac{f(x + \Delta x) - f(x)}{\Delta x} .$$

Here are the main ones:

$$f'(x) = \frac{\mathrm{d}f}{\mathrm{d}x} = \frac{\mathrm{d}}{\mathrm{d}x} f(x) = [f(x)]' . \tag{4.24}$$

The notation f' was used first by Lagrange[6]. The $\mathrm{d}f/\mathrm{d}x$ notation, due to Leibniz[7], is suggested by (4.21), and is a reminder of the fact that $f'(x)$ is the limit of the expression $\Delta f/\Delta x$ as Δx tends to zero (see Fig. 4.6). The symbol $\mathrm{d}f/\mathrm{d}x$ is called a differential *quotient*.

The symbol $\mathrm{d}/\mathrm{d}x$ in (4.24) denotes the differential *operator* which, acting on the function f, produces the derivative f'.

The derivative $f'(x_0)$ at a special point x_0 is denoted by

$$f'(x_0), \quad \frac{\mathrm{d}f}{\mathrm{d}x}(x_0) \quad \text{or} \quad \frac{\mathrm{d}f}{\mathrm{d}x}\bigg|_{x=x_0} .$$

For example, if $f(x) = x^n$, then
- the derivative $f'(x)$ is denoted by $\mathrm{d}x^n/\mathrm{d}x$, or $(\mathrm{d}/\mathrm{d}x)\,x^n$;
- the derivative $f'(x_0)$ is written $\mathrm{d}x^n/\mathrm{d}x(x_0)$, $((\mathrm{d}/\mathrm{d}x)x^n)(x_0)$, or $\mathrm{d}x^n/\mathrm{d}x|_{x=x_0}$.

MACSYMA-Session 4.5. The derivative function f' can easily be computed in MACSYMA by using `diff(expression, variable)`. For example,
c1. `diff(x^n,x)`
d1. nx^{n-1}
To compute the derivative $f'(x_0)$ at a point x_0, we use `limit(diff(f,x),x,x0)`, assuming that f' is continuous at $x = x_0$. For example, the derivative of x^3 at $x = -2$
c2. `limit(diff(x^3,x),x,-2)`
d2.   12

We end this section with a fundamental property of derivatives that they inherit from their definition as a limit.

---

6 Joseph Louis Lagrange (1736–1813).
7 Gottfried Wilhelm Leibniz (1646–1716).

Theorem 4.22 (Linearity of the derivative). If f and g are differentiable at x, and $c \in \mathbf{R}$, then $f + g$ and $c \cdot f$ are differentiable, and their derivatives are

$$(f + g)'(x) = f'(x) + g'(x) , \tag{4.25a}$$

$$(cf)'(x) = c \cdot f'(x) . \tag{4.25b}$$

Proof. This follows immediately from the linearity property of the limit operation as

$$(f + g)'(x) = \lim_{\xi \to x} \frac{(f(\xi) + g(\xi)) - (f(x) + g(x))}{\xi - x}$$

$$= \lim_{\xi \to x} \frac{f(\xi) - f(x)}{\xi - x} + \lim_{\xi \to x} \frac{g(\xi) - g(x)}{\xi - x} = f'(x) + g'(x) ,$$

and

$$(cf)'(x) = \lim_{\xi \to x} \frac{cf(\xi) - cf(x)}{\xi - x} = c \cdot \lim_{\xi \to x} \frac{f(\xi) - f(x)}{\xi - x} = c \cdot f'(x) . \qquad \square$$

Remark 4.23. a. If the derivatives f' and g' have been computed, using limits such as (4.11), then Theorem 4.22 allows writing the derivatives of $f + g$ and cf with no further effort.

b. The identity (4.25a) can be stated in words: The *derivative of a sum* is the *sum of the derivatives*.

c. The meaning of (4.25b) is that a constant can be taken out of the differentiation. For example, the derivative of $5 \sin x$ is 5 times the derivative of $\sin x$.

d. Two functions are mentioned in (4.25a), but the result holds for any number of summands. Theorem 4.22 is equivalent to:
Let the functions $f_1, f_2, \ldots, f_n$ be differentiable, and let $c_1, c_2, \ldots, c_n \in \mathbf{R}$ be given constants. Then the function $c_1 f_1 + c_2 f_2 + \cdots + c_n f_n$ is differentiable, and its derivative is

$$\left(\sum_{k=1}^{n} c_k f_k \right)'(x) = \sum_{k=1}^{n} c_k f_k'(x) . \tag{4.26}$$

MACSYMA-Session 4.6. The line tangent to a function f at a point $x = x_0$ is computed by
c1. `tangent(f,x,x0):=limit(f,x,x0)+(x-x0)*limit(diff(f,x),x,x0)$`
where `limit` is used to evaluate $f(x_0)$ and $f'(x_0)$. For example, the line tangent to $y = \sqrt{x}$ at $x_0 = 100$:

c2. `tangent(sqrt(x),x,100)`

d2. $\dfrac{x - 100}{20} + 10$

which can be expanded
c3. `expand(%)`

d3. $\dfrac{x}{20} + 5$

Another example: The line tangent to $y = \tan(x)$ at $x = \pi/4$ is:

c4. tangent(tan(x),x,%pi/4)

d4. $\qquad 2\left(x - \dfrac{\pi}{4}\right) + 1$

c5. expand(%)

d5. $\qquad 2x - \dfrac{\pi}{2} + 1$

Exercises

4.11 Show that for a differentiable function $f : I \to \mathbf{R}$ of a symmetric interval $I := [-a, a]$, the derivative f' is (a) even if f is odd, (b) odd if f is even.

4.12 The function deriv(f,x,x0) (in MACSYMA-Session 4.3), used to compute the derivative f' at a given point x_0, can be adapted for computing the derivative f' at arbitrary points x:

derivative(f,x):=lim(deriv(f,x,x0),x0,x).

    a. Use MACSYMA to compute the derivative of $f(x) = x^2$, and compare with (4.20).

    b. Calculate the derivative function for

(1) $f(x) = x^n$ (2) $f(x) = \sin x$ (3) $f(x) = 1 + 2x + 3x^2 + 4x^3$

and compare with the results of the built-in MACSYMA function diff. This exercise is not meant as a practical alternative for computing f'.

4.13 Use the function

tangent(f,x,x0):=lim(f,x,x0)+(x-x0)lim(diff(f,x),x,x0)

(Session 4.6) to compute the tangent line of f at $x_0 := 0, 1, 2$, for the following functions:

(a) $f(x) := x^3$ (b) $f(x) := \dfrac{1}{x^2 + 1}$

(c) $f(x) := \dfrac{x^3 - 2x^2 + 5x - 9}{x^2 - 3}$ (d) $f(x) := x(x - 1)(x - 2)$

(e) $f(x) := \sin \dfrac{\pi x}{4}$ (f) $f(x) := \tan \dfrac{\pi x}{6}$ (g) $f(x) := e^x$

(h) $f(x) := e^{-x}$ (i) $f(x) := \ln(x + 1)$

In each case, use MACSYMA to plot the graphs of f and the tangent line.

4.14 Write a MACSYMA function secant(f,x,x1,x2) that calculates the secant line of f, with respect to the variable x, through the points $(x_1, f(x_1))$ and $(x_2, f(x_2))$.

4.15 Use MACSYMA to find the derivative of the absolute-value function. MACSYMA's answer is signum(x). Is this correct for all $x \in \mathbf{R}$? Note that under MACSYMA's convention, signum(0) is undefined and so remains unsimplified. Now differentiate signum(x). What happens? Is this result correct?[8]

---

8 You should remember this example when working with a symbolic-computation system like MACSYMA. The moral here is that you must *always* steer the system where you want to go, rather than be guided by it. The system may assume, for any of its operations, that the functions it operates on are continuous or even differentiable. If this is not the case, as in the above example, the system may give incorrect results.

4.4 Special derivatives

In this section we use the defining limit

$$f'(x) = \lim_{\xi \to x} \frac{f(\xi) - f(x)}{\xi - x} = \lim_{\Delta x \to 0} \frac{f(x + \Delta x) - f(x)}{\Delta x}$$

to compute the derivatives of several functions.

Example 4.24 (The derivative of x^n, $n \in \mathbf{N}$). Let n be a positive integer. Then x^n is differentiable everywhere, and

$$\frac{d}{dx} x^n = nx^{n-1} . \tag{4.27}$$

Proof. We give two alternative proofs of (4.27).
   a. The sought derivative is

$$\frac{d}{dx} x^n = \lim_{\Delta x \to 0} \frac{(x + \Delta x)^n - x^n}{\Delta x} .$$

Using the *binomial expansion* of $(x + \Delta x)^n$, we compute

$$\frac{(x + \Delta x)^n - x^n}{\Delta x}$$

$$= \frac{\sum_{k=0}^{n} \binom{n}{k} x^{n-k} (\Delta x)^k - x^n}{\Delta x}$$

$$= \frac{\left(x^n + nx^{n-1}\Delta x + \binom{n}{2} x^{n-2}(\Delta x)^2 + \text{higher powers of } \Delta x \right) - x^n}{\Delta x}$$

$$= nx^{n-1} + \binom{n}{2} x^{n-2}\Delta x + \text{higher powers of } \Delta x$$

$$\therefore \quad \lim_{\Delta x \to 0} \frac{(x + \Delta x)^n - x^n}{\Delta x} = nx^{n-1} .$$

   b. Writing the derivative as

$$\lim_{\xi \to x} \frac{\xi^n - x^n}{\xi - x} = \lim_{\xi \to x} \frac{x^n - \xi^n}{x - \xi} ,$$

we observe that $(x^n - \xi^n)/(x - \xi)$ is a rational function in the variable x, and both the numerator and the denominator have a zero at the point $x = \xi$. Now use long division to cancel this common factor.

$$
\begin{array}{r}
x^{n-1} \;+\xi x^{n-2} \;+\; \cdots \;+\xi^{n-1} \\
\end{array}
$$

$$
(x-\xi)\;\big|\; x^n \hspace{5cm} -\xi^n
$$

$$
\begin{aligned}
&-\;(x^n \quad -\xi x^{n-1}) \\[2pt]
\hline
&\xi x^{n-1} \hspace{4cm} -\xi^n \\
&-\;(\xi x^{n-1} \quad -\xi^2 x^{n-2}) \\[2pt]
\hline
&\xi^2 x^{n-2} \hspace{3.6cm} -\xi^n \\
&\quad\; - \hspace{4cm} \vdots \\[2pt]
\hline
&\hspace{3cm} \xi^{n-1}x \quad -\xi^n \\
&-\hspace{3cm} (\xi^{n-1}x \quad -\xi^n) \\[2pt]
\hline
&\hspace{4.5cm} 0
\end{aligned}
$$

Therefore

$$
\frac{x^n - \xi^n}{x - \xi} = x^{n-1} + \xi x^{n-2} + \cdots + \xi^{n-1} = \sum_{k=0}^{n-1} x^{n-k}\xi^k . \tag{4.28}
$$

Finally

$$
f'(x) = \lim_{\xi \to x} \frac{x^n - \xi^n}{x - \xi} = \lim_{\xi \to x}(x^{n-1} + \xi x^{n-2} + \cdots + \xi^{n-1}) = nx^{n-1},
$$

since there are n summands, all of value x^{n-1} in the limit as $\xi \to x$.

The special case $n = 1$ gives the line $f(x) = x$. There (4.27) gives $f'(x) = 1$, a constant slope.

The formula $f'(x) = nx^{n-1}$ also holds for $n = 0$, i.e., for the constant function $f(x) = 1$. In that case we have

$$
\lim_{\xi \to x} \frac{f(\xi) - f(x)}{\xi - x} = \lim_{\xi \to x} \frac{1 - 1}{\xi - x} = 0
$$

for all $x \in \mathbf{R}$. This corresponds to the geometrical fact that the constant function $f(x) = 1$ has a horizontal tangent everywhere. We have thus proved the following theorem.

Theorem 4.25 (Derivative of powers). The monomial $f(x) = x^n$ ($n \in \mathbf{N}_0$) has the derivative $f'(x) = nx^{n-1}$.

The linearity of the derivative (see Remark 4.23 d) makes it very easy to calculate the derivative of a polynomial, such as $p(x) = -3x^5 + 12x^4 + 5x^3 - 2x + 17$. Polynomials are just linear combinations of monomials x^k, and the derivatives of monomials are known.

Corollary 4.26 (Derivatives of polynomials). The polynomial

$$p(x) := a_0 + a_1 x + \cdots + a_n x^n = \sum_{k=0}^{n} a_k x^k$$

has the derivative

$$p'(x) = a_1 + 2a_2 x + \cdots + n a_n x^{n-1} = \sum_{k=0}^{n} k a_k x^{k-1} .$$

In fact, Theorem 4.25, together with a differentiation rule given in Theorem 5.5, allows the differentiation of *rational functions*, i.e., functions which are quotients of polynomials.

Example 4.27 (Sphere). Consider a sphere of radius r. Its volume V and its surface area A are both functions of r,

$$V(r) = \tfrac{4}{3}\pi r^3, \quad A(r) = 4\pi r^2 .$$

The derivative of the volume is, by (4.25b) and (4.27),

$$V'(r) = \tfrac{4}{3}\pi 3 r^2 = 4\pi r^2 ,$$

which is the area $A(r)$! This is perhaps easier to understand if we use the differentials dV and dr,

$$dV = 4\pi r^2 \, dr = A \, dr . \tag{4.29}$$

For example, consider the problem of plating a sphere of radius r with a gold layer of thickness dr. The gold-plated sphere has radius $r + dr$ and volume $V + dV$, of which dV is the volume of the gold plate. Can you explain why dV is equal to $A \, dr$? See Fig. 4.8.

Example 4.28 (Derivative of $\sqrt{x}$).

$$(d/dx) \sqrt{x} = 1/2\sqrt{x} . \tag{4.30}$$

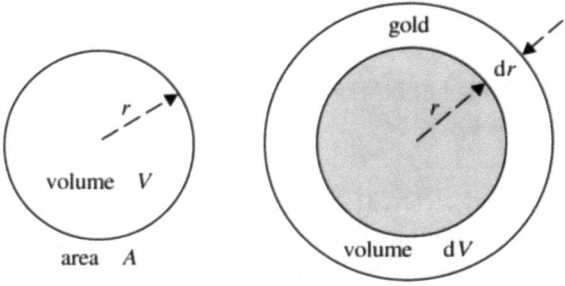

Fig. 4.8 a, b. Gold-plating a sphere. **a** Original sphere; **b** gold-plated sphere

Proof.

$$\frac{d}{dx}\sqrt{x} = \lim_{\Delta x \to 0}\frac{\sqrt{x+\Delta x}-\sqrt{x}}{\Delta x} = \lim_{\Delta x \to 0}\frac{(\sqrt{x+\Delta x}-\sqrt{x})(\sqrt{x+\Delta x}+\sqrt{x})}{\Delta x(\sqrt{x+\Delta x}+\sqrt{x})}$$

$$= \lim_{\Delta x \to 0}\frac{\Delta x}{\Delta x(\sqrt{x+\Delta x}+\sqrt{x})} = \frac{1}{2\sqrt{x}}.$$

$\square$

For the computation of $(d/dx)\sin x$ and $(d/dx)\cos x$, we need the following auxiliary result.

Lemma 4.29. If θ is measured in radians, then

$$\lim_{\theta \to 0}\frac{\sin \theta}{\theta} = 1, \tag{4.31a}$$

$$\lim_{\theta \to 0}\frac{\cos \theta - 1}{\theta} = 0. \tag{4.31b}$$

Proof. Consider the angle θ as a central angle in a circle with a radius of 1 (see Fig. 4.9). Then

a. the area of the circular sector OAB is $\theta/2$ – this is so because the radius is 1 and the angle θ is measured in radians,
b. the area of the triangle OAB is $(\sin\theta)/2$, since its basis is 1 and its height is $\sin\theta$.
c. the area of the triangle OAC is $(\tan\theta)/2$.
    Therefore

$$\frac{\sin\theta}{2} \le \frac{\theta}{2} \le \frac{\tan\theta}{2},$$

and since $\tan\theta = \sin\theta/\cos\theta$,

$$\frac{\sin\theta}{\theta} \le 1 \le \frac{\sin\theta}{\theta}\frac{1}{\cos\theta}$$

$$\therefore \quad \lim_{\theta \to 0}\frac{\sin\theta}{\theta} \le 1 \le \lim_{\theta \to 0}\frac{\sin\theta}{\theta}\lim_{\theta \to 0}\frac{1}{\cos\theta} = \lim_{\theta \to 0}\frac{\sin\theta}{\theta}\frac{1}{\lim_{\theta \to 0}\cos\theta}$$

proving (4.31a), since $\lim_{\theta \to 0}\cos\theta = 1$.

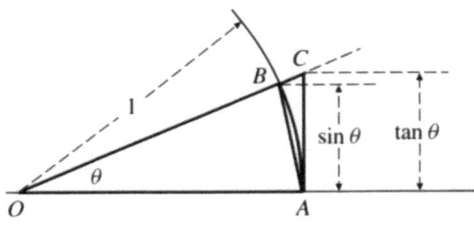

Fig. 4.9. Angle θ and its sine and tangent

Note that here we use the continuity of the cosine function. The proof of (4.31b) is from

$$\lim_{\theta \to 0} \frac{\cos\theta - 1}{\theta} = \lim_{\theta \to 0} \frac{(\cos\theta - 1)(\cos\theta + 1)}{\theta(\cos\theta + 1)}$$

$$= \lim_{\theta \to 0} \frac{\cos^2\theta - 1}{\theta(\cos\theta + 1)} = \lim_{\theta \to 0} \frac{-\sin^2\theta}{\theta(\cos\theta + 1)}$$

$$= -\lim_{\theta \to 0} \frac{\sin\theta}{\theta} \lim_{\theta \to 0} \frac{1}{\cos\theta + 1} \lim_{\theta \to 0} \sin\theta = -1 \times \frac{1}{2} \times 0 = 0$$

using the continuity of $\sin\theta$.   □

Example 4.30 (Derivative of sin x).

$$(d/dx) \sin x = \cos x . \tag{4.32}$$

Proof. We use the identity $\sin(\alpha + \beta) = \sin\alpha \cos\beta + \cos\alpha \sin\beta$ to calculate

$$\frac{d}{dx} \sin x = \lim_{\Delta x \to 0} \frac{\sin(x + \Delta x) - \sin x}{\Delta x}$$

$$= \lim_{\Delta x \to 0} \frac{\sin x \cos\Delta x + \cos x \sin\Delta x - \sin x}{\Delta x}$$

$$= \lim_{\Delta x \to 0} \sin x \frac{\cos\Delta x - 1}{\Delta x} + \cos x \lim_{\Delta x \to 0} \frac{\sin\Delta x}{\Delta x}$$

$$= \cos x ,$$

by (4.31a), (4.31b), and the continuity of the trigonometric functions.   □

Example 4.31 (Derivative of cos x).

$$(d/dx) \cos x = -\sin x . \tag{4.33}$$

Proof. We use the identity $\cos(\alpha + \beta) = \cos\alpha \cos\beta - \sin\alpha \sin\beta$ to calculate

$$\frac{d}{dx} \cos x = \lim_{\Delta x \to 0} \frac{\cos(x + \Delta x) - \cos x}{\Delta x}$$

$$= \lim_{\Delta x \to 0} \frac{\cos x \cos\Delta x - \sin x \sin\Delta x - \cos x}{\Delta x}$$

$$= \cos x \lim_{\Delta x \to 0} \frac{\cos\Delta x - 1}{\Delta x} - \sin x \lim_{\Delta x \to 0} \frac{\sin\Delta x}{\Delta x}$$

$$= -\sin x, \quad \text{by (4.31a) and (4.31b)} .$$ □

Example 4.32 (Derivative of tan x).

$$(d/dx) \tan x = \sec^2 x . \tag{4.34}$$

Proof. We use the identity

$$\tan(\alpha + \beta) = \frac{\tan \alpha + \tan \beta}{1 - \tan \alpha \tan \beta}$$

(see Proposition C.4e) to calculate

$$\frac{d}{dx} \tan x = \lim_{\Delta x \to 0} \frac{\tan(x + \Delta x) - \tan x}{\Delta x} = \lim_{\Delta x \to 0} \frac{\dfrac{\tan x + \tan \Delta x}{1 - \tan x \tan \Delta x} - \tan x}{\Delta x}$$

$$= \lim_{\Delta x \to 0} \frac{\tan \Delta x (1 + \tan^2 x)}{\Delta x (1 - \tan x \tan \Delta x)} = (1 + \tan^2 x) \lim_{\Delta x \to 0} \frac{\tan \Delta x}{\Delta x}$$

$$= (1 + \tan^2 x) \lim_{\Delta x \to 0} \frac{\sin \Delta x}{\Delta x} \lim_{\Delta x \to 0} \frac{1}{\cos \Delta x} = 1 + \tan^2 x$$

$$= 1 + \frac{\sin^2 x}{\cos^2 x} = \frac{\sin^2 x + \cos^2 x}{\cos^2 x} = \frac{1}{\cos^2 x} = \sec^2 x . \qquad \square$$

Next we compute the derivative of the exponential function $\exp x := e^x$. Recall (3.16) that

$$e := \lim_{n \to \infty} \left(1 + \frac{1}{n}\right)^n ,$$

so that

$$\exp x := e^x = \lim_{n \to \infty} \left(1 + \frac{1}{n}\right)^{nx} .$$

We also recall the limit (3.13),

$$\lim_{x \to 0} \frac{e^x - 1}{x} = 1 ,$$

which is needed below.

Example 4.33 (Derivative of e^x).

$$(d/dx)\, e^x = e^x . \qquad (4.35)$$

Proof.

$$\frac{d}{dx} e^x = \lim_{\Delta x \to 0} \frac{e^{x+\Delta x} - e^x}{\Delta x} = e^x \lim_{\Delta x \to 0} \frac{e^{\Delta x} - 1}{\Delta x} = e^x, \quad \text{by (3.13) .} \qquad \square$$

Remark 4.34. We see that the exponential function $f(x) = e^x$ is its own derivative, i.e.,

$$f'(x) = f(x) .$$

This property makes e^x natural for describing processes where the rate of change is proportional to the value of the function.

Example 4.35 (Derivative of $e^{\lambda x}$).

$$(d/dx)\, e^{\lambda x} = \lambda e^{\lambda x} \,. \tag{4.36}$$

Proof.

$$\frac{d}{dx}\, e^{\lambda x} = \lim_{\Delta x \to 0} \frac{e^{\lambda x + \lambda \Delta x} - e^{\lambda x}}{\Delta x} = e^{\lambda x} \lim_{\Delta x \to 0} \frac{e^{\lambda \Delta x} - 1}{\Delta x}$$

$$= e^{\lambda x} \lim_{\lambda \Delta x \to 0} \lambda \frac{e^{\lambda \Delta x} - 1}{\lambda \Delta x} = \lambda e^{\lambda x}, \quad \text{by (3.13)}\,. \qquad \square$$

Exercises

4.16 Compute the derivatives of the following functions $f(x)$ by forming the quotient $[f(x + h) - f(x)]/h$ and compute the limit as $h \to 0$. You may use MACSYMA to assist you in computing the limits if you are unable to do these limits by hand calculations.

(a) $f(x) := \sqrt{x}$ (b) $f(x) := \sqrt{x + 1}$ (c) $f(x) := \sqrt{x^2 + 1}$

(d) $f(x) := e^x$ (e) $f(x) := \sin x$ (f) $f(x) := \cos x$

(g) $f(x) := x^3 + ax^2 + bx + c$ (h) $f(x) := 1/x$

(i) $f(x) := 1/x^3$ (j) $f(x) := \ln x$ (k) $f(x) := 1/(x^2 + 1)$

(l) $f(x) := 1/\sin x$

4.17 Verify Eq. (4.28) with MACSYMA.

4.18 a. Compute the limits

$$\lim_{\theta \to 0} \frac{\sin \theta}{\theta} \quad \text{and} \quad \lim_{\theta \to 0} \frac{\cos \theta - 1}{\theta}$$

if θ is measured in degrees.

b. Calculate the derivatives

$$\frac{d}{dx} \sin x, \quad \frac{d}{dx} \cos x \quad \text{and} \quad \frac{d}{dx} \tan x$$

if x is measured in degrees.

The answer will convince you that radians are more natural.

4.19 a. Use (4.31a) to prove that

$$\lim_{n \to \infty} \frac{n}{2} \sin\left(\frac{2\pi}{n}\right) = \pi$$

showing that the areas of the n-polygons, inscribed in the unit circle, converge as $n \to \infty$ to the area of the circle (see Example 3.3).

b. Compute $\lim_{\theta \to 0}(\tan \theta / \theta)$.

c. Use part b to prove that $\lim_{n\to\infty} n \tan(\pi/n) = \pi$, showing that the areas of the circumscribed n-gons also converge to the area of the unit circle.

4.20 Differentiate: (a) $\sin(-x)$, (b) $\cos(-x)$, and (c) e^{-x}.

4.21 Find the derivative of $f_n(x) := |x^n + 1|$ wherever it exists and plot the graphs of the functions f_n for $n = 1, \ldots, 5$.

4.22 Prove: For any nonnegative integer n and any real number a,

$$\frac{d}{dx}(x+a)^n = n(x+a)^{n-1} . \tag{4.37}$$

Hint: The binomial expansion of $(x + a)^n$ is a polynomial in x, which can be differentiated using Corollary 4.26.

4.23 Find a function $f(x)$, not equal to e^x, which is its own derivative, i.e., $f'(x) = f(x)$ for all x.

4.24 Let $f(x) := a^x$ for some $a > 0$. Find the linear tangent of $f(x)$ at $x = 0$.
Hint: Since $a = e^{\log a}$, it follows that $a^x = e^{(\log a)x}$.

4.5 Rectilinear motion and velocity

The movement of physical bodies[9] is a natural application of derivatives. A movement (or motion) is realized when the *body* in question *changes its position* (or location) with *time*. In this section we take the simplest case, that of

- rectilinear motion, i.e., movement *along a line* (which we take as a coordinate axis), of a
- particle, i.e., a body which consists of a single *point*[10], so its
- position, at any time t, is simply its *coordinate* in time t.

An example of linear movement is given in Fig. 4.10a. The particle starts at s_0, moves right to the point s_1, then left to s_2, then right again to s_3 where it comes to a stop. Figure 4.10a gives the places visited by the particle, but not the times when these places were visited. To answer the question "when?", as well as "where?", the position s must be given as a function $s(t)$ of time t; see Fig. 4.10b, where we read that the movement started at time t_0 and ended at t_3. For each place visited by the particle, we can read from Fig. 4.10a the time (or times) of these visits. For example, the place s_1 was visited twice, and the first visit occurred at time t_1.[11]

The position s is a function of time t, which is the independent variable. For this reason we take the t-axis as horizontal and the s-axis vertical.

Time is measured in seconds [s], hours, years, etc.

Position is the s-coordinate, i.e., the distance from the origin of the s-axis. Position is therefore measured in units of *length*: meter [m], kilometer [km], foot [ft], mile, etc.

---

9 For example, planets, space vehicles, or cars on the N.J. turnpike.

10 A relatively small body, such as planet earth relative to the sun, or a space vehicle relative to earth, can be regarded as a point.

11 Read from Fig. 4.10b the time of the second visit of s_1.

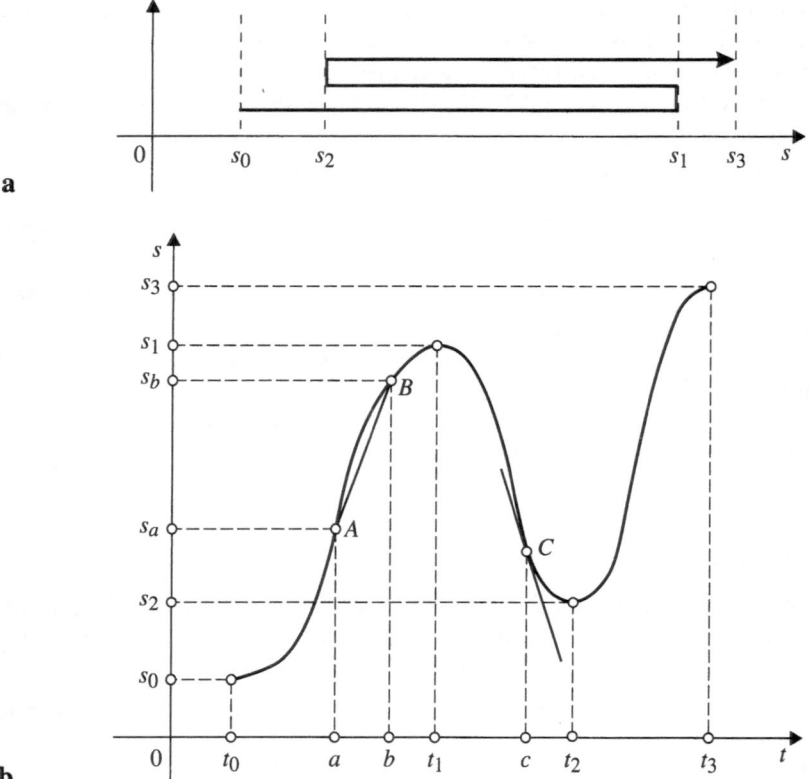

a

b

Fig. 4.10 a, b. Rectilinear motion and its position function. **a** Movement along a line. A particle moves along the s axis, from the point s_0 to the point s_3. It reverses direction at the points s_1 and s_2. **b** Position function for the movement in **a**. Position s plotted as a function of time t

With movement we associate *velocity*, the rate at which position changes with time. It is measured in units of

$$\left[\frac{\text{length}}{\text{time}}\right], \quad \text{such as} \quad \left[\frac{\text{miles}}{\text{hour}}\right] \text{(abbreviated MPH) and} \left[\frac{\text{m}}{\text{s}}\right], \text{etc.}$$

There are two notions of velocity: average and instantaneous.

If during the time interval $[t, t + \Delta t]$ the particle moved from $s(t)$ to $s(t + \Delta t)$, then its *displacement* during that interval is $\Delta s := s(t + \Delta t) - s(t)$, and its *average velocity* is

$$\bar{v}(t, t + \Delta t) := \Delta s / \Delta t , \tag{4.38}$$

which can be positive (if the displacement is positive, i.e., the net movement is in the direction of the s-axis) or negative (if $\Delta s < 0$).

Even if the displacement Δs is positive, it is not necessarily the same as the *total distance* $dist(t, t + \Delta t)$ covered during the interval $[t, t + \Delta t]$. For example, in

Fig. 4.10a the total distance covered during $[t_0, t_3]$ is $|s_1 - s_0| + |s_2 - s_1| + |s_3 - s_2|$, and the displacement is $s_3 - s_0$.

Remark 4.36. The above definition of average velocity is not intuitive. If a particle returns to its initial point, i.e., if $s(t + \Delta t) = s(t)$, then $\Delta s = 0$ and $\bar{v}(t, t + \Delta t) = 0$, regardless of how fast the particle traveled during that interval. We therefore define the *average speed* during $[t, t + \Delta t]$ as

$$\frac{dist(t, t + \Delta t)}{\Delta t}.$$

In Fig. 4.10a, the average speed during $[t_0, t_3]$ is

$$\frac{|s_1 - s_0| + |s_2 - s_1| + |s_3 - s_2|}{t_3 - t_0}.$$

The average speed is useful for making plans and schedules. For example, if, under typical conditions, a truck covers a distance of 200 miles in 4 hours, then its "average speed" is 50 MPH. Under the same conditions, a trip of 325 miles should take about about $325/50 = 6.5$ [h].

The average velocity $\bar{v}(t, t + \Delta t)$ is the slope of the secant line through the points $(t, s(t))$ and $(t + \Delta t, s(t + \Delta t))$. For example, from Fig. 4.10b we read that during the time interval $[a, b]$ the displacement is $s_b - s_a$.

$$\therefore \quad \bar{v}(a, b) = \frac{s_b - s_a}{b - a},$$

which is the slope of the secant line AB.

As $\Delta t \to 0$ we get the slope of the tangent of the position function at t, which we call the *instantaneous velocity* (or, simply, velocity) at time t. The velocity at time t is denoted by $v(t)$. Therefore

$$v(t) := \lim_{\Delta t \to 0} \bar{v}(t, t + \Delta t) = \lim_{\Delta t \to 0} \frac{\Delta s}{\Delta t} = s'(t) \tag{4.39}$$

the derivative of the position function $s(t)$.

The absolute value of the velocity is called *speed*. When we say that the speed is 55 MPH, we ignore the direction of the movement.

In Fig. 4.10b we see that the tangent, at time c, has negative slope, i.e., the velocity $v(c)$ at time c is negative. A negative velocity at any time indicates that the position s is decreasing at that time, i.e., the movement is against the direction of the s-axis.

Similarly, in any time interval when the movement is in the direction of the s-axis, we have a nonnegative derivative $v(t)$ and a nondecreasing[12] position function $s(t)$.

---

12 We allow for stops, without reversing directions, so that $s(t)$ may remain constant during certain intervals, during which $v(t) = 0$.

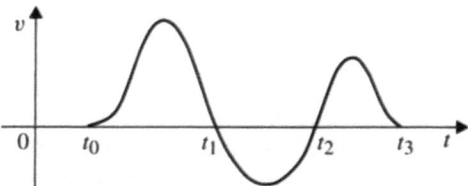

Fig. 4.11. Velocity function $v(t)$, the derivative of the position function $s(t)$ of Fig. 4.10 b. Velocity v plotted as a function of time t

The velocity is therefore positive during (t_0, t_1), negative during (t_1, t_2), and positive again during (t_2, t_3); see Fig. 4.11, where the velocity $v(t)$ is plotted. At the point t_1, the position function s has a horizontal tangent, i.e., the velocity $v(t_1) = 0$. The velocity v is also zero at

the starting time t_0, indicating a smooth start,

the direction reversal time t_2, and at

the ending time t_3, indicating a gentle stop.

Example 4.37. A circle with a radius of 1 is centered L units to the right of the origin, where $L > 2$. A point A moves on a circle; such motion is called *circular*. At any time, the position of the point A is given by the angle θ between $\overline{CA}$ and the s-axis (see Fig. 4.12).

The point A is connected, with a rigid bar of length L, to another point P which moves on the s-axis, back and forth, between $s = -1$ and $s = 1$. This connection transforms the circular motion of A to rectilinear motion of P.

At time 0 the point A is at $\theta = 0$, and the point P is at $s = 1$.

It is required to find the position, and velocity, of P for all $t > 0$, given that the point A makes one revolution each second and its speed is uniform.

Solution. The angle θ is a function of time t. Since P makes one revolution per second, at a uniform speed,

$$\theta = \theta(t) := 2\pi t, \quad \text{for all } t > 0. \tag{4.40}$$

The position s of the point P is given by

$$s = \overline{OP} = \overline{OB} - \overline{PB} = L + \overline{CB} - \overline{PB} = L + \cos\theta - L\cos\phi$$

(see Fig. 4.12) and thus depends on the two angles θ and ϕ. But these angles are not independent, indeed

$$\sin\phi = \frac{\overline{AB}}{L} = \frac{\sin\theta}{L}.$$

Now $\cos\phi$ is nonnegative, since ϕ is between $\pi/2$ and $-\pi/2$,

$$\therefore \quad \cos\phi = \sqrt{1 - \sin^2\phi} = \sqrt{1 - \frac{\sin^2\theta}{L^2}}$$

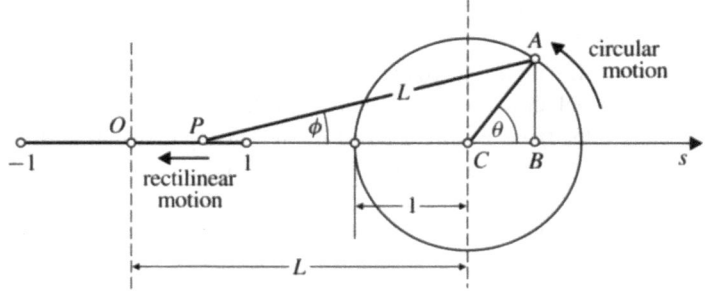

Fig. 4.12. Point A (circular motion) connected with point P (rectilinear motion)

which allows writing the position s as a function of θ alone,

$$s(\theta) = L + \cos\theta - L\sqrt{1 - \frac{\sin^2\theta}{L^2}}$$

$$= L + \cos\theta - \sqrt{L^2 - \sin^2\theta}\,, \tag{4.41a}$$

$$\therefore \quad s(t) = L + \cos(2\pi t) - \sqrt{L^2 - \sin^2(2\pi t)}, \quad \text{by (4.40)}, \tag{4.41b}$$

giving the position s as a function of time t. We will learn to differentiate such functions in the next chapter (see Example 5.23). For now we can use MACSYMA to obtain the velocity of P

$$v(t) = \frac{ds}{dt} = \sin\theta\left(\frac{\cos\theta}{\sqrt{L^2 - \sin^2\theta}} - 1\right)\frac{d\theta}{dt}$$

$$= 2\pi\sin(2\pi t)\left(\frac{\cos(2\pi t)}{\sqrt{L^2 - \sin^2(2\pi t)}} - 1\right).$$

Exercises

4.25 The velocity of a train moving along a straight length of track increases constantly until the driver sees a truck stalled across the track. At this time the driver applies the brakes and brings the train to rest just before the truck. Draw a plausible graph of velocity versus time to describe this situation.

4.26 If the distance $s(t)$ is given as a function of the time t, calculate the average velocity for the intervals $[0, 1]$, $[1/2, 1]$, $[3/4, 1]$, $[7/8, 1]$, and the instantaneous velocity at $t = 1$. Use MACSYMA to assist with the calculations.
    (a) $s(t) := t^3 + 7t$ (b) $s(t) := e^t$ (c) $s(t) := \sin t + \cos t$
    (d) $s(t) := t/(t^2 + 1)$ (e) $s(t) := (t - \pi/2)\tan t$ (f) $s(t) := \sec^2 t + t$

4.27 For the interval $[0, 1]$, find the *instantaneous* rate of change at a point $x \in [0, 1]$ and the *average* rate of change over this interval for the following functions.
    (a) $f(x) := x$ (b) $f(x) := x^2$ (c) $f(x) := e^x$

(d) $f(x) := 1/(x^2 + 1)$ (e) $f(x) := 1/(x + \varepsilon)$, where $\varepsilon > 0$

(f) $f(x) := \sin \pi x$ (g) $f(x) := \cos \pi x$ (h) $f(x) := x(x - 1)$

4.28 A truck covered 120 mi in 2 h, giving an average velocity of 60 MPH. The first 30 min were slow, with velocity not greater than 50 MPH.

Did the truck velocity necessarily exceed 70 MPH at any time during the last 90 min?

4.29 A body moves along a coordinate axis, according to $s(t) = t^3$. The movement lasts from $t = -\infty$ to $t = +\infty$.

a. What is the average velocity in the base period $[0, t]$?

b. What is the velocity $v(t)$ at any time t?

c. What can be said about the point where $v(t) = 0$?

4.30 Two cities, A and B, are connected by a straight road with a length of 150 mi. A car travels, from A to B, for two hours at a velocity of 50 MPH. How fast must it travel in the next 50 mi to average 60 MPH for the entire trip?

4.31 A car travels along a straight road at the speed $50(1 - |\sin \pi t|)$ MPH. Determine the average speed for the first hour. What is the instantaneous speed?

4.32 Consider the mechanism in Example 4.37.

a. Find the positions (angles θ_1 and θ_2) of the point P when the point P passes through the origin.

b. Consider the speed $|v|$ of the point P. Is it symmetric with respect to the origin, i.e., is it true that

$$\text{speed at the point } s = \text{speed at the point } -s, \quad \text{for all } -1 \le s \le 1 .$$

4.6 Approximations

Given a function

$$y = f(x) , \tag{4.42}$$

a point x_0 and the corresponding value $y_0 = f(x_0)$, it is required to approximate $y = f(x)$ at points close to x_0, say at $x = x_0 + \Delta x$, with small $|\Delta x|$. If f is differentiable at x_0, then the line *tangent* to f at x_0 is

$$y = f(x_0) + f'(x_0) (x - x_0) . \tag{4.43}$$

At nearby points $x_0 + \Delta x$ we expect the graph of f and the tangent line (4.43) to be close, so that $f(x_0 + \Delta x)$ can be approximated by the tangent line at $x := x_0 + \Delta x$,

$$f(x_0 + \Delta x) \approx f(x_0) + f'(x_0)\Delta x . \tag{4.44}$$

Example 4.38. Approximate $\sqrt{98.6}$.

Solution: Let $f(x) = \sqrt{x}$, $x_0 = 100$ and $\Delta x = -1.4$ (Fig. 4.13 a). Then, by Example 4.28, $f'(x) = 1/(2\sqrt{x})$, and (4.44) gives

$$\sqrt{98.6} \approx \sqrt{100} - \frac{1.4}{2\sqrt{100}} = 10 - 0.07 = 9.93 \quad \text{(Fig. 4.13 b)} .$$

The correct value is $\sqrt{98.6} = 9.929753$ (to six decimals).

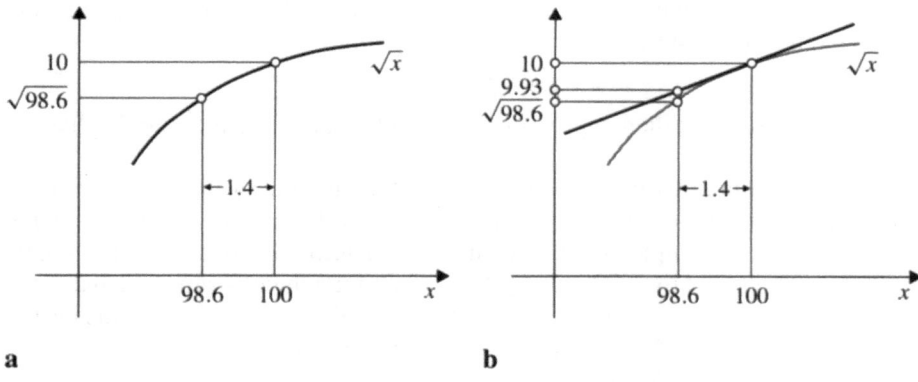

Fig. 4.13 a, b. Illustration of Example 4.38

Example 4.39. Approximate tan 46°.
Solution: We convert 46° to 0.802851 radian. Then use (4.44) and Example 4.32, with

$$f(x) = \tan x, \quad f'(x) = \sec^2 x, \quad x_0 = \frac{\pi}{4} = 0.785398 \,\text{radian} = 45°,$$

$$\Delta x = 0.0174533 \,\text{radian} = 1°,$$

to get

$$\tan(0.802851) \approx \tan\frac{\pi}{4} + \frac{0.0174533}{\cos^2(\pi/4)} = 1 + \frac{0.0174533}{(0.707107)^2}$$

$$= 1 + \frac{0.0174533}{0.5} = 1.0349066.$$

The correct value is $\tan 46° = 1.035530$ (to six decimals).

MACSYMA-Session 4.7. Recall the MACSYMA function from Session 4.6,
c1. `tangent(f,x,x0):= limit(f,x,x0)+(x-x0)*limit(diff(f,x),x,x0)$`
which computes the tangent line (4.43). The approximation (4.44) is then computed by the MACSYMA function
c2. `tangent_x(f,x,x0,x1):=limit(tangent(f,x,x0), x, x1)$`
To approximate $\sqrt{98.6}$ use
c3. `tangent_x(sqrt(x),x,100,98.6)`
d3.   9.93
in agreement with Example 4.38.
Similarly, the approximation of $\tan(46°)$, or $\tan(0.802851 \,\text{radian})$:
c4. `tangent_x(tan(x),x,%pi/4,0.802851), sfloat`
d4.   1.03491
to be compared with Example 4.39.

These ideas are used in studying measurement errors. Suppose we want to determine a quantity y which cannot be measured directly. However, another quantity, x, related to y by (4.42),

$$y = f(x) \, ,$$

can be measured. We then approximate y by substituting the measured value of x in (4.42).

Our measuring instruments are not perfect, so instead of the correct value x_0 we get $x_0 + \Delta x$, where Δx is the *measurement error*. Usually we are given a *bound* on the error. For example, an instrument for measuring length may claim that the error Δx is bounded by $-0.5 \, \text{mm} \le \Delta x \le 0.5 \, \text{mm}$, which is self-explanatory. In other cases there are bounds on the *relative error*. Thus an error of 4% means that $-0.04 \le \Delta x / x \le 0.04$. Note that the ratio $\Delta x / x$ is dimensionless.

Substituting $x_0 + \Delta x$ in (4.42) gives

$$y_0 + \Delta y = f(x_0 + \Delta x) \tag{4.45}$$

instead of the correct value y_0. Here Δy is the *resulting error* in y. How does the measurement error Δx influence the resulting error Δy? If f is differentiable at x_0, and Δx is small, then

$$f(x_0 + \Delta x) \approx f(x_0) + f'(x_0)\Delta x \, ,$$

so the resulting error is approximately linear in the measurement error,

$$\Delta y \approx f'(x_0)\Delta x \, . \tag{4.46}$$

For the relative error we get similarly,

$$\frac{\Delta y}{y_0} \approx \frac{f'(x_0)\Delta x}{f(x_0)} \, , \tag{4.47}$$

and in the special case $f(x) = cx^n$, for $n \in \mathbf{N}$,

$$\frac{\Delta y}{y_0} \approx n\frac{\Delta x}{x_0} \, , \tag{4.48}$$

independently of the coefficient c.

Example 4.40. The volume v of a sphere is estimated from its radius r whose measurement is subject to error Δr. Since $v = (4\pi/3)r^3 = cr^3$, we get from (4.48),

$$\frac{\Delta v}{v} \approx 3\frac{\Delta r}{r} \, .$$

If the radius is measured as $r \approx 10 \, \text{m}$, with relative error of 1%, then the volume is $v \approx 4000\pi/3 = 4188.79 \, \text{m}^3$ with relative error of 3%.

Exercises

4.33 Approximate each of the following values, and compare with the correct values. Use MACSYMA to compute the needed derivatives $f'(x)$.

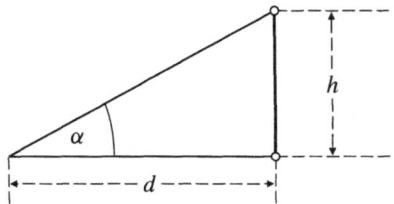

Fig. 4.14. Estimating the height h

(a) $\sqrt{17}$ (b) $\sqrt{16.1}$ (c) $\sqrt{399}$ (d) $1/2.01$

(e) $\sin 0.9\pi$ (f) $\sin 59°$ (g) $\cos 91°$ (h) $\tan 1.1\pi/4$

(i) $e^{0.001}$ (j) $\ln 1.002$ (k) $\log_{10} 102$ (l) $\ln(e^2 + 0.01)$

(m) $e^{0.01 + \ln 2}$ (n) $\log_2 65$ (o) $(1.001)^{1000}$ (p) $(1.02)^{1.02}$

(q) $\cosh 10^{-2}$ (r) $(26)^{1/3}$ (s) $(82)^{1/4}$ (t) $1/\sqrt{3.99}$

4.34 The radius of a sphere is measured as 1 ft to within an accuracy of 0.001 ft. Use differentials to estimate the error in the calculated surface area of the sphere.

4.35 The length of the side of a cube is measured as 1 m to within an accuracy of 1 mm. Use differentials to estimate the error in the calculated volume of the cube.

4.36 Why was it necessary in Example 4.39 to convert from degrees to radians?

4.37 The height h of a tree is estimated by measuring, at ground level distance d away from the tree, the angle α between ground and the tip of the tree; see Fig. 4.14. Let the correct values of d and α be $d_0 = 10$ m and $\alpha_0 = \pi/6$ radian respectively.

     a. Estimate the error in h if the measurement error of d is bounded by -0.05 m $\leq \Delta d \leq 0.05$ m.

     b. Estimate the error in h if the relative error of measuring α is bounded by 2%.

4.38 A right circular cylinder's radius is measured to within an accuracy of 2%; whereas the height of the cylinder is measured to within an accuracy of 3%. Use differentials to estimate the percentage error in the calculated lateral surface area of the cylinder.

4.39 The radius of a right circular cylinder is equal to its height. Suppose the height is measured with a ruler which has become inaccurate through heavy usage, and is known to be only accurate to 1%. Estimate the error in the calculated volume of the cylinder.

4.40 Do the previous problem for a right cirular cone.

4.7 Higher derivatives

The derivative f' of a given function f is itself a function $f'(x)$ which may or may not be differentiable. If f' is differentiable, its derivative $(f')'$ is called the *second derivative* of f and is denoted by

$$f''(x), \quad \text{or} \quad \frac{d^2}{dx^2} f \; .$$

For example, if $f(x) = x^5$, then $f'(x) = 5x^4$ and $f''(x) = 20x^3$.

     The second derivative has many applications in physics, since acceleration is the second derivative of position (see Sect. 4.8).

     In Sect. 4.2 we saw the geometrical interpretation of the first derivative $f'(x)$ as the slope of the tangent of the graph of f at the point x. The geometric interpretation of the second derivative is more complicated: It is the rate at which the slope $f'(x)$ changes, a measure of the *curvature* of the graph.

If the second derivative f'' is itself differentiable, then its derivative $(f'')'$ is called the *third derivative* of f. Proceeding in this manner we can define nth derivatives, for any positive integer n.

Definition 4.41. Let the function f be defined in a neighborhood of x, and let n be any positive integer. The *nth derivative* of f at the point x is defined by

$$f^{(n)}(x) := \begin{cases} f'(x) & \text{if } n = 1, \\ \left(f^{(n-1)}\right)'(x) & \text{otherwise,} \end{cases} \tag{4.49}$$

provided the derivatives exist. If $f^{(n)}(x)$ exists for all $x \in \mathbf{I}$, then f is called n times differentiable in $\mathbf{I}$.

Special cases: $f^{(1)}$ is the usual derivative, and $f^{(2)} = f''$. The 3rd derivative $f^{(3)}$ is sometimes denoted by f''', and the 4th derivative by f^{iv}.

We implement the nth derivative with MACSYMA.

MACSYMA-Session 4.8. The following MACSYMA function is a direct implementation of the recursive definition (4.49)
c1. dif(f,x,n) := if n=1 then diff(f,x) else diff(dif(f,x,n-1), x)$
We see that, while $n > 1$, the function dif(f,x,n) calls itself in the form
dif(f,x,n-1). As n is reduced each time dif is invoked, it will eventually come to $n = 1$, at which point MACSYMA's built-in diff function is used.
To see how dif(f,x,n) works, we trace it
c2. trace(dif)$
For example, the third derivative of x^5,
c3. dif(x^5,x,3)

$$(x^5, x, 3)$$
$$(x^5, x, 2)$$
$$(x^5, x, 1)$$
$$(5x^4)$$
$$(20x^3)$$
$$(60x^2)$$

d3. $60x^2$
The above function dif(f,x,n) serves to illustrate the recursive nature of the nth derivative but is not needed otherwise: MACSYMA computes the nth derivative by specifying an optional third argument n in the function diff(f,x,n). For example,
c4. diff(x^5,x,3)
d4. $60x^2$
Use showtime:true to compare the running time of the functions diff(f,x,n) and dif(f,x,n) for various f and n. Which is faster?

Exercises

4.41 Calculate the second and third derivative of
(a) $\sin x$, (b) $\cos x$, (c) x^2, (d) x^n $(n \in \mathbf{N})$, (e) e^x.

4.42 What will happen if the MACSYMA function `diff(f,x,n)` is invoked for some $n \notin \mathbf{N}$, e.g., for negative, fractional, or formal n?

4.8 Acceleration

For a moving particle, the acceleration $a(t)$ at any time t is the derivative of the velocity $v(t)$, or the second derivative of position,

$$a(t) := v'(t) = s''(t) . \tag{4.50}$$

Acceleration, the rate of change of velocity, is measured in units of

$$\frac{\text{velocity}}{\text{time}}, \quad \text{such as} \quad \left[\frac{\text{m}}{\text{s}^2}\right].$$

Example 4.42 (Constant acceleration). A particle begins moving at time t_0, from position s_0 (the initial position), with velocity v_0 (the initial velocity). It moves under a constant acceleration a. It is required to find (a) the velocity $v(t)$ and (b) the position $s(t)$ at any time $t \geq t_0$.
Solution: a. Since the rate of change of velocity is constant, $v'(t) = a$, we get

$$v(t) = v_0 + a(t - t_0) \tag{4.51}$$

i.e., a line through (t_0, v_0) with slope a; see Fig. 4.15b, where the initial velocity v_0 is negative, i.e., against the direction of the position axis in Fig. 4.15c.

b. The displacement $s(t) - s_0$ (covered during the time interval $[t_0, t]$) is equal to the average velocity times the duration $t - t_0$. Since the velocity is linear, the average velocity $\bar{v}(t_0, t)$ is

$$\bar{v}(t_0, t) = \frac{v_0 + v(t)}{2} = \frac{v_0 + v_0 + a(t - t_0)}{2} = v_0 + \frac{a}{2}(t - t_0) .$$

Therefore

$$s(t) - s_0 = v_0(t - t_0) + \frac{a}{2}(t - t_0)^2$$

or

$$s(t) = s_0 + v_0(t - t_0) + \frac{a}{2}(t - t_0)^2 \tag{4.52}$$

a quadratic parabola. This is illustrated in Fig. 4.15c, where the particle reverses direction, and the displacement during the time interval $[t_0, t]$ is $s(t) - s_0$.

Equations (4.51) and (4.52) were discovered by Galilei[13] , who studied the motion of falling bodies. Such bodies are subject to the (earth) *gravity acceleration*,

---

13 Galileo Galilei (1564–1642).

which at sea level is given by

$$g \approx 9.81 \, \text{m/s}^2 \,. \tag{4.53}$$

The gravity acceleration points toward the center of earth. We assume that it is constant throughout the movement.[14] For a body with initial position $s(t_0) = s_0$ and initial velocity $v(t_0) = v_0$, Eqs. (4.51) and (4.52) become

$$v(t) = v_0 + gt \tag{4.54a}$$

$$s(t) = s_0 + v_0(t - t_0) + \frac{g}{2}(t - t_0)^2 \,, \tag{4.54b}$$

where the direction of the s-axis is chosen the same as the direction of the acceleration g, i.e., the s-axis is pointing downward towards earth center.

Example 4.43 (Free fall). A ball is released in time $t = 0$ at an altitude of 10,000 m. It is required to find the velocity $v(t)$ at any time t, in particular the velocity at the moment the ball hits ground.

Solution: Put the origin at the point of release (at altitude 10,000 m, and let the s-axis point downward (Fig. 4.16a). Then the position $s(t)$ is just the s-coordinate of the body in time t, and is given by (4.54b) with $t_0 = 0$, $s_0 = 0$ and $v_0 = 0$ (the ball is just released, not thrown, so its initial velocity is zero),

$$s(t) = (g/2)t^2 \,. \tag{4.55}$$

The velocity $v(t)$ is similarly given by (4.54a) with $t_0 = 0$, $v_0 = 0$,

$$v(t) = g\,t \tag{4.56}$$

which is just the derivative of (4.55).

The body reaches earth at time t_1, when its coordinate is $s(t_1) = 10,000$ m. This time is found by solving

$$(g/2)t^2 = 10,000$$

or $t_1 = \sqrt{20,000/g} = 45.15$ [s].

The velocity at the time $t = t_1$ is, by (4.56),

$$v(t_1) = g\,t_1 = 9.81 \, \text{m/s}^2 \cdot 45.15 \, \text{s} = 442.94 \, \text{m/s} \,,$$

giving the velocity with which the ball returns to ground. The air resistance was ignored in the above analysis (which is correct in vacuum). The air resistance acts to slow the motion, so the velocity at any time is smaller than the value given by (4.56).

---

14 The gravity acceleration is inversely proportional to the square of the distance from the earth center, and so decreases as the altitude increases. We assume that the whole movement takes place sufficiently close to sea level, so there is no need to study changes in g.

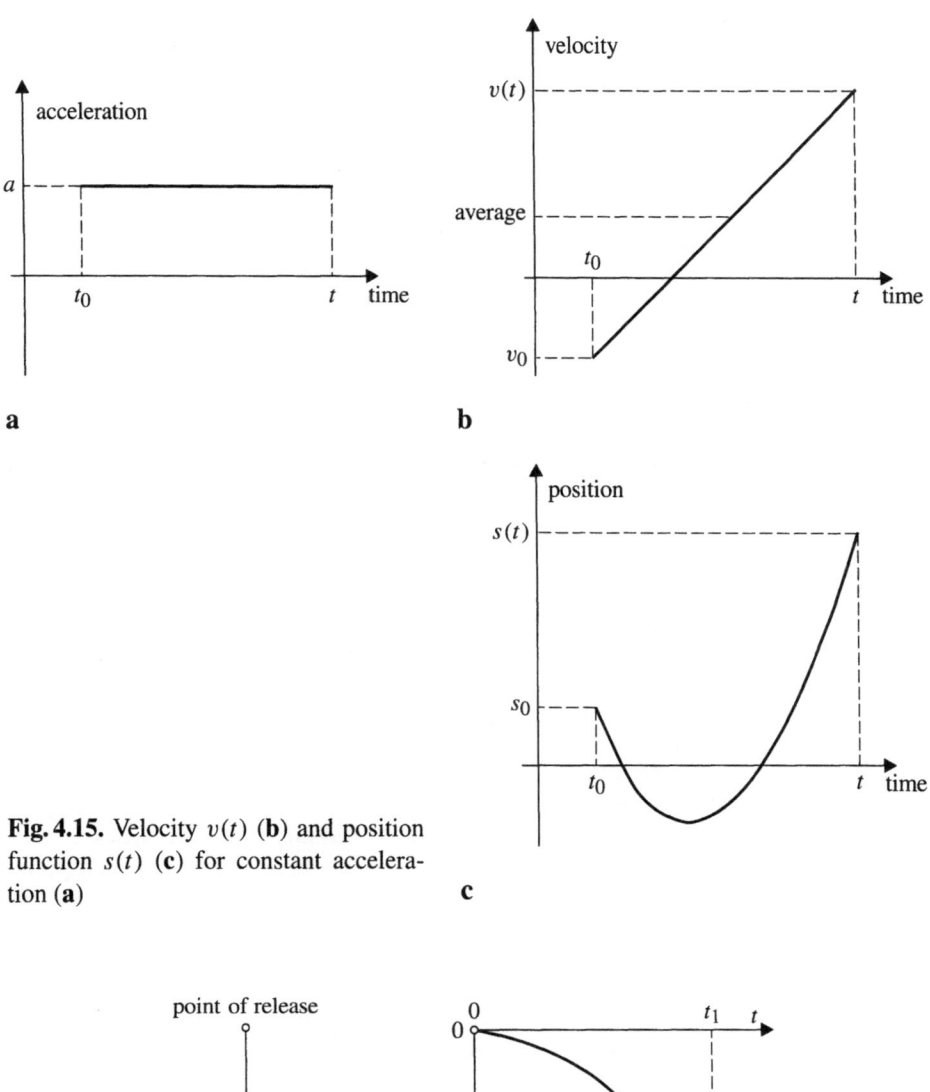

Fig. 4.15. Velocity $v(t)$ (**b**) and position function $s(t)$ (**c**) for constant acceleration (**a**)

Fig. 4.16. a A falling particle released at altitude 10,000 [m]; **b** its position function

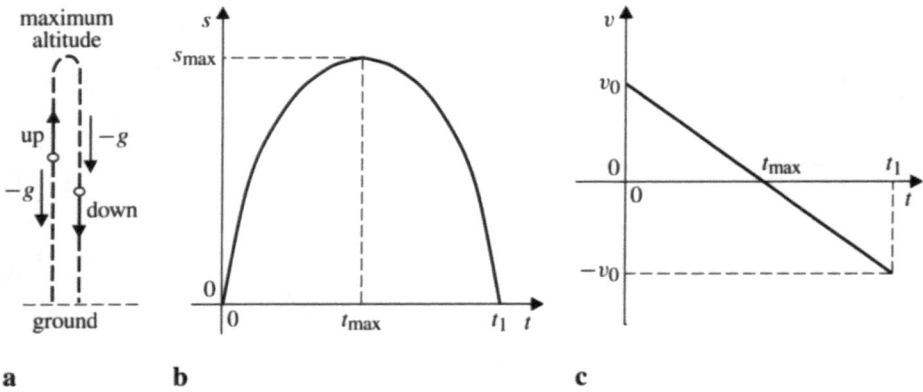

Fig. 4.17. a Trajectory (and acceleration $-g$), **b** position, and **c** velocity of a ball thrown up

Example 4.44 (Ball thrown up). At time $t = 0$ a ball is thrown up, from the surface of earth, at a velocity of $v_0 = 50$ m/s. The ball rises a certain distance, then falls until it reaches ground. Ignoring the air resistance, determine:

a. the time t_1 when the ball will hit ground;
b. the velocity at any time $t \le t_1$;
c. the velocity with which the ball hits ground;
d. the maximal altitude reached by the ball.

Solution: Place the s-axis, pointing up, at the initial point (Fig. 4.17 b).

The acceleration at any time t is still the gravity acceleration g, which points downward (toward earth center) and so against the direction of the s-axis. We therefore assign the acceleration g a negative sign; see Fig. 4.17 a, where the acceleration $-g$ "works against" the particle on its journey up, and "works with" the particle on its way down.

The position at any time $0 \le t \le t_1$ is therefore determined by (4.52) with $a = -g$, $t_0 = 0$, $s_0 = 0$, and the given initial velocity v_0,

$$s(t) = v_0 t - (g/2)t^2 \,, \tag{4.57}$$

at any time $0 \le t \le t_1$.

a. The time t_1 when the ball returns to the earth surface is found by solving the quadratic equation $s(t) = v_0 t - (g/2)t^2 = 0$, which has two solutions: $t_0 = 0$ (correct, but uninteresting, since we already know that the ball was on the ground at time $t = 0$, when the movement began), and

$$t_1 = 2v_0/g \,. \tag{4.58}$$

Here $t_1 = 2 \cdot 50/9.81 = 10.194$ [s].

b. The velocity is the derivative of $s(t)$,

$$v(t) = v_0 - g\,t \tag{4.59}$$

a line with a negative slope $-g$. Alternatively, the velocity can be determined from (4.51) with $a = -g$, $t_0 = 0$, and the given v_0. The velocity is plotted in Fig. 4.17c.

c. The velocity of the ball at the time of impact is $v(t_1)$. Substituting $t := t_1$ in (4.59) we get

$$v(t_1) = v_0 - g\frac{2v_0}{g} = -v_0 ,$$

i.e., the ball hits ground at a speed equal to its initial speed.[15]

d. The velocity $v(t)$ is
- positive when the ball moves in the direction of the s-axis, that is, up,
- negative when movement is against the direction of the s-axis, i.e., the ball falls.

Therefore the maximum altitude is reached when $v(t) = 0$. To find the time t_{max} when this happens, we solve $v(t) = 0$, or

$$t_{max} = v_0/g , \tag{4.60}$$

and in our example, $t_{max} = 50/9.81 = 5.097$ [s]. The maximum altitude s_{max} is, by (4.57) and (4.60),

$$s_{max} = s(t_{max}) = v_0 t_{max} - \frac{g}{2} t_{max}^2 = v_0\frac{v_0}{g} - \frac{g}{2}\frac{v_0^2}{g^2} = \frac{v_0^2}{2g} . \tag{4.61}$$

In our example, $s_{max} = 50^2/(2 \cdot 9.81) = 127.421$ [m].

Exercises

4.43 Prove: If a body moves with constant acceleration, then it can reverse its direction at most once.

4.44 A bomb is dropped from a height of h [ft] above its target. Show that the bomb impacts its target in $\sqrt{(2/g)h}$ [s], where we approximate g by 32 ft/s^2.

4.45 If instead of a bomb being dropped as in the previous problem, it is projected downwards with a velocity v_d [ft/s], show that the time it takes to impact its target is given by $(1/32)(\sqrt{v_d^2 + 64h} - |v_d|)$ [s].

4.46 A pigeon drops a pebble from a windowsill 100 [ft] above ground level. Ignoring the effect of air resistance, compute:
a. the velocity of the pebble when it hits the ground;
b. the time it took the pebble to reach the ground.

4.47 A rocket is fired straight upwards and its height, t minutes after launch, is given by $s(t) = (1/40)t^3$ [miles]. The rocket's engine fails after 10 min.
a. How high is the rocket and what is its velocity when its engine fails?
b. How long will it take the rocket to fall to earth?
c. How fast will it be traveling when it impacts with the earth?

---

15 Does this make sense?

5 Differentiation rules

If f is differentiable, its derivative f' can be computed using the limit (4.11),

$$f'(x) = \lim_{\xi \to x} \frac{f(\xi) - f(x)}{\xi - x} \, ,$$

which is often difficult. However, sometimes f has a special structure that allows differentiating it without evaluating the limit (4.11). For example, if u and v are differentiable functions, and if f is their product $f = uv$, then the derivative f' can be easily computed from the derivatives u' and v'. This situation is covered by a differentiation rule called the product rule (Theorem 5.1). Other rules given in this chapter are the quotient rule (Theorem 5.5) and the chain rule (Theorem 5.11).

5.1 Product and quotient rules

We saw that the derivative of a sum is the sum of the derivatives (Theorem 4.22). A natural question is whether there is also an easy way to calculate the derivative of a product of differentiable functions. The answer is given in the following theorem.

Theorem 5.1 (Product rule). If the functions u and v are differentiable at the point x, then so is their product $u \cdot v$, and

$$(u \cdot v)'(x) = u'(x)v(x) + u(x)v'(x) \, . \tag{5.1}$$

Proof.
$$
\begin{aligned}
(u \cdot v)'(x) &= \lim_{\xi \to x} \frac{u(\xi)v(\xi) - u(x)v(x)}{\xi - x} \\
&= \lim_{\xi \to x} \frac{u(\xi)v(\xi) - u(x)v(\xi) + u(x)v(\xi) - u(x)v(x)}{\xi - x} \\
&= \lim_{\xi \to x} \frac{u(\xi) - u(x)}{\xi - x} \lim_{\xi \to x} v(\xi) + u(x) \lim_{\xi \to x} \frac{v(\xi) - v(x)}{\xi - x} \\
&= u'(x)v(x) + u(x)v'(x) \, .
\end{aligned}
$$

We used here $\lim_{\xi \to x} v(\xi) = v(x)$, justified because v is continuous at x, which follows from the assumption that v is differentiable at x (Theorem 4.11). $\qquad \square$

Remark 5.2. The product rule can be stated in words as follows:

The slope of $uv = u \times$ the slope of $v + v \times$ the slope of u.

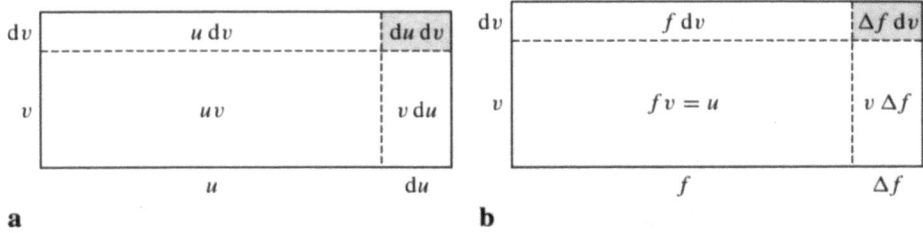

Fig. 5.1. Illustration of the product rule (**a**) and the quotient rule (**b**)

Note that the derivative of the product is not the product of the derivatives. Compare with Remark 4.23 b.

Figure 5.1 a illustrates the product rule in terms of the differentials $d(uv)$, du, and dv. We interpret the product uv as the area of the rectangle with sides[1] u and v. As the sides change from (u, v) to $(u + du, v + dv)$, the area changes to $(u + du)(v + dv) = uv + u\,dv + v\,du + du\,dv$, and the difference

$$\Delta(uv) = (u + du)(v + dv) - uv = u\,dv + v\,du + du\,dv \ .$$

The differential $d(uv)$ consists of the linear part of the difference,

$$d(uv) = u\,dv + v\,du \ , \tag{5.2}$$

omitting the product of differentials $du\,dv$, which is the shaded area in Fig. 5.1 a. A trick to remember (5.2) is to write the product rule (5.1) as

$$\frac{d(uv)}{dx} = u\frac{dv}{dx} + v\frac{du}{dx} \ ,$$

and multiply by dx to obtain

$$d(uv) = u\,dv + v\,du \ .$$

Example 5.3. It is required to compute the derivative of

$$f(x) = (1 + x + x^2)(1 - x + x^2) \ .$$

Solution 1: Expand f to get

$$f(x) = (1 + x + x^2)(1 - x + x^2) = 1 + x^2 + x^4 \ ,$$

and compute the derivative using the general formula of Corollary 4.26 for differentiating polynomials,

$$f'(x) = 2x + 4x^3 \ .$$

---

1 This interpretation is restricted to u and v with positive values, while the product rule holds for general u and v, as long as they are differentiable.

Solution 2: Use the product rule, with $u := 1 + x + x^2$ and $v := 1 - x + x^2$,

$$f'(x) = u'(x)v(x) + u(x)v'(x) = (1+2x)(1-x+x^2) + (1+x+x^2)(-1+2x),$$

giving the same answer as above.

Which solution is easier? Which solution would be easier if u and v were polynomials of degree 10?

Example 5.4. The product rule (5.1) can be applied to products of more than two functions. For example,

$$(u \cdot v \cdot w)'(x) = u'(x)v(x)w(x) + u(x)v'(x)w(x) + u(x)v(x)w'(x) . \quad (5.3)$$

Proof.

$$(u \cdot v \cdot w)'(x) = ([u \cdot v] \cdot w)'(x) = [u \cdot v]'(x)w(x) + [u \cdot v](x)w'(x)$$
$$= [u'(x)v(x) + u(x)v'(x)]w(x) + u(x)v(x)w'(x) . \quad \square$$

The derivatives of functions which are ratios u/v can be similarly expressed in terms of the derivatives of u and v.

Theorem 5.5 (Quotient rule). If u and v are differentiable at x, and if $v(x) \neq 0$, then the quotient u/v is differentiable at x, and

$$\left(\frac{u}{v}\right)'(x) = \frac{u'(x)v(x) - u(x)v'(x)}{(v(x))^2} . \quad (5.4)$$

In particular, for $u = 1$, we get the *reciprocal rule*: the reciprocal $1/v$ is differentiable, and its derivative is

$$\left(\frac{1}{v}\right)'(x) = -\frac{v'(x)}{(v(x))^2} . \quad (5.5)$$

Proof. We first prove the reciprocal rule

$$\left(\frac{1}{v}\right)'(x) = \lim_{\xi \to x} \frac{\frac{1}{v(\xi)} - \frac{1}{v(x)}}{\xi - x} = \frac{1}{v(x)} \lim_{\xi \to x} \frac{1}{v(\xi)} \lim_{\xi \to x} \frac{v(x) - v(\xi)}{\xi - x} = -\frac{v'(x)}{(v(x))^2} ,$$

where again we used the continuity of v. The quotient u/v is the product of u and the reciprocal $1/v$. Therefore

$$\left(\frac{u}{v}\right)'(x) = \left(u \cdot \frac{1}{v}\right)'(x) = \frac{u'(x)}{v(x)} + u(x)\left(\frac{1}{v(x)}\right)'$$
$$= \frac{u'(x)}{v(x)} - \frac{u(x)v'(x)}{(v(x))^2} = \frac{u'(x)v(x) - u(x)v'(x)}{(v(x))^2} ,$$

where we used both the product rule (5.1) and the reciprocal rule (5.5). $\quad \square$

It may be easier to remember the product rule, quotient rule, and the reciprocal rule in the following short form (suppressing the argument)

$$(uv)' = u'v + uv', \quad \left(\frac{u}{v}\right)' = \frac{u'v - v'u}{v^2}, \quad \left(\frac{1}{v}\right)' = -\frac{v'}{v^2},$$

$$(uvw)' = u'vw + uv'w + uvw'.$$

Example 5.6. We consider again the derivative of $\tan x$ (Example 4.32). Since $\tan x = \sin x / \cos x$ we use the quotient rule with $u := \sin x$ and $v := \cos x$,

$$\frac{d}{dx} \tan x = \frac{\cos x \dfrac{d}{dx} \sin x - \sin x \dfrac{d}{dx} \cos x}{\cos^2 x},$$

$$= \frac{\cos^2 x + \sin^2 x}{\cos^2 x}, \quad \text{by (4.32) and (4.33)},$$

$$= \frac{1}{\cos^2 x}.$$

This derivation is easier than the direct computation in Example 4.32.

Example 5.7 (Rational functions). The quotient rule allows an easy differentiation of rational functions,

$$r(x) = p(x)/q(x),$$

which are quotients of polynomials p and q. The derivative is

$$r'(x) = \frac{p'(x)q(x) - p(x)q'(x)}{q^2(x)}, \tag{5.6}$$

which is itself a rational function. For example, the derivative of

$$r(x) = \frac{1+x}{-6 + 5x + 2x^2 - x^3}$$

is computed, using (5.6) with $p(x) := 1 + x$ and $q(x) := -6 + 5x + 2x^2 - x^3$,

$$r'(x) = \frac{p'(x)q(x) - p(x)q'(x)}{(q(x))^2}$$

$$= \frac{(-6 + 5x + 2x^2 - x^3) - (5 + 4x - 3x^2)(1 + x)}{(-6 + 5x + 2x^2 - x^3)^2}.$$

We illustrate the quotient rule in Fig. 5.1b. Let $f = u/v$. As the values of (u, v) change to $(u + du, v + dv)$, the quotient changes to

$$f + \Delta f = \frac{u + du}{v + dv}$$

or

$$(f + \Delta f)(v + dv) = u + du$$

$$\therefore \quad fv + v\,\Delta f + f\,\mathrm{d}v + \Delta f\,\mathrm{d}v = u + \mathrm{d}u$$

$$\therefore \quad \Delta f = \frac{\mathrm{d}u - f\,\mathrm{d}v - \Delta f\,\mathrm{d}v}{v}, \quad \text{since } u = fv\,.$$

The differential $\mathrm{d}f$ is the linear part of the difference Δf,

$$\mathrm{d}f = \frac{\mathrm{d}u - f\,\mathrm{d}v}{v} = \frac{\mathrm{d}u - \dfrac{u}{v}\,\mathrm{d}v}{v}$$

$$= \frac{v\,\mathrm{d}u - u\,\mathrm{d}v}{v^2}\,. \tag{5.7}$$

A common calculus mistake is to differentiate the numerator and denominator separately, and to take their ratio $u'(x)/v'(x)$ as the derivative of $u(x)/v(x)$. Resist that temptation.

Example 5.8 (Derivative of x^{-n}). Consider a monomial with a negative exponent, $f(x) := x^{-n}$, where $n \in \mathbf{N}$, and $x \neq 0$. Writing $f(x) := 1/v(x)$ with $v(x) := x^n$, we can use the reciprocal rule, together with $v'(x) = nx^{n-1}$ (Theorem 4.25),

$$f'(x) = \left(\frac{1}{v}\right)'(x) = -\frac{v'(x)}{(v(x))^2} = -\frac{nx^{n-1}}{x^{2n}} = -nx^{-n-1}\,.$$

We see that this is the old power rule, now for powers with negative exponents.

Combining this result with the power rule of Theorem 4.25, we get the following theorem.

Theorem 5.9 (Power rule). Let $n \in \mathbf{Z}$. The power function $f(x) := x^n$ has the derivative $f'(x) = nx^{n-1}$ provided that $x \neq 0$ for $n < 0$.

Now we demonstrate the use of MACSYMA to handle differentiation rules.

MACSYMA-Session 5.1. If the expression f has not been assigned a value, then
c1. `diff(f,x)`
yields
d1.   0
since f is interpreted as independent of x; thus its derivative with respect to x is 0.

If f is assigned a value, say
c2. `f:x^5$`
then
c3. `diff(f,x)`
gives the correct answer
d3. $5x^4$

If $g(x)$ is an unspecified function, then
c4. `diff(g(x),x)`

gives an echo
d4.   diff$(g(x), x)$
lacking the information required to compute $g'(x)$ explicitly.

If $h(x)$ is a specified function, say
c5. h(x):=x^4$
then
c6. diff(h(x),x)
d6. $4x^3$
Similarly, the derivative of the composite function $(2x + 1)^4$ is computed as
c7. diff(h(2*x+1),x)
d7. $8(2x + 1)^3$

To illustrate the product rule, let $u(x)$, $v(x)$ and $w(x)$ be unspecified functions. Then
c8. diff(u(x)*v(x),x)
d8. $u(x) * \text{diff}(v(x), x) + v(x) * \text{diff}(u(x), x)$
Similarly,
c9. diff(u(x)*v(x)*w(x),x)
d9. $u(x)*v(x)*\text{diff}(w(x), x)+u(x)*w(x)*\text{diff}(v(x), x)+v(x)*w(x)*\text{diff}(u(x), x)$
in agreement with (5.3).
a. Compute: diff(u(x)^5,x) and diff(u(x)*u(x)*u(x)*u(x)*u(x),x)
b. Verify the quotient rule: diff(u(x)/v(x),x)
c. The 2nd derivative of a product:
c10. diff(u(x)*v(x),x,2)
d10. $u(x) * \text{diff}(v(x), x, 2) + 2 * \text{diff}(u(x), x) * \text{diff}(v(x), x) + v(x) * \text{diff}(u(x), x, 2)$
Now try the higher derivatives of the product $u(x)v(x)$ with respect to x. Can you guess
the general formula for the nth derivative? See the answer in Exercise 5.7.

Example 5.10. The function

$$f(x) := \begin{cases} x/(1 + x) & \text{if } x > 0, \\ x/(1 - x) & \text{if } x \leq 0, \end{cases}$$

is differentiable in all of **R** with derivative

$$f'(x) = \begin{cases} 1/(1 + x)^2 & \text{if } x > 0, \\ 1/(1 - x)^2 & \text{if } x \leq 0. \end{cases}$$

This is easily proved, by the quotient rule, for both $x > 0$ and $x < 0$. That the
formula also holds for $x = 0$ has to be proved separately. For $x := 0$ we use the
definition of the right and left derivative to show that

$$f'_+(0) := \lim_{\xi \to 0^+} \frac{f(\xi) - f(0)}{\xi - 0} = \lim_{\xi \to 0^+} \frac{1}{1 + \xi} = 1,$$

and

$$f'_-(0) := \lim_{\xi \to 0^-} \frac{f(\xi) - f(0)}{\xi - 0} = \lim_{\xi \to 0^-} \frac{1}{1 - \xi} = 1,$$

and as both are equal, $f'(0) = 1$. Plot the function f and its derivative f' with MACSYMA to get an idea of the situation.

Similarly one can compute the second derivative of f. Using the quotient rule we get

$$f''(x) = \begin{cases} -2/(1+x)^3 & \text{if } x > 0, \\ 2/(1-x)^3 & \text{if } x < 0. \end{cases}$$

Now it is easy to see that f' is not differentiable at $x = 0$ (see Exercise 5.9).

Exercises

5.1 Prove the rule for the derivative of the monomial $f(x) := x^m$, for $m \in \mathbf{N}$, i.e., Theorem 4.25, by induction using the derivative of $u(x) = x$ and the product rule.

5.2 Find the derivative of $f(x) := x(x-1)(x-2)(x+1)(x+2)$. Find the zeros of f'. Can you explain why these numbers were mentioned in Exercise 1.41?

5.3 Use the methods of this section to compute the derivatives of:

(a) $f(x) = \dfrac{1+x^2}{x^3 - 2x - 1}$ (b) $f(x) = \dfrac{1}{(x-2)(x-1)x(x+1)(x+2)}$

(c) $f(x) = \cot x$ (d) $f(x) = \dfrac{3x^4}{x^3 - 1}$

(e) $f(x) = \dfrac{1}{x^4 - 1}$ (f) $f(x) = e^{-x}$

Hint: In c you can use the reciprocal rule ($\cot x = 1/\tan x$) or the quotient rule ($\cot x = \cos x / \sin x$).

5.4 Show that the derivative of $f(x) := (1+x)^n$, for $n \in \mathbf{N}$, is $f'(x) = n(1+x)^{n-1}$. *Hint:* Use the binomial formula.

5.5 Find the derivative of $f(x) := |(1+x)/(1-x)|$ wherever it exists, and plot the graph of f and its derivative.

5.6 Find a formula for the second and third derivatives of the product $u(x)v(x)$. Assume that u and v have derivatives of sufficiently high order.

5.7 Generalize the result of the previous problem to obtain the *Leibniz rule*, or generalized product rule, for the nth derivative ($n \in \mathbf{N}$)

$$(u \cdot v)^{(n)} = \sum_{k=0}^{n} \binom{n}{k} u^{(k)} v^{(n-k)}, \tag{5.8a}$$

or $\quad \dfrac{d^n}{dx^n}(u(x)v(x)) = \sum_{k=0}^{n} \binom{n}{k} \left(\dfrac{d^k u}{dx^k}\right) \left(\dfrac{d^{n-k} v}{dx^{n-k}}\right).$ \tag{5.8b}

5.8 Use the Leibniz rule to calculate the fourth derivatives of the following functions. Check your results with MACSYMA

(a) $x^2 e^x$ (b) $x^3 \sin x$ (c) $x^4 \sqrt{1-x}$

(d) $x^2 y(x)$ (e) $x^3/(1-x)$ (f) $\sin x \cos x$

5.9 Define the function f of Example 5.10 with MACSYMA. Try to differentiate f. Show that f is differentiable at $x = 0$ with the functions `dif_left(f,x,x0)` and `dif_right(f,x,x0)` of MACSYMA-Session 4.4. Then differentiate f separately for positive and for negative x and build the derivative function. Plot both f and f'. Finally differentiate f' and plot also f''. For which $x \in \mathbf{R}$ is f' differentiable?

5.10 Find the even function f whose values have the form $f(x) := x^3$ for $x > 0$ and differentiate f. Declare the function with MACSYMA, and plot the graphs of f, f', and f''.

5.11 Do the same as in Exercise 5.10 for $f(x) := x/(1 + x^3)$.

5.12 Write a recursive function $\Delta(f, x, h, n)$ that calculates

(a) the difference quotient $\dfrac{f(x+h) - f(x)}{h}$ for $n = 1$, and

(b) the difference quotient of $\Delta(f, x, h, n - 1)$ for $n \in \mathbf{N}$, $n > 1$.

Calculate $\Delta(x^m, x, h, n)$ for some values of n, h, and m (try also formal values), and compare the results with the nth derivative.

5.2 Chain rule and implicit differentiation

In Exercise 5.4 we calculated, by brute force (using the binomial formula twice), the derivative of $f(x) := (1 + x)^n$, for $n \in \mathbf{N}$, which turned out to be $f'(x) = n(1+x)^{n-1}$. We see that this is exactly like the power formula $(d/du) u^n = nu^{n-1}$ if we identify u with $1 + x$.

To be precise, the function $f(x) = (1+x)^n$ is the *composition* of two functions: $f(x) = F(u(x))$ where $F(u) := u^n$ and $u(x) := 1 + x$. In such a case we can write the derivative of f in terms of the derivatives dF/du and du/dx.

Let f be the composition of u and F, denoted by,

$$f(x) = F(u(x)), \quad \text{or } f(x) = (F \circ u)(x)$$

where the function u is called the *inner function*, and F is called the *outer function* of the composed function $F \circ u = F(u)$.

The next theorem gives the derivative of a composed function.

Theorem 5.11 (Chain rule). If u is differentiable at the point x, F is differentiable at the point $u(x)$, and if $f(x) = F(u(x))$, then f is differentiable at x with

$$f'(x) = F'(u(x)) \cdot u'(x) . \tag{5.9}$$

Proof. We suppose first that $u'(x) \neq 0$. Then $u(\xi) - u(x) \neq 0$ if ξ is near enough to x, and we get

$$f'(x) = F(u(x))'(x) = \lim_{\xi \to x} \frac{F(u(\xi)) - F(u(x))}{\xi - x}$$

$$= \lim_{\xi \to x} \frac{F(u(\xi)) - F(u(x))}{u(\xi) - u(x)} \lim_{\xi \to x} \frac{u(\xi) - u(x)}{\xi - x}$$

$$= \lim_{\xi \to x} \frac{F(u(\xi)) - F(u(x))}{u(\xi) - u(x)} u'(x) .$$

Now we write $y := u(x)$ and $\eta := u(\xi)$. As u is continuous at the point x by hypothesis and by Theorem 4.11, we get that $\lim_{\xi \to x} u(\xi) = u(x)$, i.e., $\eta \to y$ as $\xi \to x$, and so

$$\lim_{\xi \to x} \frac{F(u(\xi)) - F(u(x))}{u(\xi) - u(x)} = \lim_{\eta \to y} \frac{F(\eta) - F(y)}{\eta - y} = F'(y) = F'(u(x)) ,$$

which finishes the proof in that case. Now let $u'(x) = 0$. The differentiability of F at $u(x)$ implies that

$$|F(u(\xi)) - F(u(x))| \le M|u(\xi) - u(x)|, \quad \text{for some } M \in \mathbf{R} ,$$

so that

$$|f'(x)| = \left|\left(F(u(x))\right)'(x)\right| = \left|\lim_{\xi \to x} \frac{F(u(\xi)) - F(u(x))}{\xi - x}\right|$$

$$\le M \lim_{\xi \to x} \left|\frac{u(\xi) - u(x)}{\xi - x}\right| ,$$

and so equals zero since the difference quotient of u tends to zero as $\xi \to x$. $\square$

Remark 5.12. a. The chain rule is perhaps the most important of all differentiation rules. In the given case the function u' is called the *inner derivative*, and the function $F'(u)$ is called the *outer derivative*. We can summarize the chain rule in words: The derivative of a composite function $F \circ u$ is the product of the inner derivative u' and the composition $F' \circ u$ of the outer derivative F' and u,

$$(F \circ u)'(x) = (F' \circ u)(x) u'(x) .$$

b. It is convenient, in the chain rule, to use the notation of the derivative as a differential quotient. The chain rule then reads

$$\frac{d(F(u))}{dx}(x) = \frac{dF}{du}(u(x)) \frac{du}{dx}(x) = \left.\frac{dF(u)}{du}\right|_{u=u(x)} \frac{du}{dx}(x) \tag{5.10}$$

or – omitting the variable x – even shorter

$$\frac{dF(u)}{dx} = \frac{dF}{du} \frac{du}{dx} , \tag{5.11}$$

and so the chain rule can be memorized as a *cancellation rule* for differential quotients. This is not surprising, since the chain rule was proved by multiplying, and dividing, the expression $\Delta F/\Delta x$ by $\Delta u = u(x+\Delta x) - u(x)$, and then taking the limit as $\Delta x \to 0$.

We now give some examples for the use of the chain rule.

Example 5.13 (Derivative of $\cos x$). The identity

$$\cos x = \sin(\pi/2 - x), \quad \forall x ,$$

gives the function $\cos x$ as a composition,

$$\cos x = (F \circ u)(x) \quad \text{where } F(u) := \sin u, \text{ and } u(x) := \pi/2 - x .$$

The derivative of $\cos x$ is then, by the chain rule,

$$\therefore \quad \frac{d}{dx} \cos x = \frac{d \sin u}{du}\bigg|_{u=\pi/2-x} \frac{d}{dx}(\pi/2 - x) = -\cos(\pi/2 - x) = -\sin x .$$

We used here the chain rule, and the derivative of $\sin x$, to compute the derivative of $\cos x$ without evaluating limits; compare with Example 4.31.

Example 5.14 (Derivative of a^x). Let $a > 0$. Then

$$\frac{d}{dx} a^x = a^x \ln a . \tag{5.12}$$

Proof. Since $a = e^{\ln a}$, it follows that $a^x = e^{(\ln a)x} = (F \circ u)(x)$, where $F(u) := e^u$ and $u(x) := (\ln a)x$. Therefore

$$\frac{d}{dx} a^x = \frac{dF}{du}\bigg|_{u=(\ln a)x} \frac{du}{dx} = e^u \ln a = e^{(\ln a)x} \ln a = a^x \ln a . \qquad \square$$

Example 5.15. It is required to differentiate $f(x) = ((1+x)/(1-x))^n$. If we set $u(x) := (1+x)/(1-x)$, and $F(u) := u^n$, then $f = F(u)$, and the chain rule gives

$$f'(x) = F'(u(x)) \cdot u'(x) = n(u(x))^{n-1} \cdot \frac{2}{(1-x)^2}$$

$$= n\left(\frac{1+x}{1-x}\right)^{n-1} \frac{2}{(1-x)^2} = \frac{2n(1+x)^{n-1}}{(1-x)^{n+1}} .$$

Although the given f is rational (if $n \in \mathbf{Z}$), we would have had great difficulty finding its derivative without the chain rule. Since the exponent n is symbolically given, an application of the quotient rule would require an expansion of both the numerator and the denominator of f.

An important application of the chain rule is in *implicit differentiation*. Let y be related to x through an implicit function (1.12),

$$F(x, y) = 0 .$$

It is required to compute the derivative dy/dx. One way is to solve (1.12) to obtain an explicit function

$$y = f(x), \quad \text{satisfying (1.12)} ,$$

and then differentiate f.

Implicit differentiation is a method which avoids computing the derivative of the explicit function f and usually avoids computing f itself.

For example, let the implicit function $F(x, y)$ be of the form

$$F(x, y) := g(y) + h(x) = 0 , \tag{5.13}$$

where g and h are differentiable. Assuming that y is differentiable, we rewrite (5.13) as

$$g(y(x)) + h(x) = 0$$

and differentiate, using the chain rule,

$$g'(y(x)) \, y'(x) + h'(x) = 0$$

from which

$$y'(x) = -\frac{h'(x)}{g'(y(x))} . \tag{5.14}$$

Example 5.16 (Derivative of $\sqrt{x}$). We use implicit differentiation to compute $(\mathrm{d}/\mathrm{d}x)\sqrt{x}$ for $x > 0$. The direct approach, using the definition of the derivative, was illustrated in Example 4.28.

The square root function $y := \sqrt{x}$ satisfies the implicit function

$$y^2 - x = 0 , \tag{5.15}$$

which is of the form (5.13) with $g(y) := y^2$ and $h(x) := -x$. From (5.14) we get therefore

$$y' = \frac{1}{2y} = \frac{1}{2\sqrt{x}} .$$

Example 5.17 (Derivative of powers with rational exponents). Find the derivative of the power function, with rational exponent,

$$f(x) = x^{p/q} = \sqrt[q]{x^p} ,$$

where p and q are integers, and $x > 0$.[2]

The function $y = x^{p/q}$ satisfies the implicit relation

$$y^q = x^p .$$

Therefore, by implicit differentiation,

$$qy^{q-1}y' = px^{p-1} ,$$

giving

$$y'(x) = \frac{p}{q}\frac{x^{p-1}}{y^{q-1}} = \frac{p}{q}\frac{x^{p-1}}{x^{p-p/q}} = \frac{p}{q}x^{p/q-1} .$$

We conclude that the differentiation rule for powers, Theorem 5.9, can be extended to rational exponents.

---

2 For example, if $x < 0$, then $x^{p/q}$ is not defined for $p = 1$ and q even. The restriction to $x > 0$ assures that $x^{p/q}$ is defined for all integer p, q.

Theorem 5.18 (Derivative of powers with rational exponents). For any rational number α,

$$\frac{d}{dx}x^\alpha = \alpha x^{\alpha-1}, \qquad (5.16)$$

for all $x > 0$.

Implicit differentiation should be handled with care. When we used the implicit relation (1.12), $F(x, y) = 0$, to get the derivative y', we made two (implicit) assumptions:

that there is a well-defined function y which satisfies (1.12);

that this function is differentiable.

If any of these assumptions does not hold, then the implicit derivative is meaningless, or has a meaning different from that we intended.

Thus we have not proved that (5.14) gives the derivative y' of any function y. All we claim is: If the function y satisfies $g(y)+h(x) = 0$, and if $y(x)$ is differentiable, then y' is given by (5.14).

Example 5.19 (Why implicit differentiation should be handled with care).
The implicit function

$$x^2 + y^2 = r^2, \qquad (5.17)$$

is the equation of the circle with center at $(0, 0)$ and radius r. Applying (5.14) with $g(y) = y^2$ and $h(x) = x^2 - r^2$ we get the implicit derivative

$$y' = -x/y. \qquad (5.18)$$

So far so good, but the implicit relation (5.17) *does not define* a function $y = f(x)$: As shown in Remark 1.22, there are actually two functions, f_1 representing the upper semicircle,

$$f_1(x) := +\sqrt{r^2 - x^2} \quad (-r \leq x \leq r) \quad \text{(Fig. 5.2b)},$$

and f_2 for the lower semicircle,

$$f_2(x) := -\sqrt{r^2 - x^2} \quad (-r \leq x \leq r) \quad \text{(Fig. 5.2c)}.$$

The derivatives can be computed by the chain rule. For example, f_1 is the composition of $u(x) = r^2 - x^2$ and $F(u) = \sqrt{u}$. Its derivative is, by (5.10),

$$f_1'(x) = F'(u(x))\, u'(x) = \frac{1}{2\sqrt{u}}(-2x)$$

$$= \frac{-2x}{2\sqrt{r^2 - x^2}} = -\frac{x}{\sqrt{r^2 - x^2}} \quad (-r < x < r).$$

The function f_1 is not differentiable at the endpoints $\pm r$ and does not even have

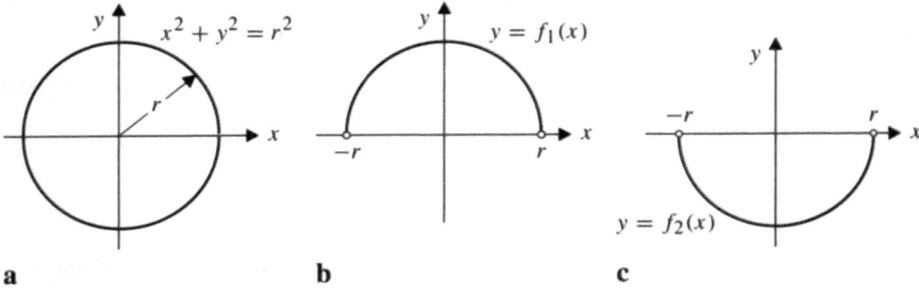

Fig. 5.2 a–c. Implicit relation $x^2 + y^2 = r^2$ defining two functions. **a** Circle with center $(0, 0)$ and radius r; **b** upper semicircle; **c** lower semicircle

one-sided derivatives there. Indeed,

$$\lim_{x \to (-r)_+} f_1'(x) = +\infty, \qquad \lim_{x \to r_-} f_1'(x) = -\infty \,.$$

Similarly

$$f_2'(x) = \frac{x}{\sqrt{r^2 - x^2}} \qquad (-r < x < r), \quad \text{and}$$

$$\lim_{x \to (-r)_+} f_2'(x) = -\infty, \qquad \lim_{x \to r_-} f_2'(x) = +\infty \,.$$

To make sense of the implicit derivative (5.18), note that the y in the denominator must be specified. If we substitute for y the upper function $f_1(x) = +\sqrt{r^2 - x^2}$, then (5.18) gives the correct value of $f_1'(x)$. Similarly, substituting $y = f_2(x)$ in (5.18) gives $f_2'(x)$. This shows that implicit differentiation cannot always avoid solving for the explicit function $y = f(x)$. Implicit differentiation only avoids computing the derivative of $f(x)$.

Note that important information about the specific circle (5.17), namely, its radius r, was lost in the implicit derivative (5.18),

$$y' = -x/y \,,$$

which does not mention r. The implicit formula (5.18) actually finds the slopes of the tangents for *all circles* with center $(0, 0)$. Through any point (x_0, y_0) there passes one such circle,

$$x^2 + y^2 = x_0^2 + y_0^2 \tag{5.19}$$

whose radius $r = \sqrt{x_0^2 + y_0^2}$ is the distance of the point (x_0, y_0) from the origin. The slope of the tangent of (5.19) at (x_0, y_0) is, by (5.18),

$$-x_0/y_0 \,.$$

In Fig. 5.3 we show the unique circle, with center at $(0, 0)$, which passes through $A(4, 3)$,

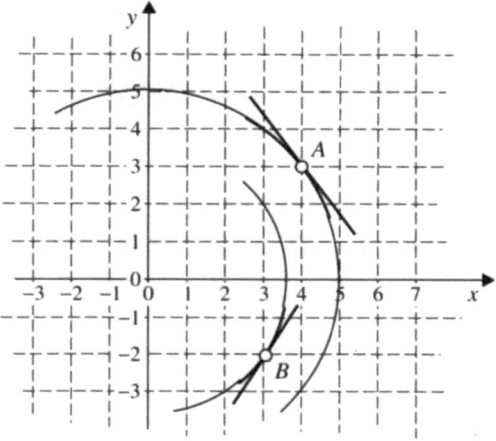

Fig. 5.3. Circles with center at $(0, 0)$ and their tangent lines

$$x^2 + y^2 = 4^2 + 3^2 ,$$

and its tangent at A. The slope of the tangent is, by (5.18), $-4/3$.

Similarly, the circle with center at $(0, 0)$ which passes through $B(3, -2)$ has a tangent at B with slope $3/2$.

MACSYMA-Session 5.2. MACSYMA implements the chain rule for differentiation. If you use unspecified functions $u(x)$ and $v(x)$, then diff(u(v(x)),x) fails to simplify further since there is no information to compute $u'(x)$ and $v'(x)$.

If $u(x)$ is a specified function, say
c1. u(x):=x^n $
then MACSYMA will use the chain rule
c2. diff(u(v(x)),x)
to obtain the correct answer
d2. $n * v(x)^{n-1} \, \text{diff}(v(x), x)$
Similarly, for
c3. u(x):=sqrt(x+1) $
c4. diff(u(v(x)),x)
d4. $\text{diff}(v(x), x)/(2 * \sqrt{v(x) + 1})$

MACSYMA also handles implicit differentiation of unspecified functions. Referring to Example 5.16, consider
c5. f(x)^2=x
d5. $f^2(x) = x$
and differentiation
c6. diff(%,x)
gives
d6. $2f(x)\dfrac{\mathrm{d}f(x)}{\mathrm{d}x} = 1$

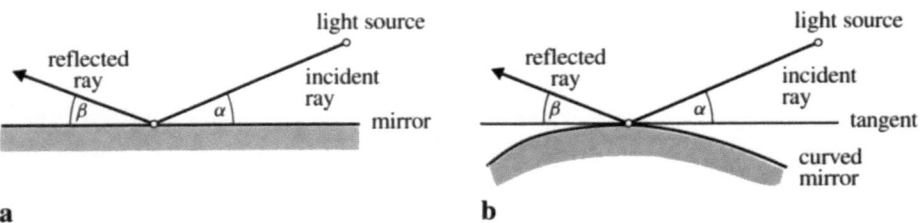

Fig. 5.4a, b. Mirrors. **a** Reflection off a planar mirror; **b** reflection off a curved mirror

We divide by $2f(x)$
c7. `%/(2*f(x))`
to get

d7. $\dfrac{\mathrm{d}f(x)}{\mathrm{d}x} = \dfrac{1}{2f(x)}$

in agreement with Example 5.16.

Example 5.20. Consider a light ray (a straight line emanating from a light source) which is reflected off a mirror. The light ray defines two angles,

α = the angle between the mirror and the incoming light,
β = the angle between the mirror and the reflected light.

We will prove in Example 6.15 that[3]

$$\alpha = \beta \qquad\qquad (5.20)$$

as illustrated in Fig. 5.4a. If the mirror is curved, then (5.20) still holds with angles measured from the tangent at the point of reflection (Fig. 5.4b).

An important property of parabolas is: Light rays parallel to the axis of the parabola are reflected by the parabola into its focus. This is illustrated in Fig. 5.5a.

Proof. Consider a parabola (Appendix D, D.3)

$$y^2 = 2px ,$$

a point $P(x_0, y_0)$ on the parabola, and a horizontal light ray which is reflected at P. We prove that the reflected ray passes through the focus $F(p/2, 0)$, by showing that the angle θ (between the horizontal line and the tangent T) is equal to the angle ϕ (between the tangent T and the line PF).

The angle θ is given by $\tan\theta = y'(x_0)$. From $y^2 = 2px$ we get

$$2yy' = 2p \quad\text{so that}\quad y' = p/y .$$

The slope of the tangent at $P(x_0, y_0)$ is therefore

$$\tan\theta = p/y_0 .$$

---

3 There we use α and β to denote the complementary angles.

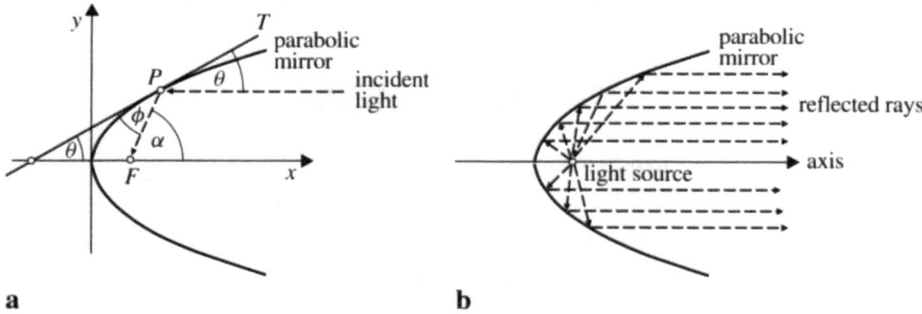

Fig. 5.5 a, b. Reflection property of parabolic mirrors. **a** Rays parallel to the parabola axis are reflected into the focus; **b** light rays emanating at the focus are reflected into parallel rays

The slope of the line FP is

$$\tan \alpha = \frac{y_0 - 0}{x_0 - p/2} .$$

The angle ϕ satisfies $\phi = \alpha - \theta$ (Fig. 5.5 a), and therefore

$$\tan \phi = \tan(\alpha - \theta) = \frac{\tan \alpha - \tan \theta}{1 + \tan \alpha \tan \theta} = \frac{\dfrac{y_0}{x_0 - p/2} - \dfrac{p}{y_0}}{1 + \dfrac{y_0}{x_0 - p/2} \dfrac{p}{y_0}}$$

$$= p/y_0, \quad \text{after some arithmetic, using the fact } y_0^2 = 2px ,$$

$$= \tan \theta, \quad \text{proving that } \phi = \theta . \qquad \square$$

This property finds many applications. Parabolic mirrors are used in telescopes to "collect" line rays from distant objects into the focus, where they are viewed. Conversely, if a light source is placed in the focus, the reflected light rays are parallel to the axis of the parabola, as illustrated in Fig. 5.5 b. For this reason parabolic reflectors are used in searchlights and car headlights, where it is important to have a concentrated light beam.

Exercises

5.13 Calculate the second and third derivatives of

    (a) $\sqrt{1 + x^2}$ (b) $\sin x^2$ (c) $\tan(e^{x+1})$

    (d) $\sqrt{1 + \sqrt{1 + x}}$ (e) $\cos\left(\dfrac{1 + x}{1 - x}\right)$ (f) $\sqrt{\dfrac{x^2}{x^2 + 1}}$

5.14 Let $f(x) := 1/(1 - x)$. Show that the nth derivative of f is given by

$$\frac{d^n f}{dx^n} := \frac{n!}{(1 - x)^{n+1}} .$$

5.15 Let $f(x) := a/(b + cx)$. Find a formula for the nth derivative of f.

5.16 Use the chain rule to prove (4.36),

$$\frac{d}{dx}e^{\lambda x} = \lambda e^{\lambda x}, \quad \text{for all real } \lambda.$$

5.17 Use the chain rule to differentiate:

(a) $(1 + x)^n$ (b) e^{2x+3} (c) e^{-x^2} (d) $\sin 2x$

(e) $e^{\sin x}$ (f) $\sin(e^x)$ (g) 2^{2x} (h) 2^{2^x}

5.18 Calculate the derivatives of the following functions.

(a) $(1 + x/n)^n$ $(n \in \mathbf{N})$ (b) $\sqrt{1 - x^2}$

(c) $\sqrt{5 + x} - \sqrt{5 - x}$ (d) $\sqrt[3]{1 + x}$

5.19 Consider the circle with center at $(0, 0)$, which passes through (x_0, y_0). Show that the equation of its tangent line, at (x_0, y_0), is $xx_0 + yy_0 = r^2$, where $r^2 = x_0^2 + y_0^2$.

5.20 In Exercise 1.62 you plotted the implicit function

$$(x^2 + y^2)^2 = 9(x^2 - y^2).$$

Find the (local) derivative $y'(x)$ by implicit differentiation, show that $P := (\sqrt{5}, 1)$ is a point of the graph, and evaluate y' at P. Plot the implicit function again, and plot its tangent at P.

5.21 Prove the following recursive version of the chain rule. If f is the composition of n functions g_k $(k = 1, \ldots, n)$,

$$f(x) := g_n(g_{n-1}(\ldots (g_1(x)\ldots))),$$

then its derivative is found by the formula

$$f'(x) = \frac{dg_n}{dg_{n-1}} \frac{dg_{n-1}}{dg_{n-2}} \cdots \frac{dg_1}{dx},$$

where we suppressed the arguments.

5.22 Use MACSYMA to calculate the derivative of $\big(f(x)g(x)\big)^n$ for arbitrary functions f and g.

5.23 Calculate the derivative of $|f(x)|$, and compare the result with MACSYMA's.

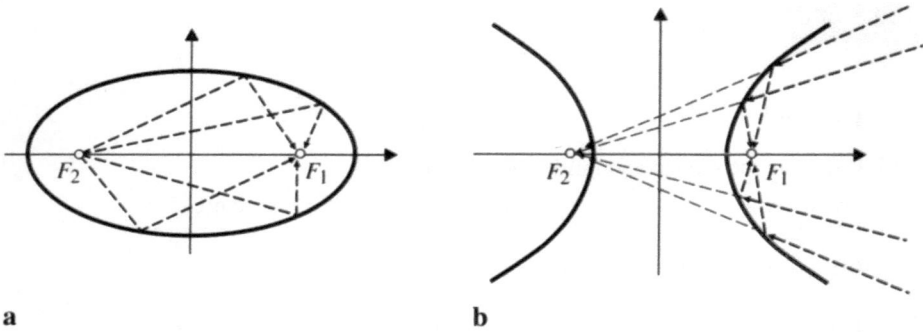

a   **b**

Fig. 5.6 a, b. Reflection property of ellipses and hyperbolas. **a** Rays emanating at F_2 are reflected to F_1; **b** rays directed to F_2 are reflected to F_1

5.24 Prove the following property of ellipses: A light ray emanating in one focus of the
     ellipse is reflected by the graph of the ellipse into the other focus (Fig. 5.6 a).
     Hint: The proof is similar to the proof of the reflection property of parabolas, in
     Example 5.20.
5.25 Prove the following property of hyperbolas: A light ray directed towards one focus of
     the hyperbola, is reflected by the graph of the hyperbola to the other focus (Fig. 5.6 b).

5.3 Rates of change

Let y be a function of time t. By the *average* rate of change of y in an interval $[\tau,
\tau + \Delta t]$ we mean the ratio

$$\Delta y / \Delta t$$

where $\Delta y = y(\tau + \Delta t) - y(\tau)$. If y is differentiable at $t := t_0$, then the *instantaneous* rate of change of y at that time is precisely the time derivative

$$\frac{dy(t_0)}{dt}, \quad \text{often denoted by } \dot{y}(t_0) \, .$$

In this section we take "rate of change" to mean "instantaneous".

Example 5.21. Let $y(t)$ be the altitude at time t of a plane taking off from a runway.
The rate of climb *per time* of the plane is given by the ratio $\Delta y / \Delta t$, or the derivative
$\dot{y}$ whenever y is differentiable.

   Another way to describe the climb is by relating the altitude y to the (ground)
distance x traveled since takeoff. The rate of climb *per distance* is given by the
ratio $\Delta y / \Delta x$, or by dy/dx wherever y is differentiable. This rate describes the
angle of climb α; indeed, $\alpha = \text{atan}(\Delta y / \Delta x)$ (Fig. 5.7).

   The two rates of change, dy/dt and dy/dx, are related by the chain rule

$$\frac{dy}{dt} = \frac{dy}{dx} \frac{dx}{dt} \, ,$$

but dx/dt is simply the (ground) *velocity* v (which is also a rate of change). Therefore

$$\dot{y}(t) = v(t) \frac{dy}{dx} \, , \tag{5.21}$$

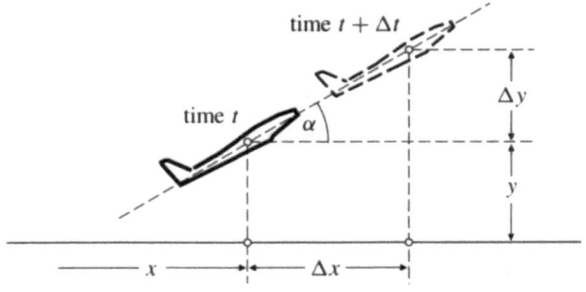

Fig. 5.7. Rate of climb

or, in words,

rate of climb per time = rate of climb per distance × velocity .

This is a relation between three rates of change, useful to express one rate in terms of other rates which are given or easier to measure.

For example, it is required to approximate the angle of climb at a moment where the plane's (ground) velocity is $v = 250$ m/s and the rate of climb per time is $\dot{y} = 100$ m/s. The angle of climb is then

$$\alpha = \text{atan}\,\frac{dy}{dx} = \text{atan}\,\frac{\dot{y}}{v} = \text{atan}\,\frac{100}{250} = 0.380506\,[\text{radian}] = 21.8014° .$$

The above example is typical of many applications where there are two variables x and y, related to each other, and both depending on a third variable t. Applying the chain rule we get *related-rates* formula

$$\frac{dy}{dt} = \frac{dy}{dx}\frac{dx}{dt} \tag{5.22}$$

relating the three rates of change dy/dt, dx/dt, and dy/dx.

Example 5.22. A conical tank, with circular cross section, height $H = 10$ m, and radius $R = 4$ m (Fig. 5.8 a), is being filled with water at the rate of $0.5\,\text{m}^3/\text{s}$. At what rate is the water level h rising when it is 6 m from the bottom.

Solution: The volume of water in the tank, when the water level is h, is

$$v = \frac{\pi r^2 h}{3} \tag{5.23}$$

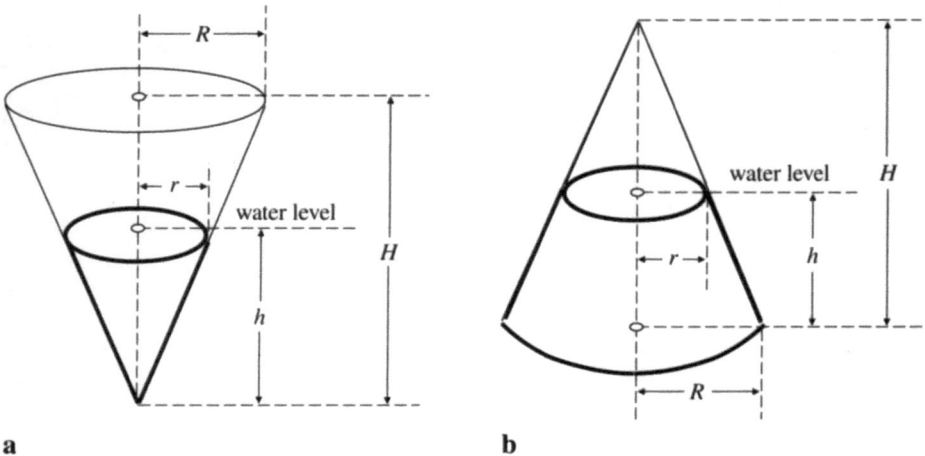

a   b

Fig. 5.8. Filling of a conical tank: **a** Example 5.22 and **b** Exercise 5.27

where r is related to h by

$$\frac{r}{h} = \frac{R}{H} = \frac{4}{10}, \quad \text{or } r = \frac{2}{5}h.$$

Substituting in (5.23) gives the water volume as a function of h alone,

$$v = v(h) = \frac{4\pi}{75}h^3.$$

Therefore

$$\frac{dv}{dt} = \frac{4\pi}{25}h^2\frac{dh}{dt}$$

and the water level rises at the rate

$$\frac{dh}{dt} = \frac{25}{4\pi h^2}\frac{dv}{dt}.$$

Given the volume change $dv/dt = 0.5\,\mathrm{m^3/s}$, we conclude that, when $h = 6\,\mathrm{m}$, it rises at the rate

$$\frac{dh}{dt} = \frac{25}{4\pi 6^2}0.5 = 0.0276311\,[\mathrm{m/s}].$$

Example 5.23. Recall the mechanism in Example 4.37, where the point A makes one revolution per second. Find the velocity $v(t)$ of the point P.

Solution: The position $s(t)$ of the point P is given as a function of θ, (4.41a),

$$s = L + \cos\theta - \sqrt{L^2 - \sin^2\theta}$$

and its velocity is

$$\begin{aligned}
v &= \frac{ds}{dt} = \frac{ds}{d\theta}\frac{d\theta}{dt} \\
&= \left(\frac{d\cos\theta}{d\theta} - \frac{d}{d\theta}\left(\sqrt{L^2 - \sin^2\theta}\right)\right)\frac{d\theta}{dt} \\
&= \left(-\sin\theta + \frac{\sin\theta\cos\theta}{\sqrt{L^2 - \sin^2\theta}}\right)\frac{d\theta}{dt}, \quad \text{(check!)} \\
&= \sin\theta\left(\frac{\cos\theta}{\sqrt{L^2 - \sin^2\theta}} - 1\right)\frac{d\theta}{dt} \\
&= \sin(2\pi t)\left(\frac{\cos(2\pi t)}{\sqrt{L^2 - \sin^2(2\pi t)}} - 1\right)(2\pi), \quad \text{since } \theta = 2\pi t \text{ (see (4.40))}.
\end{aligned}$$

Exercises

5.26 In Example 5.22 calculate the rate dh/dt when $h = 0$, i.e., when filling the tank has just begun. Explain.

5.27 Let the water tank in Example 5.22 be turned upside down (Fig. 5.8 b).

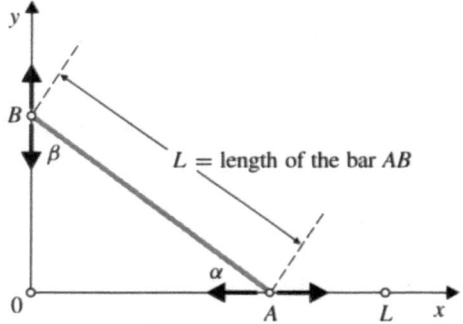

Fig. 5.9. Point A moves horizontally, between 0 and L

a. Calculate dh/dt when $h = 4\,\text{m}$

b. Calculate dh/dt when $h = 0$. Compare with Exercise 5.26

5.28 A spherical tank is used for storing water under ground. At the top there is a cap to fill the tank. If water is poured in at $4\,\text{ft}^3/\text{min}$ and the tank has a radius of $10\,\text{ft}$, at what rate is the water height raising when the height is $6\,\text{ft}$ from the bottom?

5.29 Two jet liners cross over the Atlantic. One is flying west at $30,000\,\text{ft}$ at $800\,\text{mi/h}$ and the other is flying south at $25,000\,\text{ft}$ at $700\,\text{mi/h}$.

  a. Describe the rate at which the distance between the two jets is changing.

  b. At what rate is the distance between the two jets changing when they are directly on top of each other?

  c. At what rate is the distance between the two jets changing $10\,\text{min}$ later?

5.30 A rigid bar of length L, with endpoints A and B, moves in such a way that A moves horizontally, B vertically (Fig. 5.9). You may think of a ladder, with its top B against the wall and its lower end A moving on the ground.

  a. Express the velocity dy/dt of the point B in terms of the velocity dx/dt of the point A. Explain the signs of these velocities.

  b. Express the rates of change $d\alpha/dt$ and $d\beta/dt$ in terms of dx/dt.

  c. Let A begin to move at time $t = 0$ from the point 0, and let the velocity of A be given by

$$\frac{dx}{dt} = \sin\frac{2t}{L}, \quad \text{for all } t \geq 0 .$$

  Check that this velocity is consistent with movement of A between 0 and L. Then calculate the position, and velocity, of the point B at times $t = (n + 1/4)\pi L$, for $n = 0, 1, 2, \ldots$

5.31 (This exercise has nothing to do with related rates.) Consider the rigid bar in Fig. 5.9. Describe the movement of the bar midpoint as A moves from 0 to L.

5.32 A snowball, considered a sphere of radius r (with volume $v = (4/3)\pi r^3$, surface area $A = 4\pi r^2$), is melting in a way which retains its spherical shape (but the radius decreases).

  a. Express dr/dt in terms of dv/dt

  b. Same for dr/dt and dA/dt

  c. (How does a snowball melt?) Suppose the snowball melts at a rate dv/dt propor-

before   after

Fig. 5.10. Two views of a melting snow ball

tional to the surface area A (where melting takes place),

$$dv/dt = c4\pi r^2 ,$$ (5.24)

where c is a positive constant of dimension [m/s]. Prove that the radius changes at a constant rate

$$dr/dt = c .$$ (5.25)

Conversely, if (5.25) holds[4], then the snowball melts according to (5.24).

5.33 In Example 5.23 (see also Fig. 4.12), let $L = 5$ m. Compute the velocity of P, in [m/s], for

(a) $t = \dfrac{1}{4}$ s   (b) $t = 1$ s   (c) $\theta = \dfrac{\pi}{4}$ rad   (d) $\theta = \dfrac{\pi}{2}$ rad

5.34 An ice cube is melting at the rate of 1 in^3/h. Assume that the shape remains a cube. At what rate is the diagonal changing when the cube side is 3.5 in?

5.35 A woman 5' 7" tall, is walking towards a streetlamp 20' high. At what rate is her shadow's length changing, when the woman is 6' from the lamp?

5.36 A man is riding a bicycle going west at 12 mi/h. His dog, who is 1000 ft north of him, hears his whistle and runs straight in south-west direction at 20 mi/h. At what rate is the distance between dog and master changing when it is 500 ft?

5.4 Derivatives of inverse functions

An important application of the chain rule is to compute derivatives of inverse functions. We recall Definition 1.35 that the functions f and ϕ are inverses of each other, a fact denoted by $f = \phi^{-1}$ and $\phi = f^{-1}$, if

(1.40a), $D(f) = R(\phi),$ $R(f) = D(\phi) ,$

(1.40b), $\phi(f(\xi)) = \xi,$ $\xi \in D(f) ,$

(1.40c), $f(\phi(\eta)) = \eta,$ $\eta \in R(f) .$

The inverse function $\phi = f^{-1}$, when it exists, satisfies the implicit relation

$$\phi(f(x)) = x, \quad \text{for all } x \in D(f) .$$

If ϕ is differentiable at $f(x)$, then the chain rule gives

$$\frac{d}{dx}\phi(f(x)) = \frac{d\phi(y)}{dy}\bigg|_{y=f(x)} \frac{df}{dx}(x) = 1 .$$

---

4 Suggest an experiment to check if (5.25) holds, and if it does, to measure the constant c.

We have thus obtained an identity satisfied by the derivative of the inverse function $(d/dx)f^{-1}$, if the inverse function exists and is differentiable. Here is the formal result.

Theorem 5.24 (Derivative of the inverse function). Let f have an inverse f^{-1}. If f is differentiable at x with derivative $f'(x) \neq 0$, then f^{-1} is differentiable at the point $f(x)$, and its derivative there is

$$\frac{\mathrm{d}}{\mathrm{d}y} f^{-1}(f(x)) = \frac{1}{f'(x)} . \tag{5.26}$$

Proof. Let $\phi := f^{-1}$. To show that ϕ is differentiable at $f(x)$ we must show that the limit

$$\lim_{\eta \to f(x)} \frac{\phi(\eta) - \phi(f(x))}{\eta - f(x)}$$

exists. Since f has an inverse, we can solve $f(x) = \eta$ to get a unique ξ such that

$$f(\xi) = \eta ,$$

and because f is continuous at x (it is differentiable there),

$$\xi \to x \implies f(\xi) \to f(x) .$$

Therefore, the derivative of ϕ at $f(x)$ is given by the limit

$$\lim_{\xi \to x} \frac{\phi(f(\xi)) - \phi(f(x))}{f(\xi) - f(x)} ,$$

whenever it exists. Substituting $\phi(f(\xi)) = \xi$ and $\phi(f(x)) = x$ in the last equation, we obtain

$$\lim_{\xi \to x} \frac{\xi - x}{f(\xi) - f(x)} = \lim_{\xi \to x} \frac{1}{\dfrac{f(\xi) - f(x)}{\xi - x}} = \frac{1}{\lim_{\xi \to x} \dfrac{f(\xi) - f(x)}{\xi - x}} = \frac{1}{f'(x)} ,$$

proving the theorem.   □

Remark 5.25. a. Note that (5.26) gives the derivatives f' and $(f^{-1})'$ at different points: the derivative f' at x, and $(f^{-1})'$ at $f(x)$. There is no way around this. An alternative form of (5.26) is

$$\frac{\mathrm{d}}{\mathrm{d}y} f^{-1}(y) = \frac{1}{f'(f^{-1}(y))} , \tag{5.27}$$

where we identify $f(x)$ with y.

b. Equation (5.26) can be written, using differential quotients, as

$$\frac{\mathrm{d}x}{\mathrm{d}f} = \frac{1}{\mathrm{d}f/\mathrm{d}x} ,$$

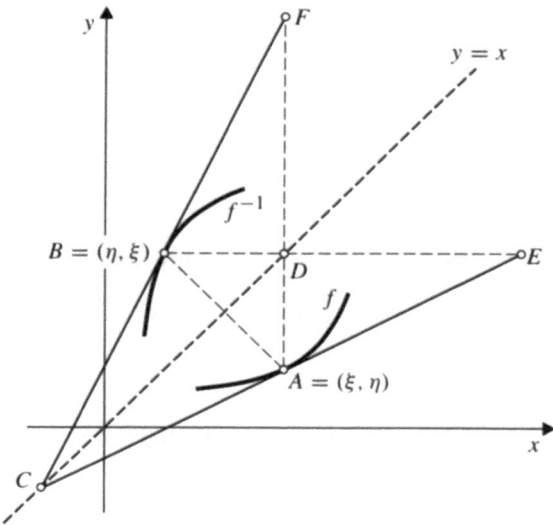

Fig. 5.11. Reciprocal slopes of tangent lines of f at A and of f^{-1} at B

which can be memorized as an "arithmetic expression" in differential quotients (see Remark 5.12b).

The geometrical interpretation of (5.26) is given in Fig. 5.11. Let $A = (\xi, \eta)$ be a point on the graph of f, and let the point $B = (\eta, \xi)$ on the graph of f^{-1} be its reflection with respect to $y = x$ (see Fig. 1.14c). Because of the symmetry, the tangent lines of f at A, and of f^{-1} at B intersect at a point C on the line $y = x$. Then:

- the line CE is the tangent of f at $A = (\xi, \eta)$, and its slope is $f'(\xi) = DA/DE$;
- the line CF is the tangent of f^{-1} at $B = (\eta, \xi)$, and its slope is $(f^{-1})'(\eta) = DF/DB$.

But $DE = DF$ and $DA = BD$ (Fig. 5.11). Therefore

$$(f^{-1})'(\eta) = (f^{-1})'(f(\xi)) = \frac{DF}{DB} = \frac{DE}{DA} = \frac{1}{f'(\xi)} \ .$$

Derivatives of $\ln x$, x^α *and logarithmic differentiation*

Example 5.26 (Derivative of $\ln x$).

$$\frac{d}{dx} \ln x = \frac{1}{x}, \quad \text{for all } x > 0 \ . \tag{5.28}$$

Proof. The function $\phi(x) = \ln x$ is the inverse of $f(x) = e^x$. Therefore, for all $y \in D(\phi) = R(f) = (0, +\infty)$,

$$\frac{d}{dy}\phi(y) = \frac{1}{f'(\phi(y))}, \quad \text{by (5.27)},$$

$$= \frac{1}{e^{\ln y}}, \quad \text{since } (e^x)' = e^x,$$

$$= 1/y.$$ □

We can finally give the derivative of x^α for any real α, and $x > 0$. We use the fact

$$x^\alpha = \left(e^{\ln x}\right)^\alpha = e^{\alpha \ln x}$$

to obtain

$$\frac{d}{dx}x^\alpha = e^{\alpha \ln x}\frac{d}{dx}(\alpha \ln x), \quad \text{by the chain rule},$$

$$= x^\alpha \alpha \frac{1}{x}, \quad \text{by (5.28)},$$

$$= \alpha x^{\alpha-1}.$$

We have thus proved the following theorem.

Theorem 5.27 (Power rule for real exponents). For any real α and $x > 0$,

$$\frac{d}{dx}x^\alpha = \alpha x^{\alpha-1}. \tag{5.29}$$

Example 5.28 (Derivative of $\ln \sqrt{(1+x)/(1-x)}$).

$$\frac{d}{dx}\ln\sqrt{\frac{1+x}{1-x}} = \tfrac{1}{2}\frac{d}{dx}(\ln(1+x) - \ln(1-x)) = \tfrac{1}{2}\left(\frac{1}{1+x} + \frac{1}{1-x}\right) = \frac{1}{1-x^2}.$$

Formula (5.28), for differentiating logarithms, can be used for computing derivatives of other functions that do not even look like logarithms. We note that, if f is differentiable at x, and $f(x) > 0$, then

$$\frac{d}{dx}\ln f(x) = \frac{f'(x)}{f(x)}, \tag{5.30}$$

as follows from the chain rule (5.10) with $F(u) = \ln u$ and $u(x) = f(x)$. Then,

$$f'(x) = f(x)\frac{d}{dx}\ln f(x), \tag{5.31}$$

expressing $f'(x)$ in terms of $(d/dx)\ln x$. This technique, called *logarithmic differentiation*, is useful if it is easier to compute $(d/dx)\ln f(x)$ than to compute $f'(x)$ directly. The following examples illustrate logarithmic differentiation.

Example 5.29 (Derivative of x^x).

$$\frac{d}{dx}x^x = x^x(1 + \ln x), \quad x > 0 .$$

Proof. We differentiate $\ln x^x = x \ln x$,

$$\frac{d}{dx}\ln x^x = \frac{d}{dx}(x \ln x)$$

$$= x\frac{d}{dx}\ln x + \ln x\frac{d}{dx}x, \quad \text{by the product rule} ,$$

$$= 1 + \ln x, \quad \text{by (5.28)} .$$

$$\therefore \quad \frac{d}{dx}x^x = x^x(1 + \ln x), \quad \text{by (5.31)} . \qquad \square$$

See also Exercise 5.37.

Example 5.30 (Product and quotient rules via logarithmic differentiation). By (5.31),

$$\frac{d}{dx}(uv) = (uv)\frac{d}{dx}(\ln u + \ln v) = (uv)\left(\frac{u'}{u} + \frac{v'}{v}\right) = u'v + uv' ,$$

$$\frac{d}{dx}\left(\frac{u}{v}\right) = \left(\frac{u}{v}\right)\frac{d}{dx}(\ln u - \ln v) = \left(\frac{u}{v}\right)\left(\frac{u'}{u} - \frac{v'}{v}\right) = \frac{u'v - uv'}{v^2} .$$

Example 5.31 (Logarithmic differentiation of products $f = f_1 f_2 \ldots f_m$). Let f be the product

$$f(x) = f_1(x) f_2(x) \ldots f_m(x)$$

of functions f_i differentiable, and positive at x. Then

$$\ln f(x) = \sum_{k=1}^{m} \ln f_k(x)$$

and therefore

$$\frac{d}{dx}\ln f(x) = \sum_{k=1}^{m}\frac{d}{dx}\ln f_k(x) = \sum_{k=1}^{m}\frac{f_k'(x)}{f_k(x)}, \quad \text{by (5.30)} .$$

$$\therefore \quad \frac{d}{dx}\prod_{k=1}^{m} f_k(x) = \sum_{k=1}^{m}\left(f_k'(x)\prod_{\substack{j=1 \\ j\neq k}}^{m} f_j(x)\right) . \tag{5.32}$$

Example 5.32. Let $a > 0$ and $b, c, \alpha, \beta, \gamma$ be real numbers, and let

$$f(x) = (x^2 + a)^{\alpha}(x - b)^{\beta}(x - c)^{\gamma}, \quad x > \max\{b, c\} . \tag{5.33}$$

Then, using (5.32),

$$f'(x) = (x^2 + a)^{\alpha-1}(x - b)^{\beta-1}(x - c)^{\gamma-1} \cdot$$
$$\cdot \left(2\alpha x(x - b)(x - c) + \beta(x^2 + a)(x - c) + \gamma(x^2 + a)(x - b)\right).$$

Derivatives of inverse trigonometric functions

We consider next the derivatives of inverse trigonometric functions.

Example 5.33 (Derivative of $\operatorname{asin} x$).

$$\frac{d}{dx} \operatorname{asin} x = \frac{1}{\sqrt{1 - x^2}}, \quad \text{for all } -1 < x < 1. \tag{5.34}$$

Proof. We restrict $f(x) = \sin x$ to the domain $D(f) := [-\pi/2, \pi/2]$. Then its range is $R(f) = [-1, 1]$. The inverse of f is the function $\phi(x) = \operatorname{asin} x$, with domain $D(\phi) = [-1, 1]$ and range $R(\phi) = [-\pi/2, \pi/2]$. The derivative of ϕ is, by (5.27),

$$\frac{d}{dx}(\operatorname{asin} x) = \frac{1}{\cos(\operatorname{asin} x)} = \frac{1}{\sqrt{1 - x^2}} \quad \text{(Fig. 5.12a)}. \qquad \square$$

Example 5.34 (Derivative of $\operatorname{atan} x$).

$$\frac{d}{dx} \operatorname{atan} x = \frac{1}{1 + x^2}, \quad \text{for all } x \in \mathbf{R}. \tag{5.35}$$

Proof. We restrict $f(x) = \tan x$ to the domain $D(f) = (-\pi/2, \pi/2)$. Then its range is $R(f) = \mathbf{R}$. The inverse of f is $\phi(x) = \operatorname{atan} x$ with domain $D(\phi) = \mathbf{R}$ and range $R(f) = (-\pi/2, \pi/2)$. Therefore,

$$\frac{d}{dx} \operatorname{atan} x = \frac{1}{\sec^2(\operatorname{atan} x)} = \cos^2(\operatorname{atan} x) = \frac{1}{1 + x^2} \quad \text{(Fig. 5.12b)}. \qquad \square$$

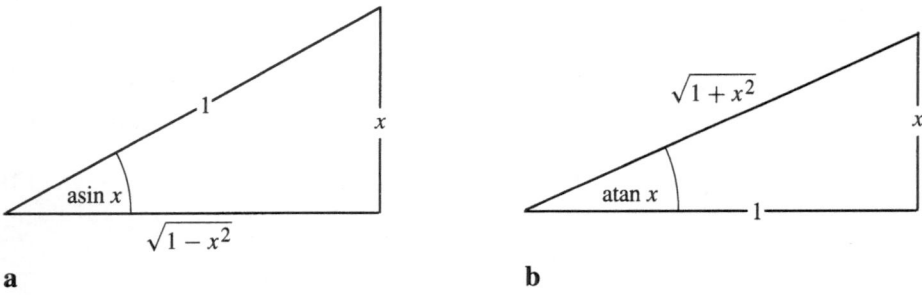

a   **b**

Fig. 5.12 a, b. Inverse trigonometric functions. **a** What is $\cos(\operatorname{asin} x)$? **b** What is $\cos^2(\operatorname{atan} x)$?

Example 5.35 (Derivative of acsc x). To differentiate the arc-cosecant, set $y = \csc x$ or $x = \mathrm{acsc}\, y$. Then for $|y| > 1$ we have the identity $x = \mathrm{acsc}(\csc x)$. As is our usual procedure we differentiate this with respect to x to obtain

$$1 = \frac{d}{dy}\,\mathrm{acsc}(y)\,\frac{d}{dx}\,\csc(x) .$$

Now

$$\frac{d}{dx}\,\csc x = -\csc x \cot x = -y\left(\pm\sqrt{y^2-1}\right)$$

and therefore

$$\frac{d}{dy}\,\mathrm{acsc}(y) = \frac{-1}{y\left(\pm\sqrt{y^2-1}\right)} .$$

We need still to determine the correct sign for the square root. Note that:

- $y > 1$ corresponds to $0 < x < \pi/2$; hence for $y > 1$ we have that $\cot x = \pm\sqrt{y^2-1} > 0$ and we *must* choose the $+$ sign.
- $y < -1$ corresponds to $-\pi/2 < x < 0$; in which case $\cot x = \pm\sqrt{y^2-1} < 0$ and we *must* choose the $-$ sign.

Combining these conclusions we get the formula

$$\frac{d}{dy}\,\mathrm{acsc}(y) = \frac{-1}{|y|\sqrt{y^2-1}} .$$

Derivatives of hyperbolic functions and their inverses

The derivatives of the hyperbolic functions (see Sect. 2.4) can now be computed. From the definitions (2.11)–(2.13) we get immediately

$$\frac{d}{dx}\,\sinh x = \frac{e^x + e^{-x}}{2} = \cosh x , \qquad\qquad (5.36a)$$

$$\frac{d}{dx}\,\cosh x = \frac{e^x - e^{-x}}{2} = \sinh x , \qquad\qquad (5.36b)$$

and using the quotient rule,

$$\frac{d}{dx}\,\tanh x = \frac{d}{dx}\,\frac{\sinh x}{\cosh x} = \frac{\cosh x\,(d/dx)(\sinh x) - \sinh x\,(d/dx)(\cosh x)}{\cosh^2 x}$$

$$= \frac{\cosh^2 x - \sinh^2 x}{\cosh^2 x} = \frac{1}{\cosh^2 x}, \quad \text{by (2.14)} ,$$

$$= \mathrm{sech}^2 x . \qquad\qquad (5.36c)$$

Also

$$\frac{d}{dx}\,\mathrm{sech}\, x = \frac{d}{dx}\left(\frac{1}{\cosh x}\right) = \frac{-\sinh x}{\cosh^2 x}$$

$$= -\,\mathrm{sech}\, x \tanh x . \qquad\qquad (5.36d)$$

The derivatives of the other hyperbolic functions, $\operatorname{csch} x := 1/\sinh x$, $\coth x := 1/\tanh x$, are left as an exercise. Next we compute the derivatives of the inverse hyperbolic functions, using Theorem 5.24 in the form (5.27).

Example 5.36. The derivative of the inverse hyperbolic sine is

$$\frac{d}{dy} \operatorname{asinh} y = \frac{1}{(d/dx)\sinh x|_{x=\operatorname{asinh} y}} = \frac{1}{\cosh x|_{x=\operatorname{asinh} y}}$$

$$= \frac{1}{\sqrt{\sinh^2(x)+1}|_{x=\operatorname{asinh} y}}, \quad \text{by (2.14)},$$

$$= \frac{1}{\sqrt{y^2+1}}, \quad -\infty < y < \infty. \quad (5.37)$$

Example 5.37. The derivative of the inverse hyperbolic cosine is similarly

$$\frac{d}{dy} \operatorname{acosh} y = \frac{1}{(d/dx)\cosh x|_{x=\operatorname{acosh} y}} = \frac{1}{\sinh x|_{x=\operatorname{acosh} y}}$$

$$= \frac{1}{\sqrt{\cosh^2(x)-1}|_{x=\operatorname{acosh} y}}, \quad \text{by (2.14)},$$

$$= \frac{1}{\sqrt{y^2-1}}, \quad 1 < y < \infty. \quad (5.38)$$

Example 5.38. The derivative of the inverse hyperbolic secant is

$$\frac{d}{dy} \operatorname{asech} y = \frac{1}{(d/dx)\operatorname{sech} x|_{x=\operatorname{asech} y}} = \frac{1}{-\operatorname{sech} x \, \tanh x|_{x=\operatorname{asech} y}} \quad (5.39)$$

by (5.36d). We use the Pythagorean relation

$$1 - \tanh^2 x = \operatorname{sech}^2 x \quad \text{(see Exercise 2.35a)},$$

to get

$$\tanh x = \pm\sqrt{1 - \operatorname{sech}^2 x} \quad (5.40)$$

which substituted in (5.39) gives

$$\frac{d}{dy} \operatorname{asech} y = \frac{-1}{y(\pm\sqrt{1 - y^2})}.$$

For the inverse hyperbolic secant to be defined, it was necessary to restrict the function $\operatorname{sech} x$ to the domain $0 \le x < \infty$, where its range is $(0, 1]$. Since the hyperbolic tangent $\tanh x > 0$ for positive x, we can select the positive sign in (5.40) to conclude

$$\frac{d}{dy} \operatorname{asech} y = \frac{-1}{y\sqrt{1 - y^2}}. \quad (5.41)$$

MACSYMA-Session 5.3. The derivative of $\sinh x$

c1. `diff(sinh(x),x)`

d1. $\cosh x$

can be written, in exponential form,

c2. `exponentialize(%)`

d2. $\dfrac{e^x + e^{-x}}{2}$

Similarly, we can reproduce (5.36c) by

c3. `diff(tanh(x),x)`

d3. $\mathrm{sech}^2 x$

c4. `factor(exponentialize(%))`

d4. $\dfrac{4e^{2x}}{(e^{2x}+1)^2}$

where the factor command reduces the exponential form in the manner we are seeking. Now consider the derivatives of the inverse hyperbolic functions. To reproduce (5.38) we do

c5. `diff(acosh(y),y)`

d5. $\dfrac{1}{\sqrt{y^2 - 1}}$

Exercises

5.37 (a) Draw the graph of $x^x, x > 0$. (b) Why the restriction to $x > 0$? (c) Use MACSYMA to calculate $\lim_{x \to 0_+} x^x$

5.38 a. The function $f(x) = x^{x^x}$ can be interpreted as $f(x) = (x^x)^x$ or as $x^{(x^x)}$. Are these equivalent for $x > 0$? If not, which do you prefer?
     b. Calculate the derivatives $(d/dx)(x^x)^x$ and $(d/dx)x^{(x^x)}$.

5.39 Differentiate the function (5.33) using the product rule for $(d/dx)(F(x)G(x)H(x))$, see MACSYMA-Session 5.1, result d9. Compare with Example 5.32.

5.40 (Derivatives of general logarithms) If a is positive and $a \neq 1$, then, for any $x > 0$, the solution y of $a^y = x$ is called the logarithm of x to the base a and denoted by $\log_a x$ (see Sect. 2.1). The most commonly used logarithms are $\log_2$ (with basis 2), the so-called common logarithms $\log_{10}$ (with basis 10), and the natural logarithms $\ln x := \log_e x$ with basis e. Prove

$$\frac{d}{dx} \log_a x = \frac{1}{x \ln a} = \frac{\log_a e}{x}. \tag{5.42}$$

5.41 Prove that $(d/dx) \mathrm{acos}\, x = -1/\sqrt{1-x^2}$.

5.42 Comparing Example 5.33 and Exercise 5.41 we see that $(d/dx)(\mathrm{asin}\, x + \mathrm{acos}\, x) = 0$. Explain.

5.43 Find the derivatives of the following functions.
     (a) $f(x) = \mathrm{acos}(\sqrt{x})$ (b) $f(x) = \mathrm{atan}(x^2 + 1)$ (c) $f(x) = \mathrm{acos}(e^x)$
     (d) $f(x) = \mathrm{asin}(x^{10})$ (e) $f(x) = \ln(\mathrm{asin}(x))$ (f) $f(x) = \mathrm{acsc}(\ln x)$
     (g) $f(x) = (\mathrm{atan}\, x)^3$ (h) $f(x) = \mathrm{asin}(\cos x)$ (i) $f(x) = \mathrm{asec}\sqrt{x^2 + 1}$

5.44 Find the derivatives of the following functions and check your results with MACSYMA.
     (a) $f(x) = \mathrm{acos}(\tan x)$ (b) $f(x) = \mathrm{atan}(\cos(\sqrt{x}))$ (c) $f(x) = \ln(\mathrm{acos}(e^x))$

(d) $f(x) = \mathrm{asin}(a^x)$ (e) $f(x) = \mathrm{asin}(x^x)$ (f) $f(x) = \mathrm{acsc}(\ln \ln x)$

(g) $f(x) = (\mathrm{atan}\, x)^x$ (h) $f(x) = \ln(\cos e^x)$ (i) $f(x) = \mathrm{asec}\,\sqrt{a^x + 1}$

5.45 Find dy/dx in the following using implicit differentiation.

(a) $\mathrm{acos}(y) + \mathrm{acos}(x) = \pi$ (b) $\mathrm{atan}(y^2 + x^2) + \mathrm{atan}\, y = \pi/2$

(c) $\mathrm{acos}(e^y) + \cos(x) = y$ (d) $\mathrm{asin}(x^2 + y^2) + \mathrm{acos}(y^2) = 0$

(e) $\ln(\mathrm{asin}(x)\,\mathrm{acos}(y)) = x + y$ (f) $(\mathrm{asin}\, x)^3 + (\mathrm{asin}\, y)^3 = \pi^3$

(g) $\mathrm{acsc}\, x + \mathrm{acsc}\, y = \pi/4$ (h) $\ln \mathrm{asin}\, x + \ln \mathrm{acos}\, y = 1$

(i) $\mathrm{asin}(x + y)\,\mathrm{asin}(x) + \sin^2(y) = \pi^2$

5.46 Find the derivatives of the following functions and check your results with MACSYMA.

(a) $f(x) = \cosh(e^x)$ (b) $f(x) = \tanh(\sqrt{x})$ (c) $f(x) = \ln(\cosh(x))$

(d) $f(x) = \sinh(a^x)$ (e) $f(x) = \sinh\sqrt{x^2 + 1}$ (f) $f(x) = \mathrm{sech}(\ln x)$

(g) $f(x) = (\tanh x)^2$ (h) $f(x) = \ln(\cosh\sqrt{x})$ (i) $f(x) = \sinh\sqrt{a^x + 1}$

5.47 Find the derivatives of the following functions and check your results with MACSYMA.

(a) $f(x) = \cosh^{-1}(x)$ (b) $f(x) = \tanh^{-1}(x)$ (c) $f(x) = \ln(\cosh^{-1}(x))$

(d) $f(x) = \sinh^{-1}(x^2)$ (e) $f(x) = \sinh^{-1}\left(\sqrt{x^2 + 1}\right)$

(f) $f(x) = \mathrm{sech}^{-1}(x)$ (g) $f(x) = \tanh^{-1}(\sqrt{x})$ (h) $f(x) = \cosh^{-1}(1/x)$

(i) $f(x) = \sinh^{-1}\sqrt{a^x + 1}$

Table 5.1. Derivatives

$\frac{d}{dx}$ function = derivative	$\frac{d}{dx}$ function = derivative	$\frac{d}{dx}$ function = derivative
$\frac{d}{dx}e^x = e^x$	$\frac{d}{dx}\ln x = \dfrac{1}{x}$	$\frac{d}{dx}\log_{10} x = \dfrac{1}{x \ln 10} = \dfrac{\log_{10} e}{x}$
$\frac{d}{dx}\sin x = \cos x$	$\frac{d}{dx}\cos x = -\sin x$	$\frac{d}{dx}\tan x = \sec^2 x$
$\frac{d}{dx}\sec x = \sec x \tan x$	$\frac{d}{dx}\csc x = \csc x \cot x$	$\frac{d}{dx}\cot x = -\csc^2 x$
$\frac{d}{dx}x^\alpha = \alpha x^{\alpha-1}$	$\frac{d}{dx}a^x = a^x \ln a$	$\frac{d}{dx}x^x = x^x(1 + \ln x)$
$\frac{d}{dx}\mathrm{asin}\, x = \dfrac{1}{\sqrt{1 - x^2}}$	$\frac{d}{dx}\mathrm{acos}\, x = -\dfrac{1}{\sqrt{1 - x^2}}$	$\frac{d}{dx}\mathrm{atan}\, x = \dfrac{1}{1 + x^2}$
$\frac{d}{dx}\mathrm{asec}\, x = \dfrac{1}{|x|\sqrt{x^2 - 1}}$	$\frac{d}{dx}\mathrm{acsc}\, x = \dfrac{-1}{|x|\sqrt{x^2 - 1}}$	$\frac{d}{dx}\mathrm{acot}\, x = \dfrac{-1}{1 + x^2}$
$\frac{d}{dx}\sinh x = \cosh x$	$\frac{d}{dx}\cosh x = \sinh x$	$\frac{d}{dx}\tanh x = \mathrm{sech}^2 x$
$\frac{d}{dx}\mathrm{sech}\, x = -\mathrm{sech}\, x \tanh x$	$\frac{d}{dx}\coth x = -\mathrm{csch}^2 x$	$\frac{d}{dx}\mathrm{csch}\, x = \mathrm{csch}\, x \coth x$
$\frac{d}{dx}\mathrm{asinh}\, x = \dfrac{1}{\sqrt{x^2 + 1}}$	$\frac{d}{dx}\mathrm{acosh}\, x = \dfrac{1}{\sqrt{x^2 - 1}}$	$\frac{d}{dx}\mathrm{atanh}\, x = \dfrac{1}{1 - x^2}$
$\frac{d}{dx}\mathrm{acoth}\, x = \dfrac{1}{1 - x^2}$	$\frac{d}{dx}\mathrm{asech}\, x = \dfrac{-1}{x\sqrt{1 - x^2}}$	$\frac{d}{dx}\mathrm{acsch}\, x = \dfrac{-1}{|x|\sqrt{1 - x^2}}$

6 Extremum problems

The following example is an illustration of the problems studied in this chapter.

Example 6.1. **You are the president of the XYZ Widget Company, which is in the business of producing and selling widgets. You must decide how many widgets, say x, to produce in the coming season. The information available to you is:**
- the company can produce no more than 10 000 widgets,
- the company can sell any quantity it produces at the price of $p := 9$ [\$/widget],
- there is a fixed cost of 6 000 [\$] (a cost such as rent, which does not depend on the decision x),
- the variable cost of producing x widgets is $2x + 0.001x^2$ [\$].

As president your job is to *maximize the profit*, which is revenue minus costs.

The sum of fixed and variable costs is the total cost $C(x) := 6000 + 2x + 0.001x^2$ [\$]. Therefore the company's profit is a function of the production quantity x,

$$f(x) := \text{revenue} - \text{costs} = 9x - (6\,000 + 2x + 0.001x^2)$$
$$= -0.001x^2 + 7x - 6\,000 \,[\$] , \tag{6.1}$$

and your decision requires solving the following mathematical problem:

Find x which maximizes the profit
$f(x) = -0.001x^2 + 7x - 6\,000$ in the interval [0, 10 000].

You try now some possible values of x and substitute in (6.1) to compute the corresponding profits. A negative profit is a loss. Some results are listed in Table 6.1. What is the best decision? This problem is solved in Example 6.13.

Table 6.1. Profit as a function of production quantity

Production quantity x	Profit	Comment
0	−6 000 [\$]	a loss, because of the fixed costs
10 000	−36 000 [\$]	producing at full capacity gives an even bigger loss
5 000	4 000 [\$]	a profit at last!
4 000	6 000 [\$]	an even higher profit

6.1 Terminology

An extremum problem (or *optimization* problem) concerns the maximization or minimization of a given function f in a given interval I. Solving extremum problems requires a complicated terminology. Some of its terms are collected in the following

Definition 6.2.

a. An *objective function* is a function $f(x)$ which has to be *maximized* or *minimized*. The argument of the objective function is called the *decision variable* of the problem.

    In Example 6.1 the objective function is the profit (6.1), and the decision variable is the production quantity x of widgets. There are problems where the objective function is minimized (for example, a non-profit company may wish to minimize its costs).

b. A *constraint* is any condition which must be satisfied by the decision variable. A problem may have one or more constraints, and any x which satisfies them is called a *feasible solution*. The set of feasible solutions is called the *feasible set*.

    In Example 6.1 the constraints are $x \geq 0$ (negative values make no sense) and $x \leq 10\,000$ (the production capacity) and the feasible set is the closed interval $[0, 10\,000]$. In Fig. 6.1 the feasible set is the closed interval $[a, b]$.

    A problem is *unconstrained* if it has no constraints, in which case its feasible set is $(-\infty, \infty)$.

c. An *extremal value* (or extremum) is a name for either the *maximum* or *minimum*[1] of the objective $f(x)$. The x where this occurs is called an *extremizer* (maximizer or minimizer), or an extremal (maximal or minimal) solution.

d. A *neighborhood* of a point x_0 is an open interval (c, d) with $c < x_0 < d$. A feasible x is an *interior feasible solution* (or interior) if it has a neighborhood of feasible solutions. A feasible x which is not interior (i.e., any neighborhood

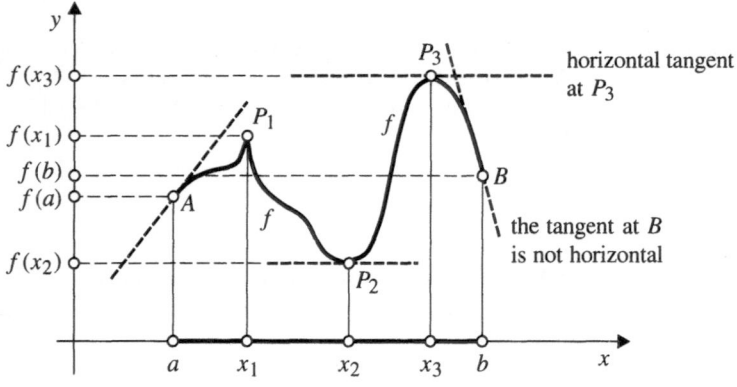

Fig. 6.1. Function f in the interval $[a, b]$

---

1 The plurals are extrema, maxima, and minima.

of x contains a nonfeasible solution) is called an *endpoint*, or a *corner* solution. In Fig. 6.1, every neighborhood of the endpoint a contains feasible as well as nonfeasible solutions. Same goes for the endpoint b.

e. A *local extremizer* is a feasible solution x_0 with a value $f(x_0)$, such that there is a neighborhood of x_0 where no feasible solution x gives a better value of f. Figure 6.1 shows two local maximizers, x_1 and x_3, and three local minimizers, a, x_2, and b.

f. A *global extremizer* is a feasible solution x_0 with a value $f(x_0)$, such that no feasible solution x gives a better value of f. Note that global is defined with respect to the feasible set. Also every global extremizer is a local extremizer. The global maximizer in Fig. 6.1 is x_3, the global minimizer is x_2.

g. A *stationary point* of f is a point x_0 where f is differentiable and $f'(x_0) = 0$. A *critical point* of f is a point which is either stationary or where f is not differentiable.

Figure 6.1 shows two stationary points, x_2 and x_3. Note that the graph of f has horizontal tangents at P_2 and P_3. The function f is not differentiable at the point x_1. The three points x_1, x_2, and x_3 are critical.

Example 6.3. The function $f(x) := x^3$ has:
a. no local extrema in $(-\infty, \infty)$ because it is increasing throughout $(-\infty, \infty)$,
b. in any closed interval $[a, b]$, with $a < b$,
    a local (and global) minimum at $x := a$, and
    a local (and global) maximum at $x := b$,
c. no local extrema in any open interval (a, b) (see also Theorem 3.52 and Exercise 3.38 a).

Example 6.4. Consider the function $f(x) := c$ (a constant). Then every point is a local (and global) minimizer, as well as a local (and global) maximizer.

Example 6.5. The function $f(x) := \sin(1/x)$ has, in the interval $[-1, 1]$, infinitely many global maxima (where its value is 1) and infinitely many global minima (where its value is -1). Find these points.
    See Fig. 3.3.

Recall the definition (Lemma 4.12) of a function *increasing* (decreasing) *at a point*.

Possible cases for a function f and a point x_0 include: (a) f has a local maximum at x_0, (b) f has a local minimum at x_0, (c) f is increasing at the point x_0, (d) f is decreasing at the point x_0 – these four cases are mutually exclusive, i.e., at most one of the cases a–d can hold for f at x_0; the next (pathological) example shows that there is a fifth case – (e) none of the above.

Example 6.6. The function

$$f(x) := \begin{cases} x^2 & \text{if } x \text{ is rational,} \\ -x^2 & \text{if } x \text{ is irrational,} \end{cases}$$

is neither increasing nor decreasing at 0 and does not have there a local extremum.

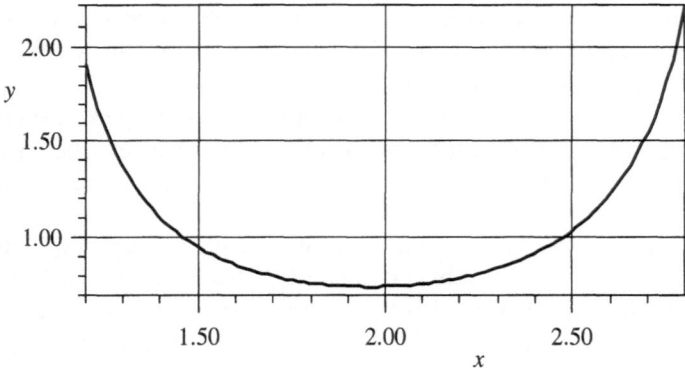

Fig. 6.2. Graph in Session 6.1, line c1

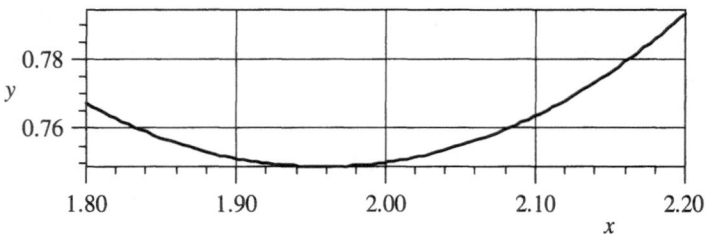

Fig. 6.3. Graph in Session 6.1, line c2

MACSYMA-Session 6.1. Plot the rational function $(1 + x)/((1 - x)(2 + x)(x - 3))$ of Session 1.12.

```
c1. plot2((1+x)/((1-x)*(2+x)*(x-3)),x,1.2,2.8)$
```

This function has poles at $x = 1$, -2, and 3. It appears to have a local minimum near the point $(2, 3/4)$ between the poles at $x = 1$ and $x = 3$.

Now reduce the range of x-values in the plot, so as to get a smaller interval around the apparent minimum, for example,

```
c2. plot2((1+x)/((1-x)*(2+x)*(x-3)),x,1.8,2.2)$
```

Continue in this fashion until you have a width of less than 0.1 around the minimum. This should enable you to get a fairly accurate estimate of its location (Figs. 6.2 and 6.3).

6.2 Necessary condition for a local extremum

The main result of this section is Corollary 6.9, which states that an interior local extremizer x_0 of a function f is necessarily a critical point of f, i.e., the derivative $f'(x_0)$ is zero or does not exist.

We deal first with the case where the function f is differentiable at x_0.

Theorem 6.7 (Necessary condition for a local extremum). Let the function f be defined in an interval I, and let x_0 be an interior point of I where f has a local extremum (maximum or minimum). If f is differentiable at x_0, then

$$f'(x_0) = 0 . \tag{6.2}$$

Proof. Suppose x_0 is a local maximum, i.e., there exists an open interval $J \subset I$ containing x_0 such that

$$f(x) \leqq f(x_0)$$

for all $x \in J$. Therefore, for $x \in J$,

$$\frac{f(x) - f(x_0)}{x - x_0} \begin{cases} \leqq 0 & \text{if } x > x_0, \\ \geqq 0 & \text{if } x < x_0. \end{cases}$$

Since f is differentiable at x_0, the limit of the above expression, as $x \to x_0$, exists. The above inequalities show this limit to be both $\leqq 0$ and $\geqq 0$, and therefore

$$f'(x_0) = \lim_{x \to x_0} \frac{f(x) - f(x_0)}{x - x_0} = 0 \,.$$

The case where x_0 is a local minimum is proved analogously. □

Remark 6.8. Theorem 6.7 states that a local extremum of a differentiable function f is necessarily a stationary point of f. The condition $f'(x_0) = 0$ is therefore *necessary* for x_0 to be a local extremum. Example 6.11a shows that this condition is *not sufficient*, i.e., it is possible for a point to be stationary, without being a local extremum.

MACSYMA-Session 6.2. The stationary points of f can be computed with MACSYMA. For example, consider

```
c1. f(x):=3*x^4-8*x^3+6*x^2-2 $
c2. factor(diff(f(x),x))
d2.     12(x − 1)²x
```

Therefore $f'(x) = 0$ has two roots, $x_1 = 0$ and $x_2 = 1$, the stationary points of f. The point $x_1 = 0$ is a local minimum, but the point $x_2 = 0$ is neither a local minimum nor a local maximum. Illustrate this by plotting the graph of $f(x)$ from $x = -1$ to 2 (Fig. 6.4)

```
c3. plot2(f(x),x,-1,2)$
```

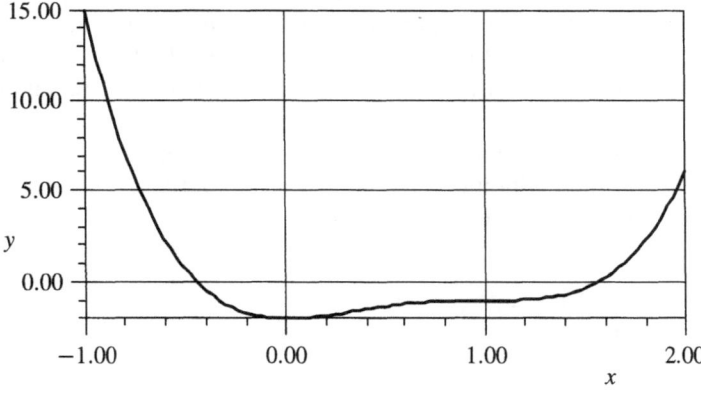

Fig. 6.4. Graph in Session 6.2, line c3

Another example is

c4. g(x):=(x^2+2*x+3)/(4*x+5) $

The derivative $g'(x)$ is found and factored by

c5. diff(g(x),x)

d5. $\dfrac{2x+2}{4x+5} - \dfrac{4(x^2+2x+3)}{(4x+5)^2}$

c6. factor(%)

d6. $\dfrac{2(2x^2+5x-1)}{(4x+5)^2}$

The stationary points of $g(x)$ are found by solving $g'(x) = 0$, or

c7. solve(%,x)

d7. $\left[x = -\dfrac{\sqrt{33}+5}{4},\ x = \dfrac{\sqrt{33}-5}{4} \right]$

The steps can be combined into one function

c8. stationary_points(f,x):=solve(diff(f,x),x) $

since solve uses factor if necessary. Calling

c9. stationary_points(g(x),x)

gives the same result

d9. $\left[x = -\dfrac{\sqrt{33}+5}{4},\ x = \dfrac{\sqrt{33}-5}{4} \right]$

See also Exercise 7.58 for the computation of a stationary point near a given point.

We now restate Theorem 6.7 as follows.

Corollary 6.9 (Necessary condition for local extremum). If x_0 is an interior local extremum of f, then x_0 is a critical point of f.

Proof. If f is differentiable at x_0, then $f'(x_0) = 0$ by Theorem 6.7. The remaining possibility is that f is not differentiable at x_0. Either way, x_0 is a critical point of f.   □

Remark 6.10. a. Corollary 6.9 is the key to solving extremum problems. It tells us that the search for local extrema can be limited to critical points. In typical problems there are few critical points, and these can be computed.

b. Once a critical point is found, it has to be tested whether it is a local *maximum*, a local *minimum*, or *neither* of these. Two such tests are studied in Sects. 6.3 and 6.4.

c. Endpoints are not covered by Corollary 6.9. For example, Fig. 6.1 shows local minima at the endpoints a and b, where the derivative is not zero, i.e., the graph has a nonhorizontal tangent.

Example 6.11. Table 6.2 gives problems a–g illustrating some of the possibilities. You should graph each function, in its feasible set, and verify all conclusions.

a. An unconstrained problem. The objective function $f(x) = x^3$ is differentiable everywhere and is increasing throughout $(-\infty, \infty)$. Therefore f cannot have local extrema, not even at $x := 0$, the only stationary point of f.

Table 6.2. Extremum problems (Example 6.11)

Prob-lem	Objective function $f(x)$	Feasible set	Local maxi-mizer(s)	Local mini-mizer(s)	Critical points Stationary point(s)	Points where f is nondif-ferentiable
a	x^3	$(-\infty, \infty)$	none	none	0	none
b	x^3	$[-1, 1]$	1	-1	0	none
c	$x^{1/3}$	$(-\infty, \infty)$	none	none	none	0
d	$x^{2/3}$, mean-ing $\left(x^{1/3}\right)^2$	$(-\infty, \infty)$	none	0	none	0
e	$x^3 - 3x$	$(-\infty, \infty)$	-1	1	$-1, 1$	none
f	$x^3 - 3x$	$[-2, 2]$	$-1, 2$	$-2, 1$	$-1, 1$	none
g	constant	any set	all feasible solutions	all feasible solutions	all feasible solutions	none

 b. Same objective as in problem a, but not the same feasible set, hence the two problems are different. The feasible set here is $[-1, 1]$, and since f is increasing throughout, the left endpoint $x := -1$ is a local (actually global) minimizer, and the right endpoint $x := 1$ is a local (and global) maximizer.

 c. Similar to problem a. The only critical point is $x := 0$ which is not a local extremum.

 d. The only critical point is $x := 0$ which is a local (and global) minimizer.

 e. An unconstrained problem. The points -1, 1 are stationary, and local extremizers. There are however no global extremizers: The function values change from $-\infty$ to ∞.

 f. Again this is a problem different from problem e, because the feasible sets are not the same. In addition to the local extremizers -1, 1 of problem e, there is a global maximum at the endpoint $x = -2$ and a global minimum at the endpoint $x = 2$.

 g. This example shows that there can be infinitely many global extremizers. Can you find a nontrivial example (i.e., f not constant) with infinitely many global maximizers, and infinitely many global minimizers, in any interval $[-\delta, \delta]$ around the origin?

 Corollary 6.9 can be adapted for endpoints of the feasible set, using the one-sided derivatives,
the *right* derivative of f at x_0:

$$f'_+(x_0) := \lim_{x \downarrow x_0} \frac{f(x) - f(x_0)}{x - x_0},$$

the *left* derivative of f at x_0:

$$f'_-(x_0) := \lim_{x \uparrow x_0} \frac{f(x) - f(x_0)}{x - x_0} \quad \text{(see Definition 4.16)} .$$

Corollary 6.12 (Necessary condition for extremum at endpoints). Let $[a, b]$ be the feasible set, f the objective function. Then:
a. if the left endpoint a is a local maximizer (minimizer) of f, and if the right derivative $f'_+(a)$ exists, then $f'_+(a) \leq 0$ $[f'_+(a) \geq 0]$;
b. if the right endpoint b is a local maximizer (minimizer) of f, and if the left derivative $f'_-(b)$ exists, then $f'_-(b) \leq 0$ $[f'_-(b) \geq 0]$.

Proof. If $f'_+(a) > 0$, then there are points $x > a$, arbitrarily close to a, with $f(x) > f(a)$, so a cannot be a local maximizer. This proves half of part a. The rest is proved similarly.   □

If the objective function f and the feasible set S are given, then the extremal problem is to find the extremal solutions in S and the extremal values of f. Often the objective function and feasible set are not given, in which case their determination is part of the problem.

We suggest a scheme for formulating and solving an extremal problem.

Step 1: Identify the objective function f.
Step 2: Identify the feasible set S.
Step 3: Find all critical points of f in S; stationary points found by solving $f'(x) = 0$, and also points where f is not differentiable .
Step 4: Determine which of the critical points are local maximizers or local minimizers.
Step 5: Compare the values of f at the endpoints with its values at the extremizers in step 4; determine the global extremizer(s) and extremum value.

In Sects. 6.3 and 6.4 we develop tests to use in step 4. For now, let us see how the scheme works for some easy examples.

Example 6.13. In Example 6.1 it was required to maximize (6.1),

$$f(x) := -0.001x^2 + 7x - 6,000 \quad \text{in the interval } [0, 10,000] .$$

Step 3: The stationary points are found by solving $f'(x) = -0.002x + 7 = 0$, giving the single point $x = 3,500$.
Step 4: We can show (see Example 6.18b) that $x = 3,500$ is a local maximizer.
Step 5: We compute the value of the objective

at the stationary point $x = 3,500$: $f(3,500) = 6,250$ [$];
at the endpoint $x = 0$: $f(0) = -6,000$ [$];
at the endpoint $x = 10,000$: $f(10,000) = -36,000$ [$]

and conclude that $x = 3,500$ is the global maximizer, giving a profit of $f(3,500) = 6,250$ [$].

Example 6.14. Given $L > 0$, it is required to calculate a rectangle having the largest possible area, and having L as its *perimeter* (= sum of lengths of sides). For example, you are given enough material to build a fence of length L, and you would like to surround the largest possible area of rectangular shape.

Solution
    Step 1: The objective, to be maximized, is the area A of the rectangle, expressed as

$$A = xy,$$

where x and y are the sides of the rectangle. However, x and y are related by

$$2x + 2y = L,$$

the given perimeter (Fig. 6.5 a). We can use the last relation to express y as a function of x,

$$y = \frac{L - 2x}{2}$$

and substitute in A to get the objective function

$$A(x) := x\left(\frac{L - 2x}{2}\right) = \frac{Lx - 2x^2}{2} \tag{6.3}$$

with one decision variable, x (Fig. 6.5 b).
    Step 2: Clearly x must be nonnegative. Also x cannot be larger than half the perimeter (why?). Therefore

$$0 \leq x \leq L/2,$$

and the feasible set is $S = [0, L/2]$.
    Step 3: The objective is differentiable everywhere. From

$$A'(x) = (L - 4x)/2 = 0$$

we get the (only) stationary point

$$x = L/4,$$

which is a feasible (interior) solution.

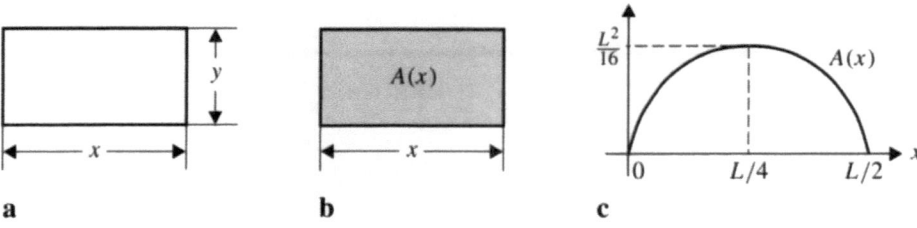

Fig. 6.5 a–c. Illustration of Example 6.14. **a** The perimeter of the rectangle is $L = 2x + 2y$. **b** The area of the rectangle is $A = A(x) = (Lx - 2x^2)/2$. **c** Graph of the area

Step 4: We can show (see Example 6.18c) that the point $x = L/4$ is a local maximizer.

Step 5: At the endpoints $0, L/2$ the objective values are

$$A(0) = 0, \quad A(L/2) = 0 .$$

Since the area $A(x)$ is nonnegative for all feasible x, we conclude that the endpoints are of no interest (they are in fact global minimizers).

Therefore the global maximizer of $A(x)$ in S is $x = L/4$ (Fig. 6.5c). The rectangle of maximal area is actually a square since $x = y = L/4$. The maximal area is

$$A\left(\frac{L}{4}\right) = \frac{L^2}{16} .$$

Example 6.15 (Reflection of light by mirror). An object, or light source (at the point A) and an observer (at B) are on the same side of a mirror. The object A and the observer B are at distances a and b, respectively, from the mirror. The horizontal separation between A and B is c.

A light ray emanating from A is reflected by the mirror (at some point C) before arriving at B (Fig. 6.6). We calculate now the exact path of that ray, in particular the point C, and the angles of reflection α and β. Our solution is based on the following principle.

Heron's[2] principle. The path of light $A \to C \to B$ is shorter than any other reflected path from A to B.

Solution

Step 1: Let x be the horizontal separation between A and C. The length L of the path $A \to C \to B$ is then a function of x

$$L(x) := \sqrt{a^2 + x^2} + \sqrt{b^2 + (c - x)^2} .$$

This, according to Hero's principle, is the objective to be minimized.

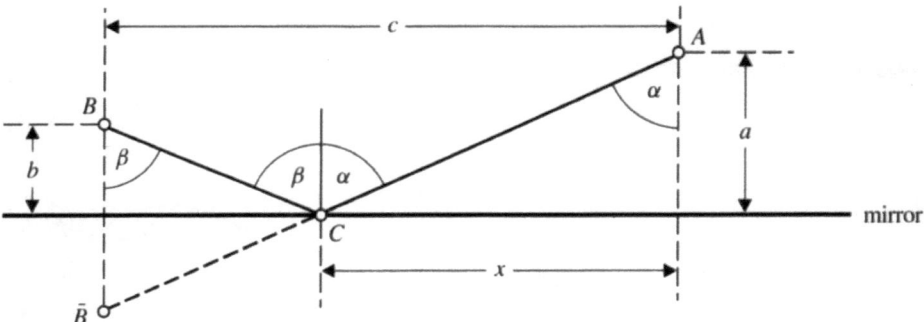

---

2 Heron of Alexandria, also called Hero (c. 125 B.C.).

Step 2: The point of reflection on the mirror, C, is between A and B. Therefore $0 \le x \le c$ and the feasible set is $[0, c]$.

Step 3: The function $L(x)$ is differentiable for all x, and

$$L'(x) = \frac{x}{\sqrt{a^2 + x^2}} - \frac{c - x}{\sqrt{b^2 + (c - x)^2}}.$$

There is a unique stationary point x^*, found by solving $L'(x) = 0$, which can be shown to be the global minimum (we omit steps 4 and 5).

Solution of $L'(x) = 0$: Writing $L'(x) = 0$ as

$$\frac{x}{\sqrt{a^2 + x^2}} = \frac{c - x}{\sqrt{b^2 + (c - x)^2}}$$

we get, by squaring both sides, a quadratic equation

$$(a^2 - b^2)x^2 - 2ca^2 x + a^2 c^2 = 0$$

whose solutions are

$$x_{1,2} = \frac{ca^2 \pm cab}{a^2 - b^2} = \frac{ca(a \pm b)}{a^2 - b^2}.$$

We reject the solution

$$x = \frac{ca(a + b)}{a^2 - b^2} = \frac{ca}{a - b}$$

because it is greater than c. Therefore

$$x^* = \frac{ca(a - b)}{a^2 - b^2} = \frac{ca}{a + b}.$$

Even without computing x^* explicitly, we can prove an interesting fact about the reflection angles: They are equal, $\alpha = \beta$ (5.20). Indeed, from Fig. 6.6,

$$\sin \alpha = \frac{x}{\sqrt{a^2 + x^2}}, \quad \sin \beta = \frac{c - x}{\sqrt{b^2 + (c - x)^2}},$$

and (5.20) follows from $L'(x) = 0$.

The reflection point C can be determined as follows. Let $\bar{B}$ be the mirror image of B, at distance b on the other side of the mirror. Then C is where the line joining A and $\bar{B}$ intersects the mirror.

Example 6.16 (Snell's law of refraction). A straight long object (e.g., pencil) placed in a glass of water is seen from the outside to be bent at the air–water boundary. This phenomenon, known as *refraction*, is described by Snell's[3] law of refraction, whose development follows.

Consider two media, say air and water, separated by a straight boundary (Fig. 6.7).

---

3 Willebrord van Roijen Snell (1591–1626)

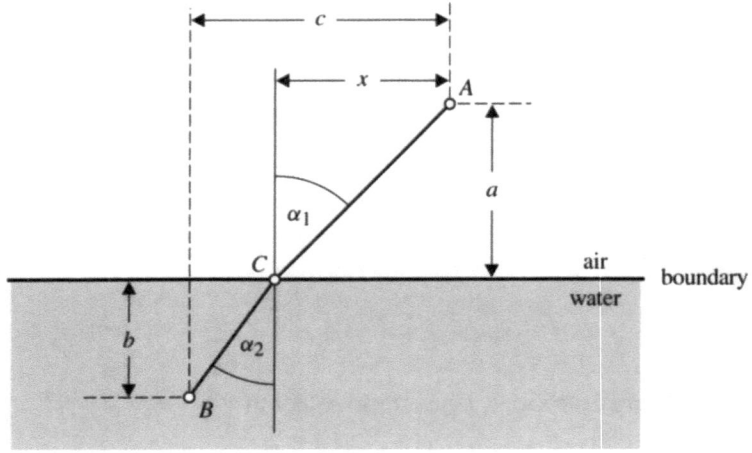

Fig. 6.7. Refraction of light

Let v_1 and v_2 be the speeds of light in air and water, respectively. There is an object (light source) at the point A, which is a distance a above the boundary. An (underwater) observer is at the point B, a distance b below the boundary. The horizontal separation between A and B is c.

How does light travel from A to B? Hero's principle (used in Example 6.15) states that light travels, between any two points, along the shortest path (i.e., straight line) joining them.

This is true if the speed of light along its path is the same.[4] When the speed of light changes along the path, then Hero's principle is no longer true. In its place, we use another principle.

Fermat's[5] principle. Light travels, between any two points, along a path with the shortest travel time.[6]

As light travels from A to B, it crosses the boundary where its velocity changes. It is clear intuitively that light would try to travel "as much as possible" in air,[7] where its speed v_1 is high, and "as little as possible in water", where the speed v_2 is lower than v_1.

Now the solution in detail.

---

4 In Example 6.15 we took the velocity of light, above the mirror, as constant.

5 Pierre de Fermat (1601–1665)

6 Since for a constant velocity v we have time being equal to the product of $(1/v)$ multiplied by distance, it follows that minimizing time (selecting path according to Fermat's principle) is the same as minimizing distance (Hero's principle).

7 Similarly, a driver who wants to get as fast as possible from city A to city B will prefer high-speed roads (such as turnpikes), although the distance along the slower roads may be shorter.

Step 1: The objective, according to Fermat's principle, is the time t_{AB} of travel from A to B. We write it as

$$t_{AB} = t_{AC} + t_{CB} ,$$

where C is the point where the light ray crosses the boundary. Let x be the horizontal distance between A and C. Then

$$\text{distance}\{A, C\} = \sqrt{a^2 + x^2}, \quad \text{and} \quad t_{AC} = \frac{\sqrt{a^2 + x^2}}{v_1} ,$$

$$\text{distance}\{C, B\} = \sqrt{b^2 + (c - x)^2}, \quad \text{and} \quad t_{CB} = \frac{\sqrt{b^2 + (c - x)^2}}{v_2} .$$

Finally the objective is

$$t_{AB} = t_{AB}(x) := \frac{\sqrt{a^2 + x^2}}{v_1} + \frac{\sqrt{b^2 + (c - x)^2}}{v_2} .$$

Step 2: The feasible set is $0 \le x \le c$.
Step 3: The objective is differentiable everywhere in $[0, c]$, and its derivative is

$$t'_{AB}(x) = \frac{1}{v_1} \frac{x}{\sqrt{a^2 + x^2}} - \frac{1}{v_2} \frac{c - x}{\sqrt{b^2 + (c - x)^2}} .$$

There is a unique stationary point x^*, which is found by solving $t'_{AB}(x) = 0$. We can then show (in steps 4 and 5, which we omit) that x^* is indeed the global minimizer.

Using $t'_{AB}(x) = 0$, and the facts

$$\frac{x}{\sqrt{a^2 + x^2}} = \sin \alpha_1, \quad \frac{c - x}{\sqrt{b^2 + (c - x)^2}} = \sin \alpha_2 \quad \text{(Fig. 6.7)},$$

we get Snell's law of refraction

$$\frac{\sin \alpha_1}{\sin \alpha_2} = \frac{v_1}{v_2}, \tag{6.4}$$

giving the angles of refraction α_1, α_2 at the bending point C.

Exercises

6.1 Determine all the stationary points and critical points for the following functions. If necessary, use MACSYMA to solve the equations which determine the location of the stationary points.

(a) $f(x) := x^2 + 5x + 8$ (b) $f(x) := 4x^2 + 16x + 1$ (c) $f(x) := x^4 + 6x^2 + 4$

(c) $f(x) := \dfrac{x^2 + 5x}{x^2 + 1}$ (d) $f(x) := \dfrac{x^2 + x}{x + 8}$ (e) $f(x) := x^{1/3}$

(f) $f(x) := (x^2 + 1)^{1/2}$ (g) $f(x) := (x^2 - 1)^{1/2}$ (h) $f(x) := (x^2 + a^2)^{3/2}$

(i) $f(x) := \sin 2x$ (j) $f(x) := \tan \dfrac{x}{3}$ (k) $f(x) := \left| \cos \dfrac{x}{2} \right|$

(l) $f(x) := x^2 \sin x$ (m) $f(x) := x^2 \tan x$ (n) $f(x) := \dfrac{\cos x}{2 + \sin x}$

(o) $f(x) := \dfrac{\cos x}{1 + \sin x}$ (p) $f(x) := \sinh x$ (q) $f(x) := \cosh x$

(r) $f(x) := \tanh x$ (s) $f(x) := \dfrac{1 + \sinh x}{1 - \cosh x}$ (t) $f(x) := \dfrac{1}{\sinh x}$

6.2 Plot the following functions f, and determine in each case the local extrema and stationary points of f in $I = (-\infty, +\infty)$, if any.

(a) $f(x) := x^{2/3}$ (b) $f(x) := |x|$ (c) $f(x) := x^3 - 3x$

(d) $f(x) := x\,e^{-x}$ (e) $f(x) := x\,e^{-x^2}$ (f) $f(x) := x^2\,e^{-x}$

6.3 Find the rectangle of largest area that can be inscribed in:

(a) a circle of radius R (b) the ellipse $\dfrac{x^2}{a^2} + \dfrac{y^2}{b^2} = 1$

6.4 A triangle is inscribed within a semicircle of radius R, such that one side corresponds to diameter of the semicircle. Find the length of the sides of the triangle having the largest area.

6.5 Let a cone be defined as a right circular cone with height h and base of radius r. Its volume is $v = \pi r^2 h/3$.

  a. Find the cone of maximum volume that can be inscribed in a sphere of radius R.
  b. Find the cone of minimum volume in which a sphere of radius R is inscribed.
  c. Is there a cone of maximum volume in which a sphere of radius R is inscribed?

6.6 Let a cylinder be defined as a right circular cylinder with height h and a base with radius r. Its volume is $v = \pi r^2 h$.

  a. Find the cylinder of maximum volume that can be inscribed in a sphere with radius R.
  b. Find the cylinder of maximum volume that can be inscribed in a circular cone with height H and radius R.

6.7 A tinsmith makes an open box out of a square piece of sheet-metal that has a side with length L. To make his box, he cuts a square with length a from each corner and folds up the sides (Fig. 6.8). What size square must he cut to make a box of maximum volume.

6.8 Repeat the above problem for a rectangular piece of sheet-metal whose sides are L_1 and L_2.

6.9 An isosceles triangle has perimeter S. Find the sides if the area is to be a maximum.

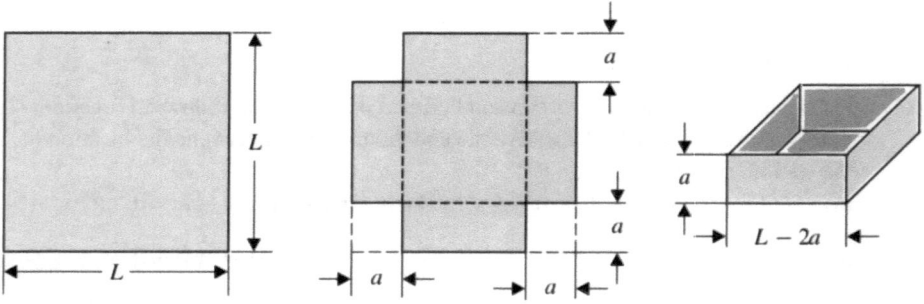

Fig. 6.8. Illustration of Exercise 6.7

6.10 Find the sides of an isosceles triangle of maximum area which is inscribed in a circle with radius R.

6.11 Find the rectangle of a given area A, which has:
(a) the smallest perimeter   (b) the smallest diameter
The diameter of a set S is the maximal distance between any two points in S.

6.12 If two sides of a triangle are given to be a and b, determine the side c which corresponds to a triangle of maximum area.

6.13 Which point of the hyperbola $x^2/4 - y^2/9 = 1$ is nearest to the point $(1, 0)$?

6.14 Find the point (x_0, y_0), with $x_0 > 0$, on the graph of $y = 1/x$, which is nearest to the origin.

6.15 Let $f(x) = \sin x$. Find the points in $[0, 2\pi]$ where the derivative is maximal.

6.16 In Example 6.16 the observer at B wants to shoot with his (underwater) gun at the object A. However, the object appears to him higher than it actually is (he sees it in the direction $\overrightarrow{BC}$). Assuming the gun shoots straight, how should the observer aim, given his view of the object A, and knowing the distances b, c and the angles α_1, α_2.

6.3 First-derivative test

We know (by Corollary 6.9) that the search for local extrema can be limited to critical points. There are two tests to determine whether a given critical point is a local *maximizer*, a local *minimizer*, or *none* of the above.

In this section we give the first-derivative test which uses only the first derivative, and in the next section the second-derivative test using the second derivative.

Let f be continuous on (a, b), and let x_0 be the only critical point of f in (a, b). Thus f is differentiable at all points of (a, b), with the possible exception of x_0. The point x_0 splits (a, b) into two open intervals, (a, x_0) and (x_0, b).

If f is increasing in (a, x_0) and decreasing in (x_0, b), then x_0 must be a local maximizer of f (actually a global maximizer) in (a, b). This and other cases are given in Table 6.3.

We recall that a differentiable function f is increasing (decreasing) at a point x_0 if its derivative $f'(x_0)$ at that point is positive (negative) (see Lemma 4.12). We will prove in Corollary 7.5 that this property extends to intervals: A differentiable function f is increasing (decreasing) in an interval if its derivative f' is positive (negative) in that interval. We can thus restate the four cases in Table 6.3 in terms of the derivative of f.

Table 6.3. Determination of extremizer of f

Case	Condition on f *	Conclusion on x_0
1	f is $\nearrow$ on (a, x_0) and $\searrow$ on (x_0, b)	x_0 is the global maximizer in (a, b)
2	f is $\searrow$ on (a, x_0), and $\nearrow$ on (x_0, b)	x_0 is the global minimizer in (a, b)
3	f is $\nearrow$ on (a, x_0), and $\nearrow$ on (x_0, b)	x_0 is not a local extremizer
4	f is $\searrow$ on (a, x_0), and $\searrow$ on (x_0, b)	x_0 is not a local extremizer

* We denote "increasing" by $\nearrow$, "decreasing" by $\searrow$.

Table 6.4. Determination of extremizer of f by first-derivative test

Case	Condition on f'	Conclusion on x_0
1	$f'(x) > 0$ on (a, x_0) and $f'(x) < 0$ on (x_0, b)	x_0 is the global maximizer of f in (a, b)
2	$f'(x) < 0$ on (a, x_0) and $f'(x) > 0$ on (x_0, b)	x_0 is the global minimizer of f in (a, b)
3	$f'(x) > 0$ on (a, x_0) and $f'(x) > 0$ on (x_0, b)	x_0 is not a local extremizer
4	$f'(x) < 0$ on (a, x_0) and $f'(x) < 0$ on (x_0, b)	x_0 is not a local extremizer

Theorem 6.17 (First-derivative test). Let f be continuous in (a, b), and differentiable in (a, x_0) and (x_0, b). Then x_0 is a global maximizer, global minimizer, or no local extremizer of f under the conditions on f' given in Table 6.4.

Example 6.18. The following examples illustrate the 1st-derivative test.
   a. The 1st-derivative test does not require the function f to be differentiable at the test point x_0. Consider the objective $f(x) = x^{2/3}$ (see Example 6.11 d). Then

$$f'(x) = \tfrac{2}{3}x^{-1/3}, \quad x \neq 0,$$

and f is not differentiable at $x_0 := 0$. However,

$$f'(x) \begin{cases} < 0 & \text{if } x < 0, \\ > 0 & \text{if } x > 0, \end{cases}$$

and $x_0 = 0$ is the global minimizer by the 1st-derivative test with $a = -\infty$, $b = \infty$.
   b. In Example 6.13, the derivative

$$f'(x) = -0.002x + 7 = \begin{cases} > 0 & \text{if } 0 \le x < 3{,}500, \\ < 0 & \text{if } 3{,}500 < x \le 10{,}000, \end{cases}$$

proving that $x = 3{,}500$ is a local maximizer.
   c. In Example 6.14, the derivative

$$A'(x) = \frac{L - 4x}{2} = \begin{cases} > 0 & \text{if } 0 \le x < L/4, \\ < 0 & \text{if } L/4 < x \le L/2, \end{cases}$$

proving that $x = L/4$ is a local maximizer.

MACSYMA-Session 6.3. We can implement an approximate version of the first-derivative test by the following MACSYMA function

```
c1. test_der_1(f,x,x0,eps) := block([prederror:false],
            case(is(subst(x0-eps,x,diff(f,x))>0),
                [unknown, "Test fails"],
                [false, case(is(limit(diff(f,x),x,x0+eps)>0),
```

```
                              [unknown,"Test fails"],
                              [false, "Function decreasing, not a
                                     local extremum"],
                              [true, "Local Minimum"]),
              [true, case(is(limit(diff(f,x),x,x0+eps)<0),
                              [unknown,"Test fails"],
                              [false, "Function increasing, not a
                                     local extremum"],
                              [true, "Local Maximum")])])$
```

`test_der_1` checks the sign of the derivative f' at the points $x_0 - \varepsilon$ and $x_0 + \varepsilon$. We
assume (with fingers crossed) that the sign of f' changes at most once between these two
points. If this assumption is correct, then the function `test_der_1` is an exact translation
of the cases in Theorem 6.17, with $a = x_0 - \varepsilon$ and $b = x_0 + \varepsilon$.

Note that "Test fails" results if MACSYMA cannot determine the sign of $f'(x_0 - \varepsilon)$ and
$f'(x_0 + \varepsilon)$ as positive or negative.

If the outer `case` is either `true` or `false`, only one of the inner `case` statements will be
used, and these correspond with the cases listed in Table 6.4. For most cases, we can give
ε a standard value such as 10^{-3}.

    a. The function $f(x) := x^2$ has a local (actually global) minimum at $x = 0$:

c2. `test_der_1(x^2,x,0,10^-3)`

`C:\MACSYMA2\share\basic.fas being loaded.`

d2.   Local Minimum

    b. For the objective function $A(x) := x(L - 2x)/2$ of Example 6.14 and $x_0 = L/4$
we find

c3. `test_der_1(x*(L-2*x)/2,x,L/4,10^-3)`

d3.   Local Maximum

Note that the parameter L is unspecified.

    c. The function $f(x) := |x|$ is not differentiable at $x = 0$, but

c4. `test_der_1(abs(x),x,0,10^-3)`

still gives the correct answer,

d4.   Local Minimum

    d. `test_der_1` is useless for testing $f(x) := \sin(1/x)$ at $x = 0$, since the function
behaves wildly, oscillating ever more rapidly as x approaches 0 (see Fig. 3.3).

    e. The result of `test_der_1` will depend on the value of ε chosen in the test. For
example,

c5. `test_der_1(sin(1/x),x,0,10^-3)`

d5.   Function decreasing, not a local extremum

but for $\varepsilon = 10^{-4}$

c6. `test_der_1(sin(1/x),x,0,10^-4)`

d6.   Function increasing, not a local extremum

Both answers are incorrect: for any $\varepsilon > 0$, the derivative of $\sin(1/x)$, $-\cos(1/x)/x^2$,
changes sign infinitely many times in the interval $(-\varepsilon, \varepsilon)$, and the function `test_der_1`
cannot be applied (to use it, we assume at most one change of sign in the tested interval).

    f. The function `test_der_1` can also fail if the tested point $x = x_0$ is too close to a
local extremizer. For example, $f(x) := x^2$ has a unique local minimum at $x = 0$. Testing
the point $x_0 = 10^{-5}$ with $\varepsilon = 10^{-3}$

c7. `test_der_1(x^2,x,10^-5,10^-3)`

gives a wrong answer, namely, that $x_0 = 10^{-5}$ is

d7.   Local Minimum

This is a worse type of failure than "Test fails". Here we get a wrong answer without any warning. However, running `test_der_1` with a smaller ε, such as 10^{-6}, gives the correct result

c8. `test_der_1(x^2,x,10^-5,10^-6)`

d9.   Function increasing, not a local extremum

since now the true extremizer lies outside the interval $(x_0 - \varepsilon, x_0 + \varepsilon)$.

   g. Explain the following contradictory answers by `test_der_1`. If necessary, graph the function in question.

c9. `test_der_1(x^2*sin(1/x),x,0,10^-3)`

d9.   Function decreasing, not a local extremum

c10. `test_der_1(x^2*sin(1/x),x,0,10^-4)`

d10.   Function increasing, not a local extremum

c11. `test_der_1(x*sin(1/x),x,0,10^-3)`

d11.   Local Maximum

c12. `test_der_1(x*sin(1/x),x,0,10^-4)`

d12.   Local Minimum

   h. Consider the function c4 of Session 6.2, `f(x):=(x^2+2*x+3)/(4*x+5)$`. As there, the stationary points are:

c13. `ans:solve(diff(f(x),x),x)`

d13. $\left[x = -\dfrac{\sqrt{33}+5}{4}, x = \dfrac{\sqrt{33}-5}{4} \right]$

c14. `x1:rhs(ans[1])$`

c15. `x2:rhs(ans[2])$`

c16. `test_der_1(f(x),x,x1,10^-3)`

d16. Local Maximum

c17. `test_der_1(f(x),x,x2,10^-3)`

d17.   Local Minimum

   i. We can apply `test_der_1` to several critical points simultaneously,

c18. `pol:x^3-2*x^2+x-11 $`

c19. `solve(diff(pol,x),x)`

d19. $\left[x = 1, x = \dfrac{1}{3} \right]$

c20. `[test_der_1(pol,x,1,10^-3),test_der_1(pol,x,1/3,10^-3)]`

d20.   [Local Minimum, Local Maximum]

Exercises

6.17 Let the feasible set be an interval $[a, b]$ and let the objective f be differentiable in (a, b). Adapt the 1st-derivative test to decide if the endpoints a, b are local extrema.

6.18 In Example 6.15 verify that the solution x^* of $L'(x) = 0$ is the global minimizer of L in the feasible set $[0, c]$.

6.19 In Example 6.16 verify that the solution x^* of $t'_{AB}(x) = 0$ is the global minimizer of t_{AB} in the feasible set $[0, c]$.

6.20 Use the 1st-derivative test to determine local maxima and minima of the following functions.

   (a) $f(x) := x^3 - 3x$ (b) $f(x) := x^4 - 4$

   (c) $f(x) := x^3 + 3x^2 - 9x + 100$ (d) $f(x) := \dfrac{x - 1}{x + 1}$

   (e) $f(x) := \dfrac{x}{x^2 + 1}$ (f) $f(x) := (x^2 - 1)^2$ (g) $f(x) := (x + 3)^6$

   (h) $f(x) := |\sin x|$ (i) $f(x) := |x|$ (j) $f(x) := |x|^{3/2}$

   (k) $f(x) := |x|^{1/3}$ (l) $f(x) := \dfrac{\sin x^2}{x}$

6.21 Find the shortest distance between a point (x_0, y_0) and the line $ax + by + c = 0$, if $c \neq 0$. Write a MACSYMA function depending on the parameters a, b, c and the point (x_0, y_0) that solves this problem.

6.22 Let $P(x_0, y_0)$ be a point in the first quadrant.
   a. Find the line through P such that the sum of the x and y intercepts is a minimum.
   b. Find the line through P such that the line segment intercepted by the x and y axes is shortest.
   If need be, use MACSYMA to compute the stationary points.

6.23 Determine a right circular cylinder, of a given volume V, which has a minimum surface area.

6.24 Determine a right circular cone, of a given volume V, which has a minimum surface area.

6.25 The illumination due to a light source is stronger the closer we get to it. Conversely, the farther we get from the source, the weaker the illumination. Denoting the intensity (measured in candle) of a light source by I, the illumination at a point, a distance d from the light source, is proportional to I/d^2.

   Two light sources of intensities I_1 and I_2, respectively, are a distance L apart. Find a point on the line segment between the two sources where the illumination is a minimum.

6.26 Let O be a light source of intensity I, let P be a point on a plane, and let θ be the angle between the line OP and the normal to the plane at P. Then the illumination at P is proportional to $(\cos \theta)\, I/d^2$, where d is the distance between O and P. In particular, the illumination is zero for $\theta = 90°$. (Explain.)

   A lamp can be moved (raised or lowered) along a vertical pole. Find the height of the lamp so that the illumination is maximal at a point on the floor 5 m away from the pole.

6.27 The strength of a beam is proportional to the product of the width of the beam and the square of the height of the beam. Determine the dimensions of the strongest beam that can be cut from a circular cylinder with radius R.

6.28 Repeat the previous problem for an elliptic cylinder, with cross section given by $x^2/a^2 + y^2/b^2 = 1$.

6.29 A square of maximum area is cut from a circle with radius R. From each of the remaining four segments of the circle a rectangle is to be cut such that each is of maximum area. Find the dimensions of these rectangles.

6.30 Two vertical poles with heights H_1 and H_2 are a distance L apart. Find the length of the shortest rope that be stretched from the top of one pole to a point on a line between them to the top of the other pole.

6.4 Second-derivative test

Let x_0 be a stationary point of f, i.e., $f'(x_0) = 0$. The first case of Theorem 6.17 is
- condition on f': $f'(x) > 0$ on (a, x_0), and $f'(x) < 0$ on (x_0, b);
- conclusion on x_0: x_0 is the global maximizer of f in (a, b).

This case surely happens if x_0 is a *stationary point* and the derivative f' is *decreasing* at x_0 (so that f' goes from *positive* values for $x < x_0$, to *zero* at x_0, then to *negative* values for $x > x_0$). If f' is differentiable at x_0, then $f''(x_0) < 0$ implies that f' is decreasing at x_0, see Lemma 4.12.

Similarly, if x_0 is a stationary point and $f''(x_0) > 0$, then f' is increasing at x_0, which gives case 2 of Theorem 6.17,
- condition on f': $f'(x) < 0$ on (a, x_0), and $f'(x) > 0$ on (x_0, b);
- conclusion on x_0: x_0 is the global minimizer of f in (a, b).

We summarize these two cases in the following theorem.

Theorem 6.19 (Second-derivative test). If f is twice differentiable at x_0, and $f'(x_0) = 0$, then

a. $f''(x_0) < 0$ implies that x_0 is a local maximizer.
b. $f''(x_0) > 0$ implies that x_0 is a local minimizer.

Example 6.20. The 2nd-derivative test gives no information in case

$$f'(x_0) = 0, \quad \text{and} \quad f''(x_0) = 0. \tag{6.5}$$

In this case, the examples (a) $f(x) = x^4$, $x_0 = 0$; (b) $f(x) = -x^4$, $x_0 = 0$; and (c) $f(x) = x^3$, $x_0 = 0$, show that x_0 can be a local minimizer, a local maximizer, or not a local extremizer.

Example 6.21. Given numbers

$$x_1, x_2, \ldots, x_n, \tag{6.6}$$

it is required to find a number x which minimizes the *sum of squares*

$$S(x) := (x_1 - x)^2 + (x_2 - x)^2 + \cdots + (x_n - x)^2 = \sum_{k=1}^{n}(x_k - x)^2.$$

This problem arises where a quantity x is unknown, cannot be measured accurately, and has to be estimated from experimental results such as (6.6). Each one of the observations x_k in (6.6) is a *deviation*

$$x_k - x$$

from the (unknown) value x. We settle for an estimate $\bar{x}$ of x which minimizes $S(x)$, the *sum of squares of deviations*.

Solution: The function $S(x)$ is differentiable for all x; its derivative

$$S'(x) = -2 \sum_{k=1}^{n} (x_k - x)$$

has a unique zero at

$$\bar{x} = \frac{\sum_{k=1}^{n} x_k}{n} \,,$$

the *average* (or arithmetic mean) of the data (6.6). The second derivative is

$$S''(x) = 2n \,,$$

and by the 2nd-derivative test, the point $\bar{x}$ is a local minimizer of $S(x)$. It is a global minimizer because it is the only critical point of $S(x)$.

Example 6.22. Let a positive number s be given, and let n nonnegative numbers $x_1, x_2, \ldots, x_n$ satisfy

$$x_1 + x_2 + \cdots + x_n = s \,.$$

Claim: The product $x_1 x_2 \ldots x_n$ is maximal if all numbers are equal, $x_1 = x_2 = \cdots = x_n = s/n$, and the maximal product is

$$x_1 x_2 \ldots x_n = (s/n)^n \,. \tag{6.7}$$

Proof. This is an optimization problem with n variables. However, using induction we can deal with these variables one at a time. Let $S(n)$ be the statement that the above claim is true for n variables.
Verification of $S(1)$: Obvious.
Proof of $S(n) \implies S(n+1)$: The problem with $n+1$ variables is

$$\begin{aligned} \max \quad & x_1 x_2 \ldots x_n x_{n+1} \\ \text{subject to} \quad & x_1 + x_2 + \cdots + x_n + x_{n+1} = s \,, \\ & x_1, x_2, \ldots, x_n, x_{n+1} \geq 0 \,. \end{aligned} \tag{6.8}$$

For any value of the variable x_{n+1} the remaining n variables $x_1, x_2, \ldots, x_n$ must be the best possible, i.e., they must solve the problem

$$\begin{aligned} \max \quad & x_1 x_2 \ldots x_n \\ \text{subject to} \quad & x_1 + x_2 + \cdots + x_n = s - x_{n+1} \,, \\ & x_1, x_2, \ldots, x_n \geq 0 \end{aligned} \tag{6.9}$$

whose optimal value is, by (6.7), equal to $((s - x_{n+1})/n)^n$. Therefore the objective function in (6.8) becomes

$$x_1 x_2 \ldots x_n x_{n+1} = \left(\frac{s - x_{n+1}}{n} \right)^n x_{n+1} = f(x_{n+1})$$

a function of a single variable! The problem (6.8) becomes

$$\begin{aligned} \max \quad & f(x_{n+1}) = \left(\frac{s - x_{n+1}}{n} \right)^n x_{n+1} \\ \text{subject to} \quad & 0 \leq x_{n+1} \leq s \end{aligned}$$

whose objective function has two stationary points

$$x_{n+1}^{(1)} = s/(n+1) \quad \text{and} \quad x_{n+1}^{(2)} = s$$

as shown by solving

$$f'(x_{n+1}) = \left(\frac{s - x_{n+1}}{n}\right)^n - n\frac{x_{n+1}}{n}\left(\frac{s - x_{n+1}}{n}\right)^{n-1}$$

$$= \left(\frac{s - x_{n+1}}{n}\right)^{n-1}\left(\frac{s - x_{n+1}}{n} - x_{n+1}\right) = 0.$$

The stationary point $x_{n+1}^{(1)}$ is a local maximizer, as shown by the 1st-derivative test or the 2nd-derivative test (verify!). The endpoints $x_{n+1} = 0$ and $x_{n+1} = s$ are not optimal (check!). Therefore the optimal solution of (6.8) is

$$x_1 = x_2 = \cdots = x_n = x_{n+1} = s/(n+1).$$ □

Example 6.23. The perimeter L of a triangle is the sum of lengths of its sides, $L = a + b + c$. An *equilateral* triangle is one with equal sides $a = b = c = L/3$. Prove: The equilateral triangle has maximal area among all triangles with the same perimeter.

Proof. The area is, by Heron's formula (Exercise C.12),

$$A = \sqrt{s(s - a)(s - b)(s - c)}, \quad \text{where } s = L/2,$$

$$= \sqrt{s}\sqrt{(s - a)(s - b)(s - c)}$$

$$= \sqrt{s}\sqrt{x_1 x_2 x_3}$$

where $x_1 := (s - a)$, $x_2 := (s - b)$, and $x_3 := (s - c)$ satisfy

$$x_1 + x_2 + x_3 = (s - a) + (s - b) + (s - c) = 3s - 2s = s.$$

The problem is therefore equivalent to finding positive numbers x_1, x_2, x_3 whose sum s is given and whose product is maximal. By Example 6.22 the solution is

$$x_1 = x_2 = x_3$$

which corresponds to the equilateral triangle $a = b = c$. □

MACSYMA-Session 6.4. The 2nd-derivative test can be carried out by the following MACSYMA function

```
cl. test_der_2(f,x,x0) := block([prederror:false],
          if limit(diff(f,x),x,x0) = 0
            then case(is(limit(diff(f,x,2),x,x0) < 0),
                      [unknown, "Test fails"],
                      [false, case(is(limit(diff(f,x,2),x,x0) > 0),
                                [[false,unknown], "Test fails"],
                                [true, "Local Minimum"])],
                      [true, "Local Maximum"])
          else "Not a stationary point")$
```

a.
c2. `test_der_2(x^2,x,0)`

   `C:\MACSYMA2\share\basic.fas being loaded.`

d2.   Local Minimum

but

c3. `test_der_2(x^3,x,0)`

d3.   Test fails

since the 2nd derivative of x^3 at $x = 0$ is zero. Here, the 1st-derivative test (see Session 6.3) gives a correct answer:

c4. `test_der_1(x^3,x,0,10^-3)`

d4.   Function increasing, not a local extremum

Similarly:

c5. `test_der_2(x^4,x,0)`

d5.   Test fails

Explain why the 2nd-derivative test failed here. Do you expect the 1st-derivative test to do better? Try.

   b. For the objective function $A(x) := x(L - 2x)/2$ of Example 6.14 and $x_0 = L/4$ we find

c6. `test_der_2(x*(L-2*x)/2,x,L/4)`

d6.   Local Maximum

a correct answer, although the parameter L is unspecified.

   c. Consider again the function $f(x) := (x^2 + 2x + 3)/(4x + 5)$ of Sessions 6.2. It has the stationary points $x_1 = -(\sqrt{33} + 5)/4$ and $x_2 = (\sqrt{33} - 5)/4$. We tested these points by the 1st-derivative test in Session 6.3h. Here are the corresponding results for the 2nd-derivative test

c7. `f(x):=(x^2+2*x+3)/(4*x+5)$`

c8. `ans:solve(diff(f(x),x),x)`

d8. $$\left[x = -\frac{\sqrt{33} + 5}{4}, \; x = \frac{\sqrt{33} - 5}{4} \right]$$

c9. `x1:rhs(ans[1])$`

c10. `x2:rhs(ans[2])$`

c11. `test_der_2(f(x),x,x1)`

d11.   Local Maximum

c12. `test_der_2(f(x),x,x2)`

d12.   Local Minimum

   d. Are the following answers correct?

c13. `test_der_2(sin(x),x,0)`

d13.   Not a stationary point

c14. `test_der_2(cos(x),x,0)`

d14.   Local Maximum

Exercises

6.31 Find a positive number x so that $x + 1/x$ is minimal.

6.32 Two electric resistors can be connected in series (Fig. 6.9a) or in parallel (Fig. 6.9b). Electric resistance is a nonnegative quantity, measured in ohm, denoted Ω.

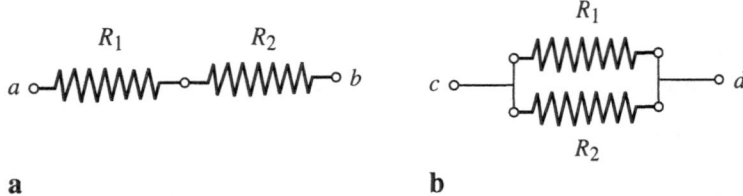

Fig. 6.9. Two resistors connected in series (**a**) or in parallel (**b**); Exercise 6.32

If two resistances R_1 and R_2 are connected in series, they form a combined resistance of $R_1 + R_2$. This is the resistance between the points a and b in Fig. 6.9 a.

The resistance R of a parallel connection (the resistance between the points c and d in Fig. 6.9 b) is zero if either $R_1 = 0$ or $R_2 = 0$, and is otherwise given by $1/R = 1/R_1 + 1/R_2$.

a. Find resistances R_1 and R_2 whose sum is 1000 Ω and which give maximal resistance when connected in parallel.

b. Same with "minimal" instead of "maximal".

c. Find resistances R_1 and R_2 whose parallel resistance is 250 Ω and which give minimal resistance when connected in series.

6.33 Rectangular boxes that are shipped by parcel post have the following size limitation: The sum of length and girth (the circumference of the cross section) must not be longer than 72 in.

Find the box, of maximum volume, that is acceptable.

6.34 Given a line L with equation $Ax + By + C = 0$ and a point $P(x_0, y_0)$, prove that the minimum distance between $P(x_0, y_0)$ and the line L is $|Ax_0 + By_0 + C|/\sqrt{A^2 + B^2}$. Compare with Appendix C, (C.21).

6.35 Prove: If f is twice differentiable in an interval (a, b), and if $f''(x) > 0$ in (a, b), then:

a. every stationary point x_0 in (a, b) is a local minimizer of f;

b. the function f does not have a local maximizer in the (open) interval (a, b);

c. if f is continuous at the endpoints a, b, then:

i. one of the points a, b is the global maximizer of f in the (closed) interval $[a, b]$;

ii. the other endpoint is either a local maximizer of f, or its global minimizer in $[a, b]$.

d. Write the analogs of a–c for $f''(x) < 0$ in (a, b).

Hint: Use Theorem 6.19.

6.36 Use the 2nd-derivative test, if possible, to determine the local maxima and minima for the functions below. If you are unable to apply the 2nd-derivative test, try to determine by other means whether the stationary point is a local extremum.

(a) $f(x) := (x + 1)(x - 2)$ (b) $f(x) := (x + 1)^2(x - 2)^2$

(c) $f(x) := (x + 1)^3(x - 2)^3$ (d) $f(x) := \sin x^2$

(e) $f(x) := \cos x^3$ (f) $f(x) := e^{x^2}$ (g) $f(x) := \sinh(x^2 - 1)$

(h) $f(x) := \ln(x^2 + 1)$ (i) $f(x) := \cos\left(\dfrac{1}{x^2 + 1}\right)$ (j) $f(x) := (x^2 - 9)^{1/3}$

(k) $f(x) := x + \cot x$ (l) $f(x) := \cos x + \sin x$

6.5 Optimal inventory

An inventory is a stock of goods held for future use. Inventory is necessary but it costs money (the money tied up in inventory could be used to earn interest). It is therefore important to set optimal inventory levels. We illustrate this with two examples.

Example 6.24 (Optimal inventory level). An assembly plant produces 500 passenger cars a day, for 360 days a year. A car requires hundreds of parts (e.g., chasis, tires, engine, doors, etc.). The total value of these items is 6,000 [$/car]. Therefore the plant uses each day parts worth $500 \times 6,000 = 3$ million [$].

To assure continuous production, the plant manager has always in stock enough parts for 30 days of production. This means that, at any time, 90 millions of the company money are tied up in inventory.

Assuming an annual interest rate of 10%, these 90 millions cost the company 9 millions [$/year], or $9,000,000/360 = 25,000$ [$/day].[8] Similarly, the *daily cost* of inventory of x working days is

$$C_I(x) = 0.1 \times \frac{3,000,000 \times x}{360} = \frac{2,500}{3}x \text{ [$/day] .}$$

An inventory of 30 production days is perhaps too high. On the other hand, an inventory which is too low (for example, 1 production day) makes it more likely that some mishap (bad weather, strike, etc.) would cause shortage of parts. The plant would then shut down until the missing parts arrive. The cost of such unscheduled shutdown is given as 1 million [$/day].

The *probability P* of shutdown depends on the inventory level. For an inventory of x working days, the shutdown probability is estimated by

$$P(x) = e^{-0.5x}, \quad x \geq 0 .$$

In particular, $P(0) = 1$ (certain shutdown for zero inventory).

The *expected daily shutdown cost*, for an inventory of x working days, is

$$C_S(x) = \text{cost} \times \text{probability} = 1,000,000 \, e^{-0.5x} \text{ [$/day] .}$$

The *expected total daily cost* is thus

$$C(x) = C_S(x) + C_I(x) = 1,000,000 \, e^{-0.5x} + \frac{2,500}{3}x \text{ [$/day] .}$$

It is required to determine the optimal inventory level, minimizing the expected total daily cost.

Solution: Solving

$$C'(x) = -500,000 \, e^{-0.5x} + \frac{2,500}{3} = 0 ,$$

---

8 In other words, a 30-day inventory adds $9,000,000/(500 \times 360) = 50$ [$] to the price of each car.

we get the stationary point (check!) $x = 2\ln(600) \approx 12.7938$ days. This is the global minimizer, by the 2nd-derivative test, since the 2nd-derivative

$$C''(x) = 250,000\,e^{-0.5x} \quad \text{is positive for all } x.$$

If x is required to be an integer, the solution is either $x = 12$ or $x = 13$. It is determined by comparing $C(12)$ and $C(13)$.

Example 6.25 (Optimal order size). A company in the widgets business has annual sales of D widgets. Demand for widgets is uniform (i.e., $D/365$ widgets are demanded every day), and no shortages (rain checks) are allowed.

The company orders the widgets from a supplier. Let x [widgets/order] denote the order size, i.e., the number of widgets in each order. We assume at least one order per year, so that

$$0 < x \le D.$$

Each order arrives when the inventory level is zero, and raises that level to x. The inventory is then depleted at a uniform rate, so the slope of the stock level is $-D$. This is illustrated in Fig. 6.10a and b, for the cases $x = D$ (one order per year) and $x = D/4$ (four orders per year).

The purchase price p [\$/widget] is assumed independent of the order size.[9]

In addition, there is a fixed cost C_O (ordering cost) of placing an order. This cost, measured in \$ per order, is also independent of the order size.[10]

Finally, there is a holding cost C_H [\$/(widget $\times$ year)], the cost of holding one widget in inventory for a whole year. This cost is prorated for quantity and time (for example, the cost of holding 7 widgets for 1 month is $(7/12)C_H$ [\$].

It is required to determine the optimal order size x, so as to minimize the total cost of the company.

Solution: If the order size is x, then the number of orders per year is D/x [orders per year]. The average inventory, throughout the year, is $x/2$ [widgets] (see Fig. 6.10b for the case $x = D/4$, with 4 orders per year).

The *total annual cost* C is the sum of the following three costs

purchase cost $= pD$ [\$/year],

orders cost $= C_O\dfrac{D}{x}$ [\$/year], since there are D/x orders per year,

inventory cost $= C_H\dfrac{x}{2}$ [\$/year], using the average inventory of $x/2$.

These costs are illustrated in Fig. 6.10c. The orders cost is a decreasing function of x, since larger orders require fewer of them. The inventory cost is an increasing function of x. The total cost is then

$$C(x) = pD + C_O\frac{D}{x} + C_H\frac{x}{2}.$$

---

9 There are no quantity discounts, i.e., the company cannot negotiate a lower price p for bigger orders. See also Exercise 6.37d.

10 For example, a truck is needed for each order, whether it consists of 1 or 100 widgets.

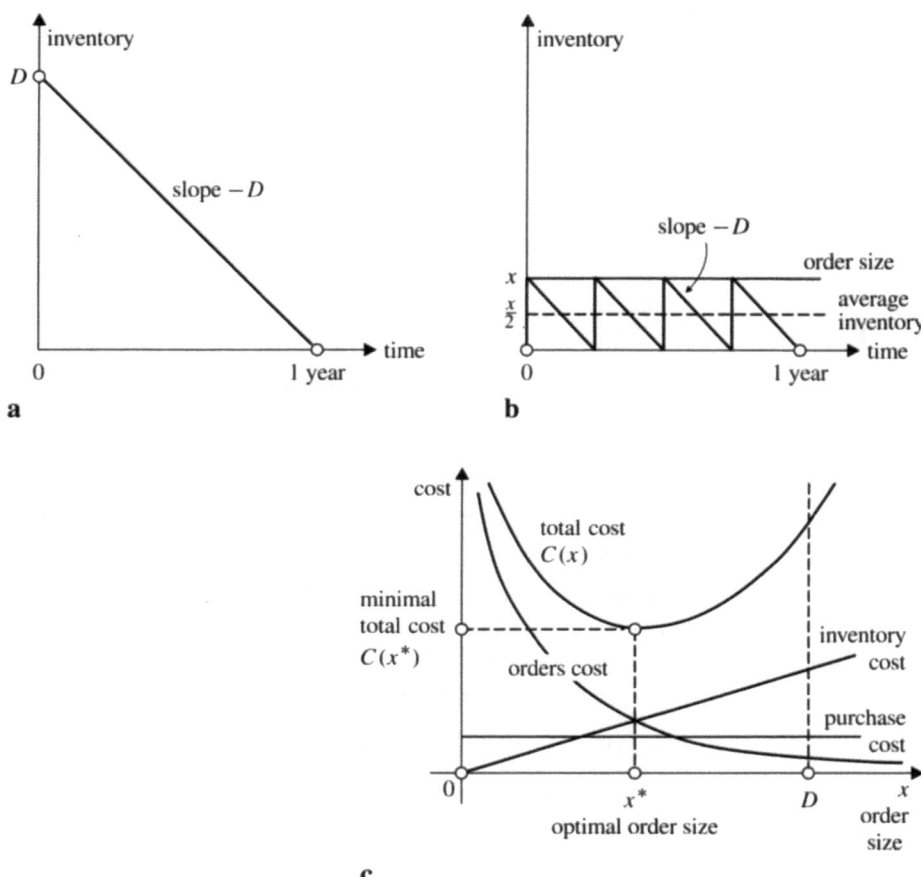

Fig. 6.10 a–c. Inventory. **a** One order per year, $x = D$; **b** four orders per year, $x = D/4$. **c** Costs and optimal solution

Solving

$$C'(x) = -C_O \frac{D}{x^2} + \frac{C_H}{2} = 0 \, ,$$

we get two stationary points $x = \pm\sqrt{(2C_O D)/C_H}$, of which the negative value is not feasible. The 2nd derivative

$$C''(x) = \frac{2C_O D}{x^3} \quad \text{is positive for all } x > 0 \, .$$

This shows (using the 2nd-derivative test) that the stationary point

$$x^* = \sqrt{\frac{2C_O D}{C_H}} \tag{6.10}$$

is the global minimizer of $C(x)$ in the half-line $(0, \infty)$. This point may not be feasible because it may be larger than D. We conclude that the optimal order size is

$$
x^* = \begin{cases} \sqrt{\dfrac{2C_O D}{C_H}} & \text{if } \sqrt{\dfrac{2C_O D}{C_H}} \leq D, \\[4mm] D & \text{if } \sqrt{\dfrac{2C_O D}{C_H}} > D. \end{cases}
$$

Exercises

6.37 Consider the problem of optimal order size in Example 6.25.

    a. Prove that at the stationary point (6.10), the orders cost and the inventory cost are equal.

    b. Determine the optimal order size, and the minimal cost, for the following data

$$p = 5 \text{ [\$ per widget]}, \quad D = 10{,}000 \text{ [widgets per year]}, \quad C_O = 50 \text{ [\$ per order]},$$
$$C_H = 20 \text{ [(percent of } p)/(\text{widget} \times \text{year})]\,.$$

    c. Show that the cost does not change much if the order size differs from the optimal size by 500 [widgets]. Conclude that the cost $C(x)$ is not sensitive to small deviations from the optimal solution (6.10).

    d. Solve b if there is a quantity discount and p depends on x as follows

$$p = p(x) = 5 - 0.0003x \quad \text{[\$ per widget]}\,.$$

6.6 Convexity*

We saw in Sect. 6.4 that, at a stationary point, the sign of the second derivative is the key for deciding whether the point is a local maximizer or local minimizer. In this section we study the case where the second derivative has the same sign throughout an interval.

If $f''(x) \geq 0$ in an interval I, then the first derivative $f'(x)$, which is the same as the *slope* of the tangent to the graph of f, is nondecreasing throughout I. This is illustrated in Fig. 6.11 a–c, where two tangents, L and M, are shown, and the slope of the tangent M is greater because it comes later.

Functions with nondecreasing first derivatives are called *convex* and may be:

- increasing (i.e., the tangents have positive slopes, as in Fig. 6.11 a),
- decreasing (tangents have negative slopes, see Fig. 6.11 b), or
- first decreasing, then increasing, as in Fig. 6.11 c.

Similarly, if the second derivative $f''(x) \leq 0$ in an interval $[a, b]$, then the tangents to f have nonincreasing slopes, as illustrated in Fig. 6.12 a–c. Such functions are called *concave*.

---

    * This section is optional. It can be skipped at first reading. The material covered here is used in few places in the following chapters.

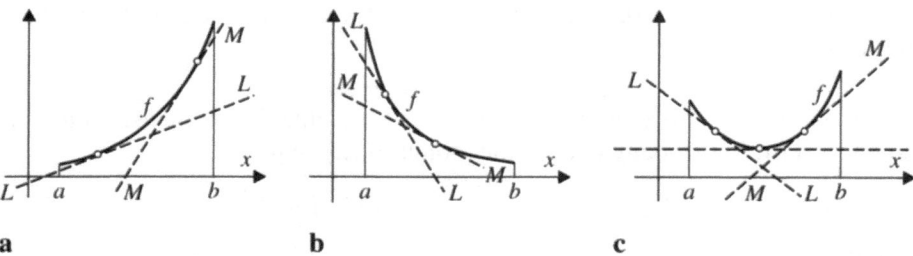

Fig. 6.11 a–c. Convex functions. **a** Increasing function, **b** decreasing function, **c** function with a global mimimum

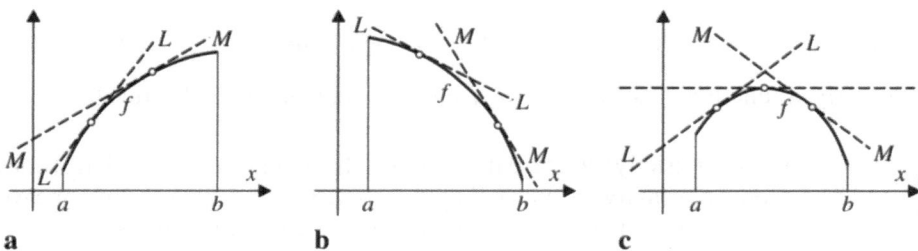

Fig. 6.12 a–c. Concave functions. **a** Increasing function, **b** decreasing function, **c** function with a global maximum

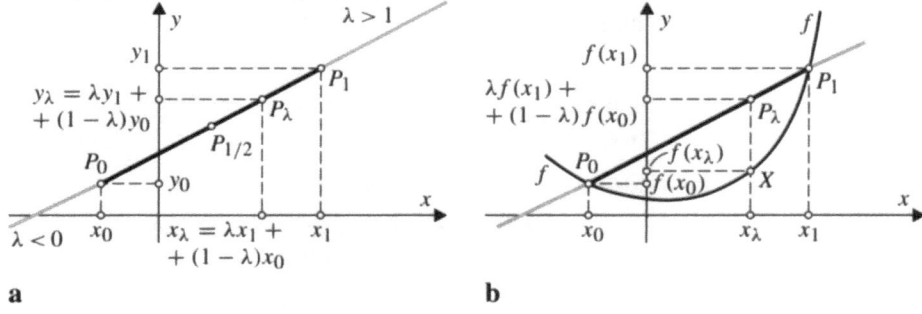

Fig. 6.13 a, b. Illustration of Definition 6.26. **a** The line segment joining two points; **b** the line segment through two points on the graph of a convex function f

Convex and concave functions have many useful properties, some of which are studied in this section.

We give now a general definition of convex functions, which does not require differentiability. First recall that the line segment $\overline{P_0 P_1}$ joining two (different) points $P_0 = (x_0, y_0)$ and $P_1 = (x_1, y_1)$ (Fig. 6.13a) is the set of points $P_\lambda := (x_\lambda, y_\lambda)$ with

$$x_\lambda := \lambda x_1 + (1 - \lambda)x_0, \quad y_\lambda := \lambda y_1 + (1 - \lambda)y_0, \quad \text{for } 0 \le \lambda \le 1. \quad (6.11)$$

In particular, for $\lambda = 0$, 1, and $1/2$ we get P_0, P_1, and their midpoint $P_{1/2}$, respectively.

Definition 6.26 (Convex and concave functions). A function f is called convex in an interval $[a, b]$ if for any two distinct points x_0, x_1 in $[a, b]$,

$$f(\lambda x_1 + (1 - \lambda)x_0) \leq \lambda f(x_1) + (1 - \lambda)f(x_0), \quad \text{for all}^{11}\ 0 < \lambda < 1 . \quad (6.12)$$

The function f is *strictly convex* if the inequality (6.12) is strict

$$f(\lambda x_1 + (1 - \lambda)x_0) < \lambda f(x_1) + (1 - \lambda)f(x_0), \quad \text{for all}\ 0 < \lambda < 1 . \quad (6.13)$$

A function f is concave[12] in $[a, b]$ if the inequality (6.12) is reversed, i.e., for all $a \leq x_0$ and $x_1 \leq b$,

$$f(\lambda x_1 + (1 - \lambda)x_0) \geq \lambda f(x_1) + (1 - \lambda)f(x_0), \quad \text{for all}\ 0 < \lambda < 1 . \quad (6.14)$$

Similarly, a function f is *strictly concave* if the inequality (6.13) is reversed.

> The function f is (strictly) concave if and only if the function $-f$ is (strictly) convex. Therefore "concavity" is covered by "convexity": Any property of concave functions follows from a corresponding property of convex functions.

Remark 6.27. Condition (6.12) has to be checked for all pairs of points x_0 and x_1 in $[a, b]$. This is impractical, except for very simple functions (see Examples 6.28 and 6.29).

If the function is differentiable, then there are more practical tests for checking convexity, see Sects. 6.5 and 6.5.

Example 6.28 (A function both convex and concave). A linear function $f(x) := mx + b$ is both convex and concave. Indeed,

$$\begin{aligned}
f(\lambda x_1 + (1 - \lambda)x_0) &= m(\lambda x_1 + (1 - \lambda)x_0) + b \\
&= m(\lambda x_1 + (1 - \lambda)x_0) + (\lambda b + (1 - \lambda)b) , \\
&\quad (\text{writing } b = \lambda b + (1 - \lambda)b) \\
&= \lambda(mx_1 + b) + (1 - \lambda)(mx_0 + b) \\
&= \lambda f(x_1) + (1 - \lambda)f(x_0) ,
\end{aligned}$$

so both inequalities (6.12) and (6.14) are satisfied (as equalities).

Example 6.29. The function $f(x) := x^2$ is strictly convex on $(-\infty, \infty)$. Indeed, for any two points $x_0 \neq x_1$ and $0 < \lambda < 1$,

---

11 The values $x_0 = x_1$, $\lambda = 0$ or $\lambda = 1$ are uninteresting since these values give a trivial equality in (6.12).

12 Some calculus books use "concave up" for "convex", and "concave down" for "concave".

$$\text{LHS } (6.13) = (\lambda x_1 + (1 - \lambda)x_0)^2$$
$$= \lambda^2 x_1^2 + 2\lambda(1 - \lambda)x_1 x_0 + (1 - \lambda)^2 x_0^2$$
$$\text{RHS } (6.13) = \lambda x_1^2 + (1 - \lambda)x_0^2$$
$$\therefore \quad \text{RHS } (6.13) - \text{LHS } (6.13) = \lambda(1 - \lambda)(x_1^2 - 2x_1 x_0 + x_0^2)$$
$$= \lambda(1 - \lambda)(x_1 - x_0)^2 > 0,$$
$$\text{since } x_1 \neq x_0, 0 < \lambda < 1.$$

Example 6.30. The function $f(x) := |x|$ is convex on $(-\infty, \infty)$. Check that the condition (6.12) is satisfied for all points x_0, x_1. Is this function strictly convex?

Example 6.31. You will prove later (in Exercise 7.34) that the function $f(x) = \log x$ is strictly concave on $(0, \infty)$.

Definition 6.26 is then equivalent to the inequality

$$\log(\lambda u + (1 - \lambda)v) > \lambda \log u + (1 - \lambda) \log v$$
$$\text{for all } u, v > 0, 0 < \lambda < 1. \tag{6.15}$$

Therefore

$$\exp(\log(\lambda u + (1 - \lambda)v)) > \exp(\lambda \log u + (1 - \lambda) \log v),$$

$$\text{since the exponential function } \exp(x) \text{ is increasing},$$

$$= \exp(\log u^\lambda + \log v^{(1-\lambda)}) = \exp(\log(u^\lambda v^{(1-\lambda)}))$$

so that,

$$\lambda u + (1 - \lambda)v \geq u^\lambda v^{1-\lambda} \quad \text{for all } u, v > 0, 0 < \lambda < 1, \tag{6.16}$$

with equality if and only if $u = v$. The inequality (6.16) is called the arithmetic geometric inequality, because the left-hand side is the *arithmetic mean* $\lambda u + (1-\lambda)v$ of u and v, and the right-hand side is their *geometric mean* $u^\lambda v^{1-\lambda}$.

See also Exercise 6.39 for the case $\lambda = 1/2$.

Remark 6.32. To explain the geometric meaning of convexity, consider two points $P_0 = (x_0, f(x_0))$ and $P_1 = (x_1, f(x_1))$ on the graph of f (Fig. 6.13b). Let $0 < \lambda < 1$, and let $x_\lambda := \lambda x_1 + (1 - \lambda)x_0$ be a point between x_0 and x_1. The corresponding point on the graph of f is $X := (x_\lambda, f(x_\lambda))$ with $f(x_\lambda) = \text{LHS } (6.12)$.

The line segment $\overline{P_0 P_1}$ is by (6.11) the set of points (x_λ, y_λ) with

$$x_\lambda := \lambda x_1 + (1 - \lambda)x_0, \quad y_\lambda := \lambda f(x_1) + (1 - \lambda)f(x_0), \quad \text{for } 0 \leq \lambda \leq 1.$$

The point corresponding to x_λ on the segment $\overline{P_0 P_1}$ is therefore

$$P_\lambda := (x_\lambda, \text{RHS}(6.12)).$$

The inequality (6.12) then says that the point X is not above P_λ (Fig. 6.13b).

For a convex function f and any two points $P_0 = (x_0, f(x_0))$ and $P_1 = (x_1, f(x_1))$ on its graph, the line segment $\overline{P_0 P_1}$ is *not below* the graph of f for all x between x_0 and x_1. For a strictly convex f, the line segment connecting any two points on the graph is *strictly above* the graph of f at all intermediate points.

An important property of convex functions is the following: If a convex function is nondecreasing, it cannot later decrease.

Lemma 6.33. Let f be convex in (a, b), let the points x_0 and u be such that

$$a < x_0 < u < b .$$

If

$$f(x_0) \leq f(u) , \tag{6.17}$$

then

$$f(x) \geq f(u) \quad \text{for all } u < x < b . \tag{6.18}$$

Proof. If there is a point x_1 in (u, b) with

$$f(x_1) < f(u) \quad \text{(Fig. 6.14)} , \tag{6.19}$$

then $u = \lambda x_1 + (1 - \lambda)x_0$, for some $0 < \lambda < 1$, since u is between x_0 and x_1. The convexity of f now implies

$$\begin{aligned} f(u) &\leq \lambda f(x_1) + (1 - \lambda) f(x_0) \\ &< f(u), \quad \text{by (6.17) and (6.19), a contradiction .} \end{aligned} \qquad \square$$

Remark 6.34. Figure 6.14 illustrates why a convex function cannot decrease after it has been nondecreasing for a while. Note that the segment $\overline{P_0 P_1}$ goes under the graph of f in the interval (x_0, x_1), violating the convexity condition (see, e.g., Remark 6.32).

Theorem 6.35. Let f be convex. Then any local minimizer of f is a global minimizer.

Proof. If x_0 be a local minimizer of f, then $f(x_0) \leq f(u)$ for some $u > x_0$. Then Lemma 6.33 shows that

$$f(x) \geq f(x_0), \quad \text{for all } x > x_0 .$$

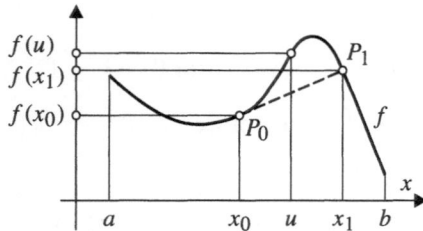

Fig. 6.14. Illustration of Lemma 6.33

The inequality

$$f(x) \geq f(x_0), \quad \text{for all } x < x_0 ,$$

is proved by the same argument, using the function $g(x) := f(-x)$, which is likewise convex. $\qquad\square$

Remark 6.36. The corresponding statement for concave functions is: Any local maximizer is a global maximizer.

The definition of convex functions (Definition 6.26) did not mention "slope" or "differentiability". We show now that the slope of a convex function is a nondecreasing function.

Lemma 6.37. Let the function f be convex, and let $P_0 := (x_0, f(x_0))$, $X := (x, f(x))$ and $P_1 := (x_1, f(x_1))$ be three points on the graph of f, with $x_0 < x < x_1$ (Fig. 6.15).
Then

$$\text{slope } \overline{P_0 X} \leq \text{slope } \overline{P_0 P_1} \leq \text{slope } \overline{X P_1} . \tag{6.20}$$

Conversely, if (6.20) holds for any three points $P_0 = (x_0, f(x_0))$, $X = (x, f(x))$, $P_1 = (x_1, f(x_1))$ on the graph of f, with $x_0 < x < x_1$, then f is convex.
This lemma is proved in Sect. 6.8.

The above facts can be used to prove the following theorem.

Theorem 6.38. If f is convex on the interval $[a, b]$, then the one-sided derivatives $f'_-(x)$ and $f'_+(x)$ exist for all x in (a, b), are nondecreasing functions on (a, b), and satisfy

$$f'_-(x) \leq f'_+(x), \quad \text{for all } x \in (a, b) . \tag{6.21}$$

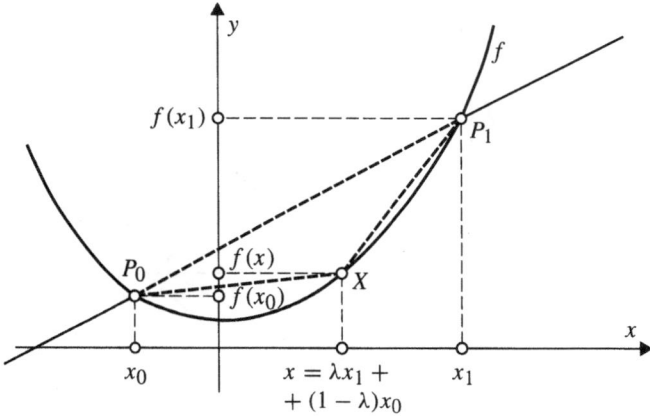

Fig. 6.15. Slopes between three points on the graph of a convex function f

If f is strictly convex on the interval $[a, b]$, then the one-sided derivatives $f'_-(x)$ and $f'_+(x)$ are strictly increasing on (a, b).

This theorem is proved in Sect. 6.8.

Let f be differentiable on (a, b), i.e. $f'_-(x) = f'_+(x) = f'(x)$ for all $a < x < b$. If f is convex (strictly convex) on (a, b), then, by Theorem 6.38, the derivative f' is nondecreasing (strictly increasing) on (a, b).

We will show later that the converse also holds: If f' is nondecreasing (strictly increasing) on (a, b), then f is convex (strictly convex) on (a, b). Summarizing these facts we get the first-derivative test for convexity.

First-derivative test for convexity

Let f be differentiable on (a, b). Then f is convex (strictly convex) on (a, b) if and only if f' is nondecreasing (strictly increasing) on (a, b).

MACSYMA-Session 6.5. To illustrate Definition 6.26, and the 1st-derivative test for convexity, use the following MACSYMA functions:

```
c1. tangent(f,x,x0) := limit(f,x,x0)+(x-x0)*limit(diff(f,x),x,x0)$
c2. secant(f,x,x1,x2) := limit(f,x,x1)+(x-x1)*(limit(f,x,x2)
                             -limit(f,x,x1))/(x2-x1)$
```

The function in c2 computes the secant line joining the two points $(x_1, f(x_1))$ and $(x_2, f(x_2))$ on the graph of f.

We now make a list of f and the secant of f joining $(a, f(a))$ and $(b, f(b))$, and concatenate it with a list of tangent lines at the $n + 1$ points

$$a,\ a + \frac{b-a}{n},\ a + \frac{2(b-a)}{n},\ \ldots,\ a + \frac{(n-1)(b-a)}{n},\ b$$

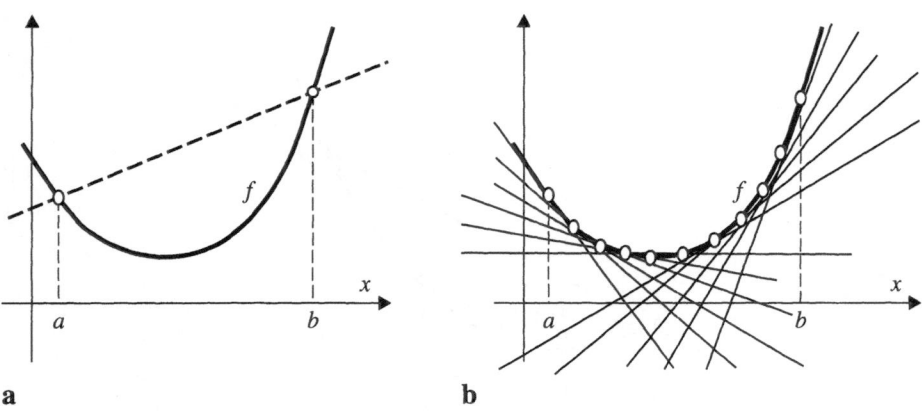

a   **b**

Fig. 6.16 a, b. Increasing slopes of tangents of a convex function. **a** Graph of f and the line joining two points on it. **b** The tangents have increasing slopes

```
c3. illustrate(f,x,a,b,n) := append([f,secant(f,x,a,b)],
                        makelist(tangent(f,x,a+k/n*(b-a)),k,0,n))$
```
If the function f is convex in $[a, b]$, plotting the list `illustrate(f,x,a,b,n)` may look like Fig. 6.16b, showing tangent lines with increasing slopes.

If the function f is not convex in $[a, b]$, we may be able to see this in the plot. We may observe, for example, that the slopes of the tangent lines are not increasing. For example,
```
c4. plot2(''(illustrate(x^3,x,0,1,10)),x,-.2,1.2)$
```
shows increasing slopes (Fig. 6.17), since x^3 is convex for $x \geq 0$. Similarly,
```
c5. plot2(''(illustrate(x^3,x,-1,0,10)),x,-1.2,.2)$
```
shows decreasing slopes (Fig. 6.18), since x^3 is concave for $x \leq 0$.

Finally, the function x^3 is neither convex nor concave in $[-1, 1]$. This is illustrated in Fig. 6.19, obtained by
```
c6. plot2(''(illustrate(x^3,x,-1,1,10)),x,-1.2,1.2)$
```
You can generate some pretty pictures by making n as large as is practically possible. See, for example, Fig. 6.20 obtained by
```
c7. plot2(''(illustrate(1/x,x,1/3,3,40)),x,1/10,4)$
```

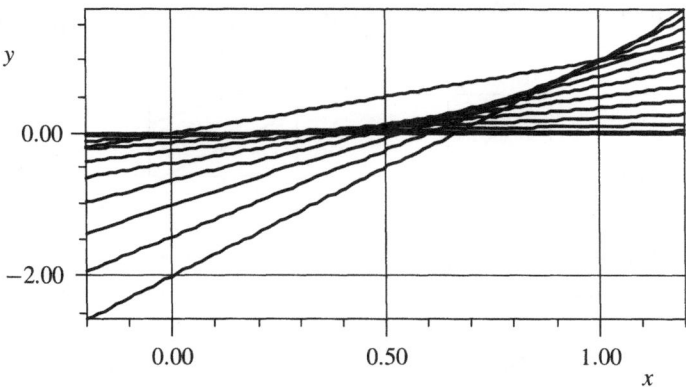

Fig. 6.17. Graph generated in Session 6.5, line c4

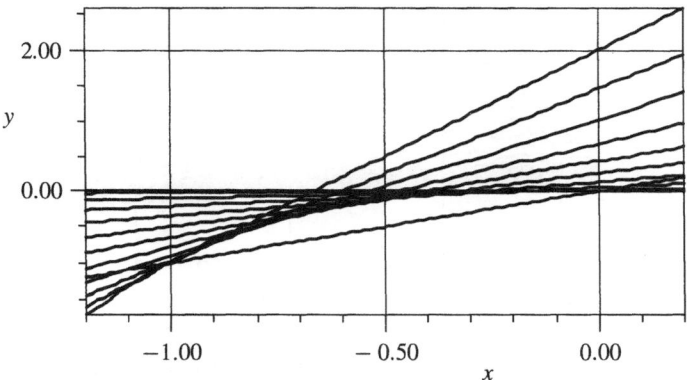

Fig. 6.18. Graph generated in Session 6.5, line c5

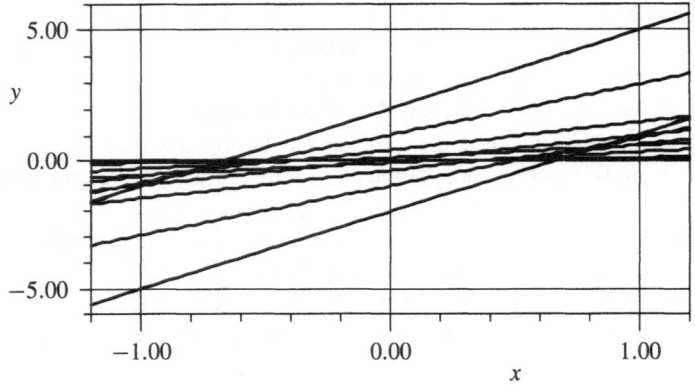

Fig. 6.19. Graph generated in Session 6.5, line c6

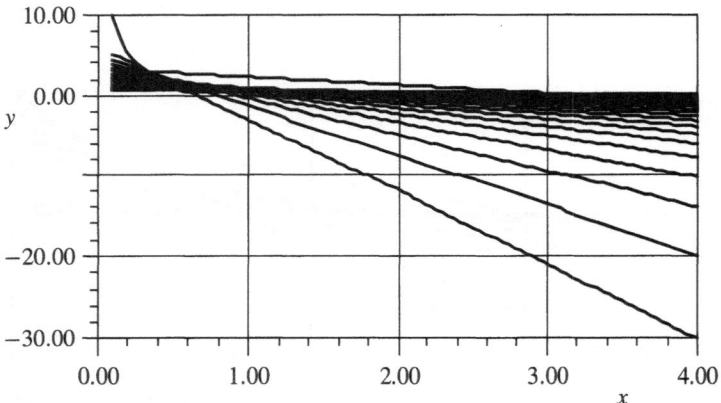

Fig. 6.20. Graph generated in Session 6.5, line c7

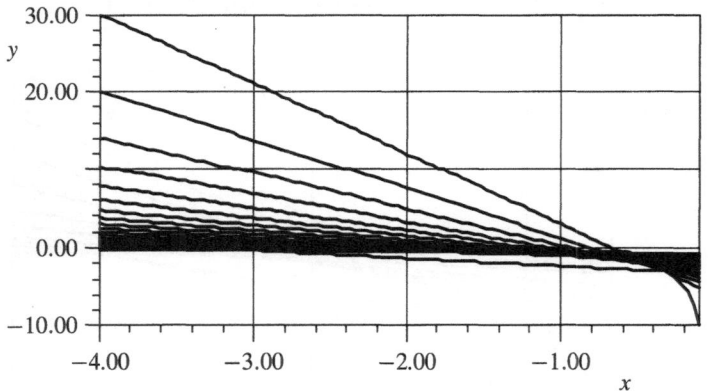

Fig. 6.21. Graph generated in Session 6.5, line c8

and Fig. 6.21 produced by
```
c8. plot2('' (illustrate(1/x,x,-3,-1/3,40)),x,-4,-1/10)$
```

Let f be twice differentiable. Then the first derivative f' is nondecreasing if and only if the second derivative f'' is nonnegative (see Corollary 7.5). We can therefore restate the above test in terms of the second derivative.

Second-derivative test for convexity

Let f be twice differentiable on (a, b). Then f is:
a.   convex on (a, b) if and only if $f''(x) \geq 0$, for all $a < x < b$;
b.   strictly convex on (a, b) if $f''(x) > 0$, for all $a < x < b$.

This test is justified in Corollary 7.32.

Remark 6.39. We cannot say more in part b since it is possible for a strictly increasing function to have derivative zero at some point(s). See Example 6.41 c, for a strictly convex function f with $f''(x_0) = 0$ at some point x_0.

Remark 6.40. Concave (strictly concave) functions have analogous statements. If f is twice differentiable on (a, b), then f is concave (strictly concave) on (a, b) if and only if $f''(x) \leq 0$ (if $f''(x) < 0$) for all $a < x < b$.

Example 6.41 (Illustrations of the 2nd-derivative test).
   a.   The linear function $f(x) = mx + b$ has second derivative $f''(x) = 0$ and therefore f is both convex and concave on $(-\infty, \infty)$.
   b.   The function $f(x) = x^2$ has $f''(x) = 2$ and therefore f is strictly convex on $(-\infty, \infty)$.
   c.   Same conclusion for $f(x) = x^n$ with n even and positive. Note however that for even integers $n > 2$ the second derivative of x^n vanishes at $x_0 = 0$, a global minimizer.
   d.   The function $f(x) = x^3$ is strictly convex (strictly concave) for all $x > 0$ $(x < 0)$.
   e.   The functions $f(x) = e^x$ and $f(x) = e^{-x}$ have $f''(x) = f(x) > 0$ for all x. Therefore e^x and e^{-x} are strictly convex on $(-\infty, \infty)$.
   f.   The function $f(x) = \log x$ has $f''(x) = -1/x^2$ and therefore f is strictly concave for all $x > 0$.

Points where a function goes from "convex" to "concave", or vice versa, are of sufficient interest to deserve a special name.

Definition 6.42. The point x_0 is called a point of *inflection* of the function f if
a. f is continuous at x_0 and
b.   for some points $a < x_0 < b$, the function f is convex in (a, x_0) and concave in (x_0, b), or vice versa.

Example 6.43.

  a. The function $f(x) = x^3$ has $x_0 = 0$ as its single point of inflection, see Example 6.41 d. The second derivative $f''(x) = 6x$ changes sign at $x_0 = 0$.

  b. The mere vanishing of $f''(x_0)$ does not imply that x_0 is an inflection point. For example, $f(x) := x^4$ has no inflection points, but $f''(0) = 0$.

  c. The function $f(x) := x^{1/3}$ is convex in $(-\infty, 0)$, concave in $(0, \infty)$, and continuous at 0 (and everywhere). Therefore $x_0 = 0$ is a point of inflection. Note that $x^{1/3}$ is not differentiable at 0.

Exercises

6.38 On any interval I, the linear functions $f(x) = mx + b$ are the only functions that are both convex and concave on I (see Example 6.28). Prove: If f is convex on (a, b), and if for any two points $a \le x_0 < x_1 \le b$ and some $0 < \lambda < 1$,

$$f(\lambda x_1 + (1 - \lambda)x_0) = \lambda f(x_1) + (1 - \lambda)f(x_0) \,,$$

then f is linear on (x_0, x_1).

6.39 For $\lambda = 1/2$ the arithmetic geometric inequality (6.16) gives, for all $u, v > 0$,

$$(u + v)/2 \ge \sqrt{uv}, \quad \text{with equality iff } u = v \,. \tag{6.22}$$

Prove (6.22).
Hint: Expand $(\sqrt{u} - \sqrt{v})^2$, and note that it is nonnegative for all positive u, v, and zero iff $u = v$.

6.40 Prove: If f and g are convex, and f is nondecreasing, then $f(g(x))$ is convex.
Hint: By the chain rule, the second derivative of the composition $f \circ g$ is

$$(f \circ g)''(x) = f''(g(x)) \, g'(x)^2 + f'(g(x)) \, g''(x) \,.$$

6.41 Determine open intervals on which the following functions are (i) increasing, or (ii) decreasing, and where they are (iii) convex, or (iv) concave

  (a) $f(x) := x^3 + 3x^2 + 3x + 1$ (b) $f(x) := e^x$

  (c) $f(x) := \ln x + 1$ (d) $f(x) := x^{3/2} - x^{1/2}$ (e) $f(x) := x^{5/4} - x^{1/4}$

  (f) $f(x) := \tan x$ (g) $f(x) := \sin x$ (h) $f(x) := \sinh x$

  (i) $f(x) := \cosh x$ (j) $f(x) := \sin^2 x$ (k) $f(x) := (x - 3)^4$

  (l) $f(x) := 1/(x^2 + 1)$ (m) $f(x) := ax^2 + bx + c$, where $a \ne 0$

  (n) $f(x) := (ax^2 + bx + c)^2$ (o) $f(x) := \begin{cases} -x & \text{for } x \le 0, \\ x^2 & \text{for } x \ge 0 \end{cases}$

  (p) $f(x) := \begin{cases} 6 - x & \text{for } x \le 3, \\ x^2 - 3 & \text{for } x \ge 3 \end{cases}$

6.42 Give an example of strictly convex functions f and g such that $f(g(x))$ is not convex.

6.43 Plotting the results of the function `illustrate(f,x,a,b)` of MACSYMA-Session 6.5 suggests that the graph of a convex (concave) function lies above (below) its tangents. Illustrate this for:

  (a) `illustrate(exp(x),x,-2,1,50)` (b) `illustrate(log(x),x,0.5,2,50)`

6.7 Analysis of graphs

"Analysis of graphs" means the study of facts and properties needed to draw the graph of a given function, or to understand the graph plotted by MACSYMA.

Although MACSYMA can plot very well, we often need information about important points and important intervals (e.g., local extrema, intervals of convexity, etc.), which is difficult to read from a graph. Such items have to be computed.

Also, the MACSYMA graphs are necessarily limited to finite intervals on the x- and y-axes, and cannot tell us about the asymptotic behavior of the function, i.e., its behavior for large values of $|x|$ and $|y|$. Here is where analysis is needed.

Important points

The important points on the graph of a function f include its

– zeros: points where $f(x) = 0$, i.e., the graph meets the x-axis;
– poles: points where f is undefined, or "blows up" (or "down");
– points of discontinuity: points where the function is not continuous;
– stationary points: points where $f'(x) = 0$;
– critical points: stationary points and points where f is not differentiable;
– local extrema: local maxima and local minima;
– global extrema: global maxima and global minima;
– points of inflection: points where the function "changes convexity" (see Definition 6.42).

These points are not present in all graphs. For example, a continuous function has no points of discontinuity, and a function monotone throughout **R** has no local extrema. Some graphs have infinitely many points from the above list. Such

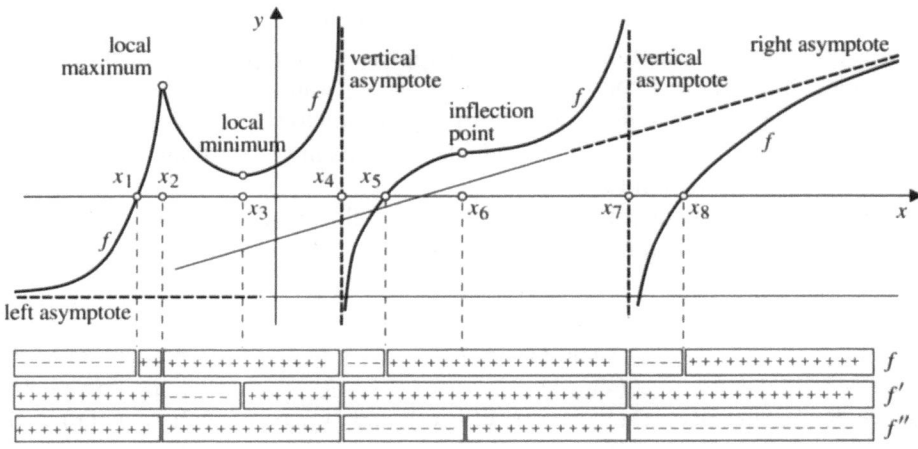

Fig. 6.22. Analysis of a graph. $+$, positive value; $-$, negative value of function f, its first derivative f', or its second derivative f''

example is $f(x) := x^2 \sin(1/x)$ which has infinitely many zeros, stationary points, local maxima, local minima, and points of inflection.

In Fig. 6.22 we have

zeros	x_1, x_5, x_8	poles	x_4, x_7
stationary points	x_3	critical points	x_2, x_3
local maxima	x_2	local minima	x_3
global extrema	none	points of inflection	x_6

Some of the "important points" are determined by solving equations. For example:
-   the zeros of f are the solutions of $f(x) = 0$. We can use here the MACSYMA function `zeros(f,x):=solve(f=0,x);`
-   the poles of a rational function $p(x)/q(x)$ are the zeros of the polynomial q (if the polynomials p and q have no common factors);
-   the stationary points of f are the solutions of $f'(x) = 0$, and can be computed by the function `stationary_points(f,x):=solve(diff(f,x)=0,x)` of Session 6.2.

Other "important points" require extra work. For example, local maxima and local minima are determined by the tests in Sects. 6.3 and 6.4.

Important intervals

The important intervals for a graph of a function f include:
-   intervals of increase, where the function f is increasing – if f is differentiable, then $f'(x) \geq 0$ on intervals of increase;
-   intervals of decrease, where f is decreasing – if f is differentiable, then $f'(x) \leq 0$ on intervals of decrease;
-   intervals of convexity, where f is convex – if f is twice differentiable, the 2nd derivative $f''(x) \geq 0$ in intervals of convexity;
-   intervals of concavity, where f is concave – if f is twice differentiable, the 2nd derivative $f''(x) \leq 0$ in intervals of concavity.

These intervals are sometimes separated by the important points of Sect. 6.7. For example, convexity and concavity intervals are separated by points of inflection. Intervals of increase and decrease are usually separated by local extrema, but not always, as shown by $f(x) := 1/x^2$, which increases in $(-\infty, 0)$ and decreases in $(0, \infty)$, but has no extremum at 0.

The "important intervals" are determined by the signs of the function f and its first derivative f' and its second derivative f'', if they exist.

In Fig. 6.22 we have

intervals of increase	$(-\infty, x_2)$, (x_3, x_4), (x_4, x_7), and (x_7, ∞)	where $f'(x) > 0$,
intervals of decrease	(x_2, x_3)	where $f'(x) < 0$,
intervals of convexity	$(-\infty, x_2)$, (x_2, x_4), and (x_6, x_7)	where $f''(x) > 0$,
intervals of concavity	(x_4, x_6) and (x_7, ∞)	where $f''(x) < 0$.

These signs can occur in all combinations. For example, $f' > 0$ and $f'' < 0$ indicate increase *and* concavity, see interval (x_4, x_6) in Fig. 6.22.

Asymptotes

Two functions f and g are said to be asymptotes at $+\infty$ if

$$\lim_{x\to\infty} (f(x) - g(x)) = 0 .$$

This means that f and g are indistinguishable as $x \to \infty$. Asymptotes at $-\infty$ are defined similarly using $\lim_{x\to-\infty}$.

Example 6.44. The functions

$$f(x) = \frac{x^7 - 3x^6 + 3x^5 - 4x^4 + 5x^3 - 3x^2 + 3x - 2}{x^4 - 2x^3 + x^2 - 4x + 4}$$

and $g(x) = x^3 - x^2 + 1$

are asymptotes at $\pm\infty$, since

$$\frac{x^7 - 3x^6 + 3x^5 - 4x^4 + 5x^3 - 3x^2 + 3x - 2}{x^4 - 2x^3 + x^2 - 4x + 4}$$
$$= x^3 - x^2 + 1 - \frac{x+3}{x^2 + x + 2}, \qquad \text{see (1.36)},$$

and

$$\lim_{x\to\pm\infty} \frac{x+3}{x^2+x+2} = \lim_{x\to\pm\infty} \frac{1/x + 3/x^2}{1 + 1/x + 2/x^2} = 0$$

Remark 6.45. Clearly, $f(x)$ and $f(x)+\varepsilon(x)$ are asymptotes at ∞ if $\lim_{x\to\infty} \varepsilon(x)$ $= 0$. Any function defined throughout **R** has therefore infinitely many asymptotes at ∞. However, a function may have at ∞ *at most* one linear asymptote

$$y = mx + b .$$

Definition 6.46. The line $y = mx + b$ is called the *right asymptote* (or asymptote at ∞) of f if

$$\lim_{x\to\infty} (f(x) - mx - b) = 0 .$$

The *left asymptote* (or asymptotes at $-\infty$) is defined similarly using the limit $\lim_{x\to-\infty}$.

Figure 6.22 shows left and right asymptotes. The left asymptote happens to be horizontal.

Lemma 6.47. The line $y = mx + b$ is the right asymptote of f if

$$m = \lim_{x\to\infty} \frac{f(x)}{x}, \quad b = \lim_{x\to\infty} (f(x) - mx) . \tag{6.23}$$

The m and b of the left asymptote are computed similarly using the limit $\lim_{x\to-\infty}$.

MACSYMA-Session 6.6. We compute the right asymptote of f, if it exists. The
MACSYMA functions

```
c1. asymptote_right_m(f,x)  := limit(f/x,x,inf)$
c2. asymptote_right_b(f,x):= limit(f-asymptote_right_m(f,x)*x,x,inf)$
```

compute the m and b using (6.23). If these limits exist and are finite, the right asymptote
$y = mx + b$ is computed by

```
c3. asymptote_right(f,x)  := asymptote_right_m(f,x)*x
                            + asymptote_right_b(f,x)$
```

The analogous expressions for the left asymptote are:

```
c4. asymptote_left_m(f,x)  := limit(f/x,x,minf)$
c5. asymptote_left_b(f,x)  := limit(f-asymptote_left_m(f,x)*x,x,minf)$
c6. asymptote_left(f,x)  := asymptote_left_m(f,x)*x
                            + asymptote_left_b(f,x)$
```

a. The right asymptote of the rational function $f(x) := (x^2 + 2x + 3)/(4x + 5)$ is
found as

```
c7. asymptote_right((x^2+2*x+3)/(4*x+5),x)
```

d7. $\dfrac{x}{4} + \dfrac{3}{16}$

which is the same as the left asymptote

```
c8. asymptote_left((x^2+2*x+3)/(4*x+5),x)
```

d8. $\dfrac{x}{4} + \dfrac{3}{16}$

see Exercise 6.45.

b. The left asymptote of $\exp(x)$ is $y = 0$:

```
c9. asymptote_left(exp(x),x)
```

d9. 0

while the right asymptote does not exist; our current definition returns

```
c10. asymptote_right(exp(x),x)
```

d10. $\infty x + \infty$

which does not make sense.

To correct such cases, we need to filter out instances where `limit` returns minf, inf,
und, or ind or where it cannot find the limit.

Re-doing the right asymptote with a test for these cases, we could write

```
c11. asymptote_right(f,x)  :=
        block([m:asymptote_right_m(f,x), b:asymptote_right_b(f,x)],
              if freeof(minf,inf,und,ind,nounify('limit),[m,b])
                  then m*x+b
              else "Asymptote does not exist or could not be found")$
```

where the function `freeof(it1,it2,...,itn,thing)` checks to see if any of the items
`it1...itn` appear in `thing`, returning "false" if any items appear and "true" if no items
appear.

Here `nounify('limit)` checks for limits that MACSYMA is unable to evaluate. As
long as both m and b do not contain any of the items, the answer $mx + b$ is returned;
otherwise either the asymptote does not exist or we cannot not determine it with
`asymptote_right`.

Until now, "asymptotes" had to do with behavior of the function as $x \to \infty$

or $x \to -\infty$. We see now another kind of "asymptotes", describing "blow-ups" of the function values.

Definition 6.48. The line $x := c$ is called a *vertical asymptote* of f if any one of the one-sided limits

$$\lim_{x \to c^-} f(x), \quad \lim_{x \to c^+} f(x)$$

is infinite.

Figure 6.22 shows vertical asymptotes $x = x_4$ and $x = x_7$.

Exercises

6.44 Compute the right and left asymptotes of
  (a) $f(x) := \sqrt{x^2}$ (b) $f(x) := \sqrt{x^2 - 1}$ (c) $f(x) := \sqrt{x^2 - 10^6}$
  and explain the results.

6.45 Consider a rational function $r(x) = p(x)/q(x)$.
  a. Prove: The function $r(x)$ has linear asymptotes at $\pm\infty$ if the degrees of the polynomials p and q satisfy

  $$\text{degree } p = \text{degree } q + 1 . \tag{$*$}$$

  b. If $(*)$ holds, are the left and right asymptotes the same?
  c. Consider the function $f(x) = 1/x$. Is it rational? Does it satisfy $(*)$? Does it have a linear asymptote at $\pm\infty$?

6.46 Prove the statement in Remark 6.45: A function f can have at most one linear asymptote $y = mx + b$ at ∞.
  Hint: Imitate the steps used to prove that a function can have at most one linear tangent at a point x_0 (see Corollary 4.5).

6.47 Find the asymptotes of the following hyperbolas
  (a) $\dfrac{x^2}{a^2} - \dfrac{y^2}{b^2} = 1$ (b) $-\dfrac{x^2}{a^2} + \dfrac{y^2}{b^2} = 1$ (c) $xy = 1$
  (d) $(x - 1)(y + 2) = -3$

6.48 Find all asymptotes and sketch the graphs of the following functions.
  (a) $f(x) := \dfrac{3x - 2}{4x - 1}$ (b) $f(x) := 2 - \dfrac{1}{x^3}$ (c) $f(x) := \dfrac{x + 1}{(x - 1)^2}$
  (d) $f(x) := \dfrac{x^3}{(x + 1)^2}$ (e) $f(x) := \dfrac{x^2 + 2x + 1}{4x + 1}$ (f) $f(x) := \dfrac{x + 1}{(x - 1)^2}$

6.49 Sketch the graphs of the following functions. Label the stationary and inflection points and show any horizontal or vertical asymptotes.
  (a) $f(x) := \dfrac{x + 1}{x - 1}$ (b) $f(x) := \dfrac{2x + 7}{(x + 5)^2}$ (c) $f(x) := \dfrac{1}{x^2 + 1}$
  (d) $f(x) := \dfrac{3x^2 + 4x + 1}{x}$ (e) $f(x) := x^2 + \dfrac{1}{x^2}$
  (f) $f(x) := \dfrac{x^4 + x^3 + 5x^2 + x + 1}{(x - 1)(x - 2)(x + 3)}$

6.50 Sketch the graphs of the following functions.

(a) $f(x) := x^{2/3}$ (b) $f(x) := \sqrt{x^2 - 1}$ (c) $f(x) := 3 + x^{4/3}$

(d) $f(x) := x - \sin x$ (e) $f(x) := \dfrac{x + \tan x}{x}$ (f) $f(x) := \sin x + \cos x$

(g) $f(x) := \sin^2 x$ (h) $f(x) := |\cos x|$ (i) $f(x) := (x - 1)^{2/3}$

(j) $f(x) := x \csc x$ (k) $f(x) := \sin |x|$ (l) $f(x) := \tan |x|$

6.51 Compare the graphs of $f(x)$, $|f(x)|$, and $f(|x|)$.

6.8 Proofs

Proof of Lemma 6.37. Since the point x is between x_0 and x_1, we have

$$x = \lambda x_1 + (1 - \lambda)x_0, \quad \text{for some } 0 < \lambda < 1,$$

(Fig. 6.15).

The slopes of the three line segments $\overline{P_0 P_1}$, $\overline{P_0 X}$, and $\overline{X P_1}$ are then:

$$\text{slope } \overline{P_0 P_1} = \frac{f(x_1) - f(x_0)}{x_1 - x_0},$$

$$\text{slope } \overline{P_0 X} = \frac{f(x) - f(x_0)}{x - x_0} = \frac{f(\lambda x_1 + (1 - \lambda)x_0) - f(x_0)}{\lambda(x_1 - x_0)},$$

$$\text{slope } \overline{X P_1} = \frac{f(x_1) - f(x)}{x_1 - x} = \frac{f(x_1) - f(\lambda x_1 + (1 - \lambda)x_0)}{(1 - \lambda)(x_1 - x_0)}.$$

Subtracting $f(x_1)$ from both sides of the inequality (6.12) we get, after some arithmetic,

$$f(x_1) - f(\lambda x_1 + (1 - \lambda)x_0) \geq f(x_1) - \lambda f(x_1) - (1 - \lambda)f(x_0),$$

or

$$\frac{f(x_1) - f(\lambda x_1 + (1 - \lambda)x_0)}{(1 - \lambda)(x_1 - x_0)} \geq \frac{f(x_1) - f(x_0)}{x_1 - x_0},$$

with strict inequality if f is strictly convex.

Similarly, subtracting $f(x_0)$ from both sides of (6.12) we get

$$\frac{f(x_1) - f(x_0)}{x_1 - x_0} \geq \frac{f(\lambda x_1 + (1 - \lambda)x_0) - f(x_0)}{\lambda(x_1 - x_0)},$$

with strict inequality if f is strictly convex. The last two inequalities prove (6.20), that

$$\text{slope } \overline{P_0 X} \leq \text{slope } \overline{P_0 P_1} \leq \text{slope } \overline{X P_1},$$

with strict inequalities if f is strictly convex.

All the above steps can be reversed. If f is a function such that for any three points $P_0 = (x_0, f(x_0))$, $X = (x, f(x))$, $P_1 = (x_1, f(x_1))$ on its graph, with $x_0 < x < x_1$, the slopes satisfy (6.20), then the function satisfies (6.12), i.e., f is convex. $\square$

Proof of Theorem 6.38. We prove only the "convex" case, leaving the "strictly convex" case as exercise. Let

$$P_0 = (x_0, f(x_0)), \ U = (u, f(u)), \ X = (x, f(x)),$$
$$V = (v, f(v)), \ P_1 = (x_1, f(x_1)),$$

be any five points on the graph of f with

$$a \le x_0 < u < x < v < x_1 \le b.$$

Then the inequality (6.20) can be used to deduce

$$\text{slope } \overline{P_0 U} \le \text{slope } \overline{P_0 X} \le \text{slope } \overline{UX} \le \text{slope } \overline{XV} \le \text{slope } \overline{XP_1} \le \text{slope } \overline{VP_1},$$

or

$$\frac{f(u) - f(x_0)}{u - x_0} \le \frac{f(x) - f(x_0)}{x - x_0} \le \frac{f(x) - f(u)}{x - u} \le \frac{f(v) - f(x)}{v - x} \le$$
$$\le \frac{f(x_1) - f(x)}{x_1 - x} \le \frac{f(x_1) - f(v)}{x_1 - v} \tag{6.24}$$

for all $a \le x_0 < u < x < v < x_1 \le b$. We extract from (6.24)

$$\frac{f(u) - f(x)}{u - x} \le \frac{f(v) - f(x)}{v - x} \quad \text{for all } a < u < x < v < b, \tag{6.25}$$

and conclude that the function $\phi(u) := \text{LHS}(6.25)$ increases as $u \uparrow x$, and the function $\psi(v) := \text{RHS}(6.25)$ decreases as $v \downarrow x$. One-sided limits of monotone functions exist by Example 3.23. Therefore

$$f'_-(x) := \lim_{u \uparrow x} \frac{f(u) - f(x)}{u - x} \quad \text{and} \quad f'_+(x) := \lim_{v \downarrow x} \frac{f(v) - f(x)}{v - x}$$

both exist. That $f'_-(x)$ and $f'_+(x)$ satisfy (6.21) follows then from (6.25).

Next we extract from (6.24) the inequality

$$\frac{f(x_0) - f(u)}{x_0 - u} \le \frac{f(u) - f(x)}{u - x} \quad \text{for all } a \le x_0 < u < x < b,$$

and take limits

$$\lim_{x_0 \uparrow u} \frac{f(x_0) - f(u)}{x_0 - u} \le \lim_{u \uparrow x} \frac{f(u) - f(x)}{u - x}$$

to conclude that $f'_-(u) \le f'_-(x)$ for all $u < x$. That $f'_+(x)$ is nondecreasing is similarly proved. $\qquad \square$

7 Mean value theorem

The derivative of a function f at a point ξ

$$f'(\xi) = \lim_{\Delta x \to 0} \frac{f(\xi + \Delta x) - f(\xi)}{\Delta x} \,,$$

is the slope of the line tangent to the graph of f at the point $P = (\xi, f(\xi))$. Restricting to $\Delta x > 0$ we see that $f'(\xi)$ is the limit of the slopes of secants PQ, as $Q \to P$ from the right (Fig. 7.1 a).

Similarly, for $\Delta x < 0$ we get $f'(\xi)$ as the limit of slopes of secants RP, as $R \to P$ from the left (Fig. 7.1 b).

These are *local* results, for points which approach P in the limit. A related question of a *global* nature is:

Let $A = (a, f(a))$ and $B = (b, f(b))$ be any two points on the graph of a differentiable function f.
Is there a point ξ, $a < \xi < b$, where the derivative $f'(\xi)$ equals the slope of the secant AB? See Fig. 7.1 c.

The answer "yes" is known as the mean value theorem. It is one of the most important results in calculus (simple questions may have profound answers).

In this chapter we study the mean value theorem, and some of its consequences and applications. The applications selected here include the rule of l'Hospital (in Sect. 7.2), Taylor's theorem (Sect. 7.3) and antiderivatives (Sect. 7.4). The mean value theorem is also used in the study of iterative methods (Sect. 7.5), in particular Newton's method (Sect. 7.6), and fixed points (Sect. 7.7).

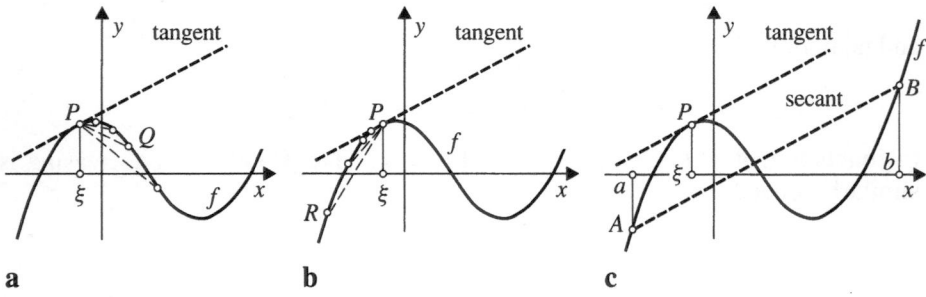

a   **b**   **c**

Fig. 7.1 a–c. Tangents and secants. **a** The derivative at P is the limit of the slopes of the secants PQ as $Q \to P$. **b** The derivative at P is the limit of the slopes of the secants RP as $R \to P$. **c** A secant AB and an "intermediate" point P where the derivative equals the slope of AB

7.1 Mean value theorem

The mean value theorem has several versions, the easiest to state and prove is the following result due to Rolle[1].

Theorem 7.1 (Rolle's theorem). If f is continuous in the closed interval $[a, b]$, differentiable in the open interval (a, b), and $f(a) = f(b)$, then there is at least one $\xi \in (a, b)$ where $f'(\xi) = 0$.

Proof. If f is constant, then the result is obviously true. Otherwise there are points $x \in (a, b)$ where either $f(x) > f(a)$ or $f(x) < f(a)$. In the first (second) case, the function f attains its maximum (minimum) in $[a, b]$ (see Theorem 3.52 at an *interior* point ξ). The necessary condition of Theorem 6.7 then guarantees that $f'(\xi) = 0$.   □

The geometrical interpretation of Rolle's theorem is: Given a closed interval, and a differentiable function f with equal values at the endpoints, there is (at least) one interior point of the interval where the graph of f has a horizontal tangent (Fig. 7.2 a).

Similarly, between any two points A and B on the graph of a differentiable function, there is at least one point of the graph where the slope of the tangent line equals the slope of the line AB (see the points ξ_1 and ξ_2 in Fig. 7.2 b). This result, due to Lagrange, is stated formally as follows.

Theorem 7.2 (Mean value theorem). If f is continuous in the closed interval $[a, b]$ and differentiable in the open interval (a, b), then there is at least one $\xi \in (a, b)$ with

$$\frac{f(b) - f(a)}{b - a} = f'(\xi) . \tag{7.1}$$

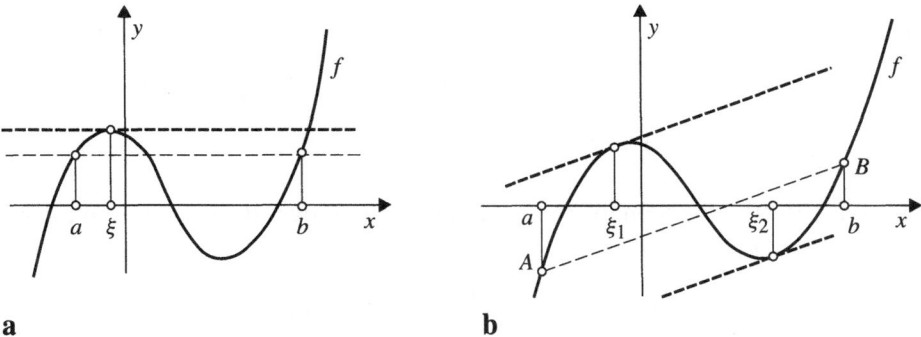

a   **b**

Fig. 7.2. Illustration of Rolle's theorem (**a**) and mean value theorem (**b**)

---

1 Michel Rolle (1652–1719).

Proof. Let

$$h(x) := f(x) - \frac{f(b) - f(a)}{b - a}(x - a) .$$

Then

$$h(a) = f(a), \quad h(b) = f(b) - \big(f(b) - f(a)\big) = f(a)$$
$$\text{and} \quad h'(x) = f'(x) - \frac{f(b) - f(a)}{b - a} .$$

By Rolle's theorem there is a point $\xi \in (a, b)$ with

$$h'(\xi) = f'(\xi) - \frac{f(b) - f(a)}{b - a} = 0$$

which proves the theorem. $\square$

The following version of the mean value theorem, due to Cauchy[2], is often useful.

Theorem 7.3 (Cauchy mean value theorem). If f and g are continuous in the closed interval $[a, b]$, differentiable in the open interval (a, b), and if g' is nonzero in (a, b), then there is at least one $\xi \in (a, b)$ where

$$\frac{f(b) - f(a)}{g(b) - g(a)} = \frac{f'(\xi)}{g'(\xi)} . \tag{7.2}$$

Proof. Since g' is nonzero in (a, b), it follows from Rolle's theorem that $g(b) - g(a) \neq 0$. Now define the function

$$h(x) := f(x) - \frac{f(b) - f(a)}{g(b) - g(a)}\big(g(x) - g(a)\big) .$$

Then

$$h(a) = f(a), \quad h(b) = f(b) - \big(f(b) - f(a)\big) = f(a)$$
$$\text{and} \quad h'(x) = f'(x) - \frac{f(b) - f(a)}{g(b) - g(a)}g'(x) .$$

Applying Rolle's theorem to the function h finally proves the existence of a point ξ in (a, b) where

$$h'(\xi) = f'(\xi) - \frac{f(b) - f(a)}{g(b) - g(a)}g'(\xi) = 0 . \qquad\qquad \square$$

---

2 Augustin Louis Cauchy (1789–1857).

Remark 7.4 (Equivalence of Theorems 7.1, 7.2 and 7.3). Comparing the above theorems, we note that
a. Theorem 7.1 $\Longrightarrow$ Theorem 7.3 ($\Longrightarrow$ denotes "implies"): indeed, Rolle's theorem was used to prove the Cauchy mean value theorem (and the mean value theorem);
b. Theorem 7.3 $\Longrightarrow$ Theorem 7.2: the mean value theorem is a special case of the Cauchy mean value theorem, with $g(x) = x$;
c. Theorem 7.2 $\Longrightarrow$ Theorem 7.1: indeed, Rolle's theorem is a special case of the mean value theorem, with $f(a) = f(b)$.
To summarize:

$$\text{Theorem 7.1} \Longrightarrow \text{Theorem 7.3} \Longrightarrow \text{Theorem 7.2} \Longrightarrow \text{Theorem 7.1},$$

and the three theorems are equivalent.

We saw in Lemma 4.12 a and Remark 4.13 a, that

$$f'(x_0) \begin{cases} > \\ < \end{cases} 0 \quad \Longrightarrow \quad f \text{ is } \begin{cases} \text{increasing} \\ \text{decreasing} \end{cases} \text{ at } x_0 \,.$$

These are local results. The corresponding global results require the mean value theorem.

Corollary 7.5. Let f be differentiable in the interval $I := (a, b)$.
a. If the derivative

$$f'(x) \begin{cases} > \\ < \end{cases} 0 \quad \text{in } I \,,$$

then the function

$$f \text{ is } \begin{cases} \text{increasing} \\ \text{decreasing} \end{cases} \text{in } I \,.$$

b. If the derivative

$$f'(x) \begin{cases} \geq \\ \leq \end{cases} 0 \quad \text{in } I \,,$$

then the function

$$f \text{ is } \begin{cases} \text{nondecreasing} \\ \text{nonincreasing} \end{cases} \text{in } I \,.$$

Proof. a. Suppose f is not "increasing in I". Then there are points $a < u < v < b$, with $f(v) \leq f(u)$, and by Theorem 7.2 there is a point ξ in (u, v) where

$$f'(\xi) = \frac{f(v) - f(u)}{v - u} \leq 0$$

proving that $f'(x) > 0$ in $I \implies f$ is increasing in I.
The remaining three statements are proved analogously. $\square$

Corollary 7.6. Let the function f be differentiable in an interval I. Then

$$f \text{ is } \begin{cases} \text{nondecreasing} \\ \text{nonincreasing} \end{cases} \text{ in } I \text{ if and only if the derivative } f' \begin{cases} \geq \\ \leq \end{cases} 0 \text{ in } I .$$

Proof. The "if" part is Corollary 7.5b. To prove "only if", let x_0 in I be a point where $f'(x_0) < 0$. Then f is decreasing at x_0 (by Remark 4.13a), so f cannot increase in I. The other statement is proved similarly.   □

Theorems 7.1–7.3 are *existence theorems*, certifying that an equation (such as (7.1)) has a solution ξ in an interval $[a, b]$. The existence of this solution allows giving useful *bounds* on some quantities of interest. This is illustrated in Corollaries 7.8 and 7.11.

First some notation. For a function g defined on $[a, b]$ we denote by
– $\inf_{a<x<b} g(x)$ the *infimum* (or greatest lower bound) of the values $\{ g(x) \mid a < x < b \}$;
– $\sup_{a<x<b} g(x)$ the *supremum* (or lowest upper bound) of the values $\{ g(x) \mid a < x < b \}$.

Remark 7.7. If g is continuous in $[a, b]$, then its minimum and maximum are attained (see Theorem 3.52). Therefore

$$\inf_{a<x<b} g(x) = \min_{a \leq x \leq b} g(x) ,$$

$$\sup_{a<x<b} g(x) = \max_{a \leq x \leq b} g(x) .$$

Corollary 7.8 (Bounds on the difference quotient). Let f and $[a, b]$ satisfy the conditions of Theorem 7.2. Then

$$\inf_{a<x<b} f'(x) \leq \frac{f(b) - f(a)}{b - a} \leq \sup_{a<x<b} f'(x) . \tag{7.3}$$

Proof. Follows from (7.1) since, for ξ in (a, b),

$$\inf_{a<x<b} f'(x) \leq f'(\xi) \leq \sup_{a<x<b} f'(x) .$$ □

Remark 7.9. If the derivative f' is continuous in an interval containing $[a, b]$, then the bounds (7.3) can be written, by Remark 7.7, as

$$\min_{a \leq x \leq b} f'(x) \leq \frac{f(b) - f(a)}{b - a} \leq \max_{a \leq x \leq b} f'(x) .$$

Example 7.10. For any $\varepsilon > 0$ there is an $M > 0$ such that

$$\ln x \leq \varepsilon x \quad \text{for all } x > M . \tag{7.4}$$

Proof. From (7.3) it follows, for any $1 < a < b$, that

$$\frac{\ln b - \ln a}{b - a} \leq \sup_{a < x < b} \frac{1}{x} = \frac{1}{a} \quad \text{(check!)}.$$

Therefore

$$\ln b \leq \ln a + \frac{b}{a}.$$

Now set

$$a := \frac{2}{\varepsilon}, \quad A := \frac{2 \ln a}{\varepsilon},$$

to get, for all $x > a$,

$$\ln x \leq \frac{\varepsilon}{2} A + \frac{\varepsilon}{2} x,$$

and (7.4) holds for $M := \max\{a, A\}$. $\qquad\square$

Since ε is the slope of the line $y = \varepsilon x$, this means that the logarithmic function grows at a rate less then any (increasing) linear function. A consequence is the following limit

$$\lim_{x \to \infty} \frac{\ln x}{x} = 0.$$

Another application of the mean value theorem is to the approximation (4.44),

$$f(x_0 + \Delta x) \approx f(x_0) + f'(x_0) \Delta x,$$

of a function f differentiable at x_0. The *approximation error* in (4.44),

$$E := f(x_0 + \Delta x) - f(x_0) - f'(x_0) \Delta x,$$

goes to 0 faster than Δx since f is differentiable at x_0. If f is twice differentiable, we can bound the error as follows.

Corollary 7.11 (Bound on the error of approximation). Let f be twice differentiable in an interval containing $[x_0, x_0 + \Delta x]$. Then

$$|f(x_0 + \Delta x) - f(x_0) - f'(x_0) \Delta x| \leq \sup_{x_0 < \xi < x_0 + \Delta x} |f''(\xi)| (\Delta x)^2. \qquad (7.5)$$

Proof. By the mean value theorem

$$f(x_0 + \Delta x) - f(x_0) = f'(\eta) \Delta x, \quad \text{for some } x_0 < \eta < x_0 + \Delta x.$$

Therefore

$$f(x_0 + \Delta x) - f(x_0) - f'(x_0) \Delta x = (f'(\eta) - f'(x_0)) \Delta x.$$

But, applying the mean value theorem again,

$$f'(\eta) - f'(x_0) = f''(\xi)(\eta - x_0), \quad \text{for some } x_0 < \xi < \eta,$$

and therefore

$$|f(x_0 + \Delta x) - f(x_0) - f'(x_0)\,\Delta x|$$
$$= |f''(\xi)|\,|\eta - x_0|\,\Delta x \leq |f''(\xi)|\,(\Delta x)^2, \quad \text{for some } x_0 < \xi < x_0 + \Delta x\,,$$

which proves (7.5).   □

Remark 7.12. a. The bound (7.5) shows that for a twice differentiable function, the approximation error E goes to 0 at least as fast as $(\Delta x)^2$.

   b. The approximation (4.44) and the error bound (7.5) hold for any real Δx positive or negative. If $\Delta x < 0$, then replace $x_0 \leq \xi \leq x_0 + \Delta x$ in (7.5) by $x_0 + \Delta x \leq \xi \leq x_0$.

   c. If the second derivative f'' is continuous in $[x_0, x_0 + \Delta x]$, then the bound (7.5) can be improved using Taylor's theorem (see (7.24)).

Example 7.13. Approximate $0.1^{4/3}$.

   Let $f(x) = x^{4/3}$ so that $f'(x) = \frac{4}{3}x^{1/3}$. At the point $x_0 = 0$, both $f(x_0) = 0$ and $f'(x_0) = 0$, so the tangent to f at 0 is the horizontal line $y = 0$. The approximation (4.44) gives $0.1^{4/3} \approx 0$, a poor approximation to the correct value of $0.1^{4/3} = 0.0464159$.

   For $x_0 = 0$ and any Δx, the approximation error

$$E = f(x_0 + \Delta x) - f(x_0) - f'(x_0)\,\Delta x = (\Delta x)^{4/3}\,,$$

which goes to zero faster than Δx, but slower than $(\Delta x)^2$. In fact,

$$\lim_{\Delta x \to 0} \frac{E}{(\Delta x)^2} = \infty\,.$$

Indeed, the function $f(x) = x^{4/3}$ is not twice differentiable at $x_0 = 0$, so the bound (7.5) does not apply.

Exercises

7.1 For the following functions f and intervals $[a, b]$, check if the hypotheses of Rolle's theorem are satisfied, in which case find all values ξ in (a, b) for which the conclusion of the theorem is satisfied.

   (a) $f(x) := x^2 - 9$, $[a, b] := [-3, 3]$

   (b) $f(x) := x^2 - 3x - 28$, $[a, b] := [-4, 7]$

   (c) $f(x) := \sin x$, $[a, b] := [0, \pi]$ (d) $f(x) := \tan x$, $[a, b] := [0, \pi]$

   (e) $f(x) := \ln(1 + \sin x)$, $[a, b] := [0, \pi]$

   (f) $f(x) := (x^2 - 1)/x$, $[a, b] := [-1, 1]$

7.2 For the following functions f and intervals $[a, b]$, verify the hypotheses of the mean value theorem. Find all values of ξ where the conclusion of the mean value theorem holds.

   (a) $f(x) := \sqrt{1 + x}$, $[a, b] := [0, 2]$ (b) $f(x) := \sqrt{1 + x^2}$, $[a, b] := [-1, 1]$

(c) $f(x) := e^x$, $[a, b] := [0, 4]$ (d) $f(x) := \sin x$, $[a, b] := [\pi/3, \pi/2]$

(e) $f(x) := \sin \pi x + \cos \pi x$, $[a, b] := [0, 1]$

(f) $f(x) := \operatorname{sech} x$, $[a, b] := [-1, 1]$

7.3   The following question is a "converse" of the question answered by the mean value theorem 7.2: Let the function f be differentiable in $(-\infty, +\infty)$, and let ξ be any point. Are there two points A and B on the graph of f, such that the slope of AB is equal to the derivative $f'(\xi)$?

7.4   Let $f : [a, b] \to \mathbf{R}$ be continuous in $[a, b]$, differentiable in (a, b), and let the absolute value of its derivative be bounded in (a, b) by two positive constants m, M,

$$0 < m \leq |f'(x)| \leq M, \quad \text{for all } x \in (a, b) .$$

Prove: For all $x, \xi \in (a, b)$,

$$m \leq \left| \frac{f(\xi) - f(x)}{\xi - x} \right| \leq M .$$

7.5   For the following functions f and intervals $[a, b]$, check the existence of a point ξ, $a < \xi < b$, satisfying (7.1). If "yes", give all such points. If "no", check which of the assumptions of Theorem 7.2 do not hold.

(a) $f(x) := x^3$, $[a, b] := [-1, 1]$ (b) $f(x) := x^3$, $[a, b] := [-1, 2]$

(c) $f(x) := x^{2/3}$, $[a, b] := [-1, 1]$ (d) $f(x) := x^{2/3}$, $[a, b] := [0, 1]$

(e) $f(x) := x^{1/3}$, $[a, b] := [-1, 1]$ (f) $f(x) := |x|$, $[a, b] := [-1, 2]$

7.6   For a function f satisfying the given hypotheses, the mean value theorem guarantees the existence of a point ξ, $a < \xi < b$, where $f'(\xi) = (f(b) - f(a))/(b - a)$ (7.1). In general there may be more than one such ξ. Give an example of a function f (not a linear function), and a finite interval $[a, b]$, where there are infinitely many points ξ satisfying (7.1).

7.7   Use Rolle's theorem to prove: If the function f is differentiable throughout $\mathbf{R}$, and if the function f has n distinct roots

$$x_1 < x_2 < \cdots < x_n ,$$

then its derivative f' has (at least) $n - 1$ distinct roots $\xi_1, \ldots, \xi_{n-1}$ such that

$$x_1 < \xi_1 < x_2 < \xi_2 < x_3 < \cdots < x_{n-1} < \xi_{n-1} < x_n .$$

7.8   Give a geometrical interpretation of the Cauchy mean value theorem.

7.9   For the following f, g and $[a, b]$, decide if the conditions of the Cauchy mean value theorem hold, in which case find a point $\xi \in (a, b)$ satisfying (7.2).

(a) $f(x) := x^2 - 2x + 3$, $g(x) := x^3 - 7x^2 + 20x - 5$, $[a, b] = [1, 4]$

(b) $f(x) := e^x$, $g(x) := \dfrac{x^2}{1 + x^2}$, $[a, b] = [-3, 3]$

7.10 Prove: The differences of square roots of successive integers, $\sqrt{n + 1} - \sqrt{n}$, are bounded by

$$\frac{1}{2\sqrt{n + 1}} < \sqrt{n + 1} - \sqrt{n} < \frac{1}{2\sqrt{n}}, \quad \text{for all positive integers } n .$$

7.11 Prove: For any integer $n > 0$, there exists an $M > 0$ such that $x^n < e^x$, for all $x > M$.

   Hint: Use Example 7.10.

7.12 Prove: For any integer $n > 0$, $\lim_{x \to \infty} (x^n / e^x) = 0$.

7.13 Use the mean value theorem to prove $|\sin x - \sin y| \le |x - y|$, for all values of x and y.

7.14 Use the Cauchy mean value theorem to prove that $|\tan x - \tan y| \ge |x - y|$, for all values of x and y in the interval $[-\pi/2, \pi/2]$.

7.15 Show that:

   a. $|\cos x - \cos y| \ge |\sin x - \sin y|$, for all values of x and y in the interval $[\pi/4, \pi/2]$;

   b. $|\sinh x - \sinh y| \ge |x - y|$, for all values of x and y;

   a. $|e^x - e^y| \ge |x - y|$, for all values of x and y.

7.16 Show that the mean value theorem fails in the interval $[-1, 1]$ for the following functions.

   (a) $f(x) := 1/x$ (b) $f(x) := |x|$ (c) $f(x) := x^{3/4}$

7.2 Rule of l'Hospital

Let the functions f and g be defined in (a, b), let x_0 be a point in (a, b), and consider the limit

$$\lim_{x \to x_0} \frac{f(x)}{g(x)} . \tag{7.6}$$

The easiest case is when the quotient function f/g is continuous at x_0, for then the limit (7.6) is just $f(x_0)/g(x_0)$. We consider here a harder, and more interesting, case where

$$\lim_{x \to x_0} f(x) = 0 \quad \text{and} \quad \lim_{x \to x_0} g(x) = 0 \tag{7.7}$$

so that

$$\lim_{x \to x_0} \frac{f(x)}{g(x)} = \frac{0}{0} = ? \, ,$$

called an *indeterminate form* of type $0/0$.

   We assume (and if necessary define or re-define)

$$f(x_0) = 0, \quad g(x_0) = 0 , \tag{7.8}$$

which makes f and g continuous at x_0.

Example 7.14. The derivative of a function f at x_0 is the limit

$$f'(x_0) = \lim_{x \to x_0} \frac{f(x) - f(x_0)}{x - x_0} ,$$

where the quotient $(f(x) - f(x_0))/(x - x_0)$ is indeterminate at x_0.

We study now a technique for computing limits, such as (7.6), due to l'Hospi-

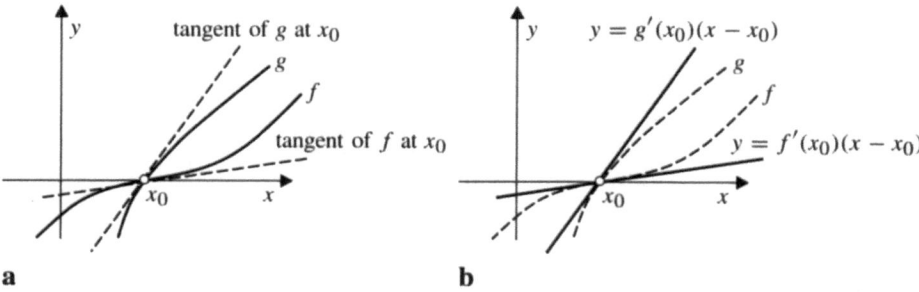

Fig. 7.3 a, b. Illustration of l'Hospital's rule. **a** The quotient f/g is indeterminate at x_0. **b** The limit of f/g as $x \to x_0$ is the quotient of the slopes of the tangents of f and g at x_0

tal[3]. This technique, called l'Hospital's rule, is based on the following computation, assuming (7.7), (7.8), and that both f and g are differentiable at x_0,

$$\lim_{x \to x_0} \frac{f(x)}{g(x)} = \lim_{x \to x_0} \frac{f(x) - f(x_0)}{g(x) - g(x_0)}$$

$$= \lim_{x \to x_0} \frac{(f(x) - f(x_0))/(x - x_0)}{(g(x) - g(x_0))/(x - x_0)}$$

$$= \frac{\displaystyle\lim_{x \to x_0} ((f(x) - f(x_0))/(x - x_0))}{\displaystyle\lim_{x \to x_0} ((g(x) - g(x_0))/(x - x_0))}, \quad \text{if } g'(x_0) \neq 0 ,$$

$$= f'(x_0)/g'(x_0) . \tag{7.9}$$

So if $g'(x_0) \neq 0$, the limit (7.6) can be computed as the ratio of derivatives $f'(x_0)/g'(x_0)$. This is illustrated in Fig. 7.3.

Example 7.15. The quotient

$$\frac{\sin x}{x} \quad \text{is indeterminate at } 0 .$$

Its limit there is, using (7.9) with $f(x) = \sin x$, $g(x) = x$ and $x_0 = 0$,

$$\lim_{x \to 0} \frac{\sin x}{x} = \frac{f'(0)}{g'(0)} = \frac{\cos(0)}{1} = 1 .$$

Example 7.16. Let $f(x) = \sqrt{5 + x} - \sqrt{5 - x}$, $g(x) = x$, and consider the limit

$$\lim_{x \to 0} \frac{f(x)}{g(x)} = \lim_{x \to 0} \frac{\sqrt{5 + x} - \sqrt{5 - x}}{x} . \tag{7.10}$$

---

3 Guillaume François Antoine Marquis de l'Hospital (1661–1704). The "s" in l'Hospital is not pronounced. In fact, the name l'Hospital is sometimes spelled l'Hôpital.

Since $f(0) = g(0) = 0$, we try the formula (7.9), with

$$f'(x) = \frac{1}{2\sqrt{5+x}} + \frac{1}{2\sqrt{5-x}}, \quad g'(x) = 1,$$

so that $f'(0) = 1/\sqrt{5}$, $g'(0) = 1$, and

$$\lim_{x\to 0} \frac{f(x)}{g(x)} = \frac{f'(0)}{g'(0)} = \frac{1}{\sqrt{5}}.$$

Remark 7.17 (One-sided limits). If the quotient $f(x)/g(x)$ is indeterminate at x_0, the one-sided limits

$$\lim_{x\to(x_0)^+} \frac{f(x)}{g(x)} \quad \text{and} \quad \lim_{x\to(x_0)^-} \frac{f(x)}{g(x)}$$

can be computed, similarly to (7.9), as the ratios of one-sided derivatives,

$$\lim_{x\to(x_0)^+} \frac{f(x)}{g(x)} = \frac{f'_+(x_0)}{g'_+(x_0)}, \quad \lim_{x\to(x_0)^-} \frac{f(x)}{g(x)} = \frac{f'_-(x_0)}{g'_-(x_0)}, \quad (7.11)$$

provided $g'_+(x_0) \neq 0$ and $g'_-(x_0) \neq 0$.

Example 7.18. The quotient

$$\frac{f(x)}{g(x)} = \frac{\tan x}{|x|} \quad \text{is indeterminate at } x = 0.$$

The numerator $f(x) = \tan x$ is differentiable at 0,

$$f'(0) = \left(\frac{d}{dx} \tan x\right)(0) = \sec^2(0) = 1,$$

where the denominator $g(x) = |x|$ has one-sided derivatives $g'_+(0) = 1$ and $g'_-(0) = -1$.
    Therefore, by (7.11),

$$\lim_{x\to 0^+} \frac{\tan x}{|x|} = \frac{f'(0)}{g'_+(0)} = \frac{\sec^2(0)}{1} = 1, \quad \text{and}$$

$$\lim_{x\to 0^-} \frac{\tan x}{|x|} = \frac{f'(0)}{g'_-(0)} = \frac{\sec^2(0)}{-1} = -1.$$

If $f(x)/g(x)$ has indeterminate form of type $0/0$ at x_0, then writing

$$\frac{f(x)}{g(x)} \quad \text{as} \quad \frac{1/g(x)}{1/f(x)}$$

we get an indeterminate form of type ∞/∞ at x_0. Therefore, the formula (7.9) is applicable to such situations as well.

The formula (7.9),

$$\lim_{x \to x_0} \frac{f(x)}{g(x)} = \frac{f'(x_0)}{g'(x_0)} \,,$$

gives no information if $g'(x_0) = 0$. However, if the quotient of derivatives $f'(x)/g'(x)$ is itself indeterminate at x_0, we replace

$$\frac{f'(x_0)}{g'(x_0)} \quad \text{in (7.9) by} \quad \lim_{x \to x_0} \frac{f'(x)}{g'(x)}$$

and get the following important result.

Theorem 7.19 (Rule of l'Hospital). Let the functions f and g be differentiable in an interval I with the possible exception of a point $x_0 \in I$ where, (7.7) and (7.8),

$$\lim_{x \to x_0} f(x) = 0, \quad \lim_{x \to x_0} g(x) = 0, \quad \text{and} \quad f(x_0) = 0, \quad g(x_0) = 0 \,,$$

and

$$g'(x) \neq 0, \quad \text{for } x_0 \neq x \in I \,. \tag{7.12}$$

Then

$$\lim_{x \to x_0} \frac{f(x)}{g(x)} \quad \text{exists if the limit} \quad \lim_{x \to x_0} \frac{f'(x)}{g'(x)} \quad \text{exists} \,,$$

in which case,

$$\lim_{x \to x_0} \frac{f(x)}{g(x)} = \lim_{x \to x_0} \frac{f'(x)}{g'(x)} \,. \tag{7.13}$$

Proof. The hypotheses of the Cauchy mean value theorem (Theorem 7.3) are satisfied in some interval $[x_0, v]$ to the right of x_0, so there exists a number $\xi \in (x_0, v)$ with

$$\frac{f(v)}{g(v)} = \frac{f(v) - f(x_0)}{g(v) - g(x_0)} = \frac{f'(\xi)}{g'(\xi)} \,.$$

As v approaches x_0, the number ξ (that lies in (x_0, v)) also approaches x_0, and therefore, the limit from the right

$$\lim_{v \to (x_0)^+} \frac{f(v)}{g(v)} = \lim_{\xi \to (x_0)^+} \frac{f'(\xi)}{g'(\xi)} \,.$$

Similarly, we get the limit from the left

$$\lim_{u \to (x_0)^-} \frac{f(u)}{g(u)} = \lim_{\xi \to (x_0)^-} \frac{f'(\xi)}{g'(\xi)} \,.$$

Taken together, these two-sided limits establish (7.13).   □

L'Hospital's rule states that a quotient f/g, which is indeterminate at x_0, has the same limit there as the quotient of their slopes f'/g'. The graphs of f and

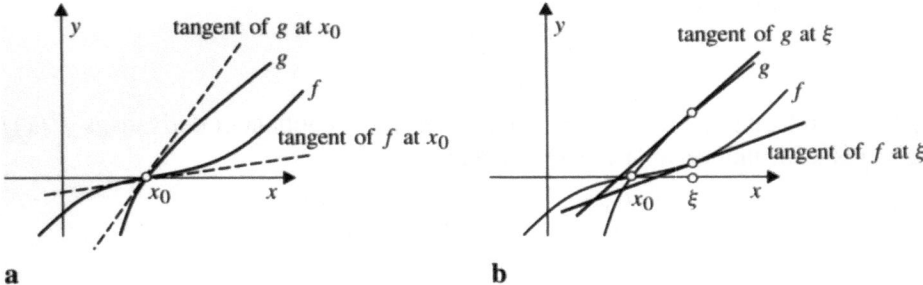

Fig. 7.4 a, b. Illustration of Theorem 7.19. **a** Functions f and g near x_0. **b** Tangents of f and g at a point near x_0

g can therefore be replaced, near $x = x_0$, by their tangents. This is illustrated in Fig. 7.4.

Example 7.20. Here is an example where $\lim_{x \to x_0}(f(x)/g(x))$ exists, but $\lim_{x \to x_0}(f'(x)/g'(x))$ does not exist. Note that this does not violate l'Hospital's rule.[4] The example is

$$f(x) := x^2 \sin(1/x), \quad g(x) := \sin x, \quad x_0 = 0 .$$

Indeed,

$$\lim_{x \to 0} \frac{x^2 \sin(1/x)}{\sin x} = \lim_{x \to 0} \frac{x}{\sin x} \cdot \lim_{x \to 0} x \sin(1/x) = 1 \cdot 0 = 0$$

but the limit of the ratio of derivatives

$$\lim_{x \to 0} \frac{2x \sin(1/x) - \cos(1/x)}{\cos x} = 0 - \lim_{x \to 0} \cos(1/x)$$

does not exist.

MACSYMA-Session 7.1. This session implements l'Hospital's rule, an exercise on our part since MACSYMA has built-in limit operators that are based on l'Hospital's rule. The following function is a translation of Theorem 7.19 into MACSYMA

```
c1. hospital(f,g,x,x0)  := if  (limit(f,x,x0)=0 and limit(g,x,x0)=0)
                    then limit(diff(f,x),x,x0)/limit(diff(g,x),x,x0)
                        else limit(f,x,x0)/limit(g,x,x0) $
```

It works as follows:
check condition (7.7), written as:

    limit(f,x,x0)=0 and limit(g,x,x0)=0

if condition (7.7) is true, apply l'Hospital's rule, returning RHS(7.13) written as

    limit(diff(f,x),x,x0)/limit(diff(g,x),x,x0)

---

4 The existence of $\lim_{x \to x_0}(f'(x)/g'(x))$ is sufficient, but not necessary, for that of $\lim_{x \to x_0}(f(x)/g(x))$.

if false, return LHS(7.13), written as
```
limit(f,x,x0)/limit(g,x,x0)
```
or an error message, "Division by zero" if $g(x_0) = 0$.

Examples:

a. The function $(1 - x^2)/(1 - x)$ simplifies to $1 + x$ for $x \neq 1$, which equals 2 at $x = 1$. Therefore:
```
c2. hospital(1-x^2,1-x,x,1)
d2.    2
```
b. An example where condition (7.7) does not hold is
```
c3. hospital(1-x^2,1-x,x,0)
d3.    1
```
since both numerator and denominator equal 1 at $x = 0$.

c. The limit

$$\lim_{x \to 0} \frac{\sqrt{1 + x + x^2} - \sqrt{1 + x}}{x^2} \tag{7.14}$$

cannot be handled by rule (7.13) alone. Indeed,
```
c4. hospital((sqrt(1+x+x^2)-sqrt(1+x)),x^2,x,0)
d4.    Division by 0
```

Remark 7.21. We proved Theorem 7.19 by establishing the following l'Hospital rule for one-sided limits:

$$\lim_{x \to (x_0)^+} \frac{f(x)}{g(x)} = \lim_{x \to (x_0)^+} \frac{f'(x)}{g'(x)}, \qquad \lim_{x \to (x_0)^-} \frac{f(x)}{g(x)} = \lim_{x \to (x_0)^-} \frac{f'(x)}{g'(x)}. \tag{7.15}$$

The rule of l'Hospital applies also for limits at infinity.

Corollary 7.22. Let f and g be differentiable throughout $\mathbf{R}$, let

$$\lim_{x \to \infty} f(x) = \lim_{x \to \infty} g(x) = 0 \quad \text{or} \quad \lim_{x \to \infty} f(x) = \lim_{x \to \infty} g(x) = \pm\infty$$

and

$$g'(x) \neq 0 \quad \text{for all } x \text{ sufficiently large} . \tag{7.16}$$

Then $\lim_{x \to \infty}(f(x)/g(x))$ exists if the limit $\lim_{x \to \infty}(f'(x)/g'(x))$ exists, in which case,

$$\lim_{x \to \pm\infty} \frac{f(x)}{g(x)} = \lim_{x \to \pm\infty} \frac{f'(x)}{g'(x)} . \tag{7.17}$$

Proof. A limit at infinity can be written as a one-sided limit at 0, using the trick

$$\lim_{x \to \infty} f(x) = \lim_{u \to 0^+} f(1/u), \quad \text{and} \quad \lim_{x \to -\infty} f(x) = \lim_{u \to 0^-} f(1/u)$$

(see Exercise 3.20). Therefore

$$\lim_{x \to \infty} \frac{f(x)}{g(x)} = \lim_{u \to 0^+} \frac{f(1/u)}{g(1/u)} = \lim_{u \to 0^+} \frac{(d/du)f(1/u)}{(d/du)g(1/u)}, \quad \text{by (7.15)}$$

$$= \lim_{u \to 0^+} \frac{-f'(1/u)u^{-2}}{-g'(1/u)u^{-2}} = \lim_{u \to 0^+} \frac{f'(1/u)}{g'(1/u)} = \lim_{x \to \infty} \frac{f'(x)}{g'(x)}. \qquad \square$$

Note that condition (7.12) of Theorem 7.19 becomes here (7.16). The rule

$$\lim_{x \to -\infty} \frac{f(x)}{g(x)} = \lim_{x \to -\infty} \frac{f'(x)}{g'(x)},$$

is proved similarly.

Example 7.23. The limit

$$\lim_{x \to \infty} \frac{2 + 2x + \sin 2x}{(2x + \sin 2x)\,\mathrm{e}^{\sin x}}$$

does not exist, since

$$\lim_{x \to \infty} \frac{2 + 2x + \sin 2x}{(2x + \sin 2x)\,\mathrm{e}^{\sin x}} = \lim_{x \to \infty} \left(1 + \frac{2}{2x + \sin 2x}\right) \mathrm{e}^{-\sin x},$$

and $\lim_{x \to \infty} \mathrm{e}^{-\sin x}$ does not exist.

However, the limit of the ratio of derivatives exists

$$\lim_{x \to \infty} \frac{2 + 2\cos 2x}{\left(2 + 2\cos 2x + (2x + \sin 2x)\cos x\right)\mathrm{e}^{\sin x}}$$

$$= \lim_{x \to \infty} \frac{4\cos^2 x}{4\cos^2 x + (2x + \sin 2x)\cos x}\,\mathrm{e}^{-\sin x}$$

$$= \lim_{x \to \infty} \frac{4\cos x}{2x + 4\cos x + \sin 2x}\,\mathrm{e}^{-\sin x}$$

$$= 0,$$

since $\displaystyle \lim_{x \to \infty} \frac{4\cos x}{2x + 4\cos x + \sin 2x} = 0$, and $\mathrm{e}^{-\sin x}$ is bounded.

Does this example contradict the l'Hospital rule given in Corollary 7.22? No, because condition (7.16) is violated here (check!), so Corollary 7.22 does not apply.

If the right-hand side of (7.13), $\lim_{x \to x_0}(f'(x)/g'(x))$, is itself indeterminate, it can sometimes be computed by using l'Hospital's rule again.

Corollary 7.24. Let the functions f and g be twice differentiable in an interval I, with the possible exception of a point $x_0 \in I$, and let f, g and their derivatives

f' and g' satisfy conditions analogous to those of Theorem 7.19. Then

$$\lim_{x \to x_0} \frac{f(x)}{g(x)} = \lim_{x \to x_0} \frac{f'(x)}{g'(x)} = \lim_{x \to x_0} \frac{f''(x)}{g''(x)}, \quad \text{etc.} \tag{7.18}$$

MACSYMA-Session 7.2. Corollary 7.24 allows using l'Hospital's rule recursively:

```
c1. hospital_recursive(f,g,x,x0)
     := if (limit(f,x,x0)=0 and limit(g,x,x0)=0)
            then hospital_recursive(diff(f,x),diff(g,x),x,x0)
        else limit(f,x,x0)/limit(g,x,x0)$
```

which (by calling itself, see 2nd line in c1) finds the first n such that the ratio of the nth derivatives $f^{(n)}(x_0)/g^{(n)}(x_0)$ is not 0/0 and then returns the result.

a. The limit (7.14) is now done easily

```
c2. hospital_recursive((sqrt(1+x+x^2)-sqrt(1+x)),x^2,x,0)
```

d2. $\dfrac{1}{2}$

b. A case requiring several iterations of `hospital_recursive` is

```
c3. hospital_recursive(exp(-1/x^2),x^3,x,0)
```

d3.   0

You can use `trace` to see how many times the l'Hospital rule was invoked here.

Example 7.25. For every positive integer n,

$$\lim_{x \to \infty} \frac{x^n}{e^x} = 0$$

i.e., the exponential function e^x grows, as $x \to \infty$, faster than x^n for all n (see also Exercise 7.12).

Proof. Use l'Hospital's rule (7.17) n times

$$\lim_{x \to \infty} \frac{x^n}{e^x} = \lim_{x \to \infty} \frac{nx^{n-1}}{e^x} \quad (\text{since } (d/dx)x^n = nx^{n-1}, (d/dx)e^x = e^x)$$

$$= \lim_{x \to \infty} \frac{n(n-1)x^{n-2}}{e^x} = \cdots = \lim_{x \to \infty} \frac{n(n-1)(n-2)\ldots 2 \cdot 1}{e^x}$$

$$= 0 . \qquad \qquad \qquad \qquad \qquad \qquad \qquad \qquad \qquad \square$$

Example 7.26. Here is an example where repeating l'Hospital's rule does not work. The limit $\lim_{x \to \pi/2}(\tan x / \sec x)$ exists and is easily computed

$$\lim_{x \to \pi/2} \frac{\tan x}{\sec x} = \lim_{x \to \pi/2} \frac{\sin x \, \cos x}{\cos x} = \lim_{x \to \pi/2} \sin x = 1 .$$

Applying l'Hospital's rule recursively,

$$\lim_{x \to \pi/2} \frac{\tan x}{\sec x} = \lim_{x \to \pi/2} \frac{\sec^2 x}{\sec x \, \tan x} = \lim_{x \to \pi/2} \frac{\sec x}{\tan x} = \lim_{x \to \pi/2} \frac{\tan x}{\sec x} = \cdots$$

keeps going in circles.

L'Hospital's rule works for limits of the indeterminate forms of types $0/0$ and ∞/∞. Sometimes we can use it to compute limits of indeterminate forms of other types, such as $\infty - \infty$, 0^0, and 1^∞, if we can transform them to $0/0$ or ∞/∞.

Example 7.27 (Indeterminate form of type $\infty - \infty$). Compute

$$\lim_{n\to\infty} \left(\sqrt[3]{n^6 + 3n^4 + n^3} - n^2\right).$$

Solution: As $n \to \infty$, the expression $\sqrt[3]{n^6 + 3n^4 + n^3} - n^2$ is an indeterminate form $\infty - \infty$. However, we can transform it to the form $0/0$ as follows

$$\lim_{n\to\infty} \left(\sqrt[3]{n^6 + 3n^4 + n^3} - n^2\right) = \lim_{n\to\infty} n^2 \left(\sqrt[3]{1 + \frac{3}{n^2} + \frac{1}{n^3}} - 1\right)$$

$$= \lim_{x\to 0^+} \frac{\sqrt[3]{1 + 3x^2 + x^3} - 1}{x^2}$$

and

$$\lim_{x\to 0^+} \frac{\sqrt[3]{1 + 3x^2 + x^3} - 1}{x^2} = \lim_{x\to 0^+} \frac{(x^2 + 2x)/(1 + 3x^2 + x^3)^{2/3}}{2x}$$

(using l'Hospital's rule)

$$= \lim_{x\to 0^+} \frac{x + 2}{2(1 + 3x^2 + x^3)^{2/3}} = 1.$$

Example 7.28 (Indeterminate form of type 0^0). Prove $\lim_{x\to 0^+} x^x = 1$.

Proof. As $x \to 0^+$, the function x^x is an indeterminate form 0^0. To transform it into ∞/∞, we define

$$y := \ln x^x = x \ln x = \frac{\ln x}{1/x} \tag{7.19}$$

and take the limit, as $x \to 0^+$

$$\lim_{x\to 0^+} \frac{\ln x}{1/x} = \lim_{x\to 0^+} \frac{1/x}{-1/x^2} = \lim_{x\to 0^+} (-x) = 0$$

and therefore

$$\lim_{x\to 0^+} x^x = \lim_{x\to 0^+} \exp(y) = 1. \qquad \square$$

Note that $y = \ln x^x$ can be written as

$$y = \frac{x}{1/\ln x}$$

instead of (7.19). Can you still compute $\lim_{x\to 0^+} y$?

Example 7.29 (Indeterminate form of type 1^∞). Recall the limit (3.16),

$$e := \lim_{n\to\infty} (1 + 1/n)^n.$$

As $x \to \infty$, the expression $(1 + 1/x)^x$ is an indeterminate form 1^∞. Let

$$y := \ln\left(\left(1 + \frac{1}{x}\right)^x\right) = x \ln\left(1 + \frac{1}{x}\right) = \frac{\ln(1 + 1/x)}{1/x} .$$

As $x \to \infty$, this is an indeterminate form $0/0$ and therefore

$$\lim_{x\to\infty} y = \lim_{x\to\infty} \frac{-\frac{1}{x^2}\left(\frac{1}{1+1/x}\right)}{-1/x^2} = \lim_{x\to\infty} \frac{1}{1 + 1/x} = 1$$

and

$$\lim_{x\to\infty}\left(1 + \frac{1}{x}\right)^x = \lim_{x\to\infty} \exp(y) = e .$$

Exercises

7.17 Use l'Hospital's rule, and the function `hospital` of MACSYMA-Session 7.1, to compute

a. $\lim_{x\to x_0}(x^n - x_0^n)/(x - x_0)$, the derivative of the monomial x^n at x_0;

b. $\lim_{x\to x_0}(f(x) - f(x_0))/(x - x_0)$, the derivative of a function f at x_0.

7.18 Determine the limits of the following functions by l'Hospital's rule.

(a) $\lim_{x\to\pi/2} \dfrac{1 - \sin x}{\cos x}$

(b) $\lim_{x\to 0} \dfrac{\ln(1 - x)}{\sin x}$

(c) $\lim_{x\to 0} \dfrac{\tan x}{x}$

(d) $\lim_{x\to 0} \dfrac{\sin 2x}{\sin 3x}$

(e) $\lim_{x\to y} \dfrac{x^{1/4} - y^{1/4}}{x - y}$

(f) $\lim_{x\to 0} \dfrac{\cosh x - 1}{x^2}$

(g) $\lim_{x\to\pi} \dfrac{\sin x}{\sqrt{x - \pi}}$

(h) $\lim_{x\to 0} \dfrac{\cosh x - 1}{\cos x - 1}$

(i) $\lim_{x\to\infty} \dfrac{\pi/2 - \operatorname{atan} x}{\ln(1 - e^{-x})}$

(j) $\lim_{x\to\infty} \dfrac{\ln\ln x}{x \ln x}$

(k) $\lim_{x\to\infty} \dfrac{\ln\ln\ln x}{\ln\ln x}$

(l) $\lim_{x\to\infty} \dfrac{4x^2 + 2}{x^2 + 5x + 1}$

(m) $\lim_{x\to 0^+} \dfrac{1 - \cos(\sqrt{x})}{x}$

(n) $\lim_{x\to 0} \dfrac{\sqrt{1 + ax} - \sqrt{1 + x}}{x}$

(o) $\lim_{x\to 0} \dfrac{\operatorname{atan} ax}{\operatorname{atan} bx}$

7.19 Use l'Hospital's rule to find the following limits.

(a) $\lim_{x\to 0^+} \sin x \ln x$

(b) $\lim_{x\to\infty} (\ln x)^{1/x}$

(c) $\lim_{x\to\infty} x^{1/\ln x}$

(d) $\lim_{x\to 0} (\cot x)^{\sin x}$

(e) $\lim_{x\to 0} x^{\tan x}$

(f) $\lim_{x\to 0} (\tan x)^x$

(g) $\lim_{x\to\infty} \dfrac{x}{\ln x}$

(h) $\lim_{x\to\infty} \left(1 + \dfrac{1}{x}\right)^x$

(i) $\lim_{x\to\infty} \left(1 + \dfrac{1}{x^2}\right)^x$

(j) $\lim_{x\to\infty} \left(1 + \dfrac{1}{x}\right)^{x^2}$

(k) $\lim_{x\to\infty} \left(1 + \dfrac{1}{x^2}\right)^{x^2}$

(l) $\lim_{x\to\pi/2} (\cos x)^{\tan x}$

(m) $\lim_{x\to 1^+} (\sin(x - 1))^{\ln x}$

(n) $\lim_{x\to\infty} \left(\dfrac{\tan x}{x}\right)^{1/x^2}$

(o) $\lim_{x\to\infty} (\sec(1/x^2))^{x^5}$

7.20 Some of the following limit problems may not require l'Hospital's rule.

(a) $\lim_{x\to\infty} (\sin(1/x^2))^{x^4}$

(b) $\lim_{x\to\infty} \left(\sqrt{x^2 + 1} - x\right)$

(c) $\lim_{x\to 0}\left(\dfrac{2}{x^2} - \dfrac{1}{e^x - 1 - x}\right)$

(d) $\lim_{x\to 0}\left(\dfrac{2}{x^2} + \dfrac{1}{\ln(1 + x) - x}\right)$

(e) $\lim\limits_{x\to 0}\left(\dfrac{1}{1-x^2/2}-\dfrac{1}{\cos x}\right)$ (f) $\lim\limits_{x\to\infty}(\ln\ln x)^{1/x}$

(g) $\lim\limits_{x\to 0}\left(\dfrac{1}{(1+x)^{1/3}}-\dfrac{1}{1+x/3}\right)$ (h) $\lim\limits_{x\to 0}\left(\dfrac{1}{(1+x)^{1/4}}-\dfrac{1}{1+x/4}\right)$

7.21 Explain why the following calculation is wrong

$$\lim_{x\to 0}\frac{x^4-x^2+x-1}{\sin^2 x}=\lim_{x\to 0}\frac{4x^3-2x^2+1}{2\sin x\cos x}=$$

$$=\lim_{x\to 0}\frac{12x^2-4x}{2\cos^2 x-2\sin^2 x}=0.$$

7.22 Use l'Hospital's rule to verify the limit $\lim_{n\to 0}(x^n-1)/n=\ln x$.
   Note: The variable here is n, not x!

7.23 Find the limits

   (a) $\lim\limits_{x\to 0}\dfrac{(1+x)^m-1}{x}$ (b) $\lim\limits_{n\to\infty}\sqrt{n}(\sqrt{1+n}-\sqrt{n})$

   (c) $\lim\limits_{x\to 0}\dfrac{1-\sqrt{(1-ax)(1-bx)}}{x}$

   by hospital or hospital_recursive.

7.24 a. Try to calculate

$$\lim_{x\to 0}\frac{(1-x)^m-(1-mx)}{x^2}$$

   by the function hospital_recursive of MACSYMA-Session 7.2. The reason it
   does not work is that MACSYMA is unable to decide if $m(m-1)=0$ is true.

   b. Rewrite the function hospital_recursive, using nested if statements, that will
   compute the limit in a.

7.25 Compute the following limits, using the l'Hospital rule twice.

   (a) $\lim\limits_{x\to x_0}\dfrac{x(f(x)-f(x_0))-x_0 f'(x_0)(x-x_0)}{(x-x_0)^2}$

   (b) $\lim\limits_{x\to x_0}\left(\dfrac{x(f(x)-f(x_0))-x_0 f'(x_0)(x-x_0)}{(f(x)-f(x_0))(x-x_0)}\right)$

7.26 Find the limits

   (a) $\lim\limits_{x\to 0}\dfrac{\sqrt[3]{1+x}-\sqrt[3]{1-x}}{x}$ (b) $\lim\limits_{n\to\infty}\dfrac{\sqrt{n^2+4}}{\sqrt{4n^2-1}}$ (c) $\lim\limits_{x\to 2}\dfrac{x^3-8}{x^4-16}$

   (d) $\lim\limits_{x\to 0}\dfrac{e^x-e^{-x}}{\ln(1+x)}$ (e) $\lim\limits_{x\to 1}\dfrac{a^{\ln x}-x}{\ln x}$ (f) $\lim\limits_{x\to 0}\dfrac{\tan x-x}{\sin x-x}$

7.27 Assuming that f is differentiable as needed, verify the limits

   (a) $\lim\limits_{h\to 0}\dfrac{f(x+h)-f(x-h)}{2h}=f'(x)$

   (b) $\lim\limits_{h\to 0}\dfrac{f(x+h)-2f(x)+f(x-h)}{h^2}=f''(x)$

7.3 Taylor theorem

The main result here is the Taylor[5] theorem, giving the value of a function $f(x)$
in terms of its value $f(x_0)$, and the values of its first n derivatives $f'(x_0)$, $f''(x_0)$,
$f'''(x_0)$, ..., $f^{(n)}(x_0)$ at some point x_0.

To explain the idea consider a function f differentiable in some interval containing x_0 and x. We conclude from the mean value theorem (Theorem 7.2) that

$$f(x) = f(x_0) + f'(\xi_1)(x - x_0) ,$$

where ξ_1 is an intermediate point between x_0 and x. If the function f is twice
differentiable, we can similarly get

$$f'(\xi_1) = f'(x_0) + f''(\xi_2)(\xi_1 - x_0) ,$$

where ξ_2 is intermediate between x_0 and ξ_1. This way we get

$$f(x) = f(x_0) + f'(x_0)(x - x_0) + f''(\xi_2)(x - x_0)(\xi_1 - x_0)$$

expressing $f(x)$ in terms of the values of f and f' at the point x_0, and a third term
which behaves like a quadratic $(x - x_0)^2$. Using this idea more precisely, and for
higher derivatives, we get the following important theorem due to Taylor.

Theorem 7.30 (Taylor theorem). Let the function f be defined and have n continuous derivatives f', f''', ..., $f^{(n)}$ on the interval $[a, b]$, and an $(n + 1)$th-derivative
$f^{(n+1)}$ on (a, b). Let x_0 be an arbitrary point in $[a, b]$. Then for every $x \in [a, b]$,
there exists a point ξ between x and x_0 such that

$$f(x) = T_n(f, x, x_0) + R_n(f, x, x_0) , \tag{7.20a}$$

where

$$T_n(f, x, x_0) := f(x_0) + f'(x_0)(x - x_0) + \frac{f''(x_0)}{2!}(x - x_0)^2$$

$$+ \cdots + \frac{f^{(n)}(x_0)}{n!}(x - x_0)^n \tag{7.20b}$$

is called the *nth-degree Taylor polynomial of f at x_0*, and

$$R_n(f, x, x_0) := \frac{f^{(n+1)}(\xi)}{(n + 1)!}(x - x_0)^{n+1} , \tag{7.20c}$$

is its *remainder*, due to Lagrange.

This theorem is proved in Sect. 7.8.

In the special case of $x_0 = 0$, a Taylor polynomial is called a Maclaurin[6] polynomial. We show now that the Taylor polynomial $T_n(f, x, x_0)$ is the polynomial
which agrees with f, and its first n derivatives, at the point x_0.

---

5 Brook Taylor (1685–1731).
6 Colin Maclaurin (1698–1746).

Example 7.31. Given a function f with n derivatives at x_0, we seek a polynomial

$$p(x) = \sum_{k=0}^{n} p_k (x - x_0)^k \tag{7.21}$$

with matching derivatives at x_0,

$$p(x_0) = f(x_0), \quad p'(x_0) = f'(x_0), \quad p''(x_0) = f''(x_0), \quad \cdots \quad ,$$
$$p^{(n)}(x_0) = f^{(n)}(x_0) . \tag{7.22}$$

Now compute the derivatives successively as

$$p'(x) = \sum_{k=1}^{n} k p_k (x - x_0)^{k-1}$$

$$p''(x) = \sum_{k=2}^{n} k(k-1) p_k (x - x_0)^{k-2}$$

$$\cdots = \cdots$$

$$p^{(m)}(x) = \sum_{k=m}^{n} k(k-1) \ldots (k - m + 1) p_k (x - x_0)^{k-m} .$$

Substituting $x = x_0$ in (7.21) we get $p_0 = p(x_0)$.

Similarly, differentiating (7.21) k times and substituting $x = x_0$ we get

$$p_k = \frac{f^{(k)}(x_0)}{k!}, \quad k = 1, 2, \ldots, n ,$$

so the sought polynomial is just the nth-degree Taylor polynomial $T_n(f, x, x_0)$, (7.20b),

$$T_n(f, x, x_0) = f(x_0) + f'(x_0)(x - x_0) + \frac{f''(x_0)}{2!}(x - x_0)^2 + \cdots$$

$$+ \frac{f^{(n)}(x_0)}{n!}(x - x_0)^n .$$

The Taylor theorem has many applications (see Chap. 13). We use it now to justify the 2nd-derivative test for convexity, given in Sect. 6.5.

Corollary 7.32. Let the function f be twice differentiable in an interval I. Then:
a. f is convex in I if and only if $f''(x) \geq 0$ for all $x \in I$;
b. f is strictly convex in I if $f''(x) > 0$ for all $x \in I$.

This corollary is proved in Sect. 7.8.

Example 7.33. We compute the 5th-degree Taylor polynomial of $\cos x$ at $x_0 = 0$. By (7.20b),

$$T_5(f, x, x_0) = f(x_0) + f'(x_0)(x - x_0) + \frac{f''(x_0)}{2!}(x - x_0)^2 + \frac{f'''(x_0)}{3!}(x - x_0)^3$$
$$+ \frac{f^{(4)}(x_0)}{4!}(x - x_0)^4 + \frac{f^{(5)}(x_0)}{5!}(x - x_0)^5,$$

and for $f(x) := \cos x$ at $x_0 = 0$,

$$f(x_0) = \cos(0) = 1, \qquad f'(x_0) = -\sin(0) = 0,$$
$$f''(x_0) = -\cos(0) = -1, \qquad f'''(x_0) = \sin(0) = 0,$$
$$f^{(4)}(x_0) = \cos(0) = 1, \qquad f^{(5)}(x_0) = -\sin(0) = 0.$$

Therefore

$$T_5(\cos x, x, 0) = 1 - \frac{x^2}{2!} + \frac{x^4}{4!}.$$

From (7.20c) it follows that the remainder $R_5(\cos x, x, 0) = \cos x - T_5(\cos x, x, 0)$ is bounded by

$$|R_5(\cos x, x, 0)| \leq \frac{x^6}{6!} = \frac{x^6}{120}.$$

Therefore, for x sufficiently close to 0, the function $\cos x$ can be approximated by the polynomial

$$\cos x \approx 1 - \frac{x^2}{2!} + \frac{x^4}{4!}. \tag{7.23}$$

Note that the 5th-degree Taylor polynomial is here a polynomial of the 4th-degree, since the coefficient of x^5 in $T_5(\cos x, x, 0)$ is zero.

MACSYMA-Session 7.3. The nth-degree Taylor polynomial of $f(x)$ at x_0 up to order $(x - x_0)^n$ can be computed by the built-in MACSYMA function `taylor(f,x,x0,n)`. MACSYMA begins the resulting display line with the letter T (for Taylor), and ends it with three dots, indicating a possibly truncated representation.

    a. The 4th-degree Taylor polynomial of x^5 at $x_0 = 1$ is

c1. `taylor(x^5,x,1,4)`
d1.   T $1 + 5(x - 1) + 10(x - 1)^2 + 10(x - 1)^3 + 5(x - 1)^4 + \ldots$

or $5x^4 - 10x^3 + 10x^2 - 5x + 1$, while the 4th-degree Taylor polynomial of x^5 about $x_0 = 0$ is

c2. `taylor(x^5,x,0,4)`
d2.   T $0 + \ldots$

which is 0. What is the 5th-degree Taylor polynomial of x^5 about $x_0 = 0$?

    b. The 1st-degree Taylor polynomial of f at x_0 is just the tangent line of f at $x = x_0$,

c3. `tangent(f,x,x0):=limit(f,x,x0)+(x-x0)*limit(diff(f,x),x,x0)$`
see Session 4.4. We illustrate this by use of an unspecified function $f(x)$
c4. `taylor(f(x),x,x0,1)-tangent(f(x),x,x0)`

d4.   T $f(x0) - \left(\lim_{x \to 0} f(x)\right) + \left(\left.\frac{df(x)}{dx}\right|_{x=x_0} - \left(\lim_{x \to 0} \frac{df(x)}{dx}\right)\right)(x - x_0) + \ldots$

which simplifies to 0.

c. Here are some examples of the use of `taylor(f,x,x0,n)`:

c5. `taylor(exp(x),x,0,5)`

d5.   T $1 + x + \dfrac{x^2}{2} + \dfrac{x^3}{6} + \dfrac{x^4}{24} + \dfrac{x^5}{120} + \cdots$

c6. `taylor(sin(x),x,0,7)`

d6.   T $x - \dfrac{x^3}{6} + \dfrac{x^5}{120} - \dfrac{x^7}{5040} + \cdots$

c7. `taylor((x+1)/(exp(x)+1),x,0,4)`

d7.   T $\dfrac{1}{2} + \dfrac{x}{4} - \dfrac{x^2}{4} + \dfrac{x^3}{48} + \dfrac{x^4}{48} + \cdots$

Try the last one by hand. It may help to recall that you can multiply the 4th-degree Taylor series of $1/(1 + \exp(x))$ about $x = 0$ by $1 + x$ to get the answer.

d. Here are some more complicated cases

c8. `taylor(sqrt(1+x),x,0,6)`

d8.   T $1 + \dfrac{x}{2} - \dfrac{x^2}{8} + \dfrac{x^3}{16} - \dfrac{5x^4}{128} + \dfrac{7x^5}{256} - \dfrac{21x^6}{1024} + \cdots$

c9. `taylor(sin(x)*cos(x),x,0,10)`

d9. T $x - \dfrac{2x^3}{3} + \dfrac{2x^5}{15} - \dfrac{4x^7}{315} + \dfrac{2x^9}{2835} + \cdots$

e. Some functions will not have a Taylor polynomial beyond a certain order because the required derivatives do not exist at that point. For example, the Taylor expansion of order 0 for $\sqrt{x}$ about $x = 0$ is 0, and higher-order expansions do not exist:

c10. `taylor(sqrt(x),x,0,0)`

d10.   T $0 + \cdots$

c11. `taylor(sqrt(x),x,0,1)`

d11.   T $\sqrt{x} + \cdots$

c12. `taylor(sqrt(x),x,0,2)`

d12.   T $\sqrt{x} + \cdots$

MACSYMA implements here a generalized notion of Taylor series, using fractional powers.

Exercises

7.28 The following problems ask for the Taylor polynomial of a product of two functions $f(x)$ and $g(x)$. Compute the Taylor polynomial at $x_0 = 0$ in two ways, (i) by computing the Taylor polynomial of the product, and (ii) by computing the Taylor polynomials of the individual functions and taking their product. Truncate all polynomials to degree 5 and compare your results.

(a) $f(x) := \sin x$, $g(x) := \cos x$ (b) $f(x) := \dfrac{1}{1+x}$, $g(x) := \ln(1 + x)$

(c) $f(x) := \sinh x$, $g(x) := \tanh x$ (d) $f(x) := \sqrt{1 - x}$, $g(x) := (1 - x)$

(e) $f(x) := e^x$, $g(x) := e^x$ (f) $f(x) := e^x$, $g(x) := e^{-x}$

(g) $f(x) := 3x + 4$, $g(x) := \dfrac{1}{1+x}$ (h) $f(x) := x^4$, $g(x) := \dfrac{1}{2x + 1}$

(i) $f(x) := \sqrt{1 - x}$, $g(x) := \sqrt{1 - x}$ (j) $f(x) := \sin x$, $g(x) := \sec x$

7.29 The Taylor theorem allows us to cut in half the bound (7.5) on the error of approximation.

Prove: Let f be twice continuously differentiable in an interval containing $[x_0, x_0 + \Delta x]$. Then

$$|f(x_0 + \Delta x) - f(x_0) - f'(x_0)\,\Delta x| \le \tfrac{1}{2} \max_{x_0 < \xi < x_0 + \Delta x} |f''(\xi)|\,(\Delta x)^2 . \tag{7.24}$$

7.30 a. Use MACSYMA to plot the graphs of $\cos x$ and of its 5th-degree Taylor polynomial at 0,

$$T_5(\cos x, x, 0) := 1 - \frac{x^2}{2!} + \frac{x^4}{4!} .$$

Conclude that (7.23) is a good approximation of $\cos x$ for x near 0.

b. Find the 5th-degree Taylor polynomial of $\cos x$ at $\pi/2$, and verify that it approximates $\cos x$ near $\pi/2$.

7.31 Calculate the nth-degree Taylor polynomial of f at x_0 (see (7.20b)) for the following f, n and x_0.

(a) $f(x) := e^x, n = 5, x_0 = 0$ (b) $f(x) := \sin x, n = 5, x_0 = 0$

(c) $f(x) := \tan x, n = 5, x_0 = 0$ (d) $f(x) := \sec x, n = 5, x_0 = 0$

(e) $f(x) := x^4, n = 3, x_0 = 1$ (f) $f(x) := (x-1)^4, n = 3, x_0 = 0$

7.32 Use MACSYMA to compute the 4th-degree Taylor polynomials at $x_0 = 0$ for the functions

(a) e^{-x^2} (b) $\dfrac{\sin x}{1 + x^2}$ (c) $\dfrac{\cos x}{\pi + x}$

Use `plot` to graph each function, and its Taylor polynomial, and observe their closeness near $x = 0$. Repeat for the Taylor polynomials of 8th degree.

7.33 Let $T_n(f, x, x_0)$ denote the nth-degree Taylor polynomial of f at x_0, and let $R_n(f, x, x_0) := f(x) - T_n(f, x, x_0)$ be its remainder, (7.20c). Prove that for all n and x_0, $R_n(x^{n+1}, x, x_0) = (x - x_0)^{n+1}$.

7.34 Use Corollary 7.32 to prove that $f(x) = \log x$ is strictly concave in $(0, \infty)$.

7.4 Antiderivatives

Until now a function was given, and we computed its derivative. In many applications the situation is reversed: The derivative is given, and one must find the original function, which is called, appropriately, an antiderivative.

Definition 7.34 (Antiderivative). Let the function f be given. A function F with derivative f,

$$F'(x) = f(x) , \tag{7.25}$$

is called an antiderivative of f.

If $F(x)$ is an antiderivative of f, so is the function,

$$G(x) := F(x) + C, \quad \text{where } C \text{ is any constant} . \tag{7.26}$$

Indeed,

$$\frac{d}{dx}(F(x) + C) = \frac{d}{dx}F(x), \quad \text{since}\frac{d}{dx}C = 0,$$
$$= f(x), \quad \text{by (7.25)}.$$

We show, in Corollary 7.36, that all antiderivatives of a function f are of the form (7.26). First a lemma, based on the mean value theorem.

Lemma 7.35. Let the functions F and G be differentiable in (a, b) and continuous at the endpoint a. Then F and G have equal derivatives in (a, b),

$$F'(x) = G'(x), \quad \text{for all } x \in (a, b), \tag{7.27}$$

if and only if the functions F and G differ, in (a, b), by a constant, $G(x) = F(x) + C$, with C as constant, (7.26).

Proof. "If": (7.27) clearly follows from (7.26).
"Only if": Let $H(x) := F(x) - G(x)$; then (7.27) implies

$$H'(x) = F'(x) - G'(x) = 0, \quad \forall x \in (a, b). \tag{7.28}$$

Let x be an arbitrary point in (a, b). Then by the mean value theorem,

$$H(x) - H(a) = H'(\xi)(x - a) \quad \text{for some point } a < \xi < x$$
$$= 0 \quad \text{by (7.28)}$$
$$\therefore \quad H(x) = H(a), \quad \text{for all } x \in (a, b),$$
$$\therefore \quad F(x) - G(x) = H(a), \quad \forall x \in (a, b),$$

i.e., F and G differ by a constant.   □

The next result shows that an antiderivative is unique, up to a constant.

Corollary 7.36 (Uniqueness, up to a constant, of antiderivative). Let F be an antiderivative of f in (a, b). Then G is an antiderivative of f in (a, b) if and only if, (7.26),

$$G(x) = F(x) + C, \quad C \text{ constant}.$$

Proof. Follows from Lemma 7.35 by substituting $F' = f$.   □

Remark 7.37. We saw that the antiderivative of a function f is a class of functions,

$$\{ F(x) + C \mid C \text{ constant} \},$$

where F is a particular antiderivative. When we talk about an antiderivative of f, we mean this whole class. The constant C is called the *constant of integration*.

We will as a rule omit the constant of integration, mentioning it only when it serves some purpose.

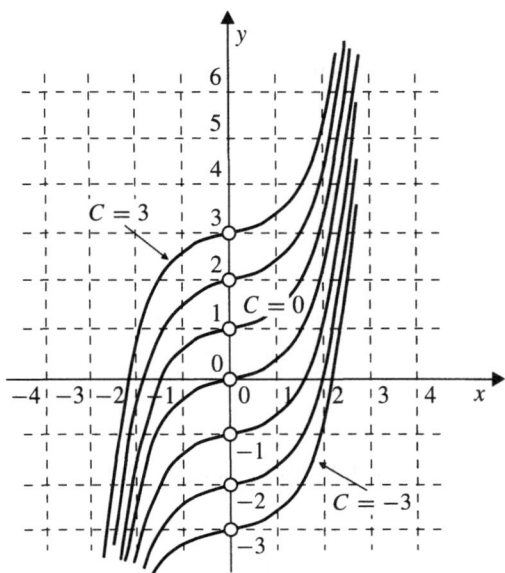

Fig. 7.5. $x^3/3 + C$ as the antiderivative of x^2, with C being an arbitrary constant

Example 7.38 (Antiderivative of x^2). An antiderivative of x^2 is $x^3/3$, as can be verified by differentiation

$$\frac{d}{dx}\frac{x^3}{3} = \frac{1}{3}\frac{d}{dx}x^3 = \frac{1}{3}3x^2 = x^2.$$

Therefore, the antiderivative of x^2 is the class of functions

$$\{x^3/3 + C \mid C \text{ constant}\},$$

some of which are plotted in Fig. 7.5. Indeed, the functions $F(x) := x^3/3 + C$ are "parallel", so they all have the same slope $F'(x) = x^2$ at any point x.

Corollary 7.39. Let F be an antiderivative of the function f in (a, b), let $a < x_0 < b$, and let y_0 be an arbitrary real number. Then there exists a unique antiderivative G of f such that

$$G(x_0) = y_0.$$

Proof. That antiderivative is $G(x) = F(x) + C$ where the constant C is determined by

$$G(x_0) = y_0, \quad \text{and therefore } C = y_0 - F(x_0). \qquad \square$$

Remark 7.40. If f has an antiderivative $F(x)$, then it has infinitely many antiderivatives $F(x) + C$ (see Remark 7.37). Corollary 7.39 shows that f has exactly one antiderivative passing through any given point (x_0, y_0). Alternatively, if two antiderivatives of f intersect, then they are the same.

Table 7.1. Antiderivatives

	Function	Antiderivative		Function	Antiderivative
a	0	C	g	e^x	e^x
b	α (constant)	αx	h	$\sin x$	$-\cos x$
c	x^n	$\dfrac{x^{n+1}}{n+1}, n \in \mathbf{N}$	i	$\cos x$	$\sin x$
d	x^α	$\dfrac{x^{\alpha+1}}{\alpha+1}, \alpha \neq -1$	j	$1/\sqrt{1-x^2}$	$\operatorname{asin} x$
e	$1/x$	$\ln x$	k	$-1/\sqrt{1-x^2}$	$\operatorname{acos} x$
f	a^x	$a^x/\ln a$	l	$1/(1+x^2)$	$\operatorname{atan} x$

The following properties of antiderivatives follow immediately from those of derivatives.

Lemma 7.41 (Properties of antiderivatives). Let the functions f and g have antiderivatives F and G respectively, and let α be a constant. Then:
a. an antiderivative of αf is αF,
b. an antiderivative of $f + g$ is $F + G$.

In Table 7.1 several antiderivatives are listed which are immediate consequences of the corresponding results on derivatives. The constant of integration C is mentioned only once. Every result in Table 7.1 can be proved by differentiation, or by consulting the corresponding entry in the table of derivatives, Table 5.1.

Exercises

7.35 Verify that both $F(x) = \sin^2 x$ and $G(x) = -\cos^2 x$ are antiderivatives of $f(x) = 2 \sin x \cos x$. Explain this in view of Corollary 7.36.

7.36 Lemma 7.41 and Table 7.1c allow computing antiderivatives of polynomials. For example, an antiderivative of $2x^3 - 5x^2 + 7x + 11$ is $\frac{1}{2}x^4 - \frac{5}{3}x^3 + \frac{7}{2}x^2 + 11x + C$.

Prove: The antiderivative of a given polynomial $p(x) = \sum_{k=0}^{n} p_k x^k$ is itself a polynomial

$$q(x) = \sum_{k=0}^{n+1} q_k x^k,$$

where q_0 is arbitrary (the constant of integration), and

$$q_k = p_{k-1}/k, \quad (k = 1, \ldots, n+1). \tag{7.29}$$

7.37 Given a function f and a point $P(x_0, y_0)$, find the antiderivative of f passing through the point P (see Corollary 7.39).
(a) $f(x) := x^2, \quad P(1, -3)$ (b) $f(x) := \sin x, \quad P(\pi/2, 2)$
(c) $f(x) := e^x, \quad P(0, -1)$

7.5 Iterative methods

There are few cases (e.g., f linear, quadratic, or polynomial of degree ≤ 4) where an equation

$$f(x) = 0 , \tag{7.30}$$

can be solved precisely and the solutions can be given in closed form. In all other cases, the most we can get is an approximate solution, usually computed by some iterative method.

This section is an introduction to iterative methods for the approximate solution of (7.30). In Sect. 7.6 we present Newton's method, the best-known iterative method for solving equations.

Definition 7.42. An iterative method uses a function $u(x)$, the *iteration function*, and a point x_0, the *initial point*, to generate a sequence

$$x_1 := u(x_0), \ x_2 := u(x_1), \ x_3 := u(x_2), \ \ldots$$

of *iteration points*, or iterations, written as

$$x_{n+1} := u(x_n), \quad (n = 0, 1, 2, \ldots) . \tag{7.31}$$

A point ξ satisfying

$$\xi = u(\xi) \tag{7.32}$$

is called a *fixed point* of u.

Figure 7.6 illustrates several iterations, starting with the initial point x_0. Note that the iterations (7.31) are obtained by a zig-zag movement, between the graph $y = u(x)$ and the line $y = x$. At the intersections of these two graphs are the points $\xi = u(\xi)$, the fixed points of u.

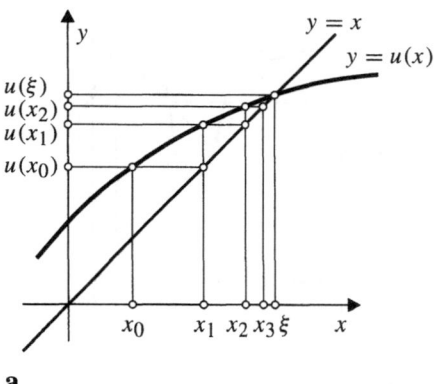
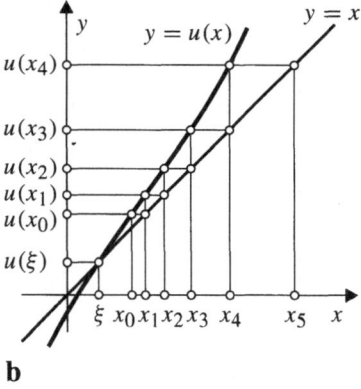

Fig. 7.6 a, b. Iterative methods and fixed points. **a** Convergence of iterations to a fixed point ξ; **b** divergence of iterations, although the initial point is near ξ

Definition 7.43. We say that a sequence of points $\{x_n \mid n = 0, 1, \ldots\}$ *converges* to a point ξ, or that ξ is a *limit* of the sequence $\{x_n\}$, if

$$\lim_{n \to \infty} |x_n - \xi| = 0 , \tag{7.33}$$

and denote this by

$$x_n \to \xi \quad \text{or} \quad \lim_{n \to \infty} x_n = \xi .$$

Note that if the sequence (7.31) converges, that is, $x_n \to \xi$, and if the function u is continuous at ξ, then ξ is a fixed point of u, $\xi = u(\xi)$.

Figure 7.6a illustrates an iterative method which converges to a fixed point ξ. In Fig. 7.6b, the initial point x_0 is close to ξ, but successive iterations move away from it.

A simple condition for the existence of a fixed point is given in the following.

Lemma 7.44. Let u be continuous in $[a, b]$, and let u take points in $[a, b]$ into $[a, b]$, i.e.,

$$\{u(x) \mid a \le x \le b\} \subset [a, b] . \tag{7.34}$$

Then u has a fixed point in $[a, b]$.

Proof. If $u(a) = a$, then it is a fixed point. Similarly, if $u(b) = b$, then b is a fixed point. The remaining case is, by (7.34),

$$u(a) > a \quad \text{and} \quad u(b) < b \quad \text{(Fig. 7.7a)} .$$

Then the function $f(x) := u(x) - x$ is continuous in $[a, b]$ and satisfies (3.23),

$$f(a)f(b) < 0 ,$$

so, by Corollary 3.57, there is a point $\xi \in [a, b]$ such that $f(\xi) = u(\xi) - \xi = 0$. $\square$

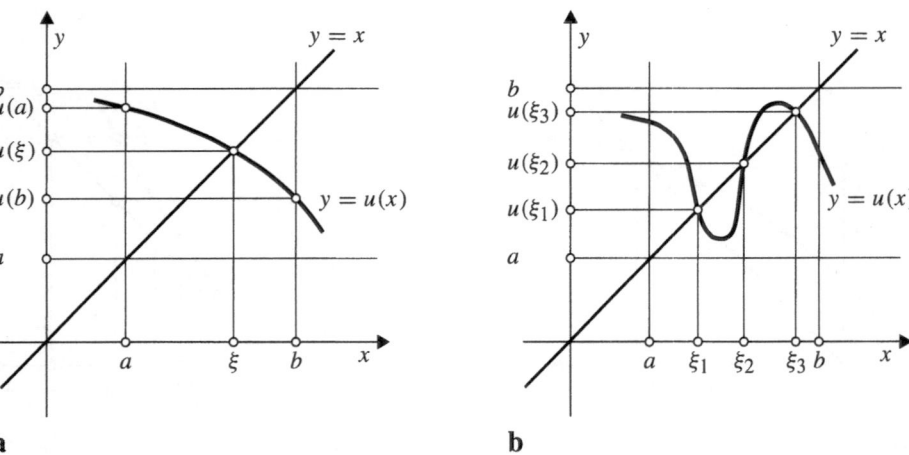

Fig. 7.7 a, b. Fixed points of u in $[a, b]$. **a** A fixed point ξ exists in $[a, b]$. **b** Several fixed points in $[a, b]$

Lemma 7.44 gives conditions for a fixed point to exist in $[a, b]$. There may how-ever be several such fixed points (see Fig. 7.7b). The next result gives conditions for the uniqueness of the fixed point. Its proof uses the mean value theorem.

Lemma 7.45. Let u and $[a, b]$ satisfy the conditions of Lemma 7.44, and in ad-dition, let u be differentiable in $[a, b]$. If there is a constant $0 \le k < 1$ such that

$$|u'(x)| \le k, \quad \text{for all } x \in (a, b) , \tag{7.35}$$

then u has a unique fixed point in $[a, b]$.

Proof. Lemma 7.44 guarantees the existence of fixed points of u in $[a, b]$. Suppose there are two distinct fixed points, ξ_1 and ξ_2 in $[a, b]$, i.e.,

$$u(\xi_1) = \xi_1, \quad u(\xi_2) = \xi_2, \quad \text{and} \quad \xi_1 \ne \xi_2 . \tag{7.36}$$

Applying the mean value theorem (Theorem 7.2) to the function u between ξ_1 and ξ_2, we conclude the existence of an intermediate point ξ where

$$\begin{aligned} u'(\xi) &= \frac{u(\xi_2) - u(\xi_1)}{\xi_2 - \xi_1} \\ &= \frac{\xi_2 - \xi_1}{\xi_2 - \xi_1}, \quad \text{by (7.36)} \\ &= 1 , \end{aligned}$$

contradicting the assumption (7.35). $\qquad\square$

Even if the function u has a fixed point ξ in $[a, b]$, it is not clear that the iterative method (7.31) will converge to ξ. It may also happen that the iterations (7.31) will converge for some initial points x_0 in $[a, b]$, and not for others. The next theorem shows that the above conditions guarantee the *global convergence* of (7.31), that is, convergence from any initial x_0 in $[a, b]$ to the unique fixed point.

Theorem 7.46. Let u and $[a, b]$ satisfy the conditions of Lemmas 7.44 and 7.45, i.e.,
a. u takes points in $[a, b]$ into $[a, b]$, (7.34),

$$\{u(x) \mid a \le x \le b\} \subset [a, b] ,$$

b. u is differentiable in $[a, b]$, and there is a constant $0 \le k < 1$ such that, (7.35),

$$|u'(x)| \le k, \quad \text{for all } x \in (a, b) .$$

Then for any $x_0 \in [a, b]$, the sequence (7.31),

$$x_{n+1} := u(x_n), \quad (n = 0, 1, 2, \dots) ,$$

converges to the unique fixed point ξ of u in $[a, b]$,

$$\lim_{n \to \infty} x_n = \xi, \quad \text{and} \tag{7.37a}$$

$$|x_{n+1} - \xi| \le k|x_n - \xi| \tag{7.37b}$$

$$\le k^{n+1}(b - a) . \tag{7.37c}$$

A proof, using the mean value theorem, is given in Sect. 7.8.

Theorem 7.46 gives conditions under which the iterative method (7.31) converges to a fixed point ξ. While the iteration points $\{x_n\}$ get closer and closer to ξ, they may never "reach" it. It is therefore necessary to decide when to stop. A reasonable *stopping rule* is

$$|x_{n+1} - x_n| < \varepsilon , \tag{7.38}$$

where $\varepsilon > 0$ is a sufficiently small number, determined in advance. Condition (7.38) means that the change in iteration n (from x_n to x_{n+1}) is negligible. Rewriting condition (7.38) as

$$|u(x_n) - x_n| < \varepsilon ,$$

we conclude that x_n is a good approximation to a fixed point.

Lemma 7.47. Let the iteration points $\{x_n\}$ satisfy (7.37b),

$$|x_{n+1} - \xi| \le k|x_n - \xi| ,$$

for some $0 \le k < 1$. Then

$$|x_{n+1} - \xi| \le \frac{k}{1-k}|x_{n+1} - x_n| . \tag{7.39}$$

Proof. Let x_n and x_{n+1} be on the same side of ξ. Therefore, by (7.37b), the point x_{n+1} is between x_n and ξ, and

$$|x_{n+1} - \xi| \le k|x_n - \xi| \le k\big(|x_n - x_{n+1}| + |x_{n+1} - \xi|\big)$$

proving (7.39). Similarly, if x_n and x_{n+1} are on opposite sides of ξ,

$$|x_{n+1} - \xi| \le \frac{k}{1+k}|x_{n+1} - x_n| , \tag{7.40}$$

this proof is left for Exercise 7.42. We can in both cases use the bound (7.39), since

$$\frac{k}{1+k} < \frac{k}{1-k}, \quad \text{for } 0 < k < 1 . \qquad \square$$

The stopping rule (7.38) guarantees, by (7.39), that x_{n+1} is sufficiently close to ξ,

$$|x_{n+1} - \xi| \le \frac{k}{1-k}\varepsilon .$$

Remark 7.48. An arbitrary equation, (7.30),

$$f(x) = 0$$

can be written as

$$x = x - f(x) \,. \tag{7.41}$$

Therefore, a number ξ is a *root* of Eq. (7.30) if, and only if, it is a *fixed point* of the iteration function

$$u(x) := x - f(x) \,. \tag{7.42}$$

For example, any solution of the equation

$$x - \cos x = 0 \tag{7.43a}$$

is a fixed point of the iteration

$$x_{n+1} := \cos(x_n) \,, \tag{7.43b}$$

and vice versa.

MACSYMA-Session 7.4. The MACSYMA function
c1. `iterates(u,x,x0,n) := block([l:[x0], numer:true, val:x0],`
        `for i thru n do`
                `(val:subst(val,x,u),`
                `l:cons(val,l)),`
                    `reverse(l))$`
computes the first $n + 1$ points: $x_0, u(x_0), u(u(x_0)), \ldots, u(u(\ldots(u(x_0))\ldots))$ of the iterative method $x_{n+1} := u(x_n)$, with initial point x_0. Here
- block introduces the variables l, numer, and val;
- l holds the list $x_0, u(x_0), u(u(x_0)), \ldots$, which grows at each iteration, and initially l: [x0], consisting of x_0;
- x0 is the initial point;
- numer:true is an instruction to carry all computations in floating point;
- val holds the value $u(x)$, updated at each iteration by val:subst(val,x,u), which substitutes the previous value of val for x into u;
- l:cons(val,l) constructs the new list l as val followed by the previous list;
- reverse(l) reverses the list upon exiting the block, resulting in the order $x_0, u(x_0), u(u(x_0)), \ldots$

For example, the first 5 iterations of (7.43b), with $x_0 = 0$ are computed by
c2. `iterates(cos(x),x,0,5)`
d2. $[0, 1.0, 0.5403, 0.85755, 0.65429, 0.79348]$

If an iterate is repeated, say $u(x_k) = x_k$, there is no reason to continue the iterations, since all successive iterates will be the same. We can instruct `iterates(u,x,x0,n)` to stop when a value is repeated. Moreover, we can make the 4th argument, n, optional, so if not given, the iterations will continue until a value is repeated. These two features are incorporated in the following MACSYMA function
c3. `iterates(u,x,x0,[n]) := block([l:[x0], numer:true, val:x0],`
                `if n=[] then`

```
                              for i do
                                 (val:subst(val,x,u),
                                  if val=first(l) then return(),
                                  l:cons(val,l))
                        else (n:first(n),
                                 for i thru n do
                                    (val:subst(val,x,u),
                                     l:cons(val,l))),
                     reverse(l))$
```

Here
- [n] indicates that the argument *n* is optional, or a list [n], containing either *n* or nothing (an empty list),
- n=[] means that *n* is not supplied, in which case the iterations continue until val=first(l), i.e., until a value is repeated,
- n:first(n) extracts *n* as the first element of the list [n], in case the list is nonempty; and the other computations are as in c1.

For example, the iterates (7.43b) with $x_0 = 0$,
c4. iterates(cos(x),x,0)
d4.   [0, 1.0, 0.5403, 0.85755, 0.65429, 0.79348, 0.70137, 0.76396, 0.7221, 0.75042,
        0.7314, 0.74424, 0.7356, 0.74142, 0.73751, 0.74015, 0.73837, 0.73957,
        0.73876, 0.7393, 0.73894, 0.73918, 0.73902, 0.73913, 0.73905, 0.73911,
        0.73907, 0.73909, 0.73908, 0.73909, 0.73908, 0.73909, 0.73908, 0.73909,
        0.73908, 0.73909, 0.73908, 0.73908, 0.73908, 0.73908, 0.73908]
producing the fixed point 0.73908. Because we carry only 6 digits, it may seem that the value 0.73908 is repeated 5 times. See what happens when you do the iterations c4 with higher precision.

If we remove the numer:true part in the iterates(u,x,x0,[n]) function, we may get *symbolic* quantities. For example, iterates(cos(x),x,0) would then give $1, \cos(1), \cos(\cos(1)), \cos(\cos(\cos(1)))$, etc.

Similarly, if the iteration function *u* is rational, and the initial x_0 is a rational number, then iterates(u,x,x0,[n]) will return rational numbers with gradually increasing numerators and denominators. As a rule, the chances of a repeated value are much smaller, showing the advantage of using floating-point arithmetic, which we assume.

Not every iteration $x_{n+1} := u(x_n)$ has a fixed point, and even if it has, a sequence of iterates $x_0, u(x_0), u(u(x_0)), \ldots$ does not necessarily converge to that point. For example,
c5. iterates(tan(x),x,1)
d5.   [1, 1.55741, 74.684, −0.86692, −1.17794, −2.41311, 0.89219, 1.24015,
        2.91337, −0.23227, −0.23653, −0.24105, −0.24583, −0.2509, −0.2563,
        −0.26206, −0.26823, −0.27486, −0.28199, −0.28971, −0.2981, −0.30726,
        −0.31731, −0.3284, −0.34074, −0.35457, −0.37022, −0.38811, −0.40885,
        −0.43327, −0.46258, −0.49867, −0.54457, −0.60566, −0.69248, −0.82951,
        −1.09236, −1.92819, 2.67789, −0.50007]
which does not *seem* to have a fixed point. We shall return to this example later. Try this iteration with another initial point, say
c6. iterates(tan(x),x,-1)

Often we need only the last iterate, which can be computed by the MACSYMA function,

```
c7. iterate(u,x,x0,[n]) := last(if n=[]
                             then iterates(u,x,x0)
                             else iterates(u,x,x0,first(n)))$
```

giving the last element of the list `iterates(u,x,x0,[n])`. A drawback of this function is the need to keep the whole list l, although only its last element is needed. This is corrected in the following MACSYMA function,

```
c8. iterate(u,x,x0,[n]) := block([numer:true, val:x0],
                         if n=[] then
                               for i do
                                   (val:subst(val,x,u),
                                   if val=x0 then return(),
                                   x0:val)
                             else (n:first(n),
                                  for i thru n do
                                     val:subst(val,x,u)),
                                  val)$
```

which dispenses with the list l altogether, and:

– if n=[] (i.e., the optional argument n is not given), the variable x_0 is assigned the current value of $u(x)$, x0:val, to be compared with the new value:
  if val=x0 then return()
– if n is supplied, then n iterations val:subst(val,x,u) are carried out, and the last value is returned.

```
c9. iterate(cos(x),x,0,5)
```
d9.   0.79348

If the argument n is not given, and if `iterates(u,x,x0)` of c3 terminates with a repeated value, then `iterate(u,x,x0)` returns this value, which is a fixed point. For example,

```
c10. iterate(cos(x),x,0)
```
d10.   0.73908

Exercises

7.38 Check the following iterations for convergence.

(a) $x_{n+1} := \dfrac{x_n}{2} - \dfrac{1}{n}, \; x_0 := 3$ (b) $x_{n+1} := x_n - \dfrac{x_n - \tan x_n}{\tan^2 x_n}, \; x_0 := 1$

(c) $x_{n+1} := \dfrac{x_n}{2} + \dfrac{1}{n}, \; x_0 := 1$ (d) $x_{n+1} := x_n - \dfrac{1 - e^{-2x_n}}{1 + e^{-2x_n}}, \; x_0 := 1$

(e) $x_{n+1} := x_n + \dfrac{\tan x_n + 1}{\tan x_n - 1}, \; x_0 := 1$ (f) $x_{n+1} := x_n + \dfrac{\tan x_n - 1}{\tan x_n + 1}, \; x_0 := 0.5$

7.39 What do you expect the following iterations have as fixed points?

(a) $x_{n+1} := \dfrac{x_n + 1}{x_n - 1}$ (b) $x_{n+1} := \dfrac{1}{2}(x_n + 1) + \dfrac{1}{x_n - 1}$

7.40 Prove: If a sequence of points $\{x_n\}$ has a limit point ξ (see Definition 7.43), then the point ξ is unique. In other words, a sequence can have at most one limit.

7.41 Give examples of sequences, (a) with a limit point, (b) with no limit point.

7.42 In the proof of Lemma 7.47, prove that (7.40) holds if x_n and x_{n+1} are on opposite
      sides of ξ.

7.6 Newton method

The Newton[7] method (also Newton–Raphson[8] method) is perhaps the best-known
iterative method for computing approximate solutions of an equation (7.30),

$$f(x) = 0 .$$

To explain this method, consider Fig. 7.8a showing the graph of a function f, a
solution ξ of $f(x) = 0$, and a nearby point x_n (called the *current approximation*).
If $|f(x_n)|$ is sufficiently small, we say that the point x_n is a good approximation
for a solution ξ. If not, we would like to get a better approximation.

The tangent of the graph of f at the point $(x_n, f(x_n))$ is given by

$$y = f(x_n) + f'(x_n)(x - x_n) . \tag{7.44}$$

If $f'(x_n) \neq 0$, then this tangent intersects the x-axis. The point of intersection is
found by solving

$$f(x_n) + f'(x_n)(x - x_n) = 0 , \tag{7.45}$$

whose solution is

$$x = x_n - \frac{f(x_n)}{f'(x_n)} ,$$

which, in the Newton method, is taken as the *next approximation* called x_{n+1}.

The Newton method is the iteration

$$x_{n+1} := x_n - \frac{f(x_n)}{f'(x_n)} \tag{7.46}$$

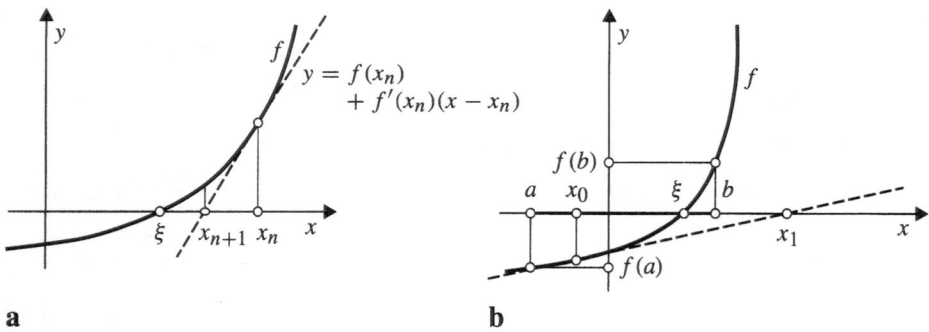

a   **b**

Fig. 7.8 a, b. Newton method. **a** The Newton method will have a happy ending. **b** A bad
start: The function f is not even defined at x_1

---

7 Isaac Newton (1642–1727).
8 Joseph Raphson (1648–1712?).

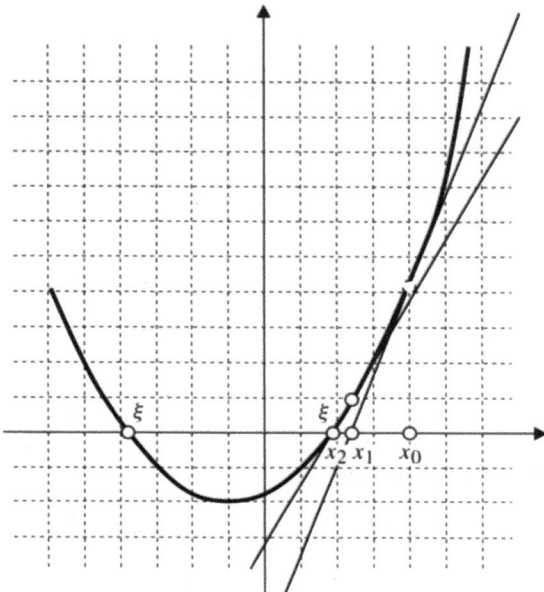

Fig. 7.9. Illustration of Example 7.49

which, at each iteration, solves a linear equation (7.45), obtained by replacing f at x_n by its tangent (7.44).

Example 7.49. Consider the equation

$$f(x) := \frac{x^2 + 2x - 7}{4} = 0 \tag{7.47}$$

(see graph in Fig. 7.9). We use Newton's method to obtain an approximate solution. The Newton iteration is here

$$x_{n+1} := x_n - \frac{f(x_n)}{f'(x_n)} = x_n - \frac{x_n^2 + 2x_n - 7}{2x_n + 2}, \tag{7.48}$$

and our first guess is $x_0 := 4$. The first iteration of (7.48) is

$$x_1 := x_0 - \frac{x_0^2 + 2x_0 - 7}{2x_0 + 2} = 4 - \frac{17}{10} = \frac{23}{10}.$$

Indeed, the tangent of f at $x_0 = 4$ is

$$y = f(x_0) + (x - x_0)f'(x_0) = \frac{17}{4} + (x - 4)\frac{5}{2} = \frac{10x - 23}{4},$$

which intersects the x-axis at $x_1 = 23/10$. The tangent of f at $x_1 = 23/10$ is

$$y = f(x_1) + (x - x_1)f'(x_1) = \frac{660x - 1229}{400}, \quad \text{which is zero for } x = \frac{1229}{660}.$$

Indeed, the second iteration of Newton's method is

$$x_2 := x_1 - \frac{x_1^2 + 2x_1 - 7}{2x_1 + 2} = \frac{1229}{660} \approx 1.8621212121212\dots .$$

These iterations are shown in Fig. 7.9. Note how close is x_2 to the solution ξ, which can be computed as

$$\xi = 2\sqrt{2} - 1 \approx 1.82842712474619\dots .$$

In this example, the Newton method (7.48) converges, to one of the two solutions of (7.47), from any initial solution x_0 except for $x_0 := -1$, where $f'(x_0) = 0$, and (7.48) is meaningless.

MACSYMA-Session 7.5. In this session we implement the Newton iteration (7.46),

$$x_{n+1} := x_n - \frac{f(x_n)}{f'(x_n)} ,$$

another exercise since MACSYMA has an efficient Newton solver. Our implementation uses the functions iterates(u,x,x0,[n]) and iterate(u,x,x0,[n]) from Session 7.4, lines c3 and c7, which must be loaded first.

The MACSYMA function
```
c1. newtons(f,x,x0,[n]) := if n=[] then
                              iterates(x-f/diff(f,x),x,x0)
                           else iterates(x-f/diff(f,x),x,x0,first(n))$
```
produces the Newton iterations (7.46), with initial x_0, and terminates if a value is repeated, or if the optional argument n is given, after n iterations, whichever comes first. Similarly, the last Newton iterate is computed by:
```
c2. newton(f,x,x0,[n]) := if n=[] then
                             iterate(x-f/diff(f,x),x,x0)
                          else iterate(x-f/diff(f,x),x,x0,first(n))$
```

    a. We solve $x - \cos(x) = 0$ by applying the Newton method to $f(x) := x - \cos$:
```
c3. newtons(x-cos(x),x,0)
```
we get convergence in only 4 iterations
```
d3.    [0, 1.0, 0.75036, 0.73911, 0.73908]
```
as compared with iterates(cos(x),x,0) in Session 7.4, line d4.

    b. We apply the Newton method to the function (7.47) of Example 7.49
```
c4. newtons((x^2+2*x-7)/4,x,4)
```
```
d4.    [4, 2.3, 1.86212, 1.82862, 1.82843]
```
obtaining a solution in 4 iterations. To get just the last iterate,
```
c5. newton((x^2+2*x-7)/4,x,4)
```
```
d5.    1.82843
```
one solution of (7.47). Starting at $x_0 = -2$ we get the second solution
```
c6. newton((x^2+2*x-7)/4,x,-2)
```
```
d6.    -3.82843
```

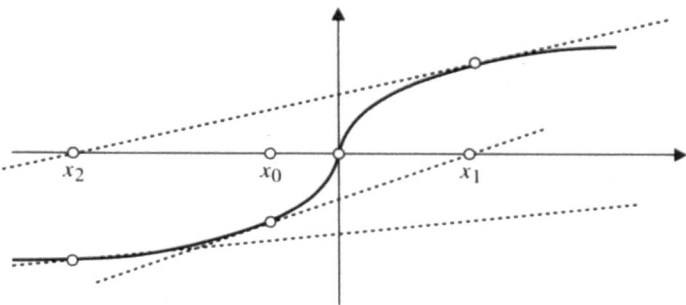

Fig. 7.10. Divergence of Newton's method for $f(x) := \sqrt[3]{x}$

Example 7.50 (Counterexample for Newton's method). An example where Newton's method does not work, for any initial point (other than the solution itself), is the equation

$$f(x) := \sqrt[3]{x} = 0 ,\tag{7.49}$$

which has a single root at $x = 0$. The derivative of f

$$f'(x) = \tfrac{1}{3}x^{-2/3}$$

exists for all $x \neq 0$, and Newton's method for solving (7.49) gives the sequence

$$x_{n+1} := x_n - \frac{f(x_n)}{f'(x_n)} = x_n - \frac{\sqrt[3]{x_n}}{\frac{1}{3}x_n^{-2/3}} = x_n - 3x_n = -2x_n .\tag{7.50}$$

Therefore

$$x_{n+1} = -2x_n = (-2)^2 x_{n-1} = \cdots = (-2)^{n+1} x_0 ,$$

showing that, for any $x_0 \neq 0$, the Newton points (7.50) move farther and farther from the solution $x = 0$. This is illustrated in Fig. 7.10.

The next theorem gives conditions on a function f and an interval I which guarantee that the Newton method works if the initial guess is good.

Theorem 7.51 (Local convergence of Newton's method). Let the function f be differentiable in an interval $[a, b]$, and assume:
a. $f(\xi) = 0$ for some interior point $a < \xi < b$;
b. $f'(\xi) \neq 0$;
c. the derivative f' is continuous at ξ.
Then there exists a subinterval $I \subset [a, b]$ with $\xi \in I$ such that the Newton sequence (7.46),

$$x_{n+1} := x_n - \frac{f(x_n)}{f'(x_n)}$$

converges to ξ for all initial values $x_0 \in I$.

Conversely, if c holds together with

d. $f'(x) \neq 0$ for all $x \in [a, b]$,

and if Newton's method converges to a point $\xi \in (a, b)$, then ξ is a zero of f.

A proof of Theorem 7.51 is given in Sect. 7.8.

Remark 7.52. i. Condition d does not require knowing ξ in advance and is there-fore more practical than the (weaker) condition b. Similarly, condition c can be replaced by the stronger condition

c′ f' is continuous in $[a, b]$

which does not require ξ.

ii. The second part of the theorem says: If Newton's method converges to a point ξ, then ξ is a zero of f. The theorem gives no information in case the Newton method diverges for certain $x_0 \in [a, b]$, in particular, it does not say that in this case there is no solution in $[a, b]$.

iii. Example 7.50 shows that condition c is essential.

iv. The proof (see Sect. 7.8) establishes that

$$|x_{n+1} - \xi| \leq k|x_n - \xi|, \quad \text{for some constant } 0 < k < 1.$$

Actually, the Newton method is better than that, and converges faster, see (7.51).

Theorem 7.53. Let f be twice differentiable on $[a, b]$, let

$$f(a)f(b) < 0$$

(3.23), and assume (i) $f'(x) \neq 0$ in $[a, b]$ and (ii) $|f''(x)|$ is bounded in $[a, b]$. Then:

a.   the interval $[a, b]$ contains a zero ξ of f which is unique;

b.   there is a subinterval I of $[a, b]$, which contains ξ, such that for any $x_0 \in I$ the Newton sequence (7.46),

$$x_{n+1} := x_n - \frac{f(x_n)}{f'(x_n)}$$

belongs to I and converges to ξ according to the rule

$$|x_{n+1} - \xi| \leq k|x_n - \xi|^2, \tag{7.51}$$

where

$$k := \frac{\max_{a \leq x \leq b} |f''(x)|}{2 \min_{a \leq x \leq b} |f'(x)|}. \tag{7.52}$$

Theorem 7.53 is proved in Sect. 7.8. Theorems 7.51 and 7.53 are *local* results in the sense that convergence is guaranteed only if the initial point x_0 is sufficiently close to the (unknown) zero ξ.

If the function f is convex in $[a, b]$, then Theorem 7.53 gives the following *global* result.

Corollary 7.54. Let the function f and the interval $[a, b]$ satisfy the assumptions of Theorem 7.53 and let f be convex on $[a, b]$.

Then, for any x_0 in $[a, b]$ such that

$$f(x_0) > 0, \qquad (7.53)$$

the Newton method (7.46),

$$x_{n+1} := x_n - \frac{f(x_n)}{f'(x_n)},$$

converges to ξ, the unique zero of f in $[a, b]$, and the convergence is monotone in the sense that, for all n, the point x_{n+1} is between x_n and ξ.

Corollary 7.54 is proved in Sect. 7.8.

Remark 7.55. If, in Corollary 7.54, the initial point x_0 is such that $f(x_0) < 0$, then the next point x_1 is on the other side of the root ξ. If $x_1 \leq b$, then, by Corollary 7.54, the Newton method will converge to ξ.

The only thing that can go wrong here is for the point x_1 to fall outside the domain of definition of f, which puts an end to the iteration. This possibility is illustrated in Fig. 7.8b.

Example 7.56 (Calculation of square roots). The square root $\xi := \sqrt{a}$ is a zero of the function $f(x) := x^2 - a$. The Newton sequence here is given by

$$x_{n+1} := \tfrac{1}{2}\left(x_n + \frac{a}{x_n}\right). \qquad (7.54)$$

Since f is convex, it follows from Corollary 7.54 that the Newton method converges for any $x_0 > \sqrt{a}$.

Remark 7.57. If a sequence $\{x_n\}$ converges to a point ξ, and if there is a positive β such that

$$|x_{n+1} - \xi| \leq \beta|x_n - \xi|^2,$$

then the convergence is called quadratic. Theorem 7.53 shows that, under some reasonable assumptions, the Newton method has quadratic convergence, (7.51),

$$|x_{n+1} - \xi| \leq k|x_n - \xi|^2.$$

To see what this means in practice, suppose that the constant

$$k := \frac{\max_{a \leq x \leq b} |f''(x)|}{2 \min_{a \leq x \leq b} |f'(x)|}$$

occurring in inequality (7.51) is approximately 1. Let the current solution x_n have an absolute error $|x_n - \xi| \approx 10^{-d}$ (roughly speaking, this means that the first d decimals of x_n are correct). Then the next iterate x_{n+1} has an absolute error of at most 10^{-2d}, and therefore at least $2d$ correct decimals.

The quadratic convergence of the Newton method thus guarantees that, sufficiently close to the sought root, each iteration roughly doubles the number of significant decimal places of the approximate solution.

This is illustrated in MACSYMA-Session 7.5, example b, where the correct decimals are

$$x_2 \approx 1.8 \,,$$

$$x_3 \approx 1.828 \,,$$

$$x_4 \approx 1.8284271 \,.$$

MACSYMA-Session 7.6. It is reasonable to stop the Newton iterations when the value of $f(x)$ is sufficiently close to 0, say $|f(x_n)| < \varepsilon$, where ε is a sufficiently small tolerance, in which case x_n can be taken as a good approximation of the sought solution. We implement this with the MACSYMA function

```
c1. newton_eps(f,x,x0,eps) := block([val:x0, u:x-f/diff(f,x),
                                      numer:true],
                         for i do
                             (if abs(subst(val,x,f)) < eps
                                 then return(),
                                 else val:subst(val,x,u))
                         val)$
```

where:
- u is the iteration function of the Newton method, $u(x) := x - f(x)/f'(x)$,
- eps is the tolerance (sufficiently small positive number),
- if abs(subst(val,x,f)) < eps is the stopping criterion $|f(x_n)| < \varepsilon$,
- val is the current iterate, updated by val:subst(val,x,u).

For example, to approximate the solution of $x - \cos(x) = 0$, with initial point $x_0 = 0$ and tolerance $\varepsilon = 0.001$,

```
c2. newton_eps(x-cos(x),x,0,.001)
```
d2.   0.73911

To obtain a better approximation of the solution, we lower the tolerance,

```
c3. newton_eps(x-cos(x),x,0,.000001)
```
d3.   0.73908

MACSYMA-Session 7.7. If the function f and its first $(p-1)$ derivatives vanish at $x = \psi$, but the pth derivative is nonzero at $x = \psi$, i.e., if

$$f(\xi) = f'(\xi) = f''(\xi) = f^{p-1}(\xi) = 0, \quad \text{and} \quad f^p(\xi) \neq 0 \,,$$

then ξ is called a *multiple root* of f, with *multiplicity* p. The Newton method may still converge to such a multiple root, but the rate of convergence is typically linear, i.e.,

$$|x_{n+1} - \xi| \le k|x_n - \xi|\,,$$

for some $0 \le k \le 1$. Indeed, the quadratic convergence was proved, in Theorem 7.53, under the condition $f'(x) \ne 0$, in $[a, b]$, which is violated by the $f'(\xi) = 0$ above.

The function $f(x) := (x - 1)^2$ has 1 as a multiple root with multiplicity 2, and the Newton method converges to it, from any initial point x_0, at a linear rate. For example, using the function `newtons(x,f,x0,n)` of Session 7.5,

c1. `newtons((x-1)^2,x,100,10)`

d1.   [100, 50.5, 25.75, 13.375, 7.1875, 4.09375, 2.54688, 1.77344, 1.38672, 1.19336, 1.09668]

the same iterations as generated by bisection.

Here is a trick to recover a quadratic convergence near a multiple root ξ. Instead of solving $f(x) = 0$, we solve

$$f(x)/f'(x) = 0\,. \tag{7.55}$$

The following MACSYMA function is an application of the Newton method to f/f',

c2. `second_newtons(f,x,x0,n):=newtons(f/diff(f,x),x,x0,n)\)`

and under reasonable assumptions will have quadratic convergence near a multiple root ξ.

For the function $f(x) := (x - 1)^2$ we have

c3. `second_newtons((x-1)^2,x,100,10)`

d3.   [100, 1, 1, 1, 1, 1, 1, 1, 1, 1, 1]

the multiple solution $\xi = 1$ was reached in just one iteration! (Compare with d1 above).

As an illustration of the strength of this procedure we return to the function $f(x) = x - \tan x$ encountered in Session 7.4.

c4. `second_newtons(x-tan(x),x,2,10)`

d4.   [2, 1.75872, 1.61498, 1.57368, 1.57081, 1.5708, 1.5708, 1.5708, 1.5708, 1.5708]

converging to the fixed point 1.5708 of $x := \tan x$ in just 6 iterations! In comparison, the Newton method:

c5. `newtons(x-tan(x),x,10,10)`

diverges (try it).

Exercises

7.43 Consider the equations
       (a) $x^{1/3} = 0$ (see Example 7.50)   (b) $x^{2/5} = 0$
       In each case, find an initial point x_0 such that the Newton method oscillates forever between the points x_0 and $-x_0$.

7.44 Use MACSYMA to reproduce the calculations of Example 7.49 and go further until x_5. Utilize the function `tangent(f,x,x0)` (see MACSYMA-Session 4.6) that produces the right-hand side of the equation of the tangent at the graph of f with respect to variable x at the point $(x_0, f(x_0))$.

7.45 Write a recursive MACSYMA function that computes the Newton sequence x_n and stops when $|f(x_n)| < \varepsilon$, for a given $\varepsilon > 0$.

7.46 Consider the sequence of polynomials

$$f_n(x) := \sum_{k=0}^{n} \frac{(-1)^k}{(2k)!} x^{2k} \ .$$

All of them have a zero between $x = 0$ and $x = 2$. Find the zeros with Newton's method for $n := 1, \ldots, 10$. Plot the functions and interpret the results.

7.47 Use MACSYMA to plot the following functions $f(x)$, and find their real zeros by Newton's method.

(a) $f(x) := x^3 - 2x + 1$ (b) $f(x) := 1 - \dfrac{x}{2} + \dfrac{x^2}{3} - \dfrac{x^3}{4} + \dfrac{x^4}{5}$

(c) $f(x) := \displaystyle\sum_{k=0}^{n} \frac{x^k}{k!}$ for $n = 1, 3, 5, 7, 9$ (d) $f(x) := x - \dfrac{x^3}{3!} + \dfrac{x^5}{5!} - \dfrac{x^7}{7!} + \dfrac{x^9}{9!}$

(e) $f(x) := 4 + 3x - 15x^2 + 5x^4$ (f) $f(x) := 1 - 10x + 7x^3$

7.48 Use MACSYMA (a) to plot the function in Example 7.50 and (b) to illustrate that Newton's method diverges for any initial $x_0 \neq 0$.

7.49 Use the Newton method (7.54) for computing square roots to find $\sqrt{k}$ for $k := 2, 3, \ldots, 10$. Repeat the computations for a precision of 6 and 12 digits.

7.50 Consider Newton's method for $f(x) := x^p - a$ ($p > 2$, $a \in \mathbf{R}^+$), and use it to calculate numerical values for

(a) $\sqrt[3]{4}$ (b) $\sqrt[5]{120}$ (c) $\sqrt[4]{2}$

(d) $\sqrt[123]{456}$ (e) $\sqrt[3]{27}$ (f) $\sqrt[10]{0.1}$

7.51 If a polynomial has a rational root, it can be computed exactly by the function newton since MACSYMA stores internally all real numbers as rationals.

Given a polynomial f, approximate the expression newton(f,x,x0) setting the bfprecision to some value n. The default precision is 20. Call the bfloat number ξ. Now substitute ξ in the polynomial f. If $f(\xi) = 0$, you are done. Otherwise, increase the precision by setting bfprecision equal to a higher value and repeat the computations.

Find the rational zeros of

(a) $f(x) := 12x^2 + 11x - 15$

(b) $f(x) := 86086415630x^2 - 34753086513x - 5555555505$

(c) $f(x) := 1245x^3 - 82236x^2 - 165717x - 254178$

(d) $f(x) := 6720x^5 - 27656x^4 + 45494x^3 - 37391x^2 + 15354x - 2520$

(e) $f(x) := \displaystyle\sum_{k=0}^{10} \binom{10}{k} x^k$

7.52 Recall Example 7.50, and give another example of a function f where the Newton method fails for any x_0 other than the solution.

7.53 Consider Newton's method for $f(x) := x^2$. Calculate the Newton sequence. Is it quadratic convergent? Explain!

7.54 Use $f(x) := 1/x^2 - a$ to give an alternative method to calculate the square root of a. Recalculate Exercise 7.49. Can you see an advantage of the new method?

7.55 If the initial point x_0 is complex, the Newton method may converge to some complex root of $f(x) = 0$. Use this to factorize the following real polynomials

(a) $f(x) := x^2 + 2x + 2$ (b) $f(x) := x^2 + 6x + 10$

(c) $f(x) := x^4 - 1$ (d) $f(x) := 4x^4 + 1$

Hint: Having found a zero ξ, divide the polynomial by the factor $x - \xi$, simplify to get a polynomial of a lower degree, then compute another root, etc.

7.56 a. Find the iteration function u of the Newton method for solving (7.55),

$$f(x)/f'(x) = 0 \quad \text{(see MACSYMA-Session 7.7)} .$$

b. Note that this method can be applied, without harm, also if $f'(x) \neq 0$ throughout $[a, b]$. Does it converge faster than the ordinary Newton method? To get an idea, compare the performance of newtons(f,x,x0,n) (MACSYMA-Session 7.5) and second_newtons(f,x,x0,n) (MACSYMA-Session 7.7) on several examples of $f(x) = 0$.

7.57 Let ξ be a root of multiplicity p of $f(x) = 0$. Then the iterative method

$$x_{n+1} := x_n - p \frac{f(x_n)}{f'(x_n)} , \tag{7.56}$$

called the Schröder method, converges to ξ (under some reasonable assumptions) and the rate of convergence is quadratic. For $p = 1$, (7.56) reduces to the Newton method.

Since the multiplicity of the root ξ must be known in advance, the Schröder method is not practical in some cases. However, if the Newton method converges slowly, then it may be worthwhile trying (7.56) with $p = 2, 3$ etc.

Use Schröder's method to solve

(a) $(x - 1)^2 = 0$ (b) $(x - 1)^3 = 0$ (c) $(x - 1)^4 = 0$ (d) $(x - \cos x)^2 = 0$

and compare with the Newton method, and the method second_newtons of MACSYMA-Session 7.7.

7.58 The Newton method can be adapted to compute the stationary points of a function f, which are the solutions of

$$f'(x) = 0 .$$

Replacing f by f' in (7.46) we get the iterative method

$$x_{n+1} := x_n - \frac{f'(x_n)}{f''(x_n)} ,$$

which approximates a stationary point of f near a given point x_0. The nth iteration is computed by the following MACSYMA function

stat_pt_newton(f,x,x0,n):=iterate(x-dif(f,x)/dif(f,x,2),x,x0,n)

In the following examples, approximate a stationary point of f near the given point x_0.

(a) $f(x) := 3x^4 - 8x^3 + 6x^2 - 2$, $x_0 = 5.0$

(b) $f(x) := 3x^4 - 8x^3 + 6x^2 - 2$, $x_0 = -5.0$

(c) $f(x) := \dfrac{x^2 + 2x + 3}{4x + 5}$, $x_0 = 0.2$ (d) $f(x) := \dfrac{x^2 + 2x + 3}{4x + 5}$, $x_0 = -2.5$

7.7 Fixed points*

The next example illustrates possible behavior of an iterative method near a fixed point.

Example 7.58. Consider the quadratic equation

$$f(x) := x^2 - 3x + 2 = 0,$$

with roots $\xi_1 = 1, \xi_2 = 2$. As in Remark 7.48 we use the iteration function

$$u(x) := x - f(x) = -x^2 + 4x - 2$$

(whose fixed points are exactly ξ_1, ξ_2) and the iterative method

$$x_{n+1} := -x_n^2 + 4x_n - 2, \quad (n = 0, 1, \dots) \tag{7.57}$$

implemented by the MACSYMA function `iterates(u,x,x0,n)` of Session 7.4, which we load first. The outcome depends on the choice of the initial point x_0. Since $u(\xi_1) = \xi_1$ and $u(\xi_2) = \xi_2$, we consider initial points different than $\xi_1 = 1$, $\xi_2 = 2$, and distinguish four cases.

Case 1: $-\infty < x_0 < 1$. In this case the iterations $\{x_n\}$ tend to $-\infty$. For example, starting with $x_0 = 0.999$, the first 17 iterations are:

c1. `iterates(-x^2+4*x-2,x,0.999,17), bfloat`

d1. [9.9899983406066894531b−1, 9.9799871444702148438b−1,
9.9599361419677734375b−1, 9.9197125434875488281b−1,
9.8387813568115234375b−1, 9.6749639511108398438b−1,
9.3393635749816894531b−1, 8.6350846290588378906b−1,
7.0838713645935058594b−1, 3.3173632621765136719b−1,
−7.8310370445251464844b−1, −5.7456645965576171875b0,
−5.799530029296875b1, −3.5974345703125b3, −1.2955924b7,
−1.67855911862272b14, −2.8175602941257177117b28,
−3.4028236692093846346b38]

so $-\infty$ cannot be too far away.

Case 2: $1 < x_0 < 2$. Here the iterations $\{x_n\}$ converge to the fixed point ξ_2. For example, starting with $x_0 = 1.001$,

c2. `iterates(-x^2+4*x-2,x,1.001,17), bfloat`

d2. [1.0009999275207519531b0, 1.0019989013671875b0,
1.0039939880371093750b0, 1.0079722404479980469b0,
1.0158810615539550781b0, 1.0315098762512207031b0,
1.0620269775390625b0, 1.1202068328857421875b0,
1.2259640693664550781b0, 1.4008684158325195313b0,
1.6410417556762695313b0, 1.8711490631103515625b0,
1.9833974838256835938b0, 1.9997243881225585938b0,
2.0b0, 2.0b0, 2.0b0, 2.0b0]

---

* This section is optional. It can be skipped at first reading. The material covered here is not used elsewhere in this book.

Case 3: $2 < x_0 \leq 3$. The sequence converges to ξ_2. For example, with $x_0 = 2.995$,

c3. `iterates(-x^2+4*x-2,x,2.995,17), bfloat`

d3.   [2.9949998855590820313b0, 1.0099754333349609375b0,
       1.0198516845703125b0, 1.0393095016479492188b0,
       1.0770740509033203125b0, 1.1482076644897460938b0,
       1.2744498252868652344b0, 1.4735770225524902344b0,
       1.7228794097900390625b0, 1.9232044219970703125b0,
       1.99410247802734375b0, 1.9999656677724609375b0,
       2.0b0, 2.0b0, 2.0b0, 2.0b0, 2.0b0, 2.0b0]

Case 4: $3 < x_0 < \infty$. The first iteration gives a point $x_1 < 1$, and the iterations diverge as in case 1.

c4. `iterates(-x^2+4*x-2,x,3.005,15), bfloat`

d4.   [3.00500011444409179688b0, 9.899749755859375b−1,
       9.7984957695007324219b−1, 9.5929312705993652344b−1,
       9.1692924499511171875b−1, 8.2695794105529785156b−1,
       6.2397241592407226563b−1, 1.0654819011688232422b−1,
       −1.5851597785949707031b0, −1.0853366851806640625b1,
       −1.63208984375b2, −2.72920078125b4, −7.4496256b8,
       −5.54969148271099904b17, −3.0799068744816352905b35,
       −3.4028236692093846346b38]

The method (7.57) thus behaves differently if the initial point x_0 is near the fixed point ξ_1 or near the fixed point ξ_2.

All sequences (7.57) which come near the fixed point ξ_2 find it irresistibly attractive. In fact, there is a neighborhood $N = (1, 3)$ of ξ_2, so that a sequence that begins in N stays forever in N, and converges to ξ_2. This is shown in cases 2 and 3 above.

The story is quite different with the fixed point ξ_1, which is repulsive in the sense that for no initial point x_0 near ξ_1 (other than ξ_1 itself) will the method (7.57) converge to ξ_1. See cases 1 and 2.

This example suggests the following classification of fixed points.

Definition 7.59. Let ξ be a fixed point of the iteration function u. Then ξ is:

a. a *stable* (also attracting) *fixed point* if there is a neighborhood N of ξ so that every sequence (7.31) which begins in N (or enters it) converges to ξ,

$$(x_0 \in N) \quad \Longrightarrow \quad (x_n \to \xi)$$

b. an *unstable* (also repelling) *fixed point* if there is a neighborhood N of ξ such that every sequence (7.31) which begins in $N \setminus \{\xi\}$ (the set N "without" the single point ξ) eventually leaves it.

The iterative method (7.31) is meant to converge to a fixed point ξ. If ξ is stable, then the method works as expected in a neighborhood of ξ. This is illustrated in Fig. 7.6a. On the other hand, if ξ is a repelling fixed point (Fig. 7.6b), then the method (7.31) is quite useless near ξ (although it may "hit" ξ accidentally, from a distance).

We see now that the attractiveness of the fixed point ξ is determined by the value of the derivative $u'(\xi)$.

Theorem 7.60. Let $u(x)$ be an iteration function defined in an interval I, and let the fixed point $\xi = u(\xi)$ be interior to I, with u differentiable at ξ, and the derivative $u'(x)$ continuous at ξ.

a. If $|u'(\xi)| < 1$, then ξ is stable, and

$$\lim_{n \to \infty} \frac{x_{n+1} - \xi}{x_n - \xi} = u'(\xi) . \tag{7.58}$$

b. If $|u'(\xi)| > 1$, then ξ is unstable.

A proof of Theorem 7.60 is given in Sect. 7.8.

Remark 7.61. Theorem 7.60 explains the behavior of the fixed points in Example 7.58. For $u(x) = -x^2 + 4x - 2$, the derivative is $u'(x) = -2x + 4$. At the fixed point $\xi_1 = 1$ we have

$$|u'(\xi_1)| = 2$$

and ξ_1 is unstable by Theorem 7.60b. At $\xi_2 = 2$ we have

$$|u'(\xi_2)| = 0$$

and, by Theorem 7.60a, ξ_2 is stable. Its interval of attraction is $(1, 3)$, but this information is not contained in Theorem 7.60.

Remark 7.62. Consider again Remark 7.48. If instead of $u(x) := x - f(x)$ we select

$$u(x) := x - \omega f(x), \quad \omega \text{ nonzero constant} , \tag{7.59}$$

then the fixed points of $u(x)$ are still roots of $f(x) = 0$, but the derivative of u is

$$u'(x) = 1 - \omega f'(x) .$$

By selecting a suitable ω we can therefore control the convergence of the iterative method

$$x_{n+1} := x_n - \omega f(x_n) , \tag{7.60}$$

making fixed points more attractive. Indeed, if we expect a fixed point in the interval I, and if

$$0 < f'(x) < M$$

for all $x \in I$, then any ω with

$$0 < \omega < 2/M$$

guarantees $|u'(x)| < 1$ for all $x \in I$, "stabilizing" any fixed point which lies in I.

Similarly, if

$$-M < f'(x) < 0$$

for all $x \in I$, then we use any ω such that

$$-2/M < \omega < 0.$$

Example 7.63. We illustrate the use of Remark 7.62. Recall Example 7.58 with $f(x) = x^2 - 3x + 2$, and the unstable fixed point $\xi_1 = 1$. Since $f'(\xi_1) = -1$, we take $M = 2$ and $\omega = -1$. The new iteration function is

$$u(x) := x - \omega f(x) = x - (-1)(x^2 - 3x + 2) = x^2 - 2x + 2$$

and now the fixed point $\xi_1 = 1$ is stable. (However, what happened to that other fixed point, $\xi_2 = 2$, which was stable before?)

In Example 7.50, the solution $x = 0$ is an unstable fixed point of the Newton method. The next result gives conditions for a solution of $f(x) = 0$ to be a stable fixed point of the Newton method.

Corollary 7.64. If ξ is a root of $f(x) = 0$, if $0 < |f'(\xi)| < \infty$ and $f''(x)$ is continuous at ξ, then ξ is a stable fixed point for the Newton method (7.46).

Proof. From

$$u(x) = x - \frac{f(x)}{f'(x)}$$

we get

$$u'(x) = 1 - \frac{f'(x)^2 - f(x)f''(x)}{f'(x)^2} = \frac{f(x)f''(x)}{f'(x)^2}$$

so that

$$u'(\xi) = \frac{f(\xi)f''(\xi)}{f'(\xi)^2} = 0$$

and, by Theorem 7.60, ξ is stable.   □

Exercises

7.59 Determine and classify the fixed points of the iteration scheme $x_{n+1} := \frac{1}{3}((x_n)^2 + 1)$.

7.60 Discuss the "rational iteration" scheme

$$x_{n+1} := \frac{ax_n + b}{cx_n + d}$$

under the conditions (a) $ad - bc = 0$, (b) $ad - bc = 1$.

7.61 Prove that if $u(x) := ax^2 + bx + c, a \neq 0$, and if u has two distinct fixed points, then at least one of them is unstable.

7.62 The equation

$$x^3 + x = 1000 \tag{7.61}$$

can be written as $x = u(x)$, where $u(x)$ has one of the forms

(a) $u(x) := 1000 - x^3$ (b) $u(x) := \dfrac{1000}{x^2} - \dfrac{1}{x}$ (c) $u(x) := \sqrt[3]{1000 - x}$

Use one of these as an iteration function for the iterative computation of the greatest positive root of (7.61), which is approximately equal to 10. Compute it to 6 decimal places.

7.63 Give an example of an iteration function u, and a fixed point ξ, which is neither stable nor unstable.

Hint: By Theorem 7.60, $|u'(\xi)| = 1$.

7.64 The iteration function

$$u(x) := \mu x (1 - x)$$

is called the logistic function. Here μ is a positive parameter, which determines the nature of the fixed points of the iterations

$$x_{n+1} := \mu x_n (1 - x_n) . \tag{7.62}$$

a.   Find the fixed points of (7.62), as functions of μ.
b.   Determine if these fixed points are stable or unstable.

7.8 Proofs

Proof of Theorem 7.30. The proof uses the Cauchy mean value theorem (Theorem 7.3). Let $F(x)$ and $G(x)$ be continuously differentiable on $[a, b]$, $F(x_0) = G(x_0) = 0$, and let $G'(x) \neq 0$ in $[a, b]$. Then for any $x \in [a, b]$

$$\frac{F(x)}{G(x)} = \frac{F(x) - F(x_0)}{G(x) - G(x_0)} = \frac{F'(\xi_0)}{G'(\xi_0)} ,$$

where ξ_0 is between x and x_0.

Similarly, if F and G have continuous second derivatives on $[a, b]$,

$$F(x_0) = G(x_0) = F'(x_0) = G'(x_0) = 0 ,$$

and $G'' \neq 0$ in $[a, b]$, then

$$\frac{F(x)}{G(x)} = \frac{F(x) - F(x_0)}{G(x) - G(x_0)} = \frac{F'(\xi_0)}{G'(\xi_0)} = \frac{F'(\xi_0) - F'(x_0)}{G'(\xi_0) - G'(x_0)} = \frac{F''(\xi_1)}{G''(\xi_1)} ,$$

for some ξ_1 between ξ_0 and x_0.

Continuing in this fashion, let F and G have n continuous derivatives in $[a, b]$, and an $(n+1)$th-derivative in (a, b), and suppose

$$F(x_0) = G(x_0) = F'(x_0) = G'(x_0) = F''(x_0) = G''(x_0) =$$
$$= \cdots = F^{(n)}(x_0) = G^{(n)}(x_0) = 0 . \tag{7.63}$$

Then there is a point ξ between x and x_0 such that

$$\frac{F(x)}{G(x)} = \frac{F^{(n+1)}(\xi)}{G^{(n+1)}(\xi)} \ . \tag{7.64}$$

We apply this now to the functions

$$F(x) := f(x) - T_n(f, x, x_0) \quad \text{and} \quad G(x) := (x - x_0)^{n+1} \ ,$$

which satisfy (7.63) (check!). We conclude, as in (7.64), that there is a point ξ between x and x_0 such that

$$\frac{f(x) - T_n(f, x, x_0)}{(x - x_0)^{n+1}} = \frac{f^{(n+1)}(\xi)}{(n+1)!}$$

so that

$$f(x) = T_n(f, x, x_0) + \frac{f^{(n+1)}(\xi)}{(n+1)!}(x - x_0)^{n+1} \ . \qquad \square$$

Proof of Corollary 7.32. a. "If": Let x_0 and x_1 be any two points in I, let $0 < \lambda < 1$, and define

$$x_\lambda := (1 - \lambda)x_0 + \lambda x_1 \ . \tag{7.65}$$

Then, by the Taylor theorem,

$$f(x_0) = f(x_\lambda) + f'(x_\lambda)(x_0 - x_\lambda) + \frac{f''(\xi_0)}{2}(x_0 - x_\lambda)^2 \ ,$$
$$\text{for some } \xi_0 \text{ between } x_0 \text{ and } x_\lambda \ ,$$

$$f(x_1) = f(x_\lambda) + f'(x_\lambda)(x_1 - x_\lambda) + \frac{f''(\xi_1)}{2}(x_1 - x_\lambda)^2 \ ,$$
$$\text{for some } \xi_1 \text{ between } x_1 \text{ and } x_\lambda \ .$$

$$\therefore \ (1 - \lambda)f(x_0) + \lambda f(x_1) = f(x_\lambda) + f'(x_\lambda)((1 - \lambda)x_0 + \lambda x_1 - x_\lambda) +$$
$$+ (1 - \lambda)\frac{f''(\xi_0)}{2}(x_0 - x_\lambda)^2 +$$
$$+ \lambda \frac{f''(\xi_1)}{2}(x_1 - x_\lambda)^2$$
$$= f(x_\lambda) + (1 - \lambda)\frac{f''(\xi_0)}{2}(x_0 - x_\lambda)^2 +$$
$$+ \lambda \frac{f''(\xi_1)}{2}(x_1 - x_\lambda)^2, \quad \text{by (7.65)} \ ,$$
$$\geq f(x_\lambda), \quad \text{since } f'' \text{ is nonnegative in } I \ . \tag{7.66}$$

$$\therefore \ (1 - \lambda)f(x_0) + \lambda f(x_1) \geq f((1 - \lambda)x_0 + \lambda x_1)$$

proving the convexity of f (see Definition 6.26).

"Only if": The convexity of f implies that the derivative f' is nondecreasing, by Theorem 6.38. Therefore, $f'' \geq 0$, by Corollary 7.6.

b. If f'' is positive in I, then there is a strict inequality in (7.66), proving the strict convexity of f.   □

Proof of Theorem 7.46. Let x_0 be an arbitrary point in $[a, b]$. Then

$$x_1 := u(x_0) \quad \text{is also in } [a, b], \quad \text{by (7.34)}.$$

For the same reason, if $x_n \in [a, b]$, then $x_{n+1} \in [a, b]$. This proves, by induction on n, that all the iterations (7.31),

$$x_{n+1} := u(x_n), \qquad (n = 0, 1, 2, \dots)$$

are in $[a, b]$. Let ξ be the unique fixed point if u in $[a, b]$, guaranteed by Lemmas 7.44 and 7.45. Then

$$
\begin{aligned}
x_{n+1} - \xi &= u(x_n) - \xi \\
&= u(x_n) - u(\xi), \quad \text{since } \xi \text{ is a fixed point}, \\
&= u'(\eta)(x_n - \xi), \quad \text{for some point } \eta \text{ between } x_n \text{ and } \xi, \\
&\qquad\qquad\qquad\quad \text{by the mean value theorem},
\end{aligned}
$$

$$
\begin{aligned}
\therefore \quad |x_{n+1} - \xi| &= |u'(\eta)|\,|x_n - \xi| \\
&\leq k|x_n - \xi|, \quad \text{by (7.35), proving (7.37b)}, \\
&\leq k^2|x_{n-1} - \xi| \leq k^3|x_{n-2} - \xi| \leq \cdots \leq k^{n+1}|x_0 - \xi| \\
&\leq k^{n+1}(b - a), \quad \text{since both } x_0 \text{ and } \xi \text{ are in } [a, b], \\
&\qquad\qquad\qquad\quad \text{proving (7.37c)}. \qquad\qquad\qquad\qquad (7.67)
\end{aligned}
$$

Since $k < 1$, it follows that $\lim_{n \to \infty} |x_n - \xi| = 0$, proving (7.37a).   □

Proof of Theorem 7.51. We assume in b that $|f'(\xi)| = m > 0$. Since f' is continuous at ξ, there is an interval $I_1 \subset [a, b]$ with $\xi \in I_1$ such that
e. $|f'(x)| > m/2$ for all $x \in I_1$.
Let an arbitrary $\varepsilon > 0$ be given. The differentiability of f at the point ξ implies that
f. $|f'(\xi) - (f(x) - f(\xi))/(x - \xi)| < \varepsilon$, for all $x \in I_2$
for some interval $I_2 \subset [a, b]$ with $\xi \in I_2$. Finally the continuity of f' at the point ξ implies that
g. $|f'(x) - f'(\xi)| < \varepsilon$, for all $x \in I_3$
and some interval $I_3 \subset [a, b]$ with $\xi \in I_3$. All the above properties hold in the interval

$$I := I_1 \cap I_2 \cap I_3 .$$

Note that I depends on the choice of ε.

For $x_n \in I$ we compute

$$|x_{n+1} - \xi| = \left| x_n - \frac{f(x_n)}{f'(x_n)} - \xi \right| = \frac{|(x_n - \xi)f'(x_n) - f(x_n)|}{|f'(x_n)|}$$

$$< \frac{2}{m}|x_n - \xi|\left| f'(x_n) - \frac{f(x_n) - f(\xi)}{x_n - \xi} \right|, \quad \text{by (a) and (e)},$$

$$= \frac{2}{m}|x_n - \xi|\left| f'(x_n) - f'(\xi) + f'(\xi) - \frac{f(x_n) - f(\xi)}{x_n - \xi} \right|$$

$$\leq \frac{2}{m}|x_n - \xi|\left(\left| f'(x_n) - f'(\xi) \right| + \left| f'(\xi) - \frac{f(x_n) - f(\xi)}{x_n - \xi} \right| \right)$$

by the triangle inequality

$$< \frac{2}{m}|x_n - \xi|(\varepsilon + \varepsilon) = \frac{4\varepsilon}{m}|x_n - \xi|, \quad \text{by (f) and (g)}.$$

We can summarize this as

$$|x_{n+1} - \xi| \leq k|x_n - \xi|, \quad \text{where } k := \frac{4\varepsilon}{m},$$

proving[9] that $\{x_n\}$ converges to ξ, provided $k < 1$, i.e., if ε is sufficiently small[10]. For example, the value $\varepsilon := m/8$ makes $k = 1/2$ which guarantees convergence.

On the other hand, if the Newton sequence x_n converges to a point ξ, then from

$$x_{n+1} - x_n = (x_{n+1} - \xi) - (x_n - \xi) \quad \text{it follows that}$$
$$|x_{n+1} - x_n| \leq |x_{n+1} - \xi| + |x_n - \xi| \quad \text{showing that}$$
$$\lim_{n \to \infty} |x_{n+1} - x_n| = 0.$$

$$\therefore \quad \lim_{n \to \infty} \left| \frac{f(x_n)}{f'(x_n)} \right| = 0, \quad \text{by (7.46), proving that } f(x_n) \to 0 \text{ as } n \to \infty. \quad \square$$

Proof of Theorem 7.53. It follows from (3.23) that f has a zero ξ in (a, b) (see Corollary 3.57). Since $f'(x) \neq 0$ in $[a, b]$, it follows from Rolle's theorem (Theorem 7.1) that ξ is the only zero of f in $[a, b]$.

Let x_n be a point in (a, b). Then by Taylor's theorem

$$0 = f(\xi) = f(x_n) + f'(x_n)(\xi - x_n) + \frac{f''(\xi_1)}{2}(\xi - x_n)^2$$

for some point ξ_1 between ξ and x_n. We rewrite this as

$$-f(x_n) = f'(x_n)(\xi - x_n) + \frac{f''(\xi_1)}{2}(\xi - x_n)^2$$

---

9 As in (7.67).

10 Small ε means small interval I.

and substitute in (7.46) to get

$$x_{n+1} := x_n - \frac{f(x_n)}{f'(x_n)} = x_n + (\xi - x_n) + \frac{1}{2}\frac{f''(\xi_1)}{f'(x_n)}(\xi - x_n)^2$$

$$\therefore \quad x_{n+1} - \xi = \frac{1}{2}\frac{f''(\xi_1)}{f'(x_n)}(\xi - x_n)^2 \tag{7.68}$$

proving that $\quad |x_{n+1} - \xi| \leq \dfrac{1}{2}\dfrac{\max_{a \leq x \leq b}|f''(x)|}{\min_{a \leq x \leq b}|f'(x)|}|x_n - \xi|^2$.

Let the point x_n be sufficiently close to the zero ξ, say x_n is in an interval

$$I := [\xi - \delta, \xi + \delta] \subset [a, b]$$

where $\delta > 0$ is sufficiently small,

$$0 < \delta < 1/k . \tag{7.69}$$

Then the point x_{n+1}, by (7.51) and (7.52),

$$|x_{n+1} - \xi| \leq k\delta^2 < \delta, \quad \text{by (7.69)} ,$$

showing that x_{n+1} is also in I.

This proves: If the initial point x_0 is in I, then the whole Newton sequence (7.46) stays in I. Therefore,

$$|x_{n+1} - \xi| \leq k|x_n - \xi|^2 \leq (k\delta)|x_n - \xi|$$
$$\leq (k\delta)^2|x_{n-1} - \xi|$$
$$\leq (k\delta)^{n+1}|x_0 - \xi| ,$$

proving that $x_n \to \xi$, since $k\delta < 1$. $\qquad\qquad\qquad\qquad\qquad\qquad\square$

Proof of Corollary 7.54. We take condition (3.23) to mean

$$f(a) < 0 \quad \text{and} \quad f(b) > 0 , \tag{7.70}$$

(the other case, $f(a) > 0$ and $f(b) < 0$, is similarly proved).

From (7.70) and the assumption

$$f'(x) \neq 0, \quad \text{in } [a, b] ,$$

we conclude that $f'(x) > 0$ in $[a, b]$.

An initial point x_0 with $f(x_0) > 0$ is then necessarily to the right of the root ξ,

$$\xi < x_0 ,$$

and monotone convergence means here

$$\xi < \cdots < x_{n+1} < x_n < \cdots < x_1 < x_0 . \tag{7.71}$$

We prove (7.71) by induction on n. For $n = 1$, it follows, since $f(x_0) > 0$ and

$f'(x_0) > 0$, that

$$x_1 := x_0 - \frac{f(x_0)}{f'(x_0)} < x_0$$

and by (7.68), with $n = 0$,

$$x_1 - \xi = \frac{1}{2} \frac{f''(\xi_1)}{f'(x_0)} (\xi - x_0)^2$$

$$\geq 0,$$

here $f'' \geq 0$ because f is convex in $[a, b]$, proving that

$$\xi \leq x_1 < x_0.$$

If $\xi < x_n$ for any n, we can show, in the same way, that the next point x_{n+1} is

$$\xi \leq x_{n+1} < x_n,$$

which proves (7.71).

Using Taylor's theorem we get

$$f(x_{n+1}) = f(x_n) + f'(x_n)(x_{n+1} - x_n) + \frac{f''(\xi_2)}{2}(x_{n+1} - x_n)^2,$$
$$\text{for some } x_{n+1} < \xi_2 < x_n$$

$$= \frac{f''(\xi_2)}{2}(x_{n+1} - x_n)^2, \quad \text{by (7.46)}.$$

$$\therefore \quad |f(x_{n+1})| \leq \frac{\max_{a \leq x \leq b} |f''(x)|}{2} |x_{n+1} - x_n|^2 \tag{7.72}$$

$$\therefore \quad |x_{n+1} - x_n| \geq \sqrt{\frac{2f(x_{n+1})}{\max_{a \leq x \leq b} |f''(x)|}}. \tag{7.73}$$

If x_{n+1} is not a good approximation of the zero ξ, say $f(x_{n+1}) > \varepsilon$, then (7.73) guarantees that

$$x_{n+1} < x_0 - (n+1)\sqrt{\frac{2\varepsilon}{\max_{a \leq x \leq b} |f''(x)|}}$$

so that eventually the sequence $\{x_n\}$ enters the interval I guaranteed by Theorem 7.53. $\square$

Proof of Theorem 7.60. a. Choose $\kappa < 1$ such that $|u'(\xi)| < \kappa$. Since $u'(x)$ is continuous at ξ, then, by the mean value theorem, there is a neighborhood $N(\xi)$ of ξ such that

$$\left| \frac{u(x) - u(\xi)}{x - \xi} \right| \leq \kappa < 1,$$

for all $\xi \neq x \in N(\xi)$. Without loss of generality we take the neighborhood $N(\xi)$ symmetric about ξ, i.e., ξ is the midpoint of $N(\xi)$.

Starting with any $x_0 \in N(\xi)$, $x_0 \neq \xi$, we get

$$\left| \frac{x_1 - \xi}{x_0 - \xi} \right| \leq \kappa < 1$$

or

$$|x_1 - \xi| \leq \kappa |x_0 - \xi| \, .$$

The point x_1 is also in the neighborhood $N(\xi)$ since it is even closer to the center ξ than is the initial x_0. Repeating this for $x_2, x_3, \ldots$, we see that the sequence $\{x_n := u(x_{n-1})\}$ is in $N(\xi)$ and satisfies the inequality

$$|x_n - \xi| \leq \kappa^n |x_0 - \xi|$$

and therefore $|x_n - \xi| \to 0$ as $n \to \infty$. Also

$$\frac{x_{n+1} - \xi}{x_n - \xi} = \frac{u(x_n) - u(\xi)}{x_n - \xi} \to u'(\xi)$$

since $x_n \to \xi$.

b. Choose $\kappa > 1$ such that $u'(\xi) > \kappa$. Therefore, there is a neighborhood $N(\xi)$ of ξ such that

$$\left| \frac{u(x) - u(\xi)}{x - \xi} \right| \geq \kappa > 1 \, ,$$

for all $x \in N(\xi)$. We assume that the neighborhood $N(\xi)$ is bounded.

As before, if $x_0 \in N(\xi)$, then

$$|x_1 - \xi| \geq \kappa |x_0 - \xi|$$

and x_1 is farther from ξ than is x_0. Each generated point x_n is therefore farther away from ξ than all its predecessors. This shows that the sequence $\{x_n\}$ eventually leaves the neighborhood $N(\xi)$.   □

Integrals

8 Definite integrals

8.1 Introduction

Calculus is based on the concept of *limit* and on two limiting operations:
- *integration*, computing *integrals* which are limits of appropriate sums;
- *differentiation*, computing *derivatives* which are limits of appropriate differences.

These operations are inverses of each other, in the sense that *subtraction* is the inverse of *addition*, and *division* is the inverse of *multiplication*. And like these inverse arithmetic operations, integration and differentiation "need each other", and we cannot achieve much using only one of them.

Integration is introduced in this chapter, and some applications of integrals are given in Chap. 11.

The main idea of integration was already known to Democritus[1] at the 5th century B.C. Democritus was the first to derive the volume of a cone or pyramid (he was also the first to speculate that matter is composed of small particles that cannot be divided any further, and which he called atoms, meaning "indivisible").

Given a region in the plane, called *basis*, and a point outside the plane, called *vertex*, a *cone* is defined as the set of all points on the line segments connecting every point in the basis with the vertex; see the line segment PQ in Fig. 8.1 a. Sometimes the shape of the basis gives the cone its name; see Fig. 8.1 b–d.

Democritus' method consists of slicing the cone, parallel to its basis, into layers. The layers get smaller as they approach the vertex, shrinking there to a point (Fig. 8.2). Democritus was puzzled by the shape and decreasing size of these layers. He asked[2]

> Are they equal or unequal? For if they are unequal, they will make the cone irregular as having many indentations, like steps, and unevenness; but, if they are equal, the cone will appear to have the property of a cylinder and to be made up of equal circles, which is very absurd.

Although he did not know about limits (they would come about 2200 years later), Democritus was able to compute the volume V of the cone

$$V = \frac{Ah}{3}, \tag{8.1}$$

---

1 Democritus of Thrace (about 460–370 B.C.).

2 This quotation is from *Method* by Archimedes, see James R. Newman, *The World of Mathematics*, vol. 1, Simon and Schuster, New York, 1956, p. 95.

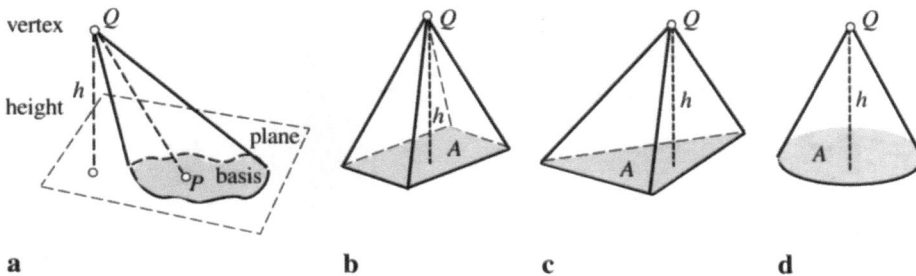

Fig. 8.1 a–d. A general cone and three familiar ones. **a** General cone defined by a basis in the plane and a vertex outside the plane. **b** Rectangular cone or pyramid. **c** Triangular cone or tetrahedron. **d** Circular cone

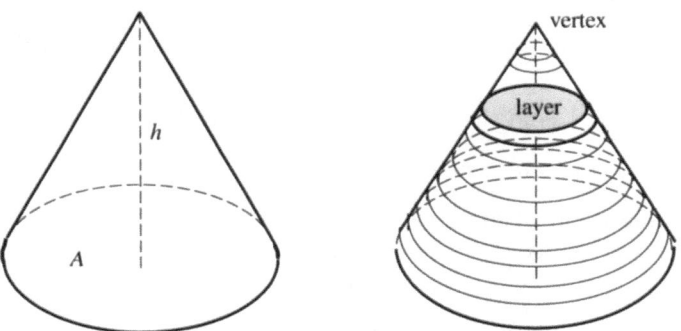

Fig. 8.2. Democritus' slicing method for computing the volume of the cone

where A is the *area* of its basis and h is its *height*, that is the distance between the vertex and the plane of the basis. If the four cones in Figure 8.1 have the same A and h, then they all have the same volume. We prove Democritus' result in Example 11.9.

Archimedes[3] used Democritus' method to compute other volumes, areas, and centroids.

Describing Democritus' method in modern language, we identify two steps:
- expressing the volume V as a *Riemann sum* (Sect. 8.2);
- computing an *integral* (the limit of the Riemann sum), which gives the volume.

Knowing limits, and how to compute them, gives us an advantage over Democritus; we can get more mileage out of his idea. We do so in this chapter, starting with the simple problem of computing an area A of a region in the plane bounded by

the graph of a given function $y = f(x)$ (above),
the x-axis (below),   (8.2)
the two vertical lines $x := a$ and $x := b$,

---

3 Archimedes of Syracuse (287–212 B.C.).

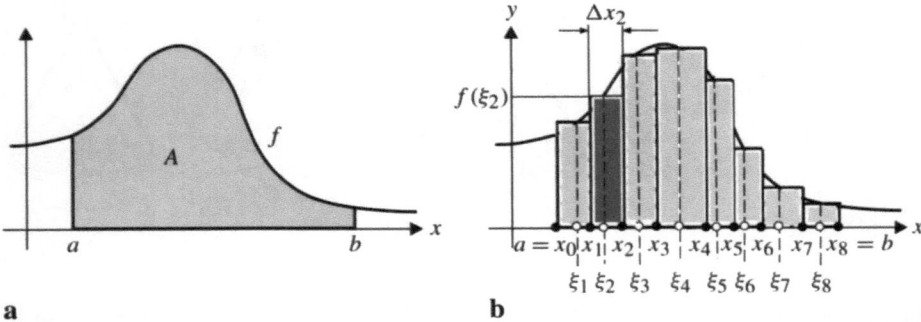

Fig. 8.3. a Region in the plane; **b** approximation of its area A as a Riemann sum with eight subintervals

such as the shaded plane region in Fig. 8.3 a. We approximate A by the following procedure.

Step 1: Select $n + 1$ points in $[a, b]$, including the endpoints,

$$x_0 := a < x_1 < x_2 < \cdots < x_{n-1} < x_n := b .$$

These points divide the interval $[a, b]$ into n subintervals $I_k := [x_{k-1}, x_k]$, for $k = 1, \ldots, n$, of lengths

$$\Delta x_k := x_k - x_{k-1} .$$

Step 2: Select a point ξ_k in each subinterval I_k

$$x_{k-1} \leq \xi_k \leq x_k \quad (k = 1, \ldots, n) .$$

Step 3: Calculate

$$A_n := \sum_{k=1}^{n} f(\xi_k) \, \Delta x_k . \tag{8.3}$$

A_n is called a *Riemann sum* of f on $[a, b]$. Its value (which depends on the selection of the points x_k and ξ_k) represents the *sum of areas of n rectangles*; the kth rectangle has *base* Δx_k (on the x-axis) and *height* $f(\xi_k)$. This is illustrated in Fig. 8.3 b.

The Riemann sum A_n is an approximation of the area A. If we make each subinterval I_k smaller, we expect A_n to be an even better approximation of A and to converge to A in the limit as all $\Delta x_k \to 0$. To guarantee that all $\Delta x_k \to 0$, we impose the condition

$$\max_{k=1,\ldots,n} \Delta x_k \to 0 ,$$

where $\max_{k=1,\ldots,n} \Delta x_k$ denotes the length of the longest subinterval I_k. In this case we expect the limit

$$\lim_{\substack{\max \Delta x_k \to 0 \\ k=1,\ldots,n}} \sum_{k=1}^{n} f(\xi_k) \, \Delta x_k = A \tag{8.4}$$

to hold, independently of the way the subintervals I_k and points ξ_k were selected. If the limit in (8.4) exists, it is called the *(definite) integral* of f from a to b and denoted by

$$\int_a^b f(x)\,dx := \lim_{\substack{\max\ \Delta x_k \to 0 \\ k=1,\dots,n}} \sum_{k=1}^n f(\xi_k)\Delta x_k . \tag{8.5}$$

The function f is called the *integrand*, and the points a and b the *limits of integration*. The symbol $\int$ in (8.28) is called the *integral sign*[4]. Finally the computation of integrals is called *integration*.

MACSYMA-Session 8.1. To compute the integral (8.5), we can use the command `integrate(f,x,a,b)`, where the first argument f is a function already known to MACSYMA and depending on the argument x (otherwise f will be interpreted as a constant).
For example,
c1. `integrate(x^2,x,0,1)`
d1. $\frac{1}{3}$
Similarly,
c2. `integrate(exp(x),x,-1,3)`
d2. $e^3 - \dfrac{1}{e}$
A numerical approximation of this value is given by
c3. `%,numer`
d3.   19.7177

The plan of this chapter is as follows. In Sect. 8.2 we consider Riemann sums, with numerical examples. In Sect. 8.3 we derive the main properties of the definite integral (Theorem 8.21), and prove that a continuous function is integrable on a finite interval $[a, b]$ (Theorem 8.34).

Two techniques of numerical integration are given in Sects. 8.4 and 8.5.

8.2 Riemann sums

In all sciences, engineering, and in every field where mathematics is applied, there is need for sums of the form[5]

$$\sum_{k=1}^n f(\xi_k)\,\Delta x_k . \tag{8.6}$$

f is a function defined and bounded on a (closed, finite) interval $[a, b]$.

---

4 The integral sign was introduced by Leibniz in 1675. It resembles a "long S", with "S" standing for "sum".

5 See the applications of integrals in Chap. 11.

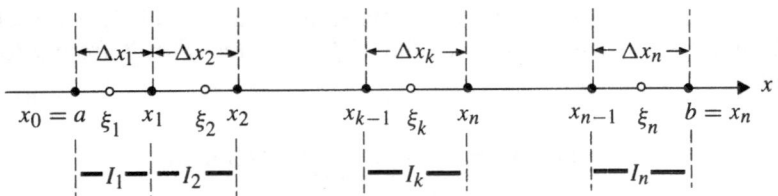

Fig. 8.4. Illustration of (8.10)

n is a positive integer.

$n + 1$ points are selected in the interval $[a, b]$, including the endpoints,

$$x_0 := a < x_1 < x_2 < \cdots < x_{n-1} < x_n := b . \tag{8.7}$$

These points partition the interval $[a, b]$ into n closed subintervals $\{I_1, \ldots, I_n\}$,

$$I_k := [x_{k-1}, x_k] , \tag{8.8}$$

of lengths

$$\Delta x_k := x_k - x_{k-1} \quad (k = 1, \ldots, n) . \tag{8.9}$$

ξ_k is a point in the kth subinterval (Fig. 8.4),

$$x_{k-1} \le \xi_k \le x_k \quad (k = 1, \ldots, n) . \tag{8.10}$$

Definition 8.1. The sum (8.6) is called a *Riemann sum*[6] of f in $[a, b]$.

Definition 8.2. Given a closed interval $[a, b]$, and a set of $n + 1$ points x_k satisfying (8.7), a *partition* $\mathcal{P}$ of $[a, b]$ is the collection of n closed subintervals $\{I_1, I_2, \ldots, I_n\}$ defined by, (8.8),

$$I_k := [x_{k-1}, x_k] \quad (k = 1, \ldots, n) .$$

The points $\{x_k\}$ are called the *partition points* of $\mathcal{P}$. The *length* of the subinterval I_k is denoted, (8.9),

$$\Delta x_k := x_k - x_{k-1} \quad (k = 1, \ldots, n) .$$

The maximum of the lengths Δx_k is called the *norm* of the partition and denoted by

$$\|\mathcal{P}\| := \max_{k=1,\ldots,n} \Delta x_k . \tag{8.11}$$

Remark 8.3. A partition $\mathcal{P} := \{I_1, I_2, \ldots, I_n\}$ of $[a, b]$ satisfies

---

6 Bernhard Riemann (1826–1866).

$$[a, b] = I_1 \cup I_2 \cup \cdots \cup I_n ,$$

$$b - a = \sum_{k=1}^{n} \Delta x_k , \tag{8.12}$$

$$\text{and} \quad I_k \cap I_{k+1} = \{x_k\} \quad (k = 1, \ldots, n) .$$

Definition 8.4 (Regular partition). A partition $\mathcal{P}$ is *regular* if all lengths Δx_k are equal. This common length is, by (8.12),

$$\Delta x := \frac{b - a}{n} . \tag{8.13}$$

Therefore, $\mathcal{P}$ is regular if the partition points (8.7) are equally spaced, i.e.,

$$x_k := x_0 + k \, \Delta x = a + k \frac{b - a}{n} \quad (k = 0, \ldots, n) , \tag{8.14}$$

$$\text{so that,} \quad I_k = [x_{k-1}, x_k] = \left[a + (k - 1)\frac{b - a}{n}, \; a + k\frac{b - a}{n} \right]. \tag{8.15}$$

For the given function f and interval $[a, b]$, the Riemann sum (8.6) depends on the selection of the partition $\mathcal{P}$ and points ξ_k ($k = 1, \ldots, n$). For each such selection, (8.6) calculates a number. In particular, for a regular partition (8.15) the Riemann sum (8.6) simplifies to

$$\sum_{k=1}^{n} f(\xi_k) \, \Delta x = \sum_{k=1}^{n} f(\xi_k) \frac{b - a}{n} \tag{8.16}$$

by (8.13).

Example 8.5 (Three special Riemann sums). Consider a regular partition of $[a, b]$. In each subinterval I_k of (8.15), select the point ξ_k in three ways.

Left endpoint. $\xi_k := x_{k-1}$, giving the Riemann sum of f in $[a, b]$ as

$$S_{\text{left}}(f, x, a, b, n) := \sum_{k=1}^{n} f\left(a + (k - 1)\frac{b - a}{n}\right)\frac{b - a}{n} . \tag{8.17}$$

Middle point. Here we select ξ_k as the midpoint of the subinterval $I_k = [x_{k-1}, x_k]$,

$$\xi_k := \frac{1}{2}\left(x_k + x_{k-1}\right) = a + \frac{2k - 1}{2} \cdot \frac{b - a}{n}, \quad \text{by (8.14)} ,$$

and the Riemann sum (8.16) becomes

$$S_{\text{middle}}(f, x, a, b, n) := \sum_{k=1}^{n} f\left(a + \frac{2k - 1}{2} \cdot \frac{b - a}{n}\right)\frac{b - a}{n} . \tag{8.18}$$

Right endpoint. $\xi_k := x_k$, giving the Riemann sum of f in $[a, b]$ as

$$S_{\text{right}}(f, x, a, b, n) := \sum_{k=1}^{n} f\left(a + k\frac{b-a}{n}\right)\frac{b-a}{n}. \tag{8.19}$$

The second argument x in $S_{\text{left}}(f, x, a, b, n)$ (and in S_{middle}, S_{right}) is the argument of the function f whose values are being summed; for example, $S_{\text{left}}(x^2, x, 0, 1, 100)$ is the left Riemann sum of $f(x) = x^2$ in $[0, 1]$ using 100 subintervals. The name we use for this variable is irrelevant; the above sum is the same as $S_{\text{left}}(t^2, t, 0, 1, 100)$.

For any given f and $[a, b]$, the three Riemann sums $S_{\text{left}}(f, x, a, b, n)$, $S_{\text{middle}}(f, x, a, b, n)$ and $S_{\text{right}}(f, x, a, b, n)$ depend only on n.

Example 8.6. Consider the function $f(x) = x^2$ in the interval $[0, 1]$. The three Riemann sums (8.17), (8.18), and (8.19) then give, omitting arithmetic details (see also Exercise 8.1),

$$S_{\text{left}}(x^2, x, 0, 1, n) = \sum_{k=1}^{n}\left(\frac{k-1}{n}\right)^2\frac{1}{n} = \frac{1}{n^3}\sum_{k=1}^{n}(k-1)^2$$

$$= \frac{(n-1)(2n-1)}{6n^2}, \tag{8.20a}$$

$$S_{\text{middle}}(x^2, x, 0, 1, n) = \sum_{k=1}^{n}\left(\frac{2k-1}{2n}\right)^2\frac{1}{n} = \frac{1}{4n^3}\sum_{k=1}^{n}(2k-1)^2 = \frac{4n^2-1}{12n^2}, \tag{8.20b}$$

$$S_{\text{right}}(x^2, x, 0, 1, n) = \sum_{k=1}^{n}\left(\frac{k}{n}\right)^2\frac{1}{n} = \frac{1}{n^3}\sum_{k=1}^{n}k^2 = \frac{(n+1)(2n+1)}{6n^2}. \tag{8.20c}$$

Note that as $n \to \infty$, the three Riemann sums (8.20a), (8.20b), and (8.20c) converge to the same limit, $1/3$, the integral of x^2 from 0 to 1.

MACSYMA-Session 8.2. A sum

$$\sum_{k=m}^{k=n} u(k) = u(m) + u(m+1) + \cdots + u(n-1) + u(n)$$

is expressed in MACSYMA as sum(u,k,m,n).
We use MACSYMA to calculate simple Riemann sums, like the three sums (8.17)–(8.19), beginning with the left Riemann sum (8.17) of f in $[a, b]$, which is computed by the MACSYMA function

```
c1. s_left(f,x,a,b,n):=
        sum(subst(a+(k-1)*(b-a)/n, x, f), k, 1, n)*(b-a)/n$
```
For example, the left Riemann sum of x^2 from 0 to 1 using 100 steps,

c2. `s_left(x^2,x,0,1,100)`

d2. $\dfrac{6567}{20000}$

The name of the argument x of the integrand f is immaterial: each of the following three expressions is equivalent to c2,

```
s_left(x^2,x,0,1,100)
s_left(u^2,u,0,1,100)
s_left(z^2,z,0,1,100)
```

The middle and right Riemann sums of f in $[a, b]$ with n steps are similarly,

c3. `s_middle(f,x,a,b,n):=`
　　　　`sum(subst(a+(2*k-1)*(b-a)/(2*n), x, f), k, 1, n)*(b-a)/n$`

c4. `s_right(f,x,a,b,n):=`
　　　　`sum(subst(a+k*(b-a)/n, x, f), k, 1, n)*(b-a)/n$`

For $f(x) = \sin x$ in $[0, \pi]$ with $n = 100$ we get:

c5. `s_left(sin(x),x,0,%pi,100),numer`

d5.　　1.99982

c6. `s_middle(sin(x),x,0,%pi,100),numer`

d6.　　2.00007

c7. `s_right(sin(x),x,0,%pi,100),numer`

d7.　　1.99982

approximating the correct value, obtained by calculating the definite integral of $\sin(x)$ from $x = 0$ to $x = \pi$

c8. `integrate(sin(x),x,0,%pi)`

d8.　　2

MACSYMA has built-in Riemann sum functions

```
riem_sum_left(f,x,a,b,n)
riem_sum_middle(f,x,a,b,n)
riem_sum_right(f,x,a,b,n)
```

which generally work better than the above functions `s_left`, `s_middle`, and `s_right`. For example,

c9. `riem_sum__left(x^2,x,0,1,n)`

d9.　$\dfrac{(n - 1)\left(\frac{n+1}{3} - \frac{1}{2}\right)}{n^2}$

c10. `factor(%)`

d10.　$\dfrac{(n - 1)(2n - 1)}{6n^2}$

in agreement with (8.20a). The limit of this expression as $n \to \infty$ is:

c11. `limit(%,n,inf)`

d11.　$\frac{1}{3}$

giving the correct value,

c12. `integrate(x^2,x,0,1)`

d12.　$\frac{1}{3}$

We now define two other important Riemann sums.

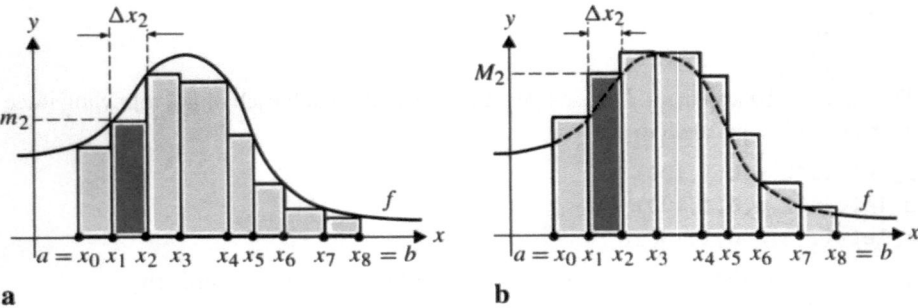

Fig. 8.5. Lower (**a**) and upper (**b**) Riemann sums bounding the area in Fig. 8.3a

Definition 8.7 (Upper and lower Riemann sums). Let $[a, b]$ be an interval, $\mathcal{P}$ a partition of $[a, b]$, and let f be a function bounded on $[a, b]$. For $k = 1, \ldots, n$, let M_k and m_k be the *supremum* and *infimum* of f in the kth subinterval I_k of $\mathcal{P}$ (see Fig. 8.5),

$$M_k := \sup_{x \in I_k} f(x) , \tag{8.21a}$$

$$m_k := \inf_{x \in I_k} f(x) . \tag{8.21b}$$

Then the upper Riemann sum (w.r.t. the partition $\mathcal{P}$) is

$$S_{\mathrm{U}}(f, \mathcal{P}) := \sum_{k=1}^{n} M_k \, \Delta x_k , \tag{8.22a}$$

and the lower Riemann sum is

$$S_{\mathrm{L}}(f, \mathcal{P}) := \sum_{k=1}^{n} m_k \, \Delta x_k . \tag{8.22b}$$

These are important because, for the given partition $\mathcal{P}$, all Riemann sums are sandwiched between $S_{\mathrm{L}}(f, \mathcal{P})$ and $S_{\mathrm{U}}(f, \mathcal{P})$.

Proposition 8.8. Let f, $[a, b]$, and $\mathcal{P}$ be as in Definition 8.7, and let

$$\sum_{k=1}^{n} f(\xi_k) \, \Delta x_k$$

be any Riemann sum, corresponding to a selection of the points ξ_k in (8.10). Then

$$S_{\mathrm{L}}(f, \mathcal{P}) \le \sum_{k=1}^{n} f(\xi_k) \, \Delta x_k \le S_{\mathrm{U}}(f, \mathcal{P}) . \tag{8.23}$$

Proof. By definition,

$$m_k \le f(\xi_k) \le M_k \quad (k = 1, \ldots, n) \, ,$$

and therefore,

$$\sum_{k=1}^{n} m_k \, \Delta x_k \le \sum_{k=1}^{n} f(\xi_k) \, \Delta x_k \le \sum_{k=1}^{n} M_k \, \Delta x_k$$

proving (8.23).   □

The lower and upper Riemann sums are used mainly in theory (theorems, proofs, etc.) and are not practical objects since they require the computation of the minimum m_k and the maximum M_k of f in each subinterval I_k. However, for the important class of monotone functions (see Definition 1.39) this computation is not necessary.

Example 8.9 (Monotone functions). Recall that a function f, on the interval $I = [a, b]$, is *monotone*, and in particular,

nondecreasing
increasing   if $\begin{cases} f(u) \le f(v) \\ f(u) < f(v) \\ f(u) \ge f(v) \\ f(u) > f(v) \end{cases}$ whenever $a \le u < v \le b$.
nonincreasing
decreasing

In each of these monotone cases the values m_k and M_k are just the values of f at the endpoints x_{k-1} and x_k. For example, if the function f is nondecreasing in I, then

$$m_k = f(x_{k-1}), \quad M_k = f(x_k) \, ,$$

with the possibility that $m_k = M_k$ if f is constant in the subinterval I_k. For nondecreasing functions, the lower Riemann sum is then the *left Riemann sum*, corresponding to the selection $\xi_k = x_{k-1}$ in (8.10)

$$S_L(f, \mathcal{P}) = \sum_{k=1}^{n} f(x_{k-1}) \, \Delta x_k \, , \tag{8.24}$$

and the upper Riemann sum is the *right Riemann sum*,

$$S_U(f, \mathcal{P}) = \sum_{k=1}^{n} f(x_k) \, \Delta x_k \, . \tag{8.25}$$

This is illustrated in Fig. 8.10b and c, where the lower Riemann sum $S_L(f, \mathcal{P})$ is the left Riemann sum and the upper Riemann sum $S_U(f, \mathcal{P})$ is the right Riemann sum.

Example 8.10. The function x^2 is increasing in $[0, 1]$. For the regular partition $\mathcal{P}$ of $[0, 1]$ with n subintervals, we get from (8.24) and (8.20a),

$$S_L(x^2, \mathcal{P}) = \frac{1}{3} - \frac{1}{2n} + \frac{1}{6n^2} \, .$$

Similarly, by (8.25) and (8.20c),

$$S_U(x^2, \mathcal{P}) = \frac{1}{3} + \frac{1}{2n} + \frac{1}{6n^2} \, .$$

Definition 8.11. Let $\mathcal{P}$ be a partition of $[a, b]$, with partition points

$$a = x_0 < x_1 < x_2 < \cdots < x_{n-1} < x_n = b \, ,$$

and let $\mathcal{Q}$ be a set of points in $[a, b]$ which contains the above $n + 1$ partition points of $\mathcal{P}$. Then $\mathcal{Q}$ is a partition of $[a, b]$ which is called *finer* than $\mathcal{P}$.

We can state this in terms of the subintervals of the partitions: If $\mathcal{P} := \{I_1, \ldots, I_n\}$ and $\mathcal{Q} := \{J_1, \ldots, J_m\}$ are partitions of $[a, b]$, then $\mathcal{Q}$ is finer than $\mathcal{P}$ if every subinterval J_k of $\mathcal{Q}$ is contained in some subinterval I_l of $\mathcal{P}$.

The following results are easy consequences of this definition.

Lemma 8.12. If $\mathcal{P}$ and $\mathcal{Q}$ are partitions of $[a, b]$, and $\mathcal{Q}$ is finer than $\mathcal{P}$, then:
a. $\|\mathcal{P}\| \geq \|\mathcal{Q}\|$;
b. $S_L(f, \mathcal{P}) \leq S_L(f, \mathcal{Q}) \leq S_U(f, \mathcal{Q}) \leq S_U(f, \mathcal{P})$ for any function f.

Exercises

8.1 Prove the following identities using MACSYMA:

(a) $\displaystyle\sum_{k=1}^{n} k = \frac{n(n+1)}{2}$ (b) $\displaystyle\sum_{k=1}^{n} k^2 = \frac{n(n+1)(2n+1)}{6}$

8.2 In the following problems use the MACSYMA command `sfloat` to approximate $S_{\text{left}}(f, x, a, b, n)$, $S_{\text{right}}(f, x, a, b, n)$, and $S_{\text{middle}}(f, x, a, b, n)$ for $n = 5, 10, 20, 100$.

(a) $f(x) := x^2 + 4x - 8, a = 0, b = 2$ (b) $f(x) := \sin x, a = 0, b = \pi$

(c) $f(x) := \cosh x, a = -1, b = 1$ (d) $f(x) := x^8 + 4x^4 + 4, a = 1, b = 3$

(e) $f(x) := 1/x, a = 1, b = 4$ (f) $f(x) := \ln x, a = 1, b = 10$

(g) $f(x) := \dfrac{x-1}{x+1}, a = 2, b = 6$ (h) $f(x) := \sqrt{x^2 + 1}, a = 0, b = 2$

(i) $f(x) := \sqrt{\dfrac{x-1}{x+1}}, a = 2, b = 6$ (j) $f(x) := \sqrt{\tan x}, a = 0, b = \dfrac{\pi}{3}$

(k) $f(x) := \sin^2 x, a = -\dfrac{\pi}{2}, b = \dfrac{\pi}{2}$ (l) $f(x) := \cos^3 x, a = 0, b = \pi$

8.3 Write a MACSYMA function to compute $S_U(f, x, a, b, n)$ and $S_L(f, x, a, b, n)$. Approximate these sums with `sfloat` for $n = 5, 10, 20, 100$.

(a) $f(x) := x^2 + 4x + 4, a = 0, b = 2$ (b) $f(x) := \sin x, a = 0, b = \pi/2$

(c) $f(x) := \cosh x, a = -1, b = 1$ (d) $f(x) := x^{10}, a = 0, b = 4$

(e) $f(x) := 1/x, a = 1, b = 10$ (f) $f(x) := \ln(1/x), a = 0.1, b = 1$

(g) $f(x) := 1/(x+1), a = 0, b = 3$ (h) $f(x) := \sqrt{x^2+1}, a = 0, b = 2$

(i) $f(x) := \sqrt{\cos x}, a = 0, b = \pi/2$ (j) $f(x) := \sqrt{\tan x}, a = 0, b = \pi/4$

(k) $f(x) := \sin^2 x, a = -\pi/2, b = \pi/2$ (l) $f(x) := \cos^3 x, a = 0, b = \pi$

8.4   Consider the following limits of Riemann sums, in the form

$$\lim_{\max \Delta x \to 0} \sum_{k=1}^{n} f(\xi_k) \Delta x_k .$$

(a) $\displaystyle\lim_{\max \Delta x \to 0} \sum_{k=1}^{n} \xi_k^4 \Delta x_k, a = 0, b = 3$

(b) $\displaystyle\lim_{\max \Delta x \to 0} \sum_{k=1}^{n} \sin(\xi_k) \Delta x_k, a = -\frac{\pi}{2}, b = \frac{\pi}{2}$

(c) $\displaystyle\lim_{\max \Delta x \to 0} \sum_{k=1}^{n} \sqrt{\xi_k^2 + 1} \Delta x_k, a = -1, b = 1$

Express these limits as integrals, but do not evaluate the integrals.

8.5   Consider partitions $\mathcal{P}$ of a bounded interval $[a, b]$, with $a < b$. Prove:

(a) $\|\mathcal{P}\| \to 0$ implies that the number of subintervals $n \to \infty$;

(b) $n \to \infty$ does not imply $\|\mathcal{P}\| \to 0$.

8.6   Prove or disprove: If f is a function bounded on an interval $[a, b]$, and $\mathcal{P}$, $\mathcal{Q}$ are any two partitions of $[a, b]$, then $S_{\mathrm{L}}(f, \mathcal{P}) \leq S_{\mathrm{U}}(f, \mathcal{Q})$.

8.7   Given two partitions $\mathcal{P}$ and $\mathcal{Q}$ of $[a, b]$, find a partition $\mathcal{R}$, with a minimal number of partition points, which is finer than both $\mathcal{P}$ and $\mathcal{Q}$.

8.8   Prove Lemma 8.12 b.

8.9   Let $a < c < b$, let $\mathcal{P}_{[a,c]}$ be any partition of $[a, c]$ and $\mathcal{P}_{[c,b]}$ any partition of $[c, b]$. Then

$$\mathcal{P} := \mathcal{P}_{[a,c]} \cup \mathcal{P}_{[c,b]} \tag{8.26}$$

is a partition of $[a, b]$, with norm

$$\|\mathcal{P}\| = \max\{\|\mathcal{P}_{[a,c]}\|, \|\mathcal{P}_{[c,b]}\|\}$$

Conversely, if $\tilde{\mathcal{P}}$ is a partition of $[a, b]$, then adjoining the point c to the partition points of $\tilde{\mathcal{P}}$ gives a partition[7] $\mathcal{P}$ of $[a, b]$ which satisfies (8.26), where now $\mathcal{P}_{[a,c]}$ $:= \{ I \cap [a, c] \mid I$ subinterval in $\mathcal{P} \}$ and $\mathcal{P}_{[c,b]} := \{ I \cap [c, b] \mid I$ subinterval in $\mathcal{P} \}$ are partitions of $[a, c]$ and $[c, b]$ respectively.

8.3 Definite integral

In the computation of area (8.4) the Riemann sums are expected to converge to the same value, as the norm of the partition, see (8.11), tends to zero. We observed this for the special Riemann sums in Example 8.6 and MACSYMA-Session 8.2. This is not always the case, see Example 8.20, but when it does, it deserves a special name.

---

7 If the point c is a partition point of $\tilde{\mathcal{P}}$, then the partitions $\tilde{\mathcal{P}}$ and $\mathcal{P}$ are the same. Otherwise, $\mathcal{P}$ is finer than $\tilde{\mathcal{P}}$, and $\mathcal{P} \neq \tilde{\mathcal{P}}$.

Definition 8.13 (Integral). Let f be a function bounded on $[a, b]$. Then f is called *integrable* on $[a, b]$ if the limit

$$\lim_{\|\mathcal{P}\| \to 0} \sum_{k=1}^{n} f(\xi_k) \, \Delta x_k \tag{8.27}$$

exists for all partitions $\mathcal{P} = \{I_1, \ldots, I_n\}$ of $[a, b]$, and for all selections of ξ_k in I_k.

If f is integrable in $[a, b]$, then the value of the limit (8.27) is called the *(definite) integral*[8] of f from a to b. It is denoted by

$$\int_a^b f(x) \, dx := \lim_{\|\mathcal{P}\| \to 0} \sum_{k=1}^{n} f(\xi_k) \, \Delta x_k \ . \tag{8.28}$$

If it exists, the integral

$$\int_a^b f(x) \, dx \tag{8.29}$$

is a number which depends only on the integrand f, and the interval $I = [a, b]$. Therefore the name of the variable used in (8.29), the *integration variable*, is of no consequence. If the letter x is needed elsewhere, we will use another letter in (8.29), say $\int_a^b f(t) \, dt$ or $\int_a^b f(u) \, du$, or suppress the variable altogether, writing the integral (8.29) as

$$\int_I f \quad \text{or} \quad \int_a^b f \ . \tag{8.30}$$

Remark 8.14 (One Riemann sum suffices if the function is integrable).
If the function f is known *to be integrable* in $[a, b]$, then its integral $\int_a^b f(x) \, dx$ can be computed with the limit of any sequence of Riemann sums whose partitions have norms $\to 0$. For example

$$\int_a^b f(x) \, dx = \lim_{n \to \infty} S_{\text{left}}(f, x, a, b, n) \ .$$

Example 8.15. $\int_{-1}^{2} x^2 \, dx = 3$. By Theorem 8.34, the function x^2 is integrable in *any* interval $[a, b]$. Therefore, using Remark 8.14, it suffices to consider the left Riemann sum of x^2 in $[-1, 2]$,

---

8 Also Riemann integral.

$$S_{\text{left}}(x^2, x, -1, 2, n) = \frac{3}{n}\sum_{k=0}^{n-1}\left(-1+\frac{3k}{n}\right)^2 = \frac{3}{n}\sum_{k=0}^{n}\left(1-\frac{6k}{n}+\frac{9k^2}{n^2}\right)$$

$$= \frac{3}{n}\left(n - \frac{3(n-1)}{2n}\right),$$

whose limit, as $n \to \infty$, is 3.

MACSYMA-Session 8.3. Up to now all integrals $\int_a^b f(x)\,dx$ had *fixed* endpoints a and b. In this session we compute integrals with *general* a and b. The fact that the endpoints a and b are arbitrary should not prevent us from using Riemann sums.

We illustrate this using the built-in MACSYMA function
`riem_sum_left(f,x,a,b,n)`; see Session 8.2. In general, $S_{\text{left}}(f, x, a, b, n)$ depends on a, b, and n. We take then the limit, as $n \to \infty$, to get the integral.

First, the Riemann sum of the constant function $f(x) = 1$ from a to b,
c1. `riem_sum_left(1,x,a,b,n)`
d1. $b - a$
This result does not depend on n, so there is no need to take the limit as $n \to \infty$.

Next, the Riemann sum of $f(x) = x$ from a to b,
c2. `riem_sum_left(x,x,a,b,n)`
gives a complicated expression, which we can factor using the command `factor(%)`, or
c3. `factor(riem_sum_left(x,x,a,b,n))`
d3. $\dfrac{(b-a)(bn+an-b+a)}{2n}$
The limit of this expression as $n \to \infty$ is the integral,
c4. `limit(%,n,inf)`
d4. $\dfrac{b^2-a^2}{2}$
Steps (c4)–(c6) can be combined in
c5. `limit(riem_sum_left(x,x,a,b,n),n,inf)`
d5. $\dfrac{b^2-a^2}{2}$

Similarly, the left Riemann sum of $f(x) = x^2$ from a to b,
c6. `riem_sum_left(x^2,x,a,b,n)`
again gives a complicated expression, but here even factoring does not help much:
c7. `factor(%)`
d7. $\dfrac{(b-a)(2b^2n^2+2abn^2+2a^2n^2-3b^2n+3a^2n+b^2-2ab+a^2)}{6n^2}$
but the limit, as $n \to \infty$, is simple
c8. `limit(%,n,inf)`
d8. $\dfrac{b^3-a^3}{3}$
Again, these steps can be combined:
c9. `limit(riem_sum_left(x^2,x,a,b,n),n,inf)`
d9. $\dfrac{b^3-a^3}{3}$
These results can be calculated directly, using the MACSYMA function

`integrate(f,x,a,b)`, see Session 8.1. For example, we verify the last result with (here `ratsimp` stands for rational simplification)

c10. `integrate(x^2,x,a,b)`, `ratsimp`

d10. $\dfrac{b^3}{3} - \dfrac{a^3}{3}$

We summarize the results of MACSYMA-Session 8.3 as follows

$$\int_a^b 1\,dx = b - a \tag{8.31a}$$

$$\int_a^b x\,dx = \frac{b^2}{2} - \frac{a^2}{2} \tag{8.31b}$$

$$\int_a^b x^2\,dx = \frac{b^3}{3} - \frac{a^3}{3} \tag{8.31c}$$

Remark 8.16 (Two Riemann sums suffice). If f is not known to be integrable in $[a, b]$, then it seems that Definition 8.13 requires checking all Riemann sums as $\|\mathcal{P}\| \to 0$. Actually it suffices to check the lower and upper Riemann sums since, for a given partition $\mathcal{P}$, all Riemann sums lie between the lower sum $S_L(f, \mathcal{P})$ and the upper sum $S_U(f, \mathcal{P})$, see Proposition 8.8, and therefore, if f is integrable,

$$S_L(f, \mathcal{P}) \le \int_a^b f(x)\,dx \le S_U(f, \mathcal{P}), \tag{8.32}$$

i.e., for any partition $\mathcal{P}$, the lower sum $S_L(f, \mathcal{P})$ is a lower bound for the integral $\int_a^b f(x)\,dx$, and the upper sum $S_U(f, \mathcal{P})$ is an upper bound.

Corollary 8.17. A bounded function f is integrable on $[a, b]$ if and only if the lower sum $S_L(f, \mathcal{P})$ and the upper sum $S_U(f, \mathcal{P})$ have a common limit as $\|\mathcal{P}\| \to 0$, for all partitions $\mathcal{P}$ of $[a, b]$,

$$\lim_{\|\mathcal{P}\| \to 0} \sum_{k=1}^n m_k\,\Delta x_k = \lim_{\|\mathcal{P}\| \to 0} \sum_{k=1}^n M_k\,\Delta x_k, \tag{8.33}$$

and the common value is the integral $\int_a^b f(x)\,dx$.

Remark 8.18. The above results can be stated formally as follows: The function f is integrable on $[a, b]$ if, and only if, for every $\varepsilon > 0$ there is a partition $\mathcal{P}$ of $[a, b]$ such that

$$S_U(f, \mathcal{P}) - S_L(f, \mathcal{P}) < \varepsilon.$$

We define the *approximation error* of the partition $\mathcal{P}$ as

$$E(f, \mathcal{P}) := S_U f, \mathcal{P}) - S_L(f, \mathcal{P}), \tag{8.34}$$

and note that its calculation does not require the value of the integral $\int_a^b f(x)\,dx$. However, if the error $E(f, \mathcal{P})$ is small, then both[9] $S_L(f, \mathcal{P})$ and $S_U(f, \mathcal{P})$ are good approximations of $\int_a^b f(x)\,dx$.

Example 8.19. Consider again the function x^2 in $[0, 1]$. Using (8.32) and Example 8.10, we bound the integral

$$S_L(f, \mathcal{P}) = \frac{1}{3} - \frac{1}{2n} + \frac{1}{6n^2} \le \int_0^1 x^2\,dx \le \frac{1}{3} + \frac{1}{2n} + \frac{1}{6n^2} = S_U(f, \mathcal{P}).$$

The approximation error, $E(f, \mathcal{P}) = 1/n$, converges to zero as $n \to 0$.

Example 8.20 (Nonintegrable functions). a. Consider the function, due to Dirichlet,

$$f_1(x) := \begin{cases} 1 & \text{if } x \text{ is rational,} \\ 0 & \text{if } x \text{ is irrational,} \end{cases}$$

in the interval $I = [0, 1]$. This function is quite artificial but serves to make the point that a bounded function need not be integrable on a finite interval. Indeed, for any partition $\mathcal{P}$ of $[0, 1]$, the lower and upper sums are

$$S_L(f_1, \mathcal{P}) = 0, \quad S_U(f_1, \mathcal{P}) = 1$$

since any subinterval, no matter how small, contains rational and irrational numbers. Therefore f_1 is not integrable on $[0, 1]$.

b. The function

$$f_2(x) := 1/x^2$$

blows up to infinity[10] as x approaches 0. We do not expect f_2 to be integrable on any interval containing 0, say $[0, 1]$. This suspicion is now confirmed by using MACSYMA.

MACSYMA-Session 8.4. We apply the three built-in Riemann sums
```
riem_sum_left(f,x,a,b,n)
riem_sum_middle(f,x,a,b,n)
riem_sum_right(f,x,a,b,n)
```
to the function $f_2(x) = 1/x^2$ of Example 8.20b, over the interval $[0, 1]$, with $n = 1000$ partition intervals.

The command for left Riemann sum `riem_sum_left(1/x^2,x,0,1,1000)` gives "error" because of division by 0 (at the leftmost endpoint 0).
The right and middle Riemann sums give

---

9 And any other Riemann sum corresponding to $\mathcal{P}$.
10 See Sect. 10.5 for integrals with unbounded integrands.

c1. `riem_sum_right(1/x^2,x,0,1,1000),numer`
d1.   1643.76
c2. `riem_sum_middle(1/x^2,x,0,1,1000),numer`
d2.   4932.98
showing a considerable lack of agreement. By increasing n to $1,000,000 = 10^6$, the
computation will take (at least a thousand times) longer, but will give similar results. In
particular, the left sum will remain infinite because of the division by 0. Indeed, the
function $1/x^2$ is not integrable in $[0, 1]$. This is confirmed in MACSYMA, which gives an
error message
c3. `integrate(1/x^2,x,0,1)`
d3.   Integral is not absolutely convergent

The following integral properties are immediate consequences of corresponding limit properties.

Theorem 8.21 (Properties of the integral). Let the functions f and g be integrable
on $[a, b]$, and let α be a constant. Then:
a.   the function αf is integrable on $[a, b]$ and

$$\int_a^b \alpha f(x)\,dx = \alpha \int_a^b f(x)\,dx \; ;$$

b.   the function $f + g$ is integrable on $[a, b]$ and

$$\int_a^b (f(x) + g(x))\,dx = \int_a^b f(x)\,dx + \int_a^b g(x)\,dx \; ;$$

c.   if $a \le c < d \le b$, then f is integrable on $[c, d]$;
d.   for any point $a \le c \le b$, the integral of f from a to b is the sum of integrals
from a to c, and from c to b,

$$\int_a^b f(x)\,dx = \int_a^c f(x)\,dx + \int_c^b f(x)\,dx \; , \tag{8.35}$$

see Fig. 8.6;
e.   if

$$f(x) \le g(x), \quad \text{for all } a \le x \le b \; , \tag{8.36}$$

then

$$\int_a^b f(x)\,dx \le \int_a^b g(x)\,dx \; . \tag{8.37}$$

This theorem is proved in Sect. 8.6.

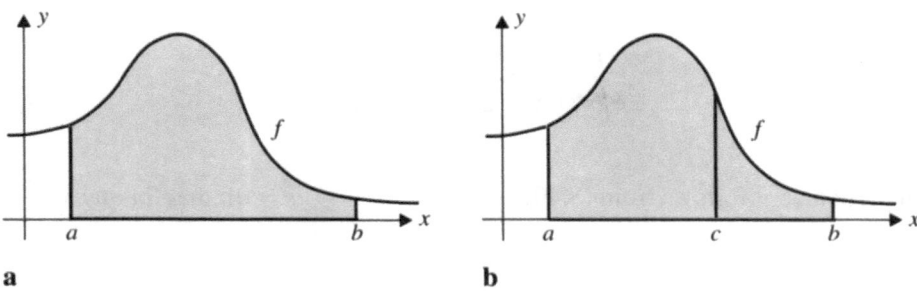

Fig. 8.6 a, b. Illustration of Theorem 8.21 d. **a** The area under f from a to b; **b** the area from a to c plus the area from c to b

Remark 8.22. Properties a and b state that integration is a *linear operation* in the sense that

$$\int_I \alpha f(x)\,dx = \alpha \int_I f(x)\,dx$$

$$\int_I (f(x) + g(x))\,dx = \int_I f(x)\,dx + \int_I g(x)\,dx$$

for any interval I where the functions are integrable. In particular, property b can be stated as: the integral of a sum is the sum of integrals.

Remark 8.23. Property c says that things cannot go wrong in a smaller interval: If f is integrable on the interval I, it is integrable on any subinterval of I. This fact is obvious for geometric reasons.

Remark 8.24. Property d states that the integral is *additive* as function of intervals. If $I_1 = [a, c]$ and $I_2 = [c, b]$, then property d gives[11]

$$\int_{I_1 \cup I_2} f(x)\,dx = \int_{I_1} f(x)\,dx + \int_{I_2} f(x)\,dx . \tag{8.38}$$

Using further subdivisions we conclude: If $I = [a, b]$ is partitioned into n subintervals I_k, then

$$\int_I f(x)\,dx = \sum_{k=1}^{n} \int_{I_k} f(x)\,dx . \tag{8.39}$$

Remark 8.25. If (8.35) is to hold for all c (not just for $a \le c \le b$), then it is

---

11 Note that I_1 and I_2 have the endpoint c in common, but even though the point c is "counted twice", its contribution to the integral is zero by (8.41).

necessary to define integrals from c to b with $b \leq c$, as follows:

$$\int_c^b f(t)\,dt = -\int_b^c f(t)\,dt \ . \tag{8.40}$$

In particular, it follows from (8.40) and ($a = -a \implies a = 0$), that for any x

$$\int_x^x f(t)\,dt = -\int_x^x f(t)\,dt = 0 \ , \tag{8.41}$$

in agreement with our intuition that the area from x to x is zero.

Remark 8.26. Finally, property e is a *monotonicity* property. If $f \leq g$, then $\int f \leq \int g$. In particular, if

$$m \leq f(x) \leq M, \quad \text{for all } a \leq x \leq b \ , \tag{8.42}$$

then, by property e,

$$\int_a^b m\,dx \leq \int_a^b f(x)\,dx \leq \int_a^b M\,dx \ ,$$

giving

$$m(b-a) \leq \int_a^b f(x)\,dx \leq M(b-a) \ . \tag{8.43}$$

Example 8.27 (Area between two graphs). Integrals can be used to calculate areas of regions more general than (8.2). Let g_1 and g_2 be continuous functions on the finite interval $[a, b]$, satisfying

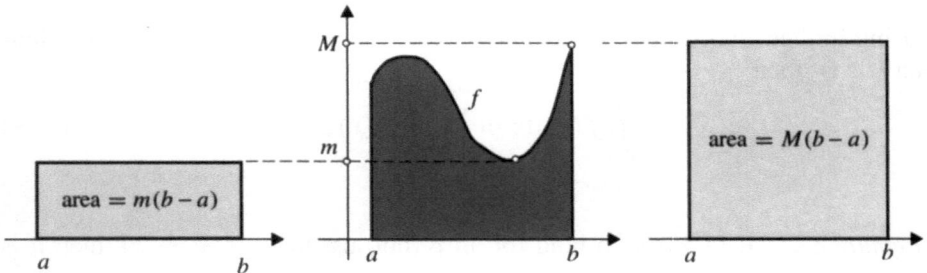

Fig. 8.7. Illustration of (8.43)

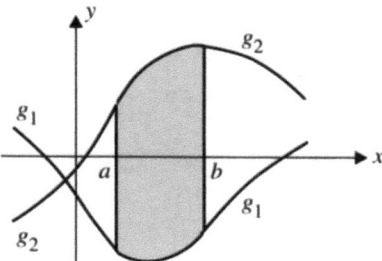

Fig. 8.8. Area between two graphs (Example 8.27)

$$g_1(x) \leq g_2(x) \quad \text{for all } a \leq x \leq b,\tag{8.44}$$

and consider the region $\mathcal{R}$ bounded by[12]

the graph of $y = g_2(x)$ (above),

the graph of $y = g_1(x)$ (below),   (8.45)

the two vertical lines $x = a$ and $x = b$ (Fig. 8.8).

The area $A(\mathcal{R})$ of $\mathcal{R}$ is the integral of the difference $g_2 - g_1$,

$$A(\mathcal{R}) = \int_a^b (g_2(x) - g_1(x))\,dx.\tag{8.46}$$

Note that the integrand, $g_2 - g_1$, is nonnegative by (8.44).

Example 8.28 (Integral of a negative function). If $f \leq 0$ on $[a, b]$, then, using Theorem 8.21 a with $\alpha = -1$, we conclude

$$\int_a^b f(x)\,dx = -\int_a^b (-f)(x)\,dx = -\int_a^b |f(x)|\,dx.\tag{8.47}$$

Equivalently, if f is negative on $[a, b]$, then $\int_a^b f(x)\,dx$ is the negative of the area of the region bounded by

the graph of $y = f(x)$ (below),

the x-axis (above),   (8.48)

the two vertical lines $x := a$ and $x := b$.

---

12 By taking $g_1(x) := 0$ (the equation of the x-axis) we see that the region (8.2) is a special case of $\mathcal{R}$.

Example 8.29.

$$\int\limits_{2-\sqrt{2}}^{2+\sqrt{2}} \left(\frac{x^2}{2} - 2x + 1\right) dx = - \int\limits_{2-\sqrt{2}}^{2+\sqrt{2}} \left|\frac{x^2}{2} - 2x + 1\right| dx$$

since $x^2/2 - 2x + 1 \leq 0$ in the interval $\left[2 - \sqrt{2}, 2 + \sqrt{2}\right]$.

Example 8.30. The integral $\int_2^1 (x - x^3)\, dx$ is

$$\int\limits_2^1 (x - x^3)\, dx \overset{(8.40)}{=\!=\!=\!=} - \int\limits_1^2 (x - x^3)\, dx \overset{(8.47)}{=\!=\!=\!=} \int\limits_1^2 (x^3 - x)\, dx \ .$$

If the function f assumes both positive and negative values on I, then its integral on I can still be expressed in terms of areas.

Corollary 8.31 (Integral of a general function in terms of areas). Let the function f be continuous on the finite interval $I = [a, b]$, and let

$$x_1 < x_2 < \cdots < x_{m-1} \tag{8.49}$$

be the points in I where f changes sign[13]. In addition to the points (8.49), take the two endpoints

$$x_0 := a, \quad x_m := b \ .$$

These points $x_0 := a < x_1 < x_2 < \cdots < x_m := b$ partition I into m subintervals $I_k := [x_{k-1}, x_k]$,

$$I = I_1 \cup I_2 \cup \cdots \cup I_m$$

such that f has the same sign (either nonnegative or nonpositive) on each subinterval I_k.

For $k = 1, \ldots, m$ let R_k be the region bounded by

the graph of $y = f(x)$,
the x-axis ,
the two vertical lines $x := x_{k-1}$ and $x := x_k$.

Then

$$\int\limits_I f(x)\, dx = \sum_{k=1}^m \int\limits_{I_k} f(x)\, dx$$

---

13 That is the set of points where the graph of f crosses the x-axis. This set is empty if f does not change sign in I. We assume for simplicity that m is finite; see also Exercise 8.23.

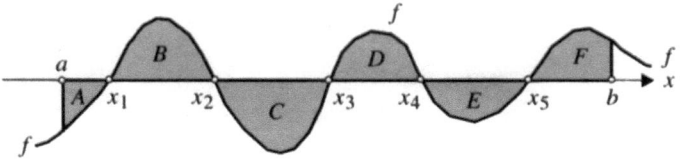

Fig. 8.9. Illustration of Corollary 8.31

is the sum of the areas of the regions R_k where f is nonnegative minus the sum of areas of R_k where f is nonpositive.

For the function f in Fig. 8.9, there are 5 sign changes in $[a, b]$, and the integral from a to b is given in terms of the areas $A, \ldots, F$ as follows:

$$\int_a^b f(x)\, dx = (B + D + F) - (A + C + E)\,.$$

Example 8.32 (Sines and cosines from 0 to 2π). Consider the integral

$$\int_0^{2\pi} \sin x\, dx\,.$$

The integrand is positive between 0 and π, and negative from π to 2π. An examination of the graph of $\sin x$ shows the "negative" area to cancel the "positive" one, and consequently,

$$\int_0^{2\pi} \sin x\, dx = 0\,.$$

For similar reasons,

$$\int_0^{2\pi} \cos x\, dx = 0\,,$$

and more generally, for any integer $n \neq 0$,

$$\int_0^{2\pi} \sin nx\, dx = 0 = \int_0^{2\pi} \cos nx\, dx\,. \tag{8.50}$$

Plot the graphs of $\sin nx$ and $\cos nx$, for different n with MACSYMA, and verify that in $[0, 2\pi]$ the "positive" areas and "negative" areas cancel each other.

Example 8.33 (Products of sines and cosines). From the trigonometric identities

$$\sin(m+n)x = \sin mx \cos nx + \cos mx \sin nx ,$$
$$\sin(m-n)x = \sin mx \cos nx - \cos mx \sin nx ,$$
$$\cos(m+n)x = \cos mx \cos nx - \sin mx \sin nx ,$$
$$\cos(m-n)x = \cos mx \cos nx + \sin mx \sin nx ,$$

it follows, by addition and subtraction, that

$$2 \sin mx \sin nx = \cos(m-n)x - \cos(m+n)x ,$$
$$2 \cos mx \cos nx = \cos(m-n)x + \cos(m+n)x ,$$
$$2 \sin mx \cos nx = \sin(m+n)x + \sin(m-n)x .$$

These identities, together with (8.50), prove that, for any integers m, n,

$$\int_0^{2\pi} \sin mx \sin nx \, dx = \int_0^{2\pi} \frac{\cos(m-n)x - \cos(m+n)x}{2} \, dx = \begin{cases} 0 & \text{if } m \neq n, \\ \pi & \text{if } m = n; \end{cases}$$

$$\int_0^{2\pi} \cos mx \cos nx \, dx = \int_0^{2\pi} \frac{\cos(m-n)x + \cos(m+n)x}{2} \, dx = \begin{cases} 0 & \text{if } m \neq n, \\ \pi & \text{if } m = n; \end{cases}$$

$$\int_0^{2\pi} \sin mx \cos nx \, dx = \int_0^{2\pi} \frac{\sin(m+n)x + \sin(m-n)x}{2} \, dx = 0 ;$$

where for $m = n$ we used the facts that $\cos 0 = 1$, and $\int_0^{2\pi} 1 \, dx = 2\pi$.

The central result in this section is that continuous functions are integrable on any finite closed interval.

Theorem 8.34 (Continuity implies integrability). *If f is continuous on a finite interval $[a, b]$, then f is integrable on $[a, b]$.*

This theorem is proved, for a special case of continuous functions f and intervals $[a, b]$, where the graph of f intersects any horizontal line in finitely many points, in Sect. 8.6.

Remark 8.35. If the continuous function f is monotone on the finite interval $[a, b]$, then the approximation error (8.34) can be computed exactly. The proof that f is integrable is then quite simple.

Let f be continuous and monotone, with minimum m and maximum M in the

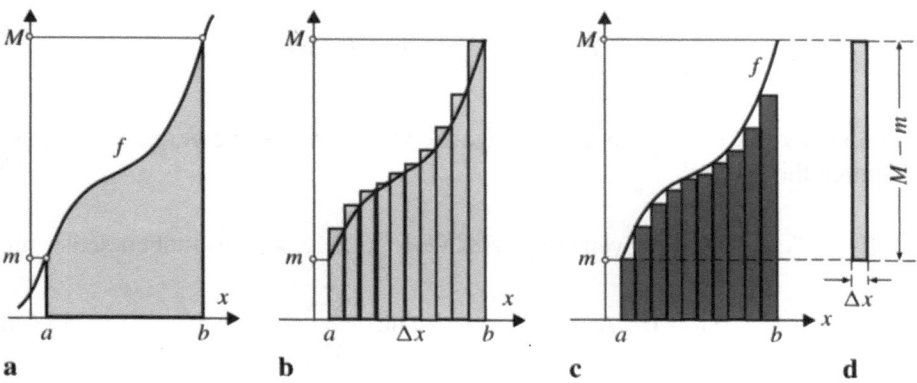

Fig. 8.10 a–d. Approximation error $E(f, \mathcal{P})$ for an increasing function. **a** An increasing function f; **b** an upper Riemann sum $S_U(f, \mathcal{P})$; **c** a lower Riemann sum $S_L(f, \mathcal{P})$; **d** error $E(f, \mathcal{P})$

finite interval $[a, b]$, and let $\mathcal{P}$ be a regular partition of $[a, b]$ into n subintervals. Then the approximation error (8.34) is

$$E(f, \mathcal{P}) = (M - m) \frac{b - a}{n}, \tag{8.51}$$

see Fig. 8.10. Since a, b, M, and m are fixed, the error $E(f, \mathcal{P})$ tends to 0 as $n \to \infty$.

In Theorem 8.34 we can allow f to have a finite number of discontinuities.

Definition 8.36 (Piecewise continuity). A function f is *piecewise continuous* on the interval $[a, b]$ if the interval is partitioned into finitely many subintervals

$$a = x_0 < x_1 < \cdots < x_n = b$$

such that, for $k = 1, \ldots, n$:
a. f is continuous in the open subinterval (x_{k-1}, x_k), and
b.   the two (one-sided) limits

$$\lim_{x \downarrow x_{k-1}} f(x), \quad \lim_{x \uparrow x_k} f(x),$$

exist and are finite.

Corollary 8.37. If f is piecewise continuous on $[a, b]$, then f is integrable on $[a, b]$.

Proof. Let $I = [a, b]$ have the partition (8.14), and let f be integrable. We rewrite (8.39) as

$$\int_I f(x)\,dx = \sum_{k=1}^n \int_{x_{k-1}}^{x_k} f(x)\,dx$$

and note, by (8.41), that we may change the values of f at the points x_k, without changing the integral.   □

If $f_1, f_2, \ldots, f_n$ are given numbers, and $\lambda_1, \lambda_2, \ldots, \lambda_n$ are numbers satisfying

$$\lambda_k \geq 0 \quad (k = 1, \ldots, n) \quad \text{and} \quad \sum_{k=1}^n \lambda_k = 1, \tag{8.52}$$

then

$$\sum_{k=1}^n \lambda_k f_k \tag{8.53}$$

is called an *average* (or *mean*) of the f_k's. The λ_k's in (8.52) are called the *weights* of the mean (8.53). The ordinary average $(1/n)\sum_{k=1}^n f_k$ is (8.53) with equal weights $\lambda_k = 1/n$.

We establish now the "mean" properties of Riemann sums and integrals. Let f be integrable on $I = [a, b]$. For any partition $\mathcal{P}$ of I, let

$$S(f, \mathcal{P}) = \sum_{k=1}^n f(\xi_k)\,\Delta x_k, \tag{8.54}$$

be a Riemann sum. Dividing by $(b - a)$ we get, by (8.12),

$$\frac{S(f, \mathcal{P})}{b - a} = \sum_{k=1}^n \frac{\Delta x_k}{\sum_{j=1}^n \Delta x_j} f(\xi_k),$$

an "average" of the values of f at the points ξ_k, with weights

$$\lambda_k := \frac{\Delta x_k}{\sum_{j=1}^n \Delta x_j}, \quad (k = 1, \ldots, n).$$

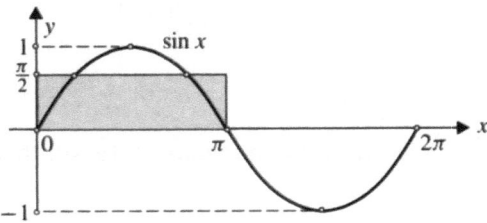

Fig. 8.11. Mean value of $\sin x$ (Example 8.38)

In the limit, as $\|\mathcal{P}\| \to 0$, we define the *average* or *mean value* of f in $I = [a, b]$ as

$$\text{mean value}(f, x, a, b) := \frac{\int_a^b f(x)\,dx}{b - a} , \qquad (8.55)$$

also denoted mean value(f, x, I).

Example 8.38. The integral of $\sin x$ from 0 to π is

$$\int_0^\pi \sin x\,dx = 2 \qquad (8.56)$$

as can be verified by MACSYMA. The mean value of $\sin x$ in the interval $[0, \pi]$ is therefore

$$\text{mean value}(\sin x, x, 0, \pi) = 2/\pi \approx 0.63662\ldots$$

In Fig. 8.11 we plot $y = \sin x$, and the horizontal line $y = 2/\pi$. From definition (8.55) it follows that the integral (8.56), the area under the graph of $\sin x$ from 0 to π, equals the shaded area.

From (8.43) it follows that

$$m \le \text{mean value}(f, x, I) \le M$$

if m and M are the minimum and maximum of f in $I = [a, b]$. Since f is continuous, it assumes in I every value between m and M. We have thus proved the following theorem.

Theorem 8.39 (Mean value theorem for integrals). Let f be continuous on $[a, b]$. Then there is a point ξ in (a, b) such that

$$\int_a^b f(x)\,dx = f(\xi)(b - a) . \qquad (8.57)$$

A point ξ, where f assumes its mean value in I, is the x-coordinate of an intersection of the graph of f and the horizontal line $y := \text{mean value}(f, x, I)$. In general, ξ is not unique; see Fig. 8.11, where in $[0, \pi]$ there are two points ξ where the mean value of $\sin x$ is attained.

Example 8.40. We show in Example 9.10 that $\int_1^3 x^2\,dx = 26/3$. Therefore the mean value of x^2 in $[1, 3]$ is

$$\text{mean value}(x^2, x, 1, 3) = \frac{\int_1^3 x^2\, dx}{3-1} = \frac{13}{3}$$

which is attained at $\xi = \sqrt{13/3}$.

MACSYMA-Session 8.5. The mean value of f in $[a, b]$ can be computed by the MACSYMA function

c1. `mean_value(f,x,a,b):=integrate(f,x,a,b)/(b-a)$`

The mean value of $\sin(x)$ in $[0, \pi]$ is

c2. `mean_value(sin(x),x,0,%pi)`

d2. $\quad \dfrac{2}{\pi}$

and the mean value of $\sin(x)$ in $[0, 2\pi]$ is

c3. `mean_value(sin(x),x,0,2*%pi)`

d3. $\quad$ 0

Explain the last result by considering the graph of $\sin(x)$ in the range $[0, 2\pi]$.

Exercises

8.10 Express the following definite integrals as Riemann sums. Approximate the Riemann sums for $n = 3, 10, 100$ with your own choice of ξ_k.

(a) $\displaystyle\int_0^\pi \sqrt{1 + \cos x}\, dx$

(b) $\displaystyle\int_0^1 \frac{x+3}{x+1}\, dx$

(c) $\displaystyle\int_{-3}^5 (x^3 + 3x - 1)\, dx$

(d) $\displaystyle\int_0^{10} \ln(x + 1)\, dx$

8.11 Recall that the Riemann sum of f on $[a, b]$ approximates the area bounded by the curve $y = f(x)$, the x-axis, and the vertical lines $x = a$ and $x = b$. In some instances it is possible to exactly compute the area with an appropriate Riemann sum. Compute the Riemann sums and show it does not matter if n is taken to be 10 or 100 if ξ_k is chosen to be the midpoint of the interval $[x_{k-1}, x_k]$ in the following problems. Why is this so?

(a) $\displaystyle\int_0^\pi (2x + 1)\, dx$

(b) $\displaystyle\int_0^1 (-3x + 6)\, dx$

(c) $\displaystyle\int_{-3}^5 (x + 18)\, dx$

(d) $\displaystyle\int_2^{10} (5x - 6)\, dx$

(e) $\displaystyle\int_{-1}^4 f(x)\, dx,$ where $f(x) := \begin{cases} 1 & \text{if } x \le 1, \\ x & \text{if } x \ge 1 \end{cases}$

(f) $\displaystyle\int_0^6 f(x)\, dx,$ where $f(x) := \begin{cases} 8, & \text{if } x \le 2, \\ 2x + 2, & \text{if } x \ge 2 \end{cases}$

8.12 Sometimes you know the area under a curve and above the x-axis from elementary geometry. What are the values of the following integrals.

(a) $\displaystyle\int_0^2 (x + 1)\, dx$

(b) $\displaystyle\int_{-1}^1 (3x + 6)\, dx$

(c) $\displaystyle\int_{-1}^1 \sqrt{1 - x^2}\, dx$

(d) $\displaystyle\int_{-2}^2 \left(3 + \sqrt{4 - x^2}\right)\, dx$

8.13 Write down the integrals expressing the area lying between the curves $y = f(x)$ and $y = g(x)$.

(a) $f(x) := x^2 + 1$, $g(x) := 5 - x^2$ (b) $f(x) := x + 2$, $g(x) := 4 - x^2$

(c) $f(x) := \sqrt{4 - x^2}$, $g(x) := 2 + x$ (d) $f(x) := x^3 + 1$, $g(x) := 1 + x$

8.14 Use elementary facts about areas of triangles and trapezoids to compute the following integrals:

(a) $\displaystyle\int_0^1 x\, dx$ (b) $\displaystyle\int_{-1}^1 x\, dx$ (c) $\displaystyle\int_{-1}^1 |x|\, dx$

(d) $\displaystyle\int_{-1}^1 (x + |x|)\, dx$ (e) $\displaystyle\int_{-4}^4 (|x| - 2)\, dx$ (f) $\displaystyle\int_{-1}^2 x\, dx$

8.15 Illustrate $\int_{-1}^2 x^2\, dx = 3$ by computing the Riemann sums $S_{\text{right}}(x^2, x, -1, 2, n)$, and $S_{\text{middle}}(x^2, x, -1, 2, n)$ and their limits as $n \to \infty$.

8.16 Recall the sign function defined by

$$\operatorname{sign} x = \begin{cases} -1, & x < 0, \\ 0, & x = 0, \\ +1, & x > 0. \end{cases}$$

Show, using elementary facts about areas of rectangles, that

$$\int_a^b \operatorname{sign} x\, dx = |b| - |a|$$

for any a, b.

8.17 Write an integral representing the area of the region bounded by
the graph of $y = 14 - 2x$ (above),
the graph of $y = -3 + x$ (below),
the two vertical lines $x := 2$ and $x := 5$.

8.18 Write an integral representing the area of the region bounded by the graphs of $y = x^2 - 1$ and $y = x + 5$.

8.19 Let f be continuous and monotone, with minimum m and maximum M in the finite interval $[a, b]$, and let $\mathcal{P}$ be a regular partition of $[a, b]$ into n subintervals. Prove (8.51).

8.20 (A generalization) Let the function f have minimum m and maximum M in the finite interval $[a, b]$, and let $\mathcal{P}$ be a regular partition of $[a, b]$ into n subintervals. Prove that the approximation error (8.34) is

$$E(f, \mathcal{P}) \le (M - m)\frac{b - a}{n}.$$

8.21 Verify, without computation: For any real α, $\int_{-\alpha}^{\alpha} \sin x\, dx = 0$.

8.22 Determine which of the following functions is integrable on $I = [0, 1]$.

(a) $f(x) := \begin{cases} \sin(1/x) & \text{if } x \ne 0, \\ 0 & \text{if } x = 0 \end{cases}$ (b) $f(x) := \begin{cases} x\sin(1/x) & \text{if } x \ne 0, \\ 0 & \text{if } x = 0 \end{cases}$

8.23 The function

$$f(x) = \begin{cases} x\sin(1/x) & \text{if } x \ne 0, \\ 0 & \text{if } x = 0, \end{cases}$$

is continuous in $\mathbf{R}$. Show that in any interval containing zero, there are infinitely many points where f changes sign. *Hint:* $f(x) = 0$ for $1/x = n\pi$, $n = \pm 1, \pm 2, \pm 3, \ldots$

8.24 Find the mean values of $\sin x$ in $[\pi, 2\pi]$ and in $[-\pi, \pi]$. Compare with Example 8.38. Is there an interval I where the mean value of $\sin x$ is greater than 1?

8.25 a. Verify, without calculation, $\int_0^{2\pi} \sin^2 x \, dx = \pi$. *Hint:* $\sin^2 x + \cos^2 x = 1$.

    b. Calculate the mean value of $f(x) = \sin^2 x$ in $[0, 2\pi]$

8.26 Compute mean value(f, x, a, b) for the following $f(x)$ and $[a, b]$. Treat u as a variable endpoint.

    (a) $\sin x$, $[0, 2\pi]$ (b) $|x|$, $[-u, u]$ (c) $\operatorname{sign} x$, $[-u, u]$

    (d) e^x, $(-\infty, u]$ (e) e^x, $[0, u]$ (f) $1/x$, $[1, u]$

8.27 Prove that

$$\lim_{b \to a} \text{mean value}(f, x, a, b) = f(a) \tag{8.58}$$

Hint: Use Theorem 8.39.

8.4 Numerical integration: trapezoid method*

Given f, a, and b, the definite integral of f in $[a, b]$ was defined above as the limit of Riemann sums, (8.5),

$$\int_a^b f(x) \, dx := \lim_{\substack{\max \Delta x_k \to 0 \\ k=1,\ldots,n}} \sum_{k=1}^n f(\xi_k) \, \Delta x_k \, .$$

Sometimes it suffices to have an approximate value of (8.29), and in such cases a suitable Riemann sum is used in lieu of the integral, and it is not necessary to compute the limit (8.5).

There are also cases where it is impossible, or impractical, to compute the above limit. One such case is when the integrand f is not given analytically, but values of f, say, $f(\xi_1)$, $f(\xi_2)$, $\ldots$, $f(\xi_n)$ are observed or measured[14]. We can then still approximate the definite integral as a Riemann sum, but we cannot take limits since only finitely many function values are given.

Computing an approximate value of a definite integral is called *numerical integration*.

The two simplest methods of numerical integration are the trapezoid method given in this section and the Simpson method in Sect. 8.5.

Both methods use a regular partition of $[a, b]$, with n subintervals of length

$$h := (b - a)/n \, . \tag{8.59}$$

The partition points here are, (8.14),

$$x_k := x_0 + kh = a + k(b - a)/n \, ,$$

and the subintervals are, (8.15),

$$I_k := [x_{k-1}, x_k] = [a + (k - 1)h, a + kh] \quad (k = 0, \ldots, n) \, .$$

---

* This section is optional. It can be skipped at first reading. The material covered here is used in Sect. 8.5.

14 This is typical in engineering and science, where functions are often obtained empirically, e.g., estimated on the basis of laboratory experiments.

Suppose we are given the $n+1$ values of f at the partition points (8.14). We denote these values by

$$f_k := f(x_k);\qquad(8.60)$$

see Fig. 8.12 a and b for illustration of given values $f_0, \ldots, f_8$ and what the function f might look like.

In each subinterval (8.15) consider the line determined by (i.e., passing through) the two points on the graph of f,

$$L = (x_{k-1}, f_{k-1}), \quad R = (x_k, f_k).\qquad(8.61)$$

The area under this line, in I_k, is the area of a *trapezoid* with *height h* (on the x-axis) and *bases* f_{k-1}, f_k,

$$A_k = \frac{f_{k-1} + f_k}{2} h = \frac{f_{k-1} + f_k}{2}\left(\frac{b-a}{n}\right).$$

This is illustrated in Fig. 8.12c for the 6th subinterval.

If n is large, then A_k is a good approximation of the area under the graph of f in I_k. Summing A_k over all subintervals and noting that the $n-1$ values $f_1, f_2, \ldots, f_{n-1}$ are counted twice, we get the following approximation of the integral (8.29),

$$\int_a^b f(x)\,dx \approx (f_0 + 2f_1 + 2f_2 + \cdots + 2f_{n-2} + 2f_{n-1} + f_n)\left(\frac{b-a}{2n}\right).\quad(8.62)$$

This is called the *trapezoid rule* (or trapezoid method). We denote the right-hand side by

$$\text{trapezoid}(f, x, a, b, n) := (f_0 + 2f_1 + 2f_2 + \cdots + 2f_{n-2}$$
$$+ 2f_{n-1} + f_n)\left(\frac{b-a}{2n}\right).\qquad(8.63)$$

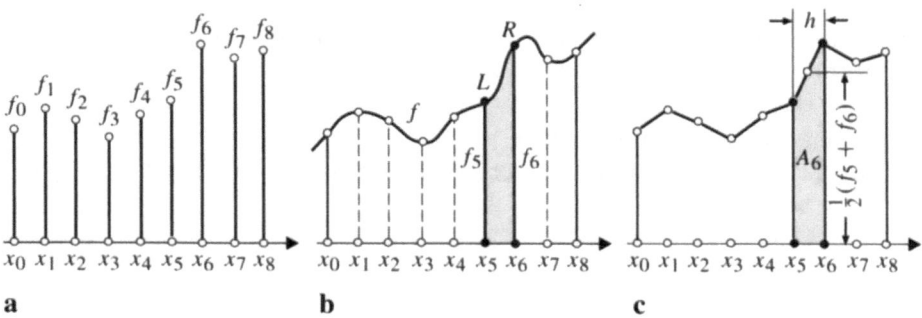

Fig. 8.12 a–c. Trapezoid method. **a** Given values $f_0, \ldots, f_8$. **b** What f might look like. **c** Trapezoidal approximation

MACSYMA-Session 8.6. The following MACSYMA function computes the trapezoid sum (8.63)

```
c1. trapezoid(f,x,a,b,n):=
        (b-a)/(2*n)*(subst(a,x,f)+subst(b,x,f))+
            riem_sum_right(f,x,a,b-(b-a)/n,n-1)$
```

which uses

`subst(a,x,f)` to compute $f(a)$ and

`riem_sum_right` to compute the $f_1 + f_2 + \cdots + f_{n-1}$ portion, from $x = a$ to $x = b - (b-a)/n$.

We check the correctness of c1 by applying it to a general function,

`c2. trapezoid(f(x),x,a,b,n)`

d2.
$$\frac{\left(-\frac{b-a}{n}+b-a\right)\sum_{k=1}^{n-1} f\left(\frac{an+(b-a)k}{n}\right)}{n-1} + \frac{(b-a)(f(b)+f(a))}{2n}$$

`c3. factor(%)`

d3.
$$\frac{(b-a)}{2n}\left(2\sum_{k=1}^{n-1} f\left(\frac{an+(b-a)k}{n}\right) + f(b) + f(a)\right)$$

which we recognize as a rewriting of the trapezoid rule (8.63).

Consider the unit circle (with center at the origin), $x^2 + y^2 = 1$, and the area A bounded by the circle, the positive x-axis and the positive y-axis,

$$A = \int_0^1 \sqrt{1 - x^2}\, dx \qquad (8.64)$$

Clearly, $A = \pi/4$ since it is one quarter of the area inside the unit circle. We can verify this by

`c4. integrate(sqrt(1-x^2),x,0,1)`

d4. $\pi/4$

Approximating A by the trapezoid method, with $n = 4$, we get

`c5. trapezoid(sqrt(1-x^2),x,0,1,4)`

d5.
$$\frac{\sqrt{15}/4 + \sqrt{7}/4 + \sqrt{3}/2}{4} + \frac{1}{8}$$

which is just under $3/4$, and still far away from $\pi/4 \approx 0.785398$.

To generate trapezoid($\sqrt{1 - x^2}, x, 0, 1, n$) for $n = 4, 8, 16, 32, 64, 128$:

```
c6. makelist(trapezoid(sqrt(1-x^2),x,0,1,n),n,[4,8,16,32,64,128]),
        numer
```

d6. $[0.74893, 0.77245, 0.78081, 0.78377, 0.78482, 0.78519]$

which makes a list of the values for the given n's. Alternatively, we can use:

`c7. makelist(trapezoid(sqrt(1-x^2),x,0,1,2^m),m,2,7),numer`

instead of c6.

To print the results in a convenient form:

```
c8. for i:2 thru 7 do print(2^i,trapezoid(sqrt(1.0-x^2),x,0,1,2^i)),
        numer
```

d8.   4   0.74893
          8   0.77245
         16   0.78081
         32   0.78377
         64   0.78482
        128   0.78519

displaying the values, but not storing them. These results show that as n increases, the values of trapezoid($\sqrt{1 - x^2}, x, 0, 1, n$) converge toward the correct result $\pi/4 \approx 0.785398\ldots$

If the 2nd derivative f'' is continuous on $[a, b]$, we can bound the error of the trapezoid method

$$\left| \int_a^b f(x)\, dx - \text{trapezoid}(f, x, a, b, n) \right|$$

as follows.

Theorem 8.41. Let f'' be continuous on $[a, b]$. Then there is a point $\xi \in [a, b]$ such that

$$\text{trapezoid}(f, x, a, b, n) - \int_a^b f(x)\, dx = \frac{(b - a)^3}{12n^2} f''(\xi) \qquad (8.65)$$

and consequently

$$\left| \int_a^b f(x)\, dx - \text{trapezoid}(f, x, a, b, n) \right| \leq \frac{(b - a)^3}{12n^2} \max_{a \leq x \leq b} |f''(x)| . \qquad (8.66)$$

The proof of this theorem, which is based on the mean value theorem, is quite tedious. You can find it in advanced books.

Remark 8.42. If f is a linear function, say

$$f(x) = p_1 x + p_0 ,$$

then $f'' = 0$ and the trapezoid rule is exact, by (8.65),

$$\int_a^b f(x)\, dx = \text{trapezoid}(f, x, a, b, n), \quad \text{for all } n \geq 1 ,$$

$$= \left(\frac{f(a) + f(b)}{2} \right)(b - a), \quad \text{in particular, for } n = 1 .$$

This can be illustrated by MACSYMA as follows: Compute the error of the trapezoid rule
c1. `trapezoid(u x^2+v x+w,x,a,b,n)-integrate(u x^2+v x+w,x,a,b)` `$`
for a quadratic function $u x^2 + v x + w$. The result after using `ratsimp` and `factor` is

d1. $\dfrac{u(b - a)^3}{6n^2}$

in agreement with (8.65), showing that the error depends on the coefficient u of x^2 but not on the coefficients v and w of the lower powers. Therefore, the trapezoid rule is exact for linear functions $vx + w$.

Exercises

8.28 Compare the trapezoid rule with the function `s_middle(f,x,a,b,n)` using the same number ($n = 4, 10, 40$) of subintervals $[x_{k-1}, x_k]$ and the functions $f(x)$ given below. Discuss the results of this numerical experiment.

(a) $\int_{-1}^{1} \sqrt{1 - x^2}\, dx$ (b) $\int_{0}^{1} \sqrt{1 + x^2}\, dx$ (c) $\int_{1}^{4} \frac{1}{x}\, dx$ (d) $\int_{0}^{\pi/2} \sin x\, dx$

8.29 Repeat the analysis of the previous problem on the integrals

(a) $\int_{\varepsilon}^{1} \frac{1}{\sqrt{x}}\, dx$ (b) $\int_{-1}^{-\varepsilon} \frac{1}{x}\, dx + \int_{\varepsilon}^{1} \frac{1}{x}\, dx$

where $\varepsilon = 0.1, 0.01$, and 0.001. Interpret your results.

8.30 Compute the following integrals to within an accuracy of 10^{-3} using the trapezoid rule.

(a) $\int_{0}^{1} (2x^2 + 1)\, dx$ (b) $\int_{-2}^{2} (x^3 + x^2 - 4x)\, dx$ (c) $\int_{-3}^{1} e^x\, dx$

(d) $\int_{2}^{4} \sinh x\, dx$ (e) $\int_{-1}^{1} f(x)\, dx$, where $f(x) := \begin{cases} |x| & \text{if } x \leq 0, \\ x^2 & \text{if } x \geq 0 \end{cases}$

(f) $\int_{0}^{20} f(x)\, dx$, where $f(x) := \begin{cases} x & \text{if } x \leq 4, \\ \sqrt{20 - x} & \text{if } 4 \leq x \leq 20 \end{cases}$

8.31 Show that the trapezoid sum (8.63) is the average of the left Riemann sum (8.17) and the right Riemann sum (8.19),

$$\text{trapezoid}(f, x, a, b, n) = \tfrac{1}{2}(S_{\text{left}}(f, x, a, b, n) + S_{\text{right}}(f, x, a, b, n)).$$

Does this formula provide an efficient method to compute the trapezoid sum?

8.32 The function f is unknown, but the following 11 values of $y = f(x)$ have been obtained empirically.

x	1.2	1.3	1.4	1.5	1.6	1.7	1.8	1.9	2.0	2.1	2.2
y	2.16	2.85	4.10	1.08	−2.14	−1.24	1.89	3.18	0.25	−1.68	−3.72

Approximate $\int_{1.2}^{2.2} f(x)\, dx$ using the trapezoid rule.

8.33 It is required to approximate the following integrals by the trapezoid rule, with an error of $\leq 10^{-4}$. Use the error bound (8.66) to determine an appropriate n, and check your results (i.e., compare the results of the trapezoid method with the correct values obtained by MACSYMA and verify that the error is as specified).

(a) $\int_{1}^{2} \frac{x^3}{6}\, dx$ (b) $\int_{\pi/2}^{\pi} \sin x\, dx$ (c) $\int_{1}^{2} \ln x\, dx$

8.5 Numerical integration: Simpson method*

In the trapezoid method the integrand f is approximated in each subinterval $[x_{k-1}, x_k]$ by the line through the two points $L = (x_{k-1}, f(x_{k-1}))$, $R = (x_k, f(x_k))$

---

    * This section is optional. It can be skipped at first reading. This section requires the results of Sect. 8.4.

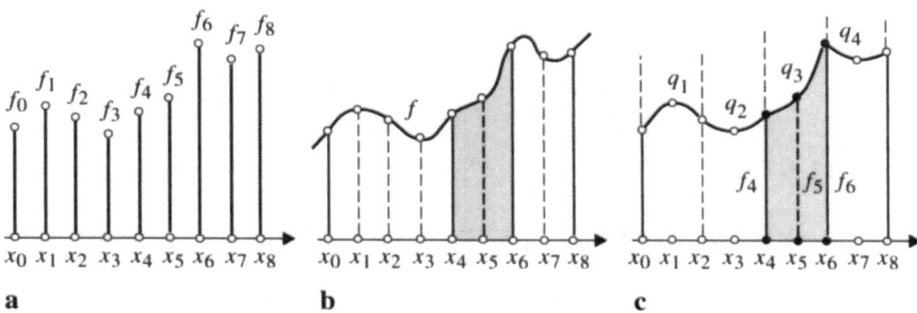

Fig. 8.13 a–c. Simpson method. **a** Given values $f_0, \ldots, f_8$. **b** What f might have looked like. **c** Approximating f by the 4 quadratics $q_1, \ldots, q_4$

on the graph of f. In the Simpson[15] method the function f is approximated by quadratic functions.

Since a quadratic function (or parabola) is determined by three points, we need 3 points, or 2 subintervals, to define each quadratic. This is illustrated in Fig. 8.13, where f is approximated by 4 quadratics $q_1, \ldots, q_4$, each requiring 3 points or 2 subintervals, giving a total number of $n = 8$ subintervals.

The Simpson method thus requires a partition with an *even number of subintervals*, n.

Given three points

$$L = (x_{k-1}, f(x_{k-1})), \quad M = (x_k, f(x_k)), \quad R = (x_{k+1}, f(x_{k+1})), \quad (8.67)$$

on the graph of f, the Simpson method finds the unique quadratic function

$$q(x) = \alpha x^2 + \beta x + \gamma, \quad (8.68)$$

passing through the three points. The function $q(x)$ is then integrated from x_{k-1} to x_{k+1}. The process is repeated for $k = 1, 3, \ldots, n-1$, i.e., for all $n/2$ parabolas used in the approximation. The work is organized in three steps.

Step 1. Compute the quadratic q passing through (8.67).
The coefficients α, β, and γ of q are determined by solving three equations, corresponding to q passing through L, M, and R. For example, the fact that q passes through the point M gives the equation

$$\alpha x_k^2 + \beta x_k + \gamma = f(x_k),$$

where x_k and $f(x_k)$ are given, and α, β, and γ are unknown.
Step 2. Integrate $q(x)$ in $[x_{k-1}, x_k]$.
Step 3. Sum over all subintervals.

The results of steps 1 and 2 depend on the three function values

$$f(x_{k-1}), \quad f(x_k), \quad f(x_{k+1}). \quad (8.69)$$

---

15 Thomas Simpson (1710–1761).

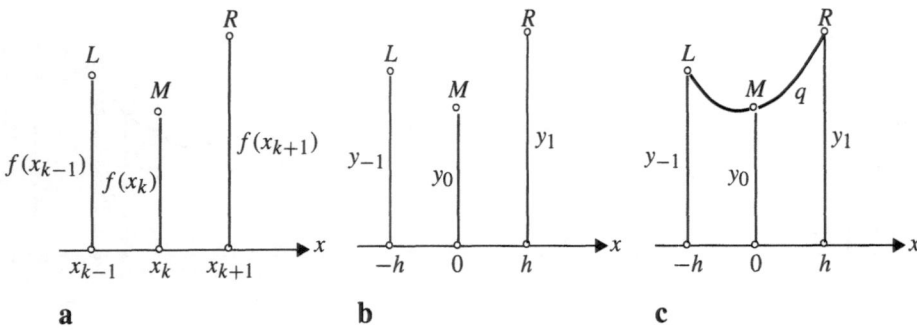

Fig. 8.14 a–c. Points L, M, and R defining a unique quadratic $q(x) = ax^2 + bx + c$. **a** Three
points L, M, and R. **b** The points centered. **c** The quadratic q

To make the arithmetic simpler we "center" the subinterval $[x_{k-1}, x_{k+1}]$ by identifying its midpoint x_k with the origin $x = 0$ and the endpoints x_{k-1} and x_{k+1} with
$x = -h$ and $x = h$, respectively, where, (8.59),

$$h := (b - a)/n \ .$$

We denote the three function values (8.69) by

$$y_{-1} := f(x_{k-1}), \quad y_0 := f(x_k), \quad y_1 := f(x_{k+1}) \ .$$

This is illustrated in Fig. 8.14b.

The quadratic q passing through

$$L := (-h, y_{-1}), \quad M := (0, y_0), \quad R := (h, y_1)$$

(Fig. 8.14c) is determined by solving the following three equations

$$\begin{aligned} \alpha h^2 - \beta h + \gamma &= y_{-1}, & & (q \text{ passes through } L) \\ \gamma &= y_0, & & (q \text{ passes through } M) \\ \alpha h^2 + \beta h + \gamma &= y_1 & & (q \text{ passes through } R) \end{aligned}$$

in the three unknowns α, β, and γ. These equations are easily solved to give

$$\gamma = y_0, \quad \beta = \frac{y_1 - y_{-1}}{2h}, \quad \alpha = \frac{1}{2h^2}(y_{-1} - 2y_0 + y_1) \ . \tag{8.70}$$

This concludes step 1. Now we integrate q from $-h$ to h,

$$\int_{-h}^{h} (\alpha x^2 + \beta x + \gamma)\, dx = \alpha \int_{-h}^{h} x^2\, dx + \beta \int_{-h}^{h} x\, dx + \gamma \int_{-h}^{h} 1\, dx$$

$$= \alpha \left(\frac{h^3}{3} - \frac{(-h)^3}{3} \right) + \beta \left(\frac{h^2}{2} - \frac{(-h)^2}{2} \right) + \gamma (h - (-h)), \quad \text{by (8.31)} \ ,$$

$$= \alpha \left(\frac{2h^3}{3} \right) + \gamma(2h), \quad \text{now substitute } \alpha \text{ and } \gamma \text{ from (8.70) to get}$$

$$= \frac{h}{3}(y_{-1} + 4y_0 + y_1) . \tag{8.71}$$

Having completed step 2, we begin step 3, the summation of the contributions (8.71) of the $n/2$ parabolas. There are altogether $n + 1$ values

$$f_0, f_1, \ldots, f_n \tag{8.72}$$

of f at the equally spaced points

$$x_k = a + kh = a + k(b - a)/n \quad (k = 0, 1, \ldots, n) .$$

The jth parabola uses three values (8.69),

$$f_{2(j-1)}, f_{2j-1}, f_{2j}, \quad j = 1, \ldots, n/2 ;$$

for example, the 3rd parabola q_3 in Fig. 8.13c uses the values f_4, f_5, f_6.
	We arrange the values (8.72) in triples, one for each parabola,

$$\{f_0, f_1, f_2\}, \{f_2, f_3, f_4\}, \ldots, \{f_{n-4}, f_{n-3}, f_{n-2}\}, \{f_{n-2}, f_{n-1}, f_n\} .$$

Summing each triple according to (8.71) we obtain an approximation of the integral

$$\int_a^b f(x)\,dx \approx \frac{h}{3}(f_0 + 4f_1 + 2f_2 + 4f_3 + 2f_4 + \cdots + 2f_{n-2}$$
$$+ 4f_{n-1} + f_n) , \tag{8.73}$$

called the *Simpson method* (Simpson rule) for approximating the integral (8.29). We denote the right-hand side of (8.73) by

$$\text{simpson}(f, x, a, b, n) := (f_0 + 4f_1 + 2f_2 + \cdots + 2f_{n-2}$$
$$+ 4f_{n-1} + f_n)\left(\frac{b - a}{3n}\right) , \tag{8.74}$$

where n is *even*.

MACSYMA-Session 8.7. The following MACSYMA function computes the Simpson sum (8.74)

```
c1. our_simpson(f,x,a,b,n):=
        (b-a)/(3*n)*(subst(a,x,f)+subst(b,x,f))
            1/3*riem_sum_right(f,x,a,b-2*(b-a)/n,n/2-1)
            2/3*riem_sum_middle(f,x,a,b,n/2)$
```

where
the second line computes $(f_0 + f_n)((b - a)/3n)$,
the third line computes $(2f_2 + 2f_4 + \cdots + 2f_{n-2})((b - a)/3n)$,
the fourth line computes $(4f_1 + 4f_3 + \cdots + f_{n-1})((b - a)/3n)$.

We named this MACSYMA function our_simpson because MACSYMA has its own
simpson function, which you should use to check the results obtained with our_simpson.

We check the correctness of `our_simpson` by applying it to a general function

c2. `our_simpson(f(x),x,a,b,n)`

The result (not shown here) is somewhat complicated, but

c3. `%,factor`

gives

d3.
$$\frac{(b-a)}{3n}\left(4\sum_{k=1}^{n/2} f\left(\frac{an+(2b-2a)k-b+a}{n}\right)\right.$$
$$\left. +2\sum_{k=1}^{(n-2)/2} f\left(\frac{an+(2b-2a)k}{n}\right)+f(b)+f(a)\right)$$

which is a rewriting of the Simpson rule (8.74).

We use `our_simpson(f(x),x,a,b,n)` to approximate the integral (8.64), see Session 8.6. For $n = 4$ we have

c4. `our_simpson(sqrt(1-x^2),x,0,1,4)`

d4.
$$\frac{\sqrt{15}/4+\sqrt{7}/4}{3}+\frac{\sqrt{3}}{12}+\frac{1}{12}$$

c5. `%,numer`

d5.   0.7709

which is closer to $\pi/4 \approx 0.785398$ than the result of the trapezoid rule for $n = 4$, see Session 8.6, line d5.

A list of values of `our_simpson(sqrt(1-x^2),x,0,1,n)` for $n = 4, 8, 16, 32,$ 64, 128 is generated by

c6. `makelist(our_simpson(sqrt(1-x^2),x,0,1,2^m),m,2,7),numer`

d6.   [0.7709, 0.7803, 0.7836, 0.78476, 0.78517, 0.78531]

To compare the precision of the Simpson and trapezoid methods, we define two lists, `simpsons` and `trapezoids`,

c7. `simpsons:makelist(our_simpson(sqrt(1-x^2),x,0,1,2^m),m,2,7),numer`

d7.   [0.7709, 0.7803, 0.7836, 0.78476, 0.78517, 0.78531]

c8. `trapezoids:makelist(trapezoid(sqrt(1-x^2),x,0,1,2^m),m,2,7),numer`

d8.   [0.74893, 0.77245, 0.78081, 0.78377, 0.78482, 0.78519]

Now subtract the correct value $\pi/4$ from each term in both lists,

          `simpsons-%pi/4, trapezoids-%pi/4`

and compute the list of ratios of corresponding terms in `simpsons-%pi/4` and `trapezoids-%pi/4`,

c9. `(simpsons-%pi/4)/(trapezoids-%pi/4),numer`

d9.   [0.39757, 0.39413, 0.39249, 0.39169, 0.39182, 0.39819]

The answer (about 0.39 for the tested n) illustrates the fact that for the same n the Simpson method gives better approximations than the trapezoid method. For this function, to get the same approximation as `our_simpson(sqrt(1-x^2),x,0,n)`, the trapezoid method needs twice as many partition points, i.e., `trapezoid(sqrt(1-x^2),x,0,1,2*n)`.

The following theorem, stated without proof, gives an error bound for the Simpson method.

Theorem 8.43. If the 4th derivative $f^{(4)}$ is continuous on $[a, b]$, then

$$\left| \int_a^b f(x)\, dx - \texttt{simpson}(f, x, a, b, n) \right| \leq \frac{(b-a)^5}{180 n^4} \max_{a \leq x \leq b} \left| f^{(4)}(x) \right| . \qquad (8.75)$$

Remark 8.44. The Simpson method approximates the integrand by a quadratic function. Therefore if f is a quadratic function to start with, say,

$$f(x) = p_2 x^2 + p_1 x + p_0 ,$$

then the Simpson rule is exact, i.e.,

$$\int_a^b f(x)\, dx = \texttt{simpson}(f, x, a, b, n), \quad \text{for all } n \geq 2 .$$

In particular, for $n = 2$,

$$\int_a^b f(x)\, dx = \left(f(a) + 4f\left(\frac{a+b}{2}\right) + f(b) \right)\left(\frac{b-a}{6}\right) . \qquad (8.76)$$

Actually the Simpson method is exact for polynomials of the third degree, by (8.75), since then the 4th derivative is zero. This can be illustrated by MACSYMA as follows: Compute the error of the Simpson method

```
c1. simpson(u x^4+v x^3+w x^2+y x+z,x,a,b,n) -
        integrate(u x^4+v x^3+w x^2+y x+z,x,a,b)$
```

for a 4th-degree polynomial $u x^4 + v x^3 + w x^2 + y x + z$. The result after using `ratsimp` and `factor` is

d1. $\dfrac{2u(b-a)^5}{15 n^4}$

in agreement with (8.75), showing that the error depends on the coefficient u of x^4 but not on the coefficients v, w, y, and z of the lower powers. Therefore the Simpson method is exact for cubic polynomials.

Exercises

8.34 The error in the Simpson method is usually better than the result (8.75). Let $h = (b-a)/n$, and let the error be written as a constant times h^4,

$$\left| \int_a^b f(x)\, dx - \texttt{simpson}(f, x, a, b, n) \right| \leq C h^4,$$

where the constant C does not depend on the partition size h. We can estimate C by computing the left-hand side using one value of $n = (b-a)/h$ and then doubling n.

Compare the coefficient C with

$$\frac{b-a}{180} \max_{a \le x \le b} |f^{(4)}(x)|$$

for the following integrals (use MACSYMA to evaluate the integrals.)

(a) $\int_0^4 x^3 \, dx$ (b) $\int_1^6 (1/x) \, dx$ (c) $\int_0^\pi |\sin x| \, dx$

(d) $\int_0^1 e^x \, dx$ (e) $\int_{-3}^6 \text{atan } x \, dx$ (f) $\int_0^2 (x^2 + 4x - 7) \, dx$

(g) $\int_{-1}^4 (x^3 + 2x^2 + 4x - 5) \, dx$ (h) $\int_0^2 \frac{\sin x}{x} \, dx$

8.35 The natural logarithm of 3 can be written as $\ln 3 = \int_1^3 (1/x) \, dx$. Compute this integral to 10^{-6} accuracy, using the Simpson rule.

8.36 Approximate the following integrals using (i) a middle Riemann sum, (ii) the trapezoid method, and (iii) the Simpson rule. Use the same number of intervals ($n = 10, 40, 100$) in the three methods, and compare your results with the correct values (obtained by MACSYMA).

(a) $\int_1^4 \sqrt{1 + x^4} \, dx$ (b) $\int_1^4 \sqrt{\frac{4-x}{x+1}} \, dx$ (c) $\int_0^{\pi/2} \sqrt{\cos x} \, dx$ (d) $\int_0^{\pi/3} \sqrt{\tan x} \, dx$

8.37 In Exercise 8.31 we expressed the trapezoid sum as an average of the left and right Riemann sums. Similarly express the Simpson sum (8.74) as an average of the left, middle, and right sums of Example 8.5. In other words, find nonnegative λ_1, λ_2, and λ_3 such that for all even n,

$$\lambda_1 S_{\text{left}}(f, x, a, b, n/2) + \lambda_2 S_{\text{middle}}(f, x, a, b, n/2) + \lambda_3 S_{\text{right}}(f, x, a, b, n/2)$$
$$= \text{simpson}(f, x, a, b, n)$$

λ_1 $+ \lambda_2$ $+ \lambda_3$
$= 1$

8.38 Given the data of Exercise 8.32, approximate $\int_{1.2}^{2.2} f(x) \, dx$ by the Simpson method.

8.39 The cross section of an airplane wing is shown in Fig. 8.15. All measurements are in meter. Estimate the area of the cross section.

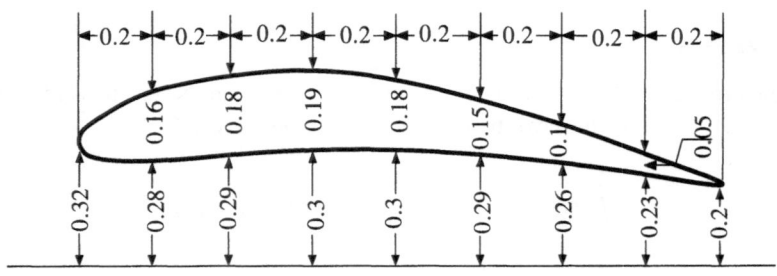

Fig. 8.15. Cross section of an airplane wing (Exercise 8.39)

8.40 Approximate the integrals

(a) $\displaystyle\int_0^1 \frac{1}{1+x}\,dx$ (b) $\displaystyle\int_0^1 \frac{1}{1+x^2}\,dx$ (c) $\displaystyle\int_{\pi/4}^{\pi/2} \frac{\sin x}{x}\,dx$

(d) $\displaystyle\int_0^1 e^{-x^2}\,dx$ (e) $\displaystyle\int_{\pi/4}^{\pi/2} \frac{x}{\sin^2 x}\,dx$ (f) $\displaystyle\int_0^{\pi/4} \ln\left(1+\tan x\right)dx$

8.41 Repeat Exercise 8.33 for the Simpson method with the error bound (8.75).

8.6 Proofs

Proof of Theorem 8.21. a. The limit of Riemann sums of the function αf,

$$\lim_{\|\mathcal{P}\|\to 0}\sum_{k=1}^{n.}\alpha f(\xi_k)\,\Delta x_k = \alpha \lim_{\|\mathcal{P}\|\to 0}\sum_{k=1}^{n} f(\xi_k)\,\Delta x_k = \alpha \int_a^b f(x)\,dx$$

if f is integrable on $[a, b]$.

b. Similarly,

$$\lim_{\|\mathcal{P}\|\to 0}\sum_{k=1}^{n}(f(\xi_k)+g(\xi_k))\,\Delta x_k = \lim_{\|\mathcal{P}\|\to 0}\sum_{k=1}^{n} f(\xi_k)\,\Delta x_k + \lim_{\|\mathcal{P}\|\to 0}\sum_{k=1}^{n} g(\xi_k)\,\Delta x_k$$

$$= \int_a^b f(x)\,dx + \int_a^b g(x)\,dx .$$

if f and g are integrable.

c. Let $[c, d]$ be a subinterval of $[a, b]$. Then for any partition $\tilde{\mathcal{P}}$ of $[a, b]$, the (perhaps finer) partition $\mathcal{P} := \tilde{\mathcal{P}} \cup \{c\} \cup \{d\}$ is the union

$$\mathcal{P} = \mathcal{P}_{[a,c]} \cup \mathcal{P}_{[c,d]} \cup \mathcal{P}_{[d,b]} ,$$

where $\mathcal{P}_{[x,y]} := \{ I \cap [x, y] \mid \text{for all subintervals } I \text{ in } \mathcal{P} \}$ is a partition of $[x, y]$. The lower Riemann sum $S_L(f, \mathcal{P})$ is therefore the sum

$$S_L(f, \mathcal{P}) = S_L(f, \mathcal{P}_{[a,c]}) + S_L(f, \mathcal{P}_{[c,d]}) + S_L(f, \mathcal{P}_{[d,b]})$$

and similarly for the upper Riemann sum $S_U(f, \mathcal{P})$. The approximation error (8.34) is therefore the sum of three errors

$$E(f, \mathcal{P}) = E(f, \mathcal{P}_{[a,c]}) + E(f, \mathcal{P}_{[c,d]}) + E(f, \mathcal{P}_{[d,b]}) \qquad (8.77)$$

due to the three subintervals $[a, c]$, $[c, d]$, and $[d, b]$. The integrability of f in $[a, b]$ means that the limit of the left-hand side (8.77), as $\|\mathcal{P}\| \to 0$, is zero. In this case, the three terms on the right also tend to zero since these terms are nonnegative. In particular,

$$\lim_{\|\mathcal{P}_{[c,d]}\|\to 0} E(f, \mathcal{P}_{[c,d]}) = 0 ,$$

showing that f is integrable on $[c, d]$.

d. This is proved similarly to c.

e. If (8.36) holds, then, for any partition $\mathcal{P}$ of $[a, b]$,

$$S_U(f, \mathcal{P}) \leq S_L(g, \mathcal{P})$$

so that

$$\lim_{\|\mathcal{P}\| \to 0} S_U(f, \mathcal{P}) \leq \lim_{\|\mathcal{P}\| \to 0} S_L(g, \mathcal{P}),$$

proving (8.37). $\square$

Proof of Theorem 8.34. We prove the theorem for a special case of continuous functions f and intervals $[a, b]$ where the graph of f intersects any horizontal line in finitely many points.

In this case, we prove that for any $\varepsilon > 0$, there is a $\delta = \delta(\varepsilon) > 0$ such that for any partition $\mathcal{P}$ with norm $\|\mathcal{P}\| \leq \delta$, the approximation error (8.34) $E(f, \mathcal{P})$ is bounded by

$$E(f, \mathcal{P}) \leq \varepsilon(b - a) . \tag{8.78}$$

Let m and M be the minimum and maximum values of f in I, which are finite since f is continuous and I is finite. Let $\varepsilon > 0$ be arbitrary, let

$$N := \left\lceil \frac{M - m}{\varepsilon/2} \right\rceil ,$$

where $\lceil \alpha \rceil$ is the smallest integer greater than or equal to α, and let

$$\varepsilon_0 = \frac{M - m}{N} \leq \frac{\varepsilon}{2} . \tag{8.79}$$

Consider the $N - 1$ horizontal lines

$$y := m + k\varepsilon_0 \quad (k = 1, \dots, N - 1) ,$$

(see illustration, with $N = 6$, in Fig. 8.16a), and the points where these lines

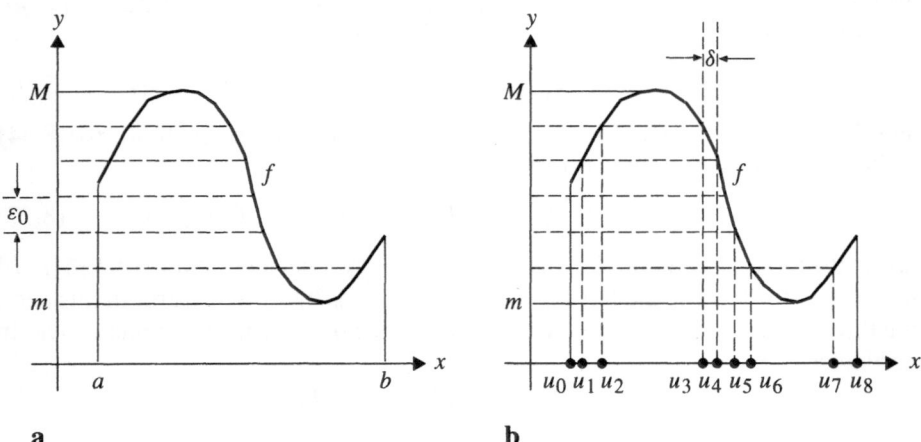

a   **b**

Fig. 8.16a, b. Partition $\mathcal{P}_\varepsilon$. **a** Horizontal lines $y = m + k\varepsilon_0$; **b** partition points

intersect the graph of f in $[a, b]$. Add to these points the two endpoints a and b, and obtain a partition $\mathcal{P}_\varepsilon$ of I[16]

$$a = u_0 < u_1 < u_2 < \cdots < u_{p-1} < u_p = b ,$$

see Fig. 8.16b. The way $\mathcal{P}_\varepsilon$ was constructed, it is obvious that in every subinterval $[u_{k-1}, u_k]$,

$$\max f(x) - \min f(x) \leq \varepsilon_0 .$$

Let δ be the minimum width in $\mathcal{P}_\varepsilon$,

$$\delta := \min_{k=1,\ldots,p} (u_k - u_{k-1}) .$$

It follows that in any subinterval of I with a width less than or equal to δ,

$$\max f(x) - \min f(x) \leq 2\varepsilon_0 \leq \varepsilon .$$

Let then $\mathcal{P}$ be any partition of I

$$a = x_0 < x_1 < x_2 < \cdots < x_{n-1} < x_n = b$$

with norm $\|\mathcal{P}\| \leq \delta$. Therefore, the values of f in each subinterval $I_k = [x_{k-1}, x_k]$ do not differ by more than ε, and consequently, for all k,

$$M_k - m_k \leq \varepsilon . \tag{8.80}$$

Summing over $k = 1, \ldots, n$, it follows that the approximation error

$$E(f, \mathcal{P}) = \sum_{k=1}^{n} (M_k - m_k) \Delta x_k \quad \text{by (8.34)}$$

$$\leq \varepsilon \sum_{k=1}^{n} \Delta x_k \quad \text{by (8.80)}$$

$$= \varepsilon(b - a) \quad \text{by (8.12)} ,$$

proving (8.78). Since ε is arbitrary, it follows that $\lim_{\|\mathcal{P}\| \to 0} E(f, \mathcal{P}) = 0$. $\qquad \square$

---

16 If we decrease ε, then the partition $\mathcal{P}_\varepsilon$ does not necessarily get any finer. For example, if f is constant on $[a, b]$, then for all $\varepsilon > 0$, we get $N = 0$ and the same (trivial) partition $\mathcal{P}_\varepsilon$ with $m = 1$ and $a = u_0 < u_1 = b$.

9 Fundamental theorem of calculus

In this chapter we study functions of the form

$$F(x) = \int_a^x f(t)\,dt$$

called *indefinite integrals* of f. If f is continuous, then F is an antiderivative of f, see Theorem 9.13. Indefinite integrals allow an easy computation of definite integrals as follows,

$$\int_a^b f(x)\,dx = F(b) - F(a)\,,$$

see Theorem 9.15. These two theorems are known jointly as the *fundamental theorem of calculus*. An application to physics is given in Sect. 9.2.

9.1 Indefinite integral

The definite integral

$$\int_a^b f(t)\,dt \tag{9.1}$$

was defined in Chap. 8 for a *fixed* interval $[a, b]$. The integral (9.1) is a number representing areas between the graph of f and the x-axis (the sum of areas above the x-axis minus the sum of areas below the x-axis). It depends on the limits of integration a and b.

MACSYMA-Session 9.1. In Session 8.3 we computed the integrals of the functions 1, t, and t^2 from a to b, for general a and b. Here we similarly compute the integrals of certain functions $f(t)$ from a to x, for general a and x.

The left Riemann sum of $\sin(t)$ from $t = a$ to $t = x$:

```
c1. riem_sum_left(sin(t),t,a,x,n)
```

$$\text{d1.} \quad \frac{\left(\sum_{k=1}^{n} \sin\left(\frac{(x-a)k - x + an + a}{n}\right)\right)(x - a)}{n}$$

an exact translation of `riem_sum_left(f,t,a,x,n)` for $f(t) = \sin t$. To get a useful answer, we use:

c2. `exponentialize(%),sum,simpsum:true`
where the `exponentialize` command converts the trigonometric functions into complex exponential form, which `simpsum:true` can simplify to a closed form. The output line d2 (not shown here) is further processed by:

c3. `tlimit(%,n,inf)`
where `tlimit` is a version of the `limit` function, using Taylor-series methods. This gives:

d3.
$$\frac{e^{-ix-ia}(e^{ix+2ia} + e^{ia}(-e^{2ix} - 1) + e^{ix})}{2}$$

which we convert back to trigonometric functions, using:

c4. `demoivre(expand(%))`

d4.
$$-\frac{i\sin x + \cos x}{2} - \frac{\cos x - i\sin x}{2} + \frac{i\sin a + \cos a}{2} + \frac{\cos a - i\sin a}{2}$$

and simplify:

c5. `ratsimp(%)`
to get finally,

d5. $\cos a - \cos x$

The same result can be obtained by using the right or the middle Riemann sums, following the same steps as in c2–c5.

Similarly, the limit of the left Riemann sum of $\cos(t)$ from $t = a$ to $t = x$ is computed as follows (we omit the display lines except for the final answer):

c6. `riem_sum_left(cos(t),t,a,x,n)`
c7. `exponentialize(%),sum,simpsum:true`
c8. `tlimit(%,n,inf)`
c9. `demoivre(expand(%))`
c10. `ratsimp(%)`
d10. $\sin(x) - \sin(a)$

The left Riemann sum of $\exp(t)$ from $t = a$ to $t = x$ is

c11. `riem_sum_left(exp(t),t,a,x,n)`
c12. `factor(%)`
c13. `tlimit(%,n,inf)`
d13. $e^x - e^a$

This approach, of computing the limit as $n \to \infty$ of a Riemann sum of f in $[a, b]$, may fail for some integrable functions f. As an example, consider the $\tan(t)$ function from $t = a$ to $t = x$

c14. `riem_sum_left(tan(t),t,a,x,n)`
c15. `exponentialize(%),sum,simpsum:true`
giving a complicated expression whose limit, as $n \to \infty$, does not give the correct integral. The integral can be computed directly:

c16. `integrate(tan(t),t,a,x)`
```
     Is  cos(a)  positive,  negative,  or zero?
     positive
     Is  cos(x)  positive,  negative,  or zero?
     positive
```
d16. $\log\cos a - \log\cos x$.

Example 9.1 (A trick).

$$\int_a^b \frac{dx}{x^2} = \frac{1}{a} - \frac{1}{b}, \quad \text{for } 0 < a < b.$$

Proof. The function $1/x^2$ is continuous in $[a, b]$ for $0 < a < b$, and therefore is integrable in $[a, b]$. By Remark 8.14 it suffices to compute the limit of one Riemann sum.

Let $\{x_0 := a, x_1, \ldots, x_{n-1}, x_n := b\}$ be any partition of $[a, b]$, and for each k let the point ξ_k, the representative of the kth subinterval $[x_{k-1}, x_k]$, be chosen as

$$\xi_k = \sqrt{x_{k-1} x_k}, \quad k = 0, \ldots, n, \quad \text{see Exercise 9.3}.$$

Then the Riemann sum becomes

$$\sum_{k=1}^n \frac{\Delta x_k}{x_{k-1} x_k} = \sum_{k=1}^n \frac{x_k - x_{k-1}}{x_{k-1} x_k} = \sum_{k=1}^n \left(\frac{1}{x_k} - \frac{1}{x_{k-1}} \right) = \frac{1}{x_0} - \frac{1}{x_n} = \frac{1}{a} - \frac{1}{b},$$

giving the correct answer without having to compute a limit! □

Although both limits of integration, a and b, are now variables, it is customary to call b a "variable name", say x, and keep a as it is. We then regard the integral

$$\int_a^x f(t) \, dt$$

as a function of x,

$$F(x) := \int_a^x f(t) \, dt, \tag{9.2}$$

where a is considered a constant.

What role does the point a (the left limit of integration) play? If we replace it by any other point, say c, we get the integral

$$F_1(x) = \int_c^x f(t) \, dt,$$

instead of the $F(x)$ of (9.2). But $\int_a^x = \int_a^c + \int_c^x$, and therefore,

$$F(x) = F_1(x) + \int_a^c f(t) \, dt,$$

$$= F_1(x) + C, \quad C = \int_a^c f(t) \, dt,$$

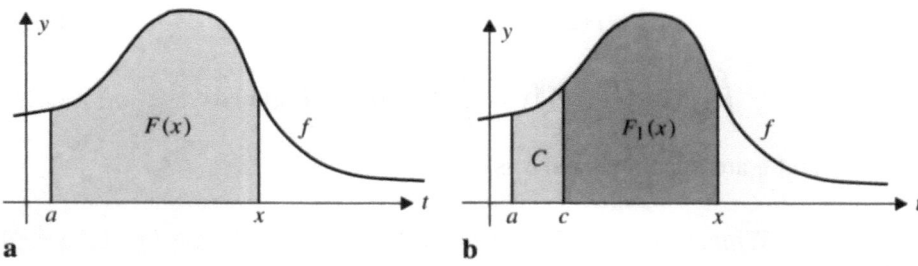

Fig. 9.1 a, b. Dependence of the indefinite integral on the left integration limit. **a** Integral from a to x; **b** integral from c to x

i.e., the two function F and F_1 differ by a constant (Fig. 9.1). Often we have no reason to distinguish between F of (9.2) and any of its "shifts" $F +$ constant. All are equally useful and have a common name: indefinite integral.

Definition 9.2 (Indefinite integral). Let f be integrable. Then, for any α, the integral

$$F(x) = \int_{\alpha}^{x} f(t)\,dt \tag{9.3}$$

is called an *indefinite integral* of f. To denote the fact that F is determined up to a constant[1], the indefinite integral is sometimes written as

$$F(x) = \int f(x)\,dx + C \tag{9.4}$$

where C is called the *constant of integration*. In (9.4) we omit the limits of integration and write the upper limit x in place of the integration variable[2] t. The notation (9.4) also serves as a reminder that the indefinite integral is a class of infinitely many functions, one for each C.

There should be no confusion between the indefinite integral $\int_{\alpha}^{x} f(t)\,dt$, which is a *function*, and a definite integral $\int_{a}^{b} f(t)\,dt$, with fixed a and b, which is a *number*. Confusion is possible because an indefinite integral is actually a definite integral with a "variable" interval $[\alpha, x]$.

Remark 9.3. From Theorem 8.21 a and b we conclude the following properties of indefinite integrals:

---

1 The constant depends on the left limit of integration α.
2 This is perhaps bad notation, but it is widely used.

$$\int \alpha f(x)\, dx = \alpha \int f(x)\, dx\ , \qquad (9.5a)$$

$$\int (f(x) + g(x))\, dx = \int f(x)\, dx + \int g(x)\, dx\ , \qquad (9.5b)$$

where f and g are integrable and α is a constant.

Remark 9.4. While there are infinitely many definite integrals, it is often required to find an indefinite integral whose graph passes through a given point, (x_0, y_0). Such an indefinite integral is

$$F(x) = \int_{x_0}^{x} f(t)\, dt + y_0\ . \qquad (9.6)$$

Indeed,

$$F(x_0) = \int_{x_0}^{x_0} f(t)\, dt + y_0 = y_0, \quad \text{by (8.41)}\ .$$

Is this the only indefinite integral passing through the given point (x_0, y_0)? The answer "yes" is proved next.

Lemma 9.5 (Uniqueness of indefinite integral passing through a given point). Let f be integrable in $I = [a, b]$ and let $x_0 \in I$. Then for any real y_0 the indefinite integral (9.6),

$$F(x) = \int_{x_0}^{x} f(t)\, dt + y_0\ ,$$

is the unique indefinite integral whose graph passes through the point (x_0, y_0).

Proof. We already saw that the graph of (9.6) passes through the point $P = (x_0, y_0)$. To prove uniqueness assume that there are two indefinite integrals, F and F_1, of f whose graphs pass through P, i.e.,

$$F(x_0) = y_0 = F_1(x_0)\ . \qquad (9.7)$$

Since any two indefinite integrals of f differ by a constant,

$$F(x) = F_1(x) + C\ ,$$

it follows from (9.7) that $C = 0$. Therefore $F(x) = F_1(x)$ for all x. $\qquad\square$

Example 9.6. We collect some results from MACSYMA-Session 9.1,

(a) $\displaystyle\int_{a}^{x} t^2\, dt = \frac{x^3}{3} - \frac{a^3}{3}$, $\qquad$ (b) $\displaystyle\int_{a}^{x} \sin t\, dt = -\cos x - (-\cos a)$,

(c) $\displaystyle\int_a^x \cos t\, dt = \sin x - \sin a$, (d) $\displaystyle\int_a^x e^t\, dt = e^x - e^a$.

Example 9.7. The indefinite integral of x^2 is, by Example 9.6 a,

$$\int x^2\, dx = \frac{x^3}{3} + C ,$$ (9.8)

consisting of infinitely many functions, one for each C. Some of these are illustrated in Fig. 7.5. The indefinite integral of x^2 passing through the point $(-2, 3)$ is then

$$F(x) = \int_{-2}^x t^2\, dt + 3 = \frac{x^3}{3} - \frac{(-2)^3}{3} + 3, \quad \text{by (9.6)} ,$$

$$= \frac{x^3}{3} + \frac{11}{3} .$$

MACSYMA-Session 9.2. The indefinite integral of $f(x)$ with respect to x is computed by
c1. `integrate(f(x),x)`
d1. $\int f(x)\, dx$
If f is not an expression depending on x, it is treated as a constant, and the result of d1 is $f \times x$.

MACSYMA uses zero for the constant of integration. For example, the indefinite integral of x^2
c2. `integrate(x^2,x)`
d2. $\dfrac{x^3}{3}$
same as (9.8) with $C = 0$. Similarly, the indefinite integral of $\sin(x)$:
c3. `integrate(sin(x),x)`
d3. $-\cos x$

An indefinite integral is said to be in *closed form* if it is given in terms of finitely many elementary operations involving polynomials, logarithms, exponentials, or trigonometric functions. Some perfectly nice functions do not have their integrals in closed forms; examples include $\sin(x^2)$, $\exp(x^2)$, and $\exp(-x^2)$. When MACSYMA is unable to perform a requested integration, it will echo the command.

Although the function $\exp(-x^2)$ does not have a closed-form integral, it is important in many applications, so a special notation is used for the integral. The error function, denoted `erf(x)`, is defined by

$$\text{erf}(x) := \frac{2}{\sqrt{\pi}} \int_0^x e^{-t^2}\, dt$$ (9.9)

as shown in:

c4. `2/sqrt(%pi)*integrate(exp(-t^2),t,0,x)`

d4. $\mathrm{erf}(x)$

The indefinite integral of e^{-t^2} is therefore

c5. `integrate(exp(-t^2),t,0,x)`

d5. $\dfrac{\sqrt{\pi}\ \mathrm{erf}(x)}{2}$

The strange coefficient $2/\sqrt{\pi}$ in (9.9) guarantees that $\int_0^\infty \mathrm{erf}(x)\,dx = 1$, see Exercise 10.31.

Since indefinite integrals are just definite integrals over a variable interval, one may wonder what is the point of studying them separately. There are several reasons why indefinite integrals are important. Here is one: They allow a painless way of computing a definite integral on any interval.

Lemma 9.8 (Use of indefinite integrals). Let f be integrable in $I = [c, d]$, and let a, b be any two points in I. Let $F(x)$ be an indefinite integral of f. Then

$$\int_a^b f(x)\,dx = F(b) - F(a) . \tag{9.10}$$

Proof. By (8.35),

$$\int_a^b f(t)\,dt = \int_\alpha^b f(t)\,dt - \int_\alpha^a f(t)\,dt , \tag{9.11}$$

$$= F(b) - F(a) . \qquad \square$$

Remark 9.9. The indefinite integral $\int_\alpha^x f(t)\,dt$ is, by Lemma 9.8,

$$\int_\alpha^x f(t)\,dt = F(x) - F(\alpha)$$

$$= F(x) + C ,$$

where C is a constant of integration. Writing now the computation (9.11) in detail

$$\int_\alpha^b f(t)\,dt - \int_\alpha^a f(t)\,dt = (F(b) - F(\alpha)) - (F(a) - F(\alpha)) = F(b) - F(a)$$

we see that the constant of integration disappeared, i.e., the computation (9.11) is independent of α. This is the reason why the definition of the indefinite integral (9.3) uses an arbitrary α.

The combination $F(b) - F(a)$ in the right-hand side of (9.10) is denoted by

$$F(x)\big|_a^b := F(b) - F(a) .\tag{9.12}$$

Example 9.10. Using (9.8) we get

$$\int_a^b x^2 \, dx = \frac{x^3}{3}\bigg|_a^b = \frac{1}{3}(b^3 - a^3) ,$$

and in particular,

$$\int_1^3 x^2 \, dx = \frac{1}{3}(3^3 - 1) = \frac{26}{3}, \quad \int_3^5 x^2 \, dx = \frac{1}{3}(5^3 - 3^3) = \frac{98}{3} ,$$

$$\int_1^5 x^2 \, dx = \frac{1}{3}(5^3 - 1) = \frac{124}{3}, \quad \text{etc.}$$

Similarly,

$$\int_0^\pi \sin x \, dx = -\cos x \big|_0^\pi, \quad \text{by MACSYMA-Session 9.2, line d3 ,}$$

$$= -\cos(\pi) - (-\cos(0)) = 2 .$$

The indefinite integral $F(x) = \int f(x) \, dx$ is better behaved than its integrand: $F(x)$ is continuous in any interval where f is merely integrable (and may have jump-discontinuities). This is illustrated in Fig. 9.2.

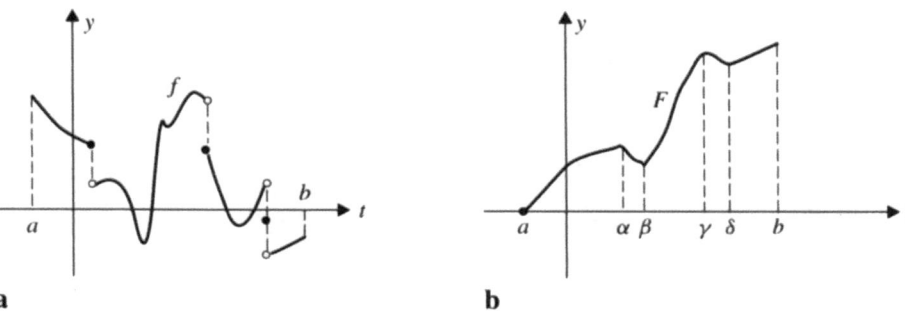

a   **b**

Fig. 9.2 a, b. Indefinite integral. **a** An integrable function f with several discontinuities. **b** The indefinite integral F passing through $(a, 0)$

Lemma 9.11 (Continuity of indefinite integrals). Let f be integrable on $I = [a, b]$, and let $F(x)$ be the indefinite integral (9.2),

$$F(x) = \int_a^x f(t)\,dt\ .$$

Then the function F is continuous on I.

Proof. Since f is integrable, it is bounded (by definition) on I, say,

$$-M \le f(x) \le M \quad \text{for all } a \le x \le b\ . \tag{9.13}$$

For any x in I and $\Delta x > 0$, we have

$$F(x + \Delta x) - F(x) = \int_a^{x+\Delta x} f - \int_a^x f = \int_x^{x+\Delta x} f\ ,$$

and therefore, by (8.43) and (9.13),

$$-M\,\Delta x \le \int_x^{x+\Delta x} f \le M\,\Delta x\ , \tag{9.14}$$

so that

$$-M\,\Delta x \le F(x + \Delta x) - F(x) \le M\,\Delta x\ ,$$

or

$$|F(x + \Delta x) - F(x)| \le M\,|\Delta x|\ . \tag{9.15}$$

If $\Delta x < 0$, then (9.15) follows similarly, using (8.40). Therefore,

$$\lim_{\Delta x \to 0} F(x + \Delta x) = F(x)$$

for all x in I, proving F continuous on I.   □

The next example shows another case where the indefinite integral $\int f(x)$ is "nicer" than its integrand f.

Example 9.12. Let the function $f(x)$ be increasing in $[a, b]$. Then its indefinite integral

$$F(x) = \int_a^x f(t)\,dt$$

is convex in $[a, b]$. This statement is proved in Sect. 9.5.

Exercises

9.1   Write the upper Riemann sum, $S_U(t^2, t, 0, x, n)$, for the integral $\int_0^x t^2 \, dt$ using a subdivision of the interval into n equal lengths $h = b/n$. Use the fact (Appendix B, (B.2))

$$\sum_{k=1}^n k^2 = 1^2 + 2^2 + 3^2 + \cdots + n^2 = \frac{n(n+1)(2n+1)}{6} \, ,$$

(see Exercise 8.1) in order to obtain

$$S_U(t^2, t, 0, x, n) = \frac{x^3}{6}\left(1 + \frac{1}{n}\right)\left(2 + \frac{1}{n}\right) .$$

By taking the limit as $n \to \infty$ obtain the well-known result $\int_0^x t^2 \, dt = x^3/3$.

9.2   Repeat the procedure of the previous problem for the integrals below. Use MACSYMA to evaluate the sums.

(a) $\displaystyle\int_0^x t^3 \, dt$ (b) $\displaystyle\int_0^x t^4 \, dt$ (c) $\displaystyle\int_0^x t^8 \, dt$ (d) $\displaystyle\int_0^x t^{11} \, dt$

9.3   Show that the selection of ξ_k in Example 9.1 is valid, i.e., show that if $0 < x_{k-1} < x_k$ and $\xi_k = \sqrt{x_{k-1}x_k}$, then $x_{k-1} < \xi_k < x_k$.

9.4   (Another trick) Verify, for $0 < a < b$ and m integer $\neq -1$,

$$\int_a^b x^m \, dx = \frac{b^{m+1} - a^{m+1}}{m+1} \, , \tag{9.16}$$

using the Riemann sum $\sum_{k=1}^n (\xi_k)^m \, \Delta x_k$ with

$$\begin{aligned} x_k &:= a(b/a)^{k/n} \quad (k = 0, 1, \ldots, n) \, , \\ \xi_k &:= x_{k-1} \quad\quad\;\; (k = 1, \ldots, n) \, . \end{aligned} \tag{9.17}$$

Here and in the next exercise, use MACSYMA to compute the Riemann sum and its limit as $n \to \infty$.

9.5   Verify, for $0 < a < b$,

$$\int_a^b (1/x) \, dx = \ln(b/a) \, , \tag{9.18}$$

using the Riemann sum $\sum_{k=1}^n \Delta x_k/\xi_k$, with x_k and ξ_k as in (9.17).

9.6   The method used in the previous problems may be generalized to take care of the integral $\int_a^b x^r \, dx$, where r is any rational number $p/q \neq -1$. Indeed, by choosing the partition points as

$$a, \; a\left(\frac{b}{a}\right)^{1/n}, \; a\left(\frac{b}{a}\right)^{2/n}, \; \ldots, \; a\left(\frac{b}{a}\right)^{(n-1)/n}, \; a\left(\frac{b}{a}\right)^{n/n} = b$$

and the evaluation points $\xi_k = x_{k-1}$, show that the Riemann sum becomes

$$\begin{aligned} & a^p(a\theta - a) + (a\theta)^r(a\theta^2 - a\theta) \\ & + (a\theta^2)^r(a\theta^3 - a\theta^2) + \cdots + (a\theta^{n-1})^p(a\theta^n - a\theta^{n-1}) \\ & = a^{p+1}(\theta - 1)\left(1 + \theta^{p+1} + \theta^{2(p+1)} + \theta^{3(p+1)} + \cdots + \theta^{(n-1)(p+1)}\right) , \end{aligned}$$

where $\theta = (b/a)^{1/n}$. Show that this series sums to

$$(b^{r+1} - a^{r+1}) \frac{\theta - 1}{\theta^{r+1} - 1} \ .$$

To evaluate $(\theta - 1)/\theta^{(p+q)/q-1}$ as $n \to \infty$, it is convenient to set $\theta^{1/q} = t$ to obtain

$$(b^{r+1} - a^{r+1}) \frac{t^q - 1}{t^{p+q} - 1} \ .$$

Evaluate this limit by MACSYMA and state the result for the definite integral.

9.7 The right Riemann sum (8.19) for the integral $\int_a^b \sin x \, dx$ is

$$S_{\text{right}}(\sin x, x, a, b, n) = h(\sin(a + h) + \sin(a + 2h) + \cdots + (\sin(a + nh)) \ ,$$

where $h := (b - a)/n$ is the common length of the n subintervals used in a regular partition of the interval $[a, b]$, see (8.13). Multiply this sum by $2\sin(h/2)$ and use the trigonometric identity

$$2 \sin \theta \sin \phi = \cos(\theta - \phi) - \cos(\theta + \phi)$$

to obtain

$$S_{\text{right}}(\sin x, x, a, b, n) = \frac{h}{2 \sin(h/2)} \left(\cos\left(a + \frac{h}{2}\right) - \cos\left(b + \frac{h}{2}\right) \right) .$$

Let $n \to \infty$ or, what is the same thing, $h \to 0$, to evaluate the integral $\int_a^b \sin x \, dx$.

9.8 Repeat a similar analysis for the integral $\int_a^b \cos x \, dx$.

9.9 a. Prove: For given α, β, and γ the indefinite integral of the quadratic
$q(x) = \alpha x^2 + \beta x + \gamma$ is

$$\int (\alpha x^2 + \beta x + \gamma) \, dx = \alpha \frac{x^3}{3} + \beta \frac{x^2}{2} + \gamma x + C \tag{9.19}$$

and consequently,

$$\int_a^b (\alpha x^2 + \beta x + \gamma) \, dx = \left(\alpha \frac{x^3}{3} + \beta \frac{x^2}{2} + \gamma x \right) \Big|_a^b$$

$$= \frac{\alpha}{3} x^3 \Big|_a^b + \frac{\beta}{2} x^2 \Big|_a^b + \gamma x \Big|_a^b$$

$$= \frac{\alpha}{3} (b^3 - a^3) + \frac{\beta}{2} (b^2 - a^2) + \gamma (b - a) . \tag{9.20}$$

b. Use MACSYMA to verify (9.19) and (9.20).

9.10 Use MACSYMA to find the indefinite integral of f whose graph passes through the point P.

(a) $f(x) := x$, $P = (1, 2)$ (b) $f(x) := -x$, $P = (1, 2)$

(c) $f(x) := x^3$, $P = (-1, 0)$ (d) $f(x) := \cos x$, $P = (\pi/2, -2)$

(e) $f(x) := 1/x$, $P = (1, 0)$ (f) $f(x) := \tan x$, $P = (0, 2)$

(g) $f(x) := e^x$, $P = (0, -1)$ (h) $f(x) := 1/(1 + x^2)$, $P = (1, \pi/2)$

9.11 Figure 9.2 a and b shows a function f and its indefinite integral $F(x) = \int_a^x f(t) \, dt$. Note that F is decreasing in the intervals $[\alpha, \beta]$ and $[\gamma, \delta]$. Explain why.

9.2 Position and distance from velocity

The velocity $v(t)$ of a particle is the derivative of its position function $s(t)$, see Sect. 4.5, (4.39),

$$v(t) := \lim_{\Delta t \to 0} \frac{\Delta s}{\Delta t} = s'(t) .$$

Consider the converse problem of recovering the position if the velocity is known. To be specific, consider a moving particle which starts at time $t = a$ in a given position $s(a)$ and moves with velocity $v(t)$ until time $t = b$. The velocity $v(t)$ is given for all $a \le t \le b$. It is required to find (a) the terminal position $s(b)$ and (b) the distance covered during the time $[a, b]$.

a. We partition the interval $[a, b]$, see Definition 8.2,

$$t_0 := a < t_1 < \cdots < t_n := b .$$

Let the kth subinterval $I_k = [t_{k-1}, t_k]$ have length Δt_k. The velocity $v(t)$ may be positive, negative, or even change sign in I_k. Let τ_k be any point in I_k. If Δt_k is small, then the velocity during the time interval $[t_{k-1}, t_k]$ can be approximated by the (constant) velocity $v(\tau_k)$. It follows that the position at time t_k is approximately

$$s(t_k) \approx s(t_{k-1}) + v(\tau_k) \Delta t_k , \tag{9.21}$$

and the approximation gets better as Δt_k gets smaller.[3] Therefore the position at time $t = b$ is approximated by the Riemann sum

$$s(b) \approx s(a) + \sum_{k=1}^{n} v(\tau_k) \Delta t_k .$$

In the limit, as max $\Delta t_k \to 0$, we get the definite integral

$$s(b) = s(a) + \int_a^b v(t) \, dt . \tag{9.22}$$

Therefore, position is the integral of velocity.

b. The *displacement* $s(b) - s(a)$ during the time interval $[a, b]$ is, by (9.22), the integral

$$s(b) - s(a) = \int_a^b v(t) \, dt . \tag{9.23}$$

If $v(t) \ge 0$ for all $t \in [a, b]$, i.e., if the movement is always in the direction of the

---

3 If the velocity $v(\tau_k)$ at the selected point τ_k is negative, then the position $s(t_k)$ is computed by (9.21) to be less than $s(t_{k-1})$, where in fact it may be that $s(t_k) > s(t_{k-1})$ because the positive velocities dominate during the subinterval I_k. The chance of selecting such an untypical point decreases if we take smaller subintervals.

s-axis, the displacement $s(b) - s(a)$ equals the *distance* covered from $t = a$ to $t = b$.

In the general case we can still use integration to calculate the distance covered. Consider again the kth subinterval I_k and an arbitrary point τ_k in it. The velocity in I_k can be positive, negative, or even change sign. The distance covered from $t = t_{k-1}$ to $t = t_k$ is approximately

$$dist(t_{k-1}, t_k) \approx |v(\tau_k)| \Delta t_k$$

so the distance covered from $t = a$ to $t = b$ is approximated by the Riemann sum

$$dist(a, b) \approx \sum_{k=1}^{n} |v(\tau_k)| \Delta t_k ,$$

and in the limit, the integral

$$dist(a, b) = \int_a^b |v(t)| \, dt . \tag{9.24}$$

Exercises

9.12 A particle, moving under a constant acceleration a, starts in time t_0 at position s_0 with velocity v_0. Use the fact that displacement is the integral of velocity to verify (4.52),

$$s(t) = s_0 + v_0(t - t_0) + \frac{a}{2}(t - t_0)^2 ,$$

in agreement with Example 4.42.

9.13 A body moves along a coordinate axis, and its coordinate at time t [s] is

$$s(t) = 10 \sin 2\pi t \ [\text{m}] .$$

Use MACSYMA (if necessary) to answer the following questions.
a. How far from the origin will the body ever get?
b. When will it return to the initial position $s(0)$?
c. What is the total distance traveled in 3 [s]?
d. What is the average velocity in the first 3 [s]?
e. What is the velocity $v(t)$ at any time t?
f. What can be said about the points where $v(t) = 0$?

9.3 Fundamental theorem of calculus

The integral of velocity $\int_a^b v(t) \, dt$ was shown in (9.23) to be the displacement $s(b) - s(a)$ which is an antiderivative[4] of v. In this section we show, for any

---

4 The antiderivative of a function f was defined, in Sect. 7.4, as the class of functions $\{F(x) + C\}$ satisfying $F'(x) = f(x)$.

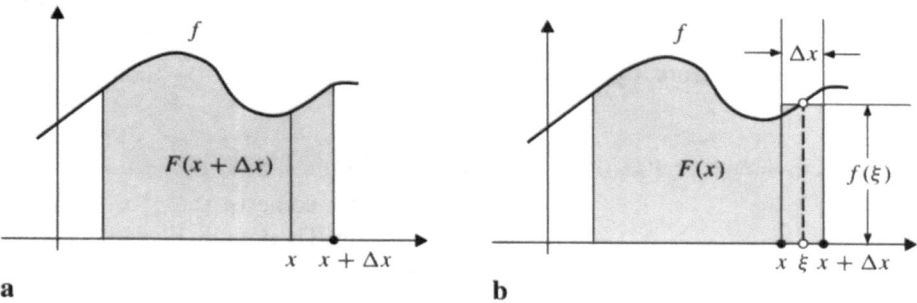

Fig. 9.3. Indefinite integrals $F(x + \Delta x)$ (**a**) and $F(x)$ (**b**)

continuous function f, an analogous result,

$$\int_a^b f(x)\,dx = F(b) - F(a), \quad \text{where } F \text{ is any antiderivative of } f .$$

A comparison with (9.10),

$$\int_a^b f(x)\,dx = F(b) - F(a), \quad \text{where } F \text{ is any indefinite integral of } f ,$$

shows that antiderivatives and indefinite integrals are quite similar. Indeed, if f is a continuous function, then its antiderivatives are exactly the same as its indefinite integrals. This is proved in Theorems 9.13 and 9.15 (known together as the fundamental theorem of calculus) and summarized in Theorem 9.16.

These results are useful because they allow using antiderivatives F in the computation (9.10) of definite integrals. Lemma 9.8 intended to use here an indefinite integral F, but antiderivatives are often easier to find.

An indefinite integral $F(x)$ is a continuous function, see Lemma 9.11. The next theorem gives a condition when F is also differentiable.

Theorem 9.13 (Fundamental theorem of calculus, part 1). Let f be integrable on $[a, b]$, and let $F(x)$ be the indefinite integral (9.2),

$$F(x) = \int_a^x f(t)\,dt, \quad a \le x \le b .$$

Then the function $F(x)$ is differentiable at any point $x_0 \in (a, b)$ where f is continuous, and its derivative is

$$F'(x_0) = f(x_0) . \tag{9.25}$$

Proof. Since f is integrable on $[a, b]$, the indefinite integral $F(x)$ is defined for $x \in (a, b)$. For sufficiently small Δx, $x + \Delta x$ is also in $[a, b]$, so $F(x + \Delta x)$ is also defined. Therefore, by Theorem 8.21 d,

$$F(x_0 + \Delta x) = F(x_0) + \int_{x_0}^{x_0 + \Delta x} f(t)\, dt = F(x_0) + f(\xi)\, \Delta x,$$

$$\text{for some } x_0 \leq \xi \leq x_0 + \Delta x,$$
$$\text{by Theorem 8.39; see Fig. 9.3.}$$

$$\therefore \quad \frac{F(x_0 + \Delta x) - F(x_0)}{\Delta x} = f(\xi),$$

$$\text{for some } x_0 \leq \xi \leq x_0 + \Delta x,$$

$$\therefore \quad F'(x_0) = \lim_{\Delta x \to 0} \frac{F(x_0 + \Delta x) - F(x_0)}{\Delta x} = f(x_0) \quad \text{if } f \text{ is continuous at } x_0.$$

$\square$

If f is continuous at x_0, its indefinite integral F is differentiable there. Again we see that an indefinite integral is better behaved than its integrand.

If the *integrand* is *not continuous*, its *indefinite integral need not be differentiable*. This is illustrated in the following example.

Example 9.14. The function sign x defined by

$$\text{sign } x = \begin{cases} -1, & x < 0, \\ 0, & x = 0, \\ +1, & x > 0, \end{cases}$$

is integrable in any finite interval $[a, b]$, see Exercise 8.16, and

$$\int_a^x \text{sign } t\, dt = |x| - |a|, \tag{9.26}$$

for all a, x. This shows that $F(x) = |x|$ is an indefinite integral of sign x. However, the absolute value function $|x|$ is not differentiable at $x = 0$.

The following theorem is a converse of Theorem 9.13, showing that antiderivatives are indefinite integrals.

Theorem 9.15 (Fundamental theorem of calculus, part 2). If f is integrable on $I = [a, b]$, F continuous on $[a, b]$, and, (7.25),

$$F'(x) = f(x), \quad \text{for all } x \in (a, b),$$

then, (9.10),

$$\int_a^b f(x)\,dx = F(b) - F(a).$$

Proof. Let $\mathcal{P} = \{x_0, x_1, \ldots, x_n\}$ be any partition (8.7) of $[a, b]$. By the mean value theorem 7.2, there is a point ξ_k in $I_k = [x_{k-1}, x_k]$ such that

$$F(x_x) - F(x_{k-1}) = F'(\xi_k)\,\Delta x_k$$
$$= f(\xi_k)\,\Delta x_k \quad \text{since } F \text{ is the antiderivative of } f.$$

Summing over $k = 1, \ldots, n$ we get

$$\sum_{k=1}^n (F(x_k) - F(x_{k-1})) = F(b) - F(a) = \sum_{k=1}^n f(\xi_k)\,\Delta x_k$$

and (9.10) follows by taking the limit as $\|\mathcal{P}\| \to 0$. $\qquad\square$

If the function f is continuous, its indefinite integrals are exactly the same as its antiderivatives. This is stated as follows.

Theorem 9.16 (Fundamental theorem of calculus). Let f be continuous on $[a, b]$, F be differentiable on $[a, b]$, and $F(a) = 0$. Then

$$F'(x) = f(x) \quad \text{for all } x \in [a, b)$$

$$\text{if and only if} \qquad F(x) = \int_a^x f(t)\,dt \quad \text{for all } x \in [a, b].$$

Proof. If: Theorem 9.13. Only if: Theorem 9.15. $\qquad\square$

Example 9.17. If n is an integer $\neq -1$, then the indefinite integral of x^n is

$$\int x^n\,dx = \frac{x^{n+1}}{n+1} + C, \tag{9.27}$$

see also Exercise 9.4.

Proof. The derivative of the right-hand side of (9.27) is x^n. $\qquad\square$

Example 9.18. If $p(x) := \sum_{k=0}^n p_k x^k = p_0 + p_1 x + p_2 x^2 + \cdots + p_{n-1}x^{n-1} + p_n x^n$ is an arbitrary polynomial of degree n, then its indefinite integral is a polynomial of degree $n + 1$,

$$\int \left(\sum_{k=0}^{n} p_k x^k \right) dx = \sum_{k=0}^{n+1} q_k x^k , \tag{9.28}$$

$$\text{where} \quad q_k := \frac{p_{k-1}}{k}, \quad (k = 1, \dots, n+1) ,$$

$$\text{and} \quad q_0 := \text{the constant of integration} ,$$

see also Exercise 7.36.

Proof. From (9.5a), (9.5b), and (9.27).   □

Remark 9.19. The fundamental theorem of calculus shows that the mean value theorem for derivatives (Theorem 7.2) is equivalent to the mean value theorem for integrals (Theorem 8.39).

Let f be continuous on $[a, b]$. Then $F(x) := \int_a^x f(t) \, dt$ is continuous on $[a, b]$ and an antiderivative of f in (a, b). By the mean value theorem 7.2 there is a point $\xi \in (a, b)$ such that

$$f(\xi) = F'(\xi) = \frac{F(b) - F(a)}{b - a}$$

$$= \frac{\int_a^b f(t) \, dt}{b - a}, \quad \text{by the fundamental theorem 9.15} ,$$

showing that Theorem 7.2 $\implies$ Theorem 8.39.

By reversing these steps we conclude that Theorem 8.39 $\implies$ Theorem 7.2.

Consider again the definite integral $\int_a^b f(t) \, dt$, and suppose the limits of integration are functions of x, say $a = a(x)$, $b = b(x)$. This makes the integral a function of x

$$I(x) := \int_{a(x)}^{b(x)} f(t) \, dt . \tag{9.29}$$

To be specific we assume that x ranges in an open interval $J = (c, d)$, and that for any $x \in J$, the integrand f is integrable on $[a(x), b(x)]$. For example, the integral

$$\int_x^{1+x^2} \frac{dt}{t^2} \tag{9.30}$$

exists for all x in the open interval $(0, \infty)$ but not for $x = 0$.

What is the derivative $I'(x)$ of the function (9.29)? If $a(x)$ and $b(x)$ are themselves differentiable, the answer turns out to be quite simple.

Corollary 9.20. Let the functions $a(x)$ and $b(x)$ be differentiable on the interval $J = (c, d)$, and let f be continuous on $[a(x), b(x)]$ for all $x \in J$. Then (9.29) is differentiable for all $x \in J$, and

$$I'(x) = \frac{d}{dx} \int_{a(x)}^{b(x)} f(t)\,dt = f(b(x))b'(x) - f(a(x))a'(x) . \tag{9.31}$$

Proof. Lemma 9.8 gives, for all $x \in J$,

$$I(x) = \int_{a(x)}^{b(x)} f(t)\,dt = F(b(x)) - F(a(x)) .$$

This is differentiable, by the chain rule, and the derivative is

$$I'(x) = F'(b(x))b'(x) - F'(a(x)a'(x) .$$

Finally, (9.31) follows since $F' = f$. $\qquad\qquad\square$

Remark 9.21. Note that (9.31) gives the derivative $I'(x)$ *without* evaluating the integral $I(x)$.

Example 9.22. We calculate the derivative of (9.30) in two different ways.
a. Integrate first, then differentiate:

$$\frac{d}{dx} \int_x^{1+x^2} \frac{dt}{t^2} = \frac{d}{dx}\left(-\frac{1}{t}\Big|_x^{1+x^2}\right) = \frac{d}{dx}\left(\frac{1}{x} - \frac{1}{1+x^2}\right)$$

$$= \frac{2x}{(1+x^2)^2} - \frac{1}{x^2} .$$

b. Use (9.31):

$$\frac{d}{dx} \int_x^{1+x^2} \frac{dt}{t^2} = \frac{1}{(1+x^2)^2} \frac{d}{dx}(1+x^2) - \frac{1}{x^2}\frac{d}{dx}x = \frac{2x}{(1+x^2)^2} - \frac{1}{x^2} .$$

Consider next the case where the integrand f in $\int_a^b f\,dt$ is a function also of a variable x, $f = f(x, t)$, making the integral a function of x,

$$I(x) := \int_a^b f(x, t)\,dt . \tag{9.32}$$

Note that I is a function only of x, since t is "integrated out". We may ask again for the derivative $I'(x)$. Here is an answer.

Corollary 9.23. Let $f = f(x, t)$ be
a. defined for all $t \in [a, b]$ and all x in an open interval J;
b. integrable (with respect to t) on $[a, b]$, for all $x \in J$;
c. continuously differentiable (with respect to x) on J, for all $t \in [a, b]$.
Then the derivative of (9.32) is

$$I'(x) = \frac{d}{dx} \int_a^b f(x, t)\, dt = \int_a^b \left[\frac{\partial}{\partial x} f(x, t)\right] dt , \qquad (9.33)$$

where $(\partial/\partial x) f(x, t)$ denotes the derivative of f with respect to x, treating t as a constant.[5]

This corollary is proved in Sect. 9.5.

Finally, if the integral $\int_a^b f\, dt$ is a function of x both through the integrand $f = f(x, t)$ and the integration limits $a = a(x)$, $b = b(x)$, then Corollaries 9.20 and 9.23 can be combined to give the derivative

$$\frac{d}{dx} \int_{a(x)}^{b(x)} f(x, t)\, dt = \int_{a(x)}^{b(x)} \left[\frac{\partial}{\partial x} f(x, t)\right] dt \qquad (9.34)$$
$$+ f(x, b(x))b'(x) - f(x, a(x))a'(x) ,$$

provided the assumptions of these corollaries hold.

Exercises

9.14 Select a function f and an interval $[a, b]$ and illustrate, (9.10), $\int_a^b f(x)\, dx = F(b) - F(a)$ by calculating the definite integral in two ways: (a) by MACSYMA (see Session 8.1) and (b) by (9.10), with antiderivative F found by MACSYMA.
9.15 Write the derivatives d/dx of

(a) $\displaystyle\int_0^x f(t)\, dt$ (b) $\displaystyle\int_x^x f(t)\, dt$ (c) $\displaystyle\int_0^{\sqrt{x}} e^{-t^2}\, dt$

(d) $\displaystyle\int_{a(x)}^{b(x)} x\, dt$ (e) $\displaystyle\int_0^x f(x - t)\, g(t)\, dt$ (f) $\displaystyle\int_{-x}^x x\, \sin t\, dt$

9.16 Calculate the stationary points of the following functions, and determine which is local maximum, local minimum, or point of inflection.

(a) $\displaystyle\int_0^x (x - t^2)\, dt$ (b) $\displaystyle\int_{-x}^x t^2\, dt$ (c) $\displaystyle\int_x^{x^2} \sqrt{x}\, dt$

---

5 $(\partial/\partial x) f(x, t)$ is called the partial derivative of f with respect to x.

9.4 List of integrals

Definite integrals are often computed by, (9.10),

$$\int_a^b f(x)\,dx = F(b) - F(a), \quad \text{denoted } F(x)\big|_a^b,$$

where F is any indefinite integral of f. Some of the most useful indefinite integrals are listed below. The constant of integration C is mentioned only in part a.

Proposition 9.24 (List of integrals).

(a) $\displaystyle\int 0\,dx = C$

(b) $\displaystyle\int \alpha\,dx = \alpha x,\ \alpha \text{ constant}$

(c) $\displaystyle\int x^\alpha\,dx = \frac{x^{\alpha+1}}{\alpha+1},\ \alpha \neq -1$

(d) $\displaystyle\int \frac{1}{x}\,dx = \ln x,\ x > 0$

(e) $\displaystyle\int e^x\,dx = e^x$

(f) $\displaystyle\int a^x\,dx = \frac{a^x}{\ln a}$

(g) $\displaystyle\int \sin x\,dx = -\cos x$

(h) $\displaystyle\int \cos x\,dx = \sin x$

(i) $\displaystyle\int \frac{1}{\sqrt{1-x^2}}\,dx = \operatorname{asin} x$

(j) $\displaystyle\int \frac{1}{1+x^2}\,dx = \operatorname{atan} x$

(k) $\displaystyle\int \sinh x\,dx = \cosh x$

(l) $\displaystyle\int \cosh x\,dx = \sinh x$

Remark 9.25. Note a difficulty in part d,

$$\int \frac{1}{x}\,dx = \ln x + C, \tag{9.35}$$

in that $\ln x$ in the right-hand side is defined only for positive x, while the left-hand side is defined for all nonzero x. So what is the integral of $1/x$ if x is negative? We can show (see Example 10.4) that for $x < 0$

$$\int \frac{1}{x}\,dx = \ln(-x) + C. \tag{9.36}$$

Many calculus books write one expression

$$\int \frac{1}{x}\,dx = \ln|x| + C \tag{9.37}$$

which seems to combine (9.35) and (9.36). This is false in the complex case and is at best questionable in the real case.[6] We recall that an indefinite integral is

---

6 See, e.g., Serge Lang, *A First Course in Calculus*, 5th edn., Springer, Berlin Heidelberg New York, 1986, pp. 269–270.

a definite integral over a variable interval, or function of intervals. The function $1/x$ is not integrable in any interval containing $x = 0$, indeed, $\ln 0$ is undefined. Therefore (9.37) cannot unify the integrals (9.35) and (9.36) with the same constant of integration C. We must deal with the cases $x > 0$ and $x < 0$ separately, using (9.35) and (9.36) respectively, and there is no need for (9.37). This is also how MACSYMA gives the indefinite integral of $1/x$.

In the complex case we have

$$\ln(-z) = \ln z - i\pi, \quad z \in \mathbf{C} \setminus \{0\},$$

so that the integration formula in Proposition 9.24 d is correct for complex z, where the constant of integration is likewise complex.

MACSYMA-Session 9.3. Consider the indefinite integral of $1/x$

c1. `integrate(1/x,x)`

d1. $\log x$

Proposition 9.24 gives different results for the indefinite integral of x^n depending on whether $n \neq -1$ (part c) or $n = -1$ (part d).

Attempting the integral $\int x^n \, dx$ in MACSYMA, it pauses the calculation to inquire:

    `Is  n+1  zero or nonzero?`

The answer `zero` results in the integral $\log(x)$, and the answer `nonzero` results in $x^{n+1}/(n+1)$.

These two answers can be unified by subtracting the constant[7] $1/(n+1)$ from $x^{n+1}/(n+1)$,

c2. `x^(n+1)/(n+1)-1/(n+1)`

d2. $\dfrac{x^{n+1}}{n+1} - \dfrac{1}{n+1}$

which is a correct indefinite integral when $n \neq -1$ and also gives the correct answer as $n \to -1$

c3. `limit(%,n,-1)`

d3. $\log x$

Exercise

9.17 For any real α and integer $n > 0$, prove that[8]

$$\int (x + \alpha)^n \, dx = \frac{(x + \alpha)^{n+1}}{n+1}. \tag{9.38}$$

Hint: Use the binomial expansion of $(x + \alpha)^n$.

---

7 Viewed as a constant of integration.
8 See also Exercise 10.1.

9.5 Proofs

Proof of Example 9.12. For any $0 < \lambda < 1$, and any two points $a < x_0 < x_1 < b$, define $x_\lambda := (1 - \lambda) x_0 + \lambda x_1$. Then

$$(1 - \lambda)F(x_0) + \lambda F(x_1) = (1 - \lambda) \int_a^{x_0} f(t) \, dt + \lambda \int_a^{x_1} f(t) \, dt$$

$$= \int_a^{x_0} f(t) \, dt + \lambda \int_{x_0}^{x_1} f(t) \, dt$$

$$\therefore \; (1 - \lambda)F(x_0) + \lambda F(x_1) - F(x_\lambda) = \int_a^{x_0} f(t) \, dt + \lambda \int_{x_0}^{x_1} f(t) \, dt - \int_a^{x_\lambda} f(t) \, dt$$

$$= \lambda \int_{x_\lambda}^{x_1} f(t) \, dt - (1 - \lambda) \int_{x_0}^{x_\lambda} f(t) \, dt$$

(check!)

$$\geq \lambda f(x_\lambda) \int_{x_\lambda}^{x_1} dt - (1 - \lambda) f(x_\lambda) \int_{x_0}^{x_\lambda} dt$$

since f is increasing and $x_0 < x_\lambda < x_1$,

$$= f(x_\lambda) \Big(\lambda \int_{x_\lambda}^{x_1} dt - (1 - \lambda) \int_{x_0}^{x_\lambda} dt \Big) = 0$$

(check!)

$$\therefore \; (1 - \lambda)F(x_0) + \lambda F(x_1) \geq F((1 - \lambda)x_0 + \lambda x_1),$$

proving that F is convex. $\qquad \square$

Proof of Corollary 9.23.

$$I'(x) = \lim_{h \to 0} \frac{I(x + h) - I(x)}{h} = \lim_{h \to 0} \frac{1}{h} \Big(\int_a^b f(x + h, t) \, dt - \int_a^b f(x, h) \, dt \Big)$$

$$= \lim_{h \to 0} \int_a^b \Big(\frac{f(x + h, t) - f(x, t)}{h} \Big) dt$$

$$= \lim_{h \to 0} \int_a^b \Big[\frac{(\partial/\partial x) f(\xi, t)h}{h} \Big] dt, \quad \text{for some } x < \xi < x + h, \text{ by the mean value theorem (regarding } t \text{ as a constant, i.e., regarding } f \text{ as a function only of } x\text{),}$$

$$= \int_a^b \lim_{h \to 0} \Big[\frac{\partial}{\partial x} f(\xi, t) \Big] dt, \quad \text{assuming that "limit" and "integral" can be interchanged (we omit this part of the proof),}$$

$$= \int_a^b \Big[\frac{\partial}{\partial x} f(x, t) \Big] dt, \quad \text{since } \frac{\partial}{\partial x} f(x, t) \text{ is continuous in } x \text{ for all } t.$$

$\qquad \square$

10 Integration techniques

We study methods for computing *antiderivatives*, methods known collectively as integration techniques. Three such methods are covered here:
- the *change of variables* (or substitution) method (Sect. 10.1),
- *integration by parts* (Sect. 10.2),
- the *partial fractions expansion* method (Sect. 10.3).

The first two are more than techniques. They are an essential part of calculus, and at the same time they will enable you to compute most integrals you are likely to encounter in practice.

The third is a technique specific to *rational functions* (see Sect. 1.4). Rational functions are important in many applications, specially in fitting curves to empirical data. The technique of Sect. 10.3 works for every rational function and is easy to compute. This is how MACSYMA, and similar packages, integrate rational functions.

In the construction of the definite integral (8.5),

$$\int_a^b f(x)\,\mathrm{d}x ,$$

we assumed a *bounded integrand* f and a *finite interval* $[a, b]$. If either one of these assumptions is violated, the integral may sometimes be salvaged by taking appropriate limits. Such integrals, called *improper*, are studied in Sect. 10.4 and 10.5.

10.1 Changing variables

Integrals, both definite $\int_a^b f(x)\,\mathrm{d}x$ and indefinite $\int f(x)\,\mathrm{d}x$, are often simplified by changing from the original variable x to a more natural variable, which is a function of x. The idea is an application of the *chain rule*

$$\frac{\mathrm{d}}{\mathrm{d}x}F(u(x)) = F'(u(x))u'(x) , \qquad (10.1)$$

and works as follows: If the integrand is an expression like the right-hand side of (10.1)

$$F'(u(x))u'(x) \qquad (10.2)$$

then by changing variables from x to u, so that $\mathrm{d}u = u'(x)\,\mathrm{d}x$, we get an immediate integral,

$$\int F'(u(x))u'(x)\,dx = \int F'(u)\,du = F(u)\,,\qquad(10.3)$$

and changing back to the original variable x,

$$\int F'(u(x))u'(x)\,dx = F(u(x)) + C\,.\qquad(10.4)$$

This is the *substitution* (or change of variables) *method*. Another way to write (10.4) is

$$\int f(u(x))u'(x)\,dx = F(u(x)) + C\,,\qquad(10.5)$$

where F is an antiderivative of f.

What makes the substitution method (10.4) useful is that in many cases the integrand, while not in the desired form (10.2), can be brought to this form with little effort, say, by multiplying or dividing by some factor. However, this method requires ingenuity: You have to "guess" a "good" function u. We illustrate this by several examples.

Example 10.1. To calculate the antiderivative of $\tan x$, recall that

$$\tan x = \sin x / \cos x\,.$$

$$\therefore\quad \int \tan x\,dx = \int \sin x / \cos x\,dx\,.\qquad(10.6)$$

Since the sine and cosine functions are related by differentiation, we substitute

$$u = \cos x,\quad \text{with } du = u'(x)\,dx = -\sin x\,dx\,,\qquad(10.7)$$

in (10.6) to obtain

$$\int \tan x\,dx = -\int du/u,\quad \text{the integrand is in the form (10.2)}\,,$$
$$= -\ln u,\quad \text{by Proposition 9.24d}\,,$$
$$= -\ln \cos x,\quad \cos x > 0\,,$$

which we add to our list of integrals (Proposition 10.14).

Note that the substitution $u = \sin x$, instead of (10.7), leads nowhere. A priori it is not obvious which substitution to use; such wisdom comes with experience.

Example 10.2. To calculate

$$\int \sin^n x \cos x\,dx\,,$$

substitute $u = \sin x$, $du = \cos x\,dx$, to get

$$\int \sin^n x \cos x \, dx = \int u^n \, du = \begin{cases} \dfrac{\sin^{n+1} x}{n+1} & \text{if } n \neq -1, \\ \ln \sin x & \text{if } n = -1. \end{cases} \qquad (10.8)$$

The last line says that

$$\int \cot x \, dx = \ln \sin x, \quad \sin x > 0,$$

which we enter in the list of integrals (Proposition 10.14).

Example 10.3. Consider the integral

$$\int \sin^3 x \, dx .$$

We use the trigonometric identity,

$$\sin^2 x + \cos^2 x = 1$$

to obtain

$$\int \sin^3 x \, dx = \int \sin x (1 - \cos^2 x) \, dx$$
$$= \int \sin x \, dx - \int \sin x \cos^2 x \, dx .$$

Of the two integrals in the right-hand side, the first is immediate,

$$\int \sin x \, dx = -\cos x ,$$

and the second is solved by substituting $u = \cos x$ and $du = -\sin x \, dx$,

$$\int \sin x \cos^2 x \, dx = -\int u^2 \, du = -\frac{\cos^3 x}{3} .$$

$$\therefore \quad \int \sin^3 x \, dx = -\cos x + \frac{\cos^3 x}{3} .$$

Example 10.4. We calculate the indefinite integral

$$\int dx/x ,$$

for $x < 0$. Now the indefinite integral is a definite integral over a variable interval,

$$\int dx/x = \int_{\alpha}^{x} dt/t$$

for some $\alpha < x$, so that $t < 0$ throughout the interval. We substitute $u = -t$ and $du = -dt$ to get

$$\int dx/x = \int_{\alpha}^{x} du/u = \ln u|_{-\alpha}^{-x}, \text{ use (9.35) since } u > 0,$$

$$= \ln(-x) - \ln(-\alpha), \text{ both make sense since } -x \text{ and } -\alpha \text{ are positive },$$
$$= \ln(-x) + C, \text{ proving (9.36) .}$$

Example 10.5. The integral

$$\int \sqrt{3x^2 + 5x}\, dx$$

is simplified by the change of variable

$$u = 3x^2 + 5, \quad \text{with } du = 6x\, dx . \tag{10.9}$$

To bring the integrand $\sqrt{3x^2 + 5x}$ to the desired form (10.2), we multiply, and divide, it by 6.

$$\therefore \quad \int \sqrt{3x^2 + 5x}\, dx = \tfrac{1}{6} \int \sqrt{3x^2 + 5} \cdot 6x\, dx$$
$$= \tfrac{1}{6} \int \sqrt{u}\, du = \tfrac{1}{6}\tfrac{2}{3} u^{3/2}$$
$$= \tfrac{1}{9}(3x^2 + 5)^{3/2} .$$

The substitution method allows replacing an indefinite integral in x by a (hopefully simpler) integral in u,

$$\int F'(u(x))u'(x)\, dx = \int F'(u)\, du = F(u) .$$

It can also be used for definite integrals

$$\int_{a}^{b} F'(u(x))\, u'(x)\, dx ,$$

to obtain

$$\int_{a}^{b} F'(u(x))u'(x)\, dx = \int_{u(a)}^{u(b)} F'(u)\, du = F(u(b)) - F(u(a)) \tag{10.10}$$

where we changed the limits of integration from $[a, b]$ to $[u(a), u(b)]$. Forgetting to do so is one of the most common errors in calculus.

Example 10.6. The integral

$$\int_{1}^{3} \sqrt{3x^2 + 5x}\, dx$$

is computed, changing variable as in Example 10.5,

$$\int\limits_{1}^{3} \sqrt{3x^2 + 5}\, x\, dx = \tfrac{1}{6} \int\limits_{8}^{32} \sqrt{u}\, du = \tfrac{1}{9} u^{3/2}\Big|_{8}^{32} = \tfrac{1}{9}(32^{3/2} - 8^{3/2}) = \frac{112\sqrt{2}}{9}\ .$$

The limits of integration were updated here by (10.9) and (10.10) with $u(1) = 3 \cdot 1 + 5 = 8$ and $u(3) = 3 \cdot 3^2 + 5 = 32$. Alternatively, we can substitute back x and keep the "old" integration limits $[1, 3]$,

$$\int\limits_{1}^{3} \sqrt{3x^2 + 5}\, x\, dx = \tfrac{1}{9}(3x^2 + 5)^{3/2}\Big|_{1}^{3} = \tfrac{1}{9}\big((3 \cdot 3^2 + 5)^{3/2} - (3 \cdot 1^2 + 5)^{3/2}\big)$$

$$= \tfrac{1}{9}(32^{3/2} - 8^{3/2}) = \frac{112\sqrt{2}}{9}\ .$$

Example 10.7 (Trigonometric substitution). In some problems the "saving" change of variable is not at all obvious. An example is the formidably looking integral

$$\int \sqrt{1 - x^2}\, dx$$

which is easily tamed with the substitution

$$x = \sin\theta, \quad \text{with } dx = \cos\theta\, d\theta\ . \tag{10.11}$$

Indeed

$$\int \sqrt{1 - x^2}\, dx = \int \cos^2\theta\, d\theta$$

$$= \frac{\theta}{2} + \frac{\sin 2\theta}{4}, \text{ by (10.24), and finally, back to } x\ ,$$

$$= \tfrac{1}{2}\operatorname{asin} x + \tfrac{1}{2}x\sqrt{1 - x^2} \tag{10.12}$$

since $\sin 2\theta = 2\sin\theta\cos\theta = 2x\sqrt{1 - x^2}$ (see Fig. 10.1).

After such success, you may try the substitution (10.11) whenever the integrand contains $\sqrt{1 - x^2}$. A method is a trick that works twice.

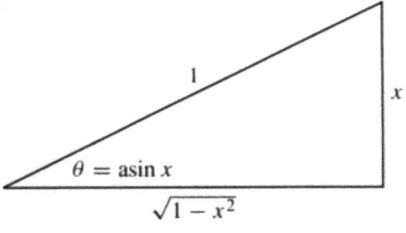

Fig. 10.1. Consequences of $x = \sin\theta$ (Examples 10.7 and 10.9). $\cos\theta = \sqrt{1 - x^2}$, $\sec\theta = 1/\sqrt{1 - x^2}$, $\tan\theta = x/\sqrt{1 - x^2}$

Example 10.8 (Another trick). Consider the integral of $\sec x$,[1]

$$\int \sec x \, dx = \int dx / \cos x \,.$$

It does not look like much until we multiply, and divide, it by $\sec x + \tan x$. Then

$$\int \sec x \, dx = \int \frac{(\sec^2 x + \sec x \tan x)}{(\sec x + \tan x)} \, dx$$

$$= \int du/u, \quad \text{where } u = \sec x + \tan x \,,$$

$$= \ln(\sec x + \tan x) \,. \tag{10.13}$$

Lest you think that $\sec x$ is too esoteric a function to worry about its integral, we hasten to put (10.13) to good use in the next example.

Example 10.9.

$$\int \frac{dx}{1-x^2} = \int \frac{\cos \theta \, d\theta}{\cos^2 \theta}, \quad \text{substituting } x = \sin \theta \,,$$

$$= \int \sec \theta \, d\theta = \ln(\sec \theta + \tan \theta), \quad \text{by (10.13)} \,,$$

$$= \ln(\sec(\operatorname{asin} x) + \tan(\operatorname{asin} x))$$

$$= \ln\left(\frac{1}{\sqrt{1-x^2}} + \frac{x}{\sqrt{1-x^2}}\right), \quad \text{see Fig. 10.1} \,,$$

$$= \ln\left(\frac{1+x}{\sqrt{1-x^2}}\right) = \ln\left(\sqrt{\frac{1+x}{1-x}}\right)$$

$$= \tfrac{1}{2} \ln\left(\frac{1+x}{1-x}\right) \,. \tag{10.14}$$

Example 10.10. In this example we denote "plus or minus" by "$\pm$", giving two results in one.

$$\int \frac{dx}{\sqrt{x^2 \pm 1}} = \int \frac{(x + \sqrt{x^2 \pm 1})}{\sqrt{x^2 \pm 1}(x + \sqrt{x^2 \pm 1})} \, dx$$

$$= \int \frac{x/\sqrt{x^2 \pm 1} + 1}{x + \sqrt{x^2 \pm 1}} \, dx = \int \frac{du}{u}, \quad \text{with } u := x + \sqrt{x^2 \pm 1} \,,$$

$$= \ln(x + \sqrt{x^2 \pm 1}) \,. \tag{10.15}$$

---

1 Recall that $\sec x = 1/\cos x$.

Example 10.11 (Translation). If the integral

$$\int_a^b f(x - \alpha)\,dx\ ,$$

is complicated by the presence of the constant α in the argument, we can translate ("move") the origin from 0 to α, by changing variable,

$$u := x - \alpha, \quad \text{with } du = dx\ , \tag{10.16}$$

resulting in a simpler integrand

$$\int_a^b f(x - \alpha)\,dx = \int_{a-\alpha}^{b-\alpha} f(u)\,du \tag{10.17}$$

but "translated" integration limits. So, e.g.,

$$\int_0^1 (x + 2)^2\,dx = \int_2^3 u^2\,du = \left.\frac{u^3}{3}\right|_2^3 = \frac{1}{3}(3^3 - 2^3) = \frac{19}{3}\ .$$

For an indefinite integral

$$\int f(x - \alpha)\,dx$$

we can still substitute (10.16), integrate with respect to u, and then substitute back x. Example:

$$\int \frac{1}{(1 - x)^2}\,dx = -\int \frac{1}{u^2}\,du = \frac{1}{u} = \frac{1}{1 - x}\ .$$

Example 10.12 (Scaling). Another reason to change variables is the need to change scale. For example, suppose we want to simplify the integrand in

$$\int_a^b f(\alpha x)\,dx\ .$$

The substitution

$$u := \alpha x, \quad \text{with } du = \alpha\,dx \tag{10.18}$$

transforms the integral into

$$\int_a^b f(\alpha x)\,dx = \frac{1}{\alpha}\int_{a/\alpha}^{b/\alpha} f(u)\,du\ . \tag{10.19}$$

So, e.g.,

$$\int_6^8 \cos 2x \, dx = \frac{1}{2} \int_3^4 \cos u \, du, \quad \text{substituting } u = 2x, \, du = 2 \, dx \, ,$$

$$= \frac{1}{2} \sin u \big|_3^4 = \frac{\sin 4 - \sin 3}{2} \, .$$

Example 10.13. By changing scale we reduce the following integral to a familiar form

$$\int \frac{dx}{\sqrt{a^2 - x^2}} = \frac{1}{a} \int \frac{dx}{\sqrt{1 - (x/a)^2}} \, ,$$

$$= \frac{1}{a} \int \frac{a \, du}{\sqrt{1 - u^2}}, \quad \text{substituting } u := x/a, \quad \text{with } dx = a \, du \, ,$$

$$= \int \frac{du}{\sqrt{1 - u^2}} = \operatorname{asin} u, \quad \text{by Proposition 9.24 i} \, ,$$

$$= \operatorname{asin}(x/a) \, .$$

Similarly,

$$\int \frac{dx}{a^2 + x^2} = \frac{1}{a^2} \int \frac{dx}{1 + (x/a)^2} = \frac{1}{a^2} \int \frac{a \, du}{1 + u^2}$$

$$= \frac{1}{a} \operatorname{atan} u, \quad \text{by Proposition 9.24 j} \, ,$$

$$= \frac{1}{a} \operatorname{atan}\left(\frac{x}{a}\right) \, .$$

Other examples:

$$\int \sqrt{a^2 - x^2} \, dx = a \int \sqrt{1 - \left(\frac{x}{a}\right)^2} \, dx = a^2 \int \sqrt{1 - u^2} \, du$$

$$= \frac{a^2}{2} \operatorname{asin} u + \frac{a^2}{2} u \sqrt{1 - u^2}, \quad \text{by (10.12)} \, ,$$

$$= \frac{a^2}{2} \operatorname{asin} \frac{x}{a} + \frac{1}{2} x \sqrt{a^2 - x^2} \, ; \tag{10.20}$$

the following "scaling" of (10.14)

$$\int \frac{dx}{a^2 - x^2} = \frac{1}{2a} \ln\left(\frac{a + x}{a - x}\right) \, ; \tag{10.21}$$

and of (10.15),

$$\int \frac{dx}{\sqrt{x^2 \pm a^2}} = \ln\left(x + \sqrt{x^2 \pm a^2}\right) \, .$$

We collect now some of the above integrals, supplementing the list given in Proposition 9.24.

Proposition 10.14 (Second list of integrals).

(a) $\displaystyle\int \tan x \, dx = -\ln\cos x, \quad \cos x > 0$

(b) $\displaystyle\int \cot x \, dx = \ln\sin x, \quad \sin x > 0$

(c) $\displaystyle\int \frac{dx}{\sqrt{a^2 - x^2}} = \operatorname{asin}\left(\frac{x}{a}\right)$ (d) $\displaystyle\int \frac{dx}{a^2 + x^2} = \frac{1}{a}\operatorname{atan}\left(\frac{x}{a}\right)$

(e) $\displaystyle\int \sqrt{a^2 - x^2}\, dx = \frac{a^2}{2}\operatorname{asin}\frac{x}{a} + \frac{x\sqrt{a^2 - x^2}}{2}$

(f) $\displaystyle\int \frac{dx}{a^2 - x^2} = \frac{1}{2a}\ln\left(\frac{a + x}{a - x}\right)$ (g) $\displaystyle\int \frac{dx}{\sqrt{x^2 \pm a^2}} = \ln\left(x + \sqrt{x^2 \pm a^2}\right)$

MACSYMA-Session 10.1. In most cases, the MACSYMA integrator does its job well, using an appropriate method for the task at hand. In this session, we "force" MACSYMA to use the substitution method, with our selection of the substitution $t = u(x)$ in the integral (10.5),

$$\int f(u(x))u'(x)\, dx = F(u(x)) + C .$$

We assume that $t = u(x)$ is one-to-one and denote its inverse by $x = u^{-1}(t)$. Writing the integrand as

$$y(x) := f(u(x))u'(x) ,$$

we get, using the rule for the derivative of the inverse function,

$$f(t) = y(x)/u'(x) = y(u^{-1}(t))/u'(u^{-1}(t)) = y(u^{-1}(t)) \frac{d}{dt} u^{-1}(t)$$

The inverse function can be computed by MACSYMA as follows:
```
c1. inverse(u,x,t) := rhs(first(solve(u=t,x)))$
```
this gives $x = u^{-1}(t)$ as the first solution x of $u(x) = t$, found by `solve`. Examples:
```
c2. inverse(cos(a*x),x,t)
```
d2. $\dfrac{\operatorname{acos} t}{a}$
```
c3. inverse(tan(x/2),x,t)
```
d3. $2\operatorname{atan} t$
```
c4. inverse(exp(a*x)+b,x,t)
```
d4. $\dfrac{\log(t - b)}{a}$
```
c5. inverse(asin(x),x,t)
```
d5. $\sin t$

c6. `inverse(atan(x),x,t)`
d6. $\tan t$

The warning message `some solutions may be lost`, which occasionally appears, indicates that `inverse(u,x,t)` is not fool-proof. Indeed, if $u(x) = t$ has several solutions, the choice of the `first` solution is not always sensible. However, when `inverse(u,x,t)` works, it can be applied, as follows, to substitute $t = u(x)$ in the integral $\int f(u(x))u'(x)\,dx$:

c7. `int_subst(y,x,u,t) := block([inv:inverse(u,x,t)],`
               `/* inv is a temporary variable, saving one computation`
                   `of inverse(u,x,t), which is used twice */`
         `subst(t=u, /* change the variable back when done */`
            `integrate(subst(x=inv,y)*diff(inv,t), t)))$`

The analogous computation of the definite integral $\int_a^b u(x)\,u'(x)\,dx$:

c8. `def_int_subst(y,x,u,t,a,b) := block([inv:inverse(u,x,t)],`
                 `integrate(subst(x=inv,y)*diff(inv,t), t,`
                      `limit(u,x,a,plus), /* lower limit */`
                      `limit(u,x,b,minus) /* upper limit */`
                      `))$%$`

Our first example is the indefinite integral of $\tan(x)$:

c9. `integrate(tan(x),x)`
d9. $\log \sec x$

Using the substitution $t = \cos(x)$ we get:

c10. `int_subst(tan(x),x,cos(x),t)`
d10. $-\log \cos x$

which is the same as d9 since $\sec(x) = 1/\cos(x)$, and therefore $\log(\sec(x)) = \log(1/\cos(x)) = -\log(\cos(x))$. When we try to substitute $t = \tan x$:

c11. `int_subst(tan(x),x,tan(x),t)`

we get

d11. $\dfrac{\log(\tan^2 x + 1)}{2}$

which agrees with d9 (verify!). Similarly, the substitution $t = \tan(x/2)$:

c12. `int_subst(tan(x),x,tan(x/2),t)`

gives

d12. $\log \sec(2 \operatorname{atan} \tan(x/2))$

which is again equivalent to d9.

Here are some additional examples:

c13. `integrate(sin(x)^n*cos(x),x)`
     `Is n+1 zero or nonzero?`
     `nonzero`

d13. $\dfrac{\sin^{n+1} x}{n+1}$

and using the substitution $t = \sin x$:

c14. `int_subst(sin(x)^n*cos(x),x,sin(x),t)`
     `Is n+1 zero or nonzero?`
     `nonzero`

d14. $\dfrac{\sin^{n+1} x}{n+1}$

The substitution $t = \cos x$ does not seem natural here:

c15. `int_subst(sin(x)^n*cos(x),x,cos(x),t)`

    `Is   n+1   zero or nonzero?`

    `nonzero`

d15. $$\dfrac{(1 - \cos^2 x)^{n/2+1/2}}{2(n/2 + 1/2)}$$

We see that the answer depends on the choice of the trigonometric function which is substituted. Using `ratsubst` to replace $\cos^2(x)$ with $1 - \sin^2(x)$

c16. `ratsubst(1-sin(x)^2,cos(x)^2,%)`

results in the absolute value function:

d16. $$\dfrac{|\sin x|^{n+1}}{n + 1}$$

which agrees with d13, as long as one does not integrate through 0.

Consider next the integral of Example 10.5,

c17. `integrate(x*sqrt(3*x^2+5),x)`

d17. $$\dfrac{(3x^2 + 5)^{3/2}}{9}$$

and using the substitution $t = 3x^2 + 5$:

c18. `int_subst(x*sqrt(3*x^2+5),x,3*x^2+5,t)`

d18. $$\dfrac{(3x^2 + 5)^{3/2}}{9}$$

The definite integral of Example 10.6 is computed similarly:

c19. `def_int_subst(x*sqrt(3*x^2+5),x,3*x^2+5,t,1,3)`

MACSYMA-Session 10.2. The function `int_subst(y,x,y,t)` in Session 10.1 computes the integral

$$\int f(u(x))u'(x)\,dx$$

by the substitution $t := u(x)$. In this session we use the inverse substitution $x := g(t)$, where g is the inverse u^{-1} of u. We implement the inverse substitution by calling the function `inverse(u,x,t)` of Session 10.1. The MACSYMA functions

c1. `int_subst_inv(y,x,g,t)  := subst(t=inverse(g,t,x),`
                      `integrate(subst(x=g,y)*diff(g,t),t))$`

c2. `def_int_subst_inv(y,x,g,t,a,b) := block([inv:inverse(g,t,x)],`
                    `integrate(subst(x=g,y)*diff(g,t), t,`
                          `limit(inv,x,a,plus),`
                          `limit(inv,x,b,minus)))$`

compute the indefinite and definite integrals of $y(x) = f(u(x))u'(x)$, respectively. As an example, consider the integral of Example 10.7, which MACSYMA integrates as

c3. `integrate(sqrt(1-x^2),x)`

d3. $$\dfrac{\operatorname{asin} x}{2} + \dfrac{x\sqrt{1 - x^2}}{2}$$

Now repeat the computation using the substitution $x = \cos(t)$:

c4. `int_subst_inv(sqrt(1-x^2),x,cos(t),t)`

d4. $$\dfrac{\pi/2 - \operatorname{acos} x}{2} + \dfrac{x\sqrt{1 - x^2}}{2}$$

which agrees with d3 (verify!). Similarly, the definite integral:
c5. `def_int_subst_inv(sqrt(1-x^2)/(1+x^2),x,sin(t),t,0,1)`

d5. $\dfrac{\sqrt{2\pi}}{2} - \dfrac{\pi}{2}$

Note that the results of this section may vary with the specific version of MACSYMA installed on your computer.

Example 10.15 (Area of disk). A disk is a set of points bounded by a circle, i.e., the points inside the circle or on it. We calculate the area A of a disk with a radius r. Consider the equation

$$x^2 + y^2 = r^2$$

of the circle with radius r and center at $(0, 0)$. The area of the quarter disk in the first quadrant is

$$\int_0^r \sqrt{r^2 - x^2}\,dx = \frac{r^2}{2}\,\mathrm{asin}\,\frac{x}{r} + \left.\frac{x\sqrt{r^2-x^2}}{2}\right|_0^r, \quad \text{by (10.20)},$$

$$= \pi r^2/4 \quad \text{since asin } 1 = \pi/2.$$

Therefore, $A = \pi r^2$.

The substitution method can also be used to prove theorems, such as:

Theorem 10.16 (Generalized integral mean value theorem). Let f be continuous, and let g be nonnegative and integrable on $I := [a, b]$. Then there is a point ξ in I such that

$$\int_a^b f(x)g(x)\,dx = f(\xi)\int_a^b g(x)\,dx.\tag{10.22}$$

Proof. Since g is nonnegative, the function

$$u(x) := \int_a^x g(t)\,dt$$

is nondecreasing. We change variable from x to $u(x)$, with $du = g(x)\,dx$. The integral (10.22) is therefore

$$\int_a^b f(x)g(x)\,dx = \int_{u(a)}^{u(b)} f(x(u))\,du,\tag{10.23}$$

where $x(u)$ is the smallest x with[2]

$$\int_a^x g(t)\, dt = u \,.$$

The mean value theorem 8.39 now guarantees a point $\xi(u)$ in I such that

$$\int_{u(a)}^{u(b)} f(x(u))\, du = f(\xi(u))(u(b) - u(a)) \,,$$

which proves (10.22) with $\xi = \xi(u)$ since $u(b) - u(a) = \int_a^b g(x)\, dx$.   □

Remark 10.17. If $g(x) \equiv 1$, this theorem reduces to the mean value theorem 8.39.

Exercises

10.1   For $n \neq -1$, calculate the integral (9.38),

$$\int (x + \alpha)^n\, dx = \frac{(x + \alpha)^{n+1}}{n + 1} \,,$$

by substituting $u = x + \alpha$. This solution is considerably simpler than the "brute-force" method suggested in Exercise 9.17.

10.2   Evaluate the integrals below. In part c, a and b are constants.

(a) $\displaystyle\int (4x + 10)^5\, dx$ (b) $\displaystyle\int x^4\, (x^5 + 3)^5\, dx$ (c) $\displaystyle\int (ax + b)^n\, dx$

(d) $\displaystyle\int x\,(x^2 + 2)^{-1/3}\, dx$ (e) $\displaystyle\int (2x + 1)^{-2}\, dx$ (f) $\displaystyle\int \frac{dx}{\sqrt{x^2 + 4}}$

(g) $\displaystyle\int (4x + 10)^5\, dx$ (h) $\displaystyle\int \frac{dt}{\sqrt{1 - 4t^2}}$ (i) $\displaystyle\int \frac{t\, dt}{t^4 + 9}$

(j) $\displaystyle\int \frac{t\, dt}{4t^4 - 16}$ (k) $\displaystyle\int (4x + 10)^5\, dx$ (l) $\displaystyle\int (1 + \sqrt{x})^5\, \frac{dx}{\sqrt{x}}$

(m) $\displaystyle\int \tan^3 x \sec x\, dx$ (n) $\displaystyle\int \tan^n x \sec x\, dx$ (o) $\displaystyle\int \cos^n \sin x\, dx$

(p) $\displaystyle\int \sec^4 x \tan x\, dx$ (q) $\displaystyle\int \frac{\ln^4 x}{x}\, dx$ (r) $\displaystyle\int \frac{\ln x^2}{x}\, dx$

(s) $\displaystyle\int \frac{\ln x^n}{x}\, dx$ (t) $\displaystyle\int (\text{atan}\, x)^2 \frac{1}{1 + x^2}\, dx$ (u) $\displaystyle\int (\text{atan}\, x)^{-1} \frac{1}{1 + x^2}\, dx$

(v) $\displaystyle\int \frac{\sec x \tan x}{(1 + \sec x)^{3/4}}\, dx$

---

2 If g is positive, then $u(x)$ is monotone increasing, and $x(u)$ is unique. If g is non-negative, then $u(x)$ is merely nondecreasing and may be constant in some intervals.

10.3 Evaluate the integrals

(a) $\displaystyle\int x \sec^2 x^2 \, dx$

(b) $\displaystyle\int x^2 \sec x^3 \tan x^3 \, dx$

(c) $\displaystyle\int \frac{e^t \, dt}{1 + e^{2t}}$

(d) $\displaystyle\int \frac{e^t \, dt}{1 + e^t}$

(e) $\displaystyle\int \frac{\sin t \, dt}{1 + \cos t}$

(f) $\displaystyle\int \frac{\sin t \, dt}{1 + \cos^2 t}$

(g) $\displaystyle\int e^{\cos t} \sin t \, dt$

(h) $\displaystyle\int e^{\operatorname{atan} x} \frac{1}{1 + x^2} \, dx$

(i) $\displaystyle\int e^{\operatorname{asin} x} \frac{1}{\sqrt{1 + x^2}} \, dx$

10.4 In each of the following integrals there is a substitution which does the job. Discover this substitution and perform the integration.

(a) $\displaystyle\int x^2 \sqrt{x + 2} \, dx$

(b) $\displaystyle\int \frac{x^2 \, dx}{\sqrt{x + 2}}$

(c) $\displaystyle\int x(x^2 + 2)\sqrt{x + 1} \, dx$

(d) $\displaystyle\int \ln(x^2 + 2) \frac{1}{x + 1} \, dx$

(e) $\displaystyle\int \frac{x \, dx}{4x^2 + 9}$

(f) $\displaystyle\int \frac{dx}{16x^2 + 4}$

(g) $\displaystyle\int \frac{e^{x+2} \, dx}{4 + e^{2x+4}}$

(h) $\displaystyle\int t\sqrt{9 - t^2} \, dt$

(i) $\displaystyle\int x^2 \sqrt{16x^2 - 9} \, dx$

(j) $\displaystyle\int \frac{1}{x} \sqrt{\ln^2 x + 2} \, dx$

(k) $\displaystyle\int \frac{1}{x} \sqrt{\ln^2 x^2 + 2} \, dx$

(l) $\displaystyle\int e^x \sqrt{e^{2x} + 4} \, dx$

10.5 Prove that the results of Example 8.33 hold if the integration is from α to $\alpha + 2\pi$, for any real α, i.e., prove that, for any integers m, n and constant α,

$$\int_\alpha^{\alpha+2\pi} \sin mx \, \sin nx \, dx = \begin{cases} 0 & \text{if } m \neq n, \\ \pi & \text{if } m = n; \end{cases}$$

$$\int_\alpha^{\alpha+2\pi} \cos mx \, \cos nx \, dx = \begin{cases} 0 & \text{if } m \neq n, \\ \pi & \text{if } m = n; \end{cases}$$

$$\int_\alpha^{\alpha+2\pi} \sin mx \, \cos nx \, dx = 0.$$

Hint: As in Example 8.33, these follow from

$$\int_\alpha^{\alpha+2\pi} \sin nx \, dx = 0, \qquad \int_\alpha^{\alpha+2\pi} \cos nx \, dx = 0,$$

for all integer n. Now we change variables in order to simplify the integration limits,

$$\int_\alpha^{\alpha+2\pi} \sin nx \, dx = \int_0^{2\pi} \sin n(u + \alpha) \, du, \quad \text{substituting (10.16)},$$

$$= \int_0^{2\pi} (\sin nu \cos n\alpha + \cos nu \sin n\alpha) \, du$$

$$= \cos n\alpha \int_0^{2\pi} \sin nu \, du + \sin n\alpha \int_0^{2\pi} \cos nu \, du$$

$$= 0, \quad \text{using (8.50)}.$$

10.6   Verify

$$\int \sin^2 x \, dx = \frac{x}{2} - \frac{\sin 2x}{4} \, ,$$

$$\int \cos^2 x \, dx = \frac{x}{2} + \frac{\sin 2x}{4} \, . \tag{10.24}$$

Hint: Obtain $\sin^2 x$ and $\cos^2 x$ by adding, or subtracting, the trigonometric identities

$$\cos^2 x + \sin^2 x = 1 \quad \text{and} \quad \cos^2 x - \sin^2 x = \cos 2x \, .$$

10.7   Use the Pythagorean relation $\sin^2 x + \cos^2 x = 1$ to simplify the following integrals into ones which are readily integrable. Check your results with MACSYMA.

(a) $\int \sin^3 x \, dx$ (b) $\int \cos^3 x \, dx$ (c) $\int \sin^3 x \cos^2 x \, dx$

(d) $\int \cos^3 x \sin^2 x \, dx$ (e) $\int \sin^5 x \, dx$ (f) $\int \cos^{2n+1} x \, dx$

(g) $\int \sin^{2n+1} x \cos^2 x \, dx$ (h) $\int \cos^{2n+1} x \sin^2 x \, dx$

10.8   Recall that another form of the Pythagorean relation is $1 + \tan^2 x = \sec^2 x$. Use this identity to integrate the following.

(a) $\int \sec^4 x \, dx$ (b) $\int \sec^6 x \, dx$ (c) $\int \tan x \sec^n x \, dx$

(d) $\int \tan^3 x \sec^2 x \, dx$ (e) $\int \tan^3 x \sec^n x \, dx$

10.9   Recall that still another form of the Pythagorean relation is $1 + \cot^2 x = \csc^2 x$. Use this identity to integrate the following.

(a) $\int \csc^4 x \, dx$ (b) $\int \csc^6 x \, dx$ (c) $\int \cot x \csc^n x \, dx$

(d) $\int \cot^3 x \csc^4 x \, dx$ (e) $\int \cot^3 x \csc^n x \, dx$

10.10 Evaluate the following integrals. Check your results with MACSYMA.

(a) $\int \sin^5 x \, dx$ (b) $\int \sin^{3/4} x \cos^3 x \, dx$ (c) $\int \tan^3 x \, dx$

(d) $\int \tan^5 x \sec^2 x \, dx$ (e) $\int \frac{\sec^2 x \, dx}{\tan^n x}$ (f) $\int \frac{\sec^4 x \, dx}{\tan^3 x}$

(g) $\int \frac{\sec x \tan x}{\sec^n x} \, dx$ (h) $\int \frac{\tan^3 x \, dx}{\sin^2 x}$ (i) $\int \frac{\cot^5 x \, dx}{\csc^4 x}$

(j) $\int \frac{\tan^{2n+1} x}{\sin^{2n} x} \, dx$ (k) $\int \frac{dx}{\sin^4 3x}$ (l) $\int \frac{dx}{\cos^4 4x}$

(m) $\int \frac{dx}{\sin^6 x}$ (n) $\int \frac{\tan x + \cos^2 x}{\sec x} \, dx$ (o) $\int \frac{\tan x + \sin x}{1 - \sin^2 x} \, dx$

(p) $\int \frac{\cos x + \cot x}{\sin^6 x} \, dx$

10.11 The following integrals contain algebraic expressions of the form $\sqrt{a^2 - x^2}$, $a^2 + x^2$, etc. These may be done by trigonometric substitutions. We list below the recommended substitutions, for each term.

term	substitute	use identity
$a^2 - x^2$	$x := a \sin \theta$	$1 - \sin^2 \theta = \cos^2 \theta$
$a^2 + x^2$	$x := a \tan \theta$	$1 + \tan^2 \theta = \sec^2 \theta$
$x^2 - a^2$	$x := a \sec \theta$	$\sec^2 \theta - 1 = \tan^2 \theta$

(a) $\displaystyle \int \sqrt{a^2 - x^2}\, dx$ (b) $\displaystyle \int \sqrt{a^2 + x^2}\, dx$ (c) $\displaystyle \int \frac{x^3\, dx}{\sqrt{1 - x^2}}$

(d) $\displaystyle \int \frac{dx}{(x^2 + a^2)^2}, a \neq 0$ (e) $\displaystyle \int \frac{x^3\, dx}{\sqrt{1 - x^2}}$ (f) $\displaystyle \int \frac{x^2\, dx}{\sqrt{1 - x^2}}$

(g) $\displaystyle \int \frac{dx}{(1 + x^2)^{3/2}}$ (h) $\displaystyle \int \frac{dx}{(1 - x^2)^{3/2}}$ (i) $\displaystyle \int (1 - x^2)^{3/2}\, dx$

(j) $\displaystyle \int (1 + x^2)^{3/2}\, dx$

Hint: In part f, use

$$\sec \theta := \frac{\cos \theta}{1 - \sin^2 \theta}, \quad \text{and then write} \quad \frac{1}{1 - \sin^2 \theta} = \frac{1}{1 - \sin \theta} + \frac{1}{1 + \sin \theta}.$$

10.12 In the following four integrals $a \neq 0$ and $b \neq 0$

(a) $\displaystyle \int \frac{dx}{(a^2 + b^2 x^2)^{3/2}}$ (b) $\displaystyle \int \frac{dx}{(a^2 - b^2 x^2)^{3/2}}$

(c) $\displaystyle \int (a^2 - b^2 x^2)^{3/2}\, dx$ (d) $\displaystyle \int (a^2 + b^2 x^2)^{3/2}\, dx$

10.13 We may also use hyperbolic substitutions to handle integrals. The recommended substitutions are shown below.

term	substitute	use identity
$a^2 - x^2$	$x := a \tanh \theta$	$1 - \tanh^2 \theta = \operatorname{sech}^2 \theta$
$a^2 + x^2$	$x := a \sinh \theta$	$1 + \sinh^2 \theta = \cosh^2 \theta$
$x^2 - a^2$	$x := a \cosh \theta$	$\cosh^2 \theta - 1 = \sinh^2 \theta$

(a) $\displaystyle \int \frac{dx}{\sqrt{a^2 + b^2 x^2}}$ (b) $\displaystyle \int \frac{\sqrt{x^2 - 1}\, dx}{x^3}$ (c) $\displaystyle \int \sqrt{x^2 - 1}\, x^n\, dx$

(d) $\displaystyle \int x^3 (x^2 - 1)^{3/2}\, dx$ (e) $\displaystyle \int \frac{dx}{(a^2 + x^2)^{3/2}}$

10.14 Find the following integrals by appropriate substitutions.

(a) $\displaystyle \int \frac{\sqrt{1 + x^2}}{1 - x^2}\, dx$ (b) $\displaystyle \int \frac{\sqrt{x^2 - 1}}{1 + x^2}\, dx$ (c) $\displaystyle \int \frac{\sqrt{1 - x^2}}{1 + x^2}\, dx$

(d) $\displaystyle \int \frac{1}{(1 + x^2)(1 + \operatorname{atan}^2 x)}\, dx$ (e) $\displaystyle \int \frac{1}{\sqrt{1 - x^2}(1 + \operatorname{asin}^2 x)}\, dx$

(f) $\displaystyle \int \frac{1}{\sqrt{(1 - x^2)(1 - \operatorname{asin}^2 x)}}\, dx$ (g) $\displaystyle \int \sqrt{x^2 + y^2}\, dx$

In part g, consider y a constant.

10.15 Given a real γ and a positive integer m, prove that

$$\int \frac{dx}{(x - \gamma)^m} = \begin{cases} -\dfrac{1}{(m - 1)(x - \gamma)^{m-1}} & \text{if } m > 1, \\ \ln(x - \gamma) & \text{if } m = 1. \end{cases} \tag{10.25}$$

10.16 Given reals A, B, α, β and a positive integer n, simplify as much as you can the
integral

$$\int \frac{A + Bx}{((x - \alpha)^2 + \beta^2)^n} \, dx \, . \tag{10.26}$$

Outline: The case $\beta = 0$ is in Exercise 10.15. In particular,

$$\int \frac{x \, dx}{(x - \gamma)^m} = \int \frac{dx}{(x - \gamma)^{m-1}} + \int \frac{\gamma \, dx}{(x - \gamma)^m} \, .$$

So assume $\beta \neq 0$. Then the substitution

$$u := \frac{x - \alpha}{\beta} \, ,$$

reduces the integral to

$$\int \frac{C + Du}{(u^2 + 1)^n} \, du = C \int \frac{du}{(u^2 + 1)^n} + D \int \frac{u \, du}{(u^2 + 1)^n} \, . \tag{10.27}$$

The second integral in the right-hand side is immediate, substituting $v := u^2 + 1$,

$$\int \frac{u \, du}{(u^2 + 1)^n} = \frac{1}{2} \int \frac{dv}{v^n} = \begin{cases} -\dfrac{v^{-n+1}}{2(n-1)} & \text{if } n \neq 1, \\[2mm] \dfrac{\ln v}{2} & \text{if } n = 1. \end{cases} \tag{10.28}$$

The first integral in the right-hand side of (10.27) is done in Exercise 10.19.

10.2 Integration by parts

In this section we use the *product rule* for two functions u and v,

$$\frac{d}{dx}(uv) = u \frac{dv}{dx} + v \frac{du}{dx} \tag{10.29}$$

to obtain another very useful technique of integration, called integration by parts.
We rewrite (10.29) as follows

$$u(x)v'(x) = \frac{d}{dx}(u(x)v(x)) - u(x)'v(x) \tag{10.30}$$

to obtain

$$\int u(x)v'(x) \, dx = \int \frac{d}{dx}(u(x)v(x)) \, dx - \int u'(x)v(x) \, dx$$

$$= u(x)v(x) - \int u'(x)v(x) \, dx \, . \tag{10.31}$$

This formula is useful if the integrand is of the form

$$u(x)v'(x) \tag{10.32}$$

where the integral $\int u'(x)v(x)\,dx$ (appearing in the right-hand side of (10.31)) is easier than the original integral $\int u(x)v'(x)\,dx$.

Formula (10.31) is called integration by parts. It is sometimes written in the form

$$\int u\,dv = uv - \int v\,du\,, \qquad (10.33)$$

which may be easier to remember.

Example 10.18. To calculate $\int \ln x\,dx$, we try $u(x) := \ln x$, $v'(x) := 1$ to get

$$\int \ln x\,dx = x\ln x - \int \frac{x}{x}\,dx$$

$$= x\ln x - x\,. \qquad (10.34)$$

Example 10.19. To calculate $\int x\cos x\,dx$, we try $u(x) := x$, $v'(x) := \cos x$:

$$\int x\cos x\,dx = x\sin x - \int \sin x\,dx \quad \text{since } u'(x) = 1,\ v(x) = \sin x$$

$$= x\sin x + \cos x\,.$$

Note that integration by parts with $u(x) := \cos x$, $v'(x) := x$ results in an integral harder than the original

$$\int x\cos x\,dx = \frac{x^2}{2}\cos x + \int \frac{x^2}{2}\sin x\,dx \quad \text{since } u'(x) = -\sin x,\ v(x) = \frac{x^2}{2}\,.$$

Example 10.20. Consider the integral

$$\int x^2 e^{-x}\,dx\,.$$

We take

$$u(x) := x^2,\quad v'(x) := e^{-x}\,, \qquad (10.35)$$

to obtain

$$\int x^2 e^{-x}\,dx = -x^2 e^{-x} + 2\int x e^{-x}\,dx \quad \text{since } u'(x) = 2x,\ v(x) = -e^{-x}\,.$$

Now use integration by parts a second time for the integral $\int x\,e^{-x}\,dx$ in the right-hand side,

$$\int x e^{-x}\,dx = -x e^{-x} + \int e^{-x}\,dx, \quad \begin{array}{l} \text{with } u(x) := x,\ v'(x) := e^{-x},\\ u'(x) = 1,\ v(x) = -e^{-x}, \end{array}$$

$$= -x e^{-x} - e^{-x}\,,$$

and finally

$$\int x^2 e^{-x}\,dx = e^{-x}(-x^2 - 2x - 2)\,.$$

Note that trying $u(x) := e^{-x}$, $v'(x) := x^2$ results in a harder integral

$$\int x^2 e^{-x} \, dx = \frac{x^3}{3} e^{-x} - \int \frac{x^3}{3} e^{-x} \, dx \ .$$

These examples suggest that whenever an integrand contains a power x^n, or a polynomial in x, then this term should be selected as $u(x)$, for then $u'(x)$ is a lower power or polynomial, hopefully giving a simpler integral in the right-hand side of (10.31), $\int u(x)v'(x) \, dx = u(x)v(x) - \int u'(x)v(x) \, dx$. If the integrand contains exponentials, sines, or cosines, then these terms could serve as either $u(x)$ or $v'(x)$.

Example 10.21. We integrate (10.8),

$$\int \sin^n x \cos x \, dx, \quad \text{(see Example 10.2)} ,$$

by parts, using:

$$u := \sin^n x, \qquad\qquad dv := \cos x \, dx ,$$
$$du = n \sin^{n-1} x \cos x \, dx, \quad v = \sin x \ .$$

Therefore,

$$\int \sin^n x \cos x \, dx = \sin^{n+1} x - n \int \sin^n x \cos x \, dx$$

$$\therefore \quad (n+1) \int \sin^n x \cos x \, dx = \sin^{n+1} x$$

$$\therefore \quad \int \sin^n x \cos x \, dx = \frac{\sin^{n+1} x}{n+1}$$

MACSYMA-Session 10.3. The integration by parts (10.33), $\int u \, dv = uv - \int v \, du$, is implemented by the following MACSYMA function:

```
c1. int_parts(u,v,x) := u*integrate(v,x) -
                        integrate(integrate(v,x)*diff(u,x),x)$
```

whose user must first decide which function is u and which is dv.

For example, consider the integral $\int \sin^3 x \cos x \, dx$.

Selecting $u = \sin^3 x$, $dv = \cos x \, dx$ we get:

```
c2. int_parts(sin(x)^3,cos(x),x)
```

d2. $\dfrac{\sin^4 x}{4}$

In this example the choice $u = \cos x$, $dv = \sin^3 x \, dx$ is unnatural. Indeed:

```
c3. int_parts(cos(x),sin(x)^3,x)
```

d3. $\qquad -\dfrac{\cos^4 x}{12} + \cos x\left(\dfrac{\cos^3 x}{3} - \cos x\right) + \dfrac{\cos^2 x}{2}$

which does not look at all like d2. However, substituting $1 - \sin^2 x$ for $\cos^2 x$ in d3 we get:

c4. `ratsubst(1-sin(x)^2,cos(x)^2,%)`

d4. $\qquad \dfrac{\sin^4 x - 1}{4}$

which differs from d2 only by the constant of integration $-1/4$. Therefore, d2 and d4 have the same derivative, $\sin^3 x \cos x$.

Consider now an integral of the form

$$\int x^n f(x)\, dx \tag{10.36}$$

for some function f. Integration by parts gives:

$$\int x^n f(x)\, dx = x^n F(x) - n \int x^{n-1} F(x)\, dx$$

where $F(x) := \int f(x)\, dx$. We can integrate (10.36) recursively as follows:

```
c5. intx_n(f,x,n) := if n=0 then integrate(f,x)
                     else block([int:integrate(f,x)],
                          x^n*int-n*intx_n(int,x,n-1))$
```

where `int` stores the value `integrate(f,x)`, which is used twice. For example, the integral $\int e^x x^2\, dx$:

c6. `intx_n(exp(x),x,2)`

d6. $\qquad x^2 e^x - 2(x\, e^x - e^x)$

We show how we may prove Taylor's expansion theorem by integration by parts.

Theorem 10.22 (Taylor's expansion theorem). Let f have $n + 1$ continuous derivatives f', f'', $\ldots$, $f^{(n)}$, $f^{(n+1)}$ on $[a, b]$. Then

$$f(b) = f(a) + f'(a)(b - a) + \frac{1}{2!} f''(a)(b - a)^2 + \cdots + \frac{1}{n!} f^{(n)}(a)(b - a)^n + R_n \tag{10.37}$$

with the remainder

$$R_n = \frac{(b - a)^{n+1}}{(n + 1)!} f^{(n+1)}(\xi) \tag{10.38}$$

for some $a \leq \xi \leq b$.

Proof. Consider first the integral

$$\int\limits_a^b g(x)(b - x)^m\, dx\ ,$$

where m is integer, and g a given function. Integration by parts gives

$$\int_a^b g(x)(b-x)^m \, dx = -g(x)\frac{(b-x)^{m+1}}{m+1}\Big|_a^b + \int_a^b g'(x)\frac{(b-x)^{m+1}}{m+1} \, dx$$

$$= g(a)\frac{(b-a)^{m+1}}{m+1} + \frac{1}{m+1}\int_a^b g'(x)(b-x)^{m+1} \, dx \, .$$

(10.39)

Applying (10.39) with $g(x) = f'(x)$, $m = 0$, we get

$$\int_a^b f'(x) \, dx = f'(a)(b-a) + \int_a^b f''(x)(b-x) \, dx \, ,$$

where the left-hand side is simply

$$\int_a^b f'(x) \, dx = f(b) - f(a) \, .$$

Applying (10.39) again to the integral in the right-hand side (taking $g(x) = f''(x)$, $m = 1$),

$$f(b) = f(a) + f'(a)(b-a) + \frac{1}{2}f''(a)(b-a)^2 + \frac{1}{2}\int_a^b f'''(x)(b-x)^2 \, dx$$

and by repeated applications of (10.39) we get

$$f(b) = f(a) + f'(a)(b-a) + \frac{1}{2!}f''(a)(b-a)^2 + \dots$$

$$+ \frac{1}{n!}f^{(n)}(a)(b-a)^n + \frac{1}{n!}\int_a^b f^{(n+1)}(x)(b-x)^n \, dx \, ,$$

which is a Taylor expansion with the remainder in integral form

$$R_n = \frac{1}{n!}\int_a^b f^{(n+1)}(x)(b-x)^n \, dx \, .$$

(10.40)

By the mean value theorem 10.16, there is a $a \le \xi \le b$ such that

$$R_n = \frac{1}{n!}f^{(n+1)}(\xi)\int_a^b (b-x)^n \, dx = \frac{1}{(n+1)!}f^{(n+1)}(\xi)(b-a)^{n+1} \, ,$$

completing the proof. $\qquad\qquad\qquad\qquad\qquad\qquad\qquad\qquad\qquad\qquad\square$

Remark 10.23. If the function f has n continuous derivatives $f', f'', \ldots, f^{(n)}$ in $[a, b]$, and the $(n+1)$th derivative $f^{(n+1)}$ is piecewise continuous on $[a, b]$, then the Taylor expansion (10.37) still holds, with remainder R_n given in integral form (10.40). If $f^{(n+1)}$ is not continuous on $[a, b]$, we cannot use the mean value theorem 10.16 to bring R_n to the form (10.38).

Exercises

10.17 Use integration by parts to compute the integrals

(a) $\int x^3 e^x \, dx$

(b) $\int x^4 e^{-x} \, dx$

(c) $\int x^2 \cos x \, dx$

(d) $\int x \sin x \, dx$

(e) $\int x \ln x \, dx$

(f) $\int x^n \ln x \, dx$

(g) $\int a\sin x \, dx$

(h) $\int a\tan x \, dx$

(i) $\int \ln x \, dx$

(j) $\int (\ln x)^2 \, dx$

(k) $\int (\ln x)^3 \, dx$

(l) $\int x \ln x \, dx$

(m) $\int x^2 \ln x \, dx$

(n) $\int x^n \ln x \, dx$

(o) $\int x(\ln x)^2 \, dx$

(p) $\int x^2 a\tan x \, dx$

(q) $\int (x^3 + 3x) a\tan x \, dx$

(r) $\int \ln(1 + x^2) \, dx$

10.18 Find the antiderivatives of the following functions

(a) $f(x) := 4x^2 + 3x - 2$

(b) $f(x) := 2/x + x^2$

(c) $f(x) := \sec^2 x + \cos x$

(d) $f(t) := \sqrt{t} + 1/\sqrt{t}$

(e) $f(x) := (x^2)^{1/3}$

(f) $f(x) := (x^2 + x^3 - 6x)/x^2$

(g) $f(x) := (4x^2 - 1)^5$

(h) $f(t) := 2t/\sqrt{t^2 + 3}$

(i) $f(t) := 4t\sqrt{t^2 + 7}$

(j) $f(x) := (2x - 1)^3/\sqrt{x}$

(k) $f(x) := \sin x^4 \cos x$

(l) $f(x) := \cos^5 x \sin x$

(m) $f(x) := \tan x \sec^2 x$

(n) $f(x) := \tan^n x \sec^2 x$

(o) $f(x) := \sec^4 x \tan x$

(p) $f(x) := \sec^n x \tan x$

(q) $f(x) := \cot^n x \csc^2 x$

(r) $f(x) := \csc^n x \cot x$

10.19 Verify, for integers $n > 1$,

$$\int \frac{dx}{(a^2 + x^2)^{n+1}} = \frac{1}{2na^2} \frac{x}{(a^2 + x^2)^n} + \frac{2n - 1}{2na^2} \int \frac{dx}{(a^2 + x^2)^n} . \qquad (10.41)$$

Hint: Integrate

$$\int \frac{dx}{(a^2 + x^2)^n}$$

by parts, with

$$u := \frac{1}{(a^2 + x^2)^n}, \quad dv := dx .$$

Equation (10.41) is an example of a *reduction formula*[3], expressing an integral of a power in terms of an integral with a lower power. What makes (10.41) useful is that it can be applied recursively, reducing the power from $n+1$ to n, then to $n-1$, etc. The process ends with $n = 1$ and an immediate integral,

$$\int \frac{1}{a^2 + x^2}\, dx = \frac{1}{a} \operatorname{atan} \frac{x}{a}.$$

10.20 *(Reduction formulas)* Prove the following reduction formulas, where n is a positive integer. Then try the formulas for small n.

(a) $\displaystyle \int (a^2 - x^2)^n\, dx = \frac{x(a^2 - x^2)^n}{2n+1} + \frac{2na^2}{2n+1} \int (a^2 - x^2)^{n-1}\, dx$

(b) $\displaystyle \int x^n\, e^{\alpha x}\, dx = \frac{1}{\alpha} x^n\, e^{\alpha x} - \frac{n}{\alpha} \int x^{n-1}\, e^{\alpha x}\, dx$

(c) $\displaystyle \int x^n \sin x\, dx = -x^n \cos x + n \int x^{n-1} \cos x\, dx$

(d) $\displaystyle \int x^n \cos x\, dx = x^n \sin x - n \int x^{n-1} \sin x\, dx$

(e) $\displaystyle \int \sin^n x\, dx = -\frac{\sin^{n-1} x \cos x}{n} + \frac{n-1}{n} \int \sin^{n-2} x\, dx \quad (n \geq 2)$

Hint: In part e prove first

$$\int \sin^n x\, dx = -\sin^{n-1} x \cos x + (n-1) \int \sin^{n-2} x \cos^2 x\, dx.$$

(f) $\displaystyle \int \frac{dx}{(x^2 + ax + b)^n} = \frac{2x + a}{(n-1)(4b - a^2)(x^2 + ax + b)^{n-1}} +$

$$+ \frac{2(2n-3)}{(n-1)(4b - a^2)} \int \frac{dx}{(x^2 + ax + b)^{n-1}}\, dx,$$

where $n \geq 2$ and $4b - a^2 \neq 0$

(g) $\displaystyle \int \frac{x\, dx}{(x^2 + ax + b)^n} = -\frac{1}{2(n-1)(x^2 + ax + b)^{n-1}} -$

$$- \frac{a}{2} \int \frac{dx}{(x^2 + ax + b)^n},$$

where $n \geq 2$ and $4b - a^2 \neq 0$

10.21 Guess a reduction formula for $\int \cos^n x\, dx$, and then check your guess.

10.22 In this exercise it is necessary to integrate by parts twice. Check your results with MACSYMA.

(a) $\displaystyle \int e^x \sin x\, dx$ (b) $\displaystyle \int e^x \cos x\, dx$

Hint: In part a integrate by parts using $u := \sin x$ and $dv := e^x\, dx$. Then integrate (the resulting integral) by parts using now $u := \cos x$ and $dv := e^x\, dx$.

---

3 Also called recurrence formula.

10.23 Evaluate the integrals

(a) $\displaystyle\int \cos(\ln x)\,dx$ (b) $\displaystyle\int e^x \cosh x\,dx$ (c) $\displaystyle\int x^2 \cosh x\,dx$

(d) $\displaystyle\int e^x \sinh x\,dx$ (e) $\displaystyle\int x^2 \sinh x\,dx$ (f) $\displaystyle\int x\,e^x \sin x\,dx$

10.24 Use the reduction formulas from Exercise 10.20 to verify

(a) $\displaystyle\int_0^{\pi/2} \sin^{2m} x\,dx = \frac{\pi}{2}\frac{1}{2}\frac{3}{4}\frac{5}{6}\cdots\frac{2m-1}{2m}$

(b) $\displaystyle\int_0^{\pi/2} \sin^{2m+1} x\,dx = \frac{2}{3}\frac{4}{5}\frac{6}{7}\cdots\frac{2m}{2m+1}$

(c) $\displaystyle\int_0^{\pi/2} \cos^{2m} x\,dx = \frac{\pi}{2}\frac{1}{2}\frac{3}{4}\frac{5}{6}\cdots\frac{2m-1}{2m}$

(d) $\displaystyle\int_0^{\pi/2} \cos^{2m+1} x\,dx = \frac{2}{3}\frac{4}{5}\frac{6}{7}\cdots\frac{2m}{2m+1}$

Can you justify the last result from the similar formula for the sine function?

10.25 Let

$$W_k := \int_0^{\pi/2} \sin^k x\,dx \quad (k = 0, 1, \dots). \tag{10.42}$$

In Exercise 10.24 we proved the *Wallis formulas*[4], giving W_k for k even and odd,

$$W_{2n} = \frac{(2n-1)(2n-3)\cdots 3\cdot 1}{2n(2n-2)\cdots 4\cdot 2}\frac{\pi}{2} \quad (n = 1, 2, \dots),$$

$$W_{2n+1} = \frac{2n(2n-2)\cdots 4\cdot 2}{(2n+1)(2n-1)\cdots 3\cdot 1} \quad (n = 0, 1, \dots). \tag{10.43}$$

Prove $\lim_{n\to\infty} W_{2n}/W_{2n+1} = 1$, and consequently

$$\frac{\pi}{2} = \prod_{n=1}^{\infty} \frac{4n^2}{4n^2-1} = \frac{2}{1}\cdot\frac{2}{3}\cdot\frac{4}{3}\cdot\frac{4}{5}\cdots\frac{6}{5}\cdot\frac{6}{7}\cdots,$$

the Wallis product formula for $\pi/2$.

Hint: For $0 < x < \pi/2$,

$$0 < \sin^{2n+1} x \le \sin^{2n} x \le \sin^{2n-1} x,$$

and therefore

$$0 < \int_0^{\pi/2} \sin^{2n+1} x\,dx \le \int_0^{\pi/2} \sin^{2n} x\,dx \le \int_0^{\pi/2} \sin^{2n-1} x\,dx.$$

Now divide by $\int_0^{\pi/2} \sin^{2n+1} x\,dx$, and use (10.43) to get $1 \le W_{2n}/W_{2n+1} \le 1 + 1/2n$.

---

4 John Wallis (1616–1703).

10.3 Rational functions

Consider an integral where the integrand is a rational function, i.e., the ratio of two polynomials,

$$f(x) = \frac{P(x)}{Q(x)} . \tag{10.44}$$

If the *degree*[5] of P, deg P, is greater than, or equal to, the degree of Q, then we divide P by Q to get a polynomial P_1 and a remainder P_2, which is a polynomial with smaller degree than Q,

$$P/Q = P_1 + P_2/Q, \quad \deg P_2 < \deg Q$$

$$\therefore \quad \int \frac{P(x)}{Q(x)} \, dx = \int P_1(x) \, dx + \int \frac{P_2(x)}{Q(x)} \, dx .$$

The integral of the polynomial P_1 presents no difficulty (see Example 9.18), so it remains to integrate the rational function $P_2(x)/Q(x)$ with deg $P_2 <$ deg Q. We therefore assume, without loss of generality, that the rational function (10.44) has

$$\deg P < \deg Q . \tag{10.45}$$

Let the polynomial $Q(x)$ have r *real roots*

$$\gamma_1, \gamma_2, \ldots, \gamma_r, \quad \text{with } \textit{multiplicities } m_1, m_2, \ldots, m_r \tag{10.46}$$

and s *pairs* of *complex roots*[6]

$$\alpha_1 \pm i\beta_1, \alpha_2 \pm i\beta_2, \ldots, \alpha_s \pm i\beta_s, \quad \text{with multiplicities } n_1, n_2, \ldots, n_s . \tag{10.47}$$

The *factorization* of Q is therefore

$$Q(x) = \prod_{k=1}^{r} (x - \gamma_k)^{m_k} \prod_{l=1}^{s} \left((x - \alpha_l)^2 + \beta_l^2\right)^{n_l} , \tag{10.48}$$

and the rational function $P(x)/Q(x)$ has the *partial fractions expansion*

$$\frac{P(x)}{Q(x)} = \frac{C_{11}}{x - \gamma_1} + \frac{C_{12}}{(x - \gamma_1)^2} + \cdots + \frac{C_{1m_1}}{(x - \gamma_1)^{m_1}} +$$

$$+ \frac{C_{21}}{x - \gamma_2} + \frac{C_{22}}{(x - \gamma_2)^2} + \cdots + \frac{C_{2m_2}}{(x - \gamma_2)^{m_2}} +$$

$$+ \cdots +$$

$$+ \frac{C_{r1}}{x - \gamma_r} + \frac{C_{r2}}{(x - \gamma_r)^2} + \cdots + \frac{C_{rm_r}}{(x - \gamma_r)^{m_r}} +$$

$$+ \frac{A_{11} + B_{11}x}{(x - \alpha_1)^2 + \beta_1^2} + \frac{A_{12} + B_{12}x}{\left((x - \alpha_1)^2 + \beta_1^2\right)^2} + \cdots + \frac{A_{1n_1} + B_{1n_1}x}{\left((x - \alpha_1)^2 + \beta_1^2\right)^{n_1}} +$$

---

5 Recall that the degree of a polynomial is its highest power.
6 If $z = a + ib$ is a complex root of the polynomial Q, so is the conjugate $\bar{z} = a - ib$.

$$+ \frac{A_{21} + B_{21}x}{(x - \alpha_2)^2 + \beta_2^2} + \frac{A_{22} + B_{22}x}{\left((x - \alpha_2)^2 + \beta_2^2\right)^2} + \cdots + \frac{A_{2n_2} + B_{2n_2}x}{\left((x - \alpha_2)^2 + \beta_2^2\right)^{n_1}} +$$

$$+ \cdots +$$

$$+ \frac{A_{s1} + B_{s1}x}{(x - \alpha_s)^2 + \beta_s^2} + \frac{A_{s2} + B_{s2}x}{\left((x - \alpha_s)^2 + \beta_s^2\right)^2} + \cdots + \frac{A_{sn_s} + B_{sn_s}x}{\left((x - \alpha_s)^2 + \beta_s^2\right)^{n_s}} .$$

$$(10.49)$$

A reason we may want to write a nice function like (10.44) in this unappealing form is that its integral can now be written as the sum of integrals of the form

$$\int \frac{C}{(x - \gamma)^k} \quad \text{and} \quad \int \frac{A + Bx}{\left((x - \alpha)^2 + \beta^2\right)^k} ,$$

which are simple; see Exercises 10.15, 10.16, and 10.19. This is the *partial fractions expansion method* for computing the integral of a rational function.

Example 10.24. Consider the integral of a rational function

$$\int \frac{2x^2 + 2x - 4}{x^4 - 2x^3 + 2x^2 - 2x + 1} \, dx .$$

Here $P(x) = 2x^2 + 2x - 4 = 2(x-1)(x+2)$ and $Q(x) = x^4 - 2x^3 + 2x^2 - 2x + 1 = (x^2 + 1)(x - 1)^2$. The integrand is therefore

$$\frac{P(x)}{Q(x)} = \frac{2x^2 + 2x - 4}{x^4 - 2x^3 + 2x^2 - 2x + 1} = \frac{2(x + 2)}{(x^2 + 1)(x - 1)} ,$$

$$\text{cancelling the common factor } (x - 1) ,$$

$$= \frac{A + Bx}{x^2 + 1} + \frac{C}{x - 1} ,$$

$$\text{the partial fractions expansion} .$$

The unknown constants A, B, and C are determined from

$$\frac{A + Bx}{x^2 + 1} + \frac{C}{x - 1} = \frac{2(x + 2)}{(x^2 + 1)(x - 1)}$$

by bringing the left-hand side to a common denominator,

$$\frac{Ax - A + Bx^2 - Bx + Cx^2 + C}{(x^2 + 1)(x - 1)} = \frac{2x + 4}{(x^2 + 1)(x - 1)} ,$$

and comparing the coefficients of the powers of x in the numerators. This gives a system of three linear equations

$$
\begin{array}{rcll}
B + C & = & 0 & \text{(coefficient of } x^2) , \\
A - B & = & 2 & \text{(coefficient of } x) , \\
-A + C & = & 4 & \text{(free coefficient) ,}
\end{array}
$$

with solution $A = -1$, $B = -3$, $C = 3$. The partial fractions expansion of the integrand is therefore

$$\frac{2x^2 + 2x - 4}{x^4 - 2x^3 + 2x^2 - 2x + 1} = \frac{-1 - 3x}{x^2 + 1} + \frac{3}{x - 1}.$$

$$\therefore \int \frac{2x^2 + 2x - 4}{x^4 - 2x^3 + 2x^2 - 2x + 1} \, dx = -\operatorname{atan} x - \frac{3}{2} \ln(x^2 + 1) + 3 \ln(x - 1).$$

MACSYMA-Session 10.4. Consider the integral of the rational function

$$\int \frac{x^4 - 4x^3 + 7x^2 - 16x + 12}{x^7 + 2x^6 - 2x^5 - 4x^4 - 7x^3 - 14x^2 - 4x - 8} \, dx$$

We enter the integrand:

c1. `(x^4-4*x^3+7*x^2-16*x+12)/`
$\qquad\qquad\qquad\qquad$`(x^7+2*x^6-2*x^5-4*x^4-7*x^3-14*x^2-4*x-8)`

d1. $\dfrac{x^4 - 4x^3 + 7x^2 - 16x + 12}{x^7 + 2x^6 - 2x^5 - 4x^4 - 7x^3 - 14x^2 - 4x - 8}$

Factoring the integrand:

c2. `factor(%)`

we get

d2. $\dfrac{(x - 3)(x - 1)(x^2 + 4)}{(x - 2)(x + 2)^2 (x^2 + 1)^2}$

and performing a partial fractions expansion:

c3. `partfrac(%,x)`

we get a convenient integrand:

d3. $\dfrac{25x - 18}{25(x^2 + 1)} + \dfrac{6x}{5(x^2 + 1)^2} - \dfrac{49}{50(x + 2)} - \dfrac{6}{5(x + 2)^2} - \dfrac{1}{50(x - 2)}$

and finally the integral:

c4. `integrate(%,x)`

d4. $\dfrac{25 \log(x^2 + 1)/2 - 18 \operatorname{atan} x}{25} - \dfrac{49 \log(x + 2)}{50} - \dfrac{\log(x - 2)}{50} - \dfrac{3}{5(x^2 + 1)} +$

$\dfrac{6}{5(x + 2)}$

If the integrand is a rational function, MACSYMA will automatically use partial fractions expansion. For example,

c5. `integrate(d2,x)`

d5. $\dfrac{\log(x^2 + 1)}{2} - \dfrac{49 \log(x + 2)}{50} - \dfrac{18 \operatorname{atan} x}{25} - \dfrac{\log(x - 2)}{50} + \dfrac{6x^2 - 3x}{5x^3 + 10x^2 + 5x + 10}$

which is equivalent to d4. Indeed:

c6. `ratsimp(%-d4)`

d6. $\quad 0$

MACSYMA-Session 10.5. This session gives the integration rules for the integrals of all the terms appearing in the partial fractions expansion (10.49). These rules show that the integral of any rational function can be given in terms of elementary functions.

First, the integrals $\int dx/(x-\gamma)$ and $\int dx/(x-\gamma)^j$, for $j \neq 1$.

c1. `integrate(1/(x-gamma),x)`

d1. $\log(x-\gamma)$

c2. `integrate(1/(x-gamma)^j,x)`

   Is j-1 zero or nonzero?

   nonzero

d2. $\dfrac{1}{(1-j)(x-\gamma)^{j-1}}$

Next, the integrals $\int dx/(x^2+ax+b)$ and $\int x/(x^2+ax+b)\,dx$

c3. `integrate(1/(x^2+a*x+b),x)`

   Is $4b-a^2$ positive or negative?

   positive

d3. $\dfrac{2\,\mathrm{atan}\left(\frac{2x+a}{\sqrt{4b-a^2}}\right)}{\sqrt{4b-a^2}}$

c4. `integrate(x/(x^2+a*x+b),x)`

   Is $4b-a^2$ positive or negative?

   negative

d4. $\dfrac{\log(x^2+ax+b)}{2} - \dfrac{a\log\big((2x-\sqrt{a^2-4b}+a)/(2x+\sqrt{a^2-4b}+a)\big)}{2\sqrt{a^2-4b}}$

Repeat c3 and c4 for the opposite signs of $4b-a^2$.

   Consider now the integrals

$$\int \frac{dx}{(x^2+ax+b)^n} \quad \text{and} \quad \int \frac{x\,dx}{(x^2+ax+b)^n}$$

for $n \geq 2$. These integrals have reduction formulas (see Exercise 10.20 f and g), which allow recursive computation.

   The integral of Exercise 10.20 f

$$\int \frac{dx}{(x^2+ax+b)^n} = \frac{2x+a}{(n-1)(4b-a^2)(x^2+ax+b)^{n-1}} + \frac{2(2n-3)}{(n-1)(4b-a^2)} \int \frac{dx}{(x^2+ax+b)^{n-1}}\,dx$$

is implemented by the recursive function

c5. `recurs_int(x,a,b,n) := block(`
```
        if n=0 then x
            else if n=1 then integrate(1/(x^2+a*x+b),x)
                else (2*x+a)/((4*b-a^2)*(n-1)*(x^2+a*x+b)^(n-1)) +
                2*(2*n-3)/((4*b-a^2)*(n-1))*recurs_int(x,a,b,n-1))$
```

Examples:

c6. `recurs_int(x,1,1,2)`

d6. $\dfrac{4\,\mathrm{atan}\big((2x+1)/\sqrt{3}\big)}{3\sqrt{3}} + \dfrac{2x+1}{3(x^2+x+1)}$

Check: Differentiate, and factor, to get back the original integrand

c7. `factor(diff(%,x))`

d7. $\dfrac{1}{(x^2+x+1)^2}$

c8. `recurs_int(x,1,2,2)`

d8. $\dfrac{4\operatorname{atan}\big((2x+1)/\sqrt{7}\big)}{7\sqrt{7}} + \dfrac{2x+1}{7(x^2+x+2)}$

Check:

c9. `factor(diff(%,x))`

d9. $\dfrac{1}{(x^2+x+2)^2}$

c10. `recurs_int(x,1,3,2)`

d10. $\dfrac{4\operatorname{atan}\big((2x+1)/\sqrt{11}\big)}{11\sqrt{11}} + \dfrac{2x+1}{11(x^2+x+3)}$

Check:

c11. `factor(diff(%,x))`

d11. $\dfrac{1}{(x^2+x+3)^2}$

The last three examples illustrate that c5 works for b^2-4a positive, zero, or negative.
  Finally, the integral of Exercise 10.20 g,

$$\int \frac{x\,dx}{(x^2+ax+b)^n} = -\frac{1}{2(n-1)(x^2+ax+b)^{n-1}} - \frac{a}{2}\int \frac{dx}{(x^2+ax+b)^n}$$

is computed recursively, using the function `recurs_int(x,a,b,n)` of c5, as follows

c12. `recurs_int_x(x,a,b,n) := block(`
    `-a/2*recurs_int(x,a,b,n) - 1/(2*(n-1)*(x^2+a*x+b)^(n-1)))$`

Example:

c13. `recurs_int_x(x,1,1,2)`

d13. $-\dfrac{\dfrac{4\operatorname{atan}((2x+1)/\sqrt{3})}{3\sqrt{3}} + \dfrac{2x+1}{3(x^2+x+1)}}{2} - \dfrac{1}{2(x^2+x+1)}$

Check:

c14. `factor(diff(%,x))`

d14. $\dfrac{x}{(x^2+x+1)^2}$

Exercises

10.26 Let the polynomial Q be factored by (10.48). Show that the degree of Q is

$$\deg Q = m_1 + m_2 + \cdots + m_r + 2(n_1 + n_2 + \cdots + n_s).$$

10.27 Prove (10.21) by the partial fractions expansion of

$$\frac{1}{a^2-x^2} = \frac{1}{(a+x)(a-x)}.$$

10.28 Evaluate the following integrals and check your results with MACSYMA.

(a) $\displaystyle\int \frac{x^3\,dx}{x+1}$ (b) $\displaystyle\int \frac{x^3\,dx}{x^2+1}$ (c) $\displaystyle\int \frac{x+1}{x-1}\,dx$

(d) $\displaystyle\int \frac{x^2\,dx}{x^2-1}$ (e) $\displaystyle\int \frac{x\,dx}{x^2+x-6}$ (f) $\displaystyle\int \frac{dx}{x^2-x}$

(g) $\int \dfrac{x^4 \, dx}{x^2 + 2}$ (h) $\int \dfrac{x^2 \, dx}{ax + b}$ (i) $\int \dfrac{x^2 + x}{x^3 + 2x^2 + x} \, dx$

(j) $\int \dfrac{x^2 \, dx}{x^2 + 2x + 1}$ (k) $\int \dfrac{x \, dx}{(x - 1)(x^2 + 1)}$ (l) $\int \dfrac{4x^2 + 2x}{(x + 1)(x^2} \, dx$

(m) $\int \dfrac{x^4 + 2x^2 + 2}{(x^2 - 1)x} \, dx$ (n) $\int \dfrac{x^2 + 2}{(x - 1)^3} \, dx$ (o) $\int \dfrac{x^2 \, dx}{x^4 - 1}$

(p) $\int \dfrac{x^3 \, dx}{ax + b}$ (q) $\int \dfrac{x^2 \, dx}{ax^2 + bx + c}$ (r) $\int \dfrac{x \, dx}{(x + a)(x^2 + b^2)}$

Here a, b, and c are constants, $a \neq 0$ and b is nonzero in part r.

10.29 Set up a system of linear equations for determining the coefficients in the partial fractions expansion given, MACSYMA-Session 10.4, line d3.

10.30 Calculate:

(a) $\int \dfrac{x^3 - 7x + 6}{x^5 - 5x^4 + 8x^3 - 8x^2 + 7x - 3} \, dx$

(b) $\int \dfrac{x^2 - 2x - 8}{x^5 + 3x^4 + 4x^3 + 8x^2 - 16} \, dx$

10.4 Improper integrals: infinite intervals

In Chap. 8 we used Riemann sums to compute definite integrals (8.5),

$$\int_a^b f(x) \, dx \, ,$$

and to prove their properties. We assumed the integrand f bounded and the integration interval $[a, b]$ finite. If either of these assumptions is violated, the integral is called *improper*. We can sometimes get around the difficulty by taking appropriate limits. In this section we handle integration on infinite intervals.

If the integration interval is not finite, say, the interval is $[a, \infty)$, and if the limit

$$\lim_{T \to \infty} \int_a^T f(x) \, dx$$

exists, then it is defined as the value of the integral,

$$\int_a^\infty f(x) \, dx := \lim_{T \to \infty} \int_a^T f(x) \, dx \, .$$

We similarly define

$$\int_{-\infty}^b f(x) \, dx := \lim_{S \to -\infty} \int_S^b f(x) \, dx$$

and

$$\int_{-\infty}^{\infty} f(x)\,dx := \lim_{S \to -\infty} \int_{S}^{a} f(x)\,dx + \lim_{T \to \infty} \int_{a}^{T} f(x)\,dx , \qquad (10.50)$$

where a is arbitrary.

If the limit exists and is finite, the corresponding integral is said to *converge*. If the limit is infinite, the integral *diverges*. The examples below show the various possibilities.

Example 10.25 (Divergence). The integral

$$\int_{1}^{\infty} \frac{dx}{x}$$

is, by definition,

$$\int_{1}^{\infty} \frac{dx}{x} = \lim_{T \to \infty} \int_{1}^{T} \frac{dx}{x}$$

$$= \lim_{T \to \infty} \ln x \big|_{1}^{T} = \lim_{T \to \infty} (\ln T - \ln 1) = \infty .$$

Example 10.26 (Convergence). For any $\varepsilon > 0$,

$$\int_{1}^{\infty} \frac{dx}{x^{1+\varepsilon}} = \lim_{T \to \infty} \int_{1}^{T} \frac{dx}{x^{1+\varepsilon}} = \lim_{T \to \infty} -\frac{x^{-\varepsilon}}{\varepsilon} \bigg|_{1}^{T} = \lim_{T \to \infty} \frac{1}{\varepsilon}(1 - T^{-\varepsilon}) = \frac{1}{\varepsilon} .$$

In particular,

$$\int_{1}^{\infty} \frac{dx}{x^2} = 1 .$$

Example 10.27 (Limit does not exist). The integral

$$\int_{0}^{\infty} \sin x \, dx$$

does not exist. Indeed,

$$\int_{0}^{T} \sin x \, dx = 1 - \cos T$$

oscillates forever between 0 and 2. Therefore the limit

$$\lim_{T \to \infty} \int_0^T \sin x \, dx$$

does not exist.

The following theorem may sometimes settle the issue of convergence.

Proposition 10.28 (Comparison theorem for improper integrals). Let a and b be finite, $a < b$, and let f be piecewise continuous for $x \geq a$.

a. If

$$\int_b^\infty g(x) \, dx \text{ converges}, \quad \text{and} \quad |f(x)| \leq g(x) \text{ for } x \geq b,$$

then $\int_a^\infty f(x) \, dx$ also converges.

b. If

$$\int_b^\infty g(x) \, dx \text{ diverges}, \quad \text{and} \quad f(x) \geq g(x) \geq 0 \text{ for } x \geq b,$$

then $\int_a^\infty f(x) \, dx$ also diverges.

Proof. Consider

$$\int_a^\infty f(x) \, dx = \int_a^b f(x) \, dx + \int_b^\infty f(x) \, dx \, .$$

The first integral in the right-hand side exists by Corollary 8.37. Convergence or divergence therefore applies only to the second integral and is determined by comparison with g. $\qquad\square$

Example 10.29. Let f be piecewise continuous for $x \geq 0$, and let

$$|f(x)| \leq K \, e^{\alpha x}, \quad \text{for all } x \geq b, \tag{10.51}$$

for constants α, b and $K > 0$. Then

$$\int_0^\infty e^{-sx} f(x) \, dx$$

converges for all $s > \alpha$. This is a straightforward application of Proposition 10.28 a since

$$\int_0^\infty e^{-sx}\, e^{\alpha x}\, dx = \int_0^\infty e^{(\alpha-s)x}\, dx = \lim_{T\to\infty} \left. \frac{e^{(\alpha-s)x}}{\alpha - s} \right|_0^T$$

$$= \frac{1}{s - \alpha} \quad \text{since } s > a \, .$$

The next example shows how factorials can be expressed as improper integrals.

Example 10.30. The integral

$$\int_0^\infty e^{-x}\, x^{n-1}\, dx \tag{10.52}$$

is called the *Gamma function* of n, denoted $\Gamma(n)$. It can be shown to converge for $n = 1, 2, \ldots$, by Example 10.29 with $s = 1$. We will compute (10.52) (and thereby establish convergence directly). For $n = 1$, the integral becomes

$$\int_0^\infty e^{-x}\, dx = \lim_{T\to\infty} \int_0^T e^{-x}\, dx$$

$$= \lim_{T\to\infty} -e^{-x}\Big|_0^T = \lim_{T\to\infty} (1 - e^{-T}) = 1 \, . \tag{10.53}$$

For $n > 1$, we integrate by parts,[7] with $u = x^{n-1}$, $v' = e^{-x}$,

$$\int e^{-x}\, x^{n-1}\, dx = -e^{-x}\, x^{n-1} + (n-1) \int e^{-x}\, x^{n-2}\, dx \, .$$

$$\therefore \int_0^\infty e^{-x}\, x^{n-1}\, dx = \lim_{T\to\infty} -e^{-x}\, x^{n-1}\Big|_0^T + \lim_{T\to\infty} (n-1) \int_0^T e^{-x}\, x^{n-2}\, dx$$

$$= (n-1) \int_0^\infty e^{-x}\, x^{n-2}\, dx \, , \tag{10.54}$$

where we used the limits

$$\lim_{x\to 0} x^n\, e^{-x} = 0, \quad \lim_{x\to\infty} x^n\, e^{-x} = 0 \, ,$$

---

7 Or use Exercise 10.20 b, with $n := n - 1$ and $\alpha := 1$.

for all $n > 0$. From (10.54) and (10.53) it follows that, for any integer $n \geq 1$,

$$\Gamma(n) := \int_0^\infty e^{-x} x^{n-1} \, dx = (n-1) \cdot (n-2) \cdots 2 \cdot 1 = (n-1)! \qquad (10.55)$$

Note that the convention $0! = 1$ agrees with (10.53).

Example 10.31 (Stirling formula). We approximate the integral

$$\int_1^n \ln x \, dx$$

using the trapezoid method (8.62) with $n - 1$ subintervals,

$$\text{trapezoid}(\ln x, x, 1, n, n-1) = \tfrac{1}{2} \ln 1 + \sum_{k=2}^{n-1} \ln k + \tfrac{1}{2} \ln n$$

$$= \ln n! - \tfrac{1}{2} \ln n \ .$$

The integral is given, by Example 10.18, as

$$\int_1^n \ln x \, dx = n \ln n - n + 1 \ .$$

Comparing these results, we get

$$\ln n! = \left(n + \tfrac{1}{2}\right) \ln n - \ln e^n + \ln e + E_{n-1} \ , \qquad (10.56)$$

from which we can approximate $n!$. A careful estimation of the error E_{n-1} actually gives[8]

$$n! \approx \sqrt{2\pi} \, n^{(n+1/2)} \, e^{-n} \qquad (10.57)$$

known as the Stirling formula[9]. This is an asymptotic approximation, meaning that

$$\lim_{n \to \infty} \frac{n!}{\sqrt{2\pi} \, n^{(n+1/2)} \, e^{-n}} = 1 \ . \qquad (10.58)$$

For large n, say, $n \geq 10$, the Stirling formula is a reasonable approximation of $n!$ and requires fewer computations. The accuracy and the relative computational advantage of the Stirling formula actually improve with n, getting better and better as $n \to \infty$. This makes the Stirling formula very useful for approximating $n!$ when n is large.

---

8 For the details, omitted here, see, e.g., R. Courant, *Differential and Integral Calculus*, vol. I, Interscience, New York, 1988, pp. 361–364.

9 Named after James Stirling (1692–1770), this formula was discovered by Abraham De Moivre (1667–1754).

MACSYMA-Session 10.6. The right-hand side of (10.57) is computed by MACSYMA as follows:

c1. `stirl(n):=sqrt(2*%pi)*n^(n+1/2)*exp(-n)$`

MACSYMA also has a built-in function `stirling(n!,n)`, which does the same. For example,

c2. `10!`

d2.   3628800

is approximated by

c3. `stirl(10)`

d3. $\sqrt{2}\sqrt{\pi}\, e^{(21 \log 10)/2} - 10$

c4. `%, numer`

d4.   3598680.0

a pretty fair approximation for d2.

The number 100! is huge. We approximate it by:

c5. `stirl(100), bfloat`

d5.   9.32484762526934324 78b157

a good approximation, as verified by dividing d5 by 100!,

c6. `%/100!`

d6.   0.9991670165679

which is close to 1.

To illustrate (10.57) for $n = 10, 20, \dots, 100$, compute:

c7. `makelist(stirl(10*n)/(10*n)!,n,1,10), bfloat`

giving values which approach 1.

For an integration interval from $-\infty$ to ∞, we also consider the special limit

$$\lim_{T \to \infty} \int_{-T}^{T} f(x)\,dx \,.$$

If the limit exists, its value is called the *Cauchy principal value* of the integral

$$\int_{-\infty}^{\infty} f(x)\,dx \tag{10.59}$$

If (10.59) converges, according to Definition (10.50), then its value equals the Cauchy principal value. However, the Cauchy principal value may exist even though the integral (10.59) diverges. An example is

$$\lim_{T \to \infty} \int_{-T}^{T} \sin x\,dx = 0, \quad \text{by Exercise 8.21} \,,$$

but

$$\int_{-\infty}^{\infty} \sin x\,dx$$

does not converge, see Example 10.27. Thus the Cauchy principal value is necessary, but not sufficient, for the convergence of the integral (10.59). It may be easier to check than the limit in (10.50).

MACSYMA-Session 10.7. Improper integrals with infinite intervals $\int_a^\infty f(x)\,dx$ and $\int_{-\infty}^b f(x)\,dx$ are computed as in Session 8.1, using inf for ∞ and minf for $-\infty$. For example, the Gamma function $\Gamma(11)$, see (10.52),

c1. integrate(exp(-x)*x^10,x,0,inf)
d1.   3628800

which is 10!, illustrating (10.55). The integration techniques of Sects. 10.1–10.3 can be used for improper integrals. For example, consider the integral

$$\int\limits_0^\infty \frac{1}{\sqrt{e^{2x}+1}}\,dx\ . \tag{10.60}$$

c2. integrate(1/sqrt(exp(2*x)+1),x,0,inf)
d2.   asinh 1

We compute this integral by the change of variable $y := \sqrt{\exp(2x)+1}$. To compute dx/dy, we have to solve for x as a function of y.

c3. solve(y=sqrt(exp(2*x)+1),x)
    Is y positive, negative or zero?
    positive

d3. $\left[x = \dfrac{\log(y^2-1)+2\log(-1)}{2}, x = \dfrac{\log(y^2-1)}{2} \right]$

we select the positive solution, i.e., the second solution in d3,

c4. x=part(%,2,2)

d4. $x = \dfrac{\log(y^2-1)}{2}$

The new integrand is therefore

c5. 1/y*diff(log(y^2-1)/2,y)

d5. $\dfrac{1}{y^2-1}$

The new integration limits corresponding to $x = 0$ and $x = \infty$ can be found by computing

c6. limit(sqrt(exp(2*x)+1),x,0,plus)
d6. $\sqrt{2}$

and

c7. limit(sqrt(exp(2*x)+1),x,inf,minus)
d7. ∞

Thus the integral (10.60) is converted to

$$\int\limits_{\sqrt{2}}^\infty \frac{dy}{y^2-1}$$

c8. integrate(1/(y^2-1),y,sqrt(2),inf)

d8. $\dfrac{\log(\sqrt{2}+1) - \log(\sqrt{2}-1)}{2}$

the same as d2 (verify!).
Another way to compute (10.60) is to take the indefinite integral
c9. `integrate(1/sqrt(exp(2*x)+1),x)`
d9. $-\operatorname{asinh} e^{-x}$
and subtract its limit as $x \to \infty$ from the limit at $x = 0$,
c10. `limit(%,x,inf)-limit(%,x,0)`
d10. $\operatorname{asinh} 1$

Exercises

10.31 Use MACSYMA to verify $\int_0^\infty e^{-x^2}\, dx = \sqrt{\pi}/2$.
10.32 (Gamma function) In Example 10.30 we showed that, for integer $n \geq 1$,

$$\int_0^\infty e^{-x} x^{n-1}\, dx = (n-1)!$$

The left-hand side is usually defined for positive reals t (not just integers n), is called
the Gamma function, and denoted by

$$\Gamma(t) := \int_0^\infty e^{-x} x^{t-1}\, dx \qquad (10.61)$$

As in Example 10.30, we have

$$\Gamma(1) = 1,$$
$$\Gamma(t) = t\Gamma(t-1), \quad \text{for all } t > 0,$$
$$\text{and in particular,} \quad \Gamma(n+1) = n!, \quad \text{for all integers } n \geq 0.$$

For what values of t does the integral (10.61) converge?
10.33 Prove $\lim_{n\to\infty} \sqrt[n]{n!}/n = 1/e$.
       Hint: Divide (10.56) by n.
10.34 Check the Stirling formula for $n = 10, 20, 30$.
10.35 The binomial coefficient

$$\binom{n}{m} := \frac{n!}{m!\,(n-m)!} = \frac{n \cdot (n-1) \cdots (n-m+1)}{1 \cdot 2 \cdots m}$$

can be computed, using MACSYMA, by the function `binomial(n,m)`. For example $\binom{50}{20}$ is computed by inputting the expression `combination(50,20)` to
get 47129212243960. The MACSYMA function `permutation(n,m)` computes $n!/$
$(n-m)!$
a. Use the Stirling formula to approximate $\binom{n}{m}$ for large n, m.
b. Check your approximation for $n = 50$, $m = 20$.
10.36 For each of the integrals below, determine if it converges (in which case compute
it), diverges, or does not exist.

(a) $\displaystyle\int_0^\infty e^{-\sqrt{x}}\,dx$ (b) $\displaystyle\int_0^\infty e^{-x}\sin x\,dx$ (c) $\displaystyle\int_1^\infty e^{-x}\log x\,dx$

(d) $\displaystyle\int_{-\infty}^0 e^x\,dx$ (e) $\displaystyle\int_{-\infty}^\infty x\,e^{-x^2}\,dx$

(f) $\displaystyle\int_{-\infty}^0 x^n\,e^x\,dx$, integer $n>0$

10.37 Determine whether the improper integrals given below converge. In the instance where they do, evaluate the integral.

(a) $\displaystyle\int_0^\infty \frac{x\,dx}{x^2+1}$ (b) $\displaystyle\int_0^\infty x^3\,e^x\,dx$ (c) $\displaystyle\int_0^\infty \frac{dx}{x^{3/2}}$

(d) $\displaystyle\int_1^\infty \frac{dx}{x^{1+\alpha}}$, where $\alpha>0$ (e) $\displaystyle\int_1^\infty \frac{dx}{x^{1-\alpha}}$, where $\alpha>0$

(f) $\displaystyle\int_1^\infty \frac{dx}{(x+1)^{3/2}}$ (g) $\displaystyle\int_0^\infty \frac{dx}{\sqrt{x}}$ (h) $\displaystyle\int_e^\infty \frac{dx}{x\ln x}$

(i) $\displaystyle\int_e^\infty \frac{dx}{x\ln^2 x}$ (j) $\displaystyle\int_{-\infty}^\infty \frac{x\,dx}{x^2+1}$ (k) $\displaystyle\int_0^\infty \frac{x^2\,dx}{x^3+6}$

(l) $\displaystyle\int_{\pi/2}^\infty \cos^2 x\,dx$ (m) $\displaystyle\int_0^\infty \frac{\varepsilon+\sin^2 x}{x}\,dx$, where $\varepsilon>0$

10.38 Calculate the Cauchy principal value for the integral $\int_{-\infty}^\infty (1+x)/(1+x^2)\,dx$. Does the integral converge?

10.5 Improper integrals: unbounded integrands

Suppose the integrand f in (8.5),

$$\int_a^b f(x)\,dx\;,$$

becomes unbounded at one of the endpoints, say,

$$\lim_{x\downarrow a} f(x) = \infty\;,$$

where $\downarrow a$ denotes approaching a from the right. Consider then the limit

$$\lim_{S\downarrow a}\int_S^b f(x)\,dx\;,$$

which if it exists is defined to be the value of the integral (8.5),

$$\int_a^b f(x)\,dx := \lim_{S\downarrow a}\int_S^b f(x)\,dx\;. \tag{10.62}$$

Similarly, if the integrand misbehaves at the right endpoint,[10]

$$\lim_{x \uparrow b} f(x) = \infty ,$$

then define

$$\int_a^b f(x)\, dx := \lim_{T \uparrow b} \int_a^T f(x)\, dx ,$$

whenever the limit exists.

Example 10.32. **The integral**

$$\int_0^1 \frac{dx}{\sqrt{x}} = \lim_{\varepsilon \downarrow 0} \int_\varepsilon^1 \frac{dx}{\sqrt{x}}, \quad \text{by Definition (10.62)} ,$$

$$= \lim_{\varepsilon \downarrow 0} 2\sqrt{x}\Big|_\varepsilon^1 = 2 \lim_{\varepsilon \downarrow 0} (1 - \sqrt{\varepsilon}) = 2 . \tag{10.63}$$

MACSYMA-Session 10.8. We verify (10.63) with MACSYMA.
c1. `integrate(1/sqrt(x),x,0,1)`
d1.   2
We can also illustrate this improper integral as the limit $\lim_{\varepsilon \to 0} \int_\varepsilon^1 1/\sqrt{x}\, dx$ where the integrals are approximated by the Simpson method (8.73) with $n = 1000$.
Define the integrand:
c2. `f(x):=1/sqrt(x)$`
For $\varepsilon = 0.1$ the Simpson method gives
c3. `simpson(f,0.1,1,1000)`
d3.   1.36749
for $\varepsilon = 0.01$,
c4. `simpson(f,0.01,1,1000)`
d4.   1.79993
for $\varepsilon = 0.001$,
c5. `simpson(f,0.001,1,1000)`
d5.   1.93683
and finally for $\varepsilon = 0.0001$,
c6. `simpson(f,0.0001,1,1000)`
d6.   1.98916
which is close to d1. Here is a tabulation of these results

ε	0.1	0.01	0.001	0.0001
simpson$\left[1/\sqrt{x}, x, \varepsilon, 1, 1000\right]$	1.36749	1.79993	1.93688	1.98916

---

10 We denote by $\uparrow b$ approaching b from the left.

Exercises

10.39 For each of the integrals below, determine if it converges (in which case compute it), diverges, or does not exist.

(a) $\displaystyle\int_0^\infty \frac{dx}{\sqrt{x}}$ (b) $\displaystyle\int_0^\infty \frac{1}{x^2}\sin x\,dx$ (c) $\displaystyle\int_0^\infty \frac{1}{x}\sin x\,dx$

10.40 If the integrand f in (8.5), $\int_a^b f(x)\,dx$, becomes unbounded at some point c between a and b, we define

$$\int_a^b f(x)\,dx := \lim_{\varepsilon\downarrow 0}\left(\int_a^{c-\varepsilon} f(x)\,dx + \int_{c+\varepsilon}^b f(x)\,dx\right)$$

whenever the limit exists. For each of the integrals below, determine if it exists (and compute it) or not.

(a) $\displaystyle\int_0^3 \frac{dx}{\sqrt{|x-1|}}$ (b) $\displaystyle\int_0^3 \frac{dx}{x-1}$ (c) $\displaystyle\int_0^3 \frac{dx}{(x-1)^2}$

10.41 Determine whether the following integrals converge

(a) $\displaystyle\int_{-\infty}^\infty \frac{dx}{x^2-1}$ (b) $\displaystyle\int_1^\infty \frac{dx}{(x-1)^{3/2}}$ (c) $\displaystyle\int_1^{10} \frac{dx}{(x-1)^{2/3}}$

(d) $\displaystyle\int_0^\infty \frac{e^{-x}}{x^2-1}\,dx$ (e) $\displaystyle\int_2^4 \frac{x^2\,dx}{x^3-27}$ (f) $\displaystyle\int_3^6 \frac{x-3}{x^2-9}\,dx$

(g) $\displaystyle\int_0^\infty \frac{e^{-\sqrt{x}}}{\sqrt{x}}\,dx$

11 Applications of integrals

What do area, length, volume, work, and hydrostatic force have in common? All of these (and many other important concepts in science and engineering) can be modelled as Riemann sums (8.6)

$$\sum_{k=1}^{n} f(\xi_k)\, \Delta x_k \, ,$$

and computed as integrals (8.28),

$$\int_a^b f(x)\, dx := \lim_{\|\mathcal{P}\| \to 0} \sum_{k=1}^{n} f(\xi_k)\, \Delta x_k \, .$$

In this chapter integrals are applied to problems of computing areas (Sects. 11.1, 11.2, and 11.6), arc lengths (Sect. 11.3), volumes (Sects. 11.4 and 11.5), moments and centroids (Sects. 11.7 and 11.8), work (Sect. 11.9), and hydrostatic force (Sect. 11.10).

11.1 Area

Area is measured in units of length squared, such as square centimeter (cm^2), square meter (m^2), square inch (in^2), square foot (ft^2).

Consider an area A bounded

above by the graph of $y = g_2(x)$	
below by the graph of $y = g_1(x)$	
on the left by the vertical line $x := a$	(11.1)
on the right by the vertical line $x := b$	

see, e.g., Fig. 8.8. We computed this area, in Example 8.27, as

$$A = \int_a^b (g_2(x) - g_1(x))\, dx \, , \qquad (11.2)$$

which follows from approximating A by Riemann sums

$$\sum_{k=1}^{n} (g_2(\xi_k) - g_1(\xi_k))\, \Delta x_k \, ,$$

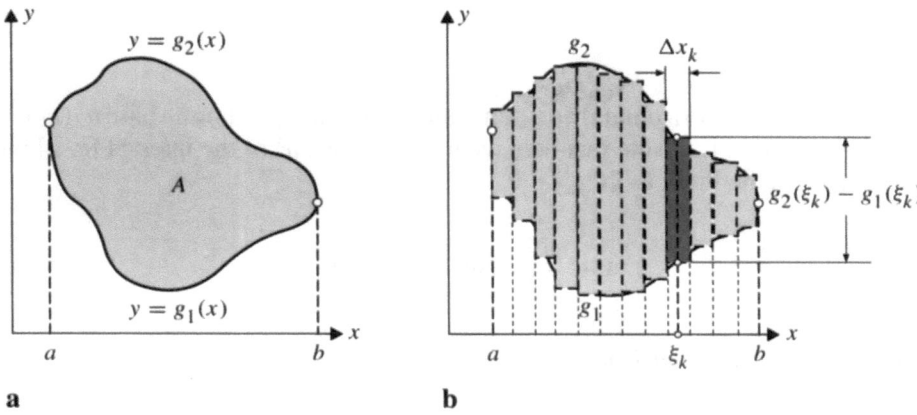

Fig. 11.1 a, b. Illustration of (11.2). **a** A plane region bounded by the graphs g_2 (above), g_1 (below), $x = a$, and $x = b$. **b** Approximating the area by vertical slices

summing areas of vertical strips, with base Δx_k and height $g_2(\xi_k) - g_1(\xi_k)$. This is called the method of *vertical slicing*.

Similarly, if the area A is bounded

on the left by the graph of $x = h_1(y)$
on the right by the graph of $x = h_2(y)$
above by the horizontal line $y := d$
below by the horizontal line $y := c$

then A can be approximated by Riemann sums

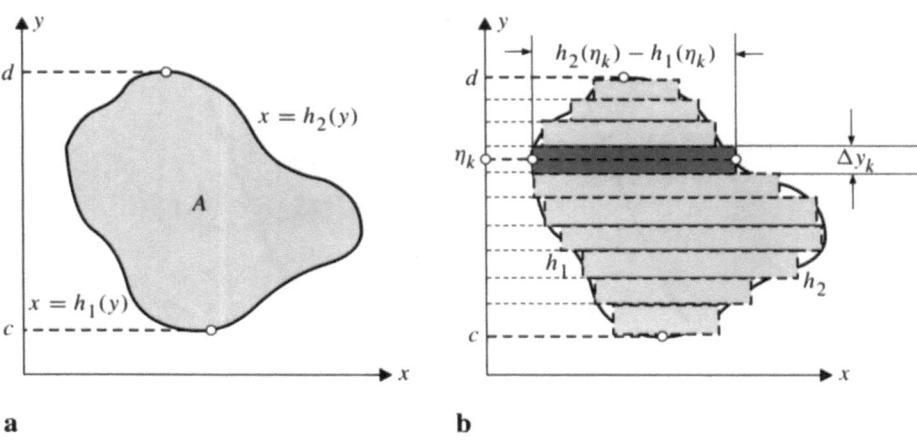

Fig. 11.2 a, b. Illustration of (11.3). **a** A plane region bounded by the graphs h_2 (right), h_1 (left), $y = c$, and $y = d$. **b** Approximating the area by horizontal slices

$$\sum_{k=1}^{n} (h_2(\eta_k) - h_1(\eta_k)) \, \Delta y_k$$

summing areas of horizontal strips, of height Δy_k (on the y-axis) and base $h_2(\eta_k) - h_1(\eta_k)$, see Fig. 11.2. Note that here we use partitions $\mathscr{P}$ of the interval $[c, d]$ on the y-axis. In the limit, as $\|\mathscr{P}\| \to 0$,

$$A = \int_{c}^{d} (h_2(y) - h_1(y)) \, dy \,, \qquad (11.3)$$

called the method of *horizontal slicing*.

Example 11.1. Calculate the area A bounded by the parabola $y = x^2$, the y-axis, and the horizontal line $y = 2$.

Vertical slicing. The area A sits on top of the x-interval $[0, \sqrt{2}]$. Then by (11.2),

$$A = \int_{0}^{\sqrt{2}} (2 - x^2) \, dx = 2x - \frac{x^3}{3} \Big|_{0}^{\sqrt{2}} = \frac{4\sqrt{2}}{3}, \quad \text{see Fig. 11.3a}.$$

Horizontal slicing. The area A is bounded on the left by the y-axis, on the right by the graph of $x = \sqrt{y}$, above by the horizontal line $y = 2$, and below by the horizontal line $y = 0$. Therefore, by (11.3),

$$A = \int_{0}^{2} \sqrt{y} \, dy = \frac{2}{3} y^{3/2} \Big|_{0}^{2} = \frac{4\sqrt{2}}{3}, \quad \text{see Fig. 11.3b}.$$

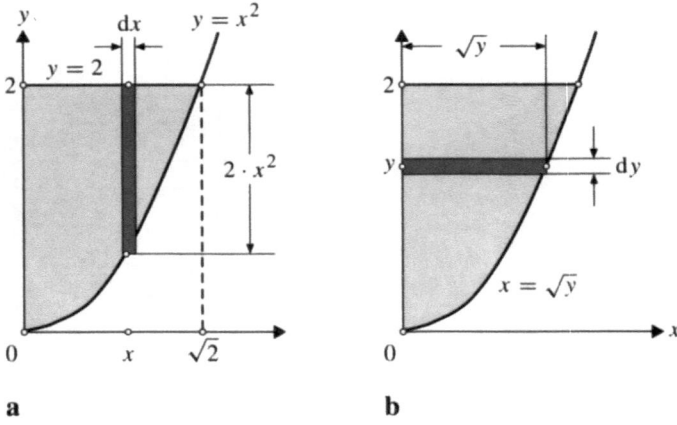

Fig. 11.3a, b. Illustration of Example 11.1 **a** Vertical slicing; **b** horizontal slicing

Example 11.2. Compute the area A bounded by the graphs $y = x^2$, $y = x^3$, the y-axis, and $x := 1$.

Vertical slicing:

$$A = \int_0^1 (x^2 - x^3)\,dx = \left(\frac{x^3}{3} - \frac{x^4}{4}\right)\bigg|_0^1 = \frac{1}{12}.$$

Horizontal slicing:

$$A = \int_0^1 (\sqrt[3]{y} - \sqrt{y})\,dy = \left(\frac{3}{4}y^{4/3} - \frac{2}{3}y^{3/2}\right)\bigg|_0^1 = \frac{1}{12}.$$

In the last two examples there was no reason to prefer any one method, since both methods (vertical slicing and horizontal slicing) gave similar integrals. There are however examples where one method may be preferred over the other.

Example 11.3. Compute the area A bounded by the graphs $y = \sin x$, $y = \cos x$, the y-axis, and $x := \pi/4$.

Using vertical slicing we get

$$A = \int_0^{\pi/4} (\cos x - \sin x)\,dx = \sin x + \cos x\big|_0^{\pi/4} = \frac{2}{\sqrt{2}} - 1 = \sqrt{2} - 1.$$

Horizontal slicing requires splitting A into two areas $A_1 + A_2$, where
- A_1 is bounded on the left by the y-axis, on the right by the graph of $x = a\sin y$ (see Fig. 11.4b), above by the horizontal line $y := 1/\sqrt{2}$;
- A_2 is bounded on the left by the y-axis, on the right by the graph of $x = a\cos y$, below by the horizontal line $y := 1/\sqrt{2}$.

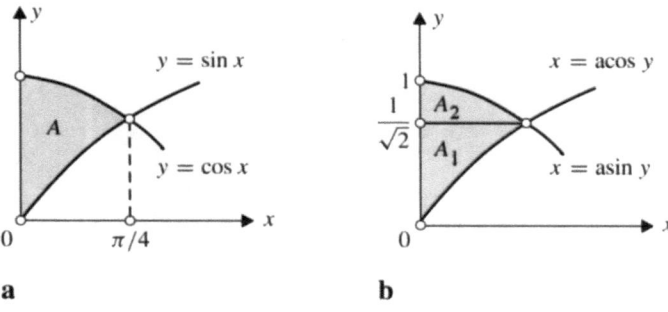

Fig. 11.4a, b. Illustration of Example 11.3. **a** Vertical slicing; **b** horizontal slicing

Therefore

$$A = \int\limits_{0}^{1/\sqrt{2}} a\sin y \, dy + \int\limits_{1/\sqrt{2}}^{1} a\cos y \, dy \,.$$

In this example vertical slicing is preferred to horizontal slicing.

Exercise

11.1 In each part are listed graphs which bound an area. Calculate each area twice, using vertical and horizontal slicing.

(a) The parabola $y^2 = x$, and the line $y = 2x - 1$

(b) The parabolas $y = 6x - x^2$ and $y = x^2 - 2x$

(c) The hyperbole $y = 1/x$, the y-axis, and the two lines $y = 1$, $y = 2$

11.2 Area by polar coordinates

In many applications, the region in question is naturally represented by *polar coordinates*

$$\begin{cases} r = \sqrt{x^2 + y^2} \\ \tan\theta = y/x \end{cases} \qquad \begin{array}{c} \text{instead of the} \\ \text{Cartesian coordinates} \end{array} \qquad \begin{cases} x = r\cos\theta \\ y = r\sin\theta \end{cases},$$

see Appendix C, C.5. The corresponding area is then an integral in the polar variable θ.

Proposition 11.4. The area A bounded by
the curve $r = f(\theta)$,
the two rays $\theta := \theta_1$ and $\theta := \theta_2$

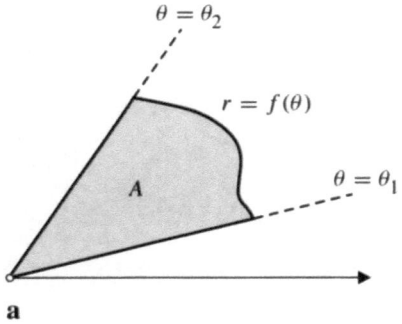
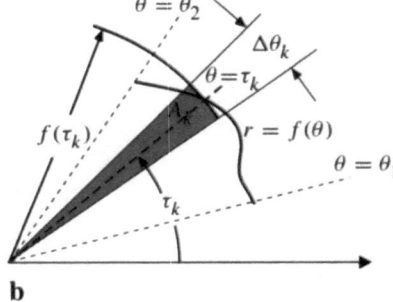

Fig. 11.5. a, b Illustration of Proposition 11.4. **a** A plane region bounded by $r = f(\theta)$, $\theta = \theta_1$, and $\theta = \theta_2$. **b** A circular sector with angle $\Delta\theta_k$ and radius $f(\tau_k)$

is

$$A = \tfrac{1}{2} \int_{\theta_1}^{\theta_2} r^2 \, d\theta = \tfrac{1}{2} \int_{\theta_1}^{\theta_2} f^2(\theta) \, d\theta \; . \tag{11.4}$$

Proof. The interval $\Theta := [\theta_1, \theta_2]$ represents a radial sector with angle $\theta_2 - \theta_1$. Let $\mathcal{P}$ be a partition of Θ into n subintervals Θ_k, the kth subinterval has the length $\Delta\theta_k$ and represents a radial sector with angle $\Delta\theta_k$. Let A_k be the part of the area A which lies in the kth sector, so that

$$A = \sum_{k=1}^{n} A_k \; .$$

If the angle $\Delta\theta_k$ is small, then A_k is approximately a circular sector with radius $r = f(\tau_k)$, for some angle $\tau_k \in \Theta_k$. The area of a circular sector with radius r and angle θ is $\tfrac{1}{2} r^2 \theta$, and therefore

$$A_k \approx \tfrac{1}{2} f^2(\tau_k) \, \Delta\theta_k \; ,$$

A is approximated by the Riemann sum

$$A \approx \tfrac{1}{2} \sum_{k=1}^{n} f^2(\tau_k) \, \Delta\theta_k \; ,$$

and in the limit, as $\|\mathcal{P}\| \to 0$,

$$A = \tfrac{1}{2} \int_{\theta_1}^{\theta_2} f^2(\theta) \, d\theta \; ,$$

completing the proof. $\qquad\square$

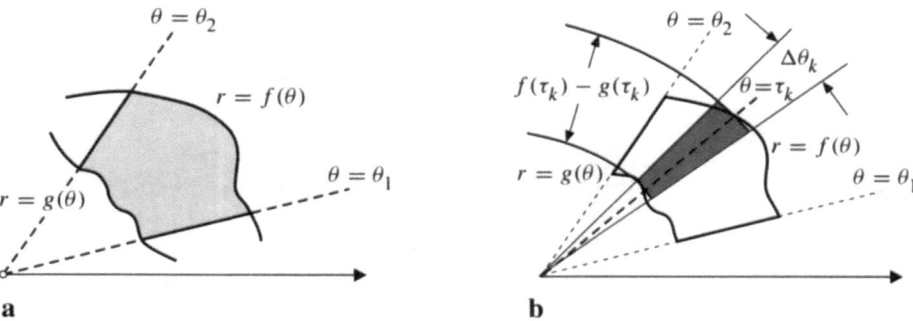

Fig. 11.6 a, b. Illustration of (11.5). **a** A plane region bounded by $r = f(\theta)$, $r = g(\theta)$, $\theta = \theta_1$, and $\theta = \theta_2$. **b** The shaded area is the difference of areas of two circular sectors with angle $\Delta\theta_k$

Similarly, the area A bounded on the outside by the curve $r = f(\theta)$, on the inside by the curve $r = g(\theta)$, and by the rays $\theta := \theta_1$ and $\theta := \theta_2$ (Fig. 11.6a) is

$$A = \tfrac{1}{2} \int_{\theta_1}^{\theta_2} \left(f^2(\theta) - g^2(\theta) \right) d\theta , \tag{11.5}$$

which is the area (11.4) minus the area of the "hole"

$$\tfrac{1}{2} \int_{\theta_1}^{\theta_2} g^2(\theta)\, d\theta .$$

This is illustrated in Fig. 11.6b, where the shaded area is approximately $\tfrac{1}{2}(f^2(\tau_k) - g^2(\tau_k))\,\Delta\theta_k$. The area A is therefore approximated by a Riemann sum

$$\tfrac{1}{2} \sum_{k=1}^{n} \left(f^2(\tau_k) - g^2(\tau_k) \right) \Delta\theta_k$$

whose limit is the integral (11.5).

Example 11.5 (Cardioid). The area A bounded by the cardioid[1]

$$r = 1 + \cos\theta \tag{11.6}$$

is, by (11.4),

$$A = \frac{1}{2} \int_0^{2\pi} (1 + \cos\theta)^2 \, d\theta = \frac{1}{2} \int_0^{2\pi} (1 + 2\cos\theta + \cos^2\theta)\, d\theta ,$$

$$= \frac{1}{2} \left(\theta + 2\sin\theta + \frac{\theta}{2} + \frac{\sin 2\theta}{4} \right) \Bigg|_0^{2\pi} , \quad \text{(see Exercise 10.6)} ,$$

$$= 3\pi/2 .$$

Example 11.6 (Spiral of Archimedes). The spiral of Archimedes is given, in polar coordinates, as

$$r = a\theta , \tag{11.7}$$

where $\theta \geq 0$ and $a > 0$ is a given parameter. The spiral makes its first revolution as θ goes from 0 to 2π. The second revolution is for θ between 2π and 4π. The nth revolution corresponds to $2(n-1)\pi \leq \theta \leq 2n\pi$.

---

1 So called because the graph of (11.6) resembles the shape of a heart.

The area inside the nth revolution of the spiral is

$$A(n) = \frac{1}{2} \int_{2(n-1)\pi}^{2n\pi} a^2\theta^2 \, d\theta = \frac{a^2}{6} \theta^3 \Big|_{2(n-1)\pi}^{2n\pi} = \frac{4a^2\pi^3}{3}(3n^2 - 3n + 1) \, .$$

The difference $A(n) - A(n-1) = 8a^2\pi^3(n-1)$ is the additional area swept by the spiral in the nth revolution. Note that $A(n) - A(n-1)$ is linear in n.

Exercises

11.2 (Which coordinate system to use?) Use $r = \sqrt{x^2 + y^2}$, $\tan\theta = y/x$, to represent in Cartesian coordinates the cardioid $r = 1 + \cos\theta$. Then attempt to calculate the area bounded by the cardioid. Which coordinate system, Cartesian or polar, is more natural for this problem?

11.3 Do Exercise 11.2 for the spiral $r = a\theta$.

11.4 Calculate the area A between the cardioid (11.6) and the circle $r = 1$.

11.3 Arc length

An arc is a (connected) part of a curve in the plane; for example, the arc in Fig. 11.7a is the part of the graph $y = f(x)$ lying between $x := a$ and $x := b$.

The length of this arc is denoted by arc_length(f, x, a, b). To compute it, we partition the interval $[a, b]$ into n subintervals $I_k := [x_{k-1}, x_k]$ and write

$$\text{arc\_length}(f, x, a, b) = \sum_{k=1}^{n} \Delta L_k \, , \tag{11.8}$$

where $\Delta L_k := \text{arc\_length}(f, x, x_{k-1}, x_k)$ is the length of the kth subarc, lying between x_{k-1} and x_k. Now denote

$$\Delta x_k := x_k - x_{k-1}, \quad \Delta y_k := f(x_k) - f(x_{k-1}) \, .$$

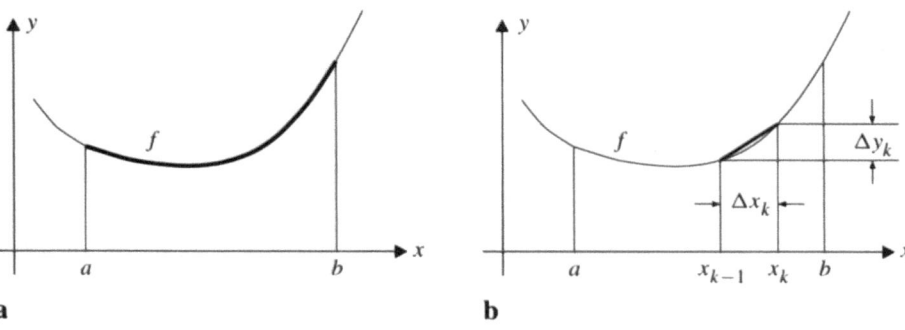

Fig. 11.7 a, b. Arc length. **a** The arc on the graph $y = f(x)$ between $x = a$ and $x = b$. **b** The arc between x_{k-1} and x_k is approximated by a line segment

If Δx_k is sufficiently small and the function f is not too wild, then the graph of f can be approximated in I_k by a line segment. This segment is the hypotenuse of a right triangle with sides Δx_k and Δy_k; see Fig. 11.7 b. Therefore, by Pythagoras's theorem,

$$(\Delta L_k)^2 \approx (\Delta x_k)^2 + (\Delta y_k)^2 ,$$

or

$$\Delta L_k \approx \sqrt{(\Delta x_k)^2 + (\Delta y_k)^2} = \sqrt{1 + \left(\frac{\Delta y_k}{\Delta x_k}\right)^2} \, \Delta x_k .$$

If the derivative $f'(x)$ is continuous, then by the mean value theorem

$$\Delta y_k / \Delta x_k = f'(\xi_k), \quad \text{for some } \xi_k \text{ in } [x_{k-1}, x_k] ,$$

and therefore

$$\Delta L_k \approx \sqrt{1 + (f'(\xi_k))^2} \, \Delta x_k . \tag{11.9}$$

Combining (11.8) and (11.9) we can thus approximate arc_length(f, x, a, b) as the Riemann sum

$$\text{arc\_length}(f, x, a, b) \approx \sum_{k=1}^{n} \sqrt{1 + (f'(\xi_k))^2} \, \Delta x_k .$$

If $\sqrt{1 + (f'(x))^2}$ is integrable on $[a, b]$, the last Riemann sum converges to the integral

$$\text{arc\_length}(f, x, a, b) = \int_a^b \sqrt{1 + f'(x)^2} \, dx . \tag{11.10}$$

Example 11.7 (Perimeter of circle). To compute the perimeter L of a circle of radius r, we center the circle at the origin, so its equation is

$$x^2 + y^2 = r^2 ,$$

an implicit relation, giving rise to two explicit functions (see Fig. 5.2),

$$y = f_1(x) := \sqrt{r^2 - x^2}, \quad \text{the upper semicircle} ,$$
$$y = f_2(x) := -\sqrt{r^2 - x^2}, \quad \text{the lower semicircle} .$$

For our purpose we may choose either one, so let

$$y = f_1(x) = \sqrt{r^2 - x^2}, \quad \text{with derivative } f_1{}'(x) = -\frac{x}{\sqrt{r^2 - x^2}} .$$

The perimeter L is therefore,

$$L = 2 \int_{-r}^{r} \sqrt{1 + (f_1')^2} \, dx = 2 \int_{-r}^{r} \sqrt{1 + \left(\frac{-x}{\sqrt{r^2 - x^2}}\right)^2} \, dx$$

$$= 2 \int_{-r}^{r} \sqrt{\frac{r^2}{r^2 - x^2}} \, dx = 2r \int_{-r}^{r} \frac{dx}{\sqrt{r^2 - x^2}}$$

$$= 2r \, \text{asin}\left(\frac{x}{r}\right)\Big|_{-r}^{r}, \quad \text{by Proposition 10.14(3)}$$

$$= 2\pi r \, .$$

Example 11.8. Compute the length of the graph $y = x^{3/2}$ from $x := 0$ to $x := 1$. Using (11.10),

$$\text{arc\_length}(x^{3/2}, x, 0, 1) = \int_{0}^{1} \sqrt{1 + \left(\frac{3}{2} x^{1/2}\right)^2} \, dx = \int_{0}^{1} \sqrt{1 + \frac{9}{4} x} \, dx$$

$$= \frac{4}{9} \int_{1}^{1+9/4} \sqrt{u} \, du = \frac{4}{9} \cdot \frac{2}{3} \cdot \left[\left(\frac{13}{4}\right)^{3/2} - 1\right].$$

This same function $f(x) = x^{3/2}$ is used in many calculus books in length examples because[2] it is one of the few f that give a decent integrand $\sqrt{1 + f'(x)^2}$. However, using MACSYMA we can easily compute lengths of less friendly graphs.

MACSYMA-Session 11.1. The MACSYMA function
```
c1. arc_length(f,x,a,b) :=
                integrate(radcan(sqrt(1+(diff(f,x))^2)), x, a, b)$
```
is a rewriting of Eq. (11.10), computing the length of the arc $y = f(x)$ between $x = a$ and $x = b$. For example, the length of a circle with a radius of 1 is twice the length of the upper semicircle:
```
c2. 2*arc_length(sqrt(1-x^2), x, -1, 1)
```
d2. 2π

The length of the graph $y = x^{3/2}$ from $x = 0$ to $x = 1$ is
```
c3. arc_length(x^(3/2),x,0,1)
```
d3. $\dfrac{26\sqrt{13}/27 - 16/27}{2}$
```
c4. factor(%)
```
d4. $\dfrac{13\sqrt{13} - 8}{27}$

in agreement with Example 11.8.
The length of the graph $y = x^{10}$ from $x = 0$ to $x = 1$ is

---

2 Also because authors copy from each other.

c5. `arc_length(x^10,x,0,1)`

d5. $\int_0^1 \sqrt{100\,x^{18} + 1}\,dx$

Since this is not given in closed form, we compute it numerically using the trapezoid method of Session 8.6. The numerical value of c5 is approximately:

c6. `trapezoid(sqrt(1+diff(x^10,x)^2), x,0,1,1000), numer`

d6.   1.75515

Exercises

11.5   Check that (11.10) gives the correct answer if $y = f(x)$ is a straight line.

11.6   Calculate the lengths of the following arcs from $x := 0$ to $x := 1$.

(a) $f(x) = \begin{cases} \sin 1/x & \text{, if } x \neq 0 \\ 0 & \text{, if } x = 0 \end{cases}$
(b) $f(x) = \begin{cases} x \sin 1/x & \text{, if } x \neq 0 \\ 0 & \text{, if } x = 0 \end{cases}$

11.7   MACSYMA-Session 11.1 suggests that arc_length$(x^n, x, 0, 1) \to 2$ as $n \to \infty$. Can you explain? Is there a power n for which arc_length$(x^n, x, 0, 1) = 2$?

11.8   Use MACSYMA to find the length of the arcs

(a) $y = \sin x$ from 0 to $\pi/2$ (b) $y = a\sin x$ from 0 to 1

(c) $y = x^2$ from 2 to 3   (d) $y = \sqrt{x}$ from 4 to 9

11.9   Prove: If f is increasing (or decreasing) on $[a, b]$ and if $f(a) = c$, $f(b) = d$, then the length of the graph $y = f(x)$ between the points $x := a$ and $x := b$ is the same as the length of $x = f^{-1}(y)$ between $y := c$ and $y := d$,

$$\text{arc\_length}(f, x, a, b) = \text{arc\_length}(f^{-1}, y, c, d) ,$$

where f^{-1} is the inverse function.

11.10 Calculate the perimeter of the ellipse $x^2/a^2 + y^2/b^2 = 1$.

11.11 The equation of the catenary is given by $y = \cosh x$. Find the length of its arc from $x = 0$ to $x = \xi$.

11.12 Calculate the arclength of the the the astroid $x^{2/3} + y^{2/3} = 1$.

11.4 Volume

Volume is measured in units of length cubed, such as cubic centimeter (cm^3), cubic meter (m^3), cubic inch (in^3), cubic foot (ft^3).

Consider a body B in $\mathbf{R}^3$, of volume V, and a plane $P := \{ (x, y, z) \mid x := \xi \}$ which is perpendicular to the x-axis. The plane P intersects the body B in an area $A(\xi)$, depending on the value $x := \xi$ where B is being sliced (see Fig. 11.8). If P and B do not intersect, then $A(\xi) = 0$.

We denote by volume(B, x, x_{k-1}, x_k) the volume of that part of B lying between the planes $x := x_{k-1}$ and $x := x_k$ (see Fig. 11.9b). If B lies between the two planes $x := a$ and $x := b$, then its volume

$$V = \text{volume}(B, x, a, b) .$$

By partitioning $[a, b]$ into n subintervals $I_k = [x_{k-1}, x_k]$, we can write

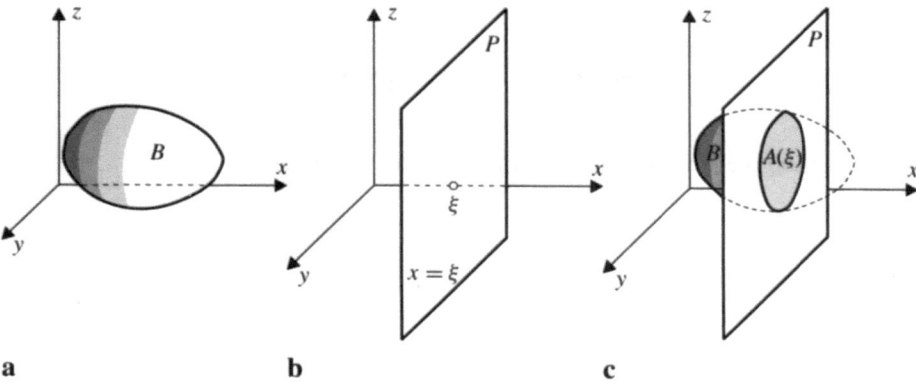

Fig. 11.8 a–c. Intersection of a body and a plane perpendicular to the x-axis. **a** Body B; **b** plane P with equation $x = \xi$. **c** The intersection of B and P has area $A(\xi)$

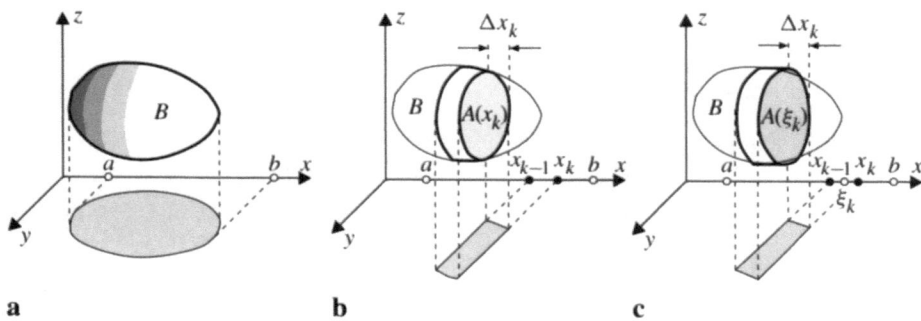

Fig. 11.9 a–c. Illustration of (11.12). **a** B lies between the planes $x = a$ and $x = b$. **b** volume(B, x, x_{k-1}, x_k). **c** A slice with basis area $A(\xi_k)$ and height Δx_k

volume(B, x, a, b) as the sum

$$\text{volume}(B, x, a, b) = \sum_{k=1}^{n} \text{volume}(B, x, x_{k-1}, x_k) \tag{11.11}$$

of volumes of its *slices*. If the length $\Delta x_k = x_k - x_{k-1}$ of I_k is small, the volume of the kth slice can be approximated by

$$\text{volume}(B, x, x_{k-1}, x_k) \approx A(\xi_k) \, \Delta x_k \tag{11.12}$$

for any ξ_k in I_k (see Fig. 11.9c). Combining (11.11) and (11.12) we approximate the volume V by the Riemann sum

$$\text{volume}(B, x, a, b) \approx \sum_{k=1}^{n} A(\xi_k) \, \Delta x_k$$

and in the limit

$$\text{volume}(B, x, a, b) = \int_a^b A(x)\, dx \ . \tag{11.13}$$

This is called the *slicing method* for computing the volume.

Example 11.9 (Volume of cone). In $\mathbf{R}^3$ consider a plane P, a region $\mathcal{R}$ in P, and a point Q outside P. By connecting Q with all points of $\mathcal{R}$, we obtain a *cone* whose *basis* is $\mathcal{R}$ and whose height is the perpendicular distance from Q to P. The volume V of the cone is, (8.1),

$$V = \frac{Ah}{3} \ ,$$

where A is the *area of the basis* and h is the *height*. We prove this by choosing Q as the origin, and P as the plane $x := h$, so the height of the cone is h; see Fig. 11.10a. We slice the cone perpendicular to the x-axis, where $A(\xi)$ is the area of the intersection of the cone with the plane $x := \xi$; see Fig. 11.10b. Therefore

$$A(0) = 0, \quad \text{and} \quad A(h) = A, \text{ the area of the basis} \ ,$$

but the following may require some thought:

$$A(\xi) = \frac{\xi^2}{h^2} A(h), \quad \text{for all } 0 \le \xi \le h \ . \tag{11.14}$$

The volume of the cone is therefore

$$V = \int_0^h A \frac{x^2}{h^2}\, dx = \frac{A}{h^2} \frac{x^3}{3}\Big|_0^h = \frac{Ah}{3} \ .$$

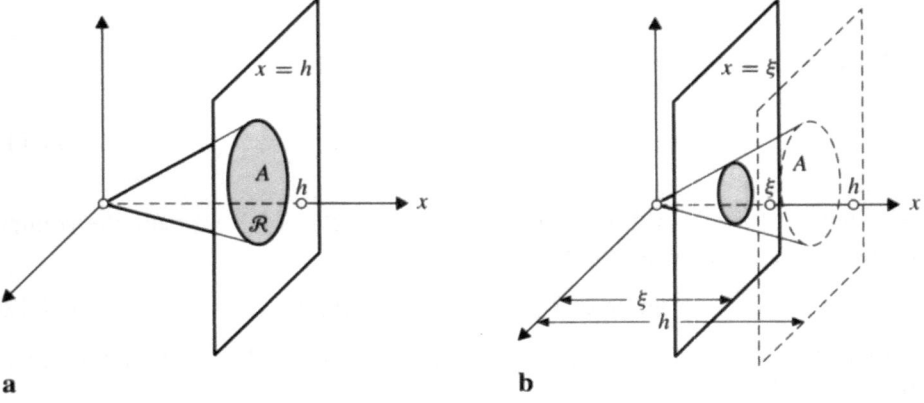

a   **b**

Fig. 11.10 a, b. Volume of a cone (Example 11.9). **a** Cone with vertex at the origin and basis in the plane $x = h$; **b** intersection of the cone with the plane $x = \xi$

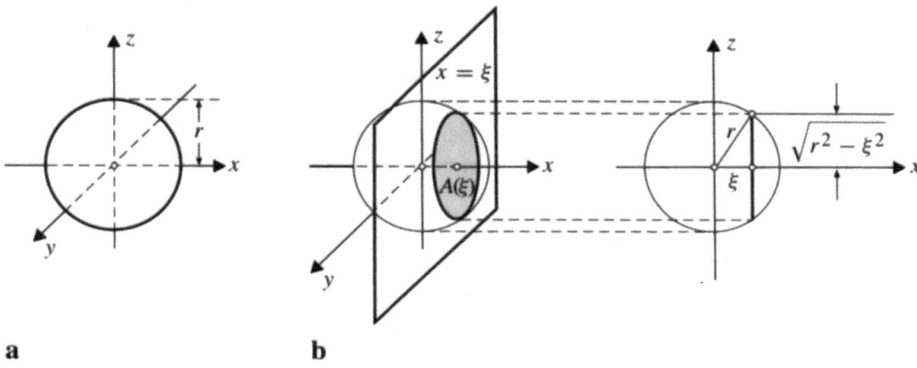

a   **b**

Fig. 11.11 a, b. Volume of a sphere (Example 11.10). **a** Sphere with radius r centered at the origin. **b** The intersection of the sphere with the plane $x = \xi$ is a disk with radius $\sqrt{r^2 - \xi^2}$

Example 11.10 (Volume of sphere). To compute the volume V of a sphere with radius r, we center it at the origin. The area of the intersection of the sphere with the plane $x := \xi$ is

$$A(\xi) = \pi(r^2 - \xi^2), \quad \text{for all } 0 \leq \xi \leq r, \tag{11.15}$$

see Fig. 11.11 b. Therefore,

$$V = \int_{-r}^{r} A(x)\,dx = \int_{-r}^{r} \pi(r^2 - x^2)\,dx = \pi \left(r^2 x - \frac{x^3}{3}\right)\Big|_{-r}^{r}$$

$$= \frac{4\pi r^3}{3}. \tag{11.16}$$

We could have simplified the arithmetic by computing the volume of the sphere as twice that of the hemisphere,

$$V = 2 \int_{0}^{r} A(x)\,dx.$$

Exercises

11.13 Prove (11.14).

11.14 Prove (11.15).

11.15 (How to slice?) Because of the spherical symmetry, we could have equally well computed V in Example 11.10 by slicing perpendicular to the y-axis or z-axis. For other bodies, there may be a preferred axis for slicing, resulting in simpler cross sections and easier integrals. Consider the volume V of a circular cylinder with radius r and height h given by $x^2 + y^2 = r^2$, for $0 \leq z \leq h$. Compute V by slicing perpendicular to

(a) the x-axis   (b) the z-axis

11.16 Let C_x be the cylinder with radius r symmetric w.r.t. the x-axis, i.e., $C_x = \{ (x, y, z) \mid y^2 + z^2 = r^2 \}$. Let C_y be the cylinder with radius r symmetric w.r.t. the y-axis. Compute the volume of the body B at the intersection of C_x and C_y.

11.5 Solids of revolution: volume

Consider a region $\mathcal{R}$ in the xy-plane, which is above (or below) the x-axis. Now rotate the region around the x-axis. By this we mean that every point $(\xi, \eta) \in \mathcal{R}$ goes into the circle

$$y^2 + z^2 = \eta^2, \quad x = \xi ,$$

in $\mathbf{R}^3$ (see Fig. 11.12 a). The body so obtained is called a *solid* (or body) *of revolution*. Its shape depends on the region $\mathcal{R}$ which is being rotated. Obviously, rotating $\mathcal{R}$ around another axis (e.g., the y-axis) results in a completely different solid of revolution.

In a normal day you come in contact with hundreds of solids of revolution, e.g., light bulbs, tires, tennis balls, and flying saucers[3].

We compute here the volume of a solid of revolution B (abbreviated volume of revolution) by the slicing method: Indeed each cross section of B is a *disk* (meaning the inside of a circle). To be specific, let the functions f be nonnegative on $[a, b]$, and let the region $\mathcal{R}$ be bounded, (8.2),

above by the graph of $y = f(x)$
below by the x-axis
on the left by the vertical line $x := a$
on the right by the vertical line $x := b$

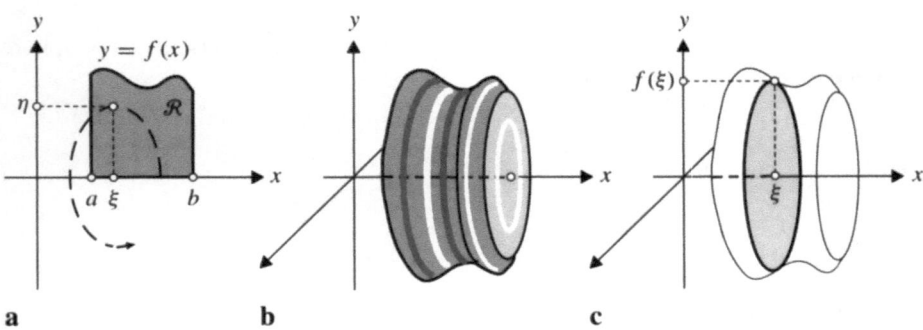

Fig. 11.12 a–c. Solid of revolution. **a** Region $\mathcal{R}$ is rotated around the x-axis; **b** resulting solid of revolution. **c** The intersection with $x = \xi$ is a disk with radius $f(\xi)$

---

3 Watch out! Here they come again.

Let B be the solid of revolution obtained by rotating $\mathcal{R}$ around the x-axis. If we slice B at $x := \xi$, the cross section is a disk with radius $f(\xi)$ and area

$$A(\xi) = \pi f(\xi)^2 ,$$

see Fig. 11.12c. Therefore, using (11.13),

$$\mathrm{volume}(B, x, a, b) = \pi \int_a^b f(x)^2 \, \mathrm{d}x , \qquad (11.17)$$

which is called the *disk method* for computing the volume of a solid of revolution.

Example 11.11 (Sphere revisited). The sphere

$$x^2 + y^2 + z^2 = r^2$$

is itself a solid of revolution, obtained by rotating around the x-axis the region (8.2) with $f(x) = \sqrt{r^2 - x^2}$, $a = -r$ and $b = r$. Therefore, the volume of the sphere can be alternatively computed as

$$V = \pi \int_{-r}^r (r^2 - x^2) \, \mathrm{d}x = \pi \left(r^2 x - \frac{x^3}{3} \right)\Big|_{-r}^r = \frac{4}{3} \pi r^3 ,$$

repeating (11.16).

Example 11.12 (Unbounded bodies). In the volume computations (11.13) or (11.17), the limits of integration a or b or both may be infinite, resulting in an improper integral (see Sect. 10.4). The body in question is unbounded, but it may still have a finite volume.

For example, consider the region $\mathcal{R}$ defined by (8.2) with $f(x) = 1/x$, $a = 1$ and $b = \infty$, illustrated in Fig. 11.13 a. Rotating $\mathcal{R}$ around the x-axis results in a body with volume

$$V = \pi \int_1^\infty \frac{1}{x^2} \, \mathrm{d}x = \pi \quad \text{(see Example 10.26)} .$$

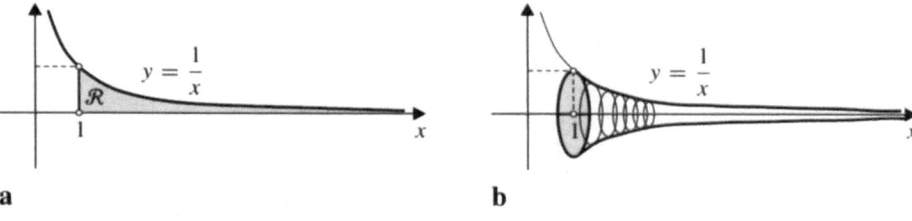

a   **b**

Fig. 11.13 a, b. Unbounded body (Example 11.12). **a** Region $\mathcal{R}$ is unbounded. **b** The solid of revolution has finite volume

Example 11.13 (Solids of revolution with holes). Consider the volume V of the body obtained by rotating, around the x-axis, the region $\mathcal{R}$ bounded above by the graph of $y = g_2(x)$, below by the graph of $y = g_1(x)$, on the left by the vertical line $x := a$, and on the right by the vertical line $x := b$, where $0 \leq g_1(x) \leq g_2(x)$ in $[a, b]$. The volume V is the difference of the volume of the body defined by g_2 and the volume of the "hole" defined by g_1.

$$V = \pi \int_a^b (g_2(x)^2 - g_1(x)^2) \, dx .$$

MACSYMA-Session 11.2. The MACSYMA function
`volume_of_revolution(f,x,a,b) := %pi*integrate(f^2,x,a,b)$`
is a rewriting of Eq. (11.17), computing the volume of the solid of revolution obtained by rotating the graph $y = f(x)$ around the x-axis, between $x = a$ and $x = b$.
For example, the volume of a sphere with radius r is
c1. `volume_of_revolution(sqrt(r^2-x^2),x,-r,r)`

d1. $\dfrac{4\pi r^3}{3}$

Similarly, the volume of the body obtained by rotating $y = \sin(x)$ around the x-axis, between $x = 0$ and $x = \pi$, is
c2. `volume_of_revolution(sin(x),x,0,%pi)`

d2. $\dfrac{\pi^2}{2}$

The volume of the body of revolution of Example 11.12 is
c3. `volume_of_revolution(1/x,x,1,inf)`

d3. π

the correct value, even though the integral is improper (infinite interval).
We clean up with
c4. `remfunction(volume_of_revolution)$`

The disk method for computing the volume of a solid of revolution, (11.17),

$$\text{volume}(B, x, a, b) = \pi \int_a^b f(x)^2 \, dx ,$$

is based on summing over the areas of the cross sections

$$A(\xi) = \pi f(\xi)^2 ,$$

which are disks. Another way of computing that volume is illustrated in Fig. 11.14. Here the solid of revolution B is obtained by rotating the region $\mathcal{R}$ which is on one side of the x-axis around the x-axis. Consider a horizontal strip of $\mathcal{R}$ and the body obtained by rotating it around the x-axis which is a *cylindrical shell* (see Fig. 11.14b). We will express the volume of B by summing the volumes of all such cylindrical shells. The result is an integral which we will write directly without

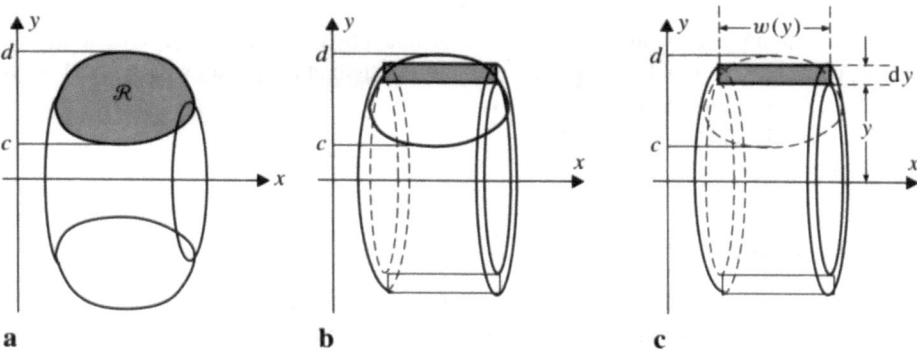

Fig. 11.14 a–c. Shell method. **a** The body obtained by rotating $\mathcal{R}$ around the x-axis. **b** The shell obtained by rotating a horizontal strip of $\mathcal{R}$. **c** The volume of the shell is approximately $2\pi yw(y)\,dy$

going through the Riemann sums. In Exercise 11.21 you are required to provide all the missing details.

The radius of the shell is y (the y-coordinate of the strip), its height is $w(y)$ (the width of the strip), and its thickness is dy. The volume of this shell is approximately $2\pi yw(y)\,dy$, as you can convince yourself by cutting the shell and spreading it open so that it becomes a rectangular box of dimensions $2\pi y$ (the circumference of the shell), $w(y)$ and dy. Summing over all these shells we get the volume formula

$$V = 2\pi \int_{c}^{d} yw(y)\,dy\,, \tag{11.18}$$

called the *shell method* for computing the volume of a solid of revolution.

MACSYMA-Session 11.3. The computation (11.18) of the volume V is implemented with the MACSYMA function
```
volume_shell(w,y,c,d) := 2*%pi*integrate(w*y,y,c,d)$
```
We now repeat the computation of the bodies of Session 11.2, where they were computed by the disk method.
The volume of a sphere with radius r is:
```
c1. volume_shell(2*sqrt(r^2-y^2),y,0,r)
    Is  r  positive or negative?
    positive
```
d1. $\dfrac{4\pi r^3}{3}$

The factor 2 in c1 is needed because for each value of y there are 2 hemispheres.
The volume of the body obtained by rotating $y = \sin(x)$ around the x-axis, between $x = 0$ and $x = \pi$, is:
```
c2. volume_shell(2*asin(y),y,0,1)
```
d2. $\dfrac{\pi^2}{2}$

Why is the factor 2 needed in c2?

Finally we compute the volume of revolution obtained by rotating $y = 1/x$ around the x-axis, between $x = 1$ and $x = \infty$. The width of the shell at the point $(x, y) = (x, 1/x)$ is $x - 1 = 1/y - 1$. Therefore, the volume is:

c3. `volume_shell(1/y-1,y,0,1)`

d3. π

Exercises

11.17 Until now we only talked about rotating plane regions around the x-axis. Write an integral representing the volume of the body obtained by rotating the region (8.2) around the horizontal line $y = \alpha$, where $f(x) \geq \alpha$ on $[a, b]$. *Hint:* The x-axis is the line $y = 0$.

11.18 Consider the region $\mathcal{R}$ defined by the three lines $y = 1 + x$, $y = 2x$, and $y = 6 - x$. Compute the volume of the body of revolution obtained by rotating $\mathcal{R}$ around the

    (a) x-axis   (b) horizontal line $y = -1$

    (c) y-axis   (d) vertical line $x = -1$

11.19 (Volume of torus)

    (a) Use the disk method (11.17) to compute the volume of the body obtained by rotating the disk $x^2 + (y - q)^2 \leq r^2$ around the x-axis, where $q \geq r > 0$. The body obtained by rotating a circular disk around a line is called a torus.

    (b) Compute the volume of the above torus using the shell method.

11.20 Compute the volume of a sphere with radius r, if a round hole with radius a, with $a < r$, is drilled through its center.

11.21 Write a Riemann sum of the volumes of shells in Fig. 11.14b and justify the shell method (11.18).

11.22 Give the shell method for the volume of a solid obtained by rotating a region $\mathcal{R}$ around the y-axis.

11.23 Compute, by the disk method and the shell method, the volumes of the following solids of revolution, obtained by rotating around the x-axis,

    (a) $x^2/a^2 - y^2/b^2 = -1$,   between $x = -h$ and $x = h$

    (b) $x^2/a^2 - y^2/b^2 = 1$,   between $x = a$ and $x = h$

11.6 Solids of revolution: surface area

In Sect. 11.5 we computed the volume of a solid of revolution. In this section we compute the surface area of such a body. Rotating the graph $y = f(x)$ around the x-axis, between $x := a$ and $x := b$, gives a *surface of revolution*, whose area is denoted by area_rev(f, x, a, b). As should by now be an automatic reaction, we partition the interval $[a, b]$ into n subintervals $I_k := [x_{k-1}, x_k]$ with lengths $\Delta x_k = x_k - x_{k-1}$. This partitions the revolution area into n strips,

$$\text{area\_rev}(f, x, a, b) = \sum_{k=1}^{n} \text{area\_rev}(f, x, x_{k-1}, x_k) .$$

If Δx_k is small, then for any ξ_k in I_k we can approximate the kth strip as a circular cylinder (without the bases) with height L_k [of Eq. (11.9)] and radius $f(\xi_k)$. This is so because the kth strip is obtained by rotating the kth length L_k around the x-axis. Therefore,

$$\text{area\_rev}(f, x, x_{k-1}, x_k) \approx 2\pi f(\xi_k)\sqrt{1 + (f'(\xi_k)^2} \, \Delta x_k$$

so that the revolution area is approximated by the Riemann sum

$$\text{area\_rev}(f, x, a, b) \approx 2\pi \sum_{k=1}^{n} f(\xi_k)\sqrt{1 + (f'(\xi_k)^2} \, \Delta x_k$$

and in the limit

$$\text{area\_rev}(f, x, a, b) = 2\pi \int_{a}^{b} f(x)\sqrt{1 + f'(x)^2} \, dx . \tag{11.19}$$

Example 11.14 (Area of sphere). To compute the area A of a sphere with radius r, we center it at the origin. This sphere is the surface of revolution obtained by rotating $y = \sqrt{r^2 - x^2}$ around the x-axis from $x := -r$ to $x := r$. Therefore,

$$A = 4\pi \int_{0}^{r} f(x)\sqrt{1 + f'(x)^2} \, dx, \quad \text{by symmetry, } A \text{ is twice the area of the hemisphere}$$

$$= 4\pi \int_{0}^{r} \sqrt{r^2 - x^2}\sqrt{1 + \left(\frac{-x}{\sqrt{r^2 - x^2}}\right)^2} \, dx$$

$$= 4\pi \int_{0}^{r} \sqrt{r^2 - x^2}\sqrt{\frac{r^2}{r^2 - x^2}} \, dx = 4\pi r \int_{0}^{r} dx$$

$$= 4\pi r^2 .$$

MACSYMA-Session 11.4. We implement Eq. (11.19) by the MACSYMA function
`surface_of_revolution(f,x,a,b):=`
$\qquad$ `2*%pi*integrate(radcan(f*sqrt(1+diff(f,x)^2)), x, a, b)$`
which computes the surface of revolution obtained by rotating the graph $y = f(x)$ around the x-axis, between $x = a$ and $x = b$.
For example, the area of a sphere with radius r is:
c1. `surface_of_revolution(sqrt(r^2-x^2),x,-r,r)`
d1. $\quad 4\pi r|r|$
rather than the well-known result $4\pi r^2$. The reason for the absolute value in d1 is that we have not told MACSYMA that r is positive, so d1 is the correct answer for the given information.

The body obtained by rotating $y = 1/x$ around the x-axis, between $x = 1$ and $x = \infty$ has infinite area. We compute it in two stages. First we compute the surface area with upper limit $x = b$ (instead of $x = \infty$)

c2. `surface_of_revolution(1/x,x,1,b)$`
    `Is    b-1    positive, negative or zero?`
    `positive`

Then we take the limit of d2 as $b \to \infty$,

c3. `limit(%,b,inf)`
d3. ∞

The area of the surface of revolution obtained by rotating the graph $y = x^{10}$ around the x-axis from $x = 0$ to $x = 1$ is computed by the MACSYMA function quanc8, an adaptive integrator.

c4. `2*%pi*quanc8(x^10*sqrt(1+100*x^18), x, 0, 1), numer`
d4.   3.22326

To compute the surface areas for $y = x^n$ where $n = 1, 10, 100$ and 1000, we use `surface_of_revolution(x,x,0,1)` for $n = 1$ and the numerical integrator quanc8 for the other cases:

c5. `cons(surface_of_revolution(x,x,0,1),`
    `2*%pi*makelist(quanc8(x^n*sqrt(1+n^2*x^(2*n-2)), x, 0, 1), n,`
                                              `[10,100,1000])), numer`

d5.   [4.44288, 3.22326, 3.14336, 3.14161]

The surface areas converge to π as $n \to \infty$ (why?)
Finally we clean up with

c6. `remfunction(surface_of_revolution)$`

Exercises

11.24 The output d5 in MACSYMA-Session 11.4 suggests that area_rev(x^n, 0, 1) $\to \pi$ as $n \to \infty$. Can you explain? Is there a power n for which area_rev(x^n, 0, 1) $= \pi$?

11.25 (Painting an infinite area with finite paint) By rotating the graph $y = 1/x$ around the x-axis from $x := 1$ to ∞ we get a body of revolution B with

finite volume: $\qquad V = \pi \int_1^\infty \frac{dx}{x^2} = \pi$ $\qquad\qquad$ see Example 11.12

infinite surface area: $\qquad A = 2\pi \int_1^\infty \frac{1}{x} \sqrt{1 + \left(\frac{-1}{x^2}\right)^2}\, dx = \infty$ $\qquad$ see MACSYMA-Session 11.4.

We can paint the infinite area A by simply filling the body B with paint. All it takes is the volume of B, or π gallons of paint.[4] Explain.

11.7 Moments and centroids

A *lever* is a simple machine used to lift loads. A typical lever is a rigid bar supported at a fixed point called the *fulcrum*. The lever can rotate around the fulcrum, and so raise the load.

---

4 You may be a good mathematician, but if you believe this, you are a dubious painter.

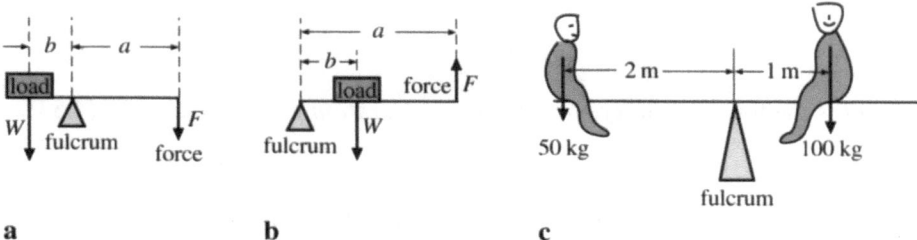

Fig. 11.15 a–c. Levers. **a** Lever with fulcrum between load and force; **b** lever with load and force on the same side. **c** The net moment is zero, and the seesaw is in balance

Figure 11.15 a and b shows two levers where a force F is applied to lift the weight W.

What makes a lever useful is that the lifting force F can be smaller than the load W. The lever is rotated by the force moments; a moment is the product of a quantity and its perpendicular distance from a reference point, here the fulcrum. In Fig. 11.15 a and b the moment of the load W is $W \times b$, and the moment of the force F is $F \times a$. The force lifts the load if its moment is bigger

$$Wb < Fa .\tag{11.20}$$

A difference between the two levers in Fig. 11.15 a and b is that in panel a the load moment tends to rotate the lever counterclockwise, and in panel b the load moment rotates the lever clockwise. To lift the load, the force must rotate the lever in an opposite direction. Since opposing moments cancel each other, we give them opposite signs. Thus a *clockwise* moment will be given a *negative* sign, and a *counterclockwise* moment a *positive* sign.

For example, in Fig. 11.15 a, the moment of W is $W \times b$ (counterclockwise) and the moment of F is $-F \times a$ (clockwise). The net moment is therefore

$$Wb - Fa .$$

If the force moment is greater than the load moment, that is, if (11.20) holds, then the net moment is negative, and the lever rotates clockwise, i.e., the force lifts the load.

Similarly, in Fig. 11.15 b, the net moment is

$$Fa - Wb .$$

Now the force lifting the load corresponds to a counterclockwise rotation of the lever, which is the case if the force moment is greater, i.e., if (11.20) holds.

Force is *mass* times *acceleration*, where mass is measured in units of gram (g) or kilogram (kg), and acceleration is measured in units of length per time squared. Two common units of force are newton (N)

$$N := \frac{\text{kg m}}{s^2}$$

and dyne (dyn),

$$\text{dyn} := \frac{\text{g cm}}{\text{s}^2}.$$

Weight is the force due to gravity, whose *acceleration* g is (at sea level), (4.53),

$$g \approx 9.81 \, [\text{m s}^{-2}],$$

see Sect. 4.8. Thus a mass of $100 \, [\text{kg}]$ weighs $100 \times 9.81 \, [\text{N}]$ on earth.[5] *Force moment* is measured in units of force times length such as Nm (newton-meter).

The seesaw in Fig. 11.15c features two masses, $100 \, [\text{kg}]$ at $1 \, [\text{m}]$ to the right of the fulcrum and $50 \, [\text{kg}]$ at $2 \, [\text{m}]$ to the left of the fulcrum. The net force moment is

$$50 \times g \times 2 - 100 \times g \times 1 = 0, \tag{11.21}$$

and the seesaw is therefore in balance (no rotation).

Since the acceleration g is common to both force moments in (11.21), we can arrive at the same conclusion (that the seesaw is balanced) by considering *mass moments* (products of mass and distance) instead of force moments. Thus the net mass moment in Fig. 11.15c is

$$50 \times 2 - 100 \times 1 = 0.$$

A point around which the net mass moment is zero is called the *center* (mass center or centroid) of the masses in question. For the two masses in Fig. 11.15c, the centroid is at the fulcrum.

In this section we compute centroids of mass systems analogous to, but more complicated than, the seesaw of Fig. 11.15c.

To begin, consider a plate of *area A* and *thickness d*, as in Fig. 11.16a. Most of the time we will find it convenient to ignore the thickness (which is the same throughout the plate), treating the plate as a planar object.

If we attach a rope to the point P on the edge of the plate and hang the plate,

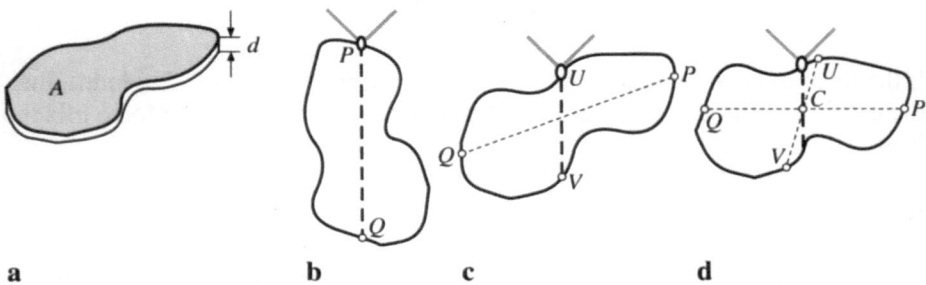

Fig. 11.16 a–d. Finding the centroid of a plate. **a** Plate of area A and thickness d. **b** Hanging the plate at the point P. **c** Hanging the plate at the point U. **d** Centroid C

---

5 But less on the moon, where the gravity acceleration is smaller.

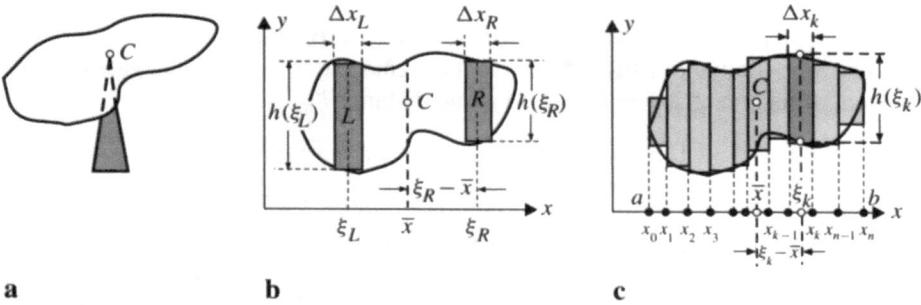

Fig. 11.17 a–c. The net moment w.r.t. the centroid being zero. **a** Balancing the plate at the centroid. **b** The moment of the strip R w.r.t. the line $x := \bar{x}$ is $-(\xi_R - \bar{x})h(\xi_R)\,\Delta x_R$. **c** The net moment of the whole plate w.r.t. $x := \bar{x}$ is approximated by a Riemann sum

it will move for a while but eventually come to rest, as in Fig. 11.16b. Let PQ be the vertical line, dropped from P. Figure 11.16c similarly describes the plate at rest, after being hung at the point U on its edge; here UV is the vertical line from U. The two lines PQ and UV intersect at the point C, which is the center of mass, or centroid, of the plate. If we hang the plate at any point on its edge, the vertical line will again pass through the centroid (Fig. 11.16d).

Let us now lay the plate flat and try to balance it by placing some sharp support directly below it. The only point of support where the plate is balanced (nobody sneeze!) is the centroid C (Fig. 11.17a). The plate is balanced for the same reason that the seesaw in Fig. 11.17b is balanced. To see this we must consider the moments of the plate with respect to the centroid.

If a mass m is placed at the point (x, y), then its moment with respect to the y-axis (or about the y-axis) is

$$M_y = -mx \tag{11.22}$$

which is positive (corresponding to counterclockwise rotation) if $x < 0$, and negative (clockwise rotation) if $x > 0$.

The *density* (or specific mass) of a given material is its mass per unit volume. The density is denoted by ρ and measured in units of mass per volume such as kg/m^3. A body with volume V and density ρ has mass $m = \rho V$.

If we compare moments of bodies with the same density, then it is possible to cancel ρ, which is the same in all mass moments $mx = \rho V x$. We then talk about the *volume moment* w.r.t. the y-axis,

$$V_y = -V x . \tag{11.23}$$

The plate in Fig. 11.16a has the same thickness d throughout, so its volume is $V = Ad$. Since d appears in all volume moments Adx, we can also cancel d and have instead the *area moment*

$$A_y = -Ax. \tag{11.24}$$

In Fig. 11.17b we denote the x-coordinate of the centroid C by $\bar{x}$. We compute now the area moments of vertical strips w.r.t. the vertical line $x := \bar{x}$.

The area moment of the strip R (to the right of $x := \bar{x}$) is the negative of the product of its area $\Delta A_R = h(\xi_R) \Delta x_R$ and distance $\xi_R - \bar{x}$,

$$(\xi_R - \bar{x})h(\xi_R) \Delta x_R = -(\xi_R - \bar{x}) \Delta A_R$$
$$= (\bar{x} - \xi_R) \Delta A_R \ .$$

Similarly the moment of the strip L in Fig. 11.17b is positive

$$(\xi_L - \bar{x})h(\xi_L) \Delta x_L = (\bar{x} - \xi_L) \Delta A_L \ ,$$

and corresponds to counterclockwise rotation. We can compare these two strips to the two seesaw riders in Fig. 11.15c: For the plate to balance at its centroid C, the net moment (of the whole plate) w.r.t. the centroid must be zero.

The area of the plate can be approximated by vertical strips like R and L in Fig. 11.17b. Therefore the net moment of the plate w.r.t. the line $x := \bar{x}$ is approximated by Riemann sums

$$A_{\{x=\bar{x}\}} \approx \sum_{k=1}^{n} (\bar{x} - \xi_k)h(\xi_k) \Delta x_k \ , \qquad (11.25)$$

$$= \sum_{k=1}^{n} (\bar{x} - \xi_k) \Delta A_k \ ,$$

corresponding to partitions $\mathcal{P}$ of the interval $I = [a, b]$ which "captures" the whole plate; see Fig. 11.17c. The area of the kth vertical strip is denoted here by ΔA_k.

In the limit, as $\|\mathcal{P}\| \to 0$, the Riemann sums (11.25) converge to the integral

$$A_{\{x=\bar{x}\}} = \int_a^b (\bar{x} - x)h(x) \, dx \ ,$$

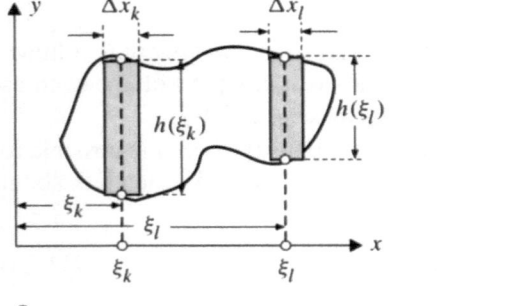

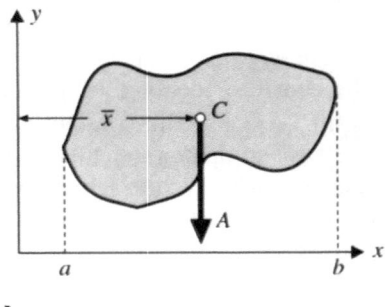

Fig. 11.18 a, b. Moments with respect to the y-axis. **a** Vertical strips and their moments w.r.t. the y-axis. **b** The moment of the whole plate w.r.t. the y-axis is the same as if the whole area were concentrated at the centroid C

which is the net area moment of the plate w.r.t. the line $x = \bar{x}$. We can rewrite this as

$$A_{\{x=\bar{x}\}} = \int_a^b (\bar{x} - x)\, dA(x)\,,$$

where $dA(x)$ denotes the *area element*,

$$dA(x) = h(x)dx\,.$$

The fact that the net area moment (w.r.t. $x = \bar{x}$) is zero gives a formula for computing $\bar{x}$,

$$0 = \int_a^b (\bar{x} - x)\, dA(x) = \int_a^b \bar{x}\, dA(x) - \int_a^b x\, dA(x)\,,$$

$$= \bar{x} \int_a^b dA(x) - \int_a^b x\, dA(x) \quad \text{since } \bar{x} \text{ is constant}\,,$$

$$= \bar{x}A - \int_a^b x\, dA(x) \quad \text{since } \int_a^b dA(x) = A\,.$$

Therefore

$$\bar{x} = \frac{\int_a^b xh(x)\, dx}{\int_a^b h(x)\, dx} = \frac{\int_a^b x\, dA(x)}{A} = -\frac{A_y}{A}\,, \tag{11.26}$$

where $A_y = -\int_a^b xh(x)\, dx$ is the area moment w.r.t. the y-axis. Another way to write (11.26) is

$$A_y = -\int_a^b xh(x)\, dx = -\bar{x}A\,,$$

i.e., the area moment A_y is the same as if the whole area A is concentrated at $x = \bar{x}$ (see Fig. 11.18).

We can analogously compute the area moment $A_{\{y=\bar{y}\}}$ of the plate w.r.t. the horizontal line $y := \bar{y}$ which passes through the centroid C. Working with horizontal strips, such as the strip H of Fig. 11.19a, we approximate the moment as a Riemann sum

$$A_{\{y=\bar{y}\}} \approx \sum_{k=1}^n (\bar{y} - \eta_k)w(\eta_k)\, \Delta y_k = \sum_{k=1}^n (\bar{y} - \eta_k)\, \Delta A_k\,,$$

where $w(\eta_k)$ is the width of the kth strip, Δy_k is its height, and its area is $\Delta A_k = w(\eta_k)\, \Delta y_k$. In the limit, the Riemann sums converge to an integral, and we get as

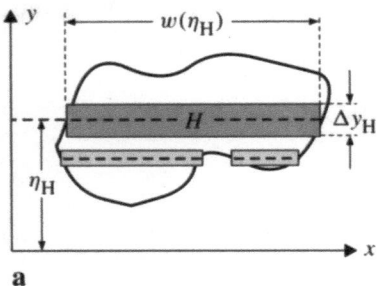

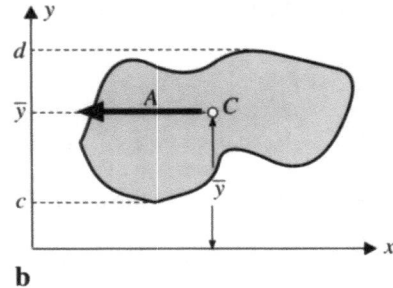

Fig. 11.19 a, b. Moments with respect to the x-axis. **a** The moment of the strip H w.r.t. the x-axis is $\eta_H w(\eta_H) \Delta y_H$. **b** The moment of the whole plate w.r.t. the x-axis is the same as if the whole area were concentrated at the centroid C

in (11.26)

$$\bar{y} = \frac{\int_c^d yw(y)\,dy}{\int_c^d w(y)\,dy} = \frac{A_x}{A}, \qquad (11.27)$$

where $[c, d]$ is an interval, on the y-axis, which "captures" the whole plate, and A_x is the area moment w.r.t. the x-axis.

Equations (11.26) and (11.27) give the coordinates $(\bar{x}, \bar{y})$ of the centroid of the plate. If the plate has some symmetry, then it is possible to correctly "guess" $\bar{x}$ or $\bar{y}$ without computations.

For example, if the plate is symmetric with respect to the vertical line $x := s$, then its centroid is on that line, i.e., $\bar{x} = s$ (see Fig. 11.20a). In this case we only have to compute the coordinate $\bar{y}$ of C. Similarly, symmetry with respect to the horizontal line $y := t$ implies that $\bar{y} = t$ (see Fig. 11.20b).

Example 11.15. Compute the centroid $(\bar{x}, \bar{y})$ of the semidisk bounded above by $x^2 + y^2 = r^2$ and below by the x-axis.

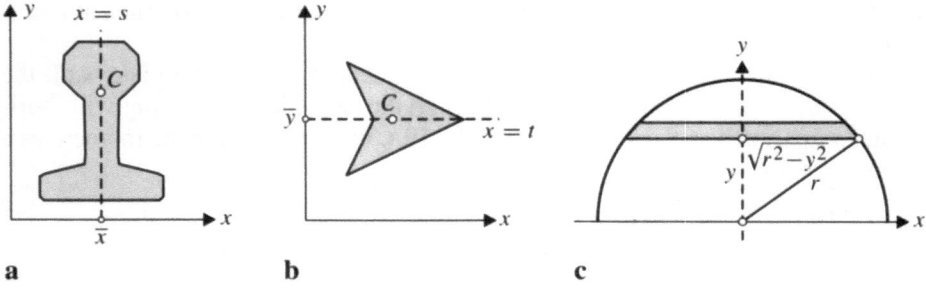

Fig. 11.20 a–c. Symmetry may give $\bar{x}$ or $\bar{y}$ without computation. **a** Plate symmetric w.r.t. the line $x = s$; **b** plate symmetric w.r.t. the line $y = t$; **c** semidisk symmetric w.r.t. the y-axis

Since the semidisk is symmetric w.r.t. the y-axis, $\bar{x} = 0$. To compute $\bar{y}$ we use (11.27) and Fig. 11.20c,

$$A_x = \int_0^r 2y\sqrt{r^2 - y^2}\, dy = \frac{2r^3}{3} \quad \text{(use MACSYMA)}.$$

$$\therefore \quad \bar{y} = \frac{A_x}{A} = \frac{2r^3/3}{\pi r^2/2} = \frac{4r}{3\pi}.$$

Example 11.16 (Second theorem of Pappus[6]). Let a region of area A, lying entirely on one side of a line, be rotated around the line. Then the volume of the solid generated is equal to the product of the area A and the distance travelled by the centroid.

Proof. Without loss of generality we take the line to be the x-axis. The volume, by the shell method, (11.18),

$$V = 2\pi \int_a^b yw(y)\, dy,$$

can be rewritten, by (11.27), as $V = 2\pi \bar{y} A$. This proves Pappus's theorem since $2\pi \bar{y}$ is the circumference of the circle travelled by the centroid. □

Exercises

11.26 Compute the centroid $(\bar{x}, \bar{y})$ of each of the areas bounded by the following curves.
   (a) the x-axis, the y-axis and the line $x + y = a$, where $a > 0$
   (b) $y^2 = 2px$, $x = h$ (c) $y = x^3$, $x = 2$, $y = 0$ (d) $y^2 = 2px$, $y = mx$
   (e) $y^2 = ax$, $x^2 = by$ (f) $y = x^2$, $y = 2x + 3$ (g) $x = 4y - y^2$, $y = x$
   (h) $y = x^2 - 2x - 3$, $y = -x^2 + 6x - 3$
   (i) $\dfrac{x^2}{a^2} - \dfrac{y^2}{b^2} = 1$, $x^2 + y^2 = 4$, first quadrant
   (j) $\dfrac{x^2}{a^2} + \dfrac{y^2}{b^2} = 1$, first quadrant   (k) $\sqrt{x} + \sqrt{y} = \sqrt{a}$, first quadrant
11.27 Compute the volume of the torus in Exercise 11.19 by Pappus's theorem (Example 11.16).
11.28 Find the volume generated by revolving the ellipse $x^2/a^2 + y^2/b^2 = 1$
   (a) around the x-axis   (b) around the y-axis

---

6 Pappus of Alexandria (4th century A.D.), the last of the great Greek geometers of antiquity.

11.8 Centroids of three-dimensional bodies

We turn now to 3-dimensional bodies and their centroids. A typical situation is the body depicted in Fig. 11.21. Its centroid C has coordinates $(\bar{x}, \bar{y}, \bar{z})$. Each is calculated by a method analogous to the calculation of $\bar{x}$ in (11.26), except that here we use mass moments, or volume moments (if the density is the same throughout the body).

We illustrate this for the coordinate $\bar{z}$ of C. The volume moment of the body, w.r.t. the xy-plane, V_{xy} is the same as if the whole volume were concentrated at C, i.e., $V_{xy} = V\bar{z}$. We now compute V_{xy} by summing the moments, w.r.t. the xy-plane, of horizontal slices as in Fig. 11.21,

$$V_{xy} \approx \sum_{k=1}^{n} \zeta_k A(\zeta_k) \, \Delta z_k \,,$$

where $A(\zeta_k)$ is the area of the horizontal cross section at $z = \zeta_k$. This is a Riemann sum over an interval $[q, r]$ along the z-axis which bounds the body. In the limit, the Riemann sum converges to the integral

$$V_{xy} = \int_{q}^{r} z A(z) \, \mathrm{d}z \,, \tag{11.28}$$

and the volume V is, by slicing perpendicularly to the z-axis,

$$V = \int_{q}^{r} A(z) \, \mathrm{d}z \,.$$

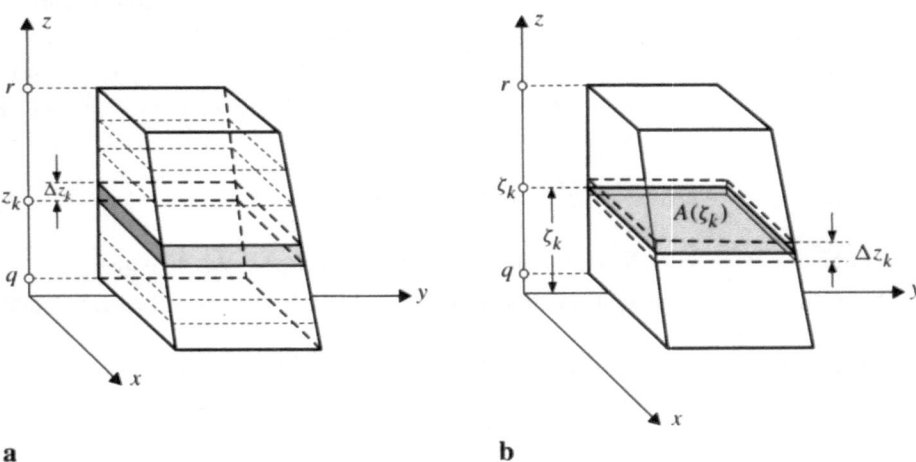

a   **b**

Fig. 11.21 a, b. Moment w.r.t. the xy-plane by horizontal slicing. **a** A horizontal slice from z_k to $z_k + \Delta z_k$ has a volume of approximately $A(\zeta_k) \, \Delta z_k$. **b** Its moment w.r.t. the xy-plane is approximately $\zeta_k A(\zeta_k) \, \Delta z_k$

Therefore

$$\bar{z} = \frac{V_{xy}}{V} = \frac{\int_q^r z A(z)\,dz}{\int_q^r A(z)\,dz}\ . \tag{11.29}$$

Example 11.17. Calculate the centroid of the semisphere bounded above by

$$x^2 + y^2 + z^2 = r^2\ ,$$

and below by the xy-plane.

The semisphere is symmetric w.r.t. the plane $x := 0$ (the yz-plane) and the plane $y := 0$ (the xz-plane). Therefore, $\bar{x} = \bar{y} = 0$.

To compute the moment V_{xy} of the semisphere w.r.t. the xy-plane, we note that the area $A(z)$ of the horizontal cross section at z is

$$A(z) = \pi(r^2 - z^2)\ .$$

Therefore, by (11.28),

$$V_{xy} = \int_0^r z\pi(r^2 - z^2)\,dz = \pi\left\{ r^2 \int_0^r z\,dz - \int_0^r z^3\,dz \right\} = \pi\left\{ \frac{r^4}{2} - \frac{r^4}{4} \right\} = \frac{\pi r^4}{4}\ .$$

The volume of the semisphere is $2\pi r^3/3$, and therefore, by (11.29),

$$\bar{z} = \frac{V_{xy}}{V} = \frac{3}{8}r\ .$$

Example 11.18. Consider a cone with its vertex at the origin and its basis with area A in the plane $x := h$ (see Fig. 11.10a. Compute the coordinate $\bar{x}$ of the centroid.

The volume moment w.r.t. the yz-plane is

$$V_{yz} = \int_0^h x \frac{x^2}{h^2} A\,dx = \frac{A}{h^2} \int_0^h x^3\,dx = \frac{A}{h^2} \frac{h^4}{4} = \frac{Ah^2}{4}\ .$$

The volume of the cone is $Ah/3$. Therefore

$$\bar{x} = \frac{V_{yz}}{V} = \frac{3}{4}h\ .$$

Example 11.19. A circular cone with basis radius of 6 [cm] and height of 10 [cm] is assembled from a tip with a height of 5 [cm], made of lead, and a copper base (see Fig. 11.22a). The density of lead is 11.3 [g cm^{-3}] and of copper 8.9 [g cm^{-3}]. Find the centroid $C = (\bar{x}, \bar{y}, \bar{z})$ of the cone.

The body in question does not have the same density throughout; therefore, we must use mass moments rather than volume moments.

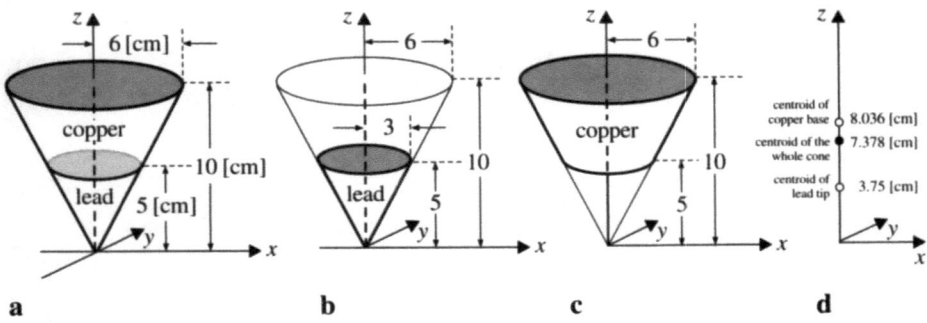

Fig. 11.22 a–d. Centroid of a cone with different densities (Example 11.19). **a** Cone with a lead tip and a copper base. **b** The lead tip has mass 532.5 [g]. **c** The copper base has mass 2935.818 [g]. **d** Three centroids

By symmetry, $\bar{x} = \bar{y} = 0$. The coordinate $\bar{z}$ is given by

$$\bar{z} = M_{xy}/M ,$$

where M_{xy} is the mass moment w.r.t. the xy-plane, and M is the total mass.

The total mass is the sum of the masses of the lead tip and of the copper base. The mass of the lead tip, M_L, is the product of volume $V_L = \pi \cdot 3^2 \cdot 5/3 \, [\text{cm}^3]$ and lead density $\rho_L = 11.3 \, [\text{g cm}^{-3}]$,

$$M_L = \pi \cdot 3 \cdot 5 \cdot 11.3 = 532.500 \, [\text{g}] .$$

The mass of the copper base M_C is the product of its volume and density ρ_C,

$$V_C = \pi \frac{6^2 \cdot 10 - 3^2 \cdot 5}{3} = 105\pi = 329.867 \, [\text{cm}^3] \quad \text{and} \quad \rho_C = 8.9 \, [\text{g cm}^{-3}] ,$$

$$M_C = \pi \cdot 105 \cdot 8.9 = 2935.818 \, [\text{g}] .$$

Therefore the mass of the cone is $M = M_L + M_C = 532.500 + 2935.818 = 3468.318 \, [\text{g}]$.

The mass moment of the lead tip w.r.t. the xy-plane is its volume moment times its density,

$$M_{L,\{xy\}} = 11.3 \int_0^5 z\pi 3^2 \frac{z^2}{5^2} \, dz = 11.3 \cdot \pi \cdot 3^2 \int_0^5 \frac{z^3}{5^2} \, dz$$

$$= 11.3 \cdot \pi \cdot 3^2 \cdot \frac{5^2}{4} = 1996.875 \, [\text{g cm}] .$$

The mass moment of the copper base w.r.t. the xy-plane is its volume moment times its density,

$$M_{C,\{xy\}} = 8.9 \int_5^{10} z\pi 6^2 \frac{z^2}{10^2}\, dz = 8.9 \cdot \pi \cdot 6^2 \int_5^{10} \frac{z^3}{10^2}\, dz$$

$$= 8.9 \cdot \pi \cdot 6^2 \cdot \frac{10^4 - 5^4}{10^2} = 8.9 \cdot \pi \cdot 6^2 \cdot 93.75 = 23591.398 \, [\text{g cm}] .$$

Therefore, the mass moment of the cone w.r.t. the xy-plane

$$M_{xy} = M_{L,\{xy\}} + M_{C,\{xy\}} = 1996.875 + 23591.398 = 25588.273 \, [\text{g cm}] .$$

Finally

$$\bar{z} = \frac{M_{xy}}{M} = \frac{25588.273}{3468.318} = 7.378 \, [\text{cm}] .$$

Note that the z-coordinate of the centroid of the lead tip, $\bar{z}_L$, is

$$\bar{z}_L = \frac{M_{L,\{xy\}}}{M_L} = \frac{1996.875}{532.500} = 3.75 \, [\text{cm}] , \tag{11.30}$$

in agreement with Example 11.18. The centroid of the copper base is similarly

$$\bar{z}_C = \frac{M_{C,\{xy\}}}{M_C} = \frac{23591.398}{2935.818} = 8.036 \, [\text{cm}] , \tag{11.31}$$

see Fig. 11.22d. Then $\bar{z}$ can be computed from

$$\bar{z} = \frac{M_L \bar{z}_L + M_C \bar{z}_C}{M_L + M_C} = \frac{532.500 \cdot 3.75 + 2935.818 \cdot 8.036}{532.500 + 2935.818} = 7.378 \, [\text{cm}] . \tag{11.32}$$

MACSYMA-Session 11.5. To compute a single coordinate of the centroid, such as $\bar{x}$, we can use the MACSYMA function

```
centroid_coordinate(h,x,a,b):=
```

$$\text{integrate}(x*h,x,a,b)/\text{integrate}(h,x,a,b)\$$$

which computes (11.26). For example, to compute the coordinate $\bar{z}$ of the semisphere of Example 11.17, we use

c1. `centroid_coordinate(%pi*(r^2-z^2),z,0,r)`

d1. $\dfrac{3r}{8}$

The coordinate $\bar{z}$ of the lead tip in Example 11.19 is found by

c2. `z_hat:centroid_coordinate(%pi*r^2/h^2*z^2,z,0,lead_height)`

d2. $\dfrac{3 \, \text{lead height}}{4}$

Similarly, the centroid of the copper base is

c3. `centroid_coordinate(%pi*r^2/h^2*z^2,z,lead_height,h)`

d3. $\dfrac{h^4/4 - \text{lead height}^4/4}{h^3/3 - \text{lead height}^3/3}$

c4. `c_hat:factor(%)`

d4. $\dfrac{3(\text{lead height} + h)(\text{lead height}^2 + h^2)}{4(\text{lead height}^2 + h \text{ lead height} + h^2)}$

and by lead height $= 5$ we get $\bar{z}$ for the lead tip as

c5. `z_hat,lead_height:5`

d5. $\dfrac{15}{4}$

c6. `%,numer`

d6.   3.75

Further, by $h = 10$ we get the centroid of the copper base as

c7. `c_hat,lead_height:5,h:10`

d7. $\dfrac{225}{28}$

c8. `%,numer`

d8.   8.03571

which agrees with (11.31).

We clean up with

c9. `(remvalue(c_hat,z_hat), remfunction(centroid_coordinate))$`

Exercises

11.29   Determine the centroid of the top half of the body B in Exercise 11.16, i.e., the part of the body which lies above the xy-plane.

11.30   Consider the cone of Example 11.19. How would the centroid change if the whole cone were made (a) of copper, (b) of lead?

11.31   Let a mass system be made of masses $M_1, M_2, \ldots, M_n$, and let the kth mass have centroid $C_k = (\bar{x}_k, \bar{y}_k, \bar{z}_k)$, for $k = 1, \ldots, n$. Then the centroid of the system is $C = (\bar{x}, \bar{y}, \bar{z})$, where

$$\bar{x} = \frac{\sum_{k=1}^{n} M_k \bar{x}_k}{\sum_{k=1}^{n} M_k}, \quad \bar{y} = \frac{\sum_{k=1}^{n} M_k \bar{y}_k}{\sum_{k=1}^{n} M_k}, \quad \bar{z} = \frac{\sum_{k=1}^{n} M_k \bar{z}_k}{\sum_{k=1}^{n} M_k}.$$

Note that Example 11.19 is a special case; see (11.32).

11.9 Work

A constant force F is applied to an object moving along a straight line and moves it a distance d. The work W done by the force is the product of the force applied and the distance travelled

$$W = F \times d.$$

Similarly, if we want to move an object a distance d against a constant force F, we must invest work in the amount of $F \times d$.

Work is defined as the product of force and length; the common units of work are joule (J) defined as newton times meter (N m), and erg defined as dyne times centimeter (dyn cm).

In most cases of interest the force is not constant but changes during the motion.

Let F be a function of the distance s from a fixed reference point,

$$F = F(s) .$$

For example, the *weight* of an object of mass m [kg] at a distance of s [m] from the center of earth is

$$F(s) = \frac{\gamma m}{s^2} \text{ [N]}, \tag{11.33}$$

where γ is a constant. Thus the weight of an object at $s = 20{,}000$ [km] is one fourth its weight at $s = 10{,}000$ [km].

To compute the work done by the force in moving the object from a to b, we partition the interval $[a, b]$ into n subintervals $I_k = [x_{k-1}, x_k]$ with lengths $\Delta x_k = x_k - x_{k-1}$. In the kth subinterval I_k the force F is approximated by its value $F(\xi_k)$ at some point $x_{k-1} \le \xi_k \le x_k$, so the work in the kth subinterval, $W(F, x_{k-1}, x_k)$, is approximately $F(\xi_k) \times \Delta x_k$. The work done in $[a, b]$ is therefore approximated by a Riemann sum

$$W(F, a, b) = \sum_{k=1}^{n} W(F, x_{k-1}, x_k) \approx \sum_{k=1}^{n} F(\xi_k) \, \Delta x_k .$$

In the limit, the Riemann sums converge to the integral,

$$W(F, a, b) = \int_a^b F(s) \, \mathrm{d}s . \tag{11.34}$$

If the force F does not change direction in $[a, b]$, then the work $W(F, a, b)$ is positive if movement is in the same direction as F, and negative otherwise. If the force changes direction in $[a, b]$, then $W(F, a, b)$ is positive if the directions of F and of the movement agree "more" than they disagree.

Example 11.20 (Springs). Consider a spring lying on the x-axis, with the left end attached to the origin. The spring can be stretched, or compressed, by applying force to the right end, which, in the absence of force, is at $x = s_0$ (we call s_0 the rest length of the spring). The force required to keep the length of the spring at s is given by *Hooke's law*[7] as

$$F(s) = k(s - s_0) , \tag{11.35}$$

where k is the *spring constant*, depending on the material and dimensions of the spring. *Stretching* corresponds to $s > s_0$ (see Fig. 11.23 b), and the force is in the direction of the x-axis. Similarly *compression* is the case $s < s_0$ (Fig. 11.23 c), and the force's direction is against the x-axis. In particular, $F(s_0) = 0$ since no force is required at $s = s_0$.

---

7 Robert Hooke (1635–1703).

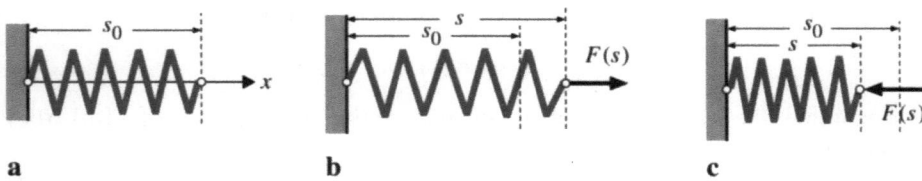

Fig. 11.23 a–c. The force required to hold the spring at length s is $F(s) = k(s - s_0)$. **a** Spring with rest length s_0; **b** spring stretched to length s; **c** spring compressed to length s

The work done in changing the spring's length from s_0 to s is stored in the spring as *potential energy*, released when the force is turned off and the spring returns to its rest length s_0. Equation (11.35) is a physical law which is valid provided s is not too far from s_0. Stretching or compressing the spring too much will cause permanent change in the length and an eventual break of the spring.

The work needed to change the spring length from a to b (assuming that Hooke's law (11.35) is valid throughout) is

$$W(F, a, b) = \int_a^b F(s)\, ds$$

$$= k \int_a^b (s - s_0)\, ds = k \int_a^b s\, ds - k s_0 \int_a^b ds$$

$$= k \left. \frac{s^2}{2} \right|_a^b - k s_0 (b - a)$$

$$= \frac{k}{2}(b^2 - a^2) - k s_0 (b - a)$$

$$= k(b - a)\left(\frac{a + b}{2} - s_0 \right). \tag{11.36}$$

A positive value of W means here energy invested in the spring, negative W is work given back by the spring ("useful work"). Equation (11.36) gives the work for all possible values of a, b and s_0 as long as Hooke's law holds.

Example 11.21 (Piston engines). In a piston engine the linear movement of pistons, inside the engine cylinders, is transformed to a rotational movement of the crankshaft. Consider the four-stroke engine illustrated in Fig. 11.24. Here a piston moves inside a cylinder, which includes two valves, intake and exhaust. Above the piston is gas.

The two extreme positions of the piston are called the *top dead center*, where the gas volume is minimal (Fig. 11.24c), and the *bottom dead center*, where gas volume is maximal (Fig. 11.25a).

The four strokes mentioned are:

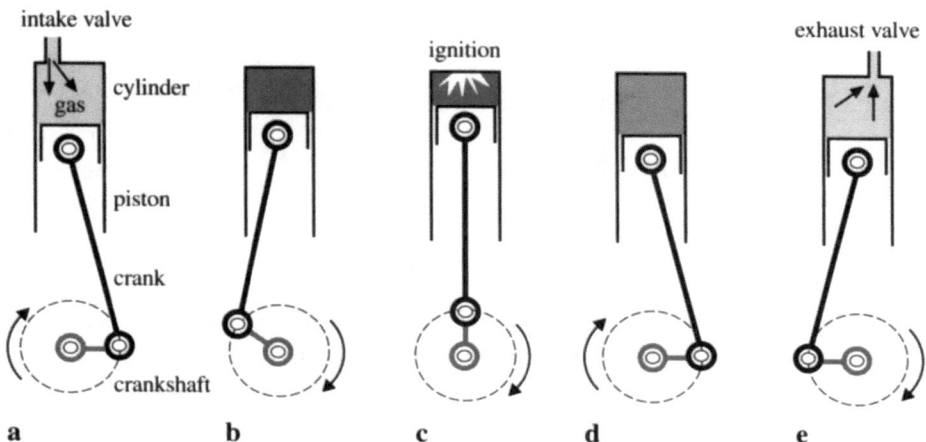

Fig. 11.24 a–e. Four-stroke engine. **a** Intake stroke; **b** compression stroke; **c** ignition; **d** power stroke; **e** exhaust stroke

stroke 1, *Intake:* The piston starts at top dead center and moves down, sucking gas[8] through the open intake valve (Fig. 11.24 a);

stroke 2, *Compression:* Both valves are closed, and the piston moves up, compressing the gas (Fig. 11.24 b); when the piston reaches top dead center, ignition occurs[9] (Fig. 11.24 c);

stroke 3, *Power:* The exploding gas pushes the piston downward (Fig. 11.24 d);

stroke 4, *Exhaust:* The exhaust valve opens, and the piston moves up, pushing the burnt gas out (Fig. 11.24 e).

The power stroke is the only one when the piston does useful work. The compression stroke requires considerable work. The intake and exhaust strokes require little work, which we neglect.

Recall that pressure is force per area, measured in units such as $N\,m^{-2}$. In gas, pressure against a surface is due to the gas molecules hitting the surface. If the gas is compressed, so its volume V decreases, then its pressure p increases according to the formula

$$pV^\alpha = C\,, \tag{11.37}$$

where C and α are constants, depending on the gas in question and its energy.

Consider now the compression stroke, illustrated in Fig. 11.25, where for convenience the cylinder is drawn horizontal. This stroke begins with the piston at bottom dead center, where the gas volume is maximal and its pressure minimal (see the point D in Fig. 11.25 a). As the gas is compressed by the piston, its pressure p

---

8 This "gas" is a mixture of gasoline and air in gasoline engines and of plain air in Diesel engines.

9 In gasoline engines, ignition is caused by an electric spark. In Diesel engines, fuel is injected into the compressed air and ignites.

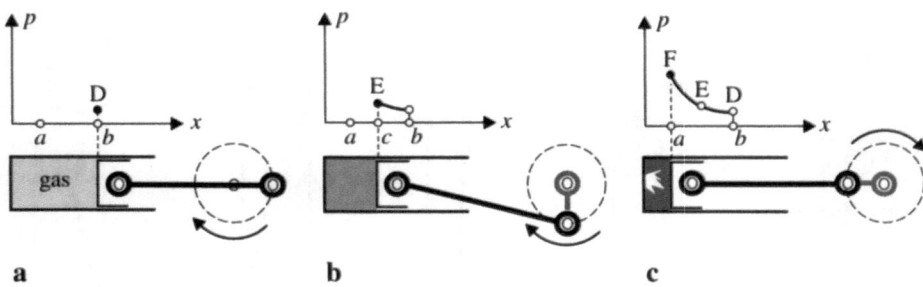

Fig. 11.25 a–c. Compression stroke. **a** Compression begins; **b** piston at an intermediate point; **c** compression ends, and ignition occurs

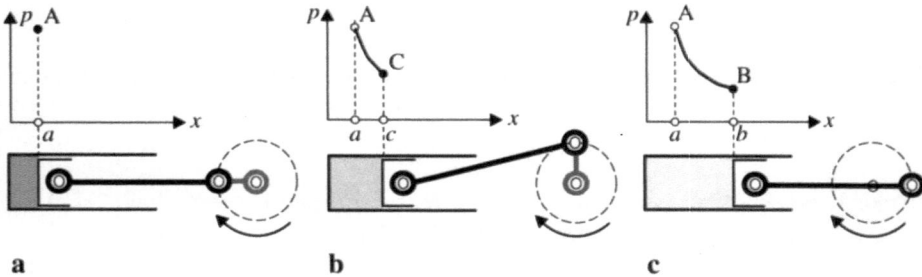

Fig. 11.26 a–c. Power stroke. **a** Piston at the top dead center; **b** piston at an intermediate point; **c** piston at bottom dead center

rises (see points E and F in Fig. 11.25 b and c). At the point F ignition occurs, and the pressure rises sharply[10], giving the point A in Fig. 11.26 a.

The power stroke begins with the piston in top dead center and gas pressure at maximum. The pressure falls as the engine moves towards the bottom dead center (see points C and B in Fig. 11.26 b and c).

Since force is pressure times area, the force on the piston is

$$F = pA \, ,$$

where A is the area of the piston, or the area of the cross section of the cylinder. The work done by the piston as it moves from a to b is therefore

$$W(F, a, b) = \int_a^b pA \, dx = A \int_a^b p \, dx \, , \tag{11.38}$$

the area under the pressure curve $p = p(x)$. The work done by the piston in the power stroke is therefore the shaded area in Fig. 11.27 a. The work invested

---

10 After ignition we have a different gas, with different parameters C and α in (11.37).

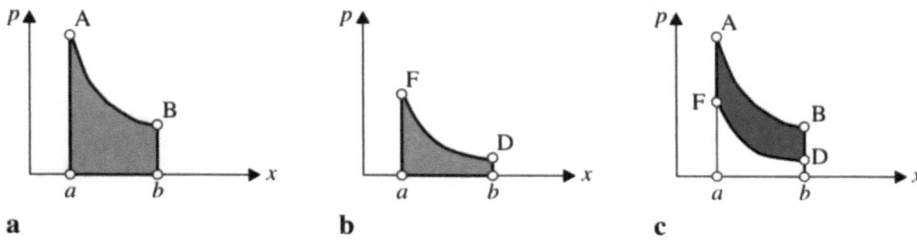

Fig. 11.27 a–c. Work during power and compression strokes. **a** Work done by the piston during power stroke; **b** work invested during compression stroke; **c** net work

during the compression stroke is the negative of the shaded area in Fig. 11.27 b. This work is negative.[11] The net work during these two strokes (which account for most of the engine work) is the difference, shown in Fig. 11.27 c.

To compute the integral (11.38), we must express the pressure p as a function of x. We already have p as a function of V, by (11.37),

$$p = p(V) = CV^{-\alpha} . \tag{11.39}$$

Since $V = Ax$, where A is the piston area, we can write the pressure p as a function of x,

$$p = p(x) = C_0 x^{-\alpha} , \tag{11.40}$$

where $C_0 = CA^{-\alpha}$. The work during the power stroke is therefore,

$$W(F, a, b) = A \cdot C_0 \int_a^b x^{-\alpha} \, dx .$$

Example 11.22 (Pumping liquid). A liquid is stored in a container; see Fig. 11.28 a. Compute the work required to pump the liquid over the top of the container, that is raise all the liquid to altitude $y = H$. The density ρ of the liquid may be a function of altitude,[12] $\rho = \rho(y)$. The area A of the cross section of the container is also a function of y, $A = A(y)$.

Consider a horizontal layer L_k of the liquid between y_k and $y_k + \Delta y_k$ (see Fig. 11.28 b), and let η_k be any point in the interval $[y_k, y_k + \Delta y_k]$. The volume V_k of L_k is approximately $V_k \approx A(\eta_k) \Delta y_k$ [m³]; see Fig. 11.28 c. Also, if Δy_k is sufficiently small, the density in L_k can be approximated by its value $\rho(\eta_k)$ at $y = \eta_k$. Therefore the mass m_k of L_k and its weight F_k are approximately

$$m_k \approx \rho(\eta_k) A(\eta_k) \Delta y_k \text{ [kg]} \quad \text{and} \quad F_k \approx g m_k = g\rho(\eta_k) A(\eta_k) \Delta y_k \text{ [N]} .$$

---

11 It is the integral of a nonnegative function p from b to a, with $b > a$.

12 As is the case when the liquid in question is composed of liquids with different densities, such as water and oil, with the heavier liquid "sinking" to the bottom, and the lighter liquids "rising" to the top. In case of a single liquid, the density is the same for all y since liquid is incompressible.

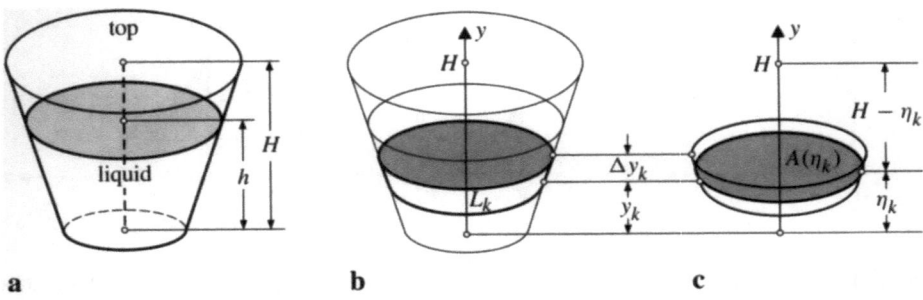

Fig. 11.28 a–c. Illustration of (11.41). **a** Liquid to be pumped over the top; **b** horizontal layer L_k; **c** volume V_k is approximately $A(\eta_k)\,\Delta y_k$

The layer L_k has to be raised a distance of approximately $H - \eta_k$ [m]; see Fig. 11.28 c. Therefore the work W_k of pumping the layer L_k is approximately,

$$W_k \approx F_k(H - \eta_k) \approx g\rho(\eta_k)A(\eta_k)\,\Delta y_k \text{ [J] .}$$

The total work W is therefore approximated by the Riemann sum

$$W = \sum_k W_k \approx g \sum_k \rho(\eta_k)A(\eta_k)(H - \eta_k)\,\Delta y_k \text{ [J] ,} \qquad (11.41)$$

and in the limit, by the integral

$$W = g \int_0^h \rho(y)A(y)(H - y)\,dy \text{ [J] .} \qquad (11.42)$$

Exercises

11.32 Consider the following six cases for (11.36).
   (a) $a < b < s_0$ (b) $b < a < s_0$ (c) $a < s_0 < b$
   (d) $b < s_0 < a$ (e) $s_0 < a < b$ (f) $s_0 < b < a$
   In each case determine whether the work $W(F, a, b)$ is positive (energy invested in the spring), negative (work given back), or that further information is required.

11.33 Explain in (11.36) why
   (a) $W(F, a, s_0) < 0$ whether $a < s_0$ or $a > s_0$
   (b) $W(F, s_0, b) > 0$ whether $b < s_0$ or $b > s_0$
   (c) $W(F, a, b) = 0$ if $s_0 = (a + b)/2$

11.34 A container has the shape of a cube with sides 10 [m] and contains 200 [m^3] water and 500 [m^3] oil. The density of water is $\rho_w = 1000$ [kg m^{-3}], and the density of oil is $\rho_o = 950$ [kg m^{-3}]. Water and oil do not mix, so the oil is above the water. Find the work required to pump all the liquids above the top of the container.

11.10 Hydrostatic force

Consider a body of liquid at rest, for example, water in a water reservoir. At any point the liquid pressure p (called *hydrostatic pressure*) can be measured by recording the force that the liquid exerts per unit area. *Pascal's law*[13] states: At any point in the liquid, the hydrostatic pressure is the same in all directions. This is illustrated in Fig. 11.29 a for a typical underwater point A.

Consider a small horizontal area A [m^2] at the point A and the *water column* and *air column* directly above the area A.

The volume V_w of the water column is $V_w = h_A A$ [m^3], where h_A is the depth of A underwater. The mass of the water column is $m_w = V_w \rho_w$ [kg]; here $\rho_w = 1000$ [kg m^{-3}] is the *density* of water. Finally the weight F_w of the water column is its mass times the gravity acceleration, $g \approx 9.81$ [m s^{-2}],

$$F_w = g\rho_w h_A A \text{ [N]} . \tag{11.43}$$

We turn now to the weight F_a of the air column[14] above the area A (at the point B in Fig. 11.29 b). This weight is

$$F_a = p_a A \text{ [N]} , \tag{11.44}$$

where p_a is the *atmospheric pressure*, measured in laboratory (at sea level) as $p_a \approx 1.013 \times 10^5$ [N m^{-2}], which is called one *atmosphere* [atm].

The sum of the water and air weights above the area A, at the point A, is therefore

$$F_A = F_w + F_a = g\rho_w h_A A + p_a A \text{ [N]}$$

and the total pressure at A is

$$p_A = \frac{F_A}{A} = g\rho_w h_A + p_a \text{ [N m}^{-2}] , \tag{11.45}$$

see Fig. 11.29 b. The hydrostatic pressure p_w is the pressure due to water alone (ignoring the atmospheric pressure p_w),

$$p_w = g\rho_w h_A \text{ [N m}^{-2}] . \tag{11.46}$$

The hydrostatic pressure is thus equal at all points with the same underwater depth. At such points along the container walls, the hydrostatic pressure against the walls is the same. The *hydrostatic force* against the container walls can be computed by a Riemann sum over horizontal strips (Fig. 11.29 c).

---

13 Blaise Pascal (1623–1662).
14 The weight F_a cannot be computed by a simple formula like (11.43) because the height of the air column is that of the earth atmosphere, several hundreds kilometers, and the density ρ_a of air decreases with altitude and is eventually zero. The weight F_a can in principle be computed as an integral $F_a = A \int_0^\infty \rho_a(y)\, dy$ but is easier to determine empirically.

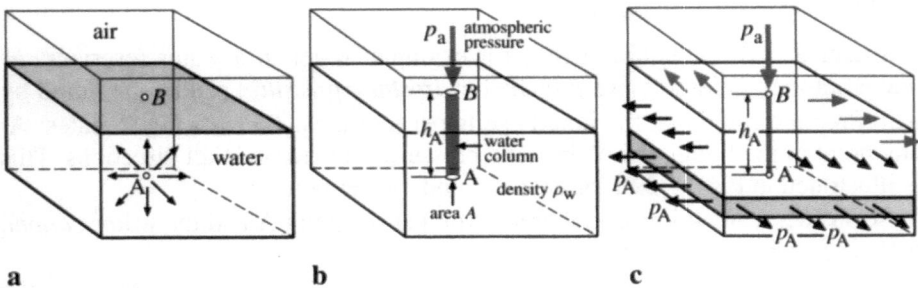

Fig. 11.29 a–c. Hydrostatic pressure. **a** The pressure at any point in the liquid is the same in all directions. **b** Pressure $P_A = g\rho_w h_A + p_a$. **c** Pressure on the container walls

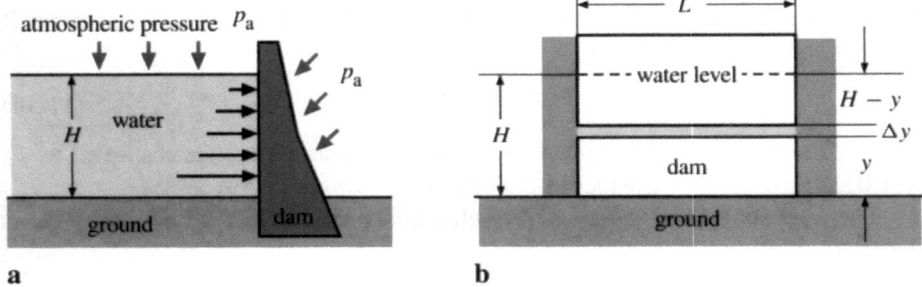

Fig. 11.30 a, b. Hydrostatic forces acting on a dam. **a** Dam holding water at level H. **b** The hydrostatic force against the lightly shaded strip is approximately $g\rho_w(H - y)L \, \Delta y$

Example 11.23 (Hydrostatic forces acting on a dam). A dam with length L and a rectangular shape holds water at level H (Fig. 11.30). Compute the hydrostatic force acting on the dam.

Consider a horizontal strip with height Δy (Fig. 11.30b). By (11.45), the hydrostatic pressure at the strip is approximately

$$g\rho_w(H - y) + p_a \, [\mathrm{N\,m^{-2}}]$$

because the underwater depth of the strip is approximately $H - y$. The area of the strip is $L \, \Delta y \, [\mathrm{m^2}]$, and the hydrostatic force on the strip is therefore approximated by

$$g\rho_w(H - y)L \, \Delta y \, [\mathrm{N}] \, .$$

The hydrostatic force on the dam can be modelled as a Riemann sum

$$F = g\rho_w \sum_{k=1}^{n}(H - \eta_k)L \, \Delta y_k \, [\mathrm{N}] \, ,$$

and in the limit, as an integral

$$F = g\rho_w \int_0^H (H - y)L\,dy \tag{11.47}$$

$$= g\rho_w L\left(\int_0^H H\,dy - \int_0^H y\,dy\right) = g\rho_w L\left(H\,y|_0^H - \frac{y^2}{2}\Big|_0^H\right)$$

$$= g\rho_w L\frac{H^2}{2}\ [\text{N}]\ . \tag{11.48}$$

In this example the dam has rectangular shape; the length L is a constant, and can be taken out of the integral (11.47). However, for dams with general shape the length L is itself a function of y, and the integral (11.47) still gives the correct result, which is different from (11.48).

Exercises

11.35 A dam has the shape of a trapezoid, with height $h = 120$ [m], lower base with length $L_1 = 100$ [m], and an upper base with length $L_2 = 300$ [m]. The dam holds water at level $H = 100$ [m]. Compute the hydrostatic force on the dam.

11.36 A rectangular gate in a vertical dam is 10 [m] wide and 6 [m] deep. Find
   a. the hydrostatic force on the gate if the water level is 8 [m] above its top
   b. how much higher the water must rise to double the hydrostatic force found in part a

11.37 Show that the hydrostatic force on any vertical surface is the product $F = \rho_w g \bar{y} A$, where A is the area of the surface, and $\bar{y}$ is the depth of its centroid.

Series and approximations

12 Sequences and series

This chapter is devoted to the study of convergence of sequences $(a_0, a_1, a_2, \ldots)$ and series $\sum_{n=0}^{\infty} a_n$ of real numbers a_n. We use the notation $\mathbf{N}_0 := \{0, 1, 2, 3, \ldots\}$ and $\mathbf{N} := \{1, 2, 3, \ldots\}$ throughout.

12.1 Sequences and convergence

Definition 12.1 (Sequence of real numbers). A sequence of real numbers is an ordered, infinite set[1] $(a_0, a_1, a_2, \ldots)$ of real numbers (a_n), $n \in \mathbf{N}_0$. The number a_n is called the nth term (or *general term*) of the sequence. The sequence is denoted as $(a_n)_{n \in \mathbf{N}_0}$, $(a_n)_n$ or $\{a_n\}_{n=0}^{\infty}$ for short.

It is sometimes convenient to number the terms of the sequence as $(a_1, a_2, a_3, \ldots)$, i.e., the first term is a_1 instead of a_0. In this case we write the sequence as $(a_n)_{n \in \mathbf{N}}$.

If the numbering is not important, we denote the sequence as $(a_n)_n$.

A *subsequence* of a sequence $(a_n)_n$ is a sequence obtained by selecting certain terms of $(a_n)_n$. We denote such a subsequence by $(a_{n_k})_k$, where n_k tells which elements of the sequence are chosen. We require that $n_k < n_{k+1}$ for all k.

Example 12.2. Here are some sequences.
a. $(0, 1, 2, 3, \ldots)$ the sequence of *nonnegative integers*, with $a_n = n$ for $n \in \mathbf{N}_0$
b. $(1, 2, 3, \ldots)$ the sequence of *positive integers*, with $a_n = n$ for $n \in \mathbf{N})$; this sequence is a subsequence of a
c. $(1, 3, 5, 7, \ldots)$ the sequence of *odd positive integers*, with $a_n = 2n - 1$ for $n \in \mathbf{N}$; this sequence is a subsequence of b or of a
d. $(1, 1, 2, 3, 5, 8, \ldots)$ the sequence $(f_n)_{n \in \mathbf{N}_0}$ of *Fibonacci numbers* defined by $f_0 = f_1 = 1$ and $f_n := f_{n-1} + f_{n-2}$ for all $n \geq 2$
e. $(1, -1/2, 1/3, -1/4, 1/5, \ldots)$ with general term $(-1)^{n+1}/n$, $n \in \mathbf{N}$; the terms of this sequence remain bounded for all $n \in \mathbf{N}$; moreover, the terms approach 0 as $n \to \infty$

Definition 12.3. A sequence $(a_n)_n$ has *limit L* as $n \to \infty$, a fact denoted by

$$\lim_{n \to \infty} a_n = L \quad \text{or} \quad a_n \to L,$$

---

1 In other words, a vector with infinitely many components.

if for any $\varepsilon > 0$ there is a positive integer N such that

$$n \geq N \quad \Longrightarrow \quad |a_n - L| \leq \varepsilon \qquad (12.1)$$

in which case the sequence is said to *converge to L*.

A sequence which converges to 0 is called a *zero sequence*.

A sequence which does not converge to any limit is said to *diverge*.

Remark 12.4. Note the similarity between this definition and Definition 3.28 of the limit at infinity of a function, $\lim_{x \to \infty} f(x) = L$. The difference is that the variable x changes continuously (over all real numbers), whereas the variable n in (12.1) takes only integer values.

Example 12.5 (Sequence of reciprocals). The sequence

$$(a_n)_n := (1/n)_{n \in \mathbb{N}} = (1, 1/2, 1/3, 1/4, \dots)$$

is a zero sequence. Clearly the reciprocals $1/n$ get arbitrarily close to zero as n increases. Using $L = 0$ in (12.1) we get, for any $\varepsilon > 0$,

$$n \geq N \quad \Longrightarrow \quad |a_n| = 1/n \leq \varepsilon,$$

which is satisfied if N is 1 more than the largest integer being less than or equal to $1/\varepsilon$. This proves that

$$\lim_{n \to \infty} 1/n = 0.$$

Example 12.6 (Alternating sequence). The sequence

$$(b_n)_n := \left((-1)^n/n\right)_{n \in \mathbb{N}} = (-1, 1/2, -1/3, 1/4, \dots)$$

is also a zero sequence. For any $\varepsilon > 0$, let N be 1 more than the largest integer being less than or equal to $1/\varepsilon$. Then for any $n \geq N$ we have

$$|b_n - 0| = |b_n| = 1/n \leq \varepsilon,$$

proving that

$$\lim_{n \to \infty} b_n = 0.$$

Example 12.7 (Subsequence). The sequence

$$(a_{n_k})_{k \in \mathbb{N}_0} = (1/n_k)_{k \in \mathbb{N}_0} = (1/2^k)_{k \in \mathbb{N}_0} = (1, 1/2, 1/4, 1/8, 1/16, \dots)$$

is a subsequence of the sequence in Example 12.5, selected according to the rule $n_k = 2^k$.

MACSYMA-Session 12.1. MACSYMA can plot a sequence of points $\{x_n, y_n\}$ in the xy-plane. In particular, given a sequence $\{a_n\}$, we can represent it graphically by plotting the points $\{(n, a_n)\}$ for $n = 1, 2, \dots, N$ with N sufficiently large.

We illustrate this for the alternating harmonic sequence $\{(-1)^n/n\}$. We first create the sequence of points $\{(n, (-1)^n/n)\}$ for $n = 1, 2, \ldots, 70$.

The first coordinates of these points are:

c1. `xlist:makelist(n,n,1,70)`

d1. $[1, 2, 3, 4, 5, 6, 7, 8, 9, 10, 11, 12, 13, \ldots\ldots, 62, 63, 64, 65, 66, 67, 68, 69, 70]$

and the second coordinates are:

c2. `ylist:makelist((-1)^n/n,n,1,70)`

d2. $\left[-1, \frac{1}{2}, -\frac{1}{3}, \frac{1}{4}, -\frac{1}{5}, \frac{1}{6}, -\frac{1}{7}, \frac{1}{8}, \ldots\ldots, \frac{1}{62}, -\frac{1}{63}, \frac{1}{64}, -\frac{1}{65}, \frac{1}{66}, -\frac{1}{67}, \frac{1}{68}, -\frac{1}{69}, \frac{1}{70}\right]$

with just the beginning and the end of the sequences shown in d1 and d2.

We now plot the sequence:

c3. `graph2(xlist,ylist)$`

obtaining the graph in Fig. 12.1. The points are connected by lines. If you want to plot just the points, use:

c4. `graph2(xlist,ylist,[99])`

where [99] tells MACSYMA to use a line style that does not draw connecting lines and to place a dot at each point of the graph.

A sequence $\{a_n\}$ is a zero sequence if for any $\varepsilon > 0$ there is an N such that

$$|a_n| < \varepsilon \quad \text{for all } n \geq N .$$

We can illustrate this by plotting the sequence of points $\{(n, a_n)\}$ in the xy-plane. If $\{a_n\}$ is a zero sequence, then for any $\varepsilon > 0$, the points $\{(n, a_n)\}$ will eventually lie in the strip between the two parallel horizontal lines $y = \varepsilon$ and $y = -\varepsilon$.

The sequence $\{y_n = (-1)^n/n\}$ is a zero sequence. In fact, for any $\varepsilon > 0$ and an integer $N > 1/\varepsilon$,

$$|y_n| < \varepsilon \quad \text{for all } n \geq N .$$

We illustrate this for $\varepsilon = 0.1$, by plotting:

c5. `graph2([xlist,[0,70],[0,70]], [ylist,[.1,.1],[-.1,-.1]], [99,0,0])`

which plots `ylist` against `xlist` as above, and adds two horizontal lines, $y = 0.1$ and $y = -0.1$. The block [99,0,0] indicates line style with unconnected points (99) for the [xlist,ylist] part and line style with points connected by lines (0) for the $y = 0.1$ and $y = -0.1$ parts.

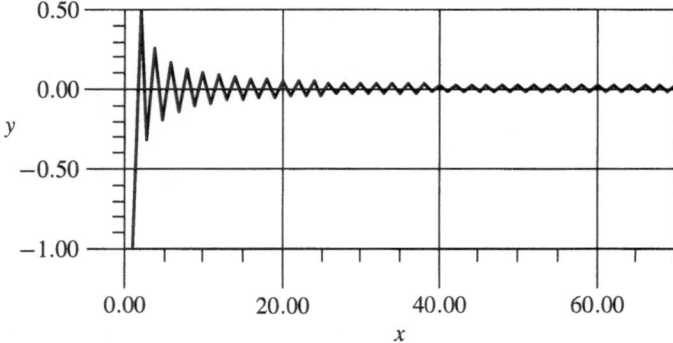

Fig. 12.1. Graph of the alternating harmonic sequence $\{(-1)^n/n\}$, for $n = 1, 2, \ldots, 70$

A sequence $(a_n)_n$ converges to L,

$$\lim_{n\to\infty} a_n = L \quad \text{if and only if} \quad \lim_{n\to\infty} (a_n - L) = 0 ,$$

i.e., the sequence $(a_n - L)_n$ (obtained by subtracting L from each element of $(a_n)_n$) is a zero sequence. To determine convergence of a sequence is therefore equivalent to recognize when a sequence is a zero sequence. The following lemma gives a *comparison rule* which sometimes helps to recognize zero sequences.

Lemma 12.8 (Comparison of sequences). Let $(a_n)_n$ and $(b_n)_n$ be real sequences such that
a. $(b_n)_n$ is a zero sequence, and
b. $|a_n| \leq C|b_n|$ for all $n \in \mathbf{N}$ and some constant $C \in \mathbf{R}$.
   Then $(a_n)_n$ is also a zero sequence. Moreover,
c. if $(a_n)_n$ and $(b_n)_n$ are zero sequences, then so is their sum $(a_n + b_n)_n$

Proof. For any given $\varepsilon > 0$ find $N \in \mathbf{N}$ such that

$$|b_n| \leq \frac{\varepsilon}{C}, \quad \text{for all } n \geq N .$$

This is possible since $(b_n)_n$ is a zero sequence[2]. Then, by assumption b,

$$|a_n| \leq C\,|b_n| \leq C\frac{\varepsilon}{C} = \varepsilon, \quad \text{for all } n \geq N ,$$

proving the $(a_n)_n$ is a zero sequence.
   To prove c, let $(a_n)_n$ and $(b_n)_n$ be zero sequences and select $N \in \mathbf{N}$ such that

$$|a_n| \leq \varepsilon/2 \quad \text{and} \quad |b_n| \leq \varepsilon/2, \quad \text{for all } n \geq N .$$

Then, for all $n \geq N$,

$$|a_n + b_n| \leq |a_n| + |b_n| \leq \varepsilon/2 + \varepsilon/2 = \varepsilon . \qquad \square$$

Remark 12.9. a. Lemma 12.8 can be written as

$$b_n \to 0 \text{ and } |a_n| \leq C|b_n| \ (n \in \mathbf{N}, C \in \mathbf{R}) \quad \Longrightarrow \quad a_n \to 0 ,$$
$$a_n \to 0 \text{ and } b_n \to 0 \quad \Longrightarrow \quad (a_n + b_n) \to 0 .$$

This notation has the advantage of being "epsilon free".
   b. In particular, the lemma states that $(a_n)_n$ is a zero sequence if there is a positive constant C such that, for any $\varepsilon > 0$, there is an $N \in \mathbf{N}$ with

$$|a_n| \leq C\varepsilon, \quad \text{for all } n \geq N .$$

As a first application we have the following

---

2 For the given ε and C, define $\bar{\varepsilon} := \varepsilon/C$. Since $(b_n)_n$ is a zero sequence, there is an $N \in \mathbf{N}$ such that $|b_n| \leq \bar{\varepsilon}$ for all $n \geq N$.

Example 12.10. Let m be any integer greater than or equal to 2. Then the sequence

$$(1/n^m)_n \quad \text{is a zero sequence .}$$

Note: This is a sequence in n, the integer m is fixed.

Proof. The result follows from Lemma 12.8 since for all $n \in \mathbf{N}$ and $m \geq 2$,

$$\frac{1}{n^m} = \frac{1}{n} \cdot \frac{1}{n^{m-1}} \leq \frac{1}{n} . \qquad \square$$

Next we show that the limit of a convergent sequence is *unique*, i.e., a sequence cannot converge to two different values.

Lemma 12.11 (Uniqueness of limit). If a sequence $(a_n)_n$ converges, then its limit is unique.

Proof. Suppose to the contrary that a sequence $(a_n)_n$ has two limits $L_1 \neq L_2$. Select $\varepsilon := |L_2 - L_1|/4$. Then there exist integers

$$N_1 \in \mathbf{N} \quad \text{such that } |a_n - L_1| \leq \varepsilon, \text{ for all } n \geq N_1 ,$$
$$N_2 \in \mathbf{N} \quad \text{such that } |a_n - L_2| \leq \varepsilon, \text{ for all } n \geq N_2 .$$

Taking $N := \max\{N_1, N_2\}$, we therefore have by the triangle inequality

$$|L_2 - L_1| \leq |L_2 - a_n| + |L_1 - a_n| \leq 2\varepsilon = |L_2 - L_1|/2 ,$$

a contradiction. $\qquad \square$

We now give some examples of convergent and divergent sequences.

Example 12.12 (Convergent and divergent sequences). Consider the sequences $(a_n)_n$ with
(a) $a_n := 1$ (b) $a_n := (-1)^n$ (c) $a_n := (-1)^n/n$ (d) $a_n := n$
    a. The constant sequence (1) obviously converges to $a = 1$ as $|a_n - a| = 0$ for all $n \in \mathbf{N}_0$, and so $|a_n - a| \leq \varepsilon$ for each choice of ε and all $n \in \mathbf{N}_0$.
    b. The values of the sequence $((-1)^n)_n$ are alternately 1 and -1. By Lemma 12.11 the sequence cannot converge, as the sequence would have to have two different limits.
    c. As $|a_n| = 1/n$, the proof given in Example 12.5 shows that $a_n \to 0$.
    d. This sequence diverges in a very specific way, namely to $+\infty$. This means that on the one hand there is no *real number* a such that $a_n \to a$, and so (a_n) diverges, but on the other hand for all $M \in \mathbf{R}^+$ there is an index $N \in \mathbf{N}$ such that $a_n \geq M$ for all numbers $n \geq N$. In this situation we say that $(a_n)_n$ *diverges to* $+\infty$ and write

$$\lim_{n \to \infty} a_n = +\infty .$$

Divergence towards $-\infty$ is similarly defined.

The last example is an *unbounded* sequence. This means that there is no real number $A \in \mathbf{R}$ such that $|a_n| \leq A$ for all $n \in \mathbf{N}_0$.

Next we show that convergent sequences always are *bounded*.

Theorem 12.13 (Convergence implies boundedness).   If the real sequence $(a_n)_{n \in \mathbf{N}_0}$ is convergent, then it is bounded, i.e., there is a real number $A \in \mathbf{R}$ such that $|a_n| \leq A$ for all $n \in \mathbf{N}_0$.

Proof. Let $\lim_{n \to \infty} a_n = L$ and assume[3] $L \neq 0$. Choose $\varepsilon := |L|$. Then there is an index $N \in \mathbf{N}$ such that

$$|a_n - L| \leq |L|, \quad \text{for all } n \geq N .$$
$$\therefore \quad |a_n| = |a_n - L + L| \leq |a_n - L| + |L| \leq 2|L|, \quad \text{for all } n \geq N .$$

Let M denote the maximum of the N numbers $|a_n|$ $(n = 0, \ldots, N - 1)$

$$M := \max \{|a_0|, |a_1|, \ldots, |a_{N-1}|\} .$$

Then $|a_n| \leq M, (n = 0, \ldots, N - 1)$, so that

$$|a_n| \leq \begin{cases} M & \text{if } n = 0, \ldots, N - 1, \\ 2|L| & \text{if } n \geq N, \end{cases}$$

and the theorem is proved for $A := \max \{M, 2\,|L|\}$.   □

Theorem 12.13 can be written for short as

$$a_n \to L \implies |a_n| \leq A, \text{ for some } A \in \mathbf{R} .$$

MACSYMA-Session 12.2.   We illustrate the limit of a sequence for the sequence $\{y_n = (2n + 1)/(3n - 2)\}$

c1. `limit((2*n+1)/(3*n-2),n,inf)`

d1. $\dfrac{2}{3}$

If we graph the points $\{(n, y_n)\}$ in the xy-plane, then the points will eventually lie between the two parallel horizontal lines $y = 2/3 + \varepsilon$ and $y = 2/3 - \varepsilon$, for any $\varepsilon > 0$. We illustrate this with the first 70 points of the sequence and $\varepsilon = 0.1$.

c2. `xlist:makelist(n,n,1,70)`

d2. $[1, 2, 3, 4, 5, 6, 7, 8, 9, 10, 11, 12, \ldots\ldots, 62, 63, 64, 65, 66, 67, 68, 69, 70]$

c3. `ylist:makelist((2*n+1)/(3*n-2),n,1,70)`

d3. $\left[3, \frac{5}{4}, 1, \frac{9}{10}, \frac{11}{13}, \frac{13}{16}, \frac{15}{19}, \frac{17}{22}, \frac{19}{25}, \frac{3}{4}, \frac{23}{31}, \frac{25}{34}, \ldots\ldots, \frac{129}{190}, \frac{131}{193}, \frac{19}{28}, \frac{135}{199}, \frac{137}{202}, \frac{139}{205}, \frac{141}{208}\right]$

Now plot:

c4. `graph2([xlist,[0,70],[0,70]],`
    `[ylist,[2/3+eps,2/3+eps],[2/3-eps,2/3-eps]],`
    `[99,0,0]),`
    `eps:0.1 $`

---

3 The case $L = 0$ is easier. Prove the theorem for $L = 0$.

The following theorem gives basic rules for computing limits. These rules state that the limit operation commutes with addition, multiplication, or division (by a nonzero).

Theorem 12.14 (Limit rules).   Let $(a_n)_n$ and $(b_n)_n$ be two sequences with $\lim_{n\to\infty} a_n = a$ and $\lim_{n\to\infty} b_n = b$. Then

a. $\lim_{n\to\infty}(C\,a_n) = C\,a$ for any constant $C \in \mathbf{R}$
b. $\lim_{n\to\infty}(a_n \pm b_n) = a \pm b$
c. $\lim_{n\to\infty} a_n b_n = ab$
d. if $b_n \neq 0$ ($n \in \mathbf{N}_0$) and $b \neq 0$, then $\lim_{n\to\infty} a_n/b_n = a/b$

This theorem is proved in Sect. 12.4.

Remark 12.15 (Properties of limits). a. Taken together, rules a and b of Theorem 12.14 are known as the *linearity property* of the limit. In particular, rule b can be stated as

the limit of a sum = the sum of the limits .

b. Theorem 12.14 is useful for computing limits of sequences $(a_n)_n$ where a_n is a rational function of n. Recall that rational functions are formed by a finite number of additions, subtractions, multiplications, and divisions. Consequently the limit of a rational expression can be found by performing the same operations on the limits; see the following three examples.

Example 12.16. Consider again the limit

$$\lim_{n\to\infty} \frac{2n+1}{3n-2},$$

see MACSYMA-Session 12.2. An application of Theorem 12.14 gives

$$\lim_{n\to\infty} \frac{2n+1}{3n-2} = \lim_{n\to\infty} \frac{2+1/n}{3-2/n} = \frac{2+\lim_{n\to\infty} 1/n}{3-\lim_{n\to\infty} 2/n} = \frac{2+0}{3-0} = \frac{2}{3}.$$

Example 12.17.

$$\lim_{n\to\infty} \frac{\sum_{k=0}^{5} kn^k}{\sum_{k=1}^{3} n^{2k-1}} = \lim_{n\to\infty} \frac{(5n^5 + 4n^4 + 3n^3 + 2n^2 + n)/n^5}{(n^5 + n^3 + n)/n^5}$$

$$= \frac{\lim_{n\to\infty} (5 + 4/n + 3/n^2 + 2/n^3 + 1/n^4)}{\lim_{n\to\infty} (1 + 1/n^2 + 1/n^4)} = \frac{5}{1} = 5.$$

Example 12.18. Here is a slightly more complicated version of the last example. Let $K \in \mathbb{N}$ be given. Then

$$\lim_{n \to \infty} \frac{\sum_{k=1}^{2K-1} kn^k}{\sum_{k=1}^{K} n^{2k-1}} = \lim_{n \to \infty} \frac{\sum_{k=1}^{2K-1} kn^{k-(2K-1)}}{\sum_{k=1}^{K} n^{2k-1-(2K-1)}}$$

$$= \frac{\lim_{n \to \infty} (1/n^{2K-2} + 2/n^{2K-3} + 3/n^{2K-4} + \cdots + (2K-2)/n + (2K-1))}{\lim_{n \to \infty} (1/n^{2K-2} + 1/n^{2K-4} + 1/n^{2K-6} + \cdots + 1/n^2 + 1)}$$

$$= 2K - 1 .$$

Example 12.19 (Limits involving square roots). The limit of the sequence $(a_n)_n$ with $a_n := \sqrt{n+1} - \sqrt{n}$,

$$\lim_{n \to \infty} \left(\sqrt{n+1} - \sqrt{n} \right) ,$$

is computed here using a special trick. We have

$$\sqrt{n+1} - \sqrt{n} = \frac{(\sqrt{n+1} - \sqrt{n})(\sqrt{n+1} + \sqrt{n})}{\sqrt{n+1} + \sqrt{n}} = \frac{(n+1) - n}{\sqrt{n+1} + \sqrt{n}}$$

$$= \frac{1}{\sqrt{n+1} + \sqrt{n}} \leq \frac{1}{2\sqrt{n}} \to 0$$

as $n \to \infty$. Other examples of this type (solvable by some trick) are given in Exercise 12.3.

The definition of convergence, Definition 12.3, assumes that the limit is known. Is it possible to determine that a sequence $(a_n)_n$ is convergent without knowing its limit? The answer is yes, but first we need the following definition.

Definition 12.20 (Cauchy sequence). A real sequence is called a Cauchy sequence if for any $\varepsilon > 0$ there is an index $N \in \mathbb{N}$ such that

$$|a_n - a_m| \leq \varepsilon, \quad \text{for all } m, n \geq N .$$

We also require the following theorem, whose proof is given in Sect. 12.4.

Theorem 12.21 (Bolzano–Weierstrass theorem[4]). Every bounded sequence has a convergent subsequence.

---

4 The Bolzano–Weierstrass theorem is very useful and important throughout mathematics.

We can finally prove a *convergence criterion* due to Cauchy which is very useful because it does not require knowledge of the actual limit.

Theorem 12.22 (Cauchy criterion). A real sequence $(a_n)_n$ is convergent if and only if it is a Cauchy sequence.

Also this theorem is proved in Sect. 12.4.

Exercises

12.1   Try to solve Example 12.18 with MACSYMA.
       Hint: MACSYMA is able to find *explicit* expressions for the sums in
       Example 12.18, however it doesn't find the limit.

12.2   Consider the following sequences of real numbers. Determine which of them are
       convergent sequences, and find the limit for the convergent sequences.

   (a) $\left(n + \dfrac{1}{n+2}\right)_{n \in \mathbb{N}_0}$ (b) $\left(\dfrac{\sin n}{n}\right)_{n \in \mathbb{N}_0}$ (c) $(n)_{n \in \mathbb{N}_0}$

   (d) $(\sin n)_{n \in \mathbb{N}_0}$

12.3   Find the limits of the following expressions as $n \to \infty$.

   (a) $\dfrac{7n^2 + 3n - 1}{n^3 + 2}$ (b) $\sqrt{n}(\sqrt{n+1} - \sqrt{n})$ (c) $\dfrac{1}{n^2} \sum_{k=1}^{n} k$

   (d) $\sqrt{n + \sqrt{n}} - \sqrt{n}$ (e) $\sqrt{9n^2 + 2n + 1} - 3n$

12.4   For each of the sequences listed below write down the first five terms

   (a) $\left\{\dfrac{k}{k+1}\right\}_{k=1}^{\infty}$ (b) $\left\{\dfrac{k!}{(k+2)!}\right\}_{k=1}^{\infty}$ (c) $\left\{\dfrac{\ln k}{k+1}\right\}_{k=1}^{\infty}$

   (d) $\left\{\dfrac{(-1)^k}{k}\right\}_{k=1}^{\infty}$ (e) $\left\{\dfrac{(2k+1)!}{(k+2)!}\right\}_{k=1}^{\infty}$ (f) $\left\{k \sin \dfrac{\pi}{k}\right\}_{k=1}^{\infty}$

   (g) $\left\{\dfrac{1}{k} - \dfrac{1}{k+1}\right\}_{k=1}^{\infty}$ (h) $\left\{\dfrac{k-1}{k} - \dfrac{k}{k+1}\right\}_{k=1}^{\infty}$

12.5   Find the limit of the sequences $\{a_k\}_{k=1}^{\infty}$ given below

   (a) $a_n = \sum_{1}^{n} \dfrac{k}{n^2}$ (b) $a_k = \dfrac{k!}{(k+2)!}$ (c) $a_k = \dfrac{\ln k}{k+1}$

   (d) $a_k = \dfrac{k^2}{e^k}$ (e) $a_n = \sum_{k=1}^{n} \dfrac{k^2}{n^3}$ (f) $a_n = \sum_{k=1}^{n} \dfrac{k^3}{n^4}$

   (g) $a_k = \dfrac{1}{k} - \dfrac{1}{k+1}$ (h) $a_k = \dfrac{k-1}{k} - \dfrac{k}{k+1}$

12.6   Below we have listed the generic term a_k in a sequence where $k = 1, 2, \ldots$ For
       each of these sequences use MACSYMA to compute the limiting value $\lim_{k \to \infty} a_k$.

   (a) $a_n = \sum_{k=1}^{n} \dfrac{1}{2 + k/n} \dfrac{1}{n}$ (b) $a_n = \dfrac{1}{n^4} \sum_{k=1}^{n} k^3$ (c) $a_n = \dfrac{1}{n^5} \sum_{k=1}^{n} k^4$

   (d) $k = \dfrac{\sin \pi/k}{k}$ (e) $a_k = \left(1 + \dfrac{1}{k}\right)^k$ (f) $a_k = (k!)^{1/k}$

   (g) $a_n = \dfrac{1 + 1/2 + \cdots + 1/n}{n}$ (h) $a_n = \dfrac{1}{n^2 + 1} + \dfrac{2}{n^2 + 2^2} + \cdots + \dfrac{n}{n^2 + n^2}$

12.7 Sometimes MACSYMA cannot evaluate a sequence, but we can help by using more information. With this in mind we state the following theorem.

Theorem 12.23. If the $a_k \to L$ as $k \to \infty$, then the *arithmetic average* tends to the same value, that is,

$$\lim_{k \to \infty} \frac{a_0 + a_1 + \cdots + a_n}{n+1} \to L.$$

Try to evaluate the following sequences first by using MACSYMA directly. Then apply Theorem 12.23 to the problem.

(a) $a_n = \dfrac{1 + 1/2 + \cdots + 1/n}{n}$

(b) $a_n = (n)^{1/n}$

(c) $a_n = \dfrac{\sum_1^n k^{1/k}}{n}$

(d) $a_n = \left(\dfrac{(n+1)^n}{n!}\right)^{1/n}$

(e) $a_k = \sum_1^n \dfrac{k}{n^2}$

(f) $a_k = \sum_1^n \left(\dfrac{k}{n}\right)^n$

(g) $a_n = \dfrac{1 + 1/2 + \cdots + 1/n}{n}$

(h) $a_n = \dfrac{1}{n}((n+1)(n+2) \cdots (n+n))$

12.8 Use MACSYMA to compute the limits of the sequences whose generic term a_k is given below.

(a) $a_n = \dfrac{1 + 1/2 + \cdots + 1/n}{\ln n}$

(b) $a_n = n(2^{1/n} - 1)$

(c) $a_n = n(2^{1/n} - 1)$

(d) $a_n = \sum_{k=1}^n \dfrac{4+k}{k^2}$

(e) $n = n(n^{1/n} - 1)$

(f) $a_n = \sum_{k=1}^\infty \left(\dfrac{k}{n}\right)^n$

(g) $a_n = \dfrac{\ln 1 + \ln 2 + \cdots + \ln n}{\ln n^n}$

(h) $a_n = 1 - \dfrac{1}{3} + \dfrac{1}{5} + \cdots + \dfrac{(-1)^{n-1}}{2n - 1}$

12.9 Find the limits

(a) $\lim_{n \to \infty} \sqrt[3]{n(n+1)^2} - \sqrt[3]{n^2(n+1)}$

(b) $\lim_{n \to \infty} \sqrt[3]{n^2}\left(\sqrt[3]{n+1} - \sqrt[3]{n}\right)$

12.10 Find $\lim_{n \to \infty} n^{(k-1)/k}\left((n+1)^{1/k} - n^{1/k}\right)$.

12.11 Show that each subsequence of a convergent sequence $(a_n)_n$ converges to the same limit.

12.12 The following sequences converge. Calculate their limits.

(a) $a_n := \left(1 + \dfrac{1}{3}\right)\left(1 + \left(\dfrac{1}{3}\right)^2\right)\left(1 + \left(\dfrac{1}{3}\right)^4\right) \cdots \left(1 + \left(\dfrac{1}{3}\right)^{2^n}\right)$

(b) $b_n := \left(1 - \dfrac{1}{4}\right)\left(1 - \dfrac{1}{9}\right)\left(1 - \dfrac{1}{16}\right) \cdots \left(1 - \dfrac{1}{n^2}\right)$

(c) (nested square root)

$c_0 := 1, c_{n+1} := \sqrt{1 + c_n}$, numerically

$c_0 := x, c_{n+1} := \sqrt{1 + c_n}$, symbolically

(d) (continued fraction)

$d_0 := 1, d_{n+1} := 1/(1 + d_n)$, numerically

$d_0 := x, d_{n+1} := 1/(1 + d_n)$, symbolically

Further calculate $c_1, c_2, \ldots, c_5$ and $d_1, d_2, \ldots, d_{10}$ in the symbolic cases of c
and d. Do you see the pattern? How will d_{11} look?
Hints: a. Multiply a_n with the term $(1 - 1/3) = 2/3$.
b. Search for factors in b_n that can be reduced.
c and d. Use MACSYMA.

12.13 Each of the following sequences (a_n) is convergent, with the given limit α, say.
Find an integer N such that if $n \geq N$, then $|a_n - \alpha| \leq 1/10$.

(a) $a_n = \dfrac{1}{n}, \quad \alpha = 0$

(b) $a_n = \dfrac{\sin n}{n}, \quad \alpha = 0$

(c) $a_n = \dfrac{n^2 + 1}{(n^2 + \sin n)}, \quad \alpha = 1$

(d) $a_n = \dfrac{\ln n}{\sqrt{n}}, \quad \alpha = 0$

Use MACSYMA to find the lowest such integer in each case.

12.2 Series

Definition 12.24 (Series). A series $(s_n)_n$ is a special sequence $(s_n)_n$

$$s_n := \sum_{k=0}^{n} a_k$$

obtained by successive sums of the terms of another sequence.

Series arise naturally in numerical work.

Example 12.25 (Series as result of long division). The fraction 10/9 is computed
by long division,

```
                    1. 11...
        9   |  10. 000...
          -    9.
               ‾‾‾‾‾
               1. 0
            -   . 9
               ‾‾‾‾‾
                . 10
            -   . 09
               ‾‾‾‾‾
                . 01
                  ⋮
```

giving a representation of 10/9 as a series

$$\frac{10}{9} = 1 + .1 + .01 + .001 + \ldots = 1 + \frac{1}{10} + \frac{1}{100} + \ldots = \sum_{k=0}^{\infty} \frac{1}{10^k}. \quad (12.2)$$

All irrational numbers, and some rational numbers (such as 10/9), are represented by infinite sums

$$\sum_{k=0}^{\infty} a_k = \sum_{k=0}^{\infty} \frac{t_k}{10^k} \,, \tag{12.3}$$

where each t_k is one of the 10 integers $0, 1, \ldots, 9$.

The ancient Greek mathematicians had difficulties with such series. Indeed, the sum (12.2) appears in the following *paradox of Zeno*[5].

Example 12.26 (Achilles and the turtle). Achilles (the fastest runner of antiquity) has a speed of 10 m/s. The tortoise (a rather slow animal) can crawl 1 m/s. When a race between Achilles and the tortoise was arranged, it was agreed that the tortoise will get 100 m head start.

The distances of the two contestants from the start line are plotted in Fig. 12.2. At time zero, when the race begins, Achilles is at the start line, and the tortoise is 100 m away.

The race begins, and Achilles covers these 100 m in 10 s. However, the tortoise has meantime advanced 10 m and is ahead. Achilles runs these 10 m in 1 s, only to find that the tortoise crawled another meter. Achilles (close to exhaustion) runs this meter in one tenth of a second, but the tortoise has not stood still and is now 1/10 m ahead, etc.

Arguing like this, Zeno wondered how will Achilles ever overtake the tortoise. He knew of course that Achilles is the eventual winner.[6]

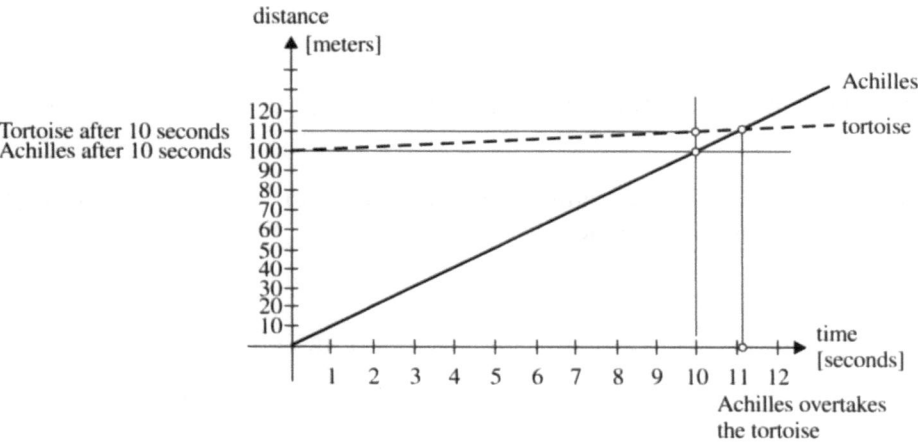

Fig. 12.2. Race between Achilles and the tortoise

5 Zeno of Elea (490–430 B.C.)

6 The race would be over in $10 + 1 + 1/10 + 1/10^2 + 1/10^3 + \ldots = 100/9$ [s] at the point in time when the lines in Fig. 12.2 intersect.

Definition 12.27 (Convergence of a series). The *limit of a series*

$$\sum_{k=0}^{\infty} a_k := \lim_{n \to \infty} \sum_{k=0}^{n} a_k$$

is the limit of its partial sums, whenever this limit exists. In this case we say that the series converges. Otherwise the series diverges.

The representations (12.3) of real numbers use series $\sum_{k=0}^{\infty} a_k$ of numbers a_k that form a zero sequence. We show next that this is a necessary condition for a series to converge.

Lemma 12.28. If the series $\sum_{k=0}^{\infty} a_k$ converges, then the sequence $(a_k)_k$ is a zero sequence.

Proof. We use the Cauchy criterion. Let $\varepsilon > 0$ be given. Then as $(s_n)_n := \left(\sum_{k=0}^{n} a_k\right)_n$ converges, by Theorem 12.22 there is some $N \in \mathbf{N}$ such that

$$|s_n - s_m| = \left| \sum_{k=m+1}^{n} a_k \right| \le \varepsilon, \quad \text{for all } m, n \ge N \;.$$

Choosing $m := n - 1$, we get

$$|a_n| \le \varepsilon$$

for all $n \ge N + 1$. As ε was arbitrary, we have $|a_n| \to 0$. $\qquad\square$

The above condition is not sufficient for convergence, i.e., it is possible for $(a_n)_n$ to be a zero sequence, but the series $\sum_{k=0}^{\infty} a_k$ diverges. Example follows.

Example 12.29 (Harmonic series). Consider the series

$$\sum_{k=1}^{\infty} \frac{1}{k}, \qquad\qquad (12.4)$$

called the harmonic series. We shall show that the partial sums

$$\sum_{k=1}^{n} \frac{1}{k} = 1 + \frac{1}{2} + \frac{1}{3} + \cdots + \frac{1}{n}$$

are not bounded, and so the series (12.4) cannot converge.

Proof. Consider the special partial sums with $n := 2^N$ for some $N \in \mathbf{N}$. For these we get

$$1 + \frac{1}{2} + \frac{1}{3} + \cdots + \frac{1}{2^N} = 1 + \frac{1}{2} + \left(\frac{1}{3} + \frac{1}{4}\right) + \left(\frac{1}{5} + \frac{1}{6} + \frac{1}{7} + \frac{1}{8}\right) + \cdots$$
$$+ \left(\frac{1}{2^{N-1}+1} + \frac{1}{2^{N-1}+2} + \cdots + \frac{1}{2^N}\right)$$
$$\geq 1 + \frac{1}{2} + \underbrace{\left(\frac{1}{4} + \frac{1}{4}\right)}_{2 \text{ terms}} + \underbrace{\left(\frac{1}{8} + \frac{1}{8} + \frac{1}{8} + \frac{1}{8}\right)}_{4 \text{ summands}} + \cdots$$
$$+ \underbrace{\left(\frac{1}{2^N} + \frac{1}{2^N} + \cdots + \frac{1}{2^N}\right)}_{2^{N-1} \text{ summands}}$$
$$= 1 + \underbrace{\frac{1}{2} + \frac{1}{2} + \cdots + \frac{1}{2}}_{N \text{ summands}} = 1 + \frac{N}{2}.$$

So, as n tends to ∞, $N \to \infty$, and the partial sums considered tend to ∞, too. Thus, the harmonic series (12.4) cannot converge by Theorem 12.13. $\square$

Example 12.30 (Geometric series). A very useful series is the geometric series

$$\sum_{k=0}^{\infty} ar^k = a + ar + ar^2 + ar^3 + \cdots .$$

The partial sums

$$s_n = a + ar + ar^2 + ar^3 + \cdots + ar^n$$

can be computed by the following trick: Subtract $r s_n$ from s_n

$$
\begin{array}{lllllllll}
s_n & = & a & + & ar & + & ar^2 & + & ar^3 & + & \cdots & + & ar^n \\
-r\, s_n & = & & & -\,(ar & + & ar^2 & + & ar^3 & + & \cdots & + & ar^n & + & ar^{n+1}), \\
\hline
(1-r)\, s_n & = & a & & & & & & & & & & & & -\,ar^{n+1}
\end{array}
$$

to obtain

$$s_n (1 - r) = a - ar^{n+1},$$

$$\text{or, if } r \neq 1, \quad s_n = a\frac{1 - r^{n+1}}{1 - r}.$$

For $|r| < 1$ these partial sums have limit

$$\lim_{n \to \infty} a\frac{1 - r^{n+1}}{1 - r} = a\frac{1 - \lim_{n \to \infty} r^{n+1}}{1 - r}$$

$$= \frac{a}{1 - r}.$$

Therefore the series

$$\sum_{k=0}^{\infty} ar^k = \begin{cases} a/(1-r) & \text{if } |r| < 1, \\ \text{diverges} & \text{if } |r| \geq 1. \end{cases} \tag{12.5}$$

As a particular example, consider again the case of Achilles and the turtle. There $a := 100\,[\text{m}]$ and $r = 1/10$. The distance traveled to "takeover" is given by the series

$$\sum_{k=0}^{\infty} 100\left(\frac{1}{10}\right)^k = \frac{100}{1 - 1/10} = \frac{1000}{9} \approx 111.111\,[\text{m}], \quad \text{see (12.2)}.$$

Example 12.31 (Partial summation). Another example where the partial sums can be computed conveniently is the series

$$\sum_{k=1}^{\infty} \frac{1}{k(k+1)} = \frac{1}{1 \cdot 2} + \frac{1}{2 \cdot 3} + \frac{1}{3 \cdot 4} + \frac{1}{4 \cdot 5} + \cdots.$$

Here we use a partial fraction expansion and get

$$s_n = \sum_{k=1}^{n} \frac{1}{k(k+1)} = \sum_{k=1}^{n}\left(\frac{1}{k} - \frac{1}{k+1}\right) = \sum_{k=1}^{n} \frac{1}{k} - \sum_{k=1}^{n} \frac{1}{k+1}.$$

Writing out the last difference

$$\sum_{k=1}^{n} \frac{1}{k} = 1 + \frac{1}{2} + \frac{1}{3} + \frac{1}{4} + \cdots + \frac{1}{n}$$

$$-\sum_{k=1}^{n} \frac{1}{k+1} = \qquad - \left(\frac{1}{2} + \frac{1}{3} + \frac{1}{4} + \cdots + \frac{1}{n} + \frac{1}{n+1}\right)$$

$$\rule{11cm}{0.4pt}$$

$$s_n = 1 \qquad\qquad\qquad\qquad\qquad\qquad\qquad - \frac{1}{n+1}$$

we get, after a massive cancellation,

$$s_n = 1 - \frac{1}{n+1}.$$

Therefore

$$\sum_{k=1}^{\infty} \frac{1}{k(k+1)} = \lim_{n\to\infty} s_n = \lim_{n\to\infty}\left(1 - \frac{1}{n+1}\right) = 1.$$

Exercises

12.14 Show that the following series converge, and find their limits

(a) $\displaystyle\sum_{k=1}^{\infty} \frac{1}{k\,(k+1)\,(k+2)}$

(b) $\displaystyle\sum_{k=1}^{\infty} \frac{1}{k\,(k+1)\,(k+2)\,(k+3)}$

(c) $\displaystyle\sum_{k=1}^{\infty} \frac{1}{4k^2 - 1}$

12.15 Calculate how many terms of the series $\sum_{k=1}^{\infty} 1/k$ are needed for the partial sum to be greater than

(a) 1   (b) 5   (c) 10

12.16 Find a number $a < 2.8$ such that $\sum_{k=0}^{n} 1/k! < a$ for all $n \in \mathbf{N}$.

12.3 Convergence criteria for series

If a series $\sum_{k=0}^{\infty} a_k$ converges, then the underlying sequence $(a_k)_k$ is a zero sequence. Example 12.29 shows that further conditions must be imposed on a zero sequence $(a_k)_k$ to assure convergence of the series $\sum_{k=0}^{\infty} a_k$. Such conditions (convergence criteria) are studied in this section. By comparing a series with another series known to converge, we can get useful criteria. The simplest such comparison test is now stated as a theorem.

Theorem 12.32 (Comparison test). Let $0 \le a_k \le b_k$, for all k. If the series $\sum_{k=1}^{\infty} b_k$ converges, so does $\sum_{k=1}^{\infty} a_k$.

Proof. The partial sums of these series satisfy

$$s_n := a_1 + \cdots + a_n \le S_n := b_1 + \cdots + b_n \,,$$

where the sequence $(S_n)_n$ is bounded (since it is convergent). The sequence $(s_n)_n$ is thus bounded, and increasing. It converges by Theorem 12.52. $\qquad\square$

Example 12.33 (Comparison test). Consider

$$\sum_{k=1}^{\infty} a_k = \sum_{k=1}^{\infty} \frac{2 + \operatorname{atan}^2(k)}{2^{k+1} + 1} \,.$$

It is not difficult to see that

$$0 \le \frac{2 + \operatorname{atan}^2(k)}{2^{k+1} + 1} \le \frac{3}{2^{k+1} + 1} \le \frac{3}{2^{k+1}} \,,$$

which compares $\sum a_k$ with a convergent geometric series.

Theorem 12.34 (Comparison test for series with positive coefficients). If the series $\sum a_k$ and $\sum b_k$ have only positive terms, and the ratio a_k/b_k tends to a nonzero limit as $k \to \infty$, then either both series are convergent or both series are divergent.

Proof. Suppose that $\lim a_k/b_k = \lambda > 0$. Then for all k sufficiently large we have the inequalities

$$\lambda/2 \le a_k/b_k \le 2\lambda ,$$

so $b_k \le (2/\lambda)a_k$ and $a_k \le 2\lambda b_k$. If the series $\sum a_k$ converges, so does $\sum (2/\lambda)a_k$ and by Theorem 12.32 so does $\sum b_k$. Also, if $\sum b_k$ converges, so does $\sum 2\lambda b_k$, and hence so does $\sum a_k$. The remarks on the simultaneous divergence follow similarly. $\qquad\square$

Note the condition that $b_k \ge a_k$ for all k in Theorem 12.32 can be relaxed to read, "if for all k larger than $n \in \mathbf{N}$, $b_k \ge a_k$."[7] We need apply the theorem only to the terms with index $k \ge n$.

Similarly, in Theorem 12.34, the condition that a_k and b_k are positive for all k can be replaced by the condition that a_k and b_k are positive for all k sufficiently large.

Here are some examples which illustrate the last two theorems.

Example 12.35. Consider the series

$$\sum_{n=1}^{\infty} \frac{n+1}{3^n} .$$

We now show for all $n \ge 4$ that

$$\frac{n+1}{3^n} \le \frac{1}{2^n}, \quad \text{or equivalently} \quad \left(\frac{2}{3}\right)^n \le \frac{1}{n+1} .$$

The last inequality is true for $n = 4$ as a calculation shows. Treating n as a continuous variable and plotting the two functions $f(n) := (2/3)^n$ and $g(n) := 1/(n+1)$ we see that $f(n)$ lies below $g(n)$ for $n \ge 4$. These estimates tell us that this series can be *compared* with the geometric series $\sum (1/2^n)$, which is convergent.

Example 12.36. Consider the series

$$\sum_{k=0}^{\infty} \frac{a^k}{k!} ,$$

where $a > 0$, and determine whether it converges or diverges. We compare this series with a convergent series of the form $\sum \lambda/2^k$. To see how to do this, break up the general term of the series as follows:

$$\frac{a^k}{k!} = \frac{a^n \cdot a^{k-n}}{n! \cdot (n+1) \cdot (n+2) \cdots (k-2) \cdot (k-1) \cdot k} .$$

---

7 Indeed, any finite number of terms in an infinite series cannot affect whether the series converges or not.

Here we suppose $k \geq n$. Now choose n to be so large that $a/n \leq 1/2$. For $k \geq n$ we have then the inequality

$$\frac{a^{k-n}}{(n+1) \cdot (n+2) \cdots (k-2) \cdot (k-1) \cdot k}$$

$$= \frac{a}{n+1} \frac{a}{n+2} \frac{a}{n+3} \cdots \frac{a}{k-1} \frac{a}{k} \leq \left(\frac{1}{2}\right)^k ,$$

which we multiply by $a^n/n!$ to obtain

$$\frac{a^k}{k!} \leq \frac{a^n}{n!} \left(\frac{1}{2}\right)^{k-n} = \frac{(2a)^n}{n!} \left(k\frac{1}{2}\right)^k ;$$

hence if we set $\lambda = (2a)^n/n!$ in our comparison series, it is seen that the series $\sum a^k/k!$ converges for any positive value of a.

Example 12.37. Consider

$$\sum_{k=1}^{\infty} \frac{k^k}{k!} .$$

Here the general term is bounded below by 1 since

$$\frac{k^k}{k!} = \frac{k \cdot k \cdots k}{1 \cdot 2 \cdots k} \geq 1 ;$$

hence, by Lemma 12.28, we see this series cannot converge.

Example 12.38. Determine whether the series

$$\sum_{k=1}^{\infty} \frac{k+2}{k!}$$

converges or diverges. For this series we take as our comparison series

$$\sum_{k=0}^{\infty} \frac{1}{k!} = \sum_{k=1}^{\infty} \frac{1}{(k-1)!} .$$

The ratio of the general terms of these series is

$$\frac{(k+2)/k!}{1/(k-1)!} = \frac{k+2}{k} \to 1$$

as $k \to \infty$. Therefore the given series converges.

The following test is probably the most important of all convergence tests for series.

Theorem 12.39 (Ratio test). Let $\sum a_k$ be a series of positive terms for which $\lim_{k \to \infty} a_{k+1}/a_k = \lambda$ exists. Then

if $\lambda < 1$, the series converges;
if $\lambda > 1$, the series diverges;
if $\lambda = 1$, the test yields no information.

Proof. If $\lambda < 1$, then there exists a positive number $r < 1$ with $\lambda < r$, such that for all $k \geq n = n(r)$ one has

$$a_{k+1}/a_k \leq r .$$

Then one has

$$a_{n+1} \leq r\, a_n, \quad a_{n+2} \leq r^2\, a_n, \quad \ldots, \quad a_{n+p} \leq r^p\, a_n .$$

Consequently the series $\sum a_k$ can be compared with the geometric series $a_n \sum r^k$, which converges since $r < 1$. This proves the first case.

If $\lambda > 1$, then there exists a positive number $r > 1$ such that for all $k \geq n = n(r)$ one has

$$a_{k+1}/a_k > r .$$

Thus $a_{n+p} \geq r^p\, a_n$ so $\lim_{k \to \infty} a_k \neq 0$, and the series does not converge. This proves the second case.

For the last case consider the series $\sum_{k=1}^{\infty} 1/k^2$ and $\sum_{k=1}^{\infty} k$. Example 12.31 shows that the first series is bounded

$$\sum_{k=1}^{n} \frac{1}{k^2} = 1 + \sum_{k=2}^{n} \frac{1}{k^2} \leq 1 + \sum_{k=2}^{n} \frac{1}{k(k-1)} = 1 + \sum_{k=1}^{n-1} \frac{1}{k(k+1)} \leq 2 ,$$

so that convergence follows from Theorem 12.52. The second series is obviously divergent. In each case, however,

$$\lim_{k \to \infty} \frac{a_{k+1}}{a_k} = 1 . \qquad \square$$

We remark that later we will see that

$$\sum_{k=1}^{\infty} \frac{1}{k^2} = \frac{\pi^2}{6} . \tag{12.6}$$

Example 12.40. The convergence of the series

$$\sum_{k=1}^{\infty} \frac{2^k (k+1)!}{2 \cdot 5 \cdot 8 \cdots (3k-1)}$$

can be decided by the ratio test. A bit of simplification yields

$$\frac{2^{k+1}(k+2)!/2 \cdot 5 \cdot 8 \cdots (3k+2)}{2^k(k+1!)/2 \cdot 5 \cdot 8 \cdots (3k-1)} = \frac{2(k+2)}{3k+2} .$$

This term tends to $2/3$ as $k \to \infty$. The series is then seen to be convergent.

Example 12.41. The series

$$\sum_{k=1}^{\infty} \frac{1}{k^p}, \quad p > 0,$$

cannot be tested by the ratio test. Indeed,

$$\lim_{k \to \infty} \frac{(k+1)^{-p}}{k^{-p}} = \lim_{k \to \infty} \left(1 + \frac{1}{k}\right)^{-p} = 1.$$

Another important test is the root test.

Theorem 12.42 (Root test). Let $\sum a_k$ be a series of positive terms for which $\lim_{k \to \infty} \sqrt[k]{a_k} = \lambda$ exists. Then
if $\lambda < 1$, the series converges;
if $\lambda > 1$, the series diverges;
if $\lambda = 1$, the test yields no information.

Note that again $\lambda \geq 0$.

Proof. If $\lambda < 1$, then there exists a positive number $r < 1$ with $\lambda < r$, such that for all $k \geq n = n(r)$ one has

$$\sqrt[k]{a_k} \leq r \quad \text{or} \quad a_k \leq r^k .$$

Consequently the series $\sum a_k$ can be compared with the geometric series $\sum r^k$, which converges since $r < 1$. This proves the first case.

If $\lambda > 1$, then there exists a positive number $r > 1$ such that for all $k \geq n = n(r)$ one has

$$\sqrt[k]{a_k} > r \quad \text{or} \quad a_k > r^k .$$

Thus $a_k > 1$ for all k, and so $\lim_{k \to \infty} a_k \neq 0$, and the series does not converge. This proves the second case.

The same examples as in Theorem 12.39 prove the last case. $\qquad\square$

The following test is quite useful for showing convergence of some series.

Theorem 12.43 (Integral test). Let (a_k), for $k \in \mathbf{N}_0$, be a zero sequence of positive terms, and let $f(x)$ be a continuous nonincreasing function on the half-infinite interval $(1, \infty)$, which takes on the values $f(k) = a_k$, for $k = 1, 2, 3, \ldots$ Then the series and the integral

$$\sum a_k \quad \text{and} \quad \int_1^{\infty} f(x) \, dx$$

either converge or diverge simultaneously (see Fig. 12.3).

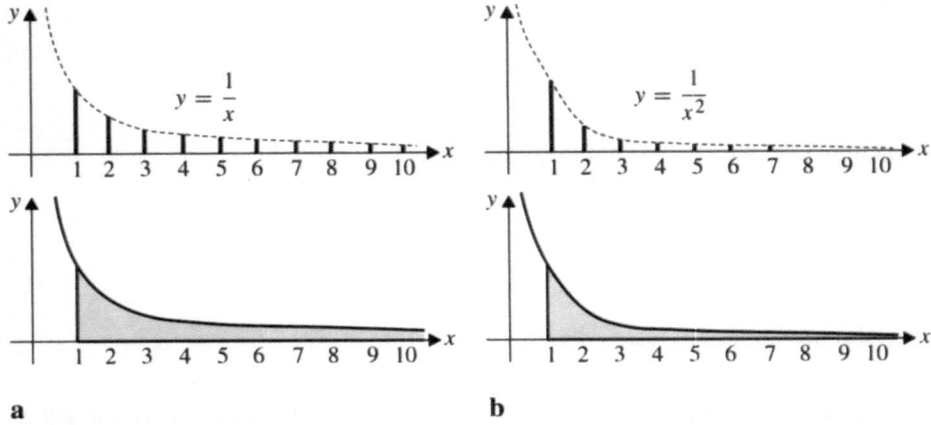

Fig. 12.3 a, b. Integral test. **a** Series $\sum_1^\infty 1/n$ and integral $\int_1^\infty 1/x\, dx$ diverge. **b** Series $\sum_1^\infty 1/n^2$ and integral $\int_1^\infty 1/x^2\, dx$ converge

The theorem can be altered to treat series which either begin with the term a_k or are used only for the remaining terms in a series after, say, the term a_{k-1}.

Proof. We recall from the hypothesis that

$$f(1) = a_1, \ f(2) = a_2, \dots, \ f(k) = a_k \dots$$

In Fig. 12.4a the rectangles are drawn so that their areas represent the terms in the summation, but lie *below* the curve $y = f(x)$. Consequently, the total area of these rectangles is less than the area under the curve. This leads to the estimate

$$a_2 + a_3 + \cdots + a_n \le \int_1^n f(x)\, dx \,,$$

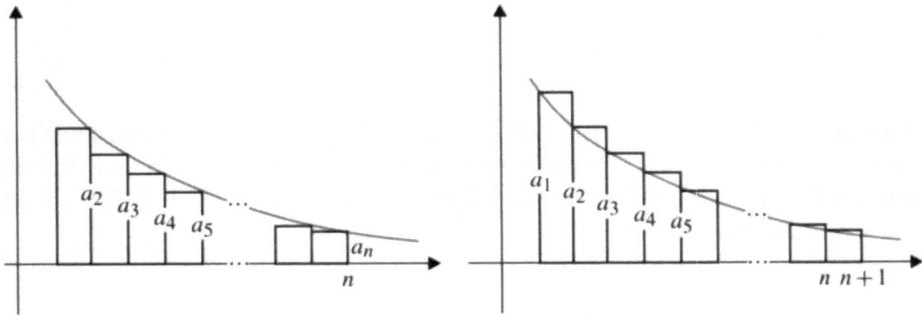

Fig. 12.4 a, b. Comparison of integral and series. **a** $\sum_1^n a_k \ \le \ a_1 + \int_1^n f(x)\, dx$. **b** $\int_1^{n+1} f(x)\, dx \le \sum_1^{n+1} a_k$

or what is the same thing

$$a_1 + a_2 + \cdots + a_n \leq a_1 + \int\limits_1^n f(x)\,dx\,,$$

which implies, if the integral converges, then the series must also converge. On the other hand, as we have drawn the rectangles in Fig. 12.4b they lie above the curve; hence, we obtain the inequality

$$\int\limits_1^{n+1} f(x)\,dx \leq a_1 + a_2 + \cdots + a_n\,.$$

This then yields an upper and lower bound for the partial sum $s_n = a_1 + \cdots + a_n$, namely,

$$\int\limits_1^{n+1} f(x)\,dx \leq s_n \leq a_1 + \int\limits_1^n f(x)\,dx\,.$$

By taking limits as $n \to \infty$, assuming the integrals exists, we have

$$\int\limits_1^\infty f(x)\,dx \leq \sum_{k=1}^\infty a_k \leq a_1 + \int\limits_1^\infty f(x)\,dx\,,$$

which is the desired result. $\qquad\qquad\qquad\qquad\qquad\qquad\qquad\qquad\qquad$ □

Example 12.44. The convergence of the *p*-series

$$\sum \frac{1}{k^p} = 1 + \frac{1}{2^p} + \frac{1}{3^p} + \cdots$$

can be decided by the integral test. We recall that if $p \neq 1$, then

$$\int\limits_1^\infty \frac{1}{x^p}\,dx = \lim_{L \to \infty} \int\limits_1^L \frac{1}{x^p}\,dx = \begin{cases} \frac{1}{p-1} & \text{if } p > 1, \\ \infty & \text{if } p \leq 1. \end{cases}$$

The case where $p = 1$ is the harmonic series, which we have already seen to be divergent.

MACSYMA-Session 12.3. We illustrate the use of the integral test.
   The series
c1. `sum(1/(k+1)/(log(k+1))^2,k,1,inf)`
d1. $\quad \displaystyle\sum_{k=1}^\infty \frac{1}{(k+1)\log^2(k+1)}$

is the sum of positive terms, so we use the integral test in the form

c2. `integrate(1/(k+1)/(log(k+1))^2,k,1,inf)`

d2. $\dfrac{1}{\log 2}$

which is finite, therefore the series converges.

    Consider the series

c3. `sum(1/k/(log(k))^p,k,2,inf)`

d3. $\displaystyle\sum_{k=2}^{\infty} \dfrac{1}{k \log^p k}$

We now

c4. `assume(p>1)$`

and perform the integration[8]

c5. `integrate(1/k/(log(k))^p,k,2,inf)`

d5. $\dfrac{\log 2}{\log^p 2 p - \log^p 2}$

c6. `factor(%)`

d6. $\dfrac{1}{(\log 2)^{p-1}(p - 1)}$

which is finite, so the series converges for $p > 1$.

    Consider next the series

c7. `sum(1/k/log(k)/(log(log(k)))^p,k,2,inf)`

d7. $\displaystyle\sum_{k=2}^{\infty} \dfrac{1}{k \log k \log^p \log k}$

Why do we start summing at $k = 2$ and not at $k = 1$? (Hint: Look for zeroes in the denominator.) With $p > 1$ already assumed, we

c8. `integrate(1/k/log(k)/(log(log(k)))^p,k,3,inf)`

d8. $\dfrac{\log \log 3}{\log^p \log 3 p - \log^p \log 3}$

c9. `factor(%)`

d9. $\dfrac{1}{(\log \log 3)^{p-1}(p - 1)}$

showing convergence for $p > 1$.

    As a final example we try the series

c10. `sum(1/k/log(k)/log(log(k))/(log(log(log(k))))^p,k,16,inf)`

d10. $\displaystyle\sum_{k=16}^{\infty} \dfrac{1}{k \log k \log \log k \log^p \log \log k}$

for $p > 1$. Why do we start summing at $k = 16$?

We apply the integral test:

c11. `integrate(1/k/log(k)/log(log(k))/(log(log(log(k))))^p,k,16,inf)`

d11. $\dfrac{\log \log(4 \log 2)}{\log^p \log(4 \log 2) p - \log^p \log(4 \log 2)}$

---

8 If we had not made assumption c4, MACSYMA would query us about the sign of $p - 1$, and we would be forced to decide.

c12. `factor(%)`

d12.
$$\frac{1}{(\log\log(4\log 2))^{p-1}(p-1)}$$
which again shows convergence for $p > 1$.

We next consider series $\sum a_k$ with both positive and negative elements. The simplest criterion for such series is due to Leibniz.

Theorem 12.45 (Leibniz criterion for decreasing alternating series). Let
$$a_0 \geq a_1 \geq a_3 \geq \cdots \geq 0 ,$$
and
$$\lim_{n\to\infty} a_n = 0 .$$
Then the alternating series
$$\sum_{k=0}^{\infty}(-1)^k a_k = a_0 - a_1 + a_2 - a_3 \pm \cdots$$
converges. Moreover, if
$$s = \sum_{k=0}^{\infty}(-1)^k a_k ,$$
then the partial sums
$$s_n = \sum_{k=0}^{n}(-1)^k a_k$$
have error
$$|s - s_n| \leq a_{n+1} , \tag{12.7}$$
i.e., the error is smaller that the absolute value of the next summand.

Proof. Consider the sequence $\left(s_n = \sum_{k=0}^{n}(-1)^k a_k\right)_n$ of partial sums and its two subsequences
$$(s_n)_{\text{even}} \quad \text{with even } n ,$$
$$(s_n)_{\text{odd}} \quad \text{with odd } n .$$

a. The sequence $(s_n)_{\text{even}}$ is nonincreasing:
$$s_{2m+2} - s_{2m} = \sum_{k=0}^{2m+2}(-1)^k a_k - \sum_{k=0}^{2m}(-1)^k a_k = a_{2m+2} - a_{2m+1} \leq 0 .$$

b. The sequence $(s_n)_{\text{odd}}$ is nondecreasing:
$$s_{2m+1} - s_{2m-1} = \sum_{k=0}^{2m+1}(-1)^k a_k - \sum_{k=0}^{2m-1}(-1)^k a_k = a_{2m} - a_{2m+1} \geq 0 .$$

c. From a and b we conclude that the intervals

$$I_k := \begin{cases} [s_{k-1}, s_k] & \text{if } k \text{ is even,} \\ [s_k, s_{k-1}] & \text{if } k \text{ is odd,} \end{cases}$$

are nested, i.e.,

$$I_1 := [s_1, s_0] \supset I_2 := [s_1, s_2] \supset I_3 := [s_3, s_2] \supset I_4 := [s_3, s_4] \supset \cdots$$

d. The intervals I_k shrink to zero. This follows from

$$s_{2m} - s_{2m-1} = \sum_{k=0}^{2m} (-1)^k a_k - \sum_{k=0}^{2m-1} (-1)^k a_k = (-1)^{2m} a_{2m} = a_{2m} \to 0,$$

and

$$s_{2m} - s_{2m+1} = \sum_{k=0}^{2m} (-1)^k a_k - \sum_{k=0}^{2m+1} (-1)^k a_k = -(-1)^{2m+1} a_{2m+1}$$

$$= -a_{2m-1} \to 0.$$

From the *nested-interval property* of the real numbers we conclude the existence of a real number s in the intersection of all the intervals I_k. Clearly $s = \lim_{n \to \infty} s_n$ and the error bound (12.7) follows from d. $\square$

Example 12.46 (Alternating harmonic series). The series

$$\sum_{k=1}^{\infty} \frac{(-1)^{k+1}}{k} = 1 - \frac{1}{2} + \frac{1}{3} - \frac{1}{4} + \cdots \qquad (12.8)$$

converges by Theorem 12.45. It converges to ln 2; see Example 13.24.

We show now one of the weirdest results in mathematics: The value of the series (12.8) depends on the order in which we sum the terms of the series. Writing

$$S := \sum_{k=1}^{\infty} \frac{(-1)^{k+1}}{k}$$

we get

$$S = 1 - \frac{1}{2} + \frac{1}{3} - \frac{1}{4} + \frac{1}{5} - \frac{1}{6} + \frac{1}{7} - \cdots$$

$$= 1 - \frac{1}{2} - \frac{1}{4} + \frac{1}{3} - \frac{1}{6} - \frac{1}{8} + \frac{1}{5} - \frac{1}{10} - \frac{1}{12} + \cdots$$

where we take one positive term and two negative ones

$$= \frac{1}{2} - \frac{1}{4} + \frac{1}{6} - \frac{1}{8} + \frac{1}{10} - \frac{1}{12} + \cdots =$$

$$= \frac{1}{2}\left(1 - \frac{1}{2} + \frac{1}{3} - \frac{1}{4} + \frac{1}{5} - \frac{1}{6} + \frac{1}{7} - \cdots\right)$$

$$\therefore \quad S = \frac{1}{2}S.$$

Before concluding that $S = 0$, we note that S is positive,

$$S = \left(1 - \frac{1}{2}\right) + \left(\frac{1}{3} - \frac{1}{4}\right) + \left(\frac{1}{5} - \frac{1}{6}\right) + \left(\frac{1}{7} - \frac{1}{8}\right) + \text{sum of positive terms} > 0.$$

Definition 12.47. The series $\sum a_k$ is said to be *absolutely convergent* if the series of absolute values $\sum |a_k|$ converges. A convergent series which is not absolutely convergent is called *conditionally convergent*.

Theorem 12.48 (Absolute convergence implies convergence).   If the series $\sum |a_n|$ converges, so does the series $\sum a_n$.

Proof. By the triangle inequality we have

$$|s_n - s_m| = |a_{m+1} + a_{m+2} + \cdots + a_n| \leq |a_{m+1}| + |a_{m+2}| + \cdots + |a_n|, \quad (12.9)$$

where $n > m$. Since $\sum |a_k|$ converges, given an $\varepsilon > 0$, there exists an integer $N = N(\varepsilon)$ such that for $n > m > N = N(\varepsilon)$ the right member of (12.9) can be made less than ε. Hence the left-hand side also is less than ε and the Cauchy criterion tells us the series $\sum a_k$ converges. $\qquad\square$

Example 12.49. a. The series

$$\sum_{k=0}^{\infty} \frac{(-1)^k}{2^k}$$

converges, by Theorem 12.48, since it is absolutely convergent:

$$\sum_{k=0}^{\infty} \frac{1}{2^k} = 2.$$

b. The alternating harmonic series (12.8) is conditionally convergent: It converges by Theorem 12.45, but it is not absolutely convergent since $\sum_{k=1}^{\infty} 1/k$ diverges. This shows that the converse of Theorem 12.48 is not true.

Example 12.50. The convergence of the series

$$\sum_{k=0}^{\infty} \frac{(-1)^k (k!)^2}{(2k)!}$$

can be decided by applying the ratio test to the series of absolute values yields

$$\lim_{k\to\infty} \frac{((k+1)!)^2/(2k+2)!}{(k!)^2/(2k)!} = \lim_{k\to\infty} \frac{(k+1)^2}{(2k+2)(2k+1)} = 1/4 .$$

Therefore the series converges absolutely and is convergent by Theorem 12.48.

Remark 12.51. a. Absolute convergence can be tested by the convergence tests given above for series with positive terms, e.g., the ratio test of Theorem 12.39.

b. Absolute convergence is important for the following reason: The sum of an absolutely convergent series $\sum^{\infty} a_k$ does not depend on the order in which the terms a_k are taken.

c. Conversely, it can be shown that for any conditionally convergent series $\sum a_k$, and *any* real number S, we can have

$$\sum a_k = S .$$

by rearranging the terms of $\sum a_k$. For example, rearranging the terms of the alternating harmonic series we can have it sum to 10^6 or to -10^6 or come within ε to the number of atoms in the universe.

Exercises

12.17 Use the ratio test to prove the convergence or divergence of the series of Examples 12.36–12.38.

12.18 Investigate the following series for convergence or divergence by any test you wish to use.

(a) $\sum_{k=0}^{\infty}(1/2)^{k-2}$ (b) $\sum_{k=0}^{\infty}(k/2)^k$ (c) $\sum_{k=0}^{\infty}(k^2/2)^k$

(d) $\sum_{k=0}^{\infty} ka^k, \ (0 < a < 1)$ (e) $\sum_{k=0}^{\infty} 2^k$ (f) $\sum_{k=0}^{\infty} \frac{2^k}{(k+1)^2}$

(g) $\sum_{k=0}^{\infty} \sin\left(\frac{2k}{k(k+1)}\right)$ (h) $\sum_{k=0}^{\infty} \frac{k^2 + \cos k}{k^2 + \sin k}$

12.19 Show that the following series are ordered so that each converges more rapidly than the preceding one.

$$\sum \frac{1}{k \ln^2 k}, \ \sum \frac{1}{k^2}, \ \sum \frac{1}{k^3}, \ \sum \frac{1}{2^k}, \ \sum \frac{1}{3^k}, \ \sum \frac{1}{k!}, \ \sum \frac{1}{k^k} .$$

Hint: Use the comparison theorem.

12.20 Show that the following series are arranged so that each series diverges less rapidly than the next.

$$\sum 2^k, \ \sum k, \ \sum 1, \ \sum \frac{1}{k}, \ \sum \frac{1}{k \ln k}, \ \sum \frac{1}{k \ln k \ln \ln k} .$$

12.21 Use the integral test to show that the series $\sum[1/k(\ln k)^p]$ diverges for all $p \le 1$.

12.22 Let

$$\ln_n k := \underbrace{\ln \ln \ldots \ln k}_{n} \ .$$

Use the integral test to show that the series

$$\sum_{3}^{\infty} \frac{1}{k \ln k (\ln_2)^p}$$

converges for $p > 1$ and diverges for $p \le 1$.

12.23 Using the notation of the preceding exercise, show (by the integral test) that

$$\sum_{16}^{\infty} \frac{1}{k \ln k \ln_2 (\ln_3)^p}$$

converges for $p > 1$ and diverges for $p \le 1$. Consider the general series

$$\sum \frac{1}{k \ln k \ln_2 \cdots \ln_{n-1}(\ln_n)^p} \ ,$$

and again show that convergence occurs for $p > 1$ and diverges for $p \le 1$.

12.24 Using the integral test and MACSYMA check the following series for convergence.

(a) $\displaystyle\sum_{k=1}^{\infty} \frac{a\sin k}{\sqrt{1 - k^2}}$ (b) $\displaystyle\sum_{k=1}^{\infty} \frac{a\tan k}{k^2 + 1}$ (c) $\displaystyle\sum_{k=1}^{\infty} \frac{a\sec k}{k\sqrt{k^2 - 1}}$ (d) $\displaystyle\sum_{k=1}^{\infty} \frac{\ln k}{k}$

(e) $\displaystyle\sum_{k=1}^{\infty} k\,e^{-k^2}$ (f) $\displaystyle\sum_{k=1}^{\infty} \frac{k^2}{k^4 + 1}$ (g) $\displaystyle\sum_{k=1}^{\infty} \left(1 + \frac{1}{k}\right)^k$ (h) $\displaystyle\sum_{k=1}^{\infty} \frac{\sqrt{k}}{k^2 + 1}$

12.25 For each of the series listed below determine whether the series converges and find its sum.

(a) $\displaystyle\sum_{k=1}^{\infty} \frac{2}{3^k}$ (b) $\displaystyle\sum_{k=1}^{\infty} \frac{1}{k(k + 1)}$ (c) $\displaystyle\sum_{k=1}^{\infty} \left(\frac{1}{k + 1} - \frac{1}{k + 2}\right)$

(d) $\displaystyle\sum_{k=1}^{\infty} \left(\frac{1}{2^k} - \frac{1}{2^k}\right)$ (e) $\displaystyle\sum_{k=2}^{\infty} \frac{1}{k^2 - 1}$ (f) $\displaystyle\sum_{k=1}^{\infty} \frac{1}{4k^2 - 1}$

(g) $\displaystyle\sum_{k=1}^{\infty} \frac{1}{4k^2 + 4k - 3}$ (h) $\displaystyle\sum_{k=1}^{\infty} \frac{\ln \sqrt{k}}{\ln k + 1}$

12.26 Give an example of a convergent series of positive terms such that the comparison test (the ratio test) shows it to be convergent.

12.27 Let $a_k > 0$ and $b_k > 0$ for $k = 0, 1, 2, \ldots$ Suppose $\sum a_k^2$ and $\sum b_k^2$ are convergent series. Show $\sum a_k b_k$ is also a convergent series.
  Hint: Expand $(a_k - b_k)^2$, which is nonnegative!

12.28 For which choice of α and β does the series $1 + \frac{1}{2}\beta + \frac{1}{3}\alpha + \frac{1}{4}\beta + \frac{1}{5}\alpha + \cdots$ converge?

12.29 We know, if $a_k > 0$ and $\sum a_k$ is convergent, then $a_k \to 0$ as $k \to \infty$. Suppose we assume also that $a_{k+1} \le a_k$ for $k \in \mathbf{N}_0$. Show that $k a_k \to 0$ as $k \to \infty$. *Hints:* Write $k = n + p$ (n and p are large positive integers). Consider $k a_k = (n + p)a_{n+p}$. Show

$$ka_k = pa_{n+p} + na_{n+p} < (a_n + a_{n+1} + \cdots + a_{n+p-1}) + na_{n+p}$$

$$\leq \left(\sum_{j=n}^{\infty} a_j \right) + na_{p+n} .$$

Now make the right-hand side of the last inequality smaller than or equal to ε by choosing n and p judiciously. Be sure you pick p and n in the proper order.

12.30 Consider a convergent series $\sum a_k$ and let $R_n = \sum_{k=n}^{\infty} a_k$ for $n > 0$. Prove the following:

a. Suppose there exist constants $M > 0$ and r $(0 \leq r < 1)$, such that $|a_k| \leq Mr^k$. Show that $|R_n| \leq Mr^n/(1 - r)$.
    Hint: Proceeding formally one has

$$|R_n| \leq \sum_{k=n}^{\infty} |a_k| \leq M \sum_{k=n}^{\infty} r^k . \tag{12.10}$$

Now complete the problem. Be sure you can justify each inequality in (12.10).

b. Repeat part a, but find an upper bound for $|R_n|$ if

$$|a_k| \leq M \frac{r^k}{k!} ,$$

with $k \geq 0$, and $0 \leq r < \infty$.
    Hint: Use $r^{n+p}/(n + p)! \leq r^n \cdot r^p/p!$ for $n, p \geq 0$.

c. Same problem but find an upper bound for $|R_n|$ if

$$|a_k| \leq M(k + 1)r^k \quad (0 \leq r < 1, \ k \geq 0) .$$

d. Same problem but find an upper bound for $|R_n|$ if

$$|a_k| \leq M \frac{r^k}{k} \quad (0 \leq r < 1, \ k \geq 1) .$$

Note these results show how to estimate the remainder in various types of convergent series.

12.31 Each of the following series is convergent with sum s, say. How large must one take n to ensure $s_n = \sum_{k=0}^{n} a_k$ approximates s with an error less than $1/100$?

(a) $\displaystyle\sum_{k=0}^{\infty} \frac{k^k}{k!}$ (b) $\displaystyle\sum_{k=0}^{\infty} k\left(\frac{1}{2}\right)^k$ (c) $\displaystyle\sum_{k=0}^{\infty} \left(k + \frac{1}{k+1}\right) \cdot \frac{k}{k!}$

12.32 Does the series

$$\sum_{k=0}^{\infty} \frac{(-1)^k}{2k + 1} = 1 - \frac{1}{3} + \frac{1}{5} - \frac{1}{7} + \cdots ,$$

converge? If it does, does it converge absolutely or conditionally?

12.4 Proofs

Proof of Theorem 12.14. We recall that by our hypothesis for each $\varepsilon > 0$ there is an $N \in \mathbf{N}$ such that for all $n \geq N$

$$|a_n - a| \le \varepsilon \quad \text{and} \quad |b_n - b| \le \varepsilon \,,$$

and moreover, by Theorem 12.13

$$|a_n| \le A \quad \text{and} \quad |b_n| \le B$$

for some $A, B \in \mathbf{R}$.

a. By Lemma 12.8 obviously $C(a_n - a) \to 0$.

b. $|(a_n \pm b_n) - (a \pm b)| = |(a_n - a) \pm (b_n - b)| \le |a_n - a| + |b_n - b| \le 2\varepsilon$, and so it follows also by Lemma 12.8 that $(a_n \pm b_n) - (a \pm b) \to 0$.

c. Similarly we have $|a_n b_n - ab| = |a_n(b_n - b) + (a_n - a)b| \le (|a_n| + |b|)\varepsilon \le (A + |b|)\varepsilon$, and so $a_n b_n \to ab$.

d. Here we use that from $b \ne 0$ it follows (choose $\varepsilon := |b|/2$) that there is some $N \in \mathbf{N}$ such that for all $n \ge N$

$$|b_n - b| \le |b|/2 \,.$$

By the triangle inequality $|b| = |(b - b_n) + b_n| \le |b - b_n| + |b_n|$, so that we get

$$|b_n| \ge |b| - |b - b_n| = |b| - |b_n - b| \ge |b| - |b|/2 = |b|/2 \,.$$

Therefore taking the reciprocal of this inequality we have

$$|1/b_n| \le 2/|b| \,.$$

Using the above we then have

$$\left| \frac{a_n}{b_n} - \frac{a}{b} \right| = \left| \frac{a_n b - b_n a}{b_n b} \right| \le \frac{|a_n(b - b_n) + b_n(a_n - a)|}{|b_n||b|}$$

$$\le \frac{A\varepsilon}{|b_n||b|} + \frac{\varepsilon}{|b|} \le \frac{2A}{|b|^2}\varepsilon + \frac{\varepsilon}{|b|} = \frac{2A + |b|}{|b|^2}\varepsilon \,,$$

which establishes our result. $\qquad\square$

Proof of Theorem 12.21. We first prove two theorems, then use them to prove Theorem 12.21.

The first is a theorem about *monotone* increasing[9] (decreasing) sequences.

Theorem 12.52 (Bounded and monotone implies convergent). If $(a_n)_n$ is a sequence which is nondecreasing and bounded above, then $(a_n)_n$ converges. If $(a_n)_n$ is nonincreasing and bounded below, then $(a_n)_n$ converges.

Proof. We prove only the first part as the proof of the second part is similar. We leave this for the reader. Since the a_n are all bounded, the set A consisting of these points on the real line has an upper bound. Indeed, from the properties of the real

---

9 If a sequence is nondecreasing, we refer to it as monotone increasing; a nonincreasing sequence is called monotone decreasing.

numbers it has a *least upper bound*.[10] Let us call this least upper bound a. We want to show that $(a_n)_n$ converges to a. By the definition of the least upper bound, given an $\varepsilon > 0$ there is an N such that $a - a_N < \varepsilon$. This means that for $n > N$, we have

$$a_N \leq a_n, \quad \text{which means} \quad a - a_n \leq a - a_N < \varepsilon .$$

Since ε may be taken as small as we like, this shows that $\lim_{n \to \infty} a_n = a$. $\square$

Using this result we can show the following theorem.

Theorem 12.53. Every bounded sequence has either a nondecreasing or a nonincreasing subsequence.[11]

Proof. To show this, it is convenient to find a subsequence which is either nondecreasing or nonincreasing. If this would be the case, then we know by the previous theorem that we would have a convergent subsequence. So let us see how we might construct such sequences. To this end we introduce the concept of a *peak point*. We say n is a peak point of the sequence $(a_n)_n$ if $a_m < a_n$ for all $m > n$. There are two such possibilities. The first is that the sequence has infinitely many peak points $n_1, n_2, n_3, n_4, \ldots$ and $a_{n_1} > a_{n_2} > a_{n_3} > a_{n_4} > \cdots$. In this case the sequence $(a_{n_k})_k$ is the sought nonincreasing subsequence. The second possibility is that there are only a finite number of peak points. Since there are only a finite number of peak points, they are bounded. Hence let n_1 be greater than any of the peak points, and it is therefore not a peak point. Since it is not a peak point, there exists an $n_2 > n_1$ such that $a_{n_2} > a_{n_1}$. Repeating this argument we find a sequence $n_1 < n_2 < n_3 < \cdots$ such that $a_{n_1} < a_{n_2} < a_{n_3} < \cdots$, i.e., we obtain a nondecreasing sequence. So in either of the two possible cases we obtain a monotone sequence; boundedness of the sequence now indicates that a limit exists. $\square$

We are now in a position to prove the Bolzano–Weierstrass theorem (every bounded sequence has a convergent subsequence).

Proof. The proof is an immediate consequence of the two proceeding results. By assuming the sequence is bounded, we may use the argument of Theorem 12.53 to extract either a monotone increasing or monotone decreasing sequence. Now we are there as a monotone bounded sequence must converge. $\square$

Proof of Theorem 12.22. Only if. Let the sequence $(a_n)_n$ converge to a. Then for

---

10 A number a is said to be a least upper bound for the set A if (a) a is an upper bound for A, and (b) if b is also an upper bound for A, then $a \leq b$. It is a property of the real numbers that if A is any nonempty set of real numbers, and A is bounded above, then A has a least upper bound. A similar statement can be said about the greatest lower bound.

11 A subsequence is a subset of the original sequence, where the sequential order is maintained, i.e., if $a_1, a_2, a_3, a_4, a_5, a_6, a_7, a_8, \ldots$, then $a_1, a_4, a_6, a_{10}, \ldots$ is a subsequence maintaining the same sequential ordering.

each $\varepsilon > 0$ there is an index $N \in \mathbf{N}$ such that

$$|a_n - a| \leq \varepsilon/2, \quad \text{for all } n \geq N .$$

If $m \geq N$, and $n \geq N$, then by the triangle inequality we have, moreover, that

$$|a_n - a_m| = |(a_n - a) + (a - a_m)| \leq |a_n - a| + |a - a_m| \leq \varepsilon ,$$

and so $(a_n)_n$ is a Cauchy sequence.

If. Conversely, let $(a_n)_n$ be a Cauchy sequence. We need to prove the *existence*[12] of some number $a \in \mathbf{R}$ such that $a_n \to a$.

a. First we show that every Cauchy sequence $(a_n)_n$ is bounded. Taking $\varepsilon := 1$ in Definition 12.20 we can find an index $N \in \mathbf{N}$ such that $|a_n - a_m| \leq 1$, for all $m, n \geq N$, and so

$$|a_n| = |a_n - a_m + a_m| \leq |a_n - a_m| + |a_m| \leq 1 + |a_m|$$

by the triangle inequality. The special choice $m := N$ gives, for $n \geq N$

$$|a_n| \leq 1 + |a_N| .$$

Let $b := \max \{|a_0|, |a_1|, \ldots, |a_N|\}$, the largest of the $N + 1$ numbers $|a_n|$ for $n = 0, \ldots, N$. Then:

$$|a_n| \leq \begin{cases} b & \text{if } n = 0, \ldots, N - 1 \\ 1 + b & \text{if } n \geq N \end{cases} \leq 1 + b ,$$

showing that the sequence $(a_n)_n$ is bounded.

b. By Theorem 12.21 there is a convergent subsequence of $(a_n)_n$. Assume now, this subsequence converges to a, i.e.,

$$\lim_{k \to \infty} a_{n_k} = a . \tag{12.11}$$

By the definition of a Cauchy sequence, for any $\varepsilon > 0$ there is some $N \in \mathbf{N}$ such that

$$|a_n - a_m| \leq \varepsilon/2, \quad \text{for all } m, n \geq N .$$

Further by (12.11) there is another index K such that

$$|a_{n_k} - a| \leq \varepsilon/2$$

for all $k \geq K$. Let now $\bar{N} := \max\{N, n_K\}$, then for all $n \geq \bar{N}$ it follows that

$$|a_n - a| \leq |a_n - a_{n_k}| + |a_{n_k} - a| \leq \varepsilon/2 + \varepsilon/2 = \varepsilon .$$

This proves that $\lim_{n \to \infty} a_n = a$. $\square$

Exercise

12.33 Prove that if $(a_n)_n$ is a nonincreasing sequence of real numbers which is bounded below, then $(a_n)_n$ converges.

---

12 This part of the proof requires the completeness of $\mathbf{R}$ studied in Appendix B, B.5.

13 Series expansions and approximations

13.1 Series of functions

Chapter 12 dealt with infinite series of numbers and criteria for their convergence. In the same way one can study infinite series of functions

$$\sum_{k=0}^{\infty} u_k(x)$$

where the functions $u_k(x)$ are all defined on a common interval.

As for series of numbers, convergence is defined here in terms of the *partial sums*

$$S_n(x) := \sum_{k=0}^{n} u_k(x) .$$

Definition 13.1. Let the $u_k(x)$ be defined on an interval I. If there is a function $f(x)$ such that

$$\lim_{n \to \infty} S_n(x) = f(x), \quad \text{for all } x \in I , \tag{13.1}$$

then the series $\sum_{k=0}^{\infty} u_k(x)$ is said to *converge to f on the interval I*, or $\sum_{k=0}^{\infty} u_k(x)$ is a *series expansion* of f in the interval I, a fact denoted by

$$\sum_{k=0}^{\infty} u_k(x) = f(x), \quad \text{for all } x \in I .$$

This can be stated as follows: For any $x \in I$ and any $\varepsilon > 0$ there is an integer $N_0 = N_0(\varepsilon, x)$ such that

$$|f(x) - S_n(x)| < \varepsilon, \quad \text{for all } n \geq N_0 . \tag{13.2}$$

Still another way to state (13.1) is

$$\lim_{n \to \infty} R_n(x) = 0, \quad \text{for all } x \in I ,$$

where $R_n(x)$ is the nth *remainder*

$$R_n(x) := f(x) - S_n(x) = \sum_{n+1}^{\infty} u_k(x) . \tag{13.3}$$

The integer N_0 in the above definition *depends* on the point x, as well as on ε. The next example illustrates this dependence.

Example 13.2. Consider the sequence of functions $u_k(x) := x^k$, for $k = 0, 1, \ldots,$ defined on the interval $[0, 1)$ and their partial sums

$$S_n(x) := \sum_{k=0}^{n} x^k = \frac{1 - x^{n+1}}{1 - x},$$

which converge for all $x \in [0, 1)$,

$$\lim_{n \to \infty} S_n(x) = \frac{1}{1 - x};$$

see Example 12.30. We may therefore write

$$\sum_{k=0}^{\infty} x^k = \frac{1}{1 - x}, \quad x \in [0, 1).$$

The nth remainder is bounded by

$$R_n(x) \leq \frac{x^{n+1}}{1 - x} \quad \text{(check!)}.$$

Therefore, in order to guarantee $R_n(x) < \varepsilon$ for some n we must have

$$n > \frac{\ln((1 - x)\varepsilon)}{\ln x} - 1. \tag{13.4}$$

For example, at $x = 1/2$

$$n > \frac{\ln(\varepsilon/2)}{\ln(1/2)} - 1.$$

As x increases (from 0 to 1), the logarithm $\ln x$ decreases, and the lower bound in (13.4) increases. As x approaches 1, that bound tends to infinity (for any $\varepsilon > 0$).

The next definition deals with an important special case of convergence, where N_0 (in condition (13.2)) does not depend on x.

Definition 13.3 (Uniform convergence). A series of functions $\sum_{k=0}^{\infty} u_k(x)$ is said to converge uniformly to a function f on the interval I if for any $\varepsilon > 0$ there is an integer N_0 such that the partial sums $S_n(x) := \sum_{k=0}^{n} u_k(x)$ satisfy the inequality

$$|f(x) - S_n(x)| < \varepsilon, \quad \text{for all } x \in I \text{ and } n \geq N_0. \tag{13.5}$$

The word *uniform* in the definition refers to the fact that the same integer N_0 "works" for all x in the interval.

Remark 13.4. Condition (13.5) states that the remainder $R_n(x)$ is bounded by

$$|R_n(x)| < \varepsilon, \quad \text{for all } x \in I .$$

Example 13.5. Consider again the series $\sum_{k=0}^{\infty} x^k$ of Example 13.2. We show that it converges uniformly on the interval $[0, 1/2]$. Indeed, the remainder

$$R_n(x) = \sum_{k=n+1}^{\infty} x^k = x^{n+1} \sum_{k=0}^{\infty} x^k = \frac{x^{n+1}}{1-x} ,$$

is uniformly bounded in $[0, 1/2]$

$$|R_n(x)| \leq \max_{[0,1/2]} x^{n+1} / \min_{[0,1/2]} (1-x) = (1/2)^n .$$

For any $\varepsilon > 0$ we can solve

$$\frac{1}{2^n} < \varepsilon \quad \text{to get} \quad n > -\frac{\ln \varepsilon}{\ln 2} .$$

Therefore, any integer $N_0 > -\ln \varepsilon / \ln 2$ will do in (13.5), uniformly for all x in $[0, 1/2]$.

We can show similarly that the series $\sum_{k=0}^{\infty} x^k$ converges uniformly on any closed interval $[0, \alpha]$ with $\alpha < 1$.

MACSYMA-Session 13.1. Use MACSYMA to draw the graphs of the sums

c1. `sum(4*sin((2*k-1)*%pi*x)/(2*k-1)/%pi,k,1,n)`

d1. $\dfrac{4 \sum_{k=1}^{n} \sin(\pi(2k-1)x)/(2k-1)}{\pi}$

for $n = 1, \ldots, 10$. To accomplish this, create a list of the functions to be plotted with

c2. `makelist(sum(4*sin((2*k-1)*%pi*x)/(2*k-1)/%pi,k,1,n), n, 1, 10)$`

Now plot the above 10 functions, using 200 points and line styles 0–9 for the ten plots

c3. `plot2('`%`,x,-4,4,'`(makelist(i,i,0,9))), plotnum:200`

Use the plots to guess the function which would appear in the limit as $n \to \infty$.

13.2 Power series

In this section we consider series of functions of the following kind.

Definition 13.6 (Power series). A power series is an infinite series

$$\sum_{k=0}^{\infty} a_k(x - x_0)^k := a_0 + a_1(x - x_0) + a_2(x - x_0)^2 + \cdots , \tag{13.6}$$

where the $a_0, a_1, a_2, \ldots$ is a sequence of real numbers, x_0 is a real number, and x

is a real variable. In particular, power series with $x_0 = 0$,

$$\sum_{k=0}^{\infty} a_k x^k := a_0 + a_1 x + a_2 x^2 + \cdots . \tag{13.7}$$

Power series are useful for working with the elementary functions of calculus (the exponential, logarithmic, trigonometric, and hyperbolic functions and their inverses). In Chap. 2 we studied these functions in an intuitive-geometric way. Certain limits, such as (3.10),

$$\lim_{x \to 0} \sin x = 0 \quad \text{and} \quad \lim_{x \to 0} \cos x = 1 ,$$

and (3.11),

$$\lim_{x \to 0} e^x = 1 ,$$

were shown by geometric arguments, which some mathematicians will not accept as valid proofs. These limits (and many others) can now be proved by simple analytical arguments, once we represent the above functions as power series.

Before using power series we must verify that they converge. Certain criteria for convergence of power series are given in this section. The following criterion is easy but useful.

Theorem 13.7. If the power series $\sum_{n=0}^{\infty} a_n x^n$ converges at the point $x = \xi \neq 0$, then the series converges uniformly and absolutely for all x such that $|x| < |\xi|$.

The proof of this theorem, which follows easily from the proof of Theorem 13.9, is assigned as exercise.

Before the next definition we recall the definition of the *least upper bound* of a set (Appendix B, Definition B.9).

Definition 13.8 (Radius of convergence). Let l be the set of all x for which the power series $\sum a_n x^n$ converges. The set l is nonempty (it contains $x = 0$). Therefore the least upper bound of l, if it exists, is a nonnegative number, which we denote R, and call the *radius of convergence* of the series. The interval $(-R, R)$ is called the *interval of convergence* of the series if $R > 0$.

These names are justified by the following theorem, the main result concerning convergence of power series.

Theorem 13.9 (Convergence of a power series). For a power series $\sum_{k=0}^{\infty} a_k x^k$ exactly one of the following statements is true:
a. the series converges only at the point 0;
b. the series converges absolutely for all values of x;

c. the series converges absolutely for all x in some finite interval $(-R, R)$ and
    diverges for $x < -R$ and for $x > R$.

A proof of this theorem is given in Sect. 13.6.

Remark 13.10. The above theorem can be adapted for power series in $(x - x_0)$,
(13.6),

$$\sum_{k=0}^{\infty} a_k(x - x_0)^k := a_0 + a_1(x - x_0) + a_2(x - x_0)^2 + \cdots ,$$

where $x_0 \neq 0$. It is only necessary to change parts a and c as follows:
a. the series converges only at the point x_0;
c. the series converges absolutely for all x in some finite interval $(x_0 - R, x_0 + R)$
    and diverges for $x < x_0 - R$ and for $x > x_0 + R$.
The value R is again referred to as the radius of convergence.
    All convergence results for a power series $\sum_{k=0}^{\infty} a_k x^k$ can be similarly adapted
for power series of the form (13.6), by translation of the origin to the point x_0. We
can therefore state all results for power series of the form (13.7).

Example 13.11. In case c of Theorem 13.9 it is possible for the series to converge,
conditionally, at the endpoints $x = -R$ or $x = R$. For example, the series

$$\sum_{k=0}^{\infty} (-1)^k \frac{x^{2k+1}}{2k+1} = x - \frac{x^3}{3} + \frac{x^5}{5} - \frac{x^7}{7} + \cdots$$

has interval of convergence $(-1, 1)$ and converges conditionally at the endpoints
$x = -1$ and $x = 1$.

Example 13.12. The series may diverge at one endpoint of its convergence interval
and converge conditionally at the other. For example, the series

$$\sum_{k=0}^{\infty} (-1)^k \frac{x^{k+1}}{k+1} = x - \frac{x^2}{2} + \frac{x^3}{3} - \frac{x^4}{4} + \cdots$$

has the interval of convergence $(-1, 1)$, diverges at the endpoint $x = -1$ but
converges (conditionally) at $x = 1$.

Theorem 13.13 (First formula for the radius of convergence). Let $\sum a_k x^k$ be a
series for which

$$\lim_{k \to \infty} \left| \frac{a_{k+1}}{a_k} \right| = \rho$$

exists. Then the radius of convergence R of the series is given by $R = 1/\rho$.

Proof. By the ratio test for series of numbers, Theorem 12.39, the power series converges for all x such that

$$\lim_{k \to \infty} \left| \frac{a_{k+1} x^{k+1}}{a_k x^k} \right| < 1 .$$

By hypothesis this implies that the series converges for all x such that $\rho|x| < 1$. The ratio test also tells us the series diverges for all x such that $\rho|x| > 1$. Hence, $R = 1/\rho$. □

Example 13.14. Consider the convergence of the power series

$$\sum_{k=0}^{\infty} \left(\frac{k!}{3 \cdot 5 \cdots (2k+1)} \right)^2 x^k .$$

Applying Theorem 13.9 to the sequence of numbers $a_k := (k!/(3 \cdot 5 \cdots (2k+1)))^2$ we obtain

$$\lim_{k \to \infty} \left| \frac{a_{k+1}}{a_k} \right| = \lim_{k \to \infty} \left(\frac{k+1}{2k+3} \right)^2 = \frac{1}{4} .$$

Therefore the power series has radius of convergence $R = 4$, and interval of convergence $(-4, 4)$.

Example 13.15. As another example consider the power series

$$\sum_{k=0}^{\infty} \frac{x^k}{k!} .$$

Applying Theorem 13.9 to the sequence of coefficients $a_k := 1/k!$ we get

$$\lim_{k \to \infty} \left| \frac{a_{k+1}}{a_k} \right| = \lim_{k \to \infty} \frac{1}{k+1} = 0 .$$

The radius of convergence is therefore $R = \infty$, i.e., the series converges for all x (case b of Theorem 13.9).

MACSYMA-Session 13.2. Theorem 13.13 allows calculating the radius of convergence of a power series $\sum a_k x^k$ by the function
```
radius_convergence(a,k) := limit(abs(a/subst(k+1,k,a)),k,inf) $
```
which computes the ratio a_k/a_{k+1}, takes the absolute value, and computes the limit as $k \to \infty$. Examples:
```
c1. radius_convergence(1/k^2,k)
d1.    1
c2. radius_convergence(1/k!,k)
d2.    ∞
```

c3. `radius_convergence(k^k/k!,k)`

d3. $\dfrac{1}{e}$

c4. `radius_convergence(binomial(2*k,k),k)`

d4. $\dfrac{1}{4}$

c5. `radius_convergence(exp(1/sin(2/k)),k)`

d5. $\dfrac{1}{\sqrt{e}}$

c6. `radius_convergence(k!,k)`

d6. 0

The last power series thus converges only at the origin.

c7. `rad:radius_convergence((k!/product(2*j+1,j,1,k))^2,k)`

   Is $\prod_{j=1}^{\infty} 2j + 1$ `zero or nonzero?`

   `nonzero`

d7. 0

MACSYMA is having trouble doing this problem as it must first deal with an identity
involving factorials and simplify this. Let us help MACSYMA by breaking up the
computation. First let us do this computation by hand. Suppose that we have the term
$1 \cdot 3 \cdot 5 \cdot 7 \cdots (2k + 1)$ that we wish to simplify. Then by multiplying and dividing by
$2 \cdot 4 \cdot 6 \cdot 8 \cdots (2k)$ we get $(2k + 1)!/(2 \cdot 4 \cdot 6 \cdot 8 \cdots (2k))$ which we simplify by factoring 2
from each term in the denominator to get k factors of 2 multiplied by $1 \cdot 2 \cdot 3 \cdot 4 \cdots k$, or
$2^k \cdot k!$

c8. `id:product(2*j+1,j,1,k)=(2*k+1)!/(2^k*k!)`

d8. $\prod_{j=1}^{k} 2j + 1 = \dfrac{(2k + 1)!}{2^k k!}$

We wish to obtain a more concise expression for the ratio in terms of factorials

c9. `b:k!/(k+1)!*subst(k+1,k,id)/id`

d9. $\dfrac{\left(\prod_{j=1}^{k+1} 2j + 1\right)k!}{\left(\prod_{j=1}^{k} 2j + 1\right)(k + 1)!} = \dfrac{k!^2(2(k + 1) + 1)!}{2(k + 1)!^2(2k + 1)!}$

c10. `factor(%)`

d10. $\dfrac{\left(\prod_{j=1}^{k+1} 2j + 1\right)k!}{\left(\prod_{j=1}^{k} 2j + 1\right)(k + 1)!} = \dfrac{k!^2(2k + 3)!}{2(k + 1)!^2(2k + 1)!}$

The `minfactorial` function finds common factorial terms and converts the expression to
a more concise form

c11. `minfactorial(%)`

d11. $\dfrac{\prod_{j=1}^{k+1} 2j + 1}{\left(\prod_{j=1}^{k} 2j + 1\right)(k + 1)} = \dfrac{(2k + 2)(2k + 3)}{2(k + 1)^2}$

and then we can take the limit successfully

c12. `%,limit`

d12. 2

The last power series thus converges in a disk with a radius of 2 about the origin.

An alternate formula for the radius of convergence may be obtained from the root test of Theorem 12.42.

Theorem 13.16 (Second formula for the radius of convergence). Let $\sum a_k x^k$ be a series for which

$$\lim_{k \to \infty} \sqrt[k]{|a_k|} = \rho$$

exists. Then the radius of convergence R of the series is given by $R = 1/\rho$.

Proof. By the root test of Theorem 12.42 the series converges for all x such that

$$\lim_{k \to \infty} \sqrt[k]{|a_k||x|^k} = |x| \lim_{k \to \infty} \sqrt[k]{|a_k|} < 1 .$$

The root test also tells us the series diverges for all x such that $\rho|x| > 1$. Hence, $R = 1/\rho$. $\qquad\square$

The next two examples illustrate working with power series expansions of functions.

Example 13.17 (Another geometric series). Consider the function

$$f(x) = \frac{1}{2x - 5} .$$

Is it possible to represent it as a power series? Recall the geometric series

$$\frac{1}{1 - x} = \sum_{k=0}^{\infty} x^k, \quad \text{for all } |x| < 1 .$$

We can bring the function f to the same form:

$$f(x) = -\left(\frac{1}{5}\right)\left(\frac{1}{1 - 2x/5}\right) = -\left(\frac{1}{5}\right) \sum_{k=0}^{\infty} \left(\frac{-2x}{5}\right)^k = \sum_{k=0}^{\infty} (-1)^{k+1} \frac{2^k x^k}{5^{k+1}} .$$

This sum converges for $2|x|/5 < 1$, or equivalently, for $|x| < 5/2$. We verify this using the function radius_convergence of MACSYMA-Session 13.2.

```
c1. radius_convergence((-1)^(k+1)*2^k/(5^(k+1)),k)
```

d1. $\dfrac{5}{2}$

Example 13.18 (Product of a series by a polynomial). Find a power series expansion for the function

$$f(x) = \frac{x^2 + 1}{1 - x^2} .$$

First we find the power series expansion

$$\frac{1}{1-x^2} = \sum_{k=0}^{\infty} x^{2k}, \quad \text{for all } |x| < 1.$$

Then by multiplying and combining terms we have

$$\frac{x^2+1}{1-x^2} = (1+x^2)\sum_{k=0}^{\infty} x^{2k} = \sum_{k=0}^{\infty} x^{2k} + \sum_{k=0}^{\infty} x^{2k+2} = 1 + 2\sum_{k=1}^{\infty} x^{2k}.$$

This method works for multiplying a series by a finite number of terms. However, if we need to multiply a power series by an infinite number of terms, the result is not clear. For example, consider the product of power series

$$\frac{1}{1-x^2} \cdot \frac{1}{2x-5} = \left(\sum_{k=0}^{\infty} x^{2k}\right) \cdot \left(\sum_{k=0}^{\infty}(-1)^{k+1}\frac{2^k}{5^{k+1}}\right),$$

which is itself a power series. This suggests the following definition.

Definition 13.19 (Multiplication of series). The formal multiplication of the two power series

$$A(x) := \sum_{k=0}^{\infty} a_k x^k, \quad \text{for all } |x| < R_1$$

and

$$B(x) := \sum_{k=0}^{\infty} b_k x^k, \quad \text{for all } |x| < R_2$$

is given by the series

$$C(x) = \sum_{k=0}^{\infty} c_k x^k := A(x) \cdot B(x) = \left(\sum_{k=0}^{\infty} a_k x^k\right) \cdot \left(\sum_{k=0}^{\infty} b_k x^k\right),$$

where

$$c_k := \sum_{j=0}^{k} a_j b_{k-j} \quad (k = 0, 1, 2, \dots), \tag{13.8}$$

which can be verified formally by multiplying $A(x) \cdot B(x)$ and collecting terms with equal powers of x,

$$\left(\sum_{k=0}^{\infty} a_k x^k\right) \cdot \left(\sum_{k=0}^{\infty} b_k x^k\right) = \sum_{k=0}^{\infty}\left(\sum_{j=0}^{k} a_j b_{k-j}\right) x^k.$$

It can be shown that the product power series converges in the smaller of the two intervals $[-R_j, R_j]$, for $j = 1, 2$.

The sequence $(c_n)_n$ (of coefficient of $C(x)$) is called the *convolution* of the sequences $(a_n)_n$ and $(b_n)_n$.

Example 13.20 (Product of two power series). The square of the power series $1/(1+x)$ is

$$\left(\frac{1}{1+x}\right)^2 = \left(\frac{1}{1+x}\right) \cdot \left(\frac{1}{1+x}\right) = \left(\sum_{k=0}^{\infty}(-1)^k x^k\right)\left(\sum_{k=0}^{\infty}(-1)^k x^k\right)$$

$$= \sum_{k=0}^{\infty} c_k x^k \, ,$$

where

$$c_k = \sum_{0}^{k}(-1)^j(-1)^{k-j} = \sum_{0}^{k}(-1)^k = (-1)^k(k+1) \, . \qquad (13.9)$$

Therefore

$$\left(\frac{1}{1+x}\right)^2 = \sum_{k=0}^{\infty}(-1)^k(k+1)x^k \, .$$

MACSYMA-Session 13.3. We attempt to calculate the coefficients of the product series (13.8),

$$c_k := \sum_{j=0}^{k} a_j b_{k-j}, \quad (k = 0, 1, 2, \dots)$$

by MACSYMA. For Example 13.20, define

c1. `c(k)=sum(a(j)*b(k-j),j,0,k)`
d1. $c(k) = \sum_{j=0}^{k} a(j)b(k-j)$
by MACSYMA. For Example 13.20, define
c2. `a(k):=(-1)^k $`
c3. `b(k):=(-1)^k $`
c4. `c(k)=sum(a(j)*b(k-j),j,0,k)`
d4. $c(k) = \sum_{j=0}^{k} a(j)b(k-j)$
Now we have $a(j) \cdot b(k-j)$ in a simple form,
c5. `a(j)*b(k-j)`
d5. $(-1)^k$
so we now have
c6. `c(k)=sum(%,j,0,k)`
d6. $c(k) = (k+1)(-1)^k$
which is in agreement with (13.9).
Another example is $a(k) := 1$ and $b(k) := k+1$, which leads to
c7. `a(j)*b(k-j)`
d7. $k - j + 1$

c8. `sum(%,j,0,k),simpsum:true,factor`

d8. $\dfrac{(k+1)(k+2)}{2}$

Some cases are too difficult for MACSYMA to handle, such as $a(k) := 1/k!$ and
$b(k) := k + 1$, which leads to a sum that MACSYMA cannot express in simpler terms.

Example 13.21 (Reciprocal of a power series). The formula (13.8) can be also
used to find the reciprocal of a power series. Suppose

$$f(x) := \sum_{k=0}^{\infty} a_k x^k \,.$$

Then what is the power series for the function $g := 1/f$? We assume that $f(0) \neq 0$
and write g as

$$g(x) = \sum_{k=0}^{\infty} b_k x^k \,.$$

Now as $f(x)g(x) \equiv 1$ for all x values, we must have

$$\left(\sum_{k=0}^{\infty} a_k x^k \right) \left(\sum_{k=0}^{\infty} b_k x^k \right) \equiv 1 \,.$$

Hence from (13.8) we have that

$$\sum_{k=0}^{\infty} \left(\sum_{j=0}^{k} a_j b_{k-j} \right) x^k = 1 \,,$$

or $a_0 b_0 = 1$, and

$$\sum_{j=0}^{k} a_j b_{k-j} = 0, \quad \text{for } k \geq 1 \,.$$

Let us try this to obtain a power series expansion for $\sec x = 1/\cos x$. The expansion for $\cos x$ will be shown, in the next section, to be

$$\cos x = \sum_{k=0}^{\infty} \frac{(-1)^k}{(2k)!} x^{2k} \,,$$

so the coefficients a_k are seen to be $a_k = (-1)^k/(2k)!$. We compute $b_0 = 1$, and

$$1 \times b_1 + \frac{-1}{2!} \times b_0 = 0 \,,$$

$$1 \times b_2 + \frac{-1}{2!} \times b_1 + \frac{(-1)^2}{(4)!} \times b_0 = 0 \,,$$

etc. So $b_1 = 1/2$, and $b_2 = 5/24$, etc.

Exercises

13.1   Prove Theorem 13.7 by the arguments of Theorem 13.9.

13.2   Prove the statements in Example 13.11.

13.3   Find intervals where the following power series are uniformly convergent.

(a) $\sum_{k=0}^{\infty} x^k$ (b) $\sum_{k=0}^{\infty} k x^k$ (c) $\sum_{k=1}^{\infty} \dfrac{x^k}{k}$

13.4   Find power series for the following functions.

(a) $\dfrac{2x - x^2}{1 - x^3}$ (b) $\dfrac{(1 + x)(3 - x)}{x + x^2}$ (c) $\dfrac{x}{1 - x} + \dfrac{1}{1 + x}$

13.5   Use the method of multiplying series to obtain the power series for

(a) $\left(\dfrac{1}{1 + x}\right)^3$ (b) $\left(\dfrac{1}{1 - x}\right)^2$ (c) $\left(\dfrac{1}{1 + x}\right)\left(\dfrac{x^2}{1 - x}\right)$

(d) $\left(\dfrac{x}{1 - x}\right)\left(\dfrac{1}{1 + 2x}\right)$

13.6   Write a MACSYMA function which produces a truncated power series for the product of two series. Use this function to check your results in the previous problem.

13.7   Determine the radius of convergence of the following power series.

(a) $\sum_{k=0}^{\infty}\left(\dfrac{k!}{3 \cdot 5 \cdots (2k + 1)}\right)^2 x^k$ (b) $\sum_{k=0}^{\infty} a^{k^2} x^k$

(c) $\sum_{k=0}^{\infty}\binom{2k}{k} x^k$ (d) $\sum_{k=0}^{\infty} k(2^{1/k} - 1)x^k$

(e) $\sum_{k=0}^{\infty}\left(1 + \dfrac{1}{2} + \dfrac{1}{3} + \cdots + \dfrac{1}{k}\right)x^k$

13.8   Give a formula for the radius of convergence for even and odd power series

(a) $\sum_{k=0}^{\infty} a_k x^{2k}$ (b) $\sum_{k=0}^{\infty} a_k x^{2k+1}$

with the aid of the ratio test.

13.9   Let the two power series

(a) $A(x) := \sum_{k=0}^{\infty} a_k x^k$ (b) $B(x) := \sum_{k=0}^{\infty} b_k x^k$

have radii of convergence R_1 and R_2, respectively. Show that the radius of convergence of the product series is at least $\min\{R_1, R_2\}$. Give examples where the radius of convergence of the product series equals this value, and where it is larger.

13.10 Determine all real numbers x for which the following power series converge. Parts a and b must converge for all $n \in \mathbf{N}$.

(a) $\sum_{k=0}^{\infty}(-1)^k \dfrac{x^{2k}}{2^{2k} k!(n + k)!}$

(b) $\sum_{k=0}^{\infty}(-1)^k \dfrac{x^{2k}}{2^{2k} k!(n + k)!}\left(\sum_{m=0}^{k}\dfrac{1}{m + 1} + \sum_{m=0}^{k}\dfrac{1}{m + n + 1}\right)$

Hint: It is easy to show $\sum_{m=1}^{p} 1/m = \ln p + O(1)$ for $p \to \infty$; read the proof of the integral test again.

(c) $\sum_{k=0}^{\infty} a_k x^k$,   with $a_0 := 1$,

$$a_k := \dfrac{\alpha(\alpha + 1) \cdots (\alpha + k - 1)\beta(\beta + 1) \cdots (\beta + k - 1)}{\gamma(\gamma + 1) \cdots (\gamma + k - 1)k!}$$

where $k \geq 0$, α, β, and γ are constants, γ is a nonnegative integer.

13.3 Differentiation of power series

Consider a convergent power series

$$\sum_{k=0}^{\infty} a_k x^k = f(x), \quad \text{for all } x \in (-R, R),$$

where the radius of convergence R is positive, possibly $R = \infty$.

Differentiating the left-hand side term by term we get another power series

$$D(f, x, 1) := \sum_{k=0}^{\infty} a_k \frac{d}{dx} x^k = \sum_{k=0}^{\infty} (k a_k) x^{k-1}. \tag{13.10}$$

This raises the following questions.

Question 1: Where does the power series $D(f, x, 1)$ converge?
Question 2: Where it converges, does it give the derivative of the function f?

Similarly, we may inquire what happens when we integrate the left-hand side term by term

$$\sum_{k=0}^{\infty} a_k \int_a^x t^k \, dt = \sum_{k=0}^{\infty} a_k \left(\frac{x^{k+1}}{k+1} - \frac{a^{k+1}}{k+1} \right), \tag{13.11}$$

where a and x are in $(-R, R)$.

Question 3: Is (13.11) equal to the integral $\int_a^x f(t) \, dt$?

The answers are very useful.

Answer 1: The term-by-term derivative $D(f, x, 1)$ of the power series has the same interval of convergence $(-R, R)$ as the original power series.
Answer 2: On that interval, the term-by-term derivative of the power series converges to the derivative $f'(x)$, so that

$$\sum_{k=0}^{\infty} (k a_k) x^{k-1} = f'(x), \quad \text{for all } x \in (-R, R). \tag{13.12}$$

Answer 3: Yes.

We begin with the last answer, showing that a power series may be integrated term by term within its interval of convergence.

Theorem 13.22. Let the function f have a power series expansion

$$f(x) := \sum_{k=0}^{\infty} a_k x^k \tag{13.13}$$

which converges for $x \in (-R, R)$. Then for any a and x in $(-R, R)$,

$$\int\limits_a^x f(t)\,\mathrm{d}t = \sum_{k=0}^\infty a_k \int\limits_a^x t^n\,\mathrm{d}t = \sum_{k=0}^\infty a_k\left(\frac{x^{k+1}}{k+1} - \frac{a^{k+1}}{k+1}\right), \qquad (13.14)$$

in other words,

$$F(x) := \sum_{k=0}^\infty \int a_k x^k\,\mathrm{d}x = \sum_{k=0}^\infty \frac{a_k}{k+1} x^{k+1} \qquad (13.15)$$

is an indefinite integral of f. $\qquad\qquad\square$

This theorem is proved in Sect. 13.6.

Example 13.23. We recall (5.35),

$$\frac{\mathrm{d}}{\mathrm{d}x} \operatorname{atan} x = \frac{1}{1+x^2}, \qquad \text{for all } x \in \mathbf{R}.$$

The right-hand side can be expressed as a geometric series

$$\frac{1}{1+x^2} = 1 - x^2 + x^4 - x^6 + \cdots = \sum_{k=0}^\infty (-1)^k x^{2k}, \quad \begin{array}{l}\text{converging} \\ \text{for } x \in (-1,1),\end{array} \qquad (13.16)$$

and integrated, definitely from 0 to x, term by term, to obtain a power series expansion

$$\operatorname{atan} x = \int\limits_0^x \frac{\mathrm{d}x}{1+x^2} = x - \frac{x^3}{3} + \frac{x^5}{5} - \frac{x^7}{7} + \cdots = \sum_{k=0}^\infty (-1)^k \frac{x^{2k+1}}{k+1}. \qquad (13.17)$$

Note that the integrated series (13.17) converges for $-1 \le x \le 1$ (see Example 13.11), while the original series (13.16) converges only for $-1 < x < 1$. Does this contradict Theorem 13.22?

We can use (13.17) to approximate π,

$$\frac{\pi}{4} = \operatorname{atan} 1 = 1 - \frac{1}{3} + \frac{1}{5} - \frac{1}{7} + \cdots ; \qquad (13.18)$$

see also Exercise 13.11.

Example 13.24. A power series expansion for $1/x$ is obtained by the trick

$$\frac{1}{x} = \frac{1}{1+(x-1)} = \sum_{k=0}^\infty (-1)^k (x-1)^k, \quad \text{converging for } 0 < x < 2.$$

Integrating, definitely from 1 to x, term by term we get a power series expansion of the natural logarithm,

$$\ln x = \int_1^x \frac{1}{x}\,dx = \sum_{k=0}^{\infty}(-1)^k \frac{(x-1)^{k+1}}{k+1}, \tag{13.19}$$

converging for $0 < x \le 2$. Indeed, this series is obtained by translation of the series

$$\sum_{k=0}^{\infty}(-1)^k \frac{x^{k+1}}{k+1}, \quad \text{which converges for } -1 < x \le 1 \text{ (see Example 13.12)}.$$

In particular, at the endpoint $x = 2$,

$$\ln 2 = \sum_{k=0}^{\infty}(-1)^k \frac{1}{k+1} = 1 - \frac{1}{2} + \frac{1}{3} - \frac{1}{4} + \frac{1}{5} - \cdots$$

We use MACSYMA to sum the first 1001 terms
```
c1. sum((-1)^k/(k+1),k,0,1000)
d1.    0.693646
```
as compared with
```
c2. ln(2), sfloat
d2.    0.693147
```

We turn now to the question of differentiating a power series term by term.

Lemma 13.25 (Derivative of a power series). Let the power series $f(x) = \sum_{k=0}^{\infty} a_k x^k$ converge for $|x| < R$ $(R > 0)$. Then the term-by-term differentiated series (13.10),

$$D(f, x, 1) := \sum_{k=1}^{\infty} k a_k x^{k-1},$$

converges for all x such that $|x| < R$.

Proof. Let x be in $(-R, R)$. Then there exists a pair of positive numbers R_0 and R_1 such that $|x| \le R_0 < R_1 < R$. Since $R_1 < R$, we know there exists a constant M such that $|a_n|R_1^{n-1} < M$ for all $n \in \mathbf{N}_0$ (compare with the proof of Theorem 13.9). Thus we have, for any x in $(-R_0, R_0)$,

$$|k a_k x^k| \le k|a_k|R_0^{k-1} = k|a_k|\left(\frac{R_0^{k-1}}{R_1^{k-1}}\right)R_1^{k-1} < kMr^{k-1},$$

where $r = R_0/R_1 < 1$. But the series

$$M\sum_{k=1}^{\infty} k r^{k-1} \quad (0 \le r < 1)$$

is a convergent series (the reader should verify this statement by using the ratio test, for example). Hence, by the comparison test (Theorem 12.32) the series

$$D(f, x, 1) = \sum_{k=1}^{\infty} k a_k x^{k-1}$$

converges in $(-R_0, R_0)$. $\square$

Lemma 13.26 (Radius of convergence of the derivative series). The power series $\sum a_k x^k$ and its term-by-term differentiated series $\sum k a_k x^{k-1}$ have the same radius of convergence.

Proof. Let the radius of convergence of $\sum a_k x^k$ and $\sum k a_k x^{k-1}$ be R and R', respectively. Theorem 13.13 shows that $R' \leq R$. If we can show $R \leq R'$, then it follows that $R = R'$. To do this, we note that for each $x \in (-R, R)$, and for $k \in \mathbf{N}$, that $|a_k x^k| \leq M_x$ for some M_x. Choose an x_0 near to R and let M be the corresponding upper bound for $|a_k x^k|$. Then for $|x| \leq \theta |x_0|$

$$|k a_k x^{k-1}| \leq |k a_k x_0^{k-1} \theta^{k-1}| \leq M k \theta^{k-1} .$$

Hence, the series $\sum |a_k x^k|$ must converge for all x where $|x| \leq \theta |x_0|$ and where the series $\sum |k \theta^{k-1}|$ converges. This latter series converges clearly for $\theta < 1$, which implies that the series $\sum k a_k x^{k-1}$ converges in $[-\theta |x_0|, \theta |x_0|] \subset [-R, R]$. As θ can be as close to 1 as wished, this means that $[-R, R]$ is also the interval of convergence for the derived series. $\square$

We summarize these lemmas in the following corollary.

Corollary 13.27. The power series $f(x) = \sum a_k x^k$ and its term-by-term differentiated series of order m,

$$D(f, x, m) := \sum_{k=m}^{\infty} \frac{k!}{(k-m)!} a_k x^{k-m} \tag{13.20}$$

have the same interval of convergence.

Proof. From Lemma 13.26 the series for $f(x)$ and $D(f, x, 1)$ have the same intervals of convergence. Applying this lemma again to the function $D(f, x, 1)$ we note that the series for $D(f, x, 1)$ and $D(f, x, 2)$ have the same interval of convergence. We can repeat this argument to see inductively that all derivatives of $f(x)$ have series representations with the same radius of convergence, the desired result. $\square$

Theorem 13.28 (Derivative of a power series). If f has a power series expansion, (13.13),

$$f(x) := \sum_{k=0}^{\infty} a_k x^k$$

which converges for $x \in (-R, R)$, then f has derivatives of all orders in $(-R, R)$, and these are obtained by differentiating the power series term by term, namely,

$$f'(x) = \sum_{k=1}^{\infty} k a_k x^{k-1} \qquad\qquad\qquad (13.21)$$

$$f''(x) = \sum_{k=2}^{\infty} k(k-1) a_k x^{k-2},$$

and the mth derivative

$$f^{(m)}(x) = \sum_{k=m}^{\infty} k(k-1) \cdots (k-m+1) a_k x^{k-m} \quad (m = 1, 2, \ldots).$$

All these series converge in the same interval $(-R, R)$.

Proof. Let, (13.13),

$$f(x) := \sum_{k=0}^{\infty} a_k x^k, \quad \text{for all } x \in (-R, R),$$

and consider the term-by-term derivative (13.10),

$$g(x) := \sum_{k=0}^{\infty} (k a_k) x^{k-1},$$

which, by Lemma 13.26, has the same interval of convergence. Integrating this series term by term we get, by Theorem 13.22,

$$\int_0^x g(t)\, dt = \sum_{k=0}^{\infty} \int_0^x k a_k t^{k-1}\, dt = \sum_{k=0}^{\infty} a_k x^k = f(x),$$

showing that g is an antiderivative of f. This proves (13.21). The results for higher derivatives are proved by repeating this argument.   □

Let the function f have a power series expansion (13.13),

$$f(x) = \sum_{k=0}^{\infty} a_k x^k$$

converging in the interval $(-R, R)$, where possibly $R = \infty$. Can the function f have a different power series expansion in the same interval? The answer is no. The power series expansion of a given function is *unique* to that function.

Theorem 13.29 (Uniqueness of power series). If two power series converge in an interval $(-R, R)$ to the same function

$$\sum_{k=0}^{\infty} a_k x^k = \sum_{k=0}^{\infty} b_k x^k = f(x),$$

then

$$a_k = b_k, \quad \text{for all } k = 0, 1, 2, \ldots \tag{13.22}$$

Proof. The nth derivative of f can be written in two ways, by differentiating $\sum_{k=0}^{\infty} a_k x^k$ or $\sum_{k=0}^{\infty} b_k x^k$

$$\begin{aligned}
f^{(n)}(x) &= n!\, a_n + \big((n+1)n(n-1)\cdots 2\big)a_{n+1}x \\
&\quad + \big((n+2)(n+1)n\cdots 3\big)a_{n+2}x^2 + \cdots \\
&= n!\, b_n + \big((n+1)n(n-1)\cdots 2\big)b_{n+1}x \\
&\quad + \big((n+2)(n+1)n\cdots 3\big)b_{n+2}x^2 + \cdots
\end{aligned}$$

The nth derivative of f at 0 is therefore $f^{(n)}(0) = n!\, a_n = n!\, b_n$, proving (13.22). $\qquad \square$

Remark 13.30. Similarly, for any $x_0 \neq 0$, the identity

$$\sum_{k=0}^{\infty} a_k (x - x_0)^k = \sum_{k=0}^{\infty} b_k (x - x_0)^k = f(x), \quad \text{for all } x \in (x_0 - R, x_0 + R),$$

implies (13.22). This follows from Theorem 13.29 by translation $\bar{x} := x - x_0$.

The next example shows an ingenious application of the above results.

Example 13.31 (Fibonacci numbers). Recall the Fibonacci numbers $(f_k)_{k=0,1,2,\ldots}$ defined recursively by

$$\begin{aligned}
f_0 &:= 0, \\
f_1 &:= 1, \\
f_n &:= f_{n-1} + f_{n-2} \quad \text{for all } n \geq 2.
\end{aligned} \tag{13.23}$$

We use power series to give an alternative proof of (13.29).
Consider the function

$$F(x) := \sum_{k=0}^{\infty} f_{k+1} x^k, \tag{13.24}$$

a power series with the Fibonacci numbers as coefficients.[1] Now

---

1 Called a generating function of the Fibonacci sequence.

$$\frac{f_{k+1}}{f_k} = \frac{f_k + f_{k-1}}{f_k} = 1 + \frac{f_{k-1}}{f_k} < 1 + 1 = 2$$

since the sequence $(f_n)_n$ is monotone increasing. This proves that

$$\lim_{n \to \infty} \frac{f_{k+1}}{f_k} \leq 2 ,$$

showing that the series (13.24) converges (at least) for $|x| < 1/2$.

Because of (13.23), the function F must satisfy the algebraic equation

$$F(x) - xF(x) - x^2 F(x) = 1 , \tag{13.25}$$

in the interval $(-1/2, 1/2)$, see Exercise 13.24. Solving for $F(x)$ we obtain

$$F(x) = \frac{-1}{x^2 + x - 1} .$$

We now use the partial fraction decomposition

$$F(x) = \frac{1}{\sqrt{5}} \left(\frac{1}{x + \frac{1+\sqrt{5}}{2}} - \frac{1}{x + \frac{1-\sqrt{5}}{2}} \right) , \tag{13.26}$$

and the identities

$$\frac{2}{1 + \sqrt{5}} = -\frac{1 - \sqrt{5}}{2} \quad \text{and} \quad \frac{2}{1 - \sqrt{5}} = -\frac{1 + \sqrt{5}}{2}$$

to obtain a geometric series expansion of F

$$F(x) = \frac{1}{\sqrt{5}} \left(\frac{\frac{2}{1+\sqrt{5}}}{1 + \frac{2}{1+\sqrt{5}}x} - \frac{\frac{2}{1-\sqrt{5}}}{1 + \frac{2}{1-\sqrt{5}}x} \right) \tag{13.27}$$

$$= \frac{1}{\sqrt{5}} \sum_{k=0}^{\infty} \left(\left(\frac{1 + \sqrt{5}}{2} \right)^{k+1} - \left(\frac{1 - \sqrt{5}}{2} \right)^{k+1} \right) x^k ,$$

which can be shown to have radius of convergence

$$R = \frac{\sqrt{5} - 1}{2} . \tag{13.28}$$

Comparing with (13.24) we get, by the uniqueness of the power series expansion of F, a proof of the identity

$$f_n = \frac{\left(\frac{1+\sqrt{5}}{2} \right)^n - \left(\frac{1-\sqrt{5}}{2} \right)^n}{\sqrt{5}} . \tag{13.29}$$

Since $(1 - \sqrt{5})/2 < 1$, it follows that $((1 - \sqrt{5})/2)^k \to 0$ as $k \to \infty$, giving the approximation

$$f_k \approx \frac{1}{\sqrt{5}} \left(\frac{1 + \sqrt{5}}{2} \right)^{k+1},$$ (13.30)

which gets better as $k \to \infty$.

Exercises

13.11 (A work-saving trick) Verify that $\operatorname{atan} 1 = \operatorname{atan} 1/4 + \operatorname{atan} 3/5$ and observe that the series (13.17) for the right-hand side

$$\sum_{k=0}^{\infty} (-1)^k \frac{(1/4)^{2k+1}}{2k+1} + \sum_{k=0}^{\infty} (-1)^k \frac{(3/5)^{2k+1}}{2k+1}$$

converges faster than (13.18), hence requires fewer terms to achieve a given accuracy.

13.12 Use MACSYMA to sum enough terms of the power series expansion (13.19), to obtain approximations of

    (a) $\ln 0.1$ (b) $\ln 0.99$ (c) $\ln 1.5$ (d) $\ln 1.99$

with error of less than 0.001. Note that more terms are needed for x near the endpoints of $(0, 2)$.

13.13 Adapt Example 13.24 to obtain approximations of $\ln x$ for x in the interval $(0, 4)$.
Hint: Use the observation

$$\ln x = \ln 2 + \ln x/2 \quad \text{where } x/2 \text{ is now in the convergence interval of (13.19).}$$

13.14 Adapt Exercise 13.13 to obtain approximations of $\ln x$ for any positive x.

13.15 Verify:

$$\ln(1 + x) = x - \frac{x^2}{2} + \frac{x^3}{3} - \frac{x^4}{4} + \cdots, \quad -1 < x < 1, \tag{13.31a}$$

$$\ln(1 - x) = -x - \frac{x^2}{2} - \frac{x^3}{3} - \frac{x^4}{4} - \cdots, \quad -1 < x < 1, \tag{13.31b}$$

$$\ln\left(\frac{1 + x}{1 - x} \right) = 2\left(x + \frac{x^3}{3} + \frac{x^5}{5} + \frac{x^7}{7} + \cdots \right), \quad -1 < x < 1. \tag{13.31c}$$

Hint: The computation of (13.31a) is analogous to the computation of (13.19) in Example 13.24.

13.16 Use $(d/dx) e^x = e^x$ and Theorem 13.29 to prove: The only possible power series expansion of e^x is

$$e^x = 1 + x + \frac{x^2}{2!} + \frac{x^3}{3!} + \cdots = \sum_{k=0}^{\infty} \frac{x^k}{k!}$$

13.17 Use the fact that $(d/dx) \operatorname{atan} x = 1/(1 + x^2)$ to obtain a Taylor expansion for $\operatorname{atan} x$.
Hint: Integrate $1/(1 + x^2)$.

13.18 Use the fact that

$$\operatorname{asin} x = \int_0^x \frac{1}{\sqrt{1 - x^2}} \, dx$$

to obtain a power series expansion for $\operatorname{asin} x$.

13.19 Using termwise integration find a power series representation for the following
functions

(a) $f(x) = \displaystyle\int_0^x \cos^2 x \, dx$ (b) $f(x) = \displaystyle\int_0^x e^{-ax^2} \, dx$

(c) $f(x) = \displaystyle\int_0^x \frac{1}{1-x^3} \, dx$ (d) $f(x) = \displaystyle\int_0^x \frac{1 - \cos x}{x^2} \, dx$

(e) $f(x) = \displaystyle\int_0^x \frac{a\sin x}{x} \, dx$ (f) $f(x) = \displaystyle\int_0^x \frac{x - a\tan x}{x^2} \, dx$

13.20 Use MACSYMA to check your results in the previous problem.

13.21 Prove the following expansions

(a) $\displaystyle\int_0^x e^{-x^2} \, dx = x - \frac{1}{3}x^3 + \frac{x^5}{5 \cdot 2!} - \frac{x^7}{7 \cdot 3!} + \cdots$

(b) $\displaystyle\int_0^x \cos x^2 \, dx = x - \frac{x^5}{5 \cdot 2!} + \frac{x^9}{13 \cdot 6!} + \cdots$

13.22 Justify the following

$$\int_0^x \frac{x^{\alpha-1}}{1+x^\beta} \, dx = \frac{1}{\alpha} - \frac{1}{\alpha+\beta} + \frac{1}{\alpha+2\beta} - \frac{1}{\alpha+3\beta} \cdots .$$

13.23 Show that the function $F(x)$ of (13.24) satisfies the second-order differential equation with constant coefficients

$$(1 - x - x^2)\frac{dF(x)}{dx} - (1 - 2x)F(x) = 0 .$$

Hint: Use term-by-term differentiation and the Fibonacci recursion (13.23).

13.24 Show that for the Fibonacci function $\sum f_{k+1} x^k$ the coefficient recursion (13.23) and the algebraic relation (13.25) are equivalent.

13.25 Show the partial fraction decomposition (13.26) with MACSYMA.

13.26 Use the exact formula (13.29), and the approximation (13.30), to calculate the Fibonacci numbers f_{10}, f_{100}, and f_{1000}.

13.4 Taylor series

In Sect. 7.3, we studied Taylor polynomials with finite degree. In this section, we study Taylor series, the series obtained when the degree of the Taylor polynomial tends to infinity.

Definition 13.32 (Taylor series). If $f(x)$ has derivatives of all orders at $x = x_0$, then the power series

$$T_\infty(f, x, x_0) := \sum_{k=0}^\infty \frac{f^{(k)}(x_0)}{k!}(x - x_0)^k \tag{13.32}$$

is called the Taylor series of f about the *center* $x = x_0$.

If the center x_0 is zero, the Taylor series

$$T_\infty(f, x, 0) := \sum_{k=0}^{\infty} \frac{f^{(k)}(0)}{k!} x^k \qquad (13.33)$$

is sometimes called the *Maclaurin series* of f.

Example 13.33 (Taylor series of a power series). We compute the Maclaurin series of a function f with a known power series expansion, (13.13),

$$f(x) = \sum_{k=0}^{\infty} a_k x^k .$$

Differentiating term by term we get the kth derivative at 0, as $f^{(k)}(0) = k! \, a_k$, or

$$a_k = \frac{f^{(k)}(0)}{k!} . \qquad (13.34)$$

The Maclaurin series is then just the original series

$$\sum_{k=0}^{\infty} \frac{f^{(k)}(0)}{k!} x^k = \sum_{k=0}^{\infty} a_k x^k ,$$

which is not surprising in view of Theorem 13.29.

Similarly, the Taylor series of the power series expansion

$$f(x) = \sum_{k=0}^{\infty} a_k (x - x_0)^k \qquad (13.35)$$

is the same series, with coefficients

$$a_k = \frac{f^{(k)}(x_0)}{k!} . \qquad (13.36)$$

Again, this is guaranteed by the uniqueness of power series; see Remark 13.30.

Definition 13.34. The coefficients a_k defined by Eq. (13.36) are called the *Taylor coefficients* of f about x_0.

In the next two examples we compute Taylor series formally, i.e., without concerning ourselves with convergence.

Example 13.35. Let

$$f(x) = \frac{1}{1 - x} .$$

Then

$$f'(x) = \frac{1}{(1-x)^2}, \quad f''(x) = \frac{2}{(1-x)^3}, \quad f'''(x) = \frac{3 \cdot 2}{(1-x)^4}$$

and the nth-derivative

$$f^{(n)}(x) := \frac{n!}{(1-x)^{n+1}}.$$

Hence, $f^n(0) = n!$ and the Taylor coefficients are given by

$$a_k = \frac{f^n(0)}{n!} = 1.$$

The Taylor series of $1/(1-x)$ is the well-known geometric series

$$T_\infty\left(\frac{1}{1-x}, x, 0\right) = \sum_{k=0}^{\infty} x^k.$$

Example 13.36. Let

$$f(x) = \frac{1}{(1+x)^3}.$$

Then

$$f'(x) = \frac{-3}{(1+x)^4}, \quad f''(x) = \frac{(-3)(-4)}{(1+x)^5}$$

and in general,

$$f^{(k)}(x) = (-1)^k \frac{3 \cdot 4 \cdot 5 \cdots (k+2)}{(1+x)^{k+2}}.$$

$$\therefore \quad f^{(k)}(0) = \frac{(-1)^k (k+2)!}{2!},$$

and the Taylor series is given by

$$T_\infty\left(\frac{1}{(1+x)^3}, x, 0\right) = \sum_{k=0}^{\infty} (-1)^k \frac{(k+2)(k+1)}{2} x^k. \tag{13.37}$$

Note that

$$\frac{d^2}{dx^2}\left(\frac{1}{1+x}\right) = \frac{2}{(1+x)^3},$$

so the power series expansion (13.37) can be obtained by differentiating (twice) the geometric series

$$\frac{1}{1+x} = \sum_{k=0}^{\infty} (-1)^k x^k.$$

Indeed,

$$\frac{d^2}{dx^2}\left(\frac{1}{1+x}\right) = \sum_{k=0}^{\infty}(-1)^k\frac{d^2}{dx^2}(x^k) = \sum_{k=2}^{\infty}(-1)^k k(k-1)x^{k-2}$$

$$= 2T_{\infty}\left(\frac{1}{(1+x)^3}, x, 0\right),$$

illustrating Theorem 13.28.

We recall (see Example 7.31) that the Taylor polynomial $T_n(f, x, x_0)$ has the property that its first n derivatives match those of the function f at the point x_0.

The Taylor polynomial provides an approximation of f near x_0. If $f(x)$ has $n+1$ continuous derivatives on an interval $I := [x_0 - c, x_0 + c]$, then the error of the approximation in I is the *Lagrange remainder* (7.20c),

$$R_n(f, x, x_0) := f(x) - T_n(f, x, x_0) = \frac{f^{(n+1)}(\xi)}{(n+1)!}(x - x_0)^{n+1},$$

for some ξ in I. This can be used to give a bound on $|R_n(f, x, x_0)|$

$$|R_n(f, x, x_0)| \le \frac{M}{(n+1)!}c^{n+1}, \quad \text{where } M := \max_{x\in I}|f^{(n+1)}(x)|.$$

The Taylor series $T_{\infty}(f, x, x_0)$ converges to f in some interval around x_0 if $|R_n(f, x, x_0)| \to 0$ as $n \to \infty$, or in other words, if

$$\lim_{n\to\infty}\left|\frac{(x - x_0)^{n+1}}{(n+1)!}f^{(n+1)}(\xi)\right| = 0,$$

for all ξ in I.

Suppose that there exist two positive constants α and K such that

$$|f^{(n)}(x)| \le \alpha K^n, \quad \text{in } I.$$

Then an upper bound on the Lagrange remainder is

$$|R_n(f, x, x_0)| \le \frac{|x - x_0|^{n+1}}{(n+1)!}|f^{(n+1)}(\xi)| \le \alpha K^{n+1}\frac{|x - x_0|^{n+1}}{(n+1)!},$$

which tends to zero as $n \to \infty$ since

$$\alpha K^{n+1}\frac{|x - x_0|^{n+1}}{(n+1)!} \to 0 \quad \text{as } n \to \infty.$$

We summarize this as follows.

Theorem 13.37. The function $f(x)$ is represented by its Taylor series $T_{\infty}(f, x, x_0)$ in the interval $(x_0 - c, x_0 + c)$ if there exist two positive constants α and K so that the inequality

$$|f^{(n)}(x)| \leq \alpha K^n \qquad (13.38)$$

holds in $(x_0 - c, x_0 + c)$.   □

Example 13.38 (Exponential series). We apply Theorem 13.37 to the *exponential function* e^x, giving the Taylor series expansion

$$e^x = \sum_{k=0}^{\infty} \frac{1}{k!} x^k = 1 + x + \frac{x^2}{2!} + \frac{x^3}{3!} + \cdots \qquad (13.39)$$

for all $x \in \mathbf{R}$. The Taylor coefficients at $x_0 = 0$ are computed by (13.36) as follows

$$a_k = \frac{1}{k!} \frac{d^k e^x}{dx^k}\Big|_{x=0} = \frac{e^x}{k!}\Big|_{x=0} = \frac{1}{k!},$$

since $e^0 = 1$. The remainder $R_n(e^x, x, 0)$ is bounded by

$$|R_n(e^x, x, 0)| \leq \frac{|x|^{n+1}}{(n+1)!} |e^\xi| \leq e^{|x|} \frac{|x|^{n+1}}{(n+1)!}.$$

For any fixed $x \in \mathbf{R}$ the remainder tends to zero as $n \to 0$, so the infinite Taylor series

$$\sum_{k=0}^{\infty} \frac{x^k}{k!} \quad \text{converges to } e^x,$$

for all x.

We prove now that e^x is continuous at $x = 0$. The first-degree Taylor polynomial gives a remainder

$$e^x - 1 - x = R_1(e^x, x, 0), \quad \text{for, say, } x \in (-1, 1)$$

bounded by

$$|R_1(e^x, x, 0)| \leq e^1 \frac{x^2}{2}.$$

Clearly $|R_1(e^x, x, 0)| \to 0$ as $x \to 0$, and from this we conclude that

$$\lim_{x \to 0} e^x = e^0 = 1,$$

which is equivalent to the continuity of the exponential function at the origin.

Example 13.39. We are frequently able to use the Taylor series for one function to develop a series for another function. For example, consider x^x which we may write using the logarithm as

$$x^x = e^{x \ln x} = \sum_{k=0}^{\infty} \frac{1}{k!} (x \ln x)^k.$$

Example 13.40 (Sine series). Let us apply Theorem 13.37 to the *sine* function $\sin x$. We shall show that

$$\sin x = \sum_{k=0}^{\infty} \frac{(-1)^k}{(2k+1)!} x^{2k+1} = x - \frac{x^3}{3!} + \frac{x^5}{5!} - \cdots \tag{13.40}$$

for all $x \in \mathbf{R}$. We want to compute the Taylor coefficients by the formula

$$a_k = \frac{1}{k!} \frac{d^k \sin x}{dx^k}\Big|_{x=0}.$$

To this end we notice that

$$\frac{d \sin x}{dx} = \cos x, \quad \frac{d^2 \sin x}{dx^2} = -\sin x, \quad \frac{d^3 \sin x}{dx^3} = -\cos x, \quad \frac{d^4 \sin x}{dx^4} = \sin x.$$

Therefore

$$\frac{d^5 \sin x}{dx^5} = \cos x, \quad \frac{d^6 \sin x}{dx^6} = -\sin x, \quad \frac{d^7 \sin x}{dx^7} = -\cos x, \quad \frac{d^8 \sin x}{dx^8} = \sin x$$

and the same derivative is repeated after 4 differentiations. Since $\sin 0 = 0$ and $\cos 0 = 1$, we conclude that only the odd terms (the 1st, 3rd, 5th, etc.) appear in the Taylor series

$$\sin x = \sum_{k=0}^{\infty} \frac{x^{2k+1}}{(2k+1)!} \frac{d^{2k+1}\sin x}{dx^{2k+1}}\Big|_{x=0} = \sum_{k=0}^{\infty} (-1)^k \frac{x^{2k+1}}{(2k+1)!} \cos x\Big|_{x=0},$$

proving (13.40).

As $|\sin x|, |\cos x| \le 1$, the remainder is bounded in the interval $(-x, x)$ by

$$R_{2n+1}(\sin x, x, 0) \le \frac{|x|^{2n+1}}{(2n+1)!} \to 0 \quad \text{as } x \to 0,$$

justifying the Taylor series (13.40).

Example 13.41 (Cosine series). As in Example 13.40 the consecutive derivatives of the cosine function follow a pattern

$$\frac{d \cos x}{dx} = -\sin x, \quad \frac{d^2 \cos x}{dx^2} = -\cos x, \quad \frac{d^3 \cos x}{dx^3} = \sin x, \quad \frac{d^4 \cos x}{dx^4} = \cos x$$

with the same derivative repeated every 4 differentiations. Since $\sin 0 = 0$ and $\cos 0 = 1$, we conclude that

$$\cos x = \sum_{k=0}^{\infty} \frac{x^k}{k!} \frac{d^k \cos x}{dx^k}\Big|_{x=0} x^k = \sum_{k=0}^{\infty} (-1)^k \frac{x^{2k}}{(2k)!} \cos x\Big|_{x=0},$$

proving

$$\cos x = \sum_{k=0}^{\infty} \frac{(-x)^{2k}}{(2k)!} = 1 - \frac{x^2}{2!} + \frac{x^4}{4!} - \frac{x^6}{6!} + \cdots$$

Example 13.42 (Binomial series). Recall the binomial formula

$$(a + b)^n = \sum_{k=0}^{n} \binom{n}{k} a^{n-k} b^k , \qquad (13.41)$$

where n is an integer; in particular,

$$(1 + x)^n = \sum_{k=0}^{n} \binom{n}{k} x^k .$$

The analogous expression for $(1 + x)^\alpha$, where α is any real number, is

$$(1 + x)^\alpha = \sum_{k=0}^{n} \binom{\alpha}{k} x^k , \qquad (13.42)$$

where we denote the *binomial coefficient*

$$\binom{\alpha}{k} := \frac{\alpha(\alpha - 1)(\alpha - 2) \cdots (\alpha - k + 1)}{k!} \qquad (k = 0, 1, 2, \ldots) .$$

Indeed,

$$\frac{d^k}{dx^k} (1 + x)^\alpha = \alpha(\alpha - 1)(\alpha - 2) \cdots (\alpha - k + 1)(1 + x)^{\alpha-k}$$

so that the Taylor series at $x = 0$,

$$(1 + x)^\alpha = \sum_{k=0}^{\infty} \left(\alpha(\alpha - 1)(\alpha - 2) \cdots (\alpha - k + 1)(1 + x)^{\alpha-k} \big|_{x=0} \right) \frac{x^k}{k!} ,$$

proving (13.42). We justify it with the following lemma.

Lemma 13.43. The power series (13.42) converges to $(1+x)^\alpha$ for all $-1 < x < 1$.

This lemma is proved in Sect. 13.6.

The next example shows that a Taylor series $T_\infty(f, x, x_0)$ may converge to something quite different than the function f near x_0.

Example 13.44 (A counterexample). Consider the function defined by

$$f(x) := \begin{cases} e^{-1/x^2} & \text{if } x \neq 0, \\ 0 & \text{if } x = 0. \end{cases} \qquad (13.43)$$

Now f has derivatives of all orders at $x = 0$, and all of them vanish

$$f^{(n)}(0) = 0, \quad \text{for all } n = 0, 1, 2, \ldots ;$$

see Exercise 13.41. Therefore all the Taylor polynomials of f at $x = 0$ are zero,

$$T_n(e^{-1/x^2}, x, 0) = 0, \quad \text{for all } n,$$

showing that the Taylor series

$$T_\infty(e^{-1/x^2}, x, 0) = 0$$

and cannot represent f near 0.

MACSYMA tells why it is impossible to find a Taylor series.

```
cl. taylor(exp(-1/x^2),x,0,4)
```

d1.   taylor encountered an essential singularity in e^{-1/x^2}

There are other forms of the remainder of the finite Taylor polynomial. One such remainder, which is a generalization of the Lagrange remainder, is the Schloemilch remainder in the following theorem.

Theorem 13.45 (Schloemilch remainder). Let $f(x)$ have $(n + 1)$ continuous derivatives on the interval $[a, b]$. Then for each fixed value of x there exists a number $0 < \theta < 1$, such that

$$f(x) = T_n(f, x, a) + \frac{f^{(n+1)}(a + \theta(b - a))}{n! \, p} (1 - \theta)^{n+1-p}(x - a)^{n+1}.$$

This theorem is proved in Sect. 13.6.

Remark 13.46 (Cauchy remainder). We remark that for $p = n + 1$ the remainder becomes the Lagrange remainder. The case $p = 1$ is known as the Cauchy remainder

$$R_n(x; a) := \frac{f^{(n+1)}(a + \theta(b - a))}{n!}(1 - \theta)^n(x - a)^{n+1}.$$

Exercises

13.27 Find formal Taylor series for the following functions.

(a) $\dfrac{1}{1 - x^2}$

(b) $\dfrac{(1 + x)}{x}$

(c) $\dfrac{x}{2 + x} + \dfrac{1}{1 + x^3}$

(d) $\sin^3 x$

(e) $(1 + x^2)^{1/4}$

(f) $\dfrac{\ln(1 + x^2)}{x^3}$

13.28 Find Taylor series of the functions

(a) e^{x^2}

(b) $\sin x \cos x$

(c) $e^{-x} \cos x$

(d) $\displaystyle\int_0^x \ln(1 + x)\, dx$

(e) $\displaystyle\int_0^x \frac{\ln(1 + x)}{x}\, dx$

(f) $\displaystyle\int_x^1 \frac{\ln x}{x}\, dx$

13.29 Calculate the radius of convergence of the Taylor series

(a) $\sin x^2$

(b) $\operatorname{atan} x$

(c) $\cosh x$

(d) $\sec x$

(e) $\ln(1 + x)$

(f) $\sqrt{1 + x}$

13.30 The following problems give a function f, an integer n, and an interval $[a, b]$. In each case, compute the error in $[a, b]$ of the Taylor polynomial at 0, $T_n(f, x, 0)$. Use MACSYMA to assist in your computations.

(a) $f(x) := \sqrt{1 + x^2}$, $\quad n = 5$, $[a, b] = [-1, 1]$

(b) $f(x) := \dfrac{\sin x - x + x^3/6}{x^2}$, $\quad n = 6$, $[a, b] = [0, 1]$

(c) $f(x) := e^x \sin x$, $\quad n = 4$, $[a, b] = [-1, 1]$

(d) $f(x) := \dfrac{x^2(1 - x^4)}{\sin x}$, $\quad n = 4$, $[a, b] = [-\pi/2, \pi/2]$

(e) $f(x) := \dfrac{\ln(1 - x)}{1 - x}$, $\quad n = 3$, $[a, b] = [-1/2, 1/2]$

(f) $f(x) := e^{\sin x}$, $\quad n = 4$, $[a, b] = [-4, 4]$

13.31 Given a constant a, let the symbol $(a)_k$ be defined as follows

$$(a)_0 = 1,$$
$$(a)_k = a(a + 1)(a + 2) \cdots (a + k - 1), \quad k = 1, 2, \ldots$$

Given three constants a, b, and c let the function $F(a, b; c; x)$ (of x) be defined by the power series

$$F(a, b; c; x) := \sum_{k=0}^{\infty} \frac{(a)_k (b)_k}{(c)_k k!} x^k.$$

Show by termwise differentiation that the power series expansion for F satisfies identically the equation

$$x(1 - x) \frac{d^2 F}{dx^2} + [c - (a + b + 1)x] \frac{dF}{dx} - abF = 0.$$

13.32 Verify the following identities by computing the power series of both in the statement

(a) $(1 + x)^a = F(-a, b; b; -x)$

(b) $\frac{1}{2}(1 + x^{1/2})^{-2a} + \frac{1}{2}(1 - x^{1/2})^{-2a} = F(a, a + 1/2; 1/2; x)$

(c) $\operatorname{asin} x = x F(1/2, 1/2; 3/2; x^2)$ $\qquad$ (d) $\operatorname{atan} x = x F(1/2, 1; 3/2; -x^2)$

(e) $\ln(x + 1) = x F(1, 1; 2; -x)$ $\qquad$ (f) $\ln \dfrac{1 + x}{1 - x} = 2x F1/2, 1; 3/2; x^2)$

13.33 Use the Lagrange form of the remainder to show that the Taylor series for the following functions converge on $(-\infty, \infty)$.

(a) $\sinh x$ $\qquad\qquad$ (b) $\displaystyle\int_0^x e^{-x^2} \, dx$ $\qquad\qquad$ (c) $\cosh x$

13.34 Obtain the series expansion

$$\frac{1}{(1 + x)^{1/3}} = 1 - \frac{1}{3}x + \frac{1 \cdot 4}{3 \cdot 6}x^2 - \frac{1 \cdot 4 \cdot 7}{3 \cdot 6 \cdot 9}x^3 + \frac{1 \cdot 4 \cdot 7 \cdot 10}{3 \cdot 6 \cdot 9 \cdot 12}x^3 - \cdots$$

13.35 Prove the following expansion, and then use it to obtain a numerical series which converges to π

$$\frac{1}{\sqrt{1 - x^2}} = 1 - \frac{1}{2}x^2 + \frac{1 \cdot 3}{2 \cdot 4}x^4 + \frac{1 \cdot 3 \cdot 5}{2 \cdot 4 \cdot 6}x^6 + \cdots.$$

13.36 Find expansions in terms of powers of x for the following functions. If you cannot find the general form of the series, give the first few terms.

(a) a^x (b) $\frac{1}{2}(e^x + e^{-x})$ (c) $\frac{1}{2}(e^x - e^{-x})$

(d) $\ln(x + \sqrt{1 + x^2})$ (e) $\cot x$ *Hint:* Expand $x \cot x$.

13.37 Use MACSYMA to obtain the following expansions, for $x^2 < \infty$,

(a) $e^{\cos x} = e\left(1 - \frac{x^2}{2!} + \frac{4x^4}{4!} - \frac{31}{6!}x^6 + \cdots\right)$

(b) $e^{\sin x} = 1 + x\frac{x^2}{2!} - \frac{3x^4}{4!} - \frac{8x^5}{5!} - \frac{3x^6}{6!} + \cdots$

Can you justify the intervals of convergence?

13.38 Prove the following estimates

(a) $e^x > 1 + x + \frac{x^2}{2!} + \cdots + \frac{x^{2n-1}}{(2n-1)!}$,   for $n \geq 1$ and all $x \neq 0$.

(b) $\ln x < (x - 1) - \frac{(x-1)^2}{2} + \frac{(x-1)^3}{3} + \cdots + \frac{(x-1)^{2n-1}}{2n-1}$,

for $n \geq 1$ and all positive $x \neq 1$.

13.39 Use power series to determine the following limits. Check your results with MAC-SYMA.

(a) $\lim_{x \to 0} \dfrac{x - \sin x}{e^x - 1 - x - x^2/2}$ (b) $\lim_{x \to 0} \dfrac{\ln^2(1+x) - \sin^2 x}{1 - e^{-x^2}}$

(c) $\lim_{x \to 0} \dfrac{e^{x^4} - 1}{(1 - \cos x)^2}$

13.40 Suppose that $f(x)$ and $g(x)$ are n times differentiable at the point $x = 0$ ($n > 0$). Moreover, let $f(x) = p_n(x) + x^n g(x)$, where $p_n(x)$ is a polynomial of degree $\leq n$. Show that $p_n(x) \equiv T_n(f, x, a)$.

13.41 Consider the function defined by

$$f(x) := \begin{cases} e^{-1/x^2} & \text{if } x \neq 0, \\ 0 & \text{if } x = 0. \end{cases}$$

Show that by using the right-sided differentiation rule $\lim_{x \to a^+} (f(x) - f(a))(x - a)$ that all the derivatives for $x > 0$ are of the form $P_n(1/x) e^{-1/x^2}$, where $P_n(t)$ is a polynomial. Conclude a similar result for $x < 0$. Use the L'Hospital rule to show that $\lim_{x \to 0} P_n(1/x) e^{-1/x^2} = 0$ for each value of $n \in \mathbf{N}$.

Plot the function with MACSYMA to see how slowly it grows near $x = 0$. Its growth is so slow that it cannot be distinguished from the zero function from the point of view of Taylor polynomial approximation.

13.42 Consider the function

$$f(x) := \begin{cases} e^{1/x} & \text{if } x < 0, \\ 0 & \text{if } x = 0, \\ e^{-1/x} & \text{if } x < 0. \end{cases}$$

Show that, like (13.43), this function is not represented well by its Taylor series at 0.

13.5 Lagrange interpolation

A line $y = mx + b$ is a polynomial of degree less than or equal to 1 (if $m = 0$, the degree is zero). It is (uniquely) determined by any two of its points (x_1, y_1) and (x_2, y_2).

Similarly, a polynomial p of degree less than or equal to n is determined by any $n + 1$ points (*data points*)

$$(x_1, y_1), \ (x_2, y_2), \ \ldots, \ (x_{n+1}, y_{n+1}), \quad \text{with different } x_1, x_2, \ldots, x_{n+1} ,$$
$$(13.44)$$

on its graph. The polynomial p passing through given data points is said to be *interpolating* such data. Finding this polynomial is called *polynomial interpolation*. The interpolating polynomial is unique by Lemma 1.30.

One of the best known methods for polynomial interpolation, due to Lagrange, is studied next.

Given the data (13.44) we first construct $n + 1$ polynomials $L_k(x)$ of degree n, such that

$$L_k(x_j) = \begin{cases} 1 & \text{if } j = k, \\ 0 & \text{if } j \neq k, \end{cases} \quad k = 1, \ldots, n + 1 , \qquad (13.45)$$

i.e., each $L_k(x)$ has the value 1 at $x := x_k$ and vanishes at other points $x = x_j$. In other words, $L_k(x)$ is the nth-degree polynomial interpolating the data

$$(x_1, 0), \ (x_2, 0), \ \ldots, \ (x_{k-1}, 0), \ (x_k, 1), \ (x_{k+1}, 0), \ \ldots, \ (x_{n+1}, 0) .$$

These *Lagrange polynomials* are given by the formula

$$
\begin{aligned}
L_k(x) &:= \frac{(x - x_1)(x - x_2) \cdots (x - x_{k-1})}{(x_k - x_1)(x_k - x_2) \cdots (x_k - x_{k-1})} \times \\
&\quad \times \frac{(x - x_{k+1})(x - x_{k+2}) \cdots (x - x_{n+1})}{(x_k - x_{k+1})(x_k - x_{k+2}) \cdots (x_k - x_{n+1})} \\
&= \frac{\prod_{j \neq k}(x - x_j)}{\prod_{j \neq k}(x_k - x_j)}, \quad \prod \text{ denotes product ;} \qquad (13.46)
\end{aligned}
$$

these polynomials satisfy (13.45) (check!). The polynomial[2] $L(x)$ interpolating the data (13.44) is finally

$$
\begin{aligned}
L(x) &:= y_1 L_1(x) + y_2 L_2 + \cdots + y_n L_n(x) + y_{n+1} L_{n+1}(x) \\
&= \sum_{k=1}^{n+1} y_k \frac{\prod_{j \neq k}(x - x_j)}{\prod_{j \neq k}(x_k - x_j)} .
\end{aligned}
\qquad (13.47)
$$

Indeed, $L(x_k) = y_k$, for $k = 1, \ldots, n + 1$, by (13.45).

Lagrange interpolation can be computed with MACSYMA, see Exercise 13.43.

---

2 L for Lagrange.

Example 13.47. **For example, to interpolate the 4 data points**

$$(-2, 3), \ (0, 3), \ (4, -2), \ (5, 1)$$

we compute the 4 Lagrange polynomials (13.46)

$$L_1(x) = \frac{(x - 0)(x - 4)(x - 5)}{(-2 - 0)(-2 - 4)(-2 - 5)} = -\frac{x^3}{84} + \frac{3x^2}{28} - \frac{5x}{21}$$

$$L_2(x) = \frac{(x + 2)(x - 4)(x - 5)}{(0 + 2)(0 - 4)(0 - 5)} = \frac{x^3}{40} - \frac{7x^2}{40} + \frac{x}{20} + 1$$

$$L_3(x) = \frac{(x + 2)(x - 0)(x - 5)}{(4 + 2)(4 - 0)(4 - 5)} = -\frac{x^3}{24} + \frac{x^2}{8} + \frac{5x}{12}$$

$$L_4(x) = \frac{(x + 2)(x - 0)(x - 4)}{(5 + 2)(5 - 0)(5 - 4)} = \frac{x^3}{35} - \frac{2x^2}{35} - \frac{8x}{35}$$

and the Lagrange interpolation is, by (13.47),

$$L(x) = 3L_1(x) + 3L_2(x) - 2L_3(x) + 1L_4(x) = \frac{127x^3}{840} - \frac{143x^2}{280} - \frac{683x}{420} + 3 \ .$$

Given a function $f(x)$ with $n + 1$ continuous derivatives in an interval $[a, b]$, the nth-degree Taylor polynomial at $x_0 = (a + b)/2$ is an approximation of f, whose error $E(x) := f(x) - T_n(f, x, x_0)$ is the Lagrange remainder $R_n(f, x, x_0)$, bounded on the interval $[a, b]$ as follows,

$$R_n(f, x, x_0) \le \frac{(b - a)^{n+1}}{(n + 1)!} \max_{x \in [a,b]} |f^{(n+1)}(x)| \ .$$

In contrast to the Taylor polynomial $T_n(f, x, x_0)$ which agrees with the function and its n derivatives at a single point, the Lagrange polynomial coincides with the function at $n + 1$ given interpolation points.

Let the function f be given at $n + 1$ points,

$$(x_1, f(x_1)), \ (x_2, f(x_2)), \ \ldots, \ (x_{n+1}, f(x_{n+1})) \ ,$$
$$\text{with different } x_1, x_2, \ldots, x_{n+1} \ .$$

We recall that the nth-degree polynomial interpolating these data points is

$$L(x) = \sum_{k=1}^{n+1} f(x_k) L_k(x) \ , \tag{13.48}$$

where, (13.46),

$$L_k(x) := \prod_{j \ne k} \frac{(x - x_j)}{(x_k - x_j)}, \quad k = 1, \ldots, n + 1 \ ;$$

see (13.47). The approximation error of Lagrange interpolation is studied in the next theorem.

Theorem 13.48 (Error of Lagrange interpolation). Let the function $f(x)$ have n continuous derivatives in the interval $[a, b]$, and let $x_1, x_2, \ldots, x_{n+1}$ be distinct points in $[a, b]$. Then there exists a point $\xi \in [a, b]$ such that

$$f(x) = \sum_{j=1}^{n+1} f(x_j) L_j(x) + \frac{f^{(n+1)}(\xi)}{(n+1)!} \prod_{j=1}^{n+1} (x - x_j) .$$

This theorem is proved in Sect. 13.6.

The error bound we get with the Lagrange interpolation is sometimes better than the error bound of the Taylor interpolation.

Example 13.49. Consider the function e^x in the interval $[-1, 1]$ and its approximation there by 2nd-degree polynomials of Lagrange and of Taylor.
    Lagrange interpolation. We use the three interpolation points

$$(-1, e^{-1}), \quad (0, e^0), \quad (1, e^1) .$$

The 2nd-degree Lagrange polynomial passing through these points is, by (13.48),

$$L(x) = e^{-1} \frac{x(x-1)}{2} - (x+1)(x-1) + e \frac{x(x+1)}{2}$$

so

$$e^x = e^{-1} \frac{x(x-1)}{2} e^{-1} - (x+1)(x-1) + e \frac{x(x+1)}{2} + \frac{(x-1)x(x+1)}{3!} e^{\xi} ,$$

where ξ is some point in $[-1, 1]$. The error term

$$E(x) = \frac{(x-1)x(x+1)}{3!} e^{\xi(x)} ,$$

can be bounded by first calculating the maximum of $(x-1)x(x+1)$ in $[-1, 1]$, namely,

$$|E(x)| \le \frac{2}{3! \, 3\sqrt{3}} \max_{x \in [-1,1]} e^x = \frac{2}{3! \, 3\sqrt{3}} e .$$

    Taylor interpolation. Using the origin as center, we get the Taylor interpolation

$$e^x = 1 + x + \frac{x^2}{2!} + \frac{x^3}{3!} e^{\xi(x)} ,$$

with error bounded by

$$|E(x)| \le \frac{e}{3!} e ,$$

which is almost three times as large as the estimate for the error in the Lagrange interpolation. This, however, does not imply that Lagrange interpolation is more exact; it just shows that the upper bound on the error term is smaller.

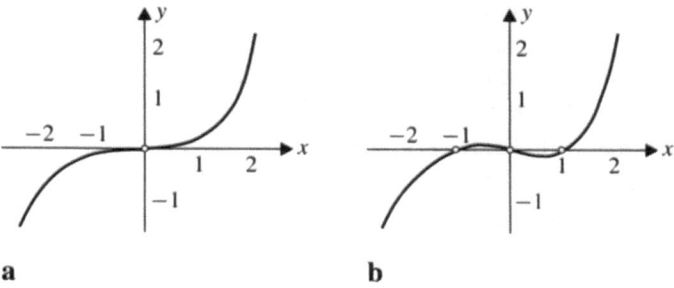

Fig. 13.1. Errors $E(x)$ of the 2nd-degree Taylor (**a**) and Lagrange (**b**) polynomials of e^x

Figure 13.1 shows the errors $E(x)$ of the Taylor polynomial of order 2 at zero and of the Lagrange polynomial with support points -1, 0, and 1. Identify which is the Taylor polynomial and which is Lagrange.

Exercises

13.43 Consider the two MACSYMA functions[3]

```
lagrange_aux(a,k,m):=
product((x-part(a,j,1))/(part(a,k,1)-part(a,j,1)),j,1,k-1)*
product((x-part(a,j,1))/(part(a,k,1)-part(a,j,1)),j,k+1,m);

lagrange(a)  := sum(part(a,k,2)*
                    lagrange_aux(a,k,length(a)),k,1,length(a));
```

a is the list of m data points, where we use m instead of $n+1$ for the number of data points in a: $[[x_1, y_1], [x_2, y_2], \ldots, [x_m, y_m]]$. part(a,k) is the MACSYMA function which produces the kth element of the vector a, but this element is itself a vector $[x_k, y_k]$:
part(a,k,1) extracts the 1st element, x_k, out of part(a,k),
part(a,k,2) extracts the 2nd element, y_k, of part(a,k).
length(a) is the MACSYMA function that calculates the dimension of the list a, i.e., the number of its elements.

a. Show that the MACSYMA function lagrange_aux(a,k,m) computes the Lagrange polynomial $L_k(x)$ of (13.46) if we identify m with $n+1$.
b. Show that the MACSYMA function lagrange(a) computes the Lagrange interpolation of the data (13.44).
c. Use the function lagrange to interpolate the data $[[-2, 3], [0, 3], [4, -2], [5, 1]]$ and compare with Example 13.47.
d. Select 4 points on a line and compute the Lagrange interpolation (13.47). What is its degree?

---

3 For better readability we wrote the first function on three lines even though you must enter it as one expression.

13.44 Use the above MACSYMA function `lagrange` to compute the Lagrange in terpola-
       tion polynomial for the following interpolation data[4]
       (a) `a:=[[0,0],[1,1],[2,2],[3,0]]`
       (b) `a:=[[-1,1],[0,0],[1,1]]`
       (c) `a:=makelist([k,1/k],k,1,5)`
       (d) `a:=[[0,0],[1,0],[2,0],[1/2,1]]`
       (e) `a:=entermatrix(i(j + 1),4,2)`
       (f) `a:=[[-4,0],[-3,1],[-2,1],[-1,1],[0,1],[1,1],[2,1],[3,1],`
           `   [4,0]]`
       (g) `a:=makelist([k,binomial(5,k)],k,0,5)`
       and plot the given data as well as the interpolating polynomials.

13.45 What does the MACSYMA function `lagrange` of Exercise 13.43 produce if you
       enter two points with the same x-value but different y-values? Explain!

13.46 Show that if the interpolation data (13.44) has even or odd symmetry, then the
       corresponding Lagrange interpolation polynomial is even or odd, respectively.

13.47 Using the function `lagrange(a)` of Exercise 13.43, write a MACSYMA function
       `polynomial_interpolation(f,x,a)` that calculates the Lagrange interpolation
       polynomial for f with respect to the variable x and the vector of x-values $a :=$
       $(x_1, x_2, \ldots, x_n)$.
       a. Use `polynomial_interpolation` to calculate the polynomial approximations
           for $f(x) = x^4$ for the vectors of x-values
           (1) $[-1, 0]$ (2) $[-1, 0, 1]$ (3) $[-1, 0, 1, 2]$ (4) $[-1, 0, 1, 2, 3]$
       b. Similarly calculate the polynomial approximations for $f(x) = \sin x$ for the
           vectors of x-values
           (1) $[-1, 0, 1]$ (2) $[-2, -1, 0, 1, 2]$
           (3) $[-3, -2, -1, 0, 1, 2, 3]$ (4) $[-4, -3, -2, -1, 0, 1, 2, 3, 4]$

13.48 Use Lagrange interpolation to approximate the integrand in each of the problems
       below. Choose your interpolating polynomial to obtain three decimal places of accu-
       racy. Use MACSYMA to perform the calculations. Compare this result with a Taylor
       polynomial of the same degree.

(a) $\displaystyle\int_0^1 \frac{\sin x}{x}\, dx$ (b) $\displaystyle\int_0^1 \frac{1 - e^x}{x}\, dx$

(c) $\displaystyle\int_0^1 \frac{1 + x - e^x}{x^2}\, dx$ (d) $\displaystyle\int_0^1 \frac{x - \sin x}{x^2}\, dx$

13.6 Proofs

Proof of Theorem 13.9. The idea behind the proof is quite simple. Moreover, from
this proof some intuition is gained about whether series converge or not. Our
argument is organized in the following way. We suppose thatthe series converges

---

4 Problems c and e show alternate ways of entering the list of data points. The list of
data points may also be entered, for example, as a column vector of row vectors $[x_k, y_k]$, or
equivalently as an $m \times 2$ matrix.

somewhere; otherwise we are in category a. We then consider the categories b and c. Suppose the series converges at some point $x_1 \neq 0$. Then it is easy to see that the power series must converge for all x where $|x| < |x_1|$. Firstly one has $|a_k x^k| < |a_k x_1^k| \cdot |x/x_1|^k$. However, from our hypothesis we know the series $\sum_{k=0}^{\infty} a_k x_1^k$ converges; hence from Theorem 12.13 we know the absolute values of the terms in the series $|a_k x_1^k|$ remain bounded for all values of k. More specifically there exists a finite, nonnegative, upper bound M for these absolute values. This fact allows us to *majorize* the power series in question, namely,

$$\left| \sum_{k=0}^{\infty} a_k x^k \right| \leq \sum_{k=0}^{\infty} |a_k| \, |x|^k \leq \sum_{k=0}^{\infty} |a_k x_1^k| \left| \frac{x}{x_1} \right|^k \leq M \sum_{k=0}^{\infty} \left| \frac{x}{x_1} \right|^k \leq \frac{M}{1 - \left| \frac{x}{x_1} \right|} \, ,$$

for all $|x| < |x_1|$. In other words, if $|x| < |x_1|$, the series is majorized by a convergent geometric series and consequently is seen to converge absolutely.

Now suppose that $\sum a_k x^k$ does not converge for *all* x. Then there exists *some* $x_1 \neq 0$ at which the series diverges. Now the point x_1 cannot be inside the open interval $(-|x_0|, |x_0|)$, that is, we cannot have $|x_1| < |x_0|$. If the point x_1 lay in the interval, our discussion above shows that x_1 is a point where the series converges. Hence x_1 is such that $|x_1| \geq |x_0|$. Now consider the set $\mathcal{S}$ of nonnegative numbers x such that $\sum a_k x^k$ converges. Clearly $\mathcal{S}$ is nonempty, as 0 is contained in $\mathcal{S}$; moreover, $\mathcal{S}$ must be bounded. Otherwise, there would be an $x_2 \in \mathcal{S}$ such that $|x_1| < x_2$. We recognize this would contradict the hypothesis that the series diverges at x_1.

Finally, if there is no point x_1 where the series diverges, we fall into category b. This concludes the discussion of the proof of the theorem. $\qquad \square$

Proof of Theorem 13.22. Suppose the function f has the power series expansion

$$f(x) = \sum_{k=0}^{\infty} a_k x^k, \quad \text{for } x \in (-R, R) \, .$$

The power series converges uniformly in any closed interval $[-\rho, \rho]$ lying in the interval of convergence. For any $\varepsilon > 0$ there exists therefore an N_0, such that for $n \geq N_0$ the remainders $R_n(x)$ are uniformly bounded by

$$|R_n(x)| \leq \varepsilon, \quad \text{for all } x \in [-\rho, \rho] \, .$$

The integral of f over an interval lying within $[-\rho, \rho]$ may now be computed as the integral of a *finite sum* of terms plus the integral of the remainder, namely,

$$\int_a^x f(x) \, \mathrm{d}x = \sum_{k=0}^{n} a_k \int_a^x x^k \, \mathrm{d}x + \int_a^x R_n(x) \, \mathrm{d}x$$

$$= \sum_{k=0}^{n} a_k \left(\frac{x^{k+1}}{k+1} - \frac{a^{k+1}}{k+1} \right) + \int_a^x R_n(x) \, \mathrm{d}x \, .$$

This shows that

$$\left| \int_a^x f(x)\,dx - \sum_{k=0}^n a_k \left(\frac{x^{k+1}}{k+1} - \frac{a^{k+1}}{k+1} \right) \right| \le \varepsilon \cdot 2\rho \, ,$$

hence, by choosing n sufficiently large the difference between the integral of the function defined by the power series and the integral of the termwise integrated partial sum can be made as small as we wish. Indeed, if we wish $\varepsilon \to 0$, then we need only let $n \to \infty$. Hence, we obtain

$$\int_a^x f(x)\,dx = \sum_{k=0}^\infty a_k \left(\frac{x^{k+1}}{k+1} - \frac{a^{k+1}}{k+1} \right) ,$$

which is the statement of our theorem. $\square$

Proof of Lemma 13.43. The finite Taylor series with the Lagrange remainder is

$$(1+x)^\alpha = \sum_{k=0}^n \binom{\alpha}{k} x^k + \binom{\alpha}{n+1}(1+\theta x)^{\alpha-n-1} ,$$

where $0 < \theta < 1$. In order to obtain the infinite series, we still need to show that the remainder

$$R_n(x,0) := \binom{\alpha}{n+1} x^n (1+\theta x)^{\alpha-n-1}$$

goes to zero for $x \in (-1,1)$. If $x \ge 0$, then for $n + 1 > \alpha$ we have

$$0 < (1+\theta x)^{\alpha-(n+1)} \le 1 .$$

So now we must just consider whether the sequence of numbers $a_n := \binom{\alpha}{n} x^n \to 0$ for $|x| < 1$, then $R_n(x,0) \to 0$. To show that the a_n tend to 0, we consider the quotient ratio

$$\frac{|a_{n+1}|}{|a_n|} = \frac{|x(\alpha-n)|}{n+1} \to |x| < 1$$

. as $n \to \infty$. We conclude that $R_n(x,0) \to 0$ as $n \to \infty$, and hence, for $0 < x < 1$ the series converges to the binomial function. The ratio test, moreover, shows us that the Taylor series representation converges for all $x \in (-1,1)$. This does not say that it converges in $(-1,0]$ to the binomial function; it shows that it converges to *some* function. To show that our Taylor series indeed converges to the binomial function on the interval $(-1,1)$, we proceed by considering the function *defined* by the power series

$$f(x) := \sum_{k=0}^\infty \binom{\alpha}{k} x^k$$

and show that it coincides with the binomial function for $x \in (-1,1)$. This is shown by demonstrating that $f(x)$ satisfies a differential identity. As the power series defining $f(x)$ converges uniformly for $|x| \le \rho < 1$, we may differentiate termwise to obtain

$$f'(x) := \sum_{k=0}^{\infty} (k+1) \binom{\alpha}{k+1} x^k = \sum_{k=0}^{\infty} \alpha \binom{\alpha-1}{k} x^k ,$$

and from this that

$$(1+x)f'(x) = \alpha + \alpha \left(\binom{\alpha-1}{1} + \binom{\alpha-1}{0} \right) x$$

$$+ \alpha \left(\binom{\alpha-1}{2} + \binom{\alpha-1}{1} \right) x^2 + \cdots$$

$$+ \alpha \left(\binom{\alpha-1}{n} + \binom{\alpha-1}{n-1} \right) x^n + \cdots .$$

We notice again that

$$\binom{\alpha-1}{n} + \binom{\alpha-1}{n-1} = \frac{(\alpha-1)\cdots(\alpha-n-1)}{(n-1)!} \left(\frac{\alpha-n}{n} + 1 \right) = \binom{\alpha}{n} ,$$

which allows the expression for $(1+x)f'(x)$ to be written as

$$(1+x)f'(x) = \alpha \sum_{k=0}^{\infty} \binom{\alpha}{n} x^n .$$

This implies that the differential relationship $(1+x)f'(x) = \alpha f(x)$ holds for all $x \in (-1, 1)$. The function $(1+x)^\alpha$ also satisfies the same differential relationship, i.e.,

$$(1+x) \frac{d(1+x)^\alpha}{dx} = \alpha (1+x)^\alpha .$$

If we assume $\alpha \neq 0$, then

$$\frac{1}{f'(x)} \frac{d(1+x)^\alpha}{dx} = \frac{(1+x)^\alpha}{f(x)} ,$$

which means

$$\left(f'(x)(1+x)^\alpha - f(x) \frac{d(1+x)^\alpha}{dx} \right) (1+x)^{-2\alpha} = \frac{d}{dx} \left(\frac{f(x)}{(1+x)^\alpha} \right) \equiv 0 .$$

This means $f(x)/(1+x)^\alpha \equiv C$ (a constant). This constant may be determined as equal to 1 by checking the values of $f(x)$ and $(1+x)^\alpha$ at $x = 0$. We conclude that $f(x)$ as defined by the power series and the binomial function $(1+x)^\alpha$ are identical on $(-1, 1)$. □

Proof of Theorem 13.45. This form of the Taylor theorem with remainder is obtained by applying the Cauchy mean value theorem to

$$\frac{F(x) - F(a)}{G(x) - G(a)} ,$$

where the functions $F(t)$ and $G(t)$ are defined by

$$F(t) := f(x) - \sum_{k=0}^{n} \frac{f^{(k)}(t)}{k!}(x-t)^k \quad \text{and} \quad G(t) := (x-t)^p .$$

We notice that $F(x) = G(x) = 0$ and that

$$F(a) = f(x) - \sum_{k=0}^{n} \frac{f^{(k)}(a)}{k!}(x-a)^k$$

is exactly the remainder which appears in the statement of the theorem. First we calculate the derivative

$$F'(t) = -\sum_{k=0}^{n} \frac{f^{(k+1)}(t)}{k!}(x-t)^k - \sum_{k=0}^{n} \frac{f^{(k)}(t)}{k!}k(x-t)^{k-1} ,$$

which by regrouping becomes

$$F'(t) = -f'(t) + \left(f'(t) - f^{(2)}(t)(x-t) \right)$$
$$+ \left(f^{(2)}(t)(x-t) - \frac{1}{2!}f^{(3)}(t)(x-t)^2 \right)$$
$$+ \left(f^{(3)}(t)(x-t)^2 - \frac{1}{3!}f^{(4)}(t)(x-t)^3 \right) + \cdots$$
$$+ \left(\frac{1}{(n-1)!}f^{(n)}(t)(x-t)^{n-1} - \frac{1}{n!}f^{(n+1)}(t)(x-t)^n \right) .$$

This is an example of an *accordion* series, since all terms but the final one cancel in pairs. Hence, we obtain the concise expression for the derivative

$$F'(t) = -\frac{f^{(n+1)}(t)}{n!}(x-t)^n \quad \text{and} \quad G'(t) = -\frac{(x-t)^n}{n!} .$$

The Cauchy mean value theorem (Theorem 7.3) now gives

$$(F(x) - F(a))G'(\xi) = (G(x) - G(a))F'(\xi)$$

for some $\xi \in (a, b)$. Since $G'(\xi) \neq 0$ (why?), and $F(x) = G(x) = 0$, we obtain after substituting in the expressions for F' and G'

$$F(a) = \frac{f^{(n+1)}(\xi)}{(n+1)!}(x-a)^{n+1} ,$$

which upon identifying the expression for $F(a)$ given above with the left-hand side is the representation we wanted to establish.   □

Proof of Theorem 13.48. The result is trivially true if x is one of the $n+1$ interpolation points since then the error term

$$E(x) := \frac{f^{(n+1)}(\xi)}{(n+1)!} \prod_{j=1}^{n+1}(x - x_j)$$

vanishes. Let us then fix a point $x \neq x_j$ and (for this fixed x) define the function $G(t)$ on $[a, b]$

$$G(t) := E(t) - \frac{P(t)}{P(x)} E(x),$$

$$\text{where} \quad P(t) := \prod_{j=1}^{n+1}(t - x_j).$$

Note that the function $G(t)$ has zeros at the points $t = x_j$, for $j = 1, 2, \ldots, n+1$, and for the point $t = x$, that is, $G(t)$ has $n + 2$ distinct zeros.

Consider the derivatives of $G(t)$. First it is easy to see that as $E(t)$ and $P(t)$ have $n + 1$ derivatives so does $G(t)$. Let us start looking at these derivatives and counting their zeros. To do this we recall that the mean value theorem 7.2 implies that between two zeros of a differentiable function the derivative has a zero. Hence, $G'(t)$ has $n + 1$ zeros on $[a, b]$, $G''(t)$ has n zeros, and proceeding inductively, $G^{(m)}(t)$ has $n + 2 - m$ zeros on $[a, b]$. Indeed, $G^{(n+1)}(t)$ has just the single zero $t = \xi$.

Now as the interpolating polynomial $P(x)$ is of degree n, its $(n+1)$st-order derivative vanishes, so $E^{(n+1)}(x) = f^{(n+1)}(x)$; moreover, $P^{(n+1)}(x) = (n+1)!$, thereby leading to an expression for $G^{(n+1)}(x)$, namely,

$$G^{(n+1)}(t) = f^{(n+1)}(t) - \frac{(n+1)!}{P(x)} E(x).$$

Upon recalling $G^{(n+1)}(\xi) = 0$, this expression provides us with the desired result

$$E(x) = \frac{P(x)}{(n+1)!} f^{(n+1)}(\xi). \qquad \square$$

Appendixes

A Introduction to MACSYMA

This appendix explains the basic features of MACSYMA, just enough to get you started. Further explanations are given, as needed, in various MACSYMA-Sessions throughout the book.

You start (assuming MACSYMA is installed on your machine), by typing `macsyma` followed by <ENTER>, or, if you use "MACSYMA for Windows," by clicking on the MACSYMA icon.

A message (depending on your MACSYMA version) will appear on the screen informing you of the software being loaded, and ending with the prompt "(c1)".

```
C:\MACSYMA2\system\init.lsp being loaded.
Batching the file C:\MACSYMA2\user\mac-init.mac
Batchload done.
(c1)
```

A MACSYMA session is an alternation of c-lines (*inputs*) and d-lines (*outputs*). For example,

```
c1. 10!
d1.     3628800
c2. trigexpand(sin(a+b))
```
d2. $\sin b \cos a + \cos b \sin a$

It is correct, but no longer necessary, to end a c-line with a semicolon. Each c-line must be followed by <ENTER> (not shown here).

To end the session use `exit` or `quit` (the difference between these two commands is explained in the MACSYMA manual).

A.1 MACSYMA inputs and outputs

An input can be:
- an *expression* to be evaluated or processed, e.g., `10!`, `sum(i^2,i,1,5)`, or `sin(a+b)`,
- a *definition*, e.g., `eq:a*x^2+b*x+c=0`, defining eq as the equation $ax^2 + bx + c = 0$,
- a *command*, e.g., `solve(eq,x)` (to solve the above equation), `trigexpand(sin(a+b))`, `assume a>0` (informing MACSYMA that a is positive), or `exit`,
- a *query*, e.g., `values` to list all variables that have been assigned values.

An output is the computer's response to the previous input, see, e.g., d1 and d2 above.

MACSYMA-Session A.1. Sometimes a d-line is just an *echo*, for example,
c1. a+b
d1. $a + b$
A d-line can be suppressed by ending the previous c-line with $. For example,
c2. 2-7$
Still, MACSYMA remembers d2, for example,
c3. 6*d2
d3. -30
You refer to a result by its number, e.g., c13 or d19. The *previous* result is denoted by %, thus in c3 we could have written 6*% (multiply the previous result by 6).
The *i*th previous result is denoted %th(i), in particular, %th(1) is the same as %.

In addition to c-lines and d-lines, there are also e-lines giving intermediate outputs.

A.2 Getting on-line help

MACSYMA is a very powerful language, with thousands of commands and options (you will be using here a small part of MACSYMA). MACSYMA is also easy to use, thanks to the excellent on-line help.

To get it, type help, or click *Help* or *MathHelp!* in your MACSYMA window. The following help options are available:
– the *MathHelp! Topic Browser*, which you must see for yourself,
– the *MathTips™ Advisor*, advising which MACSYMA command to use in a given situation,
– the *interactive primer* (to open, type primer()),
– hundreds of *executable examples* and *demos*,
– *function templates*, and
– the apropos(string) command, searching for all functions with the given string in their names. This command is useful if you remember, or guess, part of the name of the function. For example, apropos(taylor) gives a list including: taylor, taylor_ode, taylor_revert, taylor_solve, and taylorinfo (tailor-made for your taylor application).

A.3 Expressions

The basic unit of information in MACSYMA is the expression. An expression is made up of a combination of *operators*, *numbers*, *variables*, and *constants*.

MACSYMA allows expanding, simplifying, and factoring expressions, making substitutions in expressions, and extracting parts of an expression for subsequent use in other commands. These are introduced in Sects. A.10–A.14.

MACSYMA uses familiar symbols for the arithmetic operations. Table A.1 lists these operations in order of *priority*, from lowest to highest. Operations of equal

Table A.1. Arithmetic operations

Operation	Description
+	addition
−	subtraction
*	multiplication
/	division
−	negation
^	exponentiation
!	factorial
!!	double factorial

priority are performed left to right. You can use parentheses to change the order of evaluation.

Function evaluations have higher priority than arithmetic operations.

MACSYMA-Session A.2. Examples:
c1. sin(x^y)^2
d1. $\sin^2(x^y)$
If you meant $\sin((x^y)^2)$ instead of d1, use an extra pair of parentheses,
c2. sin((x^y)^2)
d2. $\sin(x^{2y})$
The "^" operator distributes over the multiplication operator "*".
c3. b^2*b^3
d3. b^5
The factorial operator "!" is defined as usual, $n! := n(n-1)(n-2)\cdots 1$. If you are not familiar with the double factorial "!!", try to guess its definition from:
c4. 3!!
d4.   3
c5. 4!!
d5.   8
c6. 5!!
d6.   15
etc.

A.4 Constants

The most commonly used constants are recognized by MACSYMA. These include e, $i = \sqrt{-1}$, and π, denoted by %e, %i and %pi, respectively.

MACSYMA-Session A.3.
c1. %e,numer
d1.   2.7182817
c2. %i^2
d2.   −1

c3. %pi,numer
d3.   3.1415927
Next we compute the unlikely expression $e^{i\pi}$,
c4. %e^(%i*%pi)
d4. -1
We have thus verified the well-known identity $e^{i\pi} + 1 = 0$, connecting the 5 most important constants in mathematics.

Other MACSYMA constants include:

inf ∞, (real) positive infinity,
minf $-\infty$, (real) negative infinity,
infinity   complex infinity,
phi $(\sqrt{5} + 1)/2 \approx 1.618034$, the reciprocal *golden section*.

A.5 Numbers

MACSYMA number types include:
- *integer*, a string of digits without a decimal point, e.g., 15934 or -523;
- *rational number*, an exact ratio of two integers, e.g., $3/2$, $-31/8$;
- *float*, or floating-point number, a number containing a decimal point; a float can be
  a *single float*, abbreviated *"sfloat"*, *(single precision)* or
  a *double float*, abbreviated *"dfloat"*, *(double precision)* or
  a *big float*, abbreviated *"bfloat"*;
- a *complex number*, for example, $\frac{1}{2} + i\frac{\sqrt{3}}{2}$ written as 1/2 + %i*sqrt(3)/2.

MACSYMA does not limit the number of digits in integers or rational numbers and can work with very large numbers, for example (see this one for yourself),
c1. 10^506 - 10^253 - 1
When computing with floating-point numbers, the results depend on the precision used – single float, double float, or big float. For an extreme example, see Sect. A.8, lines c1–d6.

A number written in scientific notation as $x10^n$, where n is integer, is represented as
- xen in single-float arithmetic, e.g., 6.023e23,
- xdn in double-float arithmetic, e.g., 8.63877340217378d85,
- xbn in big-float arithmetic, e.g., 37.567834872508325568b-98.

The digits of x are called *significant*.

Single floats have about six significant digits and cover a range of exponents from about 10^{-37} to 10^{37}. The actual range depends on your computer. You can find it by using the commands least_positive_float and most_positive_float.

Double floats have about 15 significant digits and cover a range of exponents from about 10^{-306} to 10^{306}.

Single-float computations are faster than double-float computations, which are in turn faster than big-float computations.

Fractions are preserved exactly.

MACSYMA-Session A.4.

c1. 1/121

d1. $\dfrac{1}{121}$

It will be converted to a floating-point number only on request, for example,[1]

c2. sfloat(1/121)

d2.   0.008264462

In this case the precision shown is 8, and in general it will depend on the computer used. If a higher precision is required, use

c3. bfloat(1/121)

d3.   8.2644628099173553719b−3

You can control the precision of bigfloats using the system variable bfprecision, whose default value is 20. For example, we set precision of bigfloats to 125,

c4. bfprecision:125$

and then

c5. bfloat(1/121)

d5.   8.264462809917355371900826446280991735537190082644628099173553719#
        008264462809917355371900826446280991735537190082644628099173 6b−3

where "#" indicates that the number continues on the next line.

An operation involving an integer and a floating-point number usually results in a floating-point number.

c1. 4*3.0

d1.   12.0

Complex expressions expr $= u + iv$ have specific commands, including
−   realpart(expr) $= u$, the real part,
−   imagpart(expr) $= v$, the imaginary part,
−   rectform(expr) $= u + iv$,
−   cabs(expr) $= \sqrt{u^2 + v^2}$, the absolute value of expr,
−   carg(expr) $= \mathrm{atan}(v/u)$, the argument of expr,
−   polarform(expr) $= r\,e^{i\theta}$ where $r = $ cabs(expr) and $\theta = $ carg(expr).

A.6 Assignments

We discuss the following MACSYMA operators:
:   assigns value to a variable
=   creates an equation
:=   defines a function

A variable may be *bound*, i.e., given a value, or *unbound*. When you type the name of an unbound variable, MACSYMA returns an echo:

c1. a

d1.   *a*

Binding a value to a variable is called *assignment*.

---

1 In c2 we can alternatively enter sfloat(d1) or sfloat(%).

MACSYMA-Session A.5. To assign a value to a variable, type the name of the variable, the ":" operator, and the given value, for example,

c1. a:123

d1.   123

Typing the name of a bound variable returns its value:

c2. a

d2.   123

Once bound, a stands for 123 in subsequent expressions.

c3. a+a

d3.   246

You can add bound and unbound variables:

c4. a+b

d4. $b + 123$

A single quote before a bound variable supresses evaluation:

c5. 'a

d5. a

Mistakes may occur from confusing bound and unbound variables. It is therefore safe to remove the value previously assigned to a variable,

c6. remvalue(a)

d6. a

indicating that a has become unbound. Therefore,

c7. a+b

d7. $a + b$

Compound assignments (giving values to several variables) are made by enclosing the assignments, separated by commas, in parentheses. For example,

c8. (a:5, b:15.3e5)

d8.   1530000.0

MACSYMA returns only the value of the last assignment, but both assignments in c8 were recorded:

c9. a

d9.   5

c10. b

d10.   1530000.0

The command remvalue can be used to remove the values of several variables:

c11. remvalue(a,b)

d11. $[a, b]$

or of all (bound) variables:

c12. remvalue(all)$

Expressions can be assigned values involving bound and unbound variables, for example,

c13. expr1:num+x+y

d13. $y + x + \text{num}$

If we assign to num the value 50,

c14. num:50$

the value of expr1 does not change,

c15. expr1

d15. $y + x + \text{num}$

however, when we evaluate expr1, using the command ev,

c16. ev(expr1)

d16. $y + x + 50$

Another way to write ev(expr1) is ''expr1. Also, either ev(d15) or ev(%) could be used in c16 instead of ev(expr1).

Assign the result shown in d16 to a variable named expr2:

c17. expr2:%

d17. $y + x + 50$

If we remove the value from the variable num,

c18. remvalue(num)

d18.   [num]

then evaluating expr1 gives back its original value, as in d13,

c19. ev(expr1)

d19. $y + x + num$

while expr2 does not change,

c20. ev(expr2)

d20. $y + x + 50$

More examples:

c21. expr2 + 1

d21. $y + x + 51$

c22. expr1 * 2

d22. $2(y + x + num)$

The system variable values is a list of all variables that have assigned values

c23. values

d23.   [expr1, expr2]

Now we remove the values from all variables,

c24. remvalue(all)

d24.   [expr1, expr2]

since only expr1 and expr2 are bound.

A.7 Equations

Equations are created with the "=" operator and are solved with the commands solve or, for linear equations, linsolve.

MACSYMA-Session A.6.

c1. eq:a*x^2+b*x+c=0

d1. $a x^2 + b x + c = 0$

c2. solve(eq,x)

d2. $\left[x = \dfrac{\sqrt{b^2 - 4ac} - b}{2a}, \; x = -\dfrac{\sqrt{b^2 - 4ac} + b}{2a} \right]$

Linear equations,

c3. eq1:a*x+b*y=e$

c4. eq2:c*x+d*y=f$

are solved with the linsolve command,

c5. linsolve([eq1,eq2],[x,y])

d5. $\left[x = \dfrac{de - bf}{ad - bc}, \; y = \dfrac{af - ce}{ad - bc} \right]$

The *left-hand side* and *right-hand side* of an equation are extracted by the commands lhs and rhs, respectively,

c6. lhs(d1)

d6. $a x^2 + b x + c$

c7. rhs(eq2)

d7. f

A.8 Functions

To define a function, use the function operator ":="

    function(arg$_1$, arg$_2$, ..., arg$_n$) := body

where function stands for the name of the function, arg$_i$ are the formal arguments in parentheses, and the body is any MACSYMA expression involving the variables arg$_1$ through arg$_n$.

MACSYMA-Session A.7.

c1. f(x, y) := 9*x^4 - y^4 + 2*y^2

d1. $f(x, y) := 9x^4 - y^4 + 2y^2$

The value of f at $x = 10,864$ and $y = 18,817$ (both integers) is:

c2. f(10864,18817)

d2.   1

Using floating-point numbers $x = 10,864.0$ and $y = 18,817.0$ we get different results. In single precision:

c3. f(10864.0,18817.0)

d3.   7.08155e+8

same as:

c4. f(1.0864e4,1.8817e4)

d4.   7.08155e+8

incorrect answers, since the default precision for floating-point numbers is insufficient. Double precision also gives an incorrect answer:

c5. f(1.0864d4,1.8817d4)

d5.   2.0d0+

and finally, big floats give the correct answer:

c6. f(1.0864b4,1.8817b4)

d6.   1.0b0+

We remove the definition of the function f:

c7. remfunction(f)

d7. f

MACSYMA can also handle composite functions,

c8. f(x) := x^2 + 4$

c9. g(x) := 1/x + 3$

c10. f(g(x))

d10. $\left(\dfrac{1}{x} + 3\right)^2 + 4$

c11. g(f(x))

d11. $\dfrac{1}{x^2 + 4} + 3$

c12. `f(g(2))`

d12. $\dfrac{65}{4}$

c13. `g(f(2))`

d13. $\dfrac{25}{8}$

To list the defined functions we use

c14. `functions`

d14. $[f(x), g(x)]$

and to display them:

c15. `dispfun(f,g)`

e15. $f(x) := x^2 + 4$

e16. $g(x) := \dfrac{1}{x} + 3$

d16.   done

A.9 Lists

In MACSYMA a list is an ordered set of elements, separated by commas and enclosed in square brackets. An element of a list can be any MACSYMA expression. You can assign a list as the value of a variable and then refer to its individual elements as *subscripted variables*.

Lists can be used as arguments to some commands, such as `solve` and `matrix`. Other commands return results in a list, including `remvalue`, `remfunction`, and `solve`.

c1. `list1:[a,b,c]`

d1. $[a, b, c]$

c2. `list2:[1,2,3]$`

Add the corresponding elements of the two lists:

c3. `list1+list2`

d3. $[1 + a, 2 + b, 3 + c]$

Parts of a list can be extracted with the `part` and associated commands, see Sect. A.14.

A.10 Expanding expressions

MACSYMA has several expansion-related commands, each of which expands its argument in a different way.

- `expand` expands the given expression by multiplying out products of sums and exponentiated sums at all levels of the expression.
  Use `expand(expr, p, n)` to expand the expression `expr`, multiplying out only terms with exponents between n and p.
- `multthru` multiplies a term or terms through a sum or equation.
- `distrib` expands the given expression by distributing sums over products.
- `partfrac` does a complete partial fraction decomposition, expanding an expression in partial-fractions with respect to a given main variable.

MACSYMA also provides several option variables that you can set to modify the way the expansion commands work. Two option variables, `logexpand` and `radexpand`, are discussed below.

MACSYMA-Session A.8. Examples.
c1. `expr1:(1/(a + b)^2 + x/(a - b)^3)^2`

d1. $\left(\dfrac{1}{(a+b)^2} + \dfrac{x}{(a-b)^3} \right)^2$

Expand `expr1`, assigning the result to `expr2`.
c2. `expr2:expand(expr1)`

d2. $\dfrac{x^2}{b^6 - 6ab^5 + 15a^2b^4 - 20a^3b^3 + 15a^4b^2 - 6a^5b + a^6}$
$+ \dfrac{2x}{-b^5 + ab^4 + 2a^2b^3 - 2a^3b^2 - a^4b^5 + a^5}$
$+ \dfrac{1}{b^4 + 4ab^3 + 6a^2b^2 + 4a^3b + a^4}$

c3. `expr3:expand(expr1, 2, 0)`

d3. $\dfrac{x^2}{(a-b)^6} + \dfrac{2x}{(a-b)^3(b+a)^2} + \dfrac{1}{(b+a)^2}$

The option variables `logexpand` and `radexpand` control the expansion of expressions containing logarithms and radicals, respectively.
The `log` command gives the natural log of its argument in base e.
c4. `log(%e)`
d4.　1

When `logexpand` is "true", MACSYMA does not simplify the logarithms of products and quotients.
c5. `logexpand`
d5.　true
c6. `log(a*b)`
d6.　$\log(ab)$
c7. `log(a/b)`
d7.　$\log\left(\dfrac{a}{b}\right)$

Resetting `logexpand` to `all` tells MACSYMA to simplify these logarithms.
c8. `logexpand:all$`
c9. `log(a*b)`
d9.　$\log(b) + \log(a)$
c10. `log(a/b)`
d10.　$\log(a) - \log(b)$

When `radexpand` is "true", MACSYMA does not simplify radicals containing products, quotients, and powers.
c11. `radexpand`
d11.　true
c12. `sqrt(x^y)`
d12.　$\sqrt{x^y}$
c13. `sqrt(x*y)`
d13.　$\sqrt{xy}$

c14. sqrt(x/y)

d14. $\sqrt{\frac{x}{y}}$

Resetting `radexpand` to `all` tells MACSYMA to simplify these radicals

c15. radexpand:all$

c16. sqrt(x^y)

d16. $x^{y/2}$

c17. sqrt(x*y)

d17. $\sqrt{x}\sqrt{y}$

c18. sqrt(x/y)

d18. $\dfrac{\sqrt{x}}{\sqrt{y}}$

A.11 Simplifying expressions

MACSYMA has many commands for simplifying expressions, including:

 - `ratsimp`, which simplifies an expression by combining the rational functions in the expression and then canceling out the greatest common divisor in the numerator and denominator;
 - `radcan`, which simplifies expressions containing radicals, logarithms, and exponentials;
 - `scsimp`, which implements the sequential comparative simplifier, which applies given identities to an expression in an effort to obtain a smaller expression;
 - `combine` and `rncombine`, which group the terms in a sum that have the same denominator into a single term;
 - `xthru`, which combines the terms of a sum over a common denominator without expanding them first;
 - `map`, which can apply a given function, such as a simplification command, to each term of a very large expression (this can be useful when applying the function to the entire expression would be inefficient).

MACSYMA-Session A.9. The command `ratsimp(expr)` simplifies the expression expr and all of its subexpressions. It first combines the sum of rational functions (quotients of polynomials) into one rational function and then cancels out the greatest common divisor of the numerator and denominator.

c1. -(x^5 - x)/(x - 1) + x + x^2 + x^3 + x^4 + (a + b + c)^3

d1. $-\dfrac{x^5 - x}{x - 1} + x^4 + x^3 + x^2 + x + (c + b + a)^2$

Simplify the expression in d1 with the `ratsimp` command.

c2. ratsimp(%)

d2. $c^3 + (3b + 3a)c^2 + (3b^2 + 6ab + 3a^2)c + b^3 + 3ab^2 + 3a^2b + a^3$

MACSYMA simplified the terms containing the variable x to zero, but it returned the expanded result of $(a + b + c)^3$. Another example:

c3. d*(w + a)*x + c*(w + a)*x + b*d + b*c + c*d$

c4. ratsimp(%)

d4. $((d + c)w + ad + ac)x + (c + b)d + bc$

With additional arguments, `ratsimp`(expr, var_1, var_2, ..., var_n) specifies the ordering of each variable var_i as well. For example, simplify d3, ordering d first and c second.

c5. `ratsimp(d3, c, d)`

d5. $d((w + a)x + c + b) + c((w + a)x + b)$

Simplification is controlled with the option variables `algebraic` and `ratfac`.

A.12 Factoring expressions

MACSYMA has a number of commands for factoring expressions. These include:
- `factor`, which factors an expression into factors irreducible over the integers;
- `cfactor`, which factors an expression with respect to one variable, using complex numbers, including radicals;
- `factorsum`, which tries to separate the terms in a sum into groups that can have common factors and then factors them.

MACSYMA-Session A.10. If n is a composite integer, it is factored to its prime factors, for example:

c1. `factor(1001)`

d1.   7 11 13

A prime integer is its own prime factor. For example, the 1000th prime number is

c2. `prime(1000)`

d2.   7919

c3. `factor(%)`

d3.   7919

MACSYMA can factor large integers. Consider for example the *Fermat numbers*

$$F(n) := 2^{2^n} + 1, \quad n = 1, 2, \ldots$$

The first 4 are prime. Fermat was unable to determine if $F(5)$ and higher Fermat numbers are prime. He conjectured in 1640 that all integers of the form $2^{2^n} + 1$ are primes. This conjecture lasted about 100 years: Euler showed in 1739 that $F(5)$ is composite,

$$2^{2^5} + 1 = 4294967297 = 641 \times 6700417$$

We repeat Euler's computation, and also factor $F(6)$. First define the function,

c4. `fermat(n):=2^2^n+1`

d4. $2^{2^n} + 1$

The 4th Fermat number:

c5. `fermat(4)`

d5.   65537

is prime

c6. `factor(%)`

d6.   65537

Here is a list of the first 6 Fermat numbers

c7. `makelist(concat(fermat(n)),n,1,6)`

d7.   [5, 17, 257, 65537, 4294967297, 18446744073709551617]

As Euler has shown, the 5th Fermat number is not prime,

c8. `factor(fermat(5))`
d8.   641 6700417
So is the 6th Fermat number,
c9. `factor(fermat(6))`
d9.   274177 67280421310721
The product of $F(5)$ and $F(6)$ is a big number, say
c10. `big:fermat(5)*fermat(6)`
d10.   79228162532711081671548469249
but still factors in a reasonably short time,
c11. `factor(big)`
d11.   641 274177 6700417 67280421310721
In general, it takes much time to factor a big integer[2] or determine that it is a prime
number.
We will use the `factor` command mainly for polynomials, for example:
c12. `x^28 + 1$`
c13. `factor(%)`
d13. $(x^4 + 1)(x^{24} - x^{20} + x^{16} - x^{12} + x^8 - x^4 + 1)$
Subtracting 1 from the last expression,
c14. `% - 1`
d14. $(x^4 + 1)(x^{24} - x^{20} + x^{16} - x^{12} + x^8 - x^4 + 1) - 1$
and simplifying, using the command `ratsimp`, we get:
c15. `ratsimp(%)`
d15. x^{28}
More examples:
c16. `a*x^2 - 4*a`
d16. $ax^2 - 4a$
c17. `factor(%)`
d17. $a(x - 2)(x + 2)$
c18. `a*x^2 - 2*a`
d18. $a x^2 - 2a$
c19. `factor(%)`
d19. $a\left(x^2 - 2\right)$
No more can be done with the `factor` command, since the quadratic factor $x^2 - 2$ is
irreducible over the integers. To get the linear factors of d18, we use the command
cfactor, which factors with complex numbers and radicals.
c20. `cfactor((a*x^2 - 2*a), x)`
d20. $a(x - \sqrt{2})(x + \sqrt{2})$
When you use the MACSYMA command `solve` to solve a polynomial equation, it calls
the function `factor` to factor the polynomial. If the factors are of degree 4 or less,
MACSYMA applies standard formulas to generate the solution.

---

2 This fact is used in coding and computer security.

A.13 Making substitutions

MACSYMA has several commands for substituting one expression for another in a third expression. In particular, if expr is an expression containing the argument x,
- the command ev(expr,x=a) evaluates expr for $x = a$,
- the command subst(a,x,expr) or subst(x=a,expr) replaces x by a in the expression expr,
- ratsubst is similar to subst, without restrictions on the expression being replaced.

MACSYMA-Session A.11. Examples.
c1. expr1:z*%e^-z
d1. ze^{-z}
The command
c2. ev(expr1,z=x^2)
replaces every z in d1 with x^2 by temporarily binding z to x^2 and then re-evaluates the expression,
d2. $x^2e^{-x^2}$
An equivalent way for performing this replacement is
c3. expr1, z=x^2
d3. $x^2e^{-x^2}$
The variable expr1 itself does not change:
c4. expr1
d4. ze^{-z}
The command subst(a,x,expr) works if x is an *atom* (i.e., a number, a string, or a symbol) or a complete subexpression of expr. Otherwise, use the command ratsubst(a,x,expr), which does not restrict x.
c5. subst(x^2,z,expr1)
d5. $x^2e^{-x^2}$
c6. expr2:(a+b+c)/d
d6. $\dfrac{a+b+c}{d}$
The expression $a + b$ is neither an atom nor a complete subexpression of expr2. Therefore
c7. subst(d,a+b,expr2)
does not work:
d7. $\dfrac{a+b+c}{d}$
To perform this substitution, we use
c8. ratsubst(d,a+b,expr2)
d8. $\dfrac{d+c}{d}$
Finally consider the Newton iterations:
$$x_{n+1} := x_n - \frac{f(x_n)}{f'(x_n)}, \quad n = 0, 1, \ldots \tag{A.1}$$
MACSYMA has a good *newton* program, but as an exercise we define a single iteration of (A.1) using the subst command. First we define f as a function of x:
c9. depends(f,x)$

The iteration (A.1), at the point x_0, is:

c10. `newt(f,x,x0):=subst(x0, x, x-f/diff(f,x))$`

Examples:

c11. `newt(x^3,x,1.0)`

d11.   0.66667

c12. `newt(x^(1/3),x,3.0)`

d12. -6.0

c13. `newt(x*exp(1-x),x,10.0)`

d13. -8092.08

A.14 Extracting parts of an expression

MACSYMA has many commands for extracting parts of expressions. These include:

- `part`, which returns the subexpression specified by its position in the expression or equation (the usage of `part` is summarized in Table A.2);
- `dpart`, which is similar to `part` except that it returns the entire expression, with the selected subexpression displayed inside a box;
- `substpart`, which substitutes the characters you specify for the indicated subexpression and then returns the new value of the expression;
- `pickapart`, which assigns intermediate display lines (e-lines) to all subexpressions of an expression, down to a specified depth;
- `lhs` and `rhs`, which return the left- and right-hand sides of the given equation, respectively;
- `first` and `last`, which return the first and last part of the specified expression, respectively;
- `rest`, which returns the expression with one or more of its leading elements removed;
- `coeff` and `ratcoef`, which return the coefficient of a given variable in the specified expression.

MACSYMA-Session A.12. Examples.

c1. `eq3:x^2 + 2*x + 2 = y^2 + 1$`

The second part of an equation is its right-hand side,

c2. `part(%,2)`

d2. $y^2 + 1$

Table A.2. Usage of the `part` command

Command	Returned subexpression of:	
	`expr:f(x, y, z)`	`expr:a=b`
`part(expr, 0)`	f	$=$
`part(expr, 1)`	x	a
`part(expr, 2)`	y	b
`part(expr, 3)`	z	

and the first part of d2 is

c3. part(%,1)

d3. y^2

We can combine these two steps by

c4. part(eq3, 2, 1)

d4. y^2

An example of using dpart, which works like part but displays the selected part in a box:

c5. big_expr:(x^3+3*x^2+3*x+1)/(d^(x^3+3*x^2+3*x+1)+n)

d5. $$\dfrac{x^3 + 3x^2 + 3x + 1}{n + d^{x^3+3x^2+3x+1}}$$

c6. dpart(big_expr, 2, 2, 2)

d6. $$\dfrac{x^3 + 3x^2 + 3x + 1}{n + d^{\boxed{x^3 + 3x^2 + 3x + 1}}}$$

The system variable piece holds the last expression selected with one of the part selection commands, such as part, dpart, and substpart. Currently we have:

c7. piece

d7. $x^3 + 3x^2 + 3x + 1$

We can factor the last subexpression and substitute back into the expression big_expr by the command substpart:

c8. substpart(factor(piece), big_expr, 2, 2, 2)

d8. $$\dfrac{x^3 + 3x^2 + 3x + 1}{n + d^{(x+1)^3}}$$

A.15 Trigonometric functions

The MACSYMA trigonometric (circular and hyperbolic) functions are outlined in this section. The names of these functions, and their inverses, are listed in Table A.3.

Evaluating trigonometric functions

If an argument of a trigonometric function is given as a floating-point number, MACSYMA will evaluate the function numerically. Otherwise, MACSYMA returns an exact value whenever possible.

Table A.3. Circular and hyperbolic functions and their inverses

Circular functions	Inverse circular functions	Hyperbolic functions	Inverse hyperbolic functions
sin	asin	sinh	asinh
cos	acos	cosh	acosh
tan	atan	tanh	atanh
cot	acot	coth	acoth
sec	asec	sech	asech
csc	acsc	csch	acsch

MACSYMA-Session A.13.

c1. `sin(1)`

d1. $\sin(1)$

c2. `sin(0)`

d2.   0

a number, since c2 simplifies to 0. To evaluate c1 numerically use:

c3. `sin(1), numer:true`

d3.   0.841471

MACSYMA knows the special values of trigonometric functions at the arguments $n\pi/m$, $m = 1, 2, 3, 4, 6, 12$ and n integer. For example,

c4. `cos(x)^2 - sin(x)^2`

d4. $\cos^2 x - \sin^2 x$

is evaluated at $x = \pi/3$,

c5. `%, x=%pi/3`

d5. $\frac{1}{2}$

an exact result, which can be converted to a floating-point format whenever necessary.

Expanding and simplifying trigonometric expressions

MACSYMA has several commands for expanding and simplifying trigonometric expressions, including:

—   `trigexpand`, which expands expressions containing trigonometric and hyperbolic functions of sums of angles and of multiple angles;

—   `trigreduce`, which combines products and powers of the trigonometric functions for a specified variable and tries to eliminate these functions when they occur in the denominator;

—   `trigsimp`, which converts expressions containing functions such as tan and sec to contain sin, cos, sinh, and cosh instead, so that `trigreduce` can further simplify the expressions.

MACSYMA-Session A.14. Examples.

c1. `t_expr1:sin(2*x + y) + cos(2*a)`

d1. $\sin(y + 2x) + \cos(2a)$

c2. `trigexpand(t_expr1)`

d2. $\cos(2x)\sin(y) + \sin(2x)\cos(y) - \sin^2(a) + \cos^2(a)$

Setting the option variable `trigexpand` to `true` causes the full expansion of sines and cosines in `t_expr1`

c3. `trigexpand(t_expr1), trigexpand:true`

d3. $(\cos^2(x) - \sin^2(x))\sin(y) + 2\cos(x)\sin(x)\cos(y) - \sin^2(a) + \cos^2(a)$

The option variable `trigexpandplus` controls the sum rule for `trigexpand`. By default, `trigexpandplus` is true, so MACSYMA expands sums like $\sin(x + y)$. Similarly, the option variable `trigexpandtimes` controls the product rule for `trigexpand`. By default, `trigexpandtimes` is true, so that MACSYMA expands products like $\sin(2y)$.

To simplify half angles in trigonometric expressions, set the option variable `halfangles` to `true`.

c4. `t_expr2:sin(2*x) + cosh(y-z) + tan(yz/2)`

d4. $\cosh(z - y) + \tan\left(\frac{yz}{2}\right) + \sin(2x)$

c5. `trigexpand(t_expr2)`

d5. $\quad - \sinh(y)\sinh(z) + \cosh(y)\cosh(z) + \tan\left(\frac{yz}{2}\right) + 2\cos(x)\sin(x)$

and with `trigexpandtimes` set locally to `false`,

c6. `trigexpand(t_expr2), trigexpandtimes:false`

d6. $\quad - \sinh(y)\sinh(z) + \cosh(y)\cosh(z) + \tan\left(\frac{yz}{2}\right) + \sin(2x)$

Similarly, with `trigexpandplus` bound locally to `false`,

c7. `trigexpand(t_expr2), trigexpandplus:false`

d7. $\quad \cosh(z - y) + \tan\left(\frac{yz}{2}\right) + 2\cos(x)\sin(x)$

The command `trigreduce` is the inverse of `trigexpand`, in that it converts products and powers of trigonometric functions into functions with multiple angles.

c8. `t_expr3:trigexpand(sin(2*z) + sin(2*y))`

d8. $\quad 2\cos(z)\sin(z) + 2\cos(y)\sin(y)$

c9. `trigreduce(t_expr3)`

d9. $\quad \sin(2z) + \sin(2y)$

c10. `trigreduce(t_expr3, z)`

d10. $\quad \sin(2z) + 2\cos(y)\sin(y)$

c11. `trigreduce(t_expr3, y)`

d11. $\quad 2\cos(z)\sin(z) + \sin(2y)$

The command `trigsimp` makes use of common trigonometric identities, such as

$$\sin^2 x + \cos^2 x = 1\,,$$
$$\cosh^2 x - \sinh^2 x = 1\,.$$

c12. `t_expr4:(1-sin(x))*(sec(x)+tan(x))-cos(x)+(cosh(x)^2`
`       -sinh(x)^2)^3-1`

d12. $\quad (1 - \sin(x))(\tan(x) + \sec(x)) + (\cosh^2(x) - \sinh^2(x))^3 - \cos(x) - 1$

c13. `trigsimp(t_expr4)`

d13. $\quad 0$

Alternatively, `t_expr4` can be simplified by manually substituting $\cosh(x)^2 - 1$ for $\sinh(x)^2$,

c14. `ratsubst(cosh(x)^2 - 1, sinh(x)^2, t_expr4)`

d14. $\quad (1 - \sin(x))\tan(x) - \sec(x)\sin(x) + \sec(x) - \cos(x)$

then substituting $\sin x / \cos x$ for $\tan x$,

c15. `subst(sin(x)/cos(x), tan(x),%)`

d15. $\quad \dfrac{(1 - \sin(x))\sin(x)}{\cos(x)} - \sec(x)\sin(x) + \sec(x) - \cos(x)$

and, finally

c16. `trigreduce(ratsimp(%))`

d16. $\quad 0$

Still a third way to reduce `t_expr4` is:

c17. `t_expr4, exponentialize, ratsimp`

d17. $\quad 0$

A.16 A simple program

We illustrate MACSYMA programming with the simplest possible example, deciding whether the quadratic equation

$$ax^2 + bx + c = 0$$

has a *double root, two real roots*, or *two complex roots*. A MACSYMA function
doing this is

```
c1. type_of_roots(a,b,c):=block([d:b^2-4*a*c],
                              if d=0 then "double root"
                              else if d > 0 then "two real roots"
                              else "two complex roots") $
```

The type of roots is decided according to the sign of the *discriminant* $d := b^2 - 4ac$,
in the square brackets right inside the block statement. Here block is the body
of the program. The purpose of the square brackets is to make d a *local variable*,
i.e., the assignment d:b^2-4*a*c is not valid outside type_of_roots, where the
same name d can be used for other purposes.

type_of_roots uses the MACSYMA function if in nested form:

if cond$_1$ then first else if cond$_2$ then second else third

which checks the condition cond$_1$ if true returns "first", and otherwise checks
condition cond$_2$, etc.

MACSYMA-Session A.15. Examples.
c2. type_of_roots(1,0,-1)
d2.   two real roots
c3. type_of_roots(1,1,1)
d3.   two complex roots
c4. type_of_roots(1,2,1)
d4.   double root

The function type_of_roots can be written alternatively as

```
c1. type_of_roots(a,b,c):=block([d:b^2-4*a*c],
      case(is(d=0),
          [true, "double root"],
          [false, case(is(d>0),[true, "two real roots"],
                                [false, "two complex roots"])])) $
```

using the MACSYMA function case in the form

case(is(cond$_1$), [true, first], [false, case(is(cond$_2$),

                                 [true, second], [false, third])])

MACSYMA-Session A.16. Examples.
c2. type_of_roots(1,0,1)
C:\MACSYMA2\share\basic.fas being loaded
d2.   two complex roots
c3. type_of_roots(1,5,1)

d3.   two real roots
c4. `type_of_roots(4,4,0)`
d4.   two real roots

A.17 Plotting

This is a concise introduction to the graphic capabilities of MACSYMA for both two- and three-dimensional plots.

MACSYMA-Session A.17.

Simple plots

To plot a function of one variable we use the syntax `plot(f(x), x, x-min, x-max)`, where `f(x)` is an expression in the variable x, and [`x-min`, `x-max`] is the interval of plotting.

To plot $y = x^2 - 1$ in the interval $[-4, 4]$ use:

c1. `plot(x^2-1,x,-4,4)`
d1.   done

The resulting plot is shown in Fig. A.1. The line d1 merely confirms the completion of the task. We will often suppress the d-line by ending the previous c-line with $.

Plots of user-defined functions

Consider the function

$$f(x) := \begin{cases} x^2 & \text{if } x < 0, \\ x & \text{otherwise.} \end{cases}$$

First we define the function by the if-statement

c2. `f(x) := block(if x< 0 then x^2 else x)$`

and then plot it by the command

c3. `plot(f(x), x, -2,2)$`

obtaining the graph in Fig. A.2.

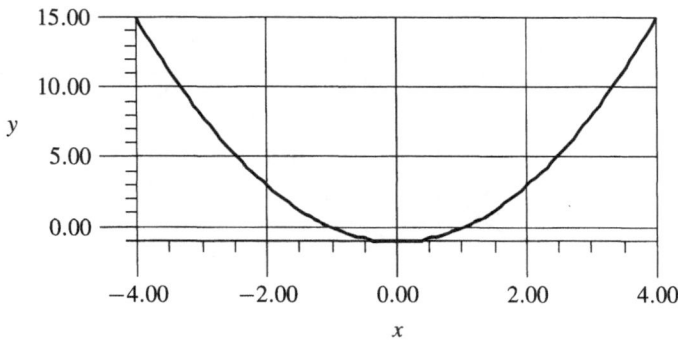

Fig. A.1. Plot of the function $x^2 - 1$

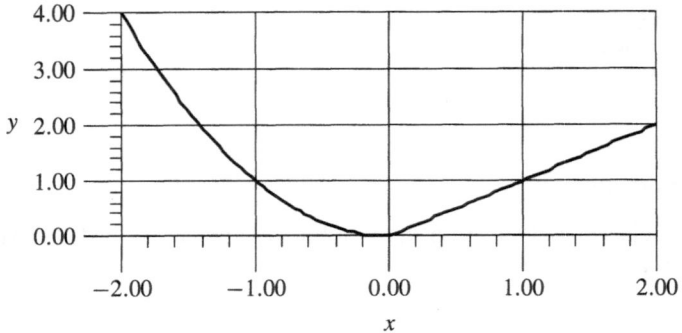

Fig. A.2. Plot of a piecewise-defined function

Combined plots
It is possible to graph more than one function on the same screen. For example,
c4. plot([sin(x), cos(x)], x, -%pi, %pi)$
gives Fig. A.3.

Parametric plots
A curve $\mathcal{C}$ in the xy-plane may be represented by a parameter t which varies over the
interval $[a, b]$,

$$
\mathcal{C} = \left\{ (x, y) \; \middle| \; \begin{cases} x = f(t), \\ y = g(t), \end{cases} \quad a \le t \le b \right\}.
$$

Such curves are plotted by the command
paramplot(f(t), g(t), t, t-min, t-max)
For example, the unit circle $\mathcal{C} := \{ (x, y) \mid x^2 + y^2 = 1 \}$ can be represented
parametrically as

$$
\mathcal{C} = \{ (\cos t, \sin t) \mid 0 \le t \le 2\pi \}
$$

We plot it by
c5. paramplot(cos(t), sin(t), t, 0, 2*%pi), equalscale:true$

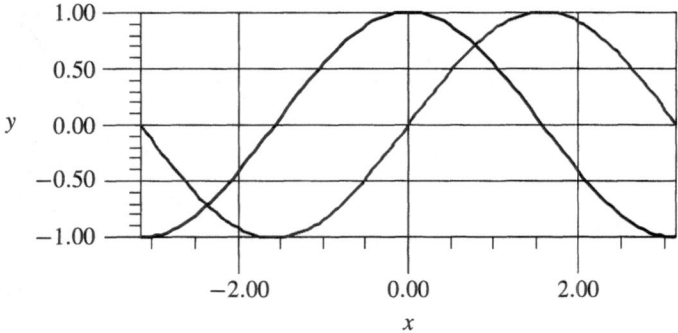

Fig. A.3. Combined plots of $\sin x$ and $\cos x$

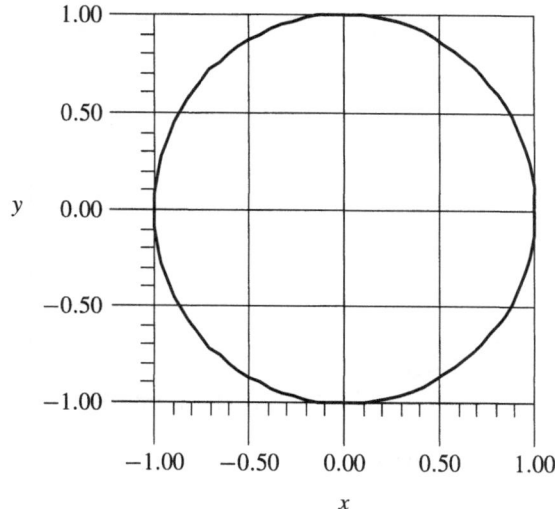

Fig. A.4. Plot of the unit circle described parametrically

giving Fig. A.4. If we forget to set `equalscale:true` in c5, then the x- and y-axes will have different scales, and the plot will appear as an ellipse rather than a circle (Fig. A.5).

Implicit plots
A curve $\mathcal{C}$ in the xy-plane may be given implicitly

$$\mathcal{C} = \{\,(x, y) \mid F(x, y) = 0\,\}$$

We plot such curves by the command
`implicit_plot(F(x,y), x, x-min, x-max, y, y-min, y-max)`
For example, the unit circle $\{\,(x, y) \mid x^2 + y^2 = 1\,\}$ is plotted by
c6. `implicit_plot(x^2+y^2-1, x, -1.5, 1.5, y, -1.5, 1.5)$`
giving Fig. A.5.

Plotting sequences of points
It is possible to plot a sequence of points in the xy-plane. We do this by first creating the x-coordinates of the points as a list,
c7. `x_list:makelist(i,i,1,10)`
d7. $[1, 2, 3, 4, 5, 6, 7, 8, 9, 10]$
then list the y-coordinates
c8. `y_list:makelist((-1)^i/i, i, 1,10)`
d8. $[-1, (1/2), -(1/3), (1/4), -(1/5), (1/6), -(1/7), (1/8), -(1/9), (1/10)]$
and then plot the sequence of points:
c9. `graph(x_list, y_list)$`
obtaining Fig. A.6.

Contour plots
A *contour* (or *level set*) of a surface $z = f(x, y)$ is the intersection of the surface with a horizontal plane $z = c$. Given sufficiently many contours, we can visualize the shape of

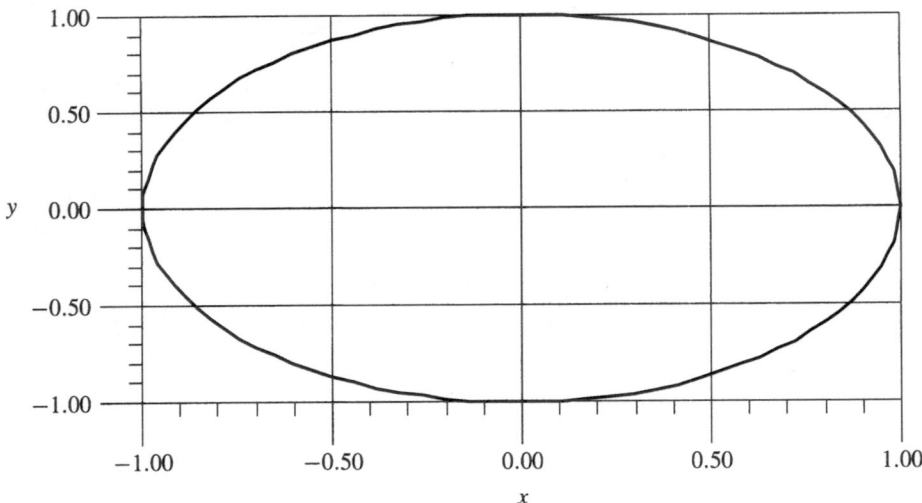

Fig. A.5. Plot of the unit circle, without specifying `equalscale:true`

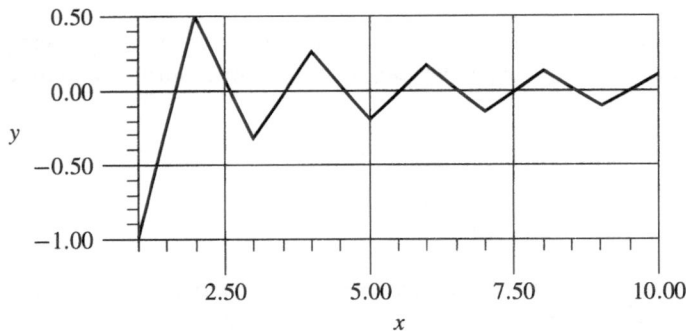

Fig. A.6. Two-dimensional plot of a set of points

the surface. The contour plot is produced by the command
`contourplot(f(x,y), x, x-min, x-max, y, y-min, y-max)`
where the number of contours is determined by the variable `contours`, with default
setting 20. To get a different number of contours, use the assignment:
　　　`contours:n.`
For example, 10 contours of the surface $\{ z = (x - 1)^2 + y^2 \mid -5 \leq x \leq 5, -5 \leq y \leq 5 \}$
are plotted by:
c10. `contours:10$`
c11. `contourplot(y^2+(x-1)^2, x, -5, 5, y, -5, 5)$`
giving Fig. A.7.

　　3-dimensional plots
A surface $z = f(x, y)$ can be plotted, over the appropriate range, by the command
`plot3d(f(x,y), x, x-min, x-max, y, y-min, y-max)`

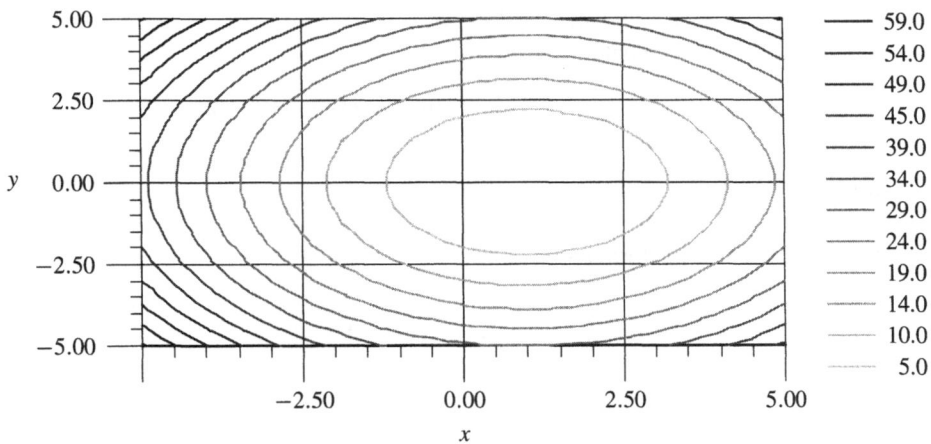

Fig. A.7. Contour plot of $z = (x-1)^2 + y^2$

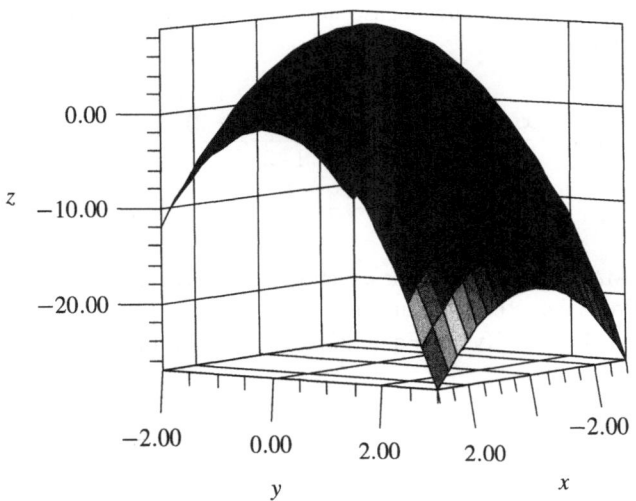

Fig. A.8. Three-dimensional plot of a surface

For example, the surface $\{z = 9 - x^2 - 3y^2 \mid -3 \leq x \leq 3, -2 \leq y \leq 3\}$ is plotted by:

```
c12. plot3d(9-x^2-3*y^2, x, -3, 3, y, -2, 3)$
```

giving Fig. A.8.

Exercises

A.1 Use MACSYMA to simplify:

(a) $\sqrt{5 + 2\sqrt{6}}$ (b) $\sqrt[4]{19601 - 13860\sqrt{2}}$ (c) $\sqrt[8]{408\sqrt{2} + 577}$

(d) $173\sqrt{34}\sqrt{2\sqrt{34} + 35} + 1394\sqrt{2\sqrt{34} + 35} - 1567\sqrt{34}$

Hint: Use nested square roots.

A.2 Factor the expressions $n^4 + 4$, and $a^{10} + a^5 + 1$.

A.3 Use MACSYMA to compute:

(a) $\displaystyle\sum_{k=1}^{n} k(k-1)$

(b) $\displaystyle\sum_{k=2}^{n} k(k-1)(k-2)$

(c) $\displaystyle\sum_{k=3}^{n} k(k-1)(k-2)(k-3)$

Use these to guess the formula for

$$\sum_{k=m}^{n} k(k-1)\cdots(k-m).$$

A.4 Recall the Stirling approximation of $n!$: $n! \approx \sqrt{2\pi}\, n^{n+1/2}\, e^{-n}$. Use MACSYMA to compute:

(a) $10!$ (b) $100!$ (c) $1000!$

and their Stirling approximations.

A.5 Show that the product of any 4 consecutive positive integers is one less than the square of an integer.

A.6 Solve the equation

$$1 - \frac{x}{1} + \frac{x(x-1)}{1\cdot 2} - \frac{x(x-1)(x-2)}{1\cdot 2\cdot 3} \pm \cdots$$
$$+ (-1)^n \frac{x(x-1)(x-2)\cdots(x-n+1)}{n!} = 0.$$

Hint: Use MACSYMA to solve the equation for $n = 1, 2, \ldots, 5$ and guess the general solution. Prove the result by induction.

A.7 Use MACSYMA to verify that Eq. (1.22) gives the roots of $ax^2 + bx + c = 0$.

B Numbers

Calculus uses numbers and functions which take numerical values. Most of the time we will use the familiar *real* numbers (Sect. B.2). However, in order to use MACSYMA intelligently we need to know a little about *complex* numbers. These are introduced in Sect. B.6.

Even if you had a good introduction to numbers, as one would normally get in a decent precalculus course, you may find this appendix useful for its MACSYMA-Sessions.

B.1 Arithmetic operations

The four operations

Arithmetic operations are defined on pairs of numbers, a, b. There are four operations

addition	$a + b,$
subtraction	$a - b,$
multiplication	written $a \cdot b, a \times b$, or $ab,$
division	a/b, or $\dfrac{a}{b}$, provided $b \neq 0.$

The number 0 (zero) has the property that

$$a + 0 = a, \quad \text{for all } a ,$$

and is called the *additive identity*. Similarly, the number 1 (one) is the *multiplicative identity*,

$$a \cdot 1 = a, \quad \text{for all } a .$$

The *reciprocal* of a nonzero number a is the number $1/a$.

Addition is *distributive*, with respect to multiplication, in the sense that

$$a \cdot (b + c) = a \cdot b + a \cdot c, \quad (a + b) \cdot c = a \cdot b + a \cdot c, \quad \text{for all } a, b, c .$$

The product of a with itself n times is denoted

$$a^n := \overbrace{a \cdot a \cdots a}^{n \text{ times}}$$

and called the nth power of a. Here a is the *base*, n is the *exponent*, and the operation a^n is called *exponentiation*.

If $a^n = b$, then a is called an nth *root* of b, denoted $a = \sqrt[n]{b}$. In particular, a *square* root of b is written as $\sqrt{b}$ (where $n = 2$ is omitted), and a *cubic* root as $\sqrt[3]{b}$.

Order conventions

The operations of addition and multiplication are *commutative*[1],

$$a + b = b + a, \quad a \cdot b = b \cdot a, \quad \text{for all } a, b \,,$$

and *associative*[2]

$$a + (b + c) = (a + b) + c, \quad a \cdot (b \cdot c) = (a \cdot b) \cdot c, \quad \text{for all } a, b, c \,,$$

where the pair, first operated on, is in parentheses. Therefore parentheses are not needed for sums $a + b + \cdots + c$ and products $a \cdot b \cdot c$ of three or more numbers.

In general, the value of an algebraic expression may depend on the order in which operations are carried out.[3] To avoid such ambiguities, mathematicians follow certain rules, called *order conventions* (the priorities of arithmetic operations):
- expressions in *parentheses* are evaluated first, e.g., $(2 + 3) \cdot 4 = 5 \cdot 4$;
- *exponentiations* are evaluated before *multiplications* or *divisions*,

$$2^3 \cdot 3^2 = 8 \cdot 9, \quad 2^3/3^2 = 8/9 \,;$$

- *multiplications* or *divisions* come before *additions* or *subtractions*, e.g., $2 + 3 \cdot 4 = 2 + 12$;
- *multiplications* and *divisions* are performed from left to right,

$$2/3 \cdot 4 = \frac{2}{3} \cdot 4, \quad 2/3/4 = \frac{2}{3}/4 = \frac{2}{3 \cdot 4} \,,$$

- *additions* and *subtractions* are performed from left to right,

$$2 + 3 - 4 = (2 + 3) - 4, \quad 2 - 3 + 4 = (2 - 3) + 4 \,;$$

- *consecutive exponentiations* are performed from right to left. Therefore

$$2^{3^4} \text{ is interpreted as } 2^{\left(3^4\right)} = 2^{81}, \quad \text{and not as } \left(2^3\right)^4 = 2^{12} \,.$$

When in doubt, use parentheses!

Exercises

B.1 Verify each of the following identities by MACSYMA.
For any numbers a, b and positive integers m, n:
(a) $a^m a^n = a^{m+n}$ (b) $a^m/a^n = a^{m-n}$ (c) $a^0 = 1$
(d) $a^n b^n = (ab)^n$ (e) $(a^m)^n = a^{mn}$ (f) $\sqrt[n]{a^m} = a^{m/n}$

---

1 Order does not matter.
2 If you add, or multiply, three numbers, it does not matter which two you handle first.
3 For example, consider $2 + 3 \cdot 4$ or $2^3/4$.

Note: If the exponent n is real, then a^n is defined, in general, only for positive a. Therefore it is necessary to specify

c1. a>0$

so that MACSYMA knows a is positive. If we want a to be real, we

c2. declare(a, real)$

B.2 Which of these is true? (a) $2^{2^{n+1}} = 2^{2^n} \times 2^2$ (b) $2^{2^{n+1}} = \left(2^{2^n}\right)^2$

B.3 Simplify:

(a) $8^{2/3}$

(b) $1000^{-1/3}$

(c) $32^{2/5}$

(d) $81^{3/4}$

(e) $125^{-2/3}$

(f) $\sqrt{36x^2 + 9}$

B.2 Real numbers

The real numbers (or reals) are the ordinary numbers of everyday life, such as 0, -3, $\sqrt{2}$, and π, which (fortunately) suffice for most of this course. The *set of all real numbers* is denoted $\mathbf{R}$.

Real line

The real numbers can be visualized as points on a line on which we define a *coordinate system* by marking one point as 0 and another point as 1. This line, called the real line, the real axis, or the x-axis, is denoted by the same letter $\mathbf{R}$.

The *origin* of $\mathbf{R}$ is at 0, its *unit* (or scale) is the distance between 0 and 1, and its *direction* is from 0 to 1 or from left to right. The direction orders the line in an obvious way, making it possible to determine for any two points, which is left and which is right (in particular, 0 is left of 1).

The *distance* between two points a and b is denoted by $dist(a, b)$. Any point x on $\mathbf{R}$ is completely described by its distance $dist(0, x)$ from 0, and its position left or right of 0. We define the *coordinate* of a point x as the number

$$- dist(0, x) \quad \text{if } x \text{ is left of } 0 ,$$
$$dist(0, x) \quad \text{if } x \text{ is right of } 0 .$$

The coordinate is unique to the point, and conversely, for any real number α, there is a unique point in $\mathbf{R}$ with α as coordinate. This shows a one-to-one correspondence between the points of the line $\mathbf{R}$ and the real numbers (considered as coordinates.)

Consider now two real numbers x and y and the corresponding points $p(x)$ and $p(y)$. We say that x is *smaller* (or less) than y, and write $x < y$ or $y > x$ (y is *greater* than x) if $p(x)$ is left of $p(y)$. We denote by $x \leq y$ (x is *less than or equal to y*) or $y \geq x$ (y is *greater than or equal to x*) the fact

$$x < y \quad \text{or} \quad x = y .$$

The *natural order* of the reals, from small to large, thus corresponds to the natural order on the real line, from left to right.

Fig. B.1. Portion of the real line

Intervals

If a, x, and b are reals and if $a \leq x \leq b$, then x is said to be between a and b.

Definition B.1 (Intervals). An interval is a set of reals between two given real numbers a and b, which are called the *endpoints* of the interval. Depending on which endpoint we include, there are four intervals:

$$[a, b] := \{ x \in \mathbf{R} \mid a \leq x \leq b \}, \quad \text{both endpoints included},$$
$$(a, b] := \{ x \in \mathbf{R} \mid a < x \leq b \}, \quad \text{right endpoint included},$$
$$[a, b) := \{ x \in \mathbf{R} \mid a \leq x < b \}, \quad \text{left endpoint included},$$
$$(a, b) := \{ x \in \mathbf{R} \mid a < x < b \}, \quad \text{both endpoints excluded}.$$

An interval of the form $[a, b]$ is called *closed*, (a, b) is called *open*, whereas $(a, b]$ and $[a, b)$ are called *half-open*.

In particular, for any real a,

$$[a, a] := \{a\}, \quad a \text{ itself},$$
$$(a, a] = [a, a) = (a, a) = \emptyset, \quad \text{the empty set}.$$

The "two points" of *infinity*[4], $-\infty$ and $+\infty$, can also serve as endpoints. Thus for any a,

$$[a, \infty) := \{ x \in \mathbf{R} \mid a \leq x \}, \qquad (a, \infty) := \{ x \in \mathbf{R} \mid a < x \},$$
$$(-\infty, a] := \{ x \in \mathbf{R} \mid x \leq a \}, \qquad (-\infty, a) := \{ x \in \mathbf{R} \mid x < a \},$$
$$(-\infty, +\infty) := \mathbf{R}, \quad \text{the entire real line}.$$

Given an interval I with endpoints $a \leq b$, the *length* of I, denoted by $|I|$, is the difference $b - a$, or the *distance* between the endpoints.

Example B.2 (Arithmetic mean). Given any two real numbers a and b, their arithmetic mean is the number $\frac{1}{2}(a + b)$. For example,

$$\frac{1}{2}(0 + 1) = \frac{1}{2}, \quad \text{and} \quad \frac{1}{2}((-1) + 1) = 0.$$

If I is the interval with endpoints a and b, then $m := \frac{1}{2}(a + b)$ is the *midpoint* of I, i.e., the point with equal distances from both endpoints. Indeed

$$dist(m, a) = \frac{1}{2}(a + b) - a = \frac{a + b - (2a)}{2} = \frac{b - a}{2},$$
$$\text{and} \quad dist(b, m) = b - \frac{1}{2}(a + b) = \frac{2b - (a + b)}{2} = \frac{b - a}{2}.$$

---

4 These are not real numbers, merely mathematical symbols.

Integers

The first numbers considered, in the history of mankind, or in personal history of every child growing up, are the natural numbers or positive integers, denoted by

$$\mathbf{N} := \{1, 2, 3, 4, \ldots\},$$

and the *nonnegative* integers, denoted by

$$\mathbf{N}_0 := \{0, 1, 2, 3, 4, \ldots\},$$

obtained from $\mathbf{N}$ by adjoining zero.

Many results in mathematics are stated for positive integers. For example,

$$\sum_{k=1}^{n} k = 1 + 2 + 3 + \cdots + n = \frac{n(n+1)}{2}, \tag{B.1}$$

$$\sum_{k=1}^{n} k^2 = 1^2 + 2^2 + 3^2 + \cdots + n^2 = \frac{n(n+1)(2n+1)}{6}, \tag{B.2}$$

$$\sum_{k=1}^{n} k^3 = 1^3 + 2^3 + 3^3 + \cdots + n^3 = \frac{n^2(n+1)^2}{4}, \tag{B.3}$$

are statements about n which are true for all $n \in \mathbf{N}$.

A number system is said to be *closed* under (or with respect to) an operation if the results of the operation stay in the number system. The set of nonnegative integers N_0 is closed under addition and multiplication (if m and n are nonnegative integers, so are their sum $m + n$ and product mn) but not under subtraction ($m - n$ is negative if $m < n$). Since N_0 is "too small" to accommodate subtraction, we extend it to the set of all *integers*, denoted by

$$\mathbf{Z} := \{\ldots, -3, -2, -1, 0, 1, 2, 3, \ldots\}.$$

While $\mathbf{Z}$ is closed under addition, subtraction, and multiplication, it is not closed under division, since the ratio m/n of two integers is in general not integer. For example, $57/19 \in \mathbf{Z}$ but $58/19 \notin \mathbf{Z}$.

Rational numbers

The last example shows that the set of integers $\mathbf{Z}$ is "too small" even for solving simple linear equations

$$nx = m, \quad \text{for given } m, n \in \mathbf{Z}, \tag{B.4}$$

whose solution, m/n, is in general "outside" $\mathbf{Z}$. This makes it necessary to extend $\mathbf{Z}$ to a bigger set which includes all solutions of (B.4).

A ratio of two integers m/n is called *rational*, and the set of all *rational numbers* is denoted by

$$\mathbf{Q} := \left\{ \frac{m}{n} \;\middle|\; m, n \in \mathbf{Z}, \; m \neq 0 \right\}.$$

In particular, every integer $n \in \mathbf{Z}$ is rational, since it can be written as $n/1$. Therefore $\mathbf{Q}$, which by definition includes the solutions of all linear equations (B.4), is an extension of $\mathbf{Z}$.

Example B.3. If a and b are rational, so is their arithmetic mean $m := \frac{1}{2}(a + b)$. Indeed, let

$$a = \frac{m_1}{n_1} \quad \text{and} \quad b = \frac{m_2}{n_2} \quad \text{with } m_1, n_1, m_2, n_2 \in \mathbf{Z}.$$

Then their arithmetic mean,

$$m = \frac{1}{2}\left(\frac{m_1}{n_1} + \frac{m_2}{n_2}\right) = \frac{m_1 \cdot n_2 + m_2 \cdot n_1}{2 \cdot n_1 \cdot n_2}$$

is rational.

Example B.3 shows that between any two rationals $a < b$ there is another rational (their arithmetic mean m), and therefore there are infinitely many rationals (by repeating the argument, taking the arithmetic mean of a and m, etc.)

In numerical computations, numbers are represented by a finite number of decimals and therefore as rational numbers.

MACSYMA-Session B.1. By increasing the precision of MACSYMA's *bigfloat* representation, we can approximate $\sqrt{2}$ with more accurate rational numbers,

```
c1. bfprecision:6 $
c2. bfloat(sqrt(2))
d2.    1.41421
c3. ratsimp(%)
```
$$d3. \quad \frac{1970}{1393}$$

approximating the bigfloat 1.41421b0 by the rational number 1970/1393.

```
c4. bfprecision:12 $
c5. bfloat(sqrt(2))
d5.    1.4142135624
c6. ratsimp(%)
```
$$d6. \quad \frac{2273378}{1607521}$$

Here `ratsimp` replaced the bigfloat d5 by the rational number d6.

```
c7. bfprecision:30 $
c8. bfloat(sqrt(2))
d8.    1.41421356237309504880168872421
c9. ratsimp(%)
```
$$d9. \quad \frac{2470433131948081}{1746860020068409}$$

The bigfloat d8 is approximated by the rational number d9.

However, not all numbers are rational.

Example B.4. The number $\sqrt{2}$ is not rational, i.e., there are no integers m and n such that
$$\sqrt{2} = m/n \, .$$
This statement was first proved by Hippasus, (ca. 500 B.C.).

The above example shows that the set of rational numbers $\mathbf{Q}$ is "too small" for solving even simple quadratic equations, such as
$$x^2 - 2 = 0 \, , \tag{B.5}$$
whose solution, $\sqrt{2}$, lies outside $\mathbf{Q}$.

Irrational numbers

A number which cannot be represented as a ratio of two integers m/n is called irrational. We saw that $\sqrt{2}$ is irrational. Other examples are $\sqrt{3}$, π, and e.

Every real number is therefore either rational or irrational. Another dichotomy is between algebraic and transcendental numbers. An *algebraic* number is a root of a polynomial with integer coefficients. All rational numbers m/n (solutions of $nx = m$) and some irrational numbers (such as $\sqrt{2}$, a solution of $x^2 - 2 = 0$) are algebraic. Real numbers which are not algebraic (examples are e and π) are called *transcendental*.

Any irrational number x can be approximated, arbitrarily close, by rational numbers. Indeed, it is possible to construct an increasing sequence of rational numbers, $\{l_0, l_1, \ldots, l_n, \ldots\}$ and a decreasing sequence $\{u_0, u_1, \ldots, u_n, \ldots\}$ such that
$$l_0 < l_1 < l_2 < \cdots < l_n < \cdots < \quad x \quad < \cdots < u_n < \cdots < u_2 < u_1 < u_0 \tag{B.6}$$
$$\text{and} \quad u_n - l_n \to 0 \quad \text{as} \quad n \to \infty \, , \tag{B.7}$$
in which case (B.6) is called a *rational approximation* of x, the rationals $\{l_0, l_1, \ldots, l_n, \ldots\}$ are called *lower approximations* and $\{u_0, u_1, \ldots, u_n, \ldots\}$ are called *upper approximations*. If we take either l_n or u_n instead of x, we make an error which is not greater than the number $u_n - l_n$ (which is called the nth *approximation error* of (B.6)).

Example B.5 (Rational approximations for $\sqrt{2}$). We show how to compute lower rational approximations $\{l_0, l_1, l_2, \ldots, l_n, \ldots\}$ and upper rational approximations $\{u_0, u_1, u_2, \ldots, u_n, \ldots\}$ for $\sqrt{2}$, with approximation error
$$u_n - l_n = 10^{-n} \, . \tag{B.8}$$
Since $1^2 < 2 < 2^2$, we conclude that $1 < \sqrt{2} < 2$, using the fact that, for positive u, $u^2 > 2$ is equivalent to $u > \sqrt{2}$. We set
$$l_0 := 1 < \sqrt{2} < u_0 := 2 \, ,$$

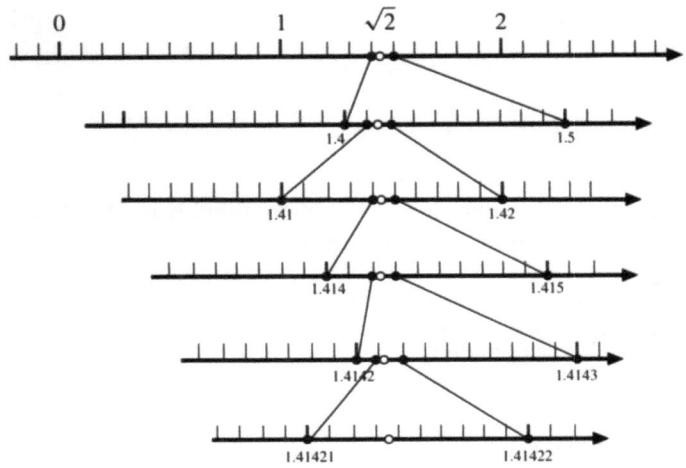

Fig. B.2. Irrational number $\sqrt{2}$ on the real line

and check the squares $1.1^2, 1.2^2, 1.3^2, \ldots, 1.9^2$ until we find the largest one which is less than 2. It is 1.4, showing that

$$l_1 := 1.4 < \sqrt{2} < u_1 := 1.5, \quad \text{with error } u_1 - l_1 = 10^{-1}.$$

Next we check the squares $\{1.41^2, 1.42^2, 1.43^2, \ldots, 1.49^2\}$ and see that $1.41^2 < 2$ but $1.42^2 > 2$. Therefore

$$l_2 := 1.41 < \sqrt{2} < u_2 := 1.42, \quad \text{with error } u_2 - l_2 = 10^{-2}.$$

Continuing in this fashion we get

$$l_3 := 1.414 < \sqrt{2} < u_3 := 1.415$$
$$l_4 := 1.4142 < \sqrt{2} < u_4 := 1.4143$$
$$l_5 := 1.41421 < \sqrt{2} < u_5 := 1.41422$$
$$l_6 := 1.414213 < \sqrt{2} < u_6 := 1.414214$$
$$l_7 := 1.4142135 < \sqrt{2} < u_7 := 1.4142136$$
$$l_8 := 1.41421356 < \sqrt{2} < u_8 := 1.41421357$$

with error $u_n - l_n = 10^{-n}, n = 1, 2, \ldots$

Exercises

B.4 (Rational approximation by bisection) Let x be an irrational number and let $l = \{l_0, l_1, \ldots, l_p\}$ and $u = \{u_0, u_1, \ldots, u_q\}$ respectively be lower and upper rational approximations of x. Since the arithmetic mean $m := (l_p + u_q)/2$ is rational (Example B.3), it cannot equal x. There are therefore two cases: either $m > x$ (in which case we can take *m* as the next upper approximation $u_{q+1} := m$) or $m < x$ (take *m* as the next lower approximation $l_{p+1} := m$.)

  a. Use this idea to obtain lower and upper rational approximations for $\sqrt{2}$, starting with $l_0 = 1$, $u_0 = 2$. Note that each arithmetic mean gives a lower approximation *or* an upper approximation, so it is possible to have several iterations without any new lower (or upper) approximation.
  b. What can be said about the error $u_n - l_n$ of this approximation?
  c. Write a MACSYMA program for lower and upper rational approximations of $\sqrt{2}$.

B.5 Long division shows that for rational numbers the decimal representation is *periodic*, i.e., there is a part of the representation which is repeatedly appended. In contrast, the decimal representations of irrational numbers are not periodic. Test this fact using MACSYMA, i.e., using the syntax bfprecision: n, set *n* sufficiently high to *observe* the repeating decimal property of the fractions

$$\frac{1}{7}, \quad \frac{2}{7}, \quad \frac{4}{13}, \quad \frac{3}{2}, \quad \frac{2}{9}, \quad \frac{100}{81}, \quad \frac{123}{456}.$$

Obtain a decimal representation of the irrational numbers $\sqrt{2}$, $\sqrt{3}$, $\sqrt{7}$, π. Try to determine which of the numbers $\sqrt{n}$, $n = 1, 2, 3, \ldots, 20$, are rational. Of course this is not a *proof* that the numbers are either rational or irrational, since one does not know in advance how many decimal digits must be computed to observe the repeating decimal property.

B.6 Use MACSYMA repeatedly to calculate $\sqrt{3}$ successively to 4 decimals and give the corresponding sequence of nested intervals. Finally let MACSYMA calculate $\sqrt{3}$ in one step with a precision of 60 digits.

B.7 Use the inequalities (3.18),

$$(1 + 1/n)^n < e < (1 + 1/n)^{n+1}, \quad n = 1, 2, \ldots,$$

to obtain lower and upper rational approximations for e.

B.8 Describe the following sets in words:
  (a) $[0, 1] \cup [2, 3]$ (b) $[a, b] \cup [b, c]$ (c) $(0, 10) \cap (-5, 5)$
  (d) $(-1, 0) \cap (0, 1)$

B.3 Absolute value

Definition B.6 (Absolute value and sign). For any real number x, the absolute *value* of x, denoted by

$$|x| := \begin{cases} x & \text{if } x \geq 0, \\ -x & \text{otherwise,} \end{cases} \tag{B.9}$$

is the distance between x and zero on the real axis. The *sign* of x (positive, negative, or zero) is indicated by the *sign function*, denoted by

$$\operatorname{sign} x := \begin{cases} 1 & \text{if } x > 0, \\ -1 & \text{if } x < 0, \\ 0 & \text{if } x = 0, \end{cases} \tag{B.10}$$

so that
$$x = \operatorname{sign} x \cdot |x|, \quad \text{for all } x \in \mathbf{R}. \tag{B.11}$$

MACSYMA-Session B.2. MACSYMA uses the absolute-value function `abs(x)` and the sign function (B.10) is `signum(x)`. We can verify (B.11) for the cases $x > 0$, $x < 0$, and $x = 0$.

```
c1. assume(x>0) $
c2. x-abs(x)*signum(x)
d2.    0
c3. forget(x>0) $
c4. assume(x<0) $
c5. x-abs(x)*signum(x)
d5.    0
c6. forget(x<0) $
c7. x:0 $
c8. x-abs(x)*signum(x)
d8.    0
```

Since the difference is 0 in all three cases, we see that (B.11) holds for all real values of x.

Some properties of the absolute value are collected below.

Lemma B.7. For any real numbers x and y,

(a) $|x \cdot y| = |x| \cdot |y|$,   (b) $\left|\dfrac{x}{y}\right| = \dfrac{|x|}{|y|}$ if $y \neq 0$,

(c) $|x + y| \leq |x| + |y|$,   (d) $|x - y| \geq \bigl||x| - |y|\bigr|$.

Proof. Parts a and b are proved by considering all possible signs of x and y. For example, if $x > 0$ and $y < 0$, then $x \cdot y < 0$, and $|x \cdot y| = -(x \cdot y) = x \cdot (-y) = |x| \cdot |y|$.

In part c we have equality if x and y have the same sign. If x and y have opposite signs, then clearly $|x + y| < |x| + |y|$. Finally, from $x = (x - y) + y$ and part c we conclude
$$|x| \leq |x - y| + |y|,$$

which can be used to prove part d.   □

Remark B.8. The inequality in part c is called the *triangle inequality*. It expresses the fact that in a triangle the *length of any one side* is *not greater* than the *sum of lengths* of the remaining two sides. This is explained in Exercise B.21 a.

Exercise

B.9 Prove, for all real numbers x and y,

(a) $\operatorname{sign}(xy) = (\operatorname{sign} x)(\operatorname{sign} y)$ (b) $\operatorname{sign}(1/x) = \operatorname{sign} x$, if $x \neq 0$

B.4 Equations and inequalities

A *real variable* is a variable which represents a real number. We often use the characters x, y, and z as real variables, but also other symbols as x_1, x_2, and x_3 will be used. For *integers* we usually use the symbols j, k, l, m, and n.

An *equation* is a statement of the form LHS $=$ RHS (left-hand side $=$ right-hand side). In this section the LHS and RHS will be algebraic expressions; however, they need not be.

Usually an equation contains variables and is true only for certain values of these variables. To *solve* an equation means finding these values. There are equations which are never true, e.g., the equation $x = x + 1$. The truth of an equation remains unchanged when adding, subtracting, multiplying, or dividing both sides of the equation by the same expression. What is to be avoided, is dividing by zero!

An *inequality* is a statement of the form LHS $\leq$ RHS, or a similar expression with $\geq$, $<$, or $>$ instead of $\leq$.

Exercises

B.10 Find those values of x for which the following hold.

(a) $|1 - x| \leq 1$ (b) $-2 \leq 2x + 4 \leq 7$ (c) $1 < 1/x < 5$

(d) $\dfrac{|1 - x|}{|1 + x|} \leq 1$ (e) $|2x - 4| > 2$ (f) $\dfrac{1}{|3x - 1|} > 4$

B.11 Prove that the identity

$$\left(x_1^2 + x_2^2\right) \cdot \left(y_1^2 + y_2^2\right) = (x_1 y_1 + x_2 y_2)^2 + (x_1 y_2 - x_2 y_1)^2$$

holds for $x_1, x_2, y_1, y_2 \in \mathbf{R}$.

B.12 Verify the identity

$$\sum_{k=1}^{n} x_k^2 \cdot \sum_{k=1}^{n} y_k^2 - \left(\sum_{k=1}^{n} x_k y_k\right)^2 = \sum_{k=1}^{n} \sum_{j=k+1}^{n} (x_k y_j - x_j y_k)^2 \tag{B.12}$$

$(x_k, y_k \in \mathbf{R}, (k = 1, \ldots, n))$ with MACSYMA for $n = 2, \ldots, 6$.
Hint: Declare x and y as functions of k by using the statement
c1. `depends([x,y],k)`
d1. $[x(k), y(k)]$
to make them arbitrary functions without predefined value. Write then $x(k)$ and $y(k)$, and $x(j)$ and $y(j)$, respectively. If you are not sure whether you have declared a functional dependence, you may use the function `dependencies` to find out
c2. `dependencies`
d2. $[x(k), y(k)]$
To remove a dependency we use the syntax
c3. `remove(k,dependency)`
d3.   done

B.13 Prove Eq. (B.12) by mathematical induction.

B.14 Prove the *Cauchy–Schwarz inequality*[5]

$$\left(\sum_{k=1}^{n} x_k y_k\right)^2 \le \sum_{k=1}^{n} x_k^2 \cdot \sum_{k=1}^{n} y_k^2, \quad \text{for } x_k, y_k \in \mathbf{R}, \ (k = 1, \ldots, n).$$

B.15 Prove *Bernoulli's inequality*[6]

$$(1 + x)^n \ge 1 + nx, \quad \text{for } n \in \mathbf{N}, \ x > -1.$$

B.16 Show that for all $x, y \in \mathbf{R}$ and $n \in \mathbf{N}_0$
  a. $x^n - y^m = (x - y)(x^{n-1} + x^{n-2}y + \cdots + xy^{n-2} + y^{n-1})$,
  and write the right-hand side with the sum sign. Also show
  b. $x^3 + y^3 = (x + y)(x^2 - xy + y^2)$.

B.5 Two fundamental properties of real numbers

In this section we will give one fundamental property, called the *least-upper-bound property*, of the real number system which will distinguish $\mathbf{R}$ from $\mathbf{Q}$. We will prove that as a consequence of the least-upper-bound property the set $\mathbf{R}$ has another important property, called the *nested-interval property*, which deals with approximations. We will need both properties at different places in this book.

To state the least-upper-bound property, we will have to give some definitions.

Definition B.9 (Upper and lower bounds, boundedness). If we have any set $S \subset \mathbf{R}$ of real numbers and if $c \in \mathbf{R}$ is a real number, then c is called an upper bound for S if the inequality $s \le c$ holds for all $s \in S$; similarly if the inequality $c \le s$ holds for all $s \in S$, then c is called a lower bound for S. If a given set S has an upper bound, then S is called bounded above, and if S has a lower bound, then S is called bounded below. If S is both bounded below and above, then it is called bounded, otherwise unbounded.

Example B.10 (Intervals). For example, if S is an interval $S = (a, b)$, then $b + 1$, e.g., is an upper bound, and $a - 10$, e.g., is a lower bound. So (a, b) is bounded. However, there are infinitely many other upper and lower bounds. In fact, if you succeed to find one upper bound for a set S, then each larger number is an upper bound for S, too. There is, however, one distinguished upper bound in the case of our interval (a, b). This is the number b, which is not only an upper bound but is the smallest of all upper bounds that exist.

Definition B.11 (Supremum and infimum). We say that a number c is the *least upper bound* of a set S if it is an upper bound such that no other upper bound of S exists which is smaller than c. Similarly c is the *greatest lower bound* of S if it is a lower bound such that no other lower bound of S exists which is larger than c.

---

5 Hermann Amandus Schwarz (1843–1921)
6 Jakob I. Bernoulli (1654–1705)

The least upper bound of a set S is also called supremum of S and is designated by sup S, and the greatest lower bound of S is also called infimum of S and is designated by inf S.

Now we are able to state the first fundamental property of the real numbers.

Least-upper-bound property. Every nonempty set of real numbers that is bounded above has a least upper bound.

It is clear that the set of rational numbers $\mathbf{Q}$ does not possess the least-upper-bound property. For example, the set

$$S := \{ x \in \mathbf{Q} \mid x^2 < 2 \}$$

does not have a least upper bound within $\mathbf{Q}$ because otherwise $\sqrt{2}$ would be rational.

Definition B.12 (Shrinking nested intervals). A collection of closed intervals $\{ I_k = [a_k, b_k] \mid k = 1, 2, \ldots \}$ is called
nested if $I_1 \supset I_2 \supset I_3 \supset \cdots$,
shrinking if the interval lengths $|I_k| = b_k - a_k$ tend to zero as k tends to infinity.

Example B.13. Let the intervals $\{ I_k \mid k = 1, 2, \ldots \}$ be defined by $I_k := [l_k, u_k]$, where the endpoints l_k and u_k are as in Example B.5.
Then $\{ I_k \mid k = 1, 2, \ldots \}$ are nested shrinking intervals, with rational endpoints, and the irrational number $\sqrt{2}$ is contained in all I_k.

Another important property of the real numbers is the nested-interval property, which follows from the least-upper-bound property as we will show in Theorem B.15.

Nested-interval property. Any sequence of shrinking nested intervals is an approximation sequence of a real number. In other words, there is a real number which is common to all intervals.

It is clear that $\mathbf{Q}$ does not possess this property as well, as in this case $\sqrt{2}$ would be rational, too. So the nested-interval property of the real numbers is another way of making the difference between $\mathbf{Q}$ and $\mathbf{R}$ precise. It makes $\mathbf{R}$ *complete*, i.e., it guarantees that all convergent sequences – as e.g., decimal representations – represent real numbers.

First we will show that shrinking nested intervals have *at most* one point in common.

Lemma B.14. Any sequence of shrinking nested intervals I_k $(k \in \mathbf{N})$, $I_k \subset \mathbf{R}$, has at most one point $c \in I_k$ $(k \in \mathbf{N})$ in common.

Proof. We assume that there are two different points $c \neq d$ common to all I_k $(k \in \mathbf{N})$. Then

$$e := |c - d| > 0 \,. \tag{B.13}$$

Moreover for all $k \in \mathbf{N}$ the relations

$$a_k \leq c \leq b_k \quad \text{and} \quad a_k \leq d \leq b_k$$

are satisfied. As $[a_k, b_k]$ is shrinking, $b_k - a_k$ tends to 0, and there must be a number $K \in \mathbf{N}$ such that $b_K - a_K \leq e/3$. Then we get the relations

$$c - a_K \leq b_K - a_K \leq e/3 \quad \text{and} \quad d - a_K \leq b_K - a_K \leq e/3 \,,$$

and finally, using the triangle inequality (Lemma B.7 c),

$$
\begin{aligned}
|c - d| &= |c - a_K + a_K - d| = |(c - a_K) + (a_K - d)| \\
&\leq |c - a_K| + |a_K - d| \leq e/3 + e/3 = 2e/3
\end{aligned}
$$

contradicting Eq. (B.13).   □

This is the first time that we proved a result by a chain of inequalities using the triangle inequality. This method is one of the main techniques in calculus.

Now we show that the nested-interval property is a consequence of the least-upper-bound property.

Theorem B.15. The least-upper-bound property implies the nested-interval property.

Proof. Let us assume the least-upper-bound property. We will have to show that a given sequence $[a_k, b_k]$ of shrinking nested intervals possesses a common real point. Lemma B.14 tells that there is at most one such point. We consider the set

$$S := \{ x \in \mathbf{R} \mid x \leq b_k \text{ for all } k \in \mathbf{N} \} \,.$$

As for all $k \in \mathbf{N}$ we have $a_k \leq b_k$, all the numbers a_k lie in S, so S is nonempty. Next we observe that for all $k \in \mathbf{N}$ the number b_k is an upper bound of S by the definition of S, so S is bounded above. By the least-upper-bound property S has a least upper bound c which is a real number. We will show that for all $k \in \mathbf{N}$

$$
\begin{aligned}
c \in [a_k, b_k] \quad &\text{or equivalently} \quad a_k \leq c \leq b_k \\
&\text{or equivalently} \quad a_k \leq c \text{ and } c \leq b_k \,, \tag{B.14}
\end{aligned}
$$

so c is the real number common to all I_k searched for, which proves the nested-interval property.

Since for all $k \in \mathbf{N}$ the number $a_k \in S$ and c is an upper bound of S, the relation

$$a_k \leq c$$

is satisfied for all $k \in \mathbf{N}$, and the first inequality of (B.14) is proved. Because for all $k \in \mathbf{N}$ the number b_k is an upper bound of S and c is the least upper bound of S, the inequality

$$c \le b_k$$

is satisfied for all $k \in \mathbf{N}$ by the definition of the least upper bound, and we are done. □

Exercises

B.17 Does $\mathbf{N}$ have upper or lower bounds? Is $\mathbf{N}$ bounded either above or below?

B.18 Find the least upper and greatest lower bounds of the sets

(a) $\{x \in \mathbf{R} \mid x^2 < 3\}$ (b) $\{x \in \mathbf{R} \mid x^2 > 3\}$ (c) $\{x \in \mathbf{R} \mid x^3 < 2\}$

(d) $\left\{ \dfrac{n-1}{n+1} \,\middle|\, n \in \mathbf{N} \right\}$ (e) $\left\{ \dfrac{(-1)^n}{n} \,\middle|\, n \in \mathbf{N} \right\}$ (f) $\left\{ \dfrac{1}{1+x^2} \,\middle|\, x \in \mathbf{Q} \right\}$

B.6 Complex numbers

Some quadratic equations such as $4x^2 = 9$ have rational solutions. Other quadratic equations, e.g., $x^2 = 2$, have real solutions (indeed the real-number system was invented in order to solve such equations). Still other quadratic equations, such as

$$x^2 = -1 , \tag{B.15}$$

have no real solution. In order to handle such equations, we must enlarge the real-number system to include solutions of (B.15).

We designate the solution of (B.15) by

$$i := \sqrt{-1} \tag{B.16}$$

called the *imaginary unit*. By its definition, i satisfies

$$i^2 = -1 . \tag{B.17}$$

A *complex number* is a number of the form

$$z := x + iy \quad (x, y \text{ real}) , \tag{B.18}$$

where x is called the *real part* of z, designated $x = \operatorname{Re} z$, and y is its *imaginary part*, designated $y = \operatorname{Im} z$. The set of complex numbers is designated

$$\mathbf{C} := \{z = x + iy \mid x, y \in \mathbf{R}\} . \tag{B.19}$$

We see that the real numbers are special complex numbers, with the imaginary part equating 0.

With (B.17) in mind, the arithmetic operations on complex numbers are quite straightforward. Let $z = x + iy$ and $w = u + iv$ be arbitrary complex numbers. Their *sum* is

$$
\begin{aligned}
z + w &= (x + iy) + (u + iv) \\
&= (x + u) + i(y + v)
\end{aligned}
\tag{B.20}
$$

and *product*:

$$zw = (x + iy)(u + iv) = xu + xiv + iyu + i^2yv$$
$$= (xu - yv) + i(xv + yu) . \qquad \text{(B.21)}$$

The number $\bar{z} := x - iy$ is called the *conjugate* of z. The following results are easy to verify:

$$\frac{1}{2}(z + \bar{z}) = \frac{1}{2}((x + iy) + (x - iy)) = x = \operatorname{Re} z, \qquad \text{(B.22a)}$$

$$\frac{1}{2i}(z - \bar{z}) = \frac{1}{2i}((x + iy) - (x - iy)) = y = \operatorname{Im} z, \qquad \text{(B.22b)}$$

$$z\bar{z} = (x + iy)(x - iy) = x^2 + y^2, \qquad \text{(B.22c)}$$

$$\overline{zw} = \bar{z}\,\bar{w} . \qquad \text{(B.22d)}$$

The last identity can be stated: the conjugate of a product is the product of conjugates.

Also, the *reciprocal* of a complex number $z := x + iy$ becomes

$$\frac{1}{z} := \frac{1}{x + iy} = \frac{x - iy}{(x + iy)(x - iy)} = \frac{x}{x^2 + y^2} - i\frac{y}{x^2 + y^2} = \frac{\bar{z}}{z\bar{z}}, \qquad \text{(B.23)}$$

where we used (B.22c) in the last step.

We can identify $\mathbf{C}$ with an xy-plane such that the x-axis represents the real part and the y-axis represents the imaginary part of the complex number $z = x + iy$. This representation for the complex numbers is called the *Gauss*[7] *number plane*. The real unit is represented by the point $(1, 0)$, and the complex unit i is represented by the point $(0, 1)$. The x-axis is called the *real* axis and the y-axis is called the *imaginary* axis.

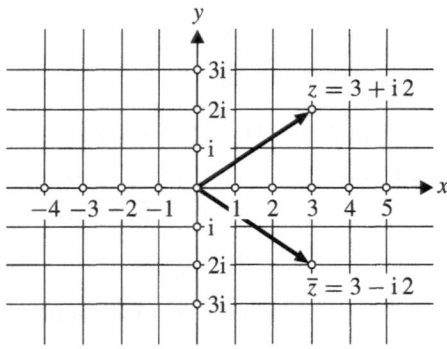

Fig. B.3. Gauss number plane

---

7 Karl Friedrich Gauss (1777–1855).

For a complex number $z = x + iy$ its *absolute value* is the distance of z from the origin, i.e., by the *Pythagoras theorem*

$$|z| = |x + iy| := \sqrt{x^2 + y^2} \, , \tag{B.24}$$

or alternatively, by (B.22c),

$$|z|^2 := z\bar{z} \, . \tag{B.25}$$

Also, for any two complex numbers, z and w,

$$|zw| = |z|\,|w| \, . \tag{B.26}$$

MACSYMA-Session B.3. The imaginary unit $\sqrt{-1}$ is designated %i
c1. z:x+%i*y
d1. $iy + x$
conjugate(expr) returns the conjugate value of its argument.
c2. conjugate(z)
d2. $x - iy$
cabs(expr) returns the complex absolute value, or *modulus* of the argument
c3. cabs(z)
d3. $\sqrt{y^2 + x^2}$
c4. cabs(1+\%i)
d4. $\sqrt{2}$
You can also use the real-absolute-value function abs(z), which will call cabs.
rectform(expr) returns an expression of the form $a + bi$, where a and b are real.
c5. rectform((2*%i+x)^3)
d5. $x^3 + i(6x^2 - 8) - 12x$
polarform(expr) returns $re^{i\theta}$, where r and θ are real.
c6. polarform(%)
d6. $\sqrt{(x^3 - 12x)^2 + (6x^2 - 8)^2} \, e^{i \, \text{atan}((6x^2-8)/(x^3-12x))}$
Notice that we return to the original expression with the next operation.
c7. rectform(%)
d7. $x^3 + i(6x^2 - 8) - 12x$
realpart(expr) and imagpart(expr) return the real part and the imaginary part, respectively, of the expression expr.
c8. realpart(%)
d8. $x^3 - 12x$
c9. imagpart(d6)
d9. $6x^2 - 8$

To check (B.25), we enter
c10. cabs(z)-sqrt(z*conjugate(z))
d11. $\sqrt{y^2 + x^2} - \sqrt{(x - iy)(iy + x)}$
and simplify:
c11. ratsimp(%)
d11.   0
We next check (B.26), written as $|zw|^2 = |z|^2|w|^2$:

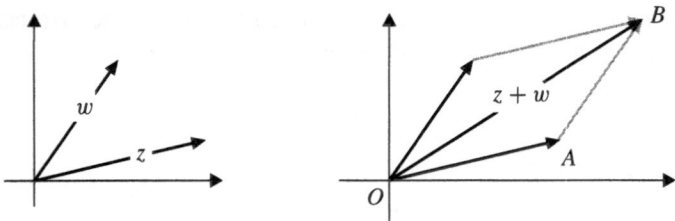

Fig. B.4. Illustration of the triangle inequality

```
c12. cabs(z*w)^2-cabs(z)^2*cabs(w)^2
```
d12. $-(v^2 + u^2)(y^2 + x^2) + (ux - vy)^2 + (uy + vx)^2$
```
c13. ratsimp(%)
```
d13.   0

Exercises

B.19 Let $z := x + iy$ and $w := u + iv$. Calculate z/w by hand.

B.20 Let $z := x + iy$. Calculate $\mathrm{Re}(1 + z)/(1 - z)$.

B.21 Prove the triangle inequality, i.e.,

$$|z + w| \leq |z| + |w| . \tag{B.27}$$

    a. Consider the triangle OAB in the Gauss plane (Fig. B.4). The lengths of its sides are $|z|$, $|w|$, and $|z + w|$. Conclude that the triangle inequality says: In a triangle, the length of any side is not greater than the sum of lengths of the remaining two sides.

    b. When does equality occur in the triangle inequality?
    Hint: Draw a triangle as above with $|z + w| = |z| + |w|$.

B.22 Prove that for all complex numbers z

    (a) $|z| = |\bar{z}|$ (b) $\left|\dfrac{1}{z}\right| = \dfrac{1}{|z|}, \quad z \neq 0$

    (c) $|z^n| = |z|^n$ for all $n \in \mathbf{Z}$

B.23 The unit circle is the set of complex numbers whose absolute value is 1,

$$U = \{ z \mid |z|^2 = 1 \} = \{ x + iy \mid x^2 + y^2 = 1 \} . \tag{B.28}$$

Find all solutions of the following equations:

    (a) $z^2 = 1$ (b) $z^3 = 1$ (c) $z^4 = 1$ (d) $z^5 = 1$

and show that they lie on the unit circle.

B.24 Find the curves, in the Gauss plane, represented by the equations:

    (a) $|z - 1| = |z + 1|$ (b) $|z - 1| = \frac{1}{2}|z + 1|$ (c) $|z - 1| = 2|z + 1|$

C Analytical geometry

C.1 Plane $\mathbf{R}^2$ with Cartesian coordinates

Real line $\mathbf{R}$

Real numbers correspond to points on a line, the real line $\mathbf{R}$. This requires turning the line into a *coordinate axis*, by selecting a *coordinate system*: an *origin*, a *direction*, and a *scale* (i.e., unit of measuring distance on the line). Every point on the line is now represented by its *coordinate*, a real number describing its position relative to the origin.

The coordinate of the origin is $x = 0$. A point with nonzero coordinate x_0 can be reached from the origin by a *move* of x_0 units *along the axis*, meaning a move of x_0 units in the direction of the axis if $x_0 > 0$,
$|x_0|$ units in the opposite direction if $x_0 < 0$.
Therefore the coordinate x_0 of a point is its *address*, giving precise instructions how to reach the point from the origin O.

Real plane $\mathbf{R}^2$

In the same way, points in the plane can be represented by pairs of real numbers, or *plane coordinates*. This requires selecting a coordinate system in the plane. The simplest such system is the *Cartesian coordinate system*[1]. It uses an (arbitrarily selected) *origin* O and two *coordinate axes*, perpendicular to each other, which intersect at O,

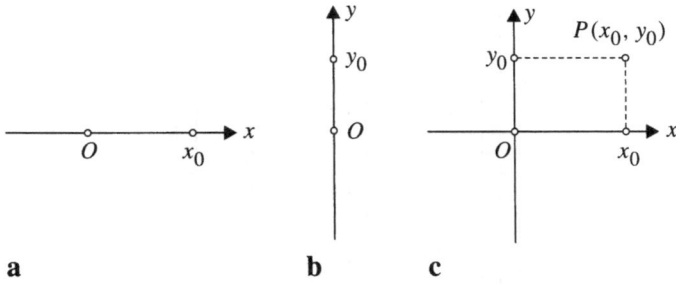

Fig. C.1 a–c. Cartesian coordinates in the plane. **a** x-axis, **b** y-axis, **c** (x, y)-plane

---

1 René Descartes (1596–1650)

a horizontal axis with direction to the right, called the *x-axis* (Fig. C.1 a), and
a vertical axis with direction up, called the *y-axis* (Fig. C.1 b).
Both coordinate axes have their origins at the point O (Fig. C.1 c).

Any point in the plane is now represented by its *Cartesian coordinates* (x, y).
The origin O has coordinates $(0, 0)$. We denote by $P(x_0, y_0)$ the fact that the
point P has coordinates (x_0, y_0). The numbers x_0 and y_0 are called the x- and
y-coordinates of P. The point $P(x_0, y_0)$ is reached from the origin by moving
x_0 units along the x-axis and
y_0 units along the y-axis (Fig. C.1 c).
The plane can therefore be identified with the set of all pairs (x, y) of real numbers,

$$\mathbf{R}^2 := \{ (x, y) \mid x, y \in \mathbf{R} \}, \qquad (C.1)$$

which is called the real plane.

Plotting in Cartesian coordinates

Any curve in the plane consists of points with coordinates (x, y). If we know a
relation, or *function*, between x and y, say,

$$y = f(x), \quad \text{for } a \le x \le b, \qquad (C.2)$$

then we can plot the curve, by computing enough points $(x_k, f(x_k))$, with $a \le x_k \le b$. The resulting curve is the *graph* of the function f for $a \le x \le b$. We
illustrate this in Fig. C.2, plotting the graph

$$y = x^2/10 - 4, \quad \text{for } -3 \le x \le 11.$$

For example, the point on the graph corresponding to $x = 3$ is $(3, -3.1)$ since
$f(3) = 3^2/10 - 4 = -3.1$.

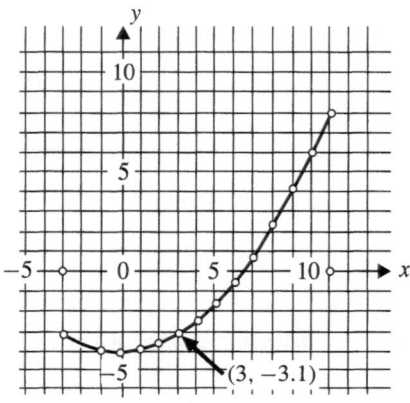

Fig. C.2. Graph of $y = x^2/10 - 4$, for $-3 \le x \le 11$

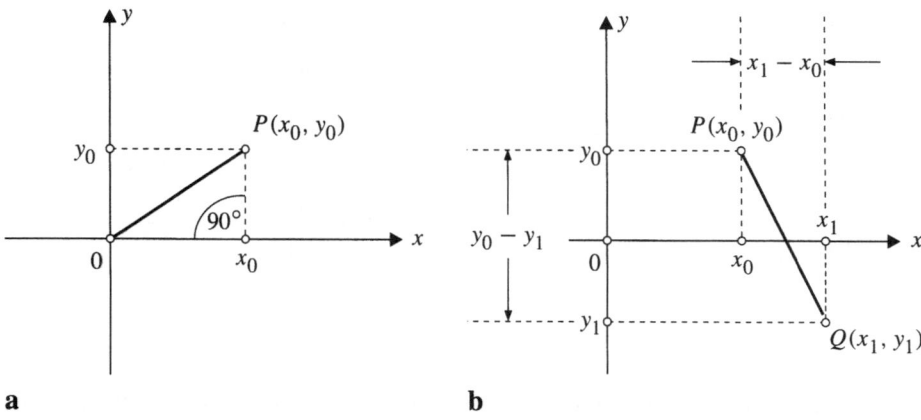

Fig. C.3 a, b. Illustration of distances in the plane. **a** Distance between 0 and P; **b** distance between P and Q

Distance

We define the *distance* between any two points P and Q in $\mathbf{R}^2$, denoted by $dist(P, Q)$, as the "aerial distance" between them, also called the *Euclidean distance*[2] between P and Q.[3]

First consider the distance $dist(O, P)$ between the origin O and a point $P(x_0, y_0)$. This distance is equal to the length $|OP|$ of the line segment OP connecting O and P (Fig. C.3 a). By Pythagoras's theorem[4],

$$|OP|^2 = x_0^2 + y_0^2, \quad \text{see Fig. C.3 a},$$

and therefore,

$$dist(O, P) = \sqrt{x_0^2 + y_0^2}. \tag{C.3}$$

Similarly, the distance $dist(P, Q)$ between two points $P(x_0, y_0)$ and $Q(x_1, y_1)$, is the length $|PQ|$ of the line segment PQ. By Pythagoras's theorem,

$$|PQ|^2 = (x_0 - x_1)^2 + (y_0 - y_1)^2, \quad \text{see Fig. C.3 b},$$

and therefore,

$$dist(P, Q) = \sqrt{(x_0 - x_1)^2 + (y_0 - y_1)^2}. \tag{C.4}$$

Translation of coordinate axes

A Cartesian coordinate system in $\mathbf{R}^2$ consists of an origin O, a horizontal x-axis, a vertical y-axis, their directions and scale.

---

2 Euclid (about 365–300 B.C.).
3 There are other ways, no less natural, to define distance in $\mathbf{R}^2$, see, e.g., Exercise C.1.
4 Pythagoras (about 570–497 B.C.).

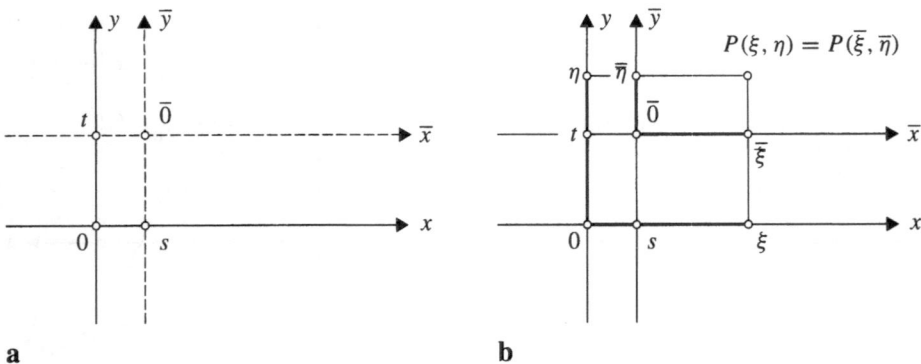

Fig. C.4 a, b. Translation of coordinates. **a** Translation of origin from 0 to $\bar{0}$, with coordinate axes $(\bar{x}, \bar{y})$ parallel to the old ones. **b** The coordinates of the point P in the two coordinate systems

If we change the origin of the coordinate system to another point, say, $\bar{O}$, and keep the directions and scale of coordinate axes as before, we get a *translated coordinate system*. We denote the new coordinate axes by $\bar{x}$ and $\bar{y}$ (Fig. C.4a).

Let the (x, y)-coordinates of the new origin $\bar{O}$ be (s, t), and let P be any point in $\mathbf{R}^2$ whose (x, y)-coordinates are (ξ, η). The $(\bar{x}, \bar{y})$-coordinates of the same point P are

$$\bar{\xi} = \xi - s, \quad \bar{\eta} = \eta - t, \quad \text{(see Fig. C.4b)} .$$

Therefore, the translated coordinates $(\bar{x}, \bar{y})$ are obtained from the (x, y)-coordinates according to the rule

$$\bar{x} = x - s, \quad \bar{y} = y - t . \tag{C.5}$$

In particular, the (x, y)-coordinates of the new origin $\bar{O}$ are (s, t), and its $(\bar{x}, \bar{y})$-coordinates are $(0, 0)$, in agreement with (C.5).

The distance between any two points P and Q in $\mathbf{R}^2$ does not change if the coordinate axes are translated. Indeed let $P(x_0, y_0)$ and $Q(x_1, y_1)$ be two such points and let their translated coordinates be $P(\bar{x}_0, \bar{y}_0)$ and $Q(\bar{x}_1, \bar{y}_1)$, respectively, where

$$\bar{x}_0 = x_0 - s, \quad \bar{y}_0 = y_0 - t, \quad \bar{x}_1 = x_1 - s, \quad \bar{y}_1 = y_1 - t .$$

Then

$$\sqrt{(\bar{x}_0 - \bar{x}_1)^2 + (\bar{y}_0 - \bar{y}_1)^2} = \sqrt{\left((x_0 - s) - (x_1 - s)\right)^2 + \left((y_0 - t) - (y_1 - t)\right)^2}$$

$$= \sqrt{(x_0 - x_1)^2 + (y_0 - y_1)^2} ,$$

showing that $dist(P, Q)$ does not depend on translations of the coordinate axes.

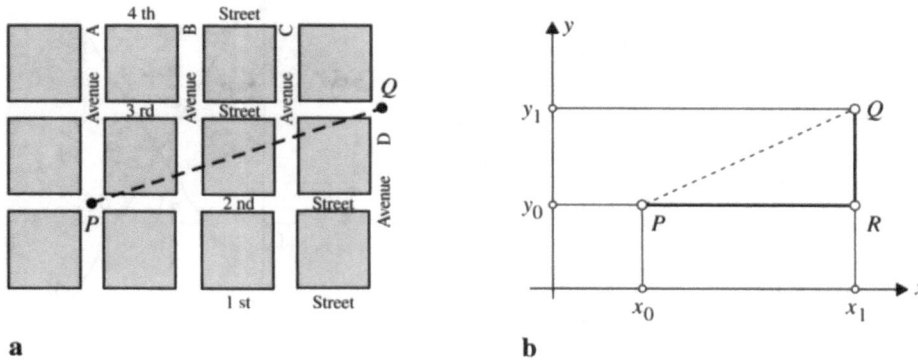

Fig. C.5 a, b. Distances in a city. **a** The aerial distance between two points is mainly for the birds. **b** The Manhattan distance between P and Q

Exercise

C.1 The map in Fig. C.5 a shows several city blocks with streets running from east to west and avenues from north to south. The distance, by car, from the point P (intersection of 2nd Street and Avenue A) to the point Q (3rd Street and Avenue D) is not the same as the aerial distance between them. The Manhattan distance[5] between any two points $P(x_0, x_1)$ and $Q(x_1, y_1)$ in $\mathbf{R}^2$ is defined as

$$dist_{\mathrm{M}}(P, Q) := |x_1 - x_0| + |y_1 - y_0| . \tag{C.6}$$

We assume that all streets and avenues allow traffic in both directions, i.e., no one-way streets or avenues. Then the Manhattan distance is the driving distance between any two points P and Q which are reachable by car, i.e., points in streets or avenues, not inside city blocks. Show:
a. For any two points P and Q, the Manhattan distance $dist_{\mathrm{M}}(P, Q) \geq$ the aerial $dist(P, Q)$.
b. There are points P and Q for which the two distances are equal. Give details.
c. For any two points P and Q, $dist_{\mathrm{M}}(P, Q) \leq \sqrt{2}\, dist(P, Q)$.

C.2 Lines

Through any two (distinct) points $P(x_0, y_0)$ and $Q(x_1, y_1)$ there passes a unique line, denoted by line(P, Q).

Slope

Let $x_0 \neq x_1$. As we move along the line from $P(x_0, y_0)$ to $Q(x_1, y_1)$, the y-coordinate changes by $\Delta y := y_1 - y_0$ and the x-coordinate by $\Delta x := x_1 - x_0$.

---

5 The borough of Manhattan, New York City, has streets and avenues as in Fig. C.5 a.

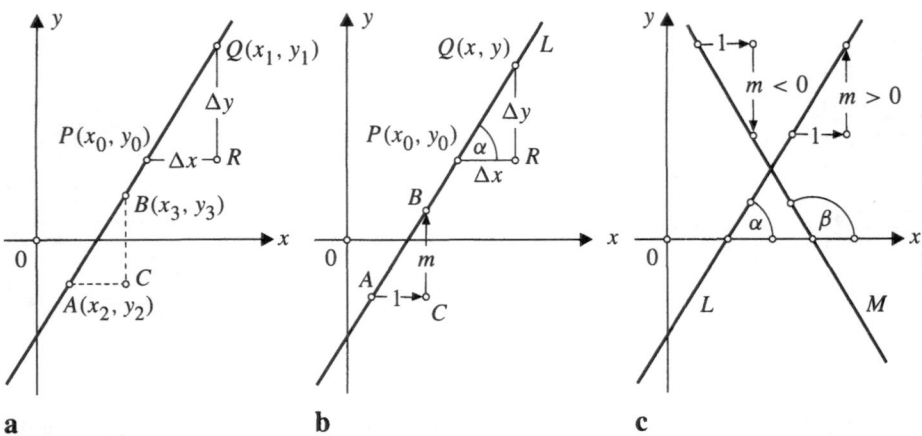

Fig. C.6 a–c. a Line through P and Q; **b** slope m of the line L; **c** positive and negative slopes

The ratio $\Delta y / \Delta x$ is called the *slope* of the line, and denoted by m,

$$m := \frac{\Delta y}{\Delta x} = \frac{y_1 - y_0}{x_1 - x_0} \, . \tag{C.7}$$

This is illustrated in Fig. C.6a, where Δx is positive (since $x_1 > x_0$) and Δy is positive ($y_1 < y_0$). Therefore the slope $m > 0$.

The slope does not depend on the particular two points P and Q used in (C.7). To see this, consider any two points on line(P, Q), say, $A(x_2, y_2)$ and $B(x_3, y_3)$ (Fig. C.6a). As we move from A to B, the coordinate changes are $\Delta y = y_3 - y_2$ and $\Delta x = x_3 - x_2$. Their ratio gives the same slope m as in (C.7):

$$\frac{y_3 - y_2}{x_3 - x_2} = \frac{y_1 - y_0}{x_1 - x_0} \, .$$

This equality follows since the triangles $\triangle(PRQ)$ and $\triangle(ACB)$ are similar, and therefore corresponding sides have the same ratios. If we move from B to A, then both Δy and Δx are negative, but their ratio is still m.

We can illustrate the slope as follows: In order to stay on a line with slope m, we must move m units along the y-axis for each unit along the x-axis. This is shown in Fig. C.6b. Moving from the point A one unit to the right and m units along the y-axis (up, since m is positive) takes us to the point B, also on the line.

Another interpretation of the slope is

$$m = \tan \alpha \, , \tag{C.8}$$

where α is the angle between the line L and a horizontal line, see the triangle $\triangle(PRQ)$ in Fig. C.6b.

A line making an acute angle α with the x-axis, such as the line L in Fig. C.6c, has a positive slope since $\tan \alpha > 0$ for $0 < \alpha < \pi/2$.

The line M in Fig. C.6c makes an obtuse angle β with the x-axis. Its slope is negative since $\tan \beta < 0$ for $\pi/2 < \beta < \pi$.

Horizontal and vertical lines

In particular, line$(P(x_0, y_0), Q(x_1, y_1))$ is
horizontal if $y_1 = y_0$, see Fig. C.7 a,
vertical if $x_1 = x_0$, see Fig. C.7 b.

The slope of a horizontal line is $m = 0$. This is shown by using $y_0 = y_1$ in (C.7). Since y is constant along a horizontal line, the equation of a horizontal line is

$$y = b, \quad \text{where } b \text{ is some constant}, \tag{C.9}$$

for example, the equation of the line in Fig. C.7 a is $y = y_0$. In particular, the x-axis is given by $y = 0$.

Similarly, x is constant along a vertical line. Therefore the equation of a vertical line is

$$x = a, \quad \text{where } a \text{ is some constant}, \tag{C.10}$$

for example, the equation of the line in Fig. C.7 b is $x = x_0$; the equation of the y-axis is $x = 0$.

The slope of a vertical line is undefined. We cannot determine it by substitut-

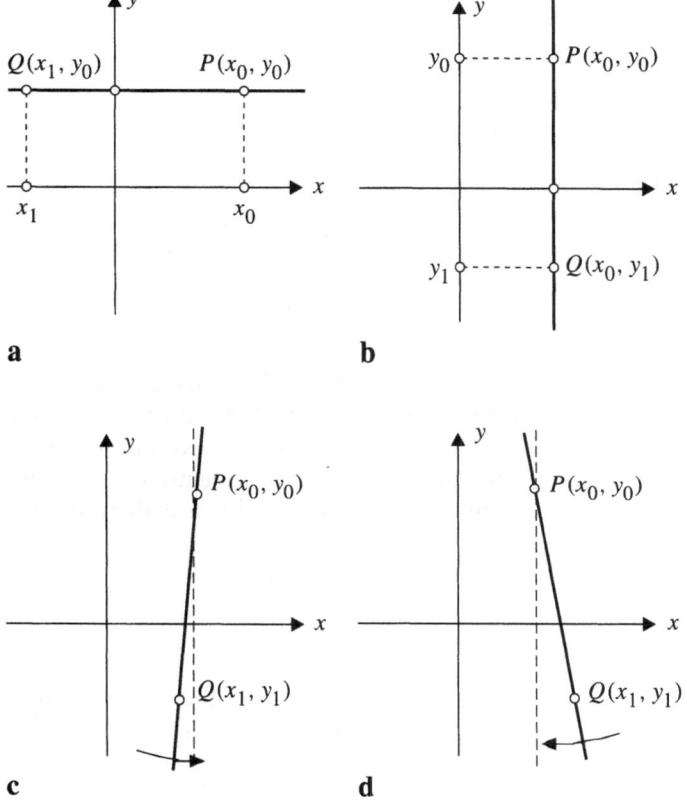

Fig. C.7 a–d. a Horizontal line, **b** vertical line, **c** "almost vertical" line with positive slope, **d** "almost vertical" line with negative slope

ing $x_0 = x_1$ in (C.7) since the denominator vanishes. Consider now the "almost vertical" line in Fig. C.7c. As we move from Q to P, $\Delta x = x_0 - x_1$ is positive and very small; $\Delta y = y_0 - y_1$ positive. Therefore the slope $m = \Delta y / \Delta x$ is positive and large. Suppose we make the line "closer" to vertical, by keeping the point P fixed and rotating the line counterclockwise around P. We see that the slope of the line increases, and in the limit, as it approaches the vertical line (the broken line in Fig. C.7c), its slope increases to $+\infty$.

Similarly, the "almost vertical" line in Fig. C.7d has negative slope, and the slope decreases to $-\infty$ as the line approaches the vertical line, by rotating it around P.

Point-slope equation of a line

When it comes to write the equation of a line, it is convenient to do so in terms of its slope and of its points.

Consider the line with slope m, passing through the point $P(x_0, y_0)$. Let $R(x, y)$ be a general point on the line (Fig. C.6c). Calculating the slope from the two points P and R we get

$$m = \frac{y - y_0}{x - x_0}, \qquad (C.11)$$

called the point-slope equation of the line.

Slope-intercept equation of a line

Solving Eq. (C.11) for y, we get

$$y = mx + (y_0 - mx_0),$$

or

$$y = mx + b, \qquad (C.12)$$

where $b = y_0 - mx_0$.

Substituting $x := 0$ in (C.12), we get $y = b$, showing that b is the coordinate of the point where the line intersects the y-axis (Fig. C.6c). The number b is called the y-*intercept* of the line. The y-intercept is undefined for a vertical line.[6]

Equation (C.12) is called the slope-intercept equation (or the slope-intercept form) of the (nonvertical) line. Every nonvertical line is therefore the graph of a function

$$y = mx + b, \quad m, b \in \mathbf{R},$$

called a *linear function*.

MACSYMA-Session C.1. The slope and y-intercept of a line given by two points $P_1 = (x_1, y_1)$ and $P_2 = (x_2, y_2)$ can be calculated by solving the two equations

$$y_1 = mx_1 + b \quad \text{and} \quad y_2 = mx_2 + b,$$

---

6 A vertical line is either the y-axis (in which case there are "infinitely many" intercepts) or a parallel to it (no intercept).

for m and b. We enter these equations
c1. `eq1:y1=m*x1+b`
d1. $\quad y_1 = mx_1 + b$
c2. `eq2:y2=m*x2+b`
d2. $\quad y_2 = mx_2 + b$
and solve
c3. `sol:solve([eq1,eq2],[m,b])`
d3. $\quad \left[m = \dfrac{y_1 - y_2}{x_1 - x_2}, \; b = -\dfrac{x_2 y_1 - x_1 y_2}{x_1 - x_2} \right]$

For example, to find the slope and y-intercept of the line $\text{line}(P_1(3, 1), P_2(1, -1))$, use
c4. `solve([m*3+b=1, m*1+b=-1], [m,b])`
d4. $\quad [m = 1, b = -2]$

Alternatively, we can substitute the values `x1=3,y1=1,x2=1,y2=-1` in `sol` (computed in d3),
c5. `sol,x1=3,y1=1,x2=1,y2=-1`
d5. $\quad [m = 1, b = -2]$

Repeat this computation for a vertical line, i.e., a line defined by two points $P_1(x_1, y_1)$ and $P_2(x_2, y_2)$ with $x_1 = x_2$, $y_1 \neq y_2$.

The equation (slope-intercept form) of the line passing through the points $P_1 = (x_1, y_1)$ and $P_2 = (x_2, y_2)$ is given by the following MACSYMA function:
c6. `line(x,x1,y1,x2,y2) := (y1-y2)/(x1-x2)*x+(x1*y2-x2*y1)/(x1-x2)`
d6. $\quad \text{line}(x, x_1, y_1, x_2, y_2) := \dfrac{(y_1 - y_2)x}{x_1 - x_2} + \dfrac{x_1 y_2 - x_2 y_1}{x_1 - x_2}$

which computes $mx + b$ using the values from `sol`. For example, the line through $P_1 = (3, 1)$ and $P_2 = (1, -1)$ is found by
c7. `line(x,3,1,1,-1)`
d7. $\quad x - 2$

in agreement with d4.

The function `line` in c6 cannot handle vertical lines, i.e., the case $x_1 = x_2$. However, it can be easily modified to work for all lines:
c8. `line(x,y,x1,y1,x2,y2) :=`
```
                if x1=x2 then x=x1
                else y=(y1-y2)/(x1-x2)*x+(x1*y2-x2*y1)/(x1-x2)$
```
For example, the line connecting $P_1(2, 1)$ and $P_2(2, 5)$ is vertical:
c9. `line(x,y,2,1,2,5)`
d9. $\quad x = 2$

and the line of c7 is:
c10. `line(x,y,3,1,1,-1)`
d10. $\quad y = x - 2$

General equation of a line

Any equation of the form

$$Ax + By + C = 0, \quad \text{with } A, B, C \in \mathbf{R}, \; A^2 + B^2 \neq 0, \qquad \text{(C.13)}$$

represents a line. We call (C.13) the general equation of the line.

We assume that $A^2 + B^2 \neq 0$, i.e., that at least one of the coefficients A and B

in (C.13) is nonzero. Otherwise Eq. (C.13) does not represent a line. To see this, suppose $A = B = 0$. If also $C = 0$, then (C.13) becomes $0 = 0$, which is satisfied by all points (x, y) and therefore represents all of $\mathbf{R}^2$. If $C \neq 0$, we get $C = 0$, a contradiction.

If the coefficient $B = 0$, then (C.13) represents the vertical line

$$x = -\frac{C}{A} \, .$$

If $A = 0$, we get the horizontal line

$$y = -\frac{C}{B} \, .$$

If $B \neq 0$, we can solve Eq. (C.13) for y,

$$y = -\frac{A}{B}x - \frac{C}{B} \, ,$$

showing that the slope is $m = -A/B$ and the y-intercept is $-C/B$.

Normal equation of a line

Consider a line which is neither horizontal nor vertical. It intersects the x-axis at a point $(a, 0)$, and the y-axis at $(0, b)$ (e.g., Fig. C.6c). It is possible to write the equation of the line in terms of its x-intercept a and its y-intercept b. This equation is

$$\frac{x}{a} + \frac{y}{b} = 1 \, , \tag{C.14}$$

called the normal equation of the line. To verify (C.14), substitute $y := 0$ and then $x := 0$ to show that the points $(a, 0)$ and $(0, b)$ satisfy the equation.

To obtain the normal equation (C.14) from the general equation of a line (C.13),

$$Ax + By + C = 0 \, ,$$

divide both sides by $-C$ to get

$$\frac{x}{-C/A} + \frac{y}{-C/B} = 1 \, ,$$

the normal form of the line equation, with x-intercept $a = -C/A$ and y-intercept $b = -C/B$.

Parallel lines

Two lines are called *parallel* if they have the same slope or if both are vertical. For example, the three lines L, M, and N in Fig. C.8a are parallel. Let m denote the common slope of these lines. Then their equations are:

Line L : $y = mx + b_1$,
Line M: $y = mx + b_2$,
Line N : $y = mx$ (the line N passes through the origin) .

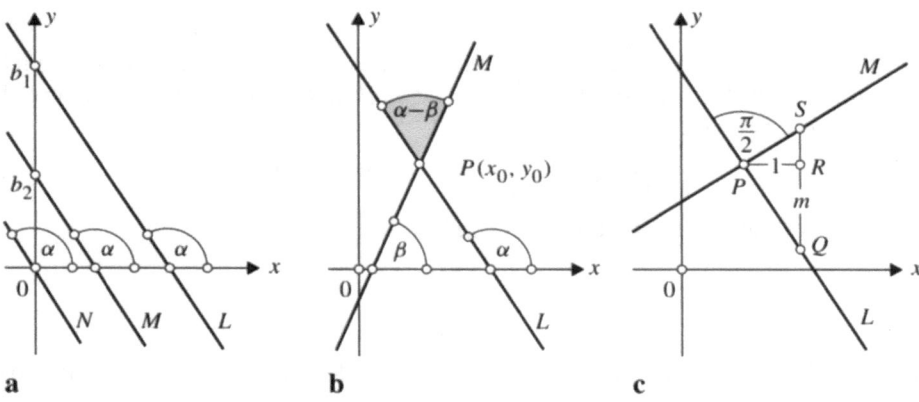

Fig. C.8 a–c. a Parallel lines, **b** two lines intersecting at a point and the angle between them, **c** perpendicular lines

Intersection

Any two nonparallel lines

$$y = m_1 x + b_1, \tag{C.15a}$$
$$y = m_2 x + b_2 \tag{C.15b}$$

intersect at a single point (e.g., Fig. C.8b). The coordinates (x_0, y_0) of the intersection point are found by solving Eqs. (C.15a) and (C.15b) for x and y.

MACSYMA-Session C.2. We use MACSYMA to compute the intersection of the lines (C.15a) and (C.15b).

```
c1. eq1:y=m1*x+b1
```
d1. $y = m_1 x + b_1$
```
c2. eq2:y=m2*x+b2
```
d2. $y = m_2 x + b_2$
```
c3. solve([eq1,eq2],[x,y])
```
d3. $\left[x = -\dfrac{b_1 - b_2}{m_1 - m_2}, \; y = \dfrac{b_2 m_1 - b_1 m_2}{m_1 - m_2} \right]$

What happens if $m_1 = m_2$, i.e., if the lines in question are parallel? The following MACSYMA function handles all possible cases (verify!)

```
c4. intersection_point(x,y,m1,m2,b1,b2):=
            if m1=m2 then if b1=b2 then "Lines coincide"
                        else "Parallel lines, do not intersect"
            else [x=(b2-b1)/(m1-m2), y=(b2*m1-b1*m2)/(m1-m2)]$
```

Examples:
```
c5. intersection_point(x,y,1,1,2,2)
```
d5.   Lines coincide
```
c6. intersection_point(x,y,1,1,2,3)
```
d6.   Parallel lines, do not intersect
```
c7. intersection_point(x,y,1,2,3,4)
```
d7. $[x = -1, y = 2]$

Angle between lines

Let the lines L and M have angles α and β, respectively, with the x-axis and let $\alpha > \beta$. Then the lines intersect at a point. The *angle between the two lines*, denoted by $\angle\{L, M\}$, is defined as

$$\angle\{L, M\} := \alpha - \beta, \quad \text{see Fig. C.8b} . \tag{C.16}$$

If the lines L and M are parallel, $\angle\{L, M\} := 0$.

Perpendicular lines

If $\angle\{L, M\} = \pi/2$, then the lines L and M are called perpendicular (or orthogonal) to each other (Fig. C.8c). The slopes of perpendicular lines are connected as follows.

Proposition C.1. Let

$$\begin{aligned} \text{Line } L : \quad & y = mx + b , \\ \text{Line } M : \quad & y = m_1 x + b_1 \end{aligned}$$

be any two lines with the slope m being finite and nonzero.[7] Then these lines are perpendicular if, and only if,

$$mm_1 = -1 . \tag{C.17}$$

We give a geometric explanation rather than a formal proof. Let P be the point where L and M intersect (Fig. C.8c). We move $\Delta x = 1$ unit along the x-axis to the point R. We now must move

m units along the y-axis in order to return to L, or
m_1 units along the y-axis in order to return to M.

Therefore m and m_1 have opposite signs,

$$mm_1 < 0 . \tag{C.18}$$

The absolute value of the slope m is the length of the segment $|RQ|$ in Fig. C.8c. Similarly, the absolute value of m_1 is the length $|RS|$. Therefore

$$\begin{aligned} |m| = \frac{|m|}{1} &= \frac{|RQ|}{|PR|} , \\ &= \frac{|PR|}{|RS|}, \quad \text{since the triangles } \triangle(PRQ) \text{ and } \triangle(PRS) \text{ are similar} , \\ &= \frac{1}{|m_1|} , \end{aligned}$$

---

7 That is, the line L is neither horizontal nor vertical.

which, together with (C.18), shows that

$$m_1 = -\frac{1}{m} .$$

Orthogonal projection

Given a point P and a line L, we construct the perpendicular line M through P. The point Q where the lines L and M intersect is called the *orthogonal projection* of P on L (Fig. C.9a). Note that $Q = P$ if P is on the line L.

Given the coordinates $P(x_0, y_0)$ and the slope-intercept form of the line L (C.12),

$$y = mx + b ,$$

we now calculate the coordinates $Q(x_1, y_1)$ of the orthogonal projection of P on L. The perpendicular line M has the equation

$$\frac{y - y_0}{x - x_0} = -\frac{1}{m} , \tag{C.19}$$

and solving the two equations (C.12) and (C.19) for x and y we get

$$x_1 = -\frac{mb - my_0 - x_0}{m^2 + 1}, \quad y_1 = \frac{b + m(my_0 + x_0)}{m^2 + 1} . \tag{C.20}$$

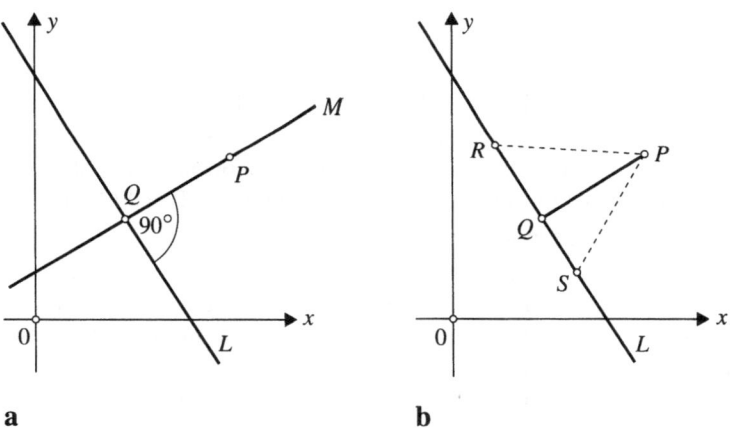

a   b

Fig. C.9 a, b. a Orthogonal projection of the point P on the line L. **b** Point Q is the closest point to P on the line L

Distance of point to line

Let Q be the orthogonal projection of P on a line L and consider any other point R on the line (Fig. C.9b). Then $dist(P, Q) < dist(P, R)$, since the triangle $\triangle(PQR)$ is a right triangle and $|PR|$ is its hypotenuse.

We conclude that of all points in L, the point Q is the closest one to P. The distance between the point P and the line L is therefore defined as

$$dist(P, L) := dist(P, Q)$$

$$= \sqrt{(x_0 - x_1)^2 + (y_0 - y_1)^2}$$

$$= \sqrt{\frac{(y_0 - mx_0 - b)^2}{m^2 + 1}}, \qquad \text{(C.21)}$$

where we substituted x_1 and y_1 from (C.20).

Exercises

C.2 Let the origin O of the coordinate system be translated to $\bar{O}(4, -1)$, so that the translated coordinates are, by (C.5), $\bar{x} = x - 4$ and $\bar{y} = y + 1$. Consider the line whose general equation, in the original coordinate system, is $2x - 3y + 5 = 0$.
   a. Find the general equation in the translated coordinate system.
   b. Find the slope-intercept equation, and normal equation, in both coordinate systems.
C.3 Verify analytically the following expressions, obtained by MACSYMA.
   (a) The intercept b in MACSYMA-Session C.1, expression d3
   (b) The solution (x, y) in MACSYMA-Session C.2, expression d3
C.4 Verify:
   (a) solution (C.20)   (b) Eq. (C.21)
C.5 Calculate the distance $dist(P, L)$ between the points P and the lines L as given below.
   (a) $P = (0, 0)$, $\quad L := \{(x, y) \in \mathbf{R}^2 \mid y = mx + b\}$
   (b) $P = (2, -3)$, $\quad L := \{(x, y) \in \mathbf{R}^2 \mid 2x - 5y + 3 = 0\}$
C.6 Give the equations of the lines determined by the following points, slopes, etc.
   (a) Points $(2, 3)$ and $(3, -1)$ (b) Points $(2, 3)$ and $(2, 4)$
   (c) Point $(2, 3)$, slope $m = -2$
   (d) Point $(-1, 2)$, parallel to $2x + 3y - 1 = 0$
C.7 Write a MACSYMA function `slopeline(m,x1,y1)` that produces the equation of the line through (x_1, y_1) which has slope m. Test the function. Then use it to plot the parallels of slope 2 through the points $(0, k)$, $k = -4, -3, \ldots, 4$, and the lines perpendicular to them which pass through the points $(k, 0)$, $k = -4, -3, \ldots, 4$.

C.3 Circles

Given a point C in the plane $\mathbf{R}^2$ and a number $r \geq 0$, the set of all points P with $dist(P, C) = r$ is called the circle with *center* C and *radius* r (Fig. C.10a). If $r = 0$, then the circle consists of one point, namely, C.

The equation of the circle with center $C(x_0, y_0)$ and radius r can be written, by (C.4), as

$$(x - x_0)^2 + (y - y_0)^2 = r^2 . \tag{C.22}$$

If the center is at the origin, the circle has the equation

$$x^2 + y^2 = r^2 . \tag{C.23}$$

Tangent

A line which intersects the circle at one point is called *tangent* to the circle at that point. The line T in Fig. C.10b is tangent at the point P.

Given a circle (C.22) and a point $P(x_1, y_1)$ on it, we now derive the equation of the tangent T to the circle at P. Recall (from plane geometry) that this tangent is perpendicular to the line segment CP connecting the center C of the circle and the point of tangency P (Fig. C.10b). The slope of the line line$(C(x_0, y_0), P(x_1, y_1))$ is

$$m = \frac{y_1 - y_0}{x_1 - x_0} .$$

Therefore the slope m_1 of the tangent T at $P(x_1, y_1)$ is, by (C.17),

$$m_1 = -\frac{x_1 - x_0}{y_1 - y_0} .$$

A general point (x, y) on T satisfies the equation

$$\frac{y - y_1}{x - x_1} = m_1 = -\frac{x_1 - x_0}{y_1 - y_0} ,$$

which we can simplify, using the fact that $(x_1 - x_0)^2 + (y_1 - y_0)^2 = r^2$,

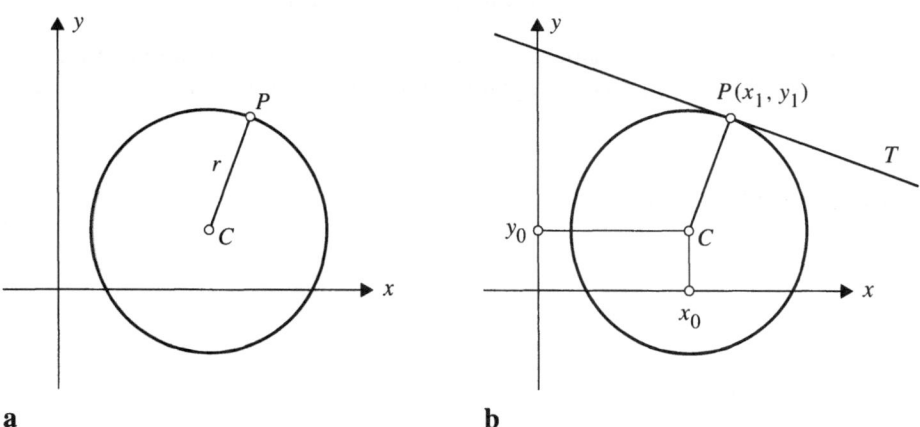

a   **b**

Fig. C.10 a, b. a Circle with center C and radius r. **b** Tangent at the point P is perpendicular to CP

$$(x_1 - x_0)x + (y_1 - y_0)y + \big(y_0(y_1 - y_0) + x_0(x_1 - x_0) - r^2\big) = 0 , \quad (C.24)$$

the general equation (C.13) of the tangent at $P(x_1, y_1)$.

If the circle is centered at the origin, i.e., if $x_0 = y_0 = 0$, then Eq. (C.24) reduces to

$$x_1 x + y_1 y = r^2 . \tag{C.25}$$

Arcs and angles

Let P and Q be any two points on a circle with center C and radius r. The *arc* from P to Q is the set of all points on the circle obtained by turning *counterclockwise* from P to Q (Fig. C.11). Note:
- the arc obtained by turning counterclockwise from P to P is the circle itself;
- the arc from P to Q is different from the arc from Q to P, the union of these arcs is the whole circle.

The angle $\angle(PCQ)$, between the radii CP and CQ, is called the *central angle* corresponding to the arc (see angle α in Fig. C.11). This is the same as the angle between the lines $\text{line}(C, Q)$ and $\text{line}(C, P)$ (Fig. C.8b).

With the radius r fixed, the length a of an arc and its central angle α are proportional

$$a = c\alpha r , \tag{C.26}$$

where c is a constant which depends on how the angles are measured. The two common units are:
- *degrees*, used in engineering, surveying, navigation, and everyday life;
- *radians*, used mainly in mathematics.

The central angle of the whole circle has 360 degrees or 2π radians. Since the length of the entire circle is $2\pi r$, this means that the constant of proportionality in (C.26),

$$c = \tfrac{2\pi}{360} \quad \text{if angles are measured in degrees} ,$$
$$c = 1 \quad \text{if angles are measured in radians} ,$$

which explains why mathematicians prefer radians. It follows that in a circle with radius $r = 1$ the arc lengths are equal to their central angles (when measured in radians),

$$a = \alpha ,$$

as shown by (C.26) with $r = 1$ and $c = 1$.

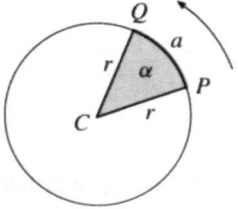

Fig. C.11. Arc of length a and its central angle α

Degrees are denoted by superscript circle and radians by the unit symbol rad. From $0° = 0\,\mathrm{rad}$ and $360° = 2\pi\,[\mathrm{rad}]$, we get in particular,

$$30° = \pi/6\,[\mathrm{rad}],\quad 45° = \pi/4\,[\mathrm{rad}],\quad 60° = \pi/3\,[\mathrm{rad}],\quad 90° = \pi/2\,[\mathrm{rad}]\,,$$
$$180° = \pi\,[\mathrm{rad}],\quad 270° = 3\pi/2\,[\mathrm{rad}],\quad \text{etc.}$$

Unit circle

The circle with center at the origin $(0, 0)$ and radius $r = 1$ is called the unit circle. Its equation is

$$x^2 + y^2 = 1\,, \tag{C.27}$$

using (C.22) with $x_0 = y_0 = 0$ and $r = 1$.

Consider the unit circle in Fig. C.12, the point $P(1, 0)$, a general point Q on the circle, and the angle $\theta = \angle\{\mathrm{line}(OP), \mathrm{line}(OQ)\}$, i.e., the central angle of the arc from P to Q. Every point Q on the unit circle thus corresponds to an angle $0 \le \theta \le 2\pi$. The point $Q(x, y)$ is said to be in the

1st quadrant	if $x \ge 0$, $y \ge 0$,	or equivalently	if $0 \le \theta \le \frac{\pi}{2}$,
2nd quadrant	if $x \le 0$, $y \ge 0$,	or equivalently	if $\frac{\pi}{2} \le \theta \le \pi$,
3rd quadrant	if $x \le 0$, $y \le 0$,	or equivalently	if $\pi \le \theta \le \frac{3\pi}{2}$,
4th quadrant	if $x \ge 0$, $y \le 0$,	or equivalently	if $\frac{3\pi}{2} \le \theta \le 2\pi$.

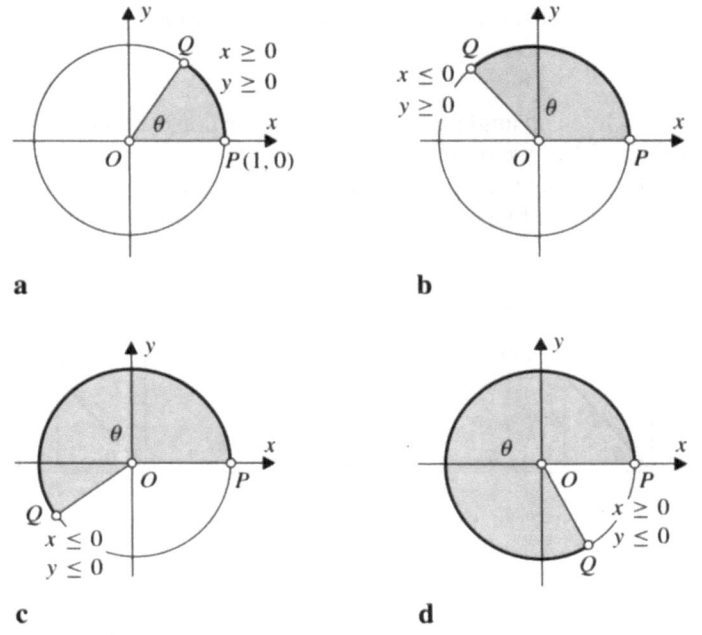

Fig. C.12 a–d. Angles and quadrants. **a** First quadrant, **b** second quadrant, **c** third quadrant, **d** fourth quadrant

The point P has $\theta = 0$. A full counterclockwise rotation, i.e., $\theta = 2\pi$, brings us back to the point P. Therefore we usually do not distinguish between the angles 0 and 2π, or for that matter between 0 and $2k\pi$, for any integer $k > 0$, since $2k\pi$ represents k counterclockwise rotations.

We agree to give negative signs to angles obtained by clockwise rotations. For example, the angle obtained in Fig. C.12d by turning clockwise from P to Q is $-(2\pi - \theta)$, or $\theta - 2\pi$, where -2π represents a clockwise rotation. k clockwise rotations are represented by the angle $-2k\pi$.

We do not distinguish between an angle θ and the angle $\theta + 2k\pi$ obtained from it by k rotations (clockwise if $k < 0$, counterclockwise if $k > 0$).

Exercises

C.8 Do the following for the circle with center $C = (2, 3)$ and radius $r = 5$.
  a. Find the equation of the tangent at $P(5, 7)$.
  b. Find the equations of the two horizontal tangents and their points of tangency.
  c. Find the equations of the two vertical tangents and their points of tangency.
  d. Find the equations and the points of tangency of the two tangents with slope 2.
C.9 Let $P(x_1, y_1)$ be a point on the circle $x^2 + y^2 = r^2$ with center at the origin. Find the equation of the tangent at P and its x- and y-intercepts.

C.4 Sine, cosine, and tangent

Recall the definition of the following trigonometric functions: sine, cosine, and tangent.

Definition C.2. Given an angle $0 \le \theta \le 2\pi$ and the corresponding point A on the unit circle (Fig. C.13a), we define:
the sine of θ, denoted by $\sin \theta$, as the y-coordinate of the point A;
the cosine of θ, denoted by $\cos \theta$, as the x-coordinate of the point A;

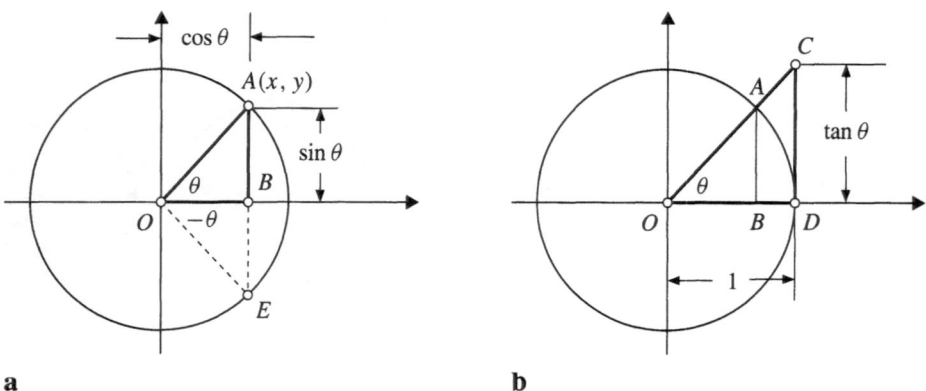

a   b

Fig. C.13a, b. Trigonometric functions: **a** sine and cosine, **b** tangent

the tangent of θ, denoted by $\tan\theta$, as the ratio

$$\tan\theta = \sin\theta/\cos\theta. \tag{C.28}$$

The sine, cosine and tangent are the main three trigonometric functions. These and other trigonometric functions are covered in Sect. 2.2.

To illustrate $\tan\theta$ for $0 \le \theta \le \pi/2$, note that,

$$\tan\theta = \frac{\sin\theta}{\cos\theta} = \frac{|AB|}{|OB|} = \frac{|CD|}{|OD|}, \quad \text{since } \triangle(OAB) \text{ and } \triangle(OCD) \text{ are similar triangles,}$$

$$= \frac{|CD|}{1}, \quad \text{see Fig. C.13b,}$$

$$= \text{the } y\text{-coordinate of the point } C.$$

Consider a triangle, its three angles α, β, and γ, and the opposite sides a, b, and c (Fig. C.14a). Two relationships between the sides and angles are given next.

Proposition C.3 (Sine and cosine laws).
Sine law:

$$\frac{a}{\sin\alpha} = \frac{b}{\sin\beta} = \frac{c}{\sin\gamma}. \tag{C.29}$$

Cosine law:

$$c^2 = a^2 + b^2 - 2ab\cos\gamma. \tag{C.30}$$

The cosine law gives the square c^2 in terms of the opposite angle γ and the other two sides a and b. Write the analogous expressions for a^2 and b^2.

In particular, if $\gamma = \pi/2$, the cosine law reduces to Pythagoras's theorem $c^2 = a^2 + b^2$.

We now draw the graph of $\sin\theta$, using Definition C.2. Let A be a general point on the unit circle and let α be the central angle of the arc PA (Fig. C.15a). The corresponding point $(\alpha, \sin\alpha)$ on the graph of $\sin\theta$, is also denoted by A (Fig. C.15b).

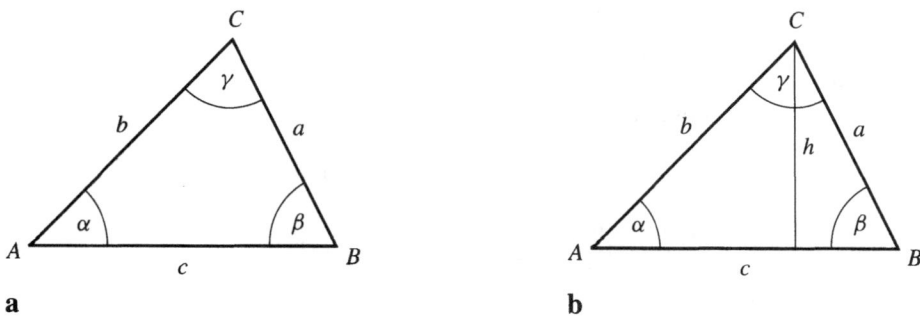

Fig. C.14 a, b. Illustration of the sine law. **a** Sides and angles of a triangle. **b** Height $h = a\sin\beta = b\sin\alpha$

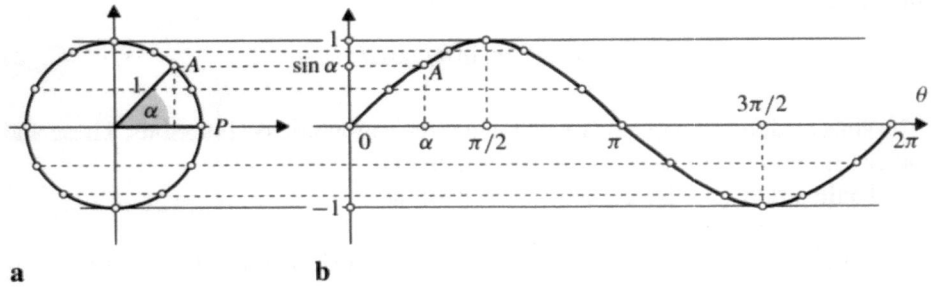

Fig. C.15 a, b. Graph of $\sin\theta$ for $0 \le \theta \le 2\pi$. **a** Points on the unit circle; **b** the corresponding points on the graph of $\sin\theta$

The following identities follow immediately from Definition C.2. For all $0 \le \theta \le 2\pi$,

$$\sin(-\theta) = -\sin\theta \,, \tag{C.31a}$$

$$\cos(-\theta) = \cos\theta \,, \tag{C.31b}$$

$$\tan(-\theta) = -\tan\theta \,, \tag{C.31c}$$

$$\cos\theta = \sin\left(\frac{\pi}{2} - \theta\right), \tag{C.31d}$$

$$\sin^2\theta + \cos^2\theta = 1 \,. \tag{C.31e}$$

The identities (C.31a), (C.31b), and (C.31c) follow from identifying negative angles with clockwise rotations. Thus in Fig. C.13 a, the point $A(x, y)$ corresponds to the angle θ, and the point $E(x, -y)$ corresponds to $-\theta$.

To illustrate (C.31d) consider Fig. C.13 a, where $\cos\theta = \cos\angle(BOA) = \sin\angle(OAB) = \sin(\pi/2 - \theta)$. The identity (C.31e) is Pythagoras's theorem for the right triangle $\triangle(OAB)$ of Fig. C.13 a.

The following identities are useful for sums and differences of angles.

Proposition C.4. For any angles α and β:
a. $\sin(\alpha + \beta) = \sin\alpha \, \cos\beta + \cos\alpha \, \sin\beta$
b. $\sin(\alpha - \beta) = \sin\alpha \, \cos\beta - \cos\alpha \, \sin\beta$
c. $\cos(\alpha + \beta) = \cos\alpha \, \cos\beta - \sin\alpha \, \sin\beta$
d. $\cos(\alpha - \beta) = \cos\alpha \, \cos\beta + \sin\alpha \, \sin\beta$
e. $\tan(\alpha + \beta) = \dfrac{\tan\alpha + \tan\beta}{1 - \tan\alpha \, \tan\beta}$ if $\tan\alpha \, \tan\beta \ne 1$

f. $\tan(\alpha - \beta) = \dfrac{\tan\alpha - \tan\beta}{1 + \tan\alpha \, \tan\beta}$ if $\tan\alpha \, \tan\beta \ne -1$

To illustrate part a, consider Fig. C.16, where $\alpha = \angle(BOA)$, $\beta = \angle(AOE)$,

$$\sin\alpha = |AB|, \quad \cos\alpha = |OB|, \quad \sin\beta = |EC|, \quad \cos\beta = |OC| \,.$$

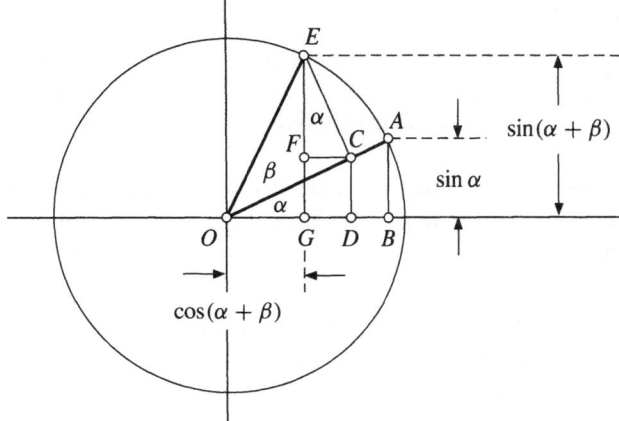

Fig. C.16. Illustration of $\sin(\alpha + \beta)$

The line segment EC is perpendicular to OA, and FC is horizontal. It follows that the angle $\angle(FEC)$ is also $= \alpha$ (check!), and therefore

$$\sin\alpha = \frac{|CF|}{|EC|}, \quad \cos\alpha = \frac{|EF|}{|EC|} \, .$$

$$\frac{|EF|}{|EC|} = \frac{|OB|}{|OA|}, \quad \text{since } \triangle(EFC) \text{ and } \triangle(OBA) \text{ are similar}$$

$$\therefore \quad \frac{|EF|}{\sin\beta} = \frac{\cos\alpha}{1} \quad \therefore \quad |EF| = \cos\alpha \, \sin\beta \, .$$

$$\frac{|CD|}{|OC|} = \frac{|AB|}{|OA|}, \quad \text{since } \triangle(ODC) \text{ and } \triangle(OBA) \text{ are similar}$$

$$\therefore \quad \frac{|CD|}{\cos\beta} = \frac{\sin\alpha}{1} \quad \therefore \quad |CD| = \sin\alpha \, \cos\beta \, .$$

Finally, $\sin(\alpha + \beta) = |EG| = |EF| + |FG| = |EF| + |CD| = \sin\alpha \, \cos\beta + \cos\alpha \, \sin\beta$.

Part c can be similarly illustrated in Fig. C.16; see Exercise C.14.

Part e follows from parts a and c, using

$$\tan(\alpha + \beta) = \frac{\sin(\alpha + \beta)}{\cos(\alpha + \beta)} = \frac{\sin\alpha \, \cos\beta + \cos\alpha \, \sin\beta}{\cos\alpha \, \cos\beta - \sin\alpha \, \sin\beta} \, ,$$

and dividing numerator and denominator by $\cos\alpha \, \cos\beta$.

The formulas for differences $(\alpha - \beta)$ follow from the corresponding sum formulas by writing

$$\alpha - \beta = \alpha + (-\beta) \, ,$$

and using (C.31a), (C.31b), and (C.31c).

Rotation of coordinate axes

Let both the x-axis and the y-axis be rotated, around the origin, by an angle α, to give new coordinate axes, denoted $\bar{x}$-axis and $\bar{y}$-axis, respectively. This is illustrated in Fig. C.17a for a positive angle α (a negative angle corresponds to a clockwise rotation).

A point in $\mathbf{R}^2$ has coordinates (x, y) in the original system and different coordinates $(\bar{x}, \bar{y})$ in the rotated system (see point P in Fig. C.17a). The coordinates change according to the following rule.

Proposition C.5 (Rotation of coordinate axes). Let the coordinate axes x and y be rotated by an angle α. If the coordinates of a point P are (x, y), its rotated coordinates are given by

$$\bar{x} = \quad x \cos \alpha + y \sin \alpha \, , \tag{C.32a}$$

$$\bar{y} = -x \sin \alpha + y \cos \alpha \, . \tag{C.32b}$$

Conversely, if $(\bar{x}, \bar{y})$ are the rotated coordinates of P, its original coordinates (x, y) are

$$x = \bar{x} \cos \alpha - \bar{y} \sin \alpha \, , \tag{C.33a}$$

$$y = \bar{x} \sin \alpha + \bar{y} \cos \alpha \, . \tag{C.33b}$$

We illustrate (C.32a) and (C.32b) with Fig. C.17b. Let AC be parallel to the $\bar{y}$-axis, so that $\angle(CAP) = \alpha$. It follows that

$$|AB| = x \sin \alpha, \quad |OB| = x \cos \alpha, \quad |AC| = y \cos \alpha \quad \text{and} \quad |CP| = y \sin \alpha \, .$$

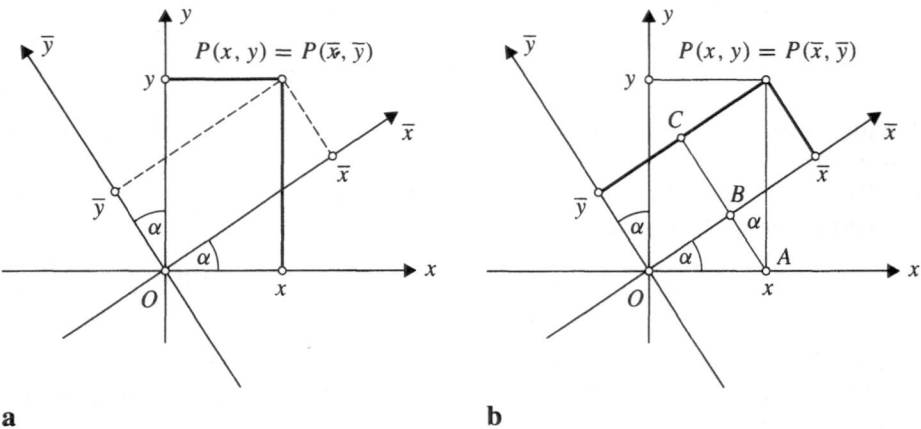

a   **b**

Fig. C.17 a, b. Rotation of coordinate axes. **a** The new coordinate axes are obtained by rotation of angle α. **b** Relating the new coordinates $(\bar{x}, \bar{y})$ to the old coordinates (x, y)

Therefore

$$\bar{x} = |OB| + |CP| = x \cos \alpha + y \sin \alpha ,$$
$$\bar{y} = |AC| - |AB| = -x \sin \alpha + y \cos \alpha .$$

Formulas (C.33a) and (C.33b) then follow, since the x- and y-axes are obtained from the $\bar{x}$- and $\bar{y}$-axes by rotating an angle $-\alpha$.

Inverse sine, cosine, and tangent

For any angle $0 \le \theta \le 2\pi$ we can compute the value of $\sin \theta$. We often ask the converse question: Given a value of the sine, what is the corresponding angle? This question is ambiguous, since for any nonzero value of $-1 < \sin \theta < 1$ there are exactly two values of $0 < \theta < 2\pi$ (see Fig. C.15). We have to pose this question more carefully, in order to assure that the answer is unambiguous. For example: Given a value of $-1 \le x \le 1$, find the angle θ in the interval $[-\pi/2, \pi/2]$, such that

$$\sin \theta = x .$$

The answer is now unique. It is called the *inverse sine* of x and denoted by asin x, where asin stands for *arc sine*, i.e., the arc (or angle) whose sine is the given x.

Similarly, the *inverse cosine* of x, acos x, answers the question: Given a value of $-1 \le x \le 1$, find the angle θ in the interval $[0, \pi]$ such that

$$\cos \theta = x .$$

Finally, for any $-\infty < x < \infty$, the *inverse tangent* of x, atan x, gives the angle θ in the open interval $(-\pi/2, \pi/2)$ such that

$$\tan \theta = x .$$

The inverse trigonometric functions are covered in Sect. 2.3.

Exercises

C.10 Verify, for any α,

(a) $\sin\left(\alpha + \dfrac{\pi}{2}\right) = \cos \alpha$ (b) $\cos\left(\alpha + \dfrac{\pi}{2}\right) = -\sin \alpha$ (c) $\tan\left(\alpha + \dfrac{\pi}{2}\right) = -\dfrac{1}{\tan \alpha}$

(d) $\sin(\alpha + \pi) = -\sin \alpha$ (e) $\cos(\alpha + \pi) = -\cos \alpha$ (f) $\tan(\alpha + \pi) = \tan \alpha$

Hint: Use Proposition C.4a and c.

C.11 Prove: (a) the sine law (C.29) (see Fig. C.14b) and (b) the cosine law (C.30)

C.12 (Heron's formula for the area of triangle) Let a triangle have sides a, b, and c. Prove Heron's formula for the area of the triangle

$$A = \sqrt{s(s-a)(s-b)(s-c)}, \quad \text{where } s = \frac{a+b+c}{2} . \tag{C.34}$$

Hint: $\qquad s(s-a) = \dfrac{(a+b)^2 - c^2}{2^2} = \dfrac{2ab(1 - \cos \theta)}{2^2}$ by the cosine rule,

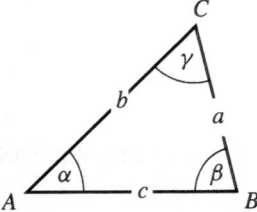

Fig. C.18. Triangle with sides a, b, and c

$$(s - b)(s - c) = \frac{c^2 - (b - a)^2}{2^2} = \frac{2ab(1 + \cos\theta)}{2^2} \quad \text{(verify!)},$$

$$\sqrt{s(s - a)(s - b)(s - c)} = \frac{ab \sin\gamma}{2}, \quad \text{which is the area (see Fig. C.18).}$$

C.13 Adapt the method of Fig. C.15 to draw the graphs of
    (a) $\cos\theta$ (b) $\tan\theta$ (c) $\sin(2\theta)$ (d) $\sin(\theta/2)$
    for $0 \le \theta \le 2\pi$. Plot the same graphs with MACSYMA and compare.
C.14 Use Fig. C.16 to illustrate the formula for $\cos(\alpha + \beta)$, Proposition C.4c.
C.15 Verify the values of trigonometric functions in Table C.1, using Definition C.2 and
    elementary geometry.
C.16 For the given angles of rotation α, find the rotated coordinates $(\bar{x}, \bar{y})$ of the points P,
    given by their (x, y) coordinates.
    (a) $\alpha = \pi/3$, $P(1, \sqrt{3})$ (b) $\alpha = -\pi/4$, $P(-1, 1)$ (c) $\alpha = \pi$, $P(-2, 5)$

Table C.1. Values of trigonometric functions

θ	$\sin\theta$	$\cos\theta$	$\tan\theta$	θ	$\sin\theta$	$\cos\theta$	$\tan\theta$
0	0	1	0	π	0	-1	0
$\dfrac{\pi}{6}$	$\dfrac{1}{2}$	$\dfrac{\sqrt{3}}{2}$	$\dfrac{1}{\sqrt{3}}$	$\dfrac{7\pi}{6}$	$-\dfrac{1}{2}$	$-\dfrac{\sqrt{3}}{2}$	$\dfrac{1}{\sqrt{3}}$
$\dfrac{\pi}{4}$	$\dfrac{1}{\sqrt{2}}$	$\dfrac{1}{\sqrt{2}}$	1	$\dfrac{5\pi}{4}$	$-\dfrac{1}{\sqrt{2}}$	$-\dfrac{1}{\sqrt{2}}$	1
$\dfrac{\pi}{3}$	$\dfrac{\sqrt{3}}{2}$	$\dfrac{1}{2}$	$\sqrt{3}$	$\dfrac{4\pi}{3}$	$-\dfrac{\sqrt{3}}{2}$	$-\dfrac{1}{2}$	$\sqrt{3}$
$\dfrac{\pi}{2}$	1	0	undefined	$\dfrac{3\pi}{2}$	-1	0	undefined
$\dfrac{2\pi}{3}$	$\dfrac{\sqrt{3}}{2}$	$-\dfrac{1}{\sqrt{2}}$	$-\sqrt{3}$	$\dfrac{5\pi}{3}$	$-\dfrac{\sqrt{3}}{2}$	$\dfrac{1}{\sqrt{2}}$	$-\sqrt{3}$
$\dfrac{3\pi}{4}$	$\dfrac{1}{\sqrt{2}}$	$-\dfrac{1}{\sqrt{2}}$	-1	$\dfrac{7\pi}{4}$	$-\dfrac{1}{\sqrt{2}}$	$\dfrac{1}{\sqrt{2}}$	-1
$\dfrac{5\pi}{6}$	$\dfrac{1}{2}$	$-\dfrac{\sqrt{3}}{2}$	$-\dfrac{1}{\sqrt{3}}$	$\dfrac{11\pi}{6}$	$-\dfrac{1}{2}$	$\dfrac{\sqrt{3}}{2}$	$-\dfrac{1}{\sqrt{3}}$

C.17 The following lines and circles are represented in (x, y)-coordinates. Find their representations in the rotated coordinates $(\bar{x}, \bar{y})$ for the given angles of rotation α.
(a) Line: $y = \sqrt{3}x + 2, \quad \alpha = \pi/3$ (b) Line: $y = -x + 1, \quad \alpha = -\pi/4$
(c) Circle: $x^2 + y^2 = r^2, \quad \alpha = $ any angle
(d) Circle: $(x - 2)^2 + (y + 1)^2 = 9, \quad \alpha = \pi/6$

C.18 Let the $\bar{x}$- and $\bar{y}$-axes be obtained from the x- and y-axes by rotation of angle α. The slope-intercept equation of a line is

$$y = mx + b, \quad \text{in the } (x, y)\text{-system},$$
$$\bar{y} = \bar{m}\bar{x} + \bar{b}, \quad \text{in the } (\bar{x}, \bar{y})\text{-system}.$$

Express $\bar{m}$ and $\bar{b}$ in terms of m, b and α.

C.19 Recall that the slope m of a line L is $m = \tan\alpha$, (C.8), where α is the angle between the line L and the x-axis. Given two lines with slopes m and $m_1, m > m_1$, show that the angle θ between them (see Fig. C.8b) is given by

$$\tan\theta = \frac{m - m_1}{1 + mm_1}. \tag{C.35}$$

Discuss the case when $1 + mm_1 = 0$.
Hint: Note that the angle $\theta = \operatorname{atan} m - \operatorname{atan} m_1$. Then use Proposition C.4 f.

C.5 Polar coordinates

A point P in $\mathbf{R}^2$ can be represented by its Cartesian coordinates (see Fig. C.19 a). Alternatively, we can represent P using
its *distance from the origin*, denoted by r,
the *angle θ between the line segment OP and the positive x-axis*, $0 \le \theta \le 2\pi$.
The pair of numbers (r, θ) are called the polar coordinates of P (see Fig. C.19 b).
The real plane is therefore identified with the set of all pairs of numbers

$$\{(r, \theta) \mid r \ge 0, 0 \le \theta < 2\pi\}, \tag{C.36}$$

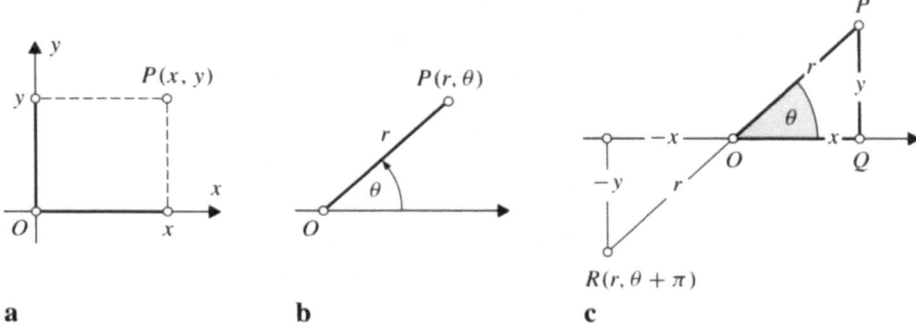

a   **b**   **c**

Fig. C.19 a–c. Two coordinate systems in the plane: **a** Cartesian coordinates, **b** polar coordinates, **c** a connection between Cartesian and polar coordinates

where, for any integer k,

$$(r, \theta + 2k\pi) \text{ represents the same point as } (r, \theta) \qquad \text{(C.37)}$$

since $2k\pi$ means k full rotations; clockwise if $k > 0$, counterclockwise if $k < 0$.

Nonuniqueness of polar coordinates

The Cartesian coordinates are unique: Every point P in the plane has a unique pair of real numbers (x, y) which are its Cartesian coordinates. This is not the case for polar coordinates, see, e.g., (C.37). Another example is the origin, whose polar coordinates are $(0, \theta)$, where θ is arbitrary.

Cartesian coordinates from polar coordinates

Figure C.19c shows how to compute the Cartesian coordinates (x, y) of a point P from its polar coordinates (r, θ):

$$x = x(r, \theta) = r \cos \theta , \qquad \text{(C.38a)}$$
$$y = y(r, \theta) = r \sin \theta . \qquad \text{(C.38b)}$$

Polar coordinates from Cartesian coordinates

Conversely, the polar coordinates (r, θ) are found from the corresponding Cartesian coordinates (x, y) as follows. The distance $r := dist(O, P)$ is, by (C.3),

$$r = r(x, y) = \sqrt{x^2 + y^2} . \qquad \text{(C.39)}$$

For $\theta \neq \pm\pi/2$ we divide (C.38b) by (C.38a), to obtain

$$\tan \theta = y/x . \qquad \text{(C.40)}$$

Solving (C.40) for θ is complicated by the fact that for any x and y,

$$y/x = -y/-x ,$$

but (x, y) and $(-x, -y)$ are in different quadrants, hence correspond to different angles; see, for example, the points $P(x, y)$ and $R(-x, -y)$ in Fig. C.19c.

The following formula gives θ for all possible values, and signs, of x and y.

$$\theta = \theta(x, y) = \begin{cases} \text{undefined} & \text{if } x = y = 0, \\ \text{atan}(y/x) & \text{if } x > 0, \\ \pi/2 & \text{if } x = 0, \ y > 0, \\ \text{atan}(y/x) + \pi & \text{if } x < 0, \\ 3\pi/2 & \text{if } x = 0, \ y < 0. \end{cases} \qquad \text{(C.41)}$$

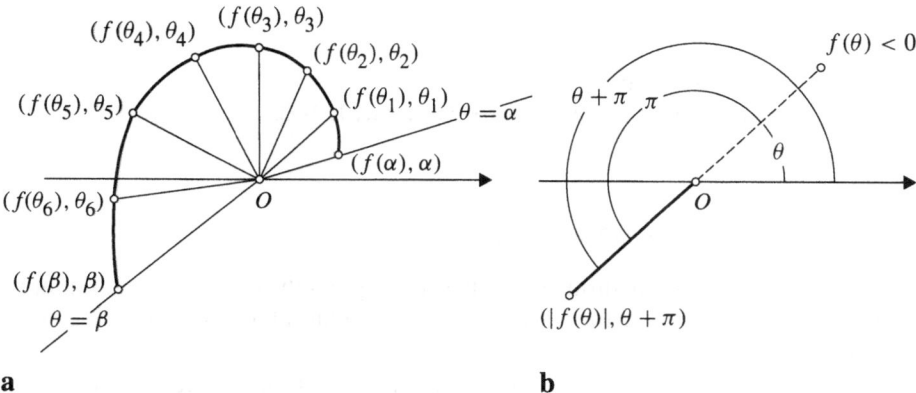

Fig. C.20 a, b. Plotting in polar coordinates. **a** Graph of $r = f(\theta)$ for $\alpha \le \theta \le \beta$. **b** Point $(f(\theta), \theta)$ with $f(\theta) < 0$ is reflected about the origin

Plotting in polar coordinates

Any curve in the plane consists of points with polar coordinates (r, θ). If we know a relation between r and θ, say,

$$r = f(\theta), \quad \text{for } \alpha \le \theta \le \beta , \tag{C.42}$$

then we can plot the curve, by computing enough points $(f(\theta_k), \theta_k)$, with $\alpha \le \theta_k \le \beta$ (see, e.g., Fig. C.20a).

Since r is the distance of the point (r, θ) from the origin, r cannot be negative. It may happen however that $f(\theta) < 0$ for some θ. In this case we take $r = |f(\theta)|$ and "reflect" about the origin, with the resulting point $(|f(\theta)|, \theta + \pi)$ (Fig. C.20b).

MACSYMA-Session C.3. Plotting in Cartesian coordinates is described in Appendix A. This session explains how to plot in polar coordinates. The plots are not shown here. To plot the spiral of Archimedes (see Exercise C.24), with $r = \theta$, $-\pi \le \theta \le \pi$, use
c1. `plot2(theta,theta,-%pi,%pi,polar)`
To plot it for $0 \le \theta \le 6\pi$,
c2. `plot2(theta,theta,0,6*%pi,polar)`
To get a floral design, make a list of sin functions
c3. `makelist(k*sin(2*k*theta)/2,k,1,4)$`
d3. $\left[\dfrac{\sin(2\theta)}{2}, \sin(4\theta), \dfrac{3\sin(6\theta)}{2}, 2\sin(8\theta)\right]$
and plot in different colors
c4. `plot2(%,theta,-%pi,%pi,polar,[0,1,2,3])`
Relate the number of petals to the argument of the sin function.

If a curve is given by an implicit relation, such as $F(r, \theta) = 0$, we solve it first for r, getting an explicit relation $r = f(\theta)$, which can be plotted. For example, the curve
c5. `r^2=theta $`
is solved
c6. `solve(%,r)`

to yield two solutions

d6. $\left[r = -\sqrt{\theta}, \; r = \sqrt{\theta} \right]$

which are spirals, as shows by

c7. `plot2([-sqrt(theta),sqrt(theta)],theta,0,2*%pi,polar,[0,1])`

Comparing polar and Cartesian coordinates

A coordinate system is "natural" for a given example if the resulting representation is simple. Examples where the Cartesian system is natural include lines (Sect. C.2) and circles (Sect. C.3).

In Example C.9 we see that the polar representation of lines is quite cumbersome. The same holds for circles, except for circles with center at the origin (Example C.6) or passing through the origin (Example C.7). On the other hand, there are examples of rather complicated curves which have a simple description in polar coordinates. A family of such curves is the *multileaved rose* given by either $r = a \cos(n\theta)$ or $r = a \sin(n\theta)$, where n is an integer. The rose has $2n$ leaves if n is even, and n leaves if $n \geq 3$ is odd.

Example C.6 (Circles with center in the origin). A circle with center at $(0, 0)$ and radius c has a very simple representation in polar coordinates. Its equation is simply

$$r = c \, . \tag{C.43}$$

Example C.7 (Circles passing through the origin). Polar coordinates are also convenient for circles which pass through the origin. Let such a circle be given in Cartesian coordinates,

$$(x - a)^2 + (y - b)^2 = a^2 + b^2 \, , \tag{C.44}$$

where (a, b) is its center and $\sqrt{a^2 + b^2}$ its radius (the circle is assumed to pass through $(0, 0)$).

Simplifying (C.44) we get

$$x^2 - 2ax + y^2 - 2by = 0 \, ,$$

and using (C.38a), (C.38b), and (C.31e),

$$r^2 = 2r(a \cos \theta + b \sin \theta) \, ,$$
$$\therefore \quad r = 2(a \cos \theta + b \sin \theta) \, .$$

We can now find constants A and α such that

$$a \cos \theta + b \sin \theta = A \sin(\theta + \alpha) \, . \tag{C.45}$$

To see this, write

$$a \cos \theta + b \sin \theta = \sqrt{a^2 + b^2} \left(\frac{a}{\sqrt{a^2 + b^2}} \cos \theta + \frac{b}{\sqrt{a^2 + b^2}} \sin \theta \right) .$$

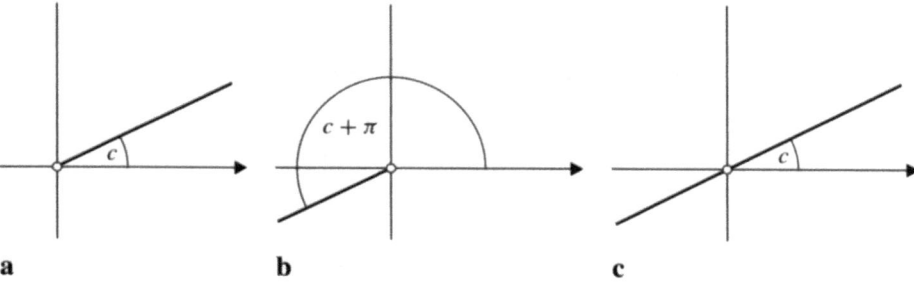

Fig. C.21 a–c. a Ray $\theta = c$; **b** ray $\theta = c + \pi$; **c** the union of the two rays is the line with
slope $m = \tan c$

By Proposition C.4(a), this reduces to (C.45) with $A = \sqrt{a^2 + b^2}$, provided α is
an angle such that

$$\sin \alpha = \frac{a}{\sqrt{a^2 + b^2}}, \quad \cos \alpha = \frac{b}{\sqrt{a^2 + b^2}}.$$

We determine α by adapting the formula (C.41),

$$\alpha = \theta(a, b) = \begin{cases} \text{atan}(b/a) & \text{if } a > 0, \\ \pi/2 & \text{if } a = 0, \ b > 0, \\ \text{atan}(b/a) + \pi & \text{if } a < 0, \\ 3\pi/2 & \text{if } a = 0, \ b < 0. \end{cases} \tag{C.46}$$

The polar representation of the circle (C.44) is therefore

$$r = 2\sqrt{a^2 + b^2} \sin(\theta + \alpha), \quad \text{where } \alpha \text{ is given by (C.46)}. \tag{C.47}$$

Example C.8 (Rays). A *ray* is a half line, emanating from the origin. Its equation
in polar coordinates is

$$\theta = c, \tag{C.48}$$

where c is a constant (Fig. C.21 a). The ray in the opposite direction is then given
by

$$\theta = c + \pi$$

(Fig. C.21 b).

Example C.9 (Lines). Let a line be given by its slope-intercept form (C.12)

$$y = mx + b.$$

If $b = 0$, i.e., if the line passes through the origin, then the line is a union of
two rays (see Fig. C.21 c).

If $b \neq 0$, we use (C.38a) and (C.38b) to rewrite (C.12) as,

$$r(\sin \theta - m \cos \theta) = b,$$

then adapt the "trick" of (C.45), with $a = -m$ and $b = 1$, to get

$$\sin\theta - m\cos\theta = \sqrt{m^2 + 1}\sin(\theta - \alpha),$$

$$\sin\alpha = \frac{m}{\sqrt{m^2 + 1}}, \quad \cos\alpha = \frac{1}{\sqrt{m^2 + 1}}.$$

We conclude that the polar representation of the line (C.12) is

$$r = \frac{b}{\sqrt{m^2 + 1}\sin(\theta - \alpha)}, \tag{C.49}$$

where α is given by (C.46) with $a = -m$ and $b = 1$.

Exercises

C.20 Use MACSYMA to plot the curves of the following roses.
  (a) $r = \sin 2\theta$ (b) $r = 3\cos 3\theta$ (c) $r = 2\sin 10\theta$

C.21 Use MACSYMA to plot the lemniscates.
  (a) $r^2 = 4\sin 2\theta$ (b) $r^2 = 9\cos 2\theta$

C.22 The following two curves are known as cardioids. Plot them.
  (a) $r = 1 + \cos\theta$ (b) $r = 1 - \sin\theta$

C.23 Plot the following limacons.
  (a) $r = 1 + 3\cos\theta$ (b) $r = 2 + \cos\theta$

C.24 Given $a \in \mathbf{R}$, the curve represented by

$$r = a\theta, \quad -\infty < \theta < \infty, \tag{C.50}$$

is a spiral of Archimedes. Plot the following spirals, for $0 \le \theta \le 10\pi$,
  (a) $r = \theta/10$ (b) $r = -\theta/10$

C.25 Given $a > 0$, the curve represented by

$$r = a^\theta, \quad 0 \le \theta < \infty, \tag{C.51}$$

is an exponential spiral. Plot the following spirals, for $-2\pi \le \theta \le 2\pi$,
  (a) $r = e^{\theta/2}$ (b) $r = -e^{\theta/2}$

C.26 Find the polar representation of the following circles, given by their centers (a, b) and radii R, then plot them.
  (a) $a = 1,\ b = 0,\ R = 1$ (b) $a = 0,\ b = 1,\ R = 1$
  (c) $a = 1,\ b = -1,\ R = \sqrt{2}$
  Hint: See Example C.7.

C.27 Find the polar representation of the following lines, given in slope-intercept form, then plot them.
  (a) $y = -x + 2$ (b) $y = 2x - 1$ (c) $y = \sqrt{3}x + 1$
  Note: Plotting the polar representations of these lines may take very long time. To avoid this, restrict θ to a small interval, say, $[-\pi/2, \pi/2]$, which will plot a segment of the line.

D Conic sections

D.1 Conic sections

Conic sections are the intersections of circular cones and planes.

Fix a vertical line (the *axis*), a point on it (the *vertex*), and a number $0 < \alpha < \pi/2$. The set of all lines through the vertex which form an angle $\leq \alpha$ with the axis is a *circular cone*[1] (Fig. D.2 a). Note that a cone has two parts, "joined" at the vertex. The angle α is called the *opening angle* of the cone.

The *generators* of the cones are the lines on its boundary, i.e., the lines forming an angle equal to α with the axis. A circular cone has infinitely many generators, two of which are shown in Fig. D.2 b).

The *angle of inclination* β *of a plane* is the smallest angle between it and the vertical axis. We have $0 \leq \beta \leq \pi/2$. In particular, $\beta = 0$ if the axis belongs to the plane, or if the axis never intersects the plane. In this case the plane is called *parallel* to the axis. Two such planes are illustrated in Fig. D.1 d.

The angle of inclination $\beta = \pi/2$ if the plane is perpendicular to the axis.

The intersection of a circular cone (with an opening angle $0 < \alpha < \pi/2$ and a plane inclined at an angle β) is always nonempty. Its shape depends on α and β.

a. If $\beta > \alpha$ (see the two planes in Fig. D.1 a), then the intersection of the plane and cone is a *point* or an *ellipse* (Fig. D.2 a). In particular, if $\beta = \pi/2$ then the ellipse becomes a *circle*.
b. If $\beta = \alpha$ (Fig. D.1 b), then the intersection is a *parabola* (Fig. D.2 b) or a *line* (Fig. D.2 d).
c. If $0 < \beta < \alpha$, then the plane intersects both parts of the cone and the intersection is a *hyperbola* (Fig. D.2 c).
d. If $\beta = 0$ and the plane contains the axis, the intersection consists of *two lines* intersecting at the vertex of the cone.
e. If $\beta = 0$ and the plane does not intersect the axis, then the intersection is again a *hyperbola*.

Definition D.1. The intersection of a circular cone (with an opening angle $0 < \alpha < \pi/2$) and a plane is called a *conic section*. A conic section is a *point*, a *line*, a *pair of intersecting lines*, an *ellipse* (in particular a *circle*), a *parabola*, or a *hyperbola*.

---

1 Also called *right circular cone*, denoting the fact that the axis is vertical.

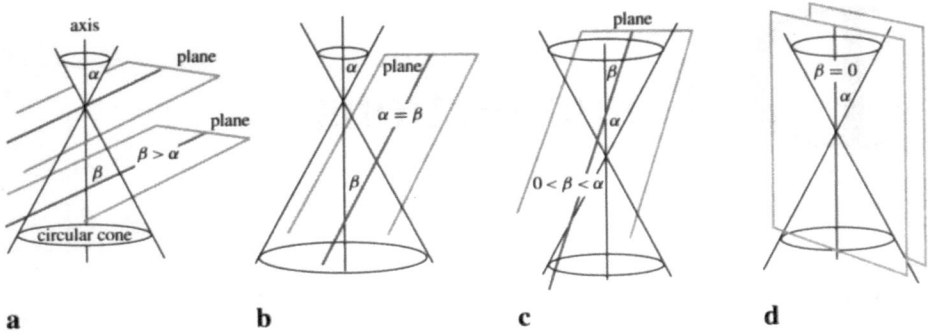

Fig. D.1 a–d. Circular cone with opening angle α and a plane with angle of inclination β.
a $\beta > \alpha$; **b** $\beta = \alpha$; **c** $0 < \beta < \alpha$; **d** $\beta = 0$

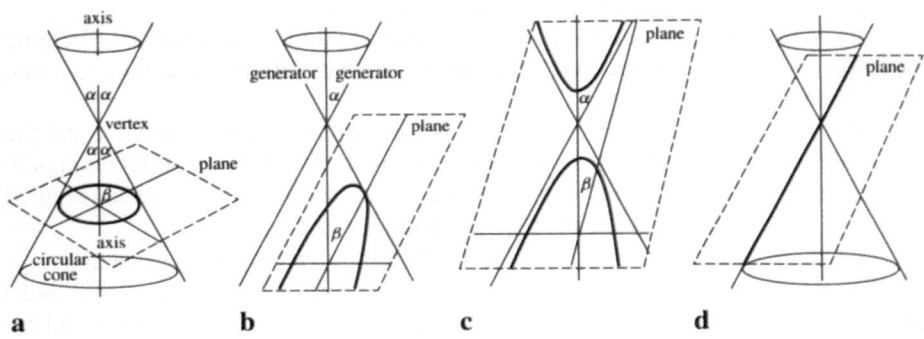

Fig. D.2 a–d. Conic sections. **a** Ellipse, **b** parabola, **c** hyperbola, **d** line

For the purposes of calculus it is convenient to approach conic sections differently. In this section we use two alternative approaches.

1. **Quadratic polynomials.** For any real constants A, B, C, D, E, and F, the graph of the quadratic equation (i.e., the set of points (x, y) satisfying the equation)

$$Ax^2 + Bxy + Cy^2 + Dx + Ey + F = 0 \,, \tag{D.1}$$

is
- a *conic section*;
- a *pair of parallel lines*, for example, if $B = C = D = E = 0$ and $AF < 0$;
- *empty*, for example, if $A = B = C = D = E = 0 \neq F$;
- the *whole plane* if $A = B = C = D = E = F = 0$.

2. **Distance properties.** Points (x, y) on the graphs of conic sections are characterized by the distances from a given point (called the *focus*), or two given points (the *foci*), and a given line (the *directrix*).

Exercises

D.1 Describe the intersection of a right circular cone with a plane that contains the axis of the cone.

D.2 Given a quadratic equation (D.1) with $A = C = 1$, $B = 0$, and $D = E = -2$,

$$x^2 + y^2 - 2x - 2y + F = 0 ,$$

determine the locus of the solutions for

(a) $F = 1$ (b) $F = 2$ (c) $F = 3$

D.2 Circle

We recall that the circle is the locus of points with equal distance r (the radius of the circle) from a fixed point (x_0, y_0), the center. The equation of such a circle (C.22) is

$$(x - x_0)^2 + (y - y_0)^2 = r^2 .$$

Expanding this equation we get (D.1) with $A = C = 1$. Conversely, given a quadratic equation (D.1) with $A = C \neq 0$,

$$Ax^2 + Ay^2 + Dx + Ey + F = 0 ,$$

we divide by A to obtain

$$\left(x^2 + \frac{D}{A}x\right) + \left(y^2 + \frac{E}{A}y\right) = -\frac{F}{A} ,$$

then complete the square (as in (1.21)) to get

$$\left(x + \frac{D}{2A}\right)^2 + \left(y + \frac{E}{2A}\right)^2 = \frac{D^2}{4A^2} + \frac{E^2}{4A^2} - \frac{F}{A} . \tag{D.2}$$

Let rhs (D.2) denote the right-hand side of (D.2). There are now two cases.

rhs (D.2) ≥ 0. In this case we designate the right-hand side by r^2, where $r \geq 0$.

If $r > 0$, then (D.2) is a *circle* with

$$\text{center} \left(-\frac{D}{2A}, -\frac{E}{2A}\right), \quad \text{and radius } r = \sqrt{\frac{D^2}{4A^2} + \frac{E^2}{4A^2} - \frac{F}{A}}$$

If $r = 0$, then (D.2) is the *point* $(D/2A, -E/2A)$.

rhs (D.2) < 0. In this case the locus is empty.

Exercises

D.3 In the following find the equation of the circle specified.

     a. Circle with center $(-1, 1)$ which passes through $(2, 5)$

     b. Circle with center $(1, 2)$ which is tangent to the line $y = -x + 4$

     c. Circle that passes through the points $(1, 1)$, $(4, 3)$, and $(-2, 6)$

d. Circle tangent to the circles $(x-1)^2 + y^2 = 1$, and $(x-2)^2 + y^2 = 4$ with center $(3/2, 0)$

e. Circles tangent to the circles $(x-1)^2 + y^2 = 1$, $(x-5)^2 + y^2 = 1$, and $(x-3)^2 + (y-6)^2 = 1$

D.4 Define the following MACSYMA functions

    a. `eqn_circle(center,radius)` that gives the equation of a circle with given center and radius

    b. `center_circle(a,d,e,f,x,y)` that gives the center (as a vector) of the circle satisfying

$$ax^2 + ay^2 + dx + ey + f = 0.$$

    c. `radius_circle(a,d,e,f,x,y)` that gives the radius of the circle satisfying the above equation.

    Use these functions to compute

    d. the equation of the circle with center at $(1, 1)$ and radius $\sqrt{2}$

    e. the center and radius of the circle satisfying the equation $x^2 + y^2 + x - y - 1 = 0$ and plot these circles.

D.5 (Steiner circles) Calculate the equation for the locus of all points having a distance to the point $(1, 0)$ that is k times larger ($k \in \mathbf{R}^+$) than the distance to the point $(-1, 0)$. Show that for $k \neq 1$ this always is a circle. What happens if $k = 1$? Plot the loci with MACSYMA for $k = j/3 (j = 1, \ldots, 6)$.

D.3 Parabola

Definition D.2. A parabola is the locus of points which are equidistant from a given line, called the *directrix*, and a given point (outside the directrix), called the *focus* of the parabola.

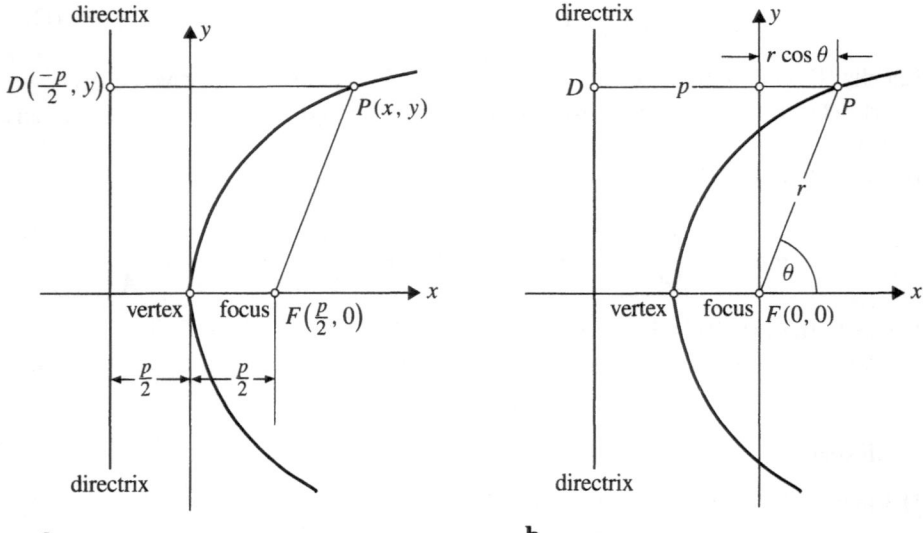

a   **b**

Fig. D.3a, b. Definition of the parabola: **a** $|PF| = |PD|$; **b** $r = p + r \cos \theta$

We derive the equation of the parabola in the special case that the directrix is vertical. For convenience we take the distance between the focus and the directrix to be p, the focus F at the point $(p/2, 0)$, and the directrix as the line $x = -p/2$ (Fig. D.3 a).

A point $P(x, y)$ on the parabola satisfies, by Definition D.2,

$$|FP| = |PD|, \quad \text{where } D \text{ is the projection of } P \text{ on the directrix}$$

$$\therefore \quad \sqrt{\left(x - \frac{p}{2}\right)^2 + y^2} = x + \frac{p}{2}$$

$$\therefore \quad \left(x - \frac{p}{2}\right)^2 + y^2 = \left(x + \frac{p}{2}\right)^2$$

$$\therefore \quad x^2 - px + \frac{p^2}{4} + y^2 = x^2 + px + \frac{p^2}{4}$$

giving the equation of the parabola as

$$y^2 = 2px . \tag{D.3}$$

Similarly, if the directrix is horizontal, given by $y = -p/2$, and the focus is at $(0, p/2)$, then the equation of the parabola is

$$x^2 = 2py . \tag{D.4}$$

The *vertex* of the parabola is the point midway between the focus and the directrix. The vertex of the parabola (D.3) is at the origin. If a parabola has a vertical directrix and a vertex at the point (h, k), we translate the coordinate axes

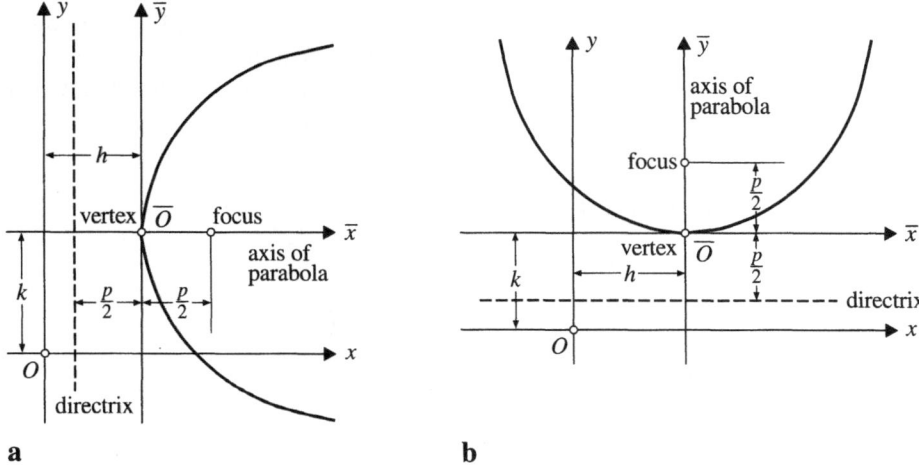

a   **b**

Fig. D.4 a, b. Parabolas with vertex (h, k) and vertical directrix (**a**), opening to the right, or horizontal directrix (**b**), opening upwards

to that point, obtaining the new coordinates

$$\bar{x} := x - h, \quad \bar{y} := y - k,$$

(Fig. D.4 a). In terms of these coordinates, the equation of the parabola is

$$\bar{y}^2 = 2p\bar{x}, \quad \text{by (D.3)},$$

or

$$(y - k)^2 = 2p(x - h). \tag{D.5}$$

The parabola (D.5) opens to the right, as illustrated in Fig. D.4 a. Similarly, the parabola

$$(y - k)^2 = -2p(x - h), \tag{D.6}$$

opens to the left. Both parabolas (D.5) and (D.6) are symmetric with respect to the horizontal line $y = k$ (the $\bar{x}$-axis), called the *axis of symmetry* of the parabola.

A parabola with a horizontal directrix and vertex at (h, k) is similarly written as

$$(x - h)^2 = 2p(y - k) \quad \text{if it opens upwards (Fig. D.4 b)}, \tag{D.7a}$$

$$(x - h)^2 = -2p(y - k) \quad \text{if it opens downwards}. \tag{D.7b}$$

The axis of symmetry of these parabolas is the vertical line $x = h$ (the $\bar{y}$-axis).

Parabola as quadratic polynomial

Equations (D.7a) and (D.7b) are of the form

$$Ax^2 + Dx + Ey + F = 0,$$

with $A \neq 0$. Similarly, Eqs. (D.5) and (D.6) are of the form

$$Cy^2 + Dx + Ey + F = 0,$$

where $C \neq 0$. These are the general forms for a parabola whose axis of symmetry is parallel to one of the coordinate axes. A particular feature of these equations is that they are linear in one of the coordinates and quadratic in the other one. Given such an equation, we can compute the parabola it represents.

Given an equation

$$Ax^2 + Dx + Ey + F = 0, \quad \text{with } A, E \neq 0,$$

divide by A to obtain

$$x^2 + \frac{D}{A}x = -\frac{E}{A}y - \frac{F}{A}$$

and complete the square to get

$$x^2 + \frac{D}{A}x + \frac{D^2}{4A} = -\frac{E}{A}y - \frac{F}{A} + \frac{D^2}{4A}$$

$$\therefore \quad \left(x + \frac{D}{2A}\right)^2 = -\frac{E}{A}\left(y + \frac{F}{E} - \frac{D^2}{4E}\right)$$

which is the form (D.7b),

$$(x - h)^2 = -2p(y - k) ,$$

with

$$h = -\frac{D}{2A}, \quad k = -\frac{F}{E} + \frac{D^2}{4E} = \frac{D^2 - 4F}{4E} ,$$

the coordinates of the vertex. The directrix is the horizontal line which is p units above the vertex (since the parabola opens downwards), where p is from

$$2p = E/A, \quad \text{giving} \quad p = E/2A .$$

Therefore the equation of the directrix is

$$y = k + \frac{p}{2} = \frac{D^2 - 4F}{4E} + \frac{E}{4A} .$$

The focus of the parabola is $p/2$ units below the vertex, at the point

$$\left(-\frac{D}{2A}, \frac{D^2 - 4F}{4E} - \frac{E}{4A} \right) .$$

Polar coordinates

Consider now the representation of a parabola in polar coordinates. For this purpose it is convenient to have the focus F at the origin $(0, 0)$. Definition D.2 then gives,

$$|FP| = |PD|, \quad \text{where } D \text{ is the projection of } P \text{ on the directrix}$$
$$\therefore \quad r = p + r \cos\theta, \quad \text{see Fig. D.3b} ,$$
$$\therefore \quad r(1 - \cos\theta) = p ,$$

giving the equation of the parabola as

$$r = p/(1 - \cos\theta) . \tag{D.8}$$

This is an equation of a parabola which opens to the right. Similarly,

$$r = p/(1 + \cos\theta) \tag{D.9}$$

is a parabola which opens to the left.

Exercises

D.6 The problems below give the vertex $V(x_0, y_0)$ and the focus $F(x_1, y_1)$. In each instance, find the equation of the parabola and use MACSYMA to graph it.
  (a) $V(0, 0)$, $F(4, 0)$ (b) $V(-1, 0)$, $F(-7, 0)$
  (c) $V(2, 3)$, $F(4, 3)$ (d) $V(-4, -5)$, $F(4, -5)$
D.7 Find the parabolas passing through three given points and having an axis of symmetry as indicated.
  (a) points: $(1, 4)$, $(3, 7)$, and $(4, 10)$;   axis: vertical

(b) points: $(1, -1)$, $(2, 0)$, and $(3, -2)$;   axis: vertical

(c) points: $(1, 4)$, $(3, 6)$, and $(0, 9)$;   axis: horizontal

D.8 Define the following MACSYMA functions

(a) `eqn_parabola(vertex, focus)` that gives the equation of a parabola with given vertex and focus

(b) `vertex_parabola(a, d, e, f, x, y)` that gives the vertex (as a vector) of the parabola satisfying the equation $ax^2 + dx + ey + f = 0$

(c) `focus_parabola(a, d, e, f, x, y)` that gives the focus (as a vector) of the above parabola

Using these functions compute

(d) the equation of the parabola with vertex at $(-1, -3)$ and focus at $(-1, -1)$

(e) the vertex, focus, and directrix of the parabola satisfying $x^2 + x - y - 1 = 0$

and plot these parabolas.

D.4 Ellipse

Definition D.3. Given two fixed points F_1 and F_2, an ellipse is the locus of all points whose sum of distances from F_1 and F_2 is constant.

The points F_1 and F_2 are called the *foci* of the ellipse. The constant sum of distances is designated by $2a$. A point P belongs to the ellipse if

$$\left| |PF_1| + |PF_2| \right| = 2a \tag{D.10}$$

(Fig. D.5 a).

To derive the equation of the ellipse, we place the foci F_1 and F_2 on the x-axis, in symmetry with respect to the origin. Let the distance between the foci be $2c$ so that the coordinates of the foci are $F_1(c, 0)$ and $F_2(-c, 0)$.

For a point $P(x, y)$, the sum of distances from the foci is

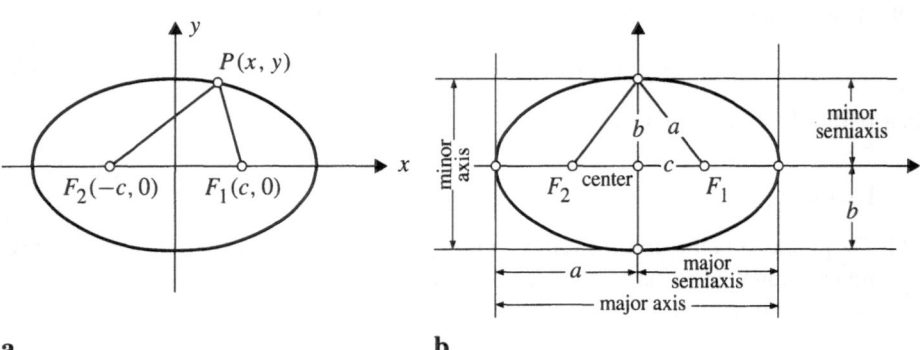

a   **b**

Fig. D.5 a, b. a Definition of the ellipse: $\left| |PF_1| + |PF_2| \right| = 2a$. **b** Characteristics of the ellipse

$$\left| |PF_1| + |PF_2| \right| = \sqrt{(x-c)^2 + y^2} + \sqrt{(x+c)^2 + y^2}$$

so condition (D.10) becomes

$$\sqrt{(x-c)^2 + y^2} + \sqrt{(x+c)^2 + y^2} = 2a \tag{D.11}$$

$$\therefore \quad \sqrt{(x+c)^2 + y^2} - a = a - \sqrt{(x-c)^2 + y^2}$$

$$\therefore \quad (x+c)^2 + y^2 + a^2 -$$
$$- 2a\sqrt{(x+c)^2 + y^2} = (x-c)^2 + y^2 + a^2 - 2a\sqrt{(x-c)^2 + y^2}$$

$$\therefore \quad 4cx = 2a\left(\sqrt{(x+c)^2 + y^2} - \sqrt{(x-c)^2 + y^2}\right)$$

$$\therefore \quad 2cx = a\left(\sqrt{(x+c)^2 + y^2} + \right.$$
$$\left. + \sqrt{(x-c)^2 + y^2} - 2\sqrt{(x-c)^2 + y^2}\right)$$

$$= a\left(2a - 2\sqrt{(x-c)^2 + y^2}\right), \quad \text{by (D.11)},$$

$$\therefore \quad cx = a\left(a - \sqrt{(x-c)^2 + y^2}\right)$$

$$\therefore \quad a\sqrt{(x-c)^2 + y^2} = a^2 - cx$$

$$\therefore \quad a^2\left((x-c)^2 + y^2\right) = a^4 - 2a^2cx + c^2x^2$$

$$\therefore \quad x^2 - 2cx + c^2 + y^2 = a^2 - 2cx + \frac{c^2}{a^2}x^2$$

$$\therefore \quad (a^2 - c^2)x^2 + a^2y^2 = a^2(a^2 - c^2)$$

$$\therefore \quad \frac{x^2}{a^2} + \frac{y^2}{a^2 - c^2} = 1$$

giving the equation of the ellipse as

$$\frac{x^2}{a^2} + \frac{y^2}{b^2} = 1, \tag{D.12}$$

where

$$b^2 := a^2 - c^2. \tag{D.13}$$

It is clear from Eq. (D.12) that the graph is symmetric with respect to both coordinate axes. Moreover, the x-intercepts are $(\pm a, 0)$, and the y-intercepts are $(0, \pm b)$. The line segment joining the two farthest intercepts is called the *major axis*, and the line segment joining the two nearest intercepts is called the *minor axis*.

The number a is the length of the *major semiaxis*, and b is the length of the *minor semiaxis* (Fig. D.5b).

The equation of an ellipse with foci $F_1(0, c)$ and $F_2(0, -c)$ (on the y-axis) is similarly given as

$$\frac{x^2}{b^2} + \frac{y^2}{a^2} = 1 \, . \tag{D.14}$$

Here the major axis is on the y-axis, and the minor axis is on the x-axis.

The *eccentricity* of the ellipse is the ratio

$$e := c/a \tag{D.15}$$

of the focal distance c to the major semiaxis a. The eccentricity is denoted by e, which should not be confused with the basis of the natural logarithms.

Remark D.4. By its definition, the eccentricity of the ellipse is a number e between 0 and 1.

Zero eccentricity corresponds to a circle. Indeed,

$$e = \frac{c}{a} = \frac{\sqrt{a^2 - b^2}}{a} = 0 \quad \Longrightarrow \quad a = b \, ,$$

and the ellipse (D.12) becomes a circle with radius a.

To show how the ellipse depends on the eccentricity, let the major semiaxis a be fixed. This fixes the "length" of the ellipse. Its "width" is the minor semiaxis b, given by (D.13) and (D.15) as

$$b = \frac{c}{e}\sqrt{1 - e^2} \, ,$$

which decreases as e increases. Therefore, as the eccentricity increases, the ellipse becomes "narrower". For $e = 1$, the ellipse has zero width, i.e., it reduces to a line segment.

The parabola was defined above as the locus of points equidistant from its focus and directrix. The ellipse can be defined analogously in terms of these distances and the eccentricity e.

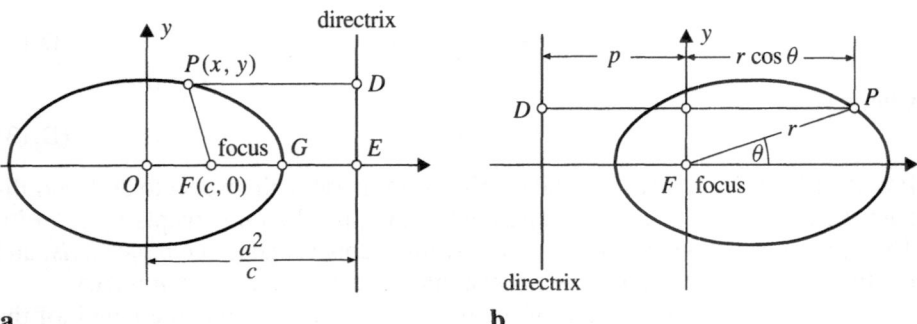

Fig. D.6 a, b. Ellipse: a focus and a directrix. **a** $|PF| = e|PD|$. **b** $r = e(p + r\cos\theta)$

Definition D.5. The ellipse is the locus of all points P with a ratio $e < 1$ between their distance from the focus F and the distance from the directrix, i.e.,

$$|PF| = e|PD|,$$ (D.16)

where D is the projection of P on the directrix (Fig. D.6 a).

For the point G on the ellipse which is closest to the directrix, it follows from (D.16) that

$$|GF| = e|GE|, \quad \text{(Fig. D.6 a)}$$
$$\therefore \quad a - c = e(|OE| - a)$$
$$\therefore \quad a - c = \frac{c}{a}(|OE| - a)$$
$$\therefore \quad |OE| = \frac{a^2}{c}$$
$$= \frac{c}{e^2}$$

showing that the equation of the directrix is

$$x = c/e^2.$$ (D.17)

Condition (D.16) then states

$$(x - c)^2 + y^2 = e^2\left(\frac{c}{e^2} - x\right)^2,$$ (D.18)

$$\therefore \quad (1 - e^2)x^2 + y^2 = c^2\left(\frac{1}{e^2} - 1\right) = \frac{c^2}{e^2}(1 - e^2)$$
$$\therefore \quad (a^2 - c^2)x^2 + a^2 y^2 = a^2(a^2 - c^2)$$

which gives (D.12), proving that Definitions D.3 and D.5 are equivalent.

Remark D.6. Definition D.5 involves a focus and a directrix. However, an ellipse has two foci and two directrices, which are symmetric with respect to the center of the ellipse. The directrix (D.17) lies to the right of the ellipse. Its counterpart on the left is

$$x = -c/e^2 = -a/e$$ (D.19)

and Definition D.5 can be used with the left focus $(-c, 0)$ and the left directrix (D.19).

Ellipse as quadratic polynomial

A translation of the origin to the point (h, k) changes the equations of the ellipses (D.12) or (D.14) into

$$\frac{(x-h)^2}{a^2} + \frac{(y-k)^2}{b^2} = 1 \quad \text{or} \quad \frac{(x-h)^2}{b^2} + \frac{(y-k)^2}{a^2} = 1, \tag{D.20}$$

respectively. Both these equations are of the form

$$Ax^2 + Cy^2 + Dx + Ey + F = 0, \tag{D.21}$$

where A and C have the same sign, which can be taken as positive.

Conversely, starting with a quadratic equation (D.21) with positive A and C, we can use completion of a square to obtain an equation (D.20). Here are the details:

$$Ax^2 + Cy^2 + Dx + Ey + F = 0,$$

$$A\left(x^2 + \frac{D}{A}x\right) + C\left(y^2 + \frac{E}{C}y\right) = -F,$$

$$A\left(x + \frac{D}{2A}\right)^2 + C\left(y + \frac{E}{2C}\right)^2 = \frac{D^2}{4A} + \frac{E^2}{4C} - F := G.$$

The left-hand side of this equation is positive. Therefore the equation has solutions (and the graph is nonempty) if the right-hand side G is nonnegative. If $G = 0$, the graph reduces to one point. If $G > 0$, the graph is an ellipse. The graph is empty if $G < 0$.

Polar coordinates

To represent an ellipse in polar coordinates, it is again convenient to identify the origin with a focus F (Fig. D.6b). We designate the distance between that focus and the directrix by p.

For a point P on the ellipse we have, by Definition D.5,

$$|FP| = e|PD|, \quad \text{where } D \text{ is the projection of } P \text{ on the directrix },$$

$$\therefore \quad r = e(p + r\cos\theta), \quad \text{see Fig. D.6b },$$

$$\therefore \quad r(1 - e\cos\theta) = pe,$$

giving the equation of the ellipse as

$$r = pe/(1 - e\cos\theta). \tag{D.22}$$

The same arithmetic with the other focus and directrix gives the equation of the ellipse as

$$r = pe/(1 + e\cos\theta). \tag{D.23}$$

Exercises

D.9   In each of the following find the equation of the ellipse indicated.
   (a) Vertices at $(\pm 3, 0)$ and $(0, \pm 5)$
   (b) Foci at $(\pm 8, 0)$ and major semiaxis 17
   (c) Foci $(-2, 5)$ and $(10, 5)$ and major axis 20

(d) Foci at (4, 3) and (4, 7) and minor axis 6

(e) Center at (0, 0), horizontal major axis 30,
     and eccentricity 0.3

D.10 Define the following MACSYMA functions

a. `eqn_ellipse(center, focus, eccentricity)` that gives the equation of an ellipse with given center, focus, and eccentricity.

b. `center_ellipse(a, c, d, e, f, x, y)` that gives the center (as a vector) of the ellipse satisfying the equation $ax^2 + cy^2 + dx + ey + f = 0$

c. `foci_ellipse(a, c, d, e, f, x, y)` that gives the foci of the ellipse satisfying the above equation

d. `eccentricity_ellipse(a, c, d, e, f, x, y)` that gives the eccentricity of the ellipse satisfying the above equation.

Use these functions to compute

e. the equation of the ellipse with center at (1, 1), focus at (1, 2), and eccentricity 0.5

f. the center, foci, and eccentricity of the ellipse satisfying the equation $4x^2 + 3y^2 - 8x - 6y - 5 = 0$

g. the center, foci, and eccentricity of the ellipse satisfying the equation $3x^2 + 4y^2 - 8x - 6y - 5 = 0$

and plot the ellipses.

D.11 Obtain the polar representation of the ellipse in Fig. D.6 b, using the directrix to the right of the ellipse, with the origin in the right focus.

D.5 Hyperbola

The hyperbola is defined analogously to the ellipse (Definition D.3), with "difference" replacing "sum".

Definition D.7. Given two fixed points F_1 and F_2, a *hyperbola* is the locus of all points whose distances from F_1 and F_2 differ by a constant.

The points F_1 and F_2 are called the *foci* of the hyperbola. The constant difference of distances is denoted by $2a$. A point P belongs to the hyperbola if

$$\left||PF_2| - |PF_1|\right| = 2a , \tag{D.24}$$

(Fig. D.7 a).

To derive the equation of the hyperbola, we place the foci F_1 and F_2 on the x-axis, in symmetry with respect to the origin. Let the distance between the foci be $2c$ so that the coordinates of the foci are $F_1(c, 0)$ and $F_2(-c, 0)$.

For a point $P(x, y)$, the difference of distances from the foci is

$$\left||PF_2| - |PF_1|\right| = \left|\sqrt{(x + c)^2 + y^2} - \sqrt{(x - c)^2 + y^2}\right| ,$$

so condition (D.24) becomes

$$\left|\sqrt{(x + c)^2 + y^2} - \sqrt{(x - c)^2 + y^2}\right| = 2a$$

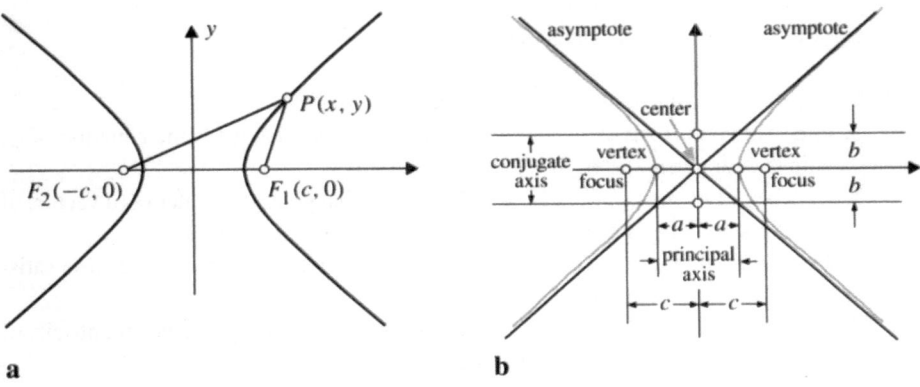

Fig. D.7 a, b. a Definition of the hyperbola: $\left||PF_2|-|PF_1|\right|=2a$. **b** Characteristics of the
hyperbola

As in the case of the ellipse this equation can be brought to the form

$$\frac{x^2}{a^2}-\frac{y^2}{c^2-a^2}=1\,,$$

giving the equation of the hyperbola as

$$\frac{x^2}{a^2}-\frac{y^2}{b^2}=1\,, \tag{D.25}$$

where

$$b^2:=c^2-a^2\,. \tag{D.26}$$

If the foci of the hyperbola are $F_1(0,c)$ and $F_2(0,-c)$, on the y-axis, then its
equation is

$$\frac{y^2}{a^2}-\frac{x^2}{b^2}=1\,, \tag{D.27}$$

with b as above.

As for the ellipse, we define the *eccentricity* of the hyperbola by (D.15),

$$e:=c/a\,.$$

The eccentricity of the hyperbola is greater than 1.

In analogy with the ellipse, Definition D.5, the hyperbola can be defined in
terms of a focus and a directrix.

Definition D.8. The hyperbola is the locus of all points P with a ratio $e>1$
between their distance from the focus F and their distance from the directrix, i.e.,
(D.16),

$$|PF|=e|PD|\,,$$

where D is the projection of P on the directrix (Fig. D.8 a).

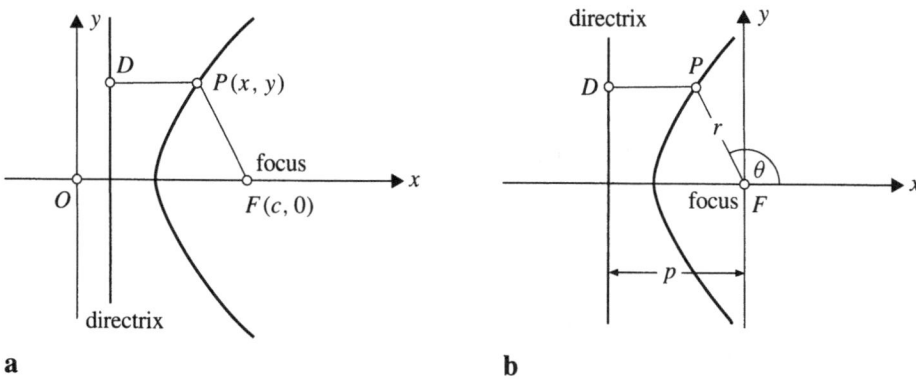

Fig. D.8 a, b. Hyperbola: a focus and a directrix. **a** $|PF| = e|PD|$. **b** $r = e(p + r\cos\theta)$

To show the equivalence of Definitions D.7 and D.8, let the focus F be at $(c, 0)$, with $c > 0$, and the directrix given by (D.17),

$$x = c/e^2 .$$

This directrix lies to the left of the focus, as illustrated in Fig. D.8 a.

A general point $P(x, y)$ on the hyperbola satisfies, by Definition D.8,

$$(x - c)^2 + y^2 = e^2(x - c/e^2)^2 ,$$

or, using the fact that $e^2 - 1 > 0$,

$$(e^2 - 1)x^2 - y^2 = c^2\left(1 - \frac{1}{e^2}\right) = \frac{c^2}{e^2}(e^2 - 1) .$$

Using $a = c/e$ and $b^2 = c^2 - a^2$, see (D.26), the last equation reduces to (D.25).

For each y the hyperbola (D.25) has two values of x,

$$x = \pm\frac{a}{b}\sqrt{y^2 + b^2} ,$$

showing that the hyperbola has two *branches* (Fig. D.7). Each branch is symmetric with respect to the *principal axis* (also transverse axis) of the hyperbola, which is the line passing through its *vertices*. The line perpendicular to the principal axis, through the *center* of the hyperbola, is called its *conjugate axis*. The conjugate axis contains no points of the hyperbola (Fig. D.7 b).

There are two lines which the hyperbola approaches in the limit as $x \to \pm\infty$. These lines are called the *asymptotes* of the hyperbola. The asymptotes of the hyperbola (D.25),

$$\frac{x^2}{a^2} - \frac{y^2}{b^2} = 1 ,$$

are

$$y = \pm\frac{b}{a}x . \tag{D.28}$$

Indeed, we solve (D.25) to get

$$y = \pm \frac{b}{a}\sqrt{x^2 - a^2}$$

which approach the lines (D.28) as $x \to \pm\infty$.

MACSYMA-Session D.1. The hyperbola (D.25) with $a = 1$ and $b = 1/2$ is
c1. x^2-4*y^2=1
d1. $x^2 - 4y^2 = 1$
and its graph is drawn by implicit_plot.
c2. implicit_plot(x^2-4*y^2=1,x,-5,5,y,-2,2)$
The asymptotes (D.28) are here the lines $y = x/2$ and $y = -x/2$. We plot them by
c3. plot([x/2,-x/2],x,-5,5,[3])$
where the last argument, [3], indicates the use of dotted lines. Now combine these plots.
c4. (implicit_plot(x^2-4*y^2=1,x,-5,5,y,-2,2,first),
              plot([x/2,-x/2],x,-5,5,[3], same, last))$
The resulting plot (not shown here) illustrates how the points on the branches of the
hyperbola get closer and closer to the asymptotes as $|x|$ becomes larger and larger, i.e., as
x approaches ∞ or $-\infty$.
We can also use the functions asymptote_right or asymptote_left of Session 6.6 to
compute the asymptotes.

Hyperbola as quadratic polynomial

A translation of the origin to the point (h, k) changes the equations of the hyperbola
(D.25) or (D.27) into

$$\frac{(x - h)^2}{a^2} - \frac{(y - k)^2}{b^2} = 1 \quad \text{or} \quad \frac{(y - k)^2}{a^2} - \frac{(x - h)^2}{b^2} = 1 \,,$$

respectively. Both these equations are of the form (D.21),

$$Ax^2 + Cy^2 + Dx + Ey + F = 0 \,,$$

where A and C have opposite signs.

Conversely, given an equation (D.21) with $AC < 0$, we may assume that $A > 0$
and $C < 0$. Completing the squares we get

$$A\left(x^2 + \frac{D}{A}x + \frac{D^2}{4A^2}\right) - |C|\left(y^2 + \frac{E}{C}y + \frac{E^2}{4C^2}\right) = -F + \frac{D^2}{4A} + \frac{E^2}{4C} \,,$$

$$A\left(x + \frac{D}{2A}\right)^2 - |C|\left(y + \frac{E}{2C}\right)^2 = -F + \frac{D^2}{4A} + \frac{E^2}{4C} \,, \quad \text{(D.29)}$$

which is a hyperbola. Its principal axis is horizontal (i.e., the hyperbola opens
up and down) if the right-hand side of (D.29) is positive, and it is vertical (i.e.,
the hyperbola opens to the left and right) if the right-hand side is negative. If the
right-hand side of (D.29) is zero, then (D.29) represents two lines intersecting at
the point $(-D/2A, -E/2C)$.

Polar coordinates

For a point P on the hyperbola, Definition D.8 states that[2]

$$|FP| = e|PD|, \quad \text{where } D \text{ is the projection of } P \text{ on the directrix}$$
$$\therefore \quad r = e(p + r\cos\theta), \quad \text{see Fig. D.8b},$$
$$\therefore \quad r(1 - e\cos\theta) = pe,$$

giving the equation of the hyperbola as

$$r = pe/(1 - e\cos\theta), \tag{D.30}$$

where now $e > 1$. Similarly, the other focus and directrix give the following equation

$$r = pe/(1 + e\cos\theta). \tag{D.31}$$

Exercises

D.12 In each of the following problems find the equations of the hyperbola described.
   (a) Foci at $(\pm3, 0)$ and vertices $(\pm2, 0)$
   (b) Foci at $(0, \pm5)$ and vertices at $(0, \pm3)$
   (c) Foci at $(\pm5, 0)$ and asymptotes $y = \pm3x/4$
   (d) Vertices at $(0, \pm3)$ and asymptotes $y = \pm4/3$
   (f) Vertices at $(\pm3, 0)$ and eccentricity $e = 5/3$
   (g) Center at $(3, 3)$, horizontal transverse axis of length 8, and eccentricity 3
D.13 Define the following MACSYMA functions
   a. `eqn_hyperbola(center, focus, eccentricity)` that gives the equation of a hyperbola with given center, focus, and eccentricity
   b. `center_hyperbola(a, c, d, e, f, x, y)` that gives the center of the hyperbola satisfying the equation $ax^2 + cy^2 + dx + ey + f = 0$ as a vector
   c. `foci_hyperbola(a, c, d, e, f, x, y)` that gives the foci of the hyperbola satisfying the equation $ax^2 + ay^2 + dx + ey + f = 0$
   d. `eccentricity_hyperbola(a, c, d, e, f, x, y)` that gives the eccentricity of the hyperbola satisfying the equation $ax^2 + ay^2 + dx + ey + f = 0$
   Using these functions let MACSYMA calculate
   e. the equation of the hyperbola with center at $(1, 1)$, focus at $(1, 2)$, and eccentricity 2
   f. the center, foci, and eccentricity of the hyperbola satisfying the equation $4x^2 - 3y^2 - 8x - 6y - 5 = 0$
   g. the center, foci, and eccentricity of the hyperbola satisfying the equation $3x^2 - 4y^2 - 8x - 6y - 5 = 0$
   and plot the hyperbolas.

---

2 Compare with the case of the ellipse.

D.6 The general quadratic $Ax^2 + Bxy + Cy^2 + Dx + Ey + F = 0$

In the previous sections we studied quadratic equations of the form

$$Ax^2 + Cy^2 + Dx + Ey + F = 0 , \qquad\qquad (D.32)$$

which have no xy term. The graph of (D.32) may be *empty*, for example,

$$x^2 + y^2 + 1 = 0 ,$$

or may be

$$\text{one point, e.g.,} \quad x^2 + y^2 = 0 ,$$
$$\text{a line, e.g.,} \quad (x - 1)^2 = 0 ,$$
$$\text{two intersecting lines, e.g.,} \quad x^2 - y^2 = 0 ,$$
$$\text{two parallel lines, e.g.,} \quad (x - 1)^2 = 1 .$$

Otherwise, the graph of (D.32) is a conic section, an ellipse (in particular a circle), a parabola, or a hyperbola.

In this section we consider the general quadratic equation (D.1),

$$Ax^2 + Bxy + Cy^2 + Dx + Ey + F = 0 .$$

If $B \neq 0$, we "get rid" of the term Bxy by rotating the coordinate axes, using Proposition C.5.

We recall that a counterclockwise rotation of the x- and y-axes, by an angle α, gives the rotated coordinates

$$\bar{x} = x \cos \alpha + y \sin \alpha \qquad\qquad \text{(C.32a)} ,$$
$$\bar{y} = -x \sin \alpha + y \cos \alpha \qquad\qquad \text{(C.32b)} .$$

Conversely, if $(\bar{x}, \bar{y})$ are the rotated coordinates of a point P, its original coordinates (x, y) are

$$x = \bar{x} \cos \alpha - \bar{y} \sin \alpha \qquad\qquad \text{(C.33a)} ,$$
$$y = \bar{x} \sin \alpha + \bar{y} \cos \alpha \qquad\qquad \text{(C.33b)} .$$

Example D.9. The simplest quadratic equation which contains a mixed term is

$$xy - 1 = 0 . \qquad\qquad (D.33)$$

If we rotate the original system by $\pi/4$ [rad], then as $\cos \pi/4 = \sin \pi/4 = 1/\sqrt{2}$, we get

$$x = \frac{1}{\sqrt{2}}(\bar{x} - \bar{y}), \quad \text{and} \quad y = \frac{1}{\sqrt{2}}(\bar{x} + \bar{y}) .$$

The quadratic (D.33) then becomes

$$\bar{x}^2 - \bar{y}^2 = 2 ,$$

which is the equation of a hyperbola.

The above example suggests that it is always possible to choose a rotated coordinate system in which there is no mixed term. This indeed is exactly the case.

Substituting (C.33a) and (C.33b) in (D.1),

$$Ax^2 + Bxy + Cy^2 + Dx + Ey + F = 0 ,$$

we get another quadratic equation,

$$\bar{A}\bar{x}^2 + \bar{B}\bar{x}\bar{y} + \bar{C}\bar{y}^2 + \bar{D}\bar{x} + \bar{E}\bar{y} + \bar{F} = 0 . \tag{D.34}$$

Proposition D.10. a. The coefficients $\bar{A}, \bar{B}, \ldots, \bar{F}$ in (D.34) are given, in terms of the original coefficients and the angle α, as follows.

$$\bar{A} = (A - C)\cos^2\alpha + B\sin\alpha\cos\alpha + C , \tag{D.35a}$$

$$\bar{B} = 2B\cos^2\alpha + 2(C - A)\sin\alpha\cos\alpha - B , \tag{D.35b}$$

$$\bar{C} = (C - A)\cos^2\alpha - B\sin\alpha\cos\alpha + A , \tag{D.35c}$$

$$\bar{D} = D\cos\alpha + E\sin\alpha , \tag{D.35d}$$

$$\bar{E} = E\cos\alpha - D\sin\alpha , \tag{D.35e}$$

$$\bar{F} = F . \tag{D.35f}$$

b. The quantity $B^2 - 4AC$ does not change with rotation

$$\bar{B}^2 - 4\bar{A}\,\bar{C} = B^2 - 4AC . \tag{D.36}$$

c. The angle of rotation α which causes $\bar{B} = 0$ is

$$\alpha = \frac{1}{2}\operatorname{atan}\left(\frac{B}{A - C}\right) . \tag{D.37}$$

Proof. Substitute (C.33a) and (C.33b) in (D.1), to get the polynomial (D.34), and identify the coefficient $\bar{A}$ of $\bar{x}^2$, the coefficient $\bar{B}$ of $\bar{x}\,\bar{y}$, etc.

The identity (D.36) is proved by simplifying $\bar{B}^2 - 4\bar{A}\bar{C}$.

Finally, the coefficient $\bar{B}$ is, by (D.35b),

$$\bar{B} = 2B\cos^2\alpha + 2(C - A)\sin\alpha\cos\alpha - B$$

$$= 2B\cos^2\alpha - 2(A - C)\sin\alpha\cos\alpha - B(\cos^2\alpha + \sin^2\alpha)$$

$$= B(\cos^2\alpha - \sin^2\alpha) - 2(A - C)\sin\alpha\cos\alpha$$

$$= B\cos(2\alpha) - (A - C)\sin(2\alpha)$$

Therefore $\bar{B} = 0$ if

$$\frac{\sin(2\alpha)}{\cos(2\alpha)} = \frac{B}{A - C} . \qquad \square$$

MACSYMA-Session D.2. The arithmetic involved in the proof of Proposition D.10 can be done with MACSYMA. We use here the MACSYMA function

```
c1. poly_coeff(u,x,n)  := 1/n!*subst(x=0,diff(u,x,n))$
```

which computes the coefficient of x^n in a polynomial u. MACSYMA has a built-in `coeff` function (which is much faster), but we use here our own version.

Next we define the 6 functions that extract the coefficients A, B, C, D, E, and F of the polynomial

```
c2. u(x,y):=A*x^2+B*x*y+C*y^2+D*x+E*y+F
```

d2. $u(x, y) := Ax^2 + Bxy + Cy^2 + Dx + Ey + F$

and these functions are

```
c3. c_A(u,x,y):=poly_coeff(u,x,2) $
c4. c_B(u,x,y):=poly_coeff(poly_coeff(u,x,1), y, 1) $
c5. c_C(u,x,y):=poly_coeff(u,y,2) $
c6. c_D(u,x,y):=subst(y=0, poly_coeff(u,x,1)) $
c7. c_E(u,x,y):=subst(x=0, poly_coeff(u,y,1)) $
c8. c_F(u,x,y):=subst([x=0,y=0],u) $
```

We will append the word "bar" to a name to represent the transformed quantities, e.g., xbar, ybar, ubar, Abar, etc. By (C.33a) and (C.33b), the coordinates (x, y) are related to the transformed coordinates $(\bar{x}, \bar{y})$ by

```
x = xbar * cos(alpha) - ybar * sin(alpha)
y = xbar * sin(alpha) + ybar * cos(alpha)
```

The quadratic polynomial in c2 is therefore, in terms of the transformed coordinates,

```
c9. ubar:u(xbar*cos(alpha)-ybar*sin(alpha),
          xbar*sin(alpha)+ybar*cos(alpha))
```

d9. $A(\cos\alpha\bar{x} - \sin\alpha\bar{y})^2 + B(\cos\alpha\bar{y} + \sin\alpha\bar{x})(\cos\alpha\bar{x} - \sin\alpha\bar{y}) +$
$+ D(\cos\alpha\bar{x} - \sin\alpha\bar{y}) + C(\cos\alpha\bar{y} + \sin\alpha\bar{x})^2 + E(\cos\alpha\bar{y} + \sin\alpha\bar{x}) + F$

a quadratic polynomial in xbar and ybar, whose coefficients $\bar{A}$, $\bar{B}$, ..., $\bar{F}$ can be determined by the functions c_A, c_B, ... c_F above:

```
c10. Abar:c_A(ubar,xbar,ybar)
```

d10. $\dfrac{2\sin^2\alpha\, C + 2\cos\alpha\sin\alpha\, B + 2A\cos^2\alpha}{2}$

```
c11. Bbar:c_B(ubar,xbar,ybar)
```

d11. $2\cos\alpha\sin\alpha\, C - \sin^2\alpha\, B + \cos^2\alpha\, B - 2A\cos\alpha\sin\alpha$

```
c12. Cbar:c_C(ubar,xbar,ybar)
```

d12. $\dfrac{2\cos^2\alpha\, C - 2\cos\alpha\sin\alpha\, B + 2A\sin^2\alpha}{2}$

```
c13. Dbar:c_D(ubar,xbar,ybar)
```

d13. $\sin\alpha\, E + \cos\alpha\, D$

```
c14. Ebar:c_E(ubar,xbar,ybar)
```

d14. $\cos\alpha\, E - \sin\alpha\, D$

```
c15. Fbar:c_F(ubar,xbar,ybar)
```

d15. F

proving the results of (D.35a)–(D.35f). To prove (D.36), we compute:

```
c16. Bbar^2-4*Abar*Cbar
```

d16. $(2\cos\alpha\sin\alpha\, C - \sin^2\alpha\, B + \cos^2\alpha\, B - 2A\cos\alpha\sin\alpha)^2 - (2\cos^2\alpha\, C -$
$- 2\cos\alpha\sin\alpha\, B + 2A\sin^2\alpha)(2\sin^2\alpha\, C + 2\cos\alpha\sin\alpha\, B + 2A\cos^2\alpha)$

which simplifies to $B^2 - 4AC$ when we replace $\cos^2(\alpha)$ by $1 - \sin^2(\alpha)$, c17. `ratsubst(1-sin(alpha)^2,cos(alpha)^2,%)`

d17. $B^2 - 4AC$

Finally, to prove (D.37), we recall

c18. `Bbar`

d18. $2\cos\alpha\sin\alpha\,C - \sin^2\alpha\,B + \cos^2\alpha\,B - 2A\cos\alpha\sin\alpha$

which simplifies to

c19. `trigreduce(%)`

d19. $\sin(2\alpha)C + \cos(2\alpha)B - A\sin(2\alpha)$

and solving for α gives (D.37).

The quantity $B^2 - 4AC$ was shown above to be invariant under rotation of the coordinate axes. We call $B^2 - 4AC$ the *discriminant* of the quadratic (D.1), $Ax^2 + Bxy + Cy^2 + Dx + Ey + F = 0$. The sign of the discriminant is the key to the type of conic section represented by (D.1).

Proposition D.11. If the graph of (D.1), $Ax^2 + Bxy + Cy^2 + Dx + Ey + F = 0$, is not empty and is not a point, a line, or two lines, then it is

$$\text{an ellipse if} \quad B^2 - 4AC < 0\,, \tag{D.38a}$$

$$\text{a parabola if} \quad B^2 - 4AC = 0\,, \tag{D.38b}$$

$$\text{a hyperbola if} \quad B^2 - 4AC > 0\,. \tag{D.38c}$$

Proof. Since the discriminant is invariant under rotation, we may assume that $B = 0$, which is guaranteed by rotating the coordinate axes by the angle α of (D.37). Therefore, the discriminant is $-AC$.

In page 585 we showed that the case $AC > 0$ corresponds to an ellipse. This proves part (D.38a).

Similarly, parts (D.38b) and (D.38c) follow from the results in pages 580 and 590, respectively. $\square$

Remark D.12. If $A = B = C = 0$ but $D^2 + E^2 > 0$, then the polynomial (D.1) represents a line

$$Dx + Ey + F = 0\,.$$

If $A = B = D = 0$ but $C \neq 0$, then the polynomial (D.1) becomes

$$Cy^2 + Ey + F = 0$$

and its graph is a horizontal line or two horizontal lines or is empty. If $B = C = E$ but $A \neq 0$, then the polynomial (D.1) becomes

$$Ax^2 + Dx + F = 0$$

and its graph is a vertical line or two vertical lines or is empty.

Proposition D.11 is a classification of conic sections according to their discriminants. The following proposition classifies them according to the eccentricity.

Proposition D.13. Let $p > 0$. Then the graph of

$$r = \frac{pe}{1 \pm e\cos\theta},$$

is

$$\text{an ellipse if} \quad e < 1, \tag{D.39a}$$
$$\text{a parabola if} \quad e = 1, \tag{D.39b}$$
$$\text{a hyperbola if} \quad e > 1. \tag{D.39c}$$

This conic section is symmetric with respect to the x-axis. It has a focus at the origin, and p is the distance between that focus and the directrix.

Proof. See (D.8) and (D.9), (D.22) and (D.23), and (D.30) and (D.31).   □

Exercises

D.14 Note that $\bar{A}$ and $\bar{C}$ (given by (D.35a) and (D.35c)) satisfy

$$\bar{A} + \bar{C} = A + C,$$

i.e., the sum $A + C$ is invariant under rotation of the coordinate axes. What information does this sum convey about the general quadratic (D.1)?

D.15 Obtain the alternative formula for the angle of rotation needed to eliminate the $\bar{x}\bar{y}$ term from a quadratic

$$\alpha = \frac{\pi}{4} - \frac{1}{2}\operatorname{atan}\left(\frac{A - C}{B}\right).$$

D.16 In the following problems classify each curve by its discriminant. Then perform the rotation needed to eliminate the mixed term. Use MACSYMA to plot the rotated curve and compare this with the plot of the original equation.
   (a) $x^2 + 2xy + y^2 = 2$
   (b) $3x^2 - 2xy + 3y^2 = 4$
   (c) $16x^2 - 24xy + 9y^2 - 60x - 80y + 100 = 0$
   (d) $73x^2 - 72xy + 52y^2 - 100 = 0$

D.17 Use the MACSYMA functions `C_A(u,x,y)`, `C_B(u,x,y)`, `C_C(u,x,y)`, `C_D(u,x,y)`, `C_E(u,x,y)`, and `C_F(u,x,y)` of Session D.2 to write the following MACSYMA functions
   a. `conic_discriminant`(g, x, y) that calculates the discriminant of the expression g using the above auxiliary functions
   b. `conic_type`(g, x, y) that answers "circle", "ellipse", "hyperbola", or "parabola" depending on its type
   c. `angle_rotation`(g, x, y) that calculates the rotation angle α of the quadratic

Use the functions to give the type and rotation angle of the following quadratics

(a) $x^2 + 2xy - y^2 + 3x + 1$

(b) $x^2 - 2xy + y^2 + 4y + 1$

(c) $x^2 + xy - 2y^2 + 2x - y - 1$

(d) $-x^2 + xy - 2y^2 + 2x - y + 10$

(e) $x^2 + xy + y^2 + x + y$

and plot the quadratics.

Subject index

SpringerComputerScience

Texts and Monographs in Symbolic Computation

Bob F. Caviness,
Jeremy R. Johnson (eds.)

Quantifier Elimination
and Cylindrical Algebraic
Decomposition

1998. XIX, 431 pages. 20 figures.
Softcover EUR 63,–
ISBN 3-211-82794-3

George Collins' discovery of Cylindrical Algebraic Decomposition (CAD) as a method for Quantifier Elimination (QE) for the elementary theory of real closed fields brought a major breakthrough in automating mathematics with recent important applications in high-tech areas (e.g. robot motion), also stimulating fundamental research in computer algebra over the past three decades.
This volume is a state-of-the-art collection of important papers on CAD and QE and on the related area of algorithmic aspects of real geometry. It contains papers from a symposium held in Linz in 1993, reprints of seminal papers from the area including Tarski's landmark paper as well as a survey outlining the developments in CAD based QE that have taken place in the last twenty years.

Dongming Wang

Elimination Methods

2001. XIII, 244 pages. 12 figures.
Softcover EUR 57,–
ISBN 3-211-83241-6

This book provides a systematic and uniform presentation of elimination methods and the underlying theories, along the central line of decomposing arbitrary systems of polynomials into triangular systems of various kinds. Highlighting methods based on triangular sets, the book also covers the theory and techniques of resultants and Gröbner bases. The methods and their efficiency are illustrated by fully worked out examples and their applications to selected problems such as from polynomial ideal theory, automated theorem proving in geometry and the qualitative study of differential equations. The reader will find the formally described algorithms ready for immediate implementation and applicable to many other problems.

SpringerWienNewYork

A-1201 Wien, Sachsenplatz 4–6, P.O. Box 89, Fax +43.1.330 24 26, e-mail: books@springer.at, Internet: www.springer.at
D-69126 Heidelberg, Haberstraße 7, Fax +49.6221.345-229, e-mail: orders@springer.de
USA, Secaucus, NJ 07096-2485, P.O. Box 2485, Fax +1.201.348-4505, e-mail: orders@springer-ny.com
Eastern Book Service, Japan, Tokyo 113, 3–13, Hongo 3-chome, Bunkyo-ku, Fax +81.3.38 18 08 64, e-mail: orders@svt-ebs.co.jp

SpringerComputerScience

Texts and Monographs in Symbolic Computation

Bernd Sturmfels

**Algorithms
in Invariant Theory**

1993. VII, 197 pages. 5 figures.
Softcover EUR 34,–*)
ISBN 3-211-82445-6

Wen-tsün Wu

**Mechanical Theorem
Proving in Geometries**

Basic Principles

Translated from the Chinese by
Xiaofan Jin and Dongming Wang
1994. XIV, 288 pages. 120 figures.
Softcover EUR 52,–*)
ISBN 3-211-82506-1

**Jochen Pfalzgraf,
Dongming Wang (eds.)**

**Automated
Practical Reasoning**

Algebraic Approaches

With a Foreword by Jim Cunningham
1995. XI, 223 pages. 23 figures.
Softcover EUR 57,–*)
ISBN 3-211-82600-9

Franz Winkler

**Polynomial Algorithms
in Computer Algebra**

1996. VIII, 270 pages. 13 figures.
Softcover EUR 47,–*)
ISBN 3-211-82759-5

Norbert Kajler (ed.)

**Computer-Human Interaction
in Symbolic Computation**

With a Foreword by D. S. Scott
1998. XI, 212 pages. 68 figures.
Softcover EUR 47,–*)
ISBN 3-211-82843-5

**Alfonso Miola,
Marco Temperini (eds.)**

**Advances in the
Design of Symbolic
Computation Systems**

1997. X, 259 pages. 39 figures.
Softcover EUR 52,–*)
ISBN 3-211-82844-3

*) Recommended retail prices
All prices are net-prices subject to local VAT.

Springer Wien New York

A-1201 Wien, Sachsenplatz 4–6, P.O. Box 89, Fax +43.1.330 24 26, e-mail: books@springer.at, Internet: www.springer.at
D-69126 Heidelberg, Haberstraße 7, Fax +49.6221.345-229, e-mail: orders@springer.de
USA, Secaucus, NJ 07096-2485, P.O. Box 2485, Fax +1.201.348-4505, e-mail: orders@springer-ny.com
Eastern Book Service, Japan, Tokyo 113, 3–13, Hongo 3-chome, Bunkyo-ku, Fax +81.3.38 18 08 64, e-mail: orders@svt-ebs.co.jp

Springer-Verlag
and the Environment

WE AT SPRINGER-VERLAG FIRMLY BELIEVE THAT AN international science publisher has a special obligation to the environment, and our corporate policies consistently reflect this conviction.

WE ALSO EXPECT OUR BUSINESS PARTNERS – PRINTERS, paper mills, packaging manufacturers, etc. – to commit themselves to using environmentally friendly materials and production processes.

THE PAPER IN THIS BOOK IS MADE FROM NO-CHLORINE pulp and is acid free, in conformance with international standards for paper permanency.